CELL DIFFERENTIATION AND NEOPLASIA

The University of Texas System Cancer Center
M. D. Anderson Hospital and Tumor Institute
30th Annual Symposium on Fundamental Cancer Research

Published for
The University of Texas System Cancer Center
M. D. Anderson Hospital and Tumor Institute
Houston, Texas, by Raven Press, New York

The University of Texas System Cancer Center
M. D. Anderson Hospital and Tumor Institute
30th Annual Symposium on Fundamental Cancer Research

Cell Differentiation and Neoplasia

Edited by

Grady F. Saunders, Ph.D.
Department of Developmental Therapeutics
The University of Texas System Cancer Center
M. D. Anderson Hospital and Tumor Institute
Houston, Texas

Raven Press ■ New York

Raven Press, 1140 Avenue of the Americas, New York, New York 10036

Made in the United States of America

Library of Congress Cataloging in Publication Data

Symposium on Fundamental Cancer Research, 30th, Anderson
Hospital and Tumor Institute, 1977.
Cell differentiation and neoplasia.

"Published for the University of Texas System Cancer Center M. D. Anderson Hospital and Tumor Institute, Houston, Texas."
Includes bibliographical references and index.
1. Carcinogenesis–Congresses. 2. Cancer cells–Congresses. 3. Cell differentiation–Congresses. 4. Genetic regulation–Congresses. I. Saunders, Grady F. II. Anderson Hospital and Tumor Institute, Houston, Tex. III. Title. [DNLM: 1. Cell differentiation–Congresses. 2. Neoplasms–Etiology–Congresses. 3. Cell transformation, Neoplastic–Congresses.
W3 SY5177 30th 1977c / QZ202 S986 1977c]
RC268.5.S95 1977 616.9'94'07 77–17694
ISBN 0–89004–200–4

The material contained herein was submitted as previously unpublished material, except in the instances in which credit has been given to the source from which some of the illustrative material was derived.

Editor's Foreword

One view of cancer is that it results from an impairment of the cell differentiation process. The 30th Annual Symposium on Fundamental Cancer Research dealt with the relationship between cell differentiation and neoplasia. Differentiation is defined as a process of unidirectional and irreversible transformation under normal physiological conditions. This process is accomplished through the ordered control of gene expression. The objective of the Symposium was to examine the viewpoint of cancer as a disease in cell differentiation.

In this Symposium, a number of tumor cell types originating at various stages of differentiation were considered. These included germ cell tumors, tumors of cells of the nervous system, and tumors of the hematopoietic system. The second part of the Symposium was devoted to the control of differentiation and neoplasia. Particular emphasis was placed on the molecular mechanisms of gene expression and the application of the techniques of molecular biology to the differentiation process. The rapid development in technology during the past few years has led to remarkable progress in our knowledge of the regulation of gene expression. The view of cancer resulting from an impairment of the cell differentiation process now can be examined in considerable detail. It was hoped that this Symposium would result in generation of new ideas and greater insight into the role of cell differentiation in human cancer.

Grady F. Saunders, Ph.D.

Acknowledgments

We wish to thank the many individuals whose knowledge and support made possible the 30th Annual Symposium on Fundamental Cancer Research at M. D. Anderson Hospital and Tumor Institute. We are especially grateful to the National Cancer Institute and the Texas Division of the American Cancer Society for their continued support. We also wish to thank The University of Texas Health Science Center at Houston, Graduate School of Biomedical Sciences for its assistance.

We offer special thanks to the members of the Symposium Committee, Grady F. Saunders (Chairman), Ralph B. Arlinghaus, Lubomir S. Hnilica, T. C. Hsu, Alfred G. Knudson, Jr., Kenneth B. McCredie, Marvin L. Meistrich, Jan van Eys, and David A. Wright; and the Advisory Committee, Donald D. Brown, Stanley Cohen, James German, Clement Markert, Paul A. Marks, Ernest A. McCulloch, Beatrice Mintz, Susumo Ohno, Bert O'Malley, John Paul, and William J. Rutter.

For the compilation of the volume, we acknowledge the excellent editorial work of the Publications Office of the Department of Information and Publications, and especially the assistance of Leslie Wildrick, Symposium Editor, and Walter Pagel, of that office.

Contents

Contributors

G. I. Abelev
Laboratory of Tumor Immunochemistry and Diagnosis
N. F. Gamaleya Institute of Epidemiology and Microbiology
USSR Academy of Medical Sciences
Moscow, USSR

Terence D. Allen
Paterson Laboratories
Christie Hospital and Holt Radium Institute
Manchester M20 9BX England

Thomas H. Alton
Department of Biology
Massachusetts Institute of Technology
Cambridge, Massachusetts 02139
Present Address: Department of Biochemistry
Stanford University School of Medicine
Stanford, California 94305

Ralph B. Arlinghaus
Department of Biology
The University of Texas System Cancer Center
M. D. Anderson Hospital and Tumor Institute
Houston, Texas 77030

Arthur Bank
Cancer Research Center, and Departments of Medicine and of Human Genetics and Development
Columbia University
New York, New York 10032

Perry F. Bartlett
Biology Department
The Johns Hopkins University
Baltimore, Maryland 21218

John E. Bergmann
Department of Biology
Massachusetts Institute of Technology
Cambridge, Massachusetts 02139

Mary R. Bordelon
Department of Cell Biology
Baylor College of Medicine
Houston, Texas 77030

B. R. Brinkley
Department of Human Biological Chemistry and Genetics
Division of Cell Biology
The University of Texas Medical Branch
Galveston, Texas 77550
Present Address: Department of Cell Biology
Baylor College of Medicine
Houston, Texas 77030

William A. Brock
Section of Experimental Radiotherapy
The University of Texas System Cancer Center
M. D. Anderson Hospital and Tumor Institute
Houston, Texas 77030

Donald D. Brown
Department of Embryology
Carnegie Institution of Washington
Baltimore, Maryland 21210

R. N. Buick
Institute of Medical Science
University of Toronto, and the Ontario Cancer Institute
Toronto, Ontario, Canada

Michael Cancro
Department of Pathology
University of Pennsylvania
Philadelphia, Pennsylvania 19104

John M. Chirgwin
Department of Biochemistry and Biophysics
University of California at San Francisco
San Francisco, California 94143

Jen-Fu Chiu
Department of Biochemistry
Vanderbilt University School of Medicine
Nashville, Tennessee 37232

W. F. Cox, Jr.
Department of Pathology
University of Colorado Medical Center
Denver, Colorado 80262

B. A. Cunningham
Department of Developmental and Molecular Biology
The Rockefeller University
New York, New York 10021

James E. Darnell, Jr.
The Rockefeller University
New York, New York 10021

E. M. De Robertis
MRC Laboratory of Molecular Biology
Cambridge CB2 2QH England

T. Michael Dexter
Paterson Laboratories
Christie Hospital and Holt Radium Institute
Manchester M20 9BX England

K. A. Dicke
Department of Developmental Therapeutics
The University of Texas System Cancer Center
M. D. Anderson Hospital and Tumor Institute
Houston, Texas 77030

G. M. Edelman
Department of Developmental and Molecular Biology
The Rockefeller University
New York, New York 10021

Michael Edidin
Biology Department
The Johns Hopkins University
Baltimore, Maryland 21218

Nina V. Fedoroff
Department of Embryology
Carnegie Institution of Washington
Baltimore, Maryland 21210

Eitan Fibach
Cancer Research Center, and Departments of Medicine and of Human Genetics and Development
Columbia University
New York, New York 10032

J. W. Fuseler
Department of Human Biological Chemistry and Genetics
Division of Cell Biology
The University of Texas Medical Branch
Galveston, Texas 77550
Present Address: Department of Cell Biology
The University of Texas Health Science Center
Dallas, Texas 75235

Hideo Fujitani
Department of Biochemistry
Vanderbilt University School of Medicine
Nashville, Tennessee 37232

Yair Gazitt
Cancer Research Center, and Departments of Medicine and of Human Genetics and Development
Columbia University
New York, New York 10032

Sidney R. Grimes, Jr.
Department of Biochemistry
The University of Texas System Cancer Center
M. D. Anderson Hospital and Tumor Institute
Houston, Texas 77030
Present Address: Veterans Administration Hospital
Shreveport, Louisiana 71130

J. B. Gurdon
MRC Laboratory of Molecular Biology
Cambridge CB2 2QH England

John D. Harding
Department of Biochemistry and Biophysics
University of California at San Francisco
San Francisco, California 94143

Kenneth Hardy
Department of Biochemistry
Vanderbilt University School of Medicine
Nashville, Tennessee 37232

R. Henning
Department of Developmental and Molecular Biology
The Rockefeller University
New York, New York 10021

Mary M. Herman
Department of Pathology (Neuropathology)
Stanford University School of Medicine
Stanford, California 94305

Lubomir S. Hnilica
Department of Biochemistry
Vanderbilt University School of Medicine
Nashville, Tennessee 37232

T. C. Hsu
Department of Biology
The University of Texas System Cancer Center
M. D. Anderson Hospital and Tumor Institute
Houston, Texas 77030

Ronald M. Humphrey
Department of Physics
The University of Texas System Cancer Center
M. D. Anderson Hospital and Tumor Institute
Houston, Texas 77030

Norman N. Iscove
Basel Institute for Immunology
CH4058 Basel, Switzerland

Judith A. Jaehning
Department of Biological Chemistry
Division of Biology and Biomedical Sciences
Washington University
St. Louis, Missouri 63110

Ghazi A. Jamjoom
Department of Biology
The University of Texas System Cancer Center
M. D. Anderson Hospital and Tumor Institute
Houston, Texas 77030

Alfred G. Knudson, Jr.
Institute for Cancer Research
Fox Chase Cancer Center
Philadelphia, Pennsylvania 19111

John Kopchick
Department of Biology
The University of Texas System Cancer Center
M. D. Anderson Hospital and Tumor Institute
Houston, Texas 77030

Sau-Ping Kwan
Department of Biological Chemistry
University of Cincinnati Medical Center
Cincinnati, Ohio 45267
Present Address: Albert Einstein College of Medicine
Yeshiva University
Bronx, New York 10461

Laszlo G. Lajtha
Paterson Laboratories
Christie Hospital and Holt Radium Institute
Manchester M20 9BX England

R. A. Laskey
MRC Laboratory of Molecular Biology
Cambridge CB2 2QH England

Jerry B. Lingrel
Department of Biological Chemistry
University of Cincinnati Medical Center
Cincinnati, Ohio 45267

Harvey F. Lodish
Department of Biology
Massachusetts Institute of Technology
Cambridge, Massachusetts 02139

Brian I. Lord
Paterson Laboratories
Christie Hospital and Holt Radium Institute
Manchester M20 9BX England

Raymond J. MacDonald
Department of Biochemistry and Biophysics
University of California at San Francisco
San Francisco, California 94143

Clement L. Markert
Department of Biology
Yale University
New Haven, Connecticut 06520

Paul A. Marks
Cancer Research Center, and Departments of Medicine and of Human Genetics and Development
Columbia University
New York, New York 10032

K. B. McCredie
Department of Developmental Therapeutics
The University of Texas System Cancer Center
M. D. Anderson Hospital and Tumor Institute
Houston, Texas 77030

E. A. McCulloch
Institute of Medical Science
University of Toronto, and the Ontario Cancer Institute
Toronto, Ontario, Canada

Anna T. Meadows
The Children's Hospital of Philadelphia
Philadelphia, Pennsylvania 19104

Marvin L. Meistrich
Section of Experimental Radiotherapy
The University of Texas System Cancer Center
M. D. Anderson Hospital and Tumor Institute
Houston, Texas 77030

J. E. Mertz
MRC Laboratory of Molecular Biology
Cambridge CB2 2QH England
Present Address: McArdle Laboratory for Cancer Research
University of Wisconsin
Madison, Wisconsin 53706

C. L. Miller
Department of Human Biological Chemistry and Genetics
Division of Cell Biology
The University of Texas Medical Branch
Galveston, Texas 77550
Present Address: Department of Biology
University of Pennsylvania
Philadelphia, Pennsylvania 19104

R. J. Milner
Department of Developmental and Molecular Biology
The Rockefeller University
New York, New York 10021

Beatrice Mintz
Institute for Cancer Research
Fox Chase Cancer Center
Philadelphia, Pennsylvania 19111

Robert B. Naso
Department of Biology
The University of Texas System Cancer Center
M. D. Anderson Hospital and Tumor Institute
Houston, Texas 77030

Sun-Yu Ng
Department of Biological Chemistry
Division of Biology and Biomedical Sciences
Washington University
St. Louis, Missouri 63110

Uri Nudel
Cancer Research Center, and Departments of Medicine and of Human Genetics and Development
Columbia University
New York, New York 10032

Bert W. O'Malley
Department of Cell Biology
Baylor College of Medicine
Houston, Texas 77030

Suzanne Ostrand-Rosenberg
Biology Department
The Johns Hopkins University
Baltimore, Maryland 21218

Joseph T. Painter
Administration
The University of Texas System Cancer Center
M. D. Anderson Hospital and Tumor Institute
Houston, Texas 77030

Carl S. Parker
Department of Biological Chemistry
Division of Biology and Biomedical Sciences
Washington University
St. Louis, Missouri 63110

G. A. Partington
MRC Laboratory of Molecular Biology
Cambridge CB2 2QH England

J. Paul
Beatson Institute for Cancer Research
Wolfson Laboratory for Molecular Pathology
Glasgow G61 1BD Scotland

D. A. Pepper
Department of Human Biological Chemistry and Genetics
Division of Cell Biology
The University of Texas Medical Branch
Galveston, Texas 77550
Present Address: Department of Cell Biology
Baylor College of Medicine
Houston, Texas 77030

Raymond L. Pictet
Department of Biochemistry and Biophysics
University of California at San Francisco
San Francisco, California 94143

G. B. Pierce
Department of Pathology
University of Colorado Medical Center
Denver, Colorado 80262

Robert D. Platz
Section of Experimental Radiotherapy
The University of Texas System Cancer Center
M. D. Anderson Hospital and Tumor Institute
Houston, Texas 77030
Present Address: Frederick Cancer Research Center
Frederick, Maryland 21701

Michael Potter
National Cancer Institute
Laboratory of Cell Biology
Bethesda, Maryland 20014

Kedar N. Prasad
Department of Radiology
University of Colorado Medical Center
Denver, Colorado 80262

Alan E. Przybyla
Department of Biochemistry and Biophysics
University of California at San Francisco
San Francisco, California 94143

K. Reske
Department of Developmental and Molecular Biology
The Rockefeller University
New York, New York 10021

Roberta Reuben
Cancer Research Center, and Departments of Medicine and of Human Genetics and Development
Columbia University
New York, New York 10032

Richard A. Rifkind
Cancer Research Center, and Departments of Medicine and of Human Genetics and Development
Columbia University
New York, New York 10032

Robert G. Roeder
Department of Biological Chemistry
Division of Biology and Biomedical Sciences
Washington University
St. Louis, Missouri 63110

Paul Rosteck, Jr.
Department of Biological Chemistry
University of Cincinnati Medical Center
Cincinnati, Ohio 45267

William J. Rutter
Department of Biochemistry and Biophysics
University of California at San Francisco
San Francisco, California 94143

Leo Sachs
Department of Genetics
Weizmann Institute of Science
Rehovot, Israel

Jane Salmon
Cancer Research Center, and Departments of Medicine and of Human Genetics and Development
Columbia University
New York, New York 10032

Grady F. Saunders
Department of Developmental Therapeutics
The University of Texas System Cancer Center
M. D. Anderson Hospital and Tumor Institute
Houston, Texas 77030

J. W. Schrader
Department of Developmental and Molecular Biology

The Rockefeller University
New York, New York 10021

Michael J. Siciliano
Department of Biology
The University of Texas System Cancer Center
M. D. Anderson Hospital and Tumor Institute
Houston, Texas 77030

Pramod K. Sinha
Department of Radiology
University of Colorado Medical Center
Denver, Colorado 80262

Virgil E. F. Sklar
Department of Biological Chemistry
Division of Biology and Biomedical Sciences
Washington University
St. Louis, Missouri 63110

Kate Smith
Department of Biological Chemistry
University of Cincinnati Medical Center
Cincinnati, Ohio 45267

G. Spitzer
Department of Developmental Therapeutics
The University of Texas System Cancer Center
M. D. Anderson Hospital and Tumor Institute
Houston, Texas 77030

Walter G. Sterling
Board of Regents
The University of Texas System
Houston, Texas 77030

Masaaki Terada
Cancer Research Center, and Departments of Medicine and of Human Genetics and Development
Columbia University
New York, New York 10032

J. E. Till
Institute of Medical Science
University of Toronto, and the Ontario Cancer Institute
Toronto, Ontario, Canada

Howard Towle
Department of Cell Biology
Baylor College of Medicine
Houston, Texas 77030
Present Address: Department of Medicine
University of Minnesota
Minneapolis, Minnesota 55455

Ming-Jer Tsai
Department of Cell Biology
Baylor College of Medicine
Houston, Texas 77030

Sophia Y. Tsai
Department of Cell Biology
Baylor College of Medicine
Houston, Texas 77030

Scott R. VandenBerg
Department of Pathology (Neuropathology)
Stanford University School of Medicine
Stanford, California 94305

D. Verma
Department of Developmental Therapeutics
M. D. Anderson Hospital and Tumor Institute
Houston, Texas 77030

L. J. Wible
Department of Human Biological Chemistry and Genetics
Division of Cell Biology
The University of Texas Medical Branch
Galveston, Texas 77550
Present Address: Department of Cell Biology
Baylor College of Medicine
Houston, Texas 77030

T. Gordon Wood
Department of Biological Chemistry
University of Cincinnati Medical Center
Cincinnati, Ohio 45267
Present Address: The University of Texas System Cancer Center
M. D. Anderson Hospital and Tumor Institute
Houston, Texas 77030

Eric G. Wright
Paterson Laboratories

Christie Hospital and Holt Radium Institute
Manchester M20 9BX England

David A. Wright
Department of Biology
The University of Texas System Cancer Center
M. D. Anderson Hospital and Tumor Institute
Houston, Texas 77030

A. D. Wyllie
MRC Laboratory of Molecular Biology
Cambridge CB2 2QH England

J. A. Ziffer
Department of Developmental and Molecular Biology
The Rockefeller University
New York, New York 10021

SYMPOSIUM COMMITTEE

Grady F. Saunders, Chairman
Ralph B. Arlinghaus
Lubomir S. Hnilica
T. C. Hsu
Alfred G. Knudson, Jr.
Kenneth B. McCredie
Marvin L. Meistrich
Jan van Eys
David A. Wright

Ex Officio Members

George Blumenschein
Benjamin Drewinko
Frances Goff
A. Clark Griffin
Felix Haas
Ronald M. Humphrey
Glenn R. Knotts
Charles R. Shaw
Stephen C. Stuyck

SESSION CHAIRMEN

Ralph B. Arlinghaus
Department of Biology, The University of Texas System Cancer Center M. D. Anderson Hospital and Tumor Institute, Houston, Texas

T. C. Hsu
Department of Biology, The University of Texas System Cancer Center M. D. Anderson Hospital and Tumor Institute, Houston, Texas

Alfred G. Knudson, Jr.
Institute for Cancer Research, Fox Chase Cancer Center, Philadelphia, Pennsylvania

Clement L. Markert
Yale University, New Haven, Connecticut

Kenneth B. McCredie
Department of Developmental Therapeutics, The University of Texas System Cancer Center M. D. Anderson Hospital and Tumor Institute, Houston, Texas

Jan van Eys
Department of Pediatrics, The University of Texas System Cancer Center M. D. Anderson Hospital and Tumor Institute, Houston, Texas

ADVISORY COMMITTEE

Donald D. Brown
Carnegie Institution of Washington, Baltimore, Maryland

Stanley Cohen
Vanderbilt University, Nashville, Tennessee

James German
New York Blood Center, New York, New York

Clement L. Markert
Yale University, New Haven, Connecticut

Paul A. Marks
Columbia University, New York, New York

Ernest A. McCulloch
The Ontario Cancer Institute, Toronto, Ontario, Canada

Beatrice Mintz
Institute for Cancer Research, Fox Chase Cancer Center, Philadelphia, Pennsylvania

Susumo Ohno
City of Hope Medical Center, Duarte, California

Bert W. O'Malley
Baylor College of Medicine, Houston, Texas

John Paul
Beatson Institute for Cancer Research, Glasgow, Scotland

William J. Rutter
University of California at San Francisco, San Francisco, California

Cell Differentiation and Neoplasia, edited by Grady F. Saunders. Raven Press, New York
© 1978.

Welcome Address

Walter G. Sterling

Board of Regents, The University of Texas System, Houston, Texas 77030

It is a privilege for me to have the opportunity to represent The University of Texas System Board of Regents here today. On behalf of the members of the board, I would like to extend a cordial welcome to the 30th Annual Symposium on Fundamental Cancer Research. I hope that you will find the scientific presentations informative and trust that you will return to your homes with a renewed sense of commitment to your important work.

For 30 years, The University of Texas System Cancer Center has sponsored an annual meeting to bring together scientists to discuss the crucial issues related to the fundamental problems associated with cancer. More that 700 scientists and physicians have assembled once again to discuss and debate these critical issues related to cancer research.

From the Symposium's small beginnings, the meeting has developed an international reputation and now attracts scientists from throughout the world. Only in 1970 was the Symposium not held. And that year, the Symposium was replaced by the Tenth International Cancer Congress held in Houston, and sponsored by the National Academy of Sciences and The University of Texas System Cancer Center.

I would like to extend my congratulations and best wishes to Dr. Beatrice Mintz of Philadelphia, who receives the 26th Annual Bertner Memorial Award for her contributions to cancer research. Dr. Mintz receives this award in recognition of her outstanding scientific career as a developmental biologist and geneticist. The award honors the late Dr. E. W. Bertner, who served as M. D. Anderson Hospital's first acting director and who was the first president of the Texas Medical Center in Houston.

I would also like to congratulate Dr. Bosco Wang, who receives the Wilson S. Stone Memorial Award. This award—presented to a student for outstanding achievement in the biomedical sciences—honors the late Dr. Wilson S. Stone. Dr. Stone, a brilliant geneticist, was, at the time of his death, vice chancellor of The University of Texas System.

Finally, let me express our gratitude and thanks to the National Cancer Institute and the Texas Division of the American Cancer Society. Both organizations—as our partners in the fight against cancer—have been long-time, enthusiastic cosponsors of this important scientific meeting.

On behalf of The University of Texas System, I commend each of you for your dedication to science and diligent efforts toward the control of cancer. Because the ultimate solution to the mysteries surrounding cancer will be found in the research laboratory, I pray the very best for each of you.

Cell Differentiation and Neoplasia, edited by Grady F. Saunders. Raven Press, New York

Introductory Remarks

Joseph T. Painter

Administration, The University of Texas System Cancer Center M. D. Anderson Hospital and Tumor Institute, Houston, Texas 77030

On behalf of Dr. R. Lee Clark, President of The University of Texas System Cancer Center, let me extend a personal welcome to each of you to the 30th Annual Symposium on Fundamental Cancer Research. The topic, "Cell Differentiation and Neoplasia," is an exciting and challenging one that promises to lead us to some of the answers we seek concerning what activities within the cell respond to the factors that may cause cell transformation, and how they respond, with the resulting modification of cellular activities and structure.

The first section of this year's Symposium, the stage of cellular differentiation in tumor cell origin, contains papers that review work with germ cells and with neoplastic cells of the neural, hematopoietic, and immunologic systems. The second section, control of differentiation and neoplasia, provides papers that review molecular mechanisms of gene expression in normal and neoplastic cells. Dr. Clark planned to review the status of knowledge regarding cell differentiation taken from previous M. D. Anderson symposia to illustrate the dramatic changes that have taken place as new technology has provided us with ever more discrete tools with which to develop theories and pursue their verification.

Eighteen years ago, during the 13th Annual Symposium in 1959, entitled *Genetics and Cancer,* a paper submitted by George and Eva Klein stated: "It is generally realized that malignancy is a composite phenomenon, including several different biological properties or, in the terminology of Foulds (1954, 1958), unit characters. Consideration of the natural history of various forms of cancer indicates that tumors often develop by a series of independent changes of several unit characters. This process, termed 'tumor progression,' is reminiscent of the development of certain microorganismal populations by a series of adaptive changes. In the view of Huxley (1956), 'all autonomous neoplasms can be regarded as the equivalents of new biological species.'

"[The field of tumor cytogenetics is in its embryonic stage at best. The] opinion has often been voiced that progression may have a classical mutation-selection basis. The situation . . . may be more complicated in reality. Unfortunately, differentiation and related phenomena are still among the most obscure in biology, but it is probable that they are not mutational in nature; i.e., are not based on random changes of structurally coded information transmitted from cell to cell in connection with division.

"Since somatic cells are presently unavailable for sexual recombination or other forms of genetic transfer, their straightforward genetic analysis is not possible. Indirect methods . . . may nevertheless yield certain information. In the particular case of tumor cells, the two main approaches hitherto employed can be registered as the study of phenotypic marker characteristics and the detailed examination of chromosome morphology."

In 1965, at the 19th Annual Symposium, entitled *Developmental and Metabolic Control Mechanisms and Neoplasia,* Dr. Clark stated in his introduction: "We may define the process of differentiation as the orderly interplay of developmental and metabolic control mechanisms and their resulting effects on cell form and function. We acknowledge that much is yet unknown. . . . Still less is known about their interrelationships whereby tissue and organ differentiation may be achieved." Dr. Clark also mentioned Erwin Chargaff's early studies of nucleic acids and the concept of base-pairing and referred to the Watson-Crick model for DNA structure (1953). "To signify this new awareness of the biologist that ultimately he may be dealing with biochemicals and their physical and chemical interrelationships, terms such as 'molecular biology' have been introduced.

"The chairman of our first session [Dr. Marshall Nirenberg] made the initial exciting discovery [of nucleic acid metabolism] involving 'poly-U.' . . . The classical enzymologists have done much to advance our ideas of metabolic control. The concepts of feedback control by metabolic end products [were] established. The study of hormone-enzyme interaction has led to the appreciation of still another facet, . . . one for which Jacob recently coined the phrase 'allosteric effects.' It appears likely that histones and other nuclear proteins serve some regulatory role, and we should learn more of this."

In a paper presented by Ulrich Clever during that 1965 Symposium, the following appeared: "The genetic information is now generally believed to be coded in the deoxyribonucleic acid (DNA) of the chromosomes. This information, then, is assumed to be transcribed into some special type of ribonucleic acid (RNA), the so-called messenger RNA (mRNA), and the base sequence of this RNA is finally translated into the amino acid sequence of a protein. While during the first years after Jacob and Monod (1961) had developed their suggestive model on regulation of protein synthesis in bacteria, many students of development liked to discuss their findings in terms of regulation of gene activity, recently a more critical consideration has set in. There is now some evidence that a stabilization of mRNA may occur under certain conditions of development, and whether protein synthesis may be regulated on the ribosomes is still a matter of discussion."

Also during that Symposium, Carl Frieden mentioned "three aspects of . . . enzymology which have come under intensive examination in the last few years . . . (1) the question of the kinetics of 'regulatory' or 'allosteric' (Monod, Changeaux, and Jacob, 1963) enzymes, (2) the relation between protein structure and enzyme activity, and (3) the role of these enzymes in metabolic control."

Rutter and Weber stated in their paper entitled "Specific Proteins in Cytodifferentiation," "In view of the evidence that differentiation is genetically regulated, analysis of the differentiation process may proceed fruitfully at the genetic level. At this time, however, many aspects of gene action are most readily investigated by changes in the levels of specific proteins."

Four years later, at the 23rd Annual Symposium in 1969, *Genetic Concepts and Neoplasia,* Dr. Clark said in his introduction: "The program this year will explore the current state of knowledge in four primary areas: (1) disorders predisposing to cancer that have a genetic etiology of either abnormal genes or chromosomes, (2) interactions at both molecular and cellular levels between the genetic apparatus and exogenous agents, namely, radiation, chemicals, and viruses, (3) the exploitable potential of studies in somatic cell genetics in relation to neoplastic growth, and (4) conceptual advances in molecular biology involving the nature of gene repair."

At this Symposium, Dr. Boris Ephrussi was honored in absentia for his work with somatic hybridization and was presented the Bertner Foundation Award. In his published manuscript, he stated: "[We cannot], at present, bypass the painstaking genetic analysis and seek the solution of the problems of cell differentiation and neoplasia by a direct attack at the molecular level. I believe that, because of the complexity of these interconnections [between a neoplastic cell and its normal ancestor], the mechanisms of differentiation and neoplasia are, and for some time will be, more accessible to genetic analysis than to analysis of any other type."

During that same Symposium, Dr. Weinstein stated: "The biochemical mechanisms underlying the normal control of gene expression during differentiation are poorly understood, but it is clear that they do not operate via somatic mutation (Gurdon, 1967). It is important, therefore, in considering mechanisms by which exogenous agents produce cancer, to keep our minds open to the possibility that they do so by changing the cytoplasmic environment of a cell in a way which induces an altered and self-sustaining pattern of gene expression."

These remarks are indicative of opinions and concepts that were prevalent prior to this year's Symposium. I can safely say that during the last five years, the scene has shifted rather dramatically.

In 1973, in the journal *Differentiation* (1:3–10, 1973), Paul Weiss stated: "[There is a] semantic duplicity of the word [differentiation], which may refer either to a state or to the process that has led up to it—an ambiguity [which] also [is] inherent in such terms as 'determination,' 'adaptation,' 'organization,' and so forth. Clearly, to recognize 'differentiation' as a process, advancing in time, makes the all-or-none description of cells as either 'differentiated' or 'undifferentiated' utterly inappropriate; . . . the complex of cells and the matrix in which they lie embedded in the body, represent a dynamic continuum, defying the notion that either of the two conjugated systems may be viewed or treated without regard to the other and their mutual interactive context. The cell . . . [is] simply an 'interactor' with its surroundings, including of course other cells and their products. Logically, therefore, no cell or cell group can be envisioned as 'differentiating' on its own; and the fact that its course always deviates from absolute stereotypy must be accounted for in terms of both the diversity of cells (e.g., mutations) and the variability of their environments (e.g., in modulation).

"We [must] recogniz[e] the inadequacy of a purely atomistic concept of piecemeal assembly of living systems and to acknowledge certain rules of order which define the patterns of collective interactions among the component units of the given system. And what do we actually know about such patterned collective group dynamics? Practically nothing.

"The crux of *modulation* is the reversibility of the observed excursions of a cell from a mean state, a fact which is in sharp contrast to the irreversibility of the progressive changes during development for which the term differentiation should be reserved. *'Differentiation,'* applied to cell strains, refers to their being in a process of unidirectional and irreversible transformation, which engenders true constitutional differentials among the emerging daughter generations; whereas, *'modulation'* merely signifies the range of variation and plasticity open to cells at any given level of that unidirectional course.

"Claims of de-differentiation were founded mostly on observations on the *regression* of microscopic features. However, if a 'regressing' cell has simply withdrawn its [identification] flag without losing its identity, this is just an ordinary modulation. . . .

"It is clear that purely visual aspects of 'regression,' whether microscopic or ultramicroscopic, are unfit to serve as diagnostic witnesses for or against irreversibility of cell character. A test of true de-differentiation would have to prove that a cell of a given type could, after regression, re-develop into any of the other cell types.

"Cytodifferentiation, that is, the translation of covert properties of an individual cell into characteristic overt form and behavior, is the simplest among the phenomena sheltered under the term 'differentiation.' The study of cytodifferentiation is yielding important data on the molecular basis of the covert-to-overt transactions in single cells. But since those conversions occur in cells which have already acquired type-specificity, they are more in the nature of modulations, hence inadequate as potential models for strain differentiation.

"It is self-evident that strain differentiation can be tested only in organisms with *somatic cell strains* of different type-specific characters. This qualification immediately disqualifies direct generalizations about strain differentiation from studies confined to microbes (prokaryotes), whose somatic and reproductive cells are one and the same.

"But having at our disposal now the magnificent advances of our knowledge of molecular biology; the great enrichment of our technical instrumentarium; the powerful method of cell hybridization; the growing awareness of the need to supplement the wealth of information about the biochemical components and processes in the cell content, . . . we can certainly expect to bring the operation [of the progressive divergent differentiation of the cell population of the embryo] . . . to light."

Present investigational model systems include one-to-eight cell mammalian embryos, spermatogenesis, the chick oviduct system, somatic cell hybridization, cultured hemopoietic (marrow) cells, and various other cultured normal and neoplastic mammalian cells used to observe and measure the synthesis of proteins in the nucleus and in the cytoplasm as the cells begin to differentiate, or if they transform from the normal to the malignant state. More and more, the genes that are believed to be responsible for the initiation and regulation of the highly discrete and specialized proteins that characterize differentiated cells are being located on chromosomes and are being histochemically manipulated to increase our knowledge of normal and abnormal mechanisms. The discoveries and theories discussed during this Symposium appear to be of special relevance to our attempts to understand the mechanisms of carcinogenesis and to design less empirical therapeutic control over the malignant processes.

We wish to acknowledge the successful efforts of our Symposium chairman, Dr. Grady Saunders, and his committee to bring from around the world scientists who are working in this new and rapidly expanding field of biology. We are grateful that the Board of Regents of The University of Texas has once again given us the opportunity to convene such an august group and that our invitations were so willingly accepted.

The cosponsors of this Symposium are The University of Texas M. D. Anderson Hospital and Tumor Institute, the National Cancer Institute, the American Cancer Society, Texas Division, Inc., and The University of Texas Health Science Center, Graduate School of Biomedical Sciences. We are especially grateful to them for their continuing support.

Cell Differentiation and Neoplasia, edited by Grady F. Saunders. Raven Press, New York

Cancer: The Survival of the Fittest*

Clement L. Markert

Department of Biology, Yale University, New Haven, Connecticut 06520

The title of this keynote address, "Cancer: The survival of the fittest," was chosen deliberately to emphasize that cancer cells represent biological successes, not failures, and to suggest that cancer cells might usefully be viewed as independent organisms living and multiplying in the various ecological niches found in the host organism. Too often, we adopt a rather subjective un-biological view of cancer. We tend to regard cancers as disordered, defective, diseased growths that threaten to kill us. The malignant nature of the cancer attack is clear enough, but the cancer cells themselves are efficient, well-organized, tough-living systems. They are versatile and adaptive, and very difficult to subdue or to eradicate from any host. We know their clinical consequences, but their origin and true nature still remain obscure.

In the preface to a recent comprehensive treatise on cancer, the editor, F. F. Becker (1975), correctly points out that despite all our experience with cancer and the truly vast amount of information we have accumulated, we still are unable to define the malignant cell. I believe our troubles in defining cancer or, more generally, neoplasia stem from our clinical obsession with cancer. We tend to view cancer as a disease akin to infectious diseases, rather than as an inherent consequence of our cellular organization, more akin to such phenomena as mutation or aging. The organization of metazoan life is not perfect, and one of the inherent imperfections is cancer. That doesn't mean we are helpless in the face of this malady. Not at all. The better we understand the origin, nature, and behavior of the innumerable kinds of neoplasms, the better we can protect ourselves.

I believe that a keynote address for a symposium on *Cell Differentiation and Neoplasia* should aim to present a unifying theme, clearly and simply if possible, to serve as a focus for attack and for defense from the other symposium participants. Thus motivated, I present the following definition:

Neoplasias are diseases of cell differentiation stemming from a mis-programming of normal gene function. This mis-programming can have many causes, but all cancer must act by changing the function of regulatory DNA.

This dogmatic definition at least provides a stable platform from which to examine the origin, development, and characteristics of neoplastic cells, as well

* Keynote Address

as the interactions between neoplastic cells and the host organism. Evaluations of various programs for the prevention or cure of malignant neoplasms may also be sharpened by this definition. At the outset, I should admit that the point of view I have adopted is general, flexible, and inclusive enough to explain virtually everything about cancer (and therefore to explain almost nothing that is helpful). Moreover, this definition is somewhat discouraging because it leads us to conclude that simple, generally applicable therapies seem most unlikely.

DISTINCTIONS AMONG CELLS

Neoplastic cells are, from the point of view of developmental biology, just special kinds of differentiated cells carrying on normal metabolic activities, albeit in unique patterns not found in any single normal cell. Describing and understanding any particular kind of neoplastic cell, whether benign or malignant, presents just the same problems that must be solved in describing and understanding any particular normal cell. Every kind of cell, normal or abnormal, is unique, but the characteristics of each of them are produced by the same basic cellular mechanisms. *Abnormal* is, after all, only a statistical concept with little relevance to understanding cell biology.

In practice, how do we distinguish cells from one another, normal from normal, or normal from neoplastic? Our present knowledge of molecular biology and molecular genetics persuades us that the genetic information in each of our cells is virtually the same. Cellular differentiation is based on a selective activation of only a very small part of the genome for each type of cell. The vast majority of the genes remain silent in each cell. Thus, from one cell genotype we obtain many different cell phenotypes by the process of differential gene activation. This leads to a specific protein or enzymatic makeup for each kind of cell. For example, we distinguish liver from muscle cells, pigment cells from pancreas cells, or mammary gland cells from adrenal gland cells by their specific enzymatic composition and by their distinctive metabolic contributions to the overall economy of the organism. Liver cells produce many enzymes and proteins not found in muscle cells, but muscle cells produce actin and myosin; pigment cells, tyrosinase; and pancreas cells, insulin. Unique cell products are dramatic evidence of cell differentiation, but most differences among cells are merely quantitative, though no less significant for cell structure and function.

The numerous biochemical differences distinguishing one differentiated cell from another can all be traced back to different repertories of functioning genes or to genes functioning to different quantitative extents. With regard to quantitative variation let me cite a single example from work in my own laboratory. An ubiquitous enzyme present in every vertebrate cell is lactate dehydrogenase, LDH. Each kind of differentiated cell has its own characteristic amount of this enzyme expressed in a characteristic pattern of five isozymes. This pattern is achieved by a random combination in tetramers of two different kinds of polypeptide subunits, designated A and B, each of which is encoded in a separate

TABLE 1. *Subunits of LDH per cell in adult mouse tissues*

	Total subunits	A subunits	B subunits
Heart	1.13×10^8	2.56×10^7	8.71×10^7
Muscle	3.60×10^8	3.34×10^8	2.56×10^7
Liver	3.94×10^7	3.78×10^7	1.60×10^6
Kidney	3.91×10^7	1.11×10^7	2.80×10^7

gene. The synthetic activity of these two genes as expressed in corresponding numbers of polypeptide subunits can vary greatly and is characteristic for each kind of cell (Table 1). Differential synthetic activity, therefore, determines the proportion of A and B subunits and hence the pattern of relative abundance of the five LDH isozymes. However, all five isozymes are degraded at the same rate in each cell—a rate that is characteristic for that cell type. The relative rates of synthesis and degradation must specify the total amount of enzyme for each cell. For example, as reported (Table 2) by Nadal-Ginard (1975), liver cells of the mouse synthesize 71 A subunits per cell per second but only 3 B subunits. Heart muscle, on the other hand, synthesizes only 25 A subunits per cell per second but 85 B subunits. Thus, the quantitative activity of each gene is cell specific. The half-life of skeletal muscle LDH is 43 days, but in liver it is only 4.3 days, one-tenth as long. Thus, muscle cells contain about 10 times as much LDH as do liver cells (see Table 1).

Evidently, some molecular mechanism must exist in each cell, normal or abnormal, benign or malignant, not only for activating or silencing structural genes but also for regulating the quantitative degree of their expression. Ultimately, it is the pattern of gene activity that defines each kind of differentiated cell and, I assert, also defines each kind of neoplastic cell. Neoplastic cells make no molecules not found also in some normal cell, adult or embryonic. The proteins encoded in the genome of oncogenic viruses and present in infected neoplastic cells may be regarded as an exception to this rule, but, in fact, they do not disturb the general hypothesis of neoplasia as set forth here. I shall take up the mechanisms of viral oncogenesis later in this report.

TABLE 2. *Half-lives, degradation coefficients, and rate of synthesis of LDH in different mouse tissues*

	Half-life of LDH in days	Fraction of LDH tetramers degraded per second	LDH molecules synthesized per cell per second: Total subunits	A subunits	B subunits
Heart	8.18	0.084	110	25	85
Muscle	43.0	0.016	67	62	5
Liver	4.3	0.161	74	71	3
Kidney	6.1	0.113	52	15	37

CHARACTERISTICS OF NEOPLASMS

If neoplastic cells do not synthesize any new kinds of molecules nor establish any unique metabolic pathways, then how do we distinguish them from normal cells? Obviously by their behavior in the organism and by their own characteristic structural and functional organization—an organization that is based entirely on cell properties, each of which is found in some normal cell. The organization is abnormal, but the basic units of the organization are individually all normal. Moreover, the deviation in metabolic or cellular organization from the normal pattern can be very small or so gross as to make impossible the identification of the normal cell progenitor of the neoplastic cell.

This finely graded spectrum of deviations from normal that characterizes the gamut of neoplastic cells could be achieved by a corresponding degree of reprogramming gene function, either to activate dormant genes, repress active genes, or by changing the relative amount of transcription from functioning genes. Such changes should be adequate to generate all of the neoplastic patterns of metabolism so far described. Moreover, so I hypothesize, all such changes in the function of structural genes would be generated by prior changes in the regulatory DNA. The changes in the regulatory DNA could involve alterations of base sequence but, more likely, would involve combination with regulating proteins to control transcription. By this view the elementary events in initiating neoplasia would be generically the same as the events responsible for producing normal cellular differentiation. Unfortunately, we cannot describe the process of normal cell differentiation in molecular terms, so we can scarcely expect at this time to provide a molecular description of neoplastic differentiation. Although an inevitable consequence of the multicellular organization of life, neoplastic transformation is a capricious, stochastic aberration that may be escaped, circumvented, neutralized, or resisted successfully, in any particular case. But there is no hope for abolishing its occurrence.

Let me emphasize again that neoplastic differentiation is in terms of mechanism *not* abnormal. We might analogize the genes of a cell to the musical instruments in a symphony orchestra. The same orchestra can play Beethoven's Fifth Symphony or a dissonant piece by Schönberg—the latter a neoplastic development, benign or malignant, according to your musical taste. Likewise, the same genes, by differential function, can produce any cell in an organism—normal, benign, or malignant.

So far we have not dealt with the detailed ways in which neoplastic cells may differ from normal cells. The variety of these differences is enormous and is doubtless much greater than the variety of differences found among normal cells. The magnitude of such variety precludes detailed or comprehensive description, but some useful generalizations can still be made. Let us consider cell division, cell mobility, and cell metabolism—all of which may be involved in neoplastic transformation (cf. Markert 1968).

Cell Division

All neoplastic cells must divide in order to amplify their effects to a level that can be recognized. A single, nondividing neoplastic cell, even of the most malignant type, would pass unnoticed. Perhaps we all harbor many such cells. Thus, in a practical clinical sense, cell division is an indispensable attribute of neoplastic cells, but the rates of cell division among neoplastic cells covers the same spectrum of rates found among normal cells. Some human tumors double their mass in less than two weeks; others require more than a year to double. The median period is approximately two months. Animal tumors generally grow much more rapidly (Steel 1973) and so do human embryonic cells.

Metastasis

Those neoplasms we commonly describe as malignant are characterized by their ability to invade adjacent tissues and to migrate via the lymphatics and blood vessels to other sites in the body, there to establish new foci for tumor development. Such mobility per se is not biologically abnormal. During embryonic life, many kinds of cells migrate or "invade" adjacent tissues. In fact, such behavior is essential to normal development. In the adult, cell migration is not common, but leukocytes, for example, must invade many tissues in discharging their normal biological role. Wound healing requires cell migration, and restricted local movements of cells also occur continuously in the epithelial lining of the gut, the formation of skin epidermis, and, of course, in the uterus during pregnancy. Invasion and migration, like cell division, are not abnormal biological properties of cells. They become abnormal only when they occur in certain kinds of cells at particular times and places in the organism.

Metabolism

To some degree, almost by definition, neoplastic cells must differ metabolically from their normal progenitors. Yet these differences range in magnitude from the undetectably small to very large indeed. Differentiated metabolic characteristics, glandular function, for example, may be retained or lost, diminished or greatly enhanced, in different neoplasms even of the same general type. New patterns of metabolic activity compounded of bits and pieces found in a variety of normal cells (but not in any one of them) may be generated during neoplastic development (Uriel 1975, Odell and Wolfsen 1975).

It was long ago recognized (Greenstein 1945) that the enzymatic activities of tumor tissues resemble one another even though the enzymatic patterns of their normal cell counterparts differ greatly. More recently, Schapira *et al.* (1963) recognized that embryonic isozyme patterns may reappear in tumor tissues even though these patterns differ quantitatively and even qualitatively from those found in the adult tissue of tumor origin. Many isozyme systems have now been shown to behave this way (Criss 1971, Schapira 1973, Weinhouse 1973). Not only isozymes but other embryonic gene products such as alpha-fetoprotein (Abelev *et al.* 1963) and a wide variety of carcinoembryonic antigens are produced by neoplastic cells. This resurgence of embryonic metabolic activity clearly represents a reactivation of genes that were repressed in the adult tissues. These are interesting and important observations, but the significance of the embryonic nature of the gene products may only reflect the similarities in lifestyle between embryonic and neoplastic cells—rapid growth, migration, etc.—and may not be of any help in uncovering the nature of the neoplastic process.

Perhaps more instructive is the fact that many tumors, even though not derived from endocrine glands, secrete peptide or protein hormones (Odell and Wolfsen 1975). These gene products were never part of the metabolic history of the normal cells that gave rise to these tumors. These hormones represent the activation of genes that never functioned in the antecedent normal cell line. Their synthesis probably confers no metabolic advantage on the tumor cell, but their presence demonstrates that new programs of gene function do arise during neoplastic transformation. Some of these programs must confer advantages on the tumor cell if it is to multiply and spread throughout the host. These new patterns of metabolism in tumors have great clinical significance and are the target of many therapies. But the individual components of these novel metabolic patterns are all normal. Only the pattern is abnormal.

ECOLOGICAL LIFE HISTORY OF NEOPLASMS: SURVIVAL OF THE FITTEST

From the foregoing analysis, one might conclude that neoplastic cells, including the malignant subset, are healthy cells, and indeed many of them are. Only their presence in our bodies is regarded as abnormal, and justifiably so. Nevertheless, these cancer cells represent the "survival of the fittest." They out-compete our normal cells for ecological space, for the requirements for survival and reproduction. Their metabolic versatility exceeds that of all normal cells, and, consequently, they can flourish in many cellular environments, each of which would be fatal to nearly all normal cells—all but those few (perhaps only one type) adapted to that particular environment.

Let us trace out the origin and evolutionary progression of a population of neoplastic cells that become malignant and kill the host. We will skip over the details of the initial carcinogenic event for now and begin with a single neoplastic cell. It arose from a cell that could divide (possessed the metabolic machinery for cell division) but was normally constrained in its division by associated cells. Cancer cells are usually said to exhibit uncontrolled growth. There is some truth in this assertion, but a more accurate statement would be that the newly arisen cancer cell can divide under conditions that would block cell division in its normal counterpart. Requirements for neoplastic cell division are not so fastidious, but they are extensive, complicated, and probably different for each kind of neoplastic cell. Nondividing normal cells may be inhibited from dividing by external or internal factors or may lack an internal stimulus, but in either case, cell division may be initiated by internal metabolic activity newly generated during neoplastic transformation. This could be achieved either by repressing or activating the appropriate genes. Probably most of the cells that are pushed onto the neoplastic pathway by a reprogramming of their gene function soon falter and die. Everything must be just right for success, external to the cell and internally as well. The rarity of success suggests that a precise sequence of *multiple* changes is needed to initiate successful, continued multiplication. Once a neoplastic population has developed, however, the restrictions and barriers begin to crumble.

The cells of the neoplastic population by their own metabolic activity modify their environment in self-serving ways. The hostility of the surrounding environment of normal cells is diminished or at least pushed farther away. The new patterns of metabolic activity within the neoplastic cells also change the molecular environment of their chromosomes during replication, leading to new programs of gene activation and repression. Similar events can be produced by placing normal cells in culture where they grow and divide under what is for them very abnormal conditions. Many cells in culture do become neoplastic or are "transformed" (Gey 1955, Sanford *et al.* 1950).

Most newly generated patterns of metabolism are surely deleterious to the incipient neoplastic cells. Most die; but occasionally one of the neoplastic variants survives, and its descendants become even better fitted to grow, multiply, and to invade adjacent tissues. Acquiring invasive properties means that the surface membranes must be altered to allow the neoplastic cell to escape from the constraining influences of adjacent normal cells. Now the neoplastic cell is truly malignant. It continues to multiply, usually more rapidly, and descendant cells migrate to other sites in the body where new environments are encountered. These environments pose different problems, jeopardizing survival and multiplication. Again, most of the cancer cells must prove inadequate to the new challenges and die or cease to divide and disappear clinically. But a constant interaction is occurring between the cancer cell and its new environments, which generates and selects for new patterns of gene function in the cancer population. By a kind of natural selection, the best adapted cancer cells survive, multiply, and migrate to still new environments where they encounter new differentiating influences, so that yet new types of cancer cells arise (Cairns 1975a). This process continues until the host dies and with it all cellular environments—a kind of self-imposed ecological extinction analogous to what is faced by any species that destroys and exhausts its environment. (The human species, with respect to the world we inhabit today, is perhaps currently in a position analogous to that of an early cancer cell population in its aggressive, destructive extension throughout the host.)

Why is the cancer fatal to the host? There is no malice involved here. Cancer cells are just trying to "make a living," like any other organism. They are blind to their long-range interests as a species (as *we* are) and only do what is possible and advantageous at the moment—metabolize, change, multiply, and colonize new territory. It's an old story. The central process of evolution: inherited variation, natural selection, and survival of the fittest, at least until the resources for living are exhausted.

The host may be killed by starvation, the malignant cells consuming too much of the available nutrients, perhaps exhausting a single essential amino acid. The host may die from mechanical damage to a critical organ because vital space is occupied by the burgeoning cancer cell population or because essential tissue or organ integrity is destroyed by the invading cells. Blood supplies may be reduced or vessels ruptured to produce fatal hemorrhages. The immune system may be so damaged that accidental infection by exogenous organisms leads to death of the host. Or the by-products of cancer metabolic activity may prove lethal to the host. Endocrine tumors such as insulinomas, thyroid tumors, or pheochromocytomas all can kill by deranging the metabolism of the host by secreting excessive amounts of normal hormones.

ORIGIN OF NEOPLASMS

Since cancer cells do seem to represent the survival of the fittest in a population of cells, why is neoplastic transformation apparently so rare, particularly in its malignant form? From the point of view of the individual human being, it is of course not so "rare." Unless there is a change for the better, perhaps one fourth of us present at this Symposium are destined to die from cancer, and, of course, everyone develops numerous benign neoplasms, most of which are scarcely noticed. However, viewed as an event in a population of cells, recognizable neoplasia is in fact extraordinarily rare. Only a few cells out of the several trillion that make up our bodies ever exhibit neoplastic transformation of any sort, and none of the cells in most individuals ever produce clinically recognized cancer cell populations.

Thus, in view of the large number of normal cells in a human being, malignant transformation of a cell must be very difficult indeed—at least, it is very rare. And no wonder. We as human beings are the product of millions of years of evolution, not just as representatives of the human species but as representatives of complex, long-lived metazoans. This long evolution occurred in the face of the potential for fatal neoplastic transformation of the normal cell. Our evolutionary success testifies that either the essential features of development and cellular organization virtually preclude successful neoplastic deviation, or, as some believe, a special organ system was developed, the immune system, for monitoring and policing host cells and destroying all deviants (Burnet 1969, Nossal 1974). Either arrangement seems possible, but I prefer to place my bets on the former as the principal line of defense. I doubt that the immune system is very important in combating cancer.

Multiple Steps to Malignancy: Genetic and Epigenetic

Metazoans can avoid neoplasias or reduce their frequency or delay their onset by achieving such a tight, interdependent cellular organization that almost every single aberrant step by any cell would prove fatal to that cell, or at least prevent proliferation. I believe that this is what has occurred. Each type of normal cell at every stage in development from the early embryo to the adult is highly dependent upon other cells or, to use evolutionary terminology, perfectly adapted to the society of cells of which it is a part. Moreover, during embryonic life each cell must change and differentiate in perfect harmony with the developing cell community. Any single deviation, getting out of step, is likely to make the cell less adapted and therefore to prove fatal. In fact, cell death is very common throughout our lives and especially during embryonic development (Saunders 1966).

However, since cancer cells do exist, we know they are biologically possible. The significant fact is that they differ in *multiple* ways from their normal progenitors. The simultaneous occurrence of these multiple changes renders the neoplastic transformation very improbable, and each of the required changes by itself, particularly out of order, is likely to prove damaging to the cell. The course of neoplastic development is, in fact, now generally recognized to be a multistep process. This tumor progression (Foulds 1949 a, b, 1954, 1964, 1969, 1975) following an initiating event involves many changes in the nature of the cell. Some of these may be genetic, that is, involve changes in base sequence in the DNA or gross changes in chromosomal structure or ploidy. However, essential changes during progression may be epigenetic and act to alter the regulation of gene function rather than to change the structure of the genes. Be that as it may, the multiple events required for neoplastic transformation account for its rarity and for the long period of time required to accumulate several changes within a single cell. In fact, most neoplastic development occurs late in life during the post-reproductive years when its occurrence escapes the evolutionary screen. Graphical representation of the incidence of various kinds of cancer with age show that the logarithm of the incidence is linearly related to the logarithm of age. This relationship suggests that several events must accumulate in a cell before it crosses a threshold into neoplasia. The number of events required varies with the kind of cancer but ranges from two to about seven. These events, at least some of them, may be mutations, and the genetic etiology of certain specific cancers such as retinoblastoma (Knudson 1975) and colon carcinoma (Ashley 1969) are clearly in accord with the multiple "hit" hypothesis.

CARCINOGENIC AGENTS

This brings us back to the problem of the definition of cancer and to the elementary causes of the malignant transformation. Recalling my dogmatic definition, I readily acknowledge that this definition cannot now be proved to be true, but can it be shown to be false or to fail to accommodate any important well-documented fact about cancers?

Mutation

Somatic mutations have often been invoked as a cause of cancer. And, in fact, single mutations behaving according to Mendel's laws are known to be involved in the etiology of a few cancers, and multiple genetic loci are clearly involved in many more, particularly in those cancers studied in experimental animals. However, none of these mutated loci has been identified by an abnormal protein product, and I doubt that any will be. Probably all of these mutations are in the regulatory DNA and act to alter the pattern of gene activation either qualitatively or quantitatively. It is especially noteworthy that these mutant "genes," although present in every cell of the body, encourage neoplastic transformation in only a single cell type or, in some cases, in a few cell types. I say "encourage" because nearly all of the cells in any affected tissue remain indifferent to the mutation. Only one out of a million or so cells at risk is ever transformed to the neoplastic type. Of course, if a large enough population is at risk, then malignant transformation of at least one cell becomes a virtual certainty.

It must be acknowledged that a change in a structural gene could lead to altered metabolic activity in a cell which might in turn divert that cell at a particular stage in differentiation onto a new pathway leading to malignancy. Such a feedback from cytoplasmic activity to gene activation and repression must occur during normal cell differentiation, but the programming of these interactions is normally very precise. Signals at the wrong time and place are ignored or are injurious. Perhaps an example of altered patterns of metabolic activity (quantitative not qualitative) leading to a higher frequency of neoplastic differentiation is provided by the association of trisomy 21 with leukemia. Individuals with three chromosomes number 21, instead of the normal two, exhibit a complex syndrome of abnormalities. The genetic information in their genomes is qualitatively normal; they simply have too much of the genetic information contained in chromosome 21. One consequence is a higher incidence of leukemia. This result may be attributed to abnormal patterns of metabolism in trisomy 21 cells—metabolic patterns that are more readily converted to neoplastic patterns.

It seems obvious, to me at least, that structural changes in the regulatory DNA could easily lead to a misprogramming of gene function during cellular differentiation. One well-recognized category of mutational or chromosomal change reinforces this belief. These are the mutants affecting the capacity of cells to repair lesions in the DNA. Xeroderma pigmentosum is the most prominent example (Setlow *et al.* 1969). Afflicted individuals are deficient in the ability to repair chromosomal lesions induced by ultraviolet light. Consequently, they develop multiple neoplastic growths on their skins. Such UV-induced damage would mostly occur in the regulatory DNA (because it is the most abundant) and could easily lead to misprogramming of gene function in many cells to produce many tumors. Moreover, only those cells that had reached that stage of differentiation requiring appropriate function of the mutated regulatory DNA would be affected. All other cells would be indifferent, even those cells of the target cell line before and after they reached the sensitive stage of differentiation. Only at a critical stage in differentiation would their defective regulatory DNA make them vulnerable. Mutant genes that enhance the frequency of neoplastic transformation do indeed exhibit an extreme specificity in the cell types affected.

Epigenetic Events: Carcinogenic Chemicals and Physiological Stress

Chemical carcinogens, of which there are many in our environment (cf. Cairns 1975b), could likewise affect the regulatory DNA either by mutating it or by combining with it. Investigators have long recognized at least two steps in chemical carcinogenesis: initiation and promotion (Foulds 1949 a, b, Berenblum 1954). Initiation may be a mutation—at least it seems to be a permanent change in the heredity of the cell. Promotion is brought about by chemicals that stimulate cell division and elicit the neoplastic phenotype, perhaps by chemical combination with regulatory DNA (cf. Heidelberger 1975). The combination could be transitory and still be effective by irreversibly diverting cells into a neoplastic pathway at a sensitive stage in their differentiation. It is well known in developmental biology that transient inductive stimuli have permanent effects on cell differentiation long after the initiating stimulus has disappeared. Carcinogens and imposed changes in metabolic activity (as, for example, by intensive prolonged hormonal stimulation) could act this way in transforming cells.

Oncogenic Viruses

And then there are viruses—DNA viruses and RNA viruses. Representatives of both types have been shown to cause cancers, although the great majority of cancers reveal no evidence of infectious viruses, even after the most diligent search. The simplest interpretation of the mode of action of oncogenic viruses, in accord with the definition of neoplasia that I offered, begins with the fact that viral DNA may be incorporated into the host cell genome (Dulbecco 1976). Certain families of RNA viruses replicate through a DNA intermediate, and others do not. Only those producing a DNA intermediate (and not all of them) can cause cancer. None of those RNA viruses that replicate through an RNA intermediate are known to be oncogenic. Among the DNA viruses, several are known to cause cancer but at a much lower frequency than the most oncogenic RNA viruses (cf. Temin 1977, 1974a, 1974b). The differences in oncogenicity among RNA tumor viruses and between them and DNA viruses apparently is related to the site specificity of integration into the host cell genome (Battula and Temin 1977). Lack of specificity in site integration would result in weak transforming ability because only rarely would the viral DNA be inserted at just that site in the genome that would so disturb the regulatory DNA as to generate a neoplastic program of gene function. At most sites in the genome, integrated viral DNA would be ineffective or at least not oncogenic. However, if the site for integration was highly specific and also led to oncogenesis, then the virus would be oncogenic nearly 100% of the time. This is true for the most strongly transforming RNA tumor viruses. In any event, the role of viruses in inducing neoplastic transformation is basically like that of other neoplastic agents—the regulatory DNA of the host cell genome must be altered so as to reprogram gene function in a neoplastic pattern.

CONCLUSIONS

Once cells have become fully neoplastic, must the condition be permanent? In principle, no more so or less so than for any differentiated cell with its genome intact. The external environmental stimuli required to transform one fully differentiated normal cell into another must be very complex, and such an event rarely occurs. However, there are examples (Braun 1974, 1975). Regeneration of the lens in the eye of a salamander from the pigmented retina is a clear example that the differentiated state can be changed (Yamada and McDevitt 1974, Yamada 1976). The most dramatic reversal of malignant cells to normal states of differentiation has been achieved by subjecting teratocarcinoma cells to embryonic environments (Mintz and Illmensee 1975). When injected into the mouse blastocyst, these teratocarcinoma cells multiplied and contributed cells to many tissues and organs of the developing mouse (Illmensee and Mintz 1976). Such chimeric mice have developed to term, reached adulthood, and even reproduced, with gametes derived from the originally implanted teratocarcinoma cells.

I have spoken at length about misprogrammed gene activity leading to neoplasia, but I should like to emphasize that most cases of misprogramming probably are fatal to the cell, or the cell does not divide and is, therefore, of no concern to the host organism. Moreover, neoplastic programs of gene function are probably only a very special subset of a larger generic set of cell aberrations that result from faulty regulatory DNA. I will hazard the suggestion that inherited teratogenic development, congenital abnormalities, are a nonmalignant expression of the same general kind of misprogrammed gene function stemming from changes in the structure or function of regulatory DNA at critical stages in embryonic development.

In summary, the difficulty with defining and understanding neoplastic change is the same difficulty that developmental geneticists have in understanding normal cell differentiation. When we understand one, we shall surely understand the other. No one can predict whether progress will be more rapid with a focus on normal cells or on abnormal cells. I believe it is all the same problem—perhaps the most important practical and theoretical problem facing us today in the immense area of cell biology. This Symposium, with its galaxy of speakers and participants, promises to lead us at least a short distance along the path toward the complete molecular description of cell differentiation, both normal and neoplastic. Even at the end of that long path, I doubt that we shall be able to prevent all neoplasia, but perhaps we can master and control it and save ourselves from much of the misery of cancer.

ACKNOWLEDGMENT

The preparation of this review was assisted by support from Grant R01 HD07741–03 from the National Institutes of Health.

REFERENCES

Abelev, G. I., S. D. Perova, N. I. Khramkova, Z. A. Postnikova, and I. S. Irlin. 1963. Production of embryonal α-globulin by transplantable mouse hepatomas. Transplantation 1:174–180.

Ashley, D. J. B. 1969. The two "hit" and multiple "hit" theories of carcinogenesis. Br. J. Cancer 23:313–328.

Battula, N., and H. M. Temin. 1977. A single site for integration of infectious spleen necrosis virus DNA in the DNA of chronically infected chicken fibroblasts. Proc. Natl. Acad. Sci. USA 74:281–285.

Becker, F. F., ed. 1975–1977. Cancer, A Comprehensive Treatise. Vols. 1–4. Plenum Press, New York.

Berenblum, I. 1954. A speculative review; the probable nature of promoting action and its significance in the understanding of the mechanism of carcinogenesis. Cancer Res. 14:471–477.

Braun, A. C. 1974. The Biology of Cancer. Addison-Wesley, Reading, Mass. 169 pp.

Braun, A. C. 1975. Differentiation and dedifferentiation, *in* Cancer, A Comprehensive Treatise, F. F. Becker, ed. Vol. 3. Biology of Tumors: Cellular Biology and Growth. Plenum Press, New York, pp. 3–20.

Burnet, M. 1969. Self and Not-Self. Melbourne University Press, Melbourne, 318 pp.

Cairns, J. 1975a. Mutation selection and the natural history of cancer. Nature 255:197–200.

Cairns, J. 1975b. The cancer problem. Sci. Am. 233:64–78.

Criss, W. E. 1971. A review of isozymes in cancer. Cancer Res. 31:1523–1542.

Dulbecco, R. 1976. From the molecular biology of oncogenic DNA viruses to cancer. Science 192:437–440.

Foulds, L. 1949a. Mammary tumors in hybrid mice: Hormone responses of transplanted tumors. Br. J. Cancer 3:240–248.

Foulds, L. 1949b. Mammary tumors in hybrid mice: Growth and progression of spontaneous tumors. Br. J. Cancer 3:345–375.

Foulds, L. 1954. Tumor progression: A review. Cancer Res. 14:327–339.

Foulds, L. 1964. Tumor progression and neoplastic development, *in* Cellular Control Mechanisms and Cancer, P. Emmelot and O. Mudlbock, eds., Elsevier, Amsterdam, pp. 242–258.

Foulds, L. 1969. Neoplastic Development. Vol. I. Academic Press, New York, 439 pp.

Foulds, L. 1975. Neoplastic Development. Vol. II. Academic Press, London, 729 pp.

Gey, G. O. 1955. Some aspects of the constitution and behavior of normal and malignant cells maintained in continuous culture, in Harvey Lectures No. 50, Academic Press, New York, pp. 154–229.

Greenstein, J. P. 1945. Enzymes in normal and neoplastic tissues, *in* AAAS Research Conference on Cancer, F. R. Moulton, ed., AAAS, Washington, D.C., pp. 191–215.

Heidelberger, C. 1975. Chemical carcinogenesis. Annu. Rev. Biochem. 44:79–121.

Illmensee, K., and B. Mintz 1976. Totipotency and normal differentiation of single teratocarcinoma cells cloned by injection into blastocysts. Proc. Natl. Acad. Sci. USA 73:549–553.

Knudson, A. G., Jr. 1975. Symposium No. 18: Oncogenetics. Genetics of human cancer. Genetics 79:305–316.

Markert, C. L. 1968. Neoplasia: A disease of cell differentiation. Cancer Res. 28:1908–1914.

Mintz, B., and K. Illmensee. 1975. Normal genetically mosaic mice produced from maligant teratocarcinoma cells. Proc. Natl. Acad. Sci. USA 72:3585–3589.

Nadal-Ginard, B. 1975. Post-transcriptional processes in the production of cell phenotypes: Quantitative regulation of the lactate dehydrogenase genes. Ph.D. Thesis. Yale University.

Nossal, G. J. V. 1974. Lymphocyte differentiation and immune surveillance against cancer, *in* Developmental Aspects of Carcinogenesis. 32nd Symposium of the Society for Developmental Biology, T. J. King, ed., Academic Press, Inc., New York, pp. 205–213.

Odell, D. O., and A. Wolfsen. 1975. Ectopic hormone secretion by tumors, *in* Cancer, A Comprehensive Treatise, F. F. Becker, ed. Vol. 3. Biology of Tumors: Cellular Biology and Growth, Plenum Press, New York, pp. 81–97.

Sanford, K. K., W. R. Earle, E. Shelton, E. L. Schilling, E. M. Duchesne, G. D. Likely, and M. M. Becker. 1950. Production of malignancy in vitro. XII. Further transformations of mouse fibroblasts to sarcomatous cells. J. Natl. Cancer Inst. 11:351–375.

Saunders, J. W., Jr. 1966. Death in embryonic systems. Science 154:604–612.

Schapira, F. 1973. Isozymes and cancer. Adv. Cancer Res. 18:77–153.

Schapira, F., J. C. Dreyfus, and G. Schapira. 1963. Aldolase in primary liver cancer. Nature 200:995–997.

Setlow, R. B., J. D. Regan, J. German, and W. L. Carrier. 1969. Evidence that xeroderma pigmentosum cells do not perform the first step in the repair of ultraviolet damage to their DNA. Proc. Natl. Acad. Sci. USA 64:1035–1041.

Steel, G. G. 1973. Cytokinetics of neoplasia, *in* Cancer Medicine, J. F. Holland and E. Frei, eds. Lea and Febiger, Philadelphia, pp. 125–140.

Temin, H. M. 1974a. On the origin of the genes for neoplasia: G. H. A. Clowes Memorial Lecture. Cancer Res. 34:2835–2841.

Temin, H. M. 1974b. On the origin of RNA tumor viruses. Annu. Rev. Genet. 8:155–177.

Temin, H. M. 1977. The relationship of tumor virology to an understanding of nonviral cancers. BioScience 27:170–176.

Uriel, J. 1975. Fetal characteristics of cancer, *in* Cancer, A Comprehensive Treatise, F. F. Becker, ed. Vol. 3. Biology of Tumors: Cellular Biology and Growth. Plenum Press, New York, pp. 21–55.

Weinhouse, S. 1973. Metabolism and isozyme alterations in experimental hepatomas. Fed. Proc. 32:2162–2167.

Yamada, T., and D. S. McDevitt. 1974. Direct evidence for transformation of differentiated iris epithelial cells into lens cells. Dev. Biol. 38:104–118.

Yamada, T. 1976. Dedifferentiation associated with cell-type conversion in the newt lens regenerating system: A review, *in* Progress in Differentiation Research, N. Muller-Berat *et al.*, eds., North-Holland Publishing Co., Amsterdam, pp. 355–360.

The Ernst W. Bertner Memorial Award Lecture

Cell Differentiation and Neoplasia, edited by Grady F. Saunders. Raven Press, New York

Introduction of The Ernst W. Bertner Memorial Award Recipient

T. C. Hsu

Department of Biology, The University of Texas System Cancer Center M. D. Anderson Hospital and Tumor Institute, Houston, Texas 77030

In the mid-1960s, I served for four years as a member of the Cell Biology Study Section of NIH to review grant applications relating to the fields of cell biology and developmental biology. At that time, the study section reviewed an application from Dr. Beatrice Mintz, a research proposal more or less in the classic approach of developmental biology. The experiments outlined were certainly well thought out, but the members of the study section were not excited with the anticipated results.

During the discussion period concerning this application, one member commented that he had just visited Bea Mintz and learned of some fascinating results she had obtained. We were told that she was able to fuse early mouse embryos and implant the combined embryos into the uterus of a pseudopregnant female who served as a foster mother. These embryos, each representing two combined embryos, would develop normally as one individual. Thus, when she used embryos of genetically different mouse strains for her experiments, e.g., one with a white coat and the other with a black coat, the combined embryos would develop into mice with a striped coat.

This informal report not unnaturally excited the members of the study section, and many of us could see the possibility of tackling numerous problems in genetics and development by this new technique. The study section decided to table the application in consideration and to communicate with Mintz for her to submit a new application to develop this new approach.

Since that time, Dr. Mintz has produced one significant contribution after another using the technique of inducing the "allophenic mice." The interplay of two mouse embryos artificially combined offers a marvelous tool for investigations in numerous problems in development, because in these mice different genetic markers can be introduced, including susceptibility to tumorigenesis. Because of these contributions, she was elected, in 1973, as a member of the National Academy of Sciences.

Dr. Mintz was born and raised in New York City and received her baccalaureate from Hunters College. She obtained her Master's degree at New York University and then went to the University of Iowa for her Ph.D. degree. She taught at Iowa and Chicago before she joined the Institute for Cancer Research in Philadelphia, where she has been a Senior member since 1965. She also has a professorial appointment at the University of Pennsylvania.

The Ernst W. Bertner Memorial Award is given annually to investigators with outstanding contributions to cancer research. Dr. Mintz' contributions to developmental biology most certainly deserve such recognition, especially in view of her recent work on teratocarcinomas. Using the allophenic mice technique, she has been able to demonstrate that her teratoma cells, carried in serial transplantations for a number of years, were still able to participate in embryonic differentiation, including tissues in which the normal counterpart of teratoma cells are not supposed to exist. These findings suggest that neoplastic conversion does not necessarily require a profound genetic change of a cell.

In behalf of Dr. Clark, Dr. Hickey, the Symposium Committee, and The University of Texas System Cancer Center, I take great pleasure in presenting the 1977 Bertner Award to Dr. Beatrice Mintz.

Cell Differentiation and Neoplasia, edited by Grady F. Saunders. Raven Press, New York

Genetic Mosaicism and In Vivo Analyses of Neoplasia and Differentiation

Beatrice Mintz

Institute for Cancer Research, Fox Chase Cancer Center, Philadelphia, Pennsylvania 19111

The most challenging and least understood aspects of mammalian development, and of many genetic diseases, involve "higher" levels of biological organization, including cell interactions, not attainable in vitro. Neoplasia, as it occurs in the organism, may be essentially an aberration of differentiation; this also is not observable in culture, however invaluable the numerous studies of cell transformation in vitro have been. It was with the hope of fashioning new in vivo possibilities for investigating these phenomena that I undertook, over a dozen years ago (Mintz 1962, 1965), experiments involving the early introduction of cellular genetic markers into intact animals. The expectation was that the coexistence of genotypically different populations of cells throughout life in "mosaic" individuals would provide ways of detecting in vivo the inceptions and consequences of specialized gene function in cell diversification, in genetic diseases, and in neoplasia.

These experiments have been done largely in mice, in which various defined strains and mutant genes are readily available. The procedure for synthesizing a composite embryo, and ultimately a mosaic animal, from cells of two embryos is shown in Figure 1 (Mintz 1971a). Four-parent *allophenic* animals are thus produced; in them, the disparate cellular phenotypes within any given tissue are ascribable to genotypic differences. However, the two strains of cells are not necessarily found in all tissues; their tissue distribution may vary widely. Experimental permutations thus fortuitously result, and they are especially useful in defining the primary tissue in which a gene is expressed or in which a clinically complex disease actually originates (Mintz 1971c, 1974). The cells retain their individuality and do not fuse, except in the differentiation of multinucleated skeletal muscle fibers. There, uninucleated myoblasts unite and form true heterokaryons (Mintz and Baker 1967).

Allophenic animals are immunologically competent to reject foreign proteins, but they remain permanently tolerant of any immunogenetic disparities in their respective cell strains (Mintz and Silvers 1967, Mintz and Palm 1969). Skin grafts from the parental pure strains are retained despite any strong histocompatibility differences, e.g., at the *H-2* locus, and graft-vs.-host disease is absent. As a result, virtually any cell strains may be developmentally associated. This

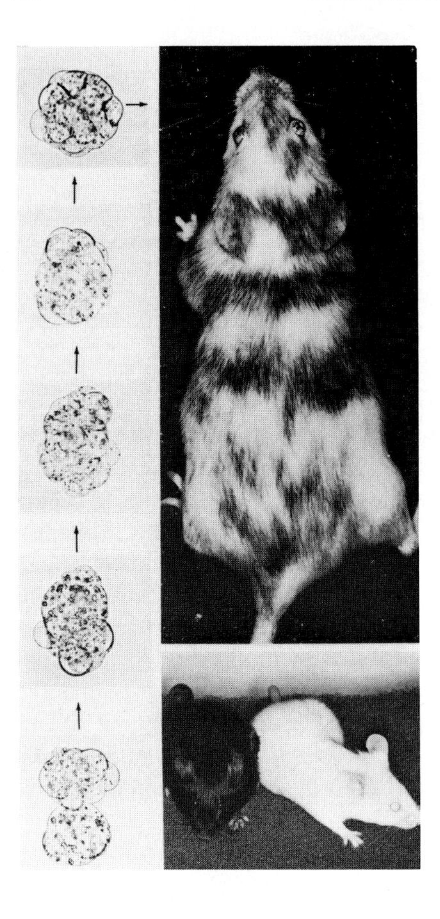

FIG. 1. A striped, four-parent allophenic female mouse produced by aggregating two cleavage-stage embryos (lower left), one of pigmented (*C/C*) and the other of albino (c/c) genotype. After enzymatic removal of the envelope, the embryos are placed in contact at 37°C. In successive photographs, a single spherical composite is formed. After transfer to the uterus of a foster mother, a normal mouse is born. The stripes represent transversely migrating clones of pigment cells of the separate genotypes. The presence of two separate strains of germ cells is evident from the colors of the progeny (lower right), after mating to a pure-strain albino male. (From Methods in Mammalian Embryology, edited by Joseph C. Daniel, Jr., W. H. Freeman and Company. Copyright © 1971.)

"intrinsic" (as opposed to experimentally "acquired") tolerance exhibited by allophenic mice seems to have furnished a valid model of the ontogeny of normal self-tolerance: New proteins introduced prior to the maturation of immune competence are not later recognized as foreign by the mature lymphocytes.

Other kinds of investigations to which these animals have lent themselves have included the detection of cell lineages and movements, cell interactions, and cell selection (Mintz 1970b,1971b,1971c,1974). Examples have been described for the clonal origin of melanocytes (see Figure 1) (Mintz 1967), the development of the neural retina (Mintz and Sanyal 1970), the formation of the vertebral column (Moore and Mintz 1972), and the genesis of certain anemias (Mintz 1971c, 1974).

The emphasis in the present discussion will be chiefly on those of our studies that bear on the developmental nature of neoplastic disease. We have utilized two experimental approaches to the problem. In one, we have availed ourselves of the existence of strain-specific malignancies in mice in order to obtain allophenic individuals in which tumor-susceptible and nonsusceptible cell strains coexist throughout life. In such associations, the question of cellular autonomy of expression of "susceptibility" may be analyzed; the primary tissue involved in heritable tumor susceptibility may be identified, and the cellular lineages contributing to tumor formation may be revealed.

In another experimental approach involving slightly different techniques, the aim has been to bring frankly malignant stem cells into contact with normal, early stem cells in young embryos. This has provided a means of testing the potential of the malignant cells for recovery of orderly control of gene expression, hence of return to normalcy.

LOCALIZED VS. SYSTEMIC CONTROL OF TUMOR SUSCEPTIBILITY

Allophenic mice composed of both tumor-susceptible and nonsusceptible cell strains, with respect to a particular kind of neoplasm, offer an opportunity to identify the tissue(s) in which "susceptibility" genes are expressed. This is a first step toward defining more precisely the cell phenotypes leading to malignancy. The fact that the two cellular genotypes in allophenics vary in their tissue distribution from one individual to another creates a series of in vivo tests of the involvement of specific tissues. Under these circumstances, the formation of a tumor and the tumor-cell genotype may be evaluated in relation to the genotypes of that animal's other tissues. For example, if only susceptible-strain cells in the ultimate target organ become malignant, irrespective of the genotypes of the individual's other tissues, the relevant genes must be acting autonomously in the cells of that organ. If, on the other hand, transformation of susceptible-strain cells occurs only in those allophenic mice that have cells of that genetic strain in another *specific* tissue, it is in the latter that these genes must first be expressed, and tumor formation is then the result of indirect or systemic events. Nonsusceptible cells might also become malignant in the allophenic situation, either because of transmission of viral or other tumor information from nearby transformed cells of the susceptible strain or through an indirect effect of gene action in susceptible cells of another tissue.

Among the possible indirect or systemic effects that might contribute to growth or suppression of an incipient tumor, immunological defense mechanisms have received special attention. The hypothesis of immunological surveillance (Burnet 1974) holds that tumors are generally immunogenic to their hosts and that immune lymphocytes tend to react by preventing tumor cell growth. Experiments (reviewed by Prehn 1976) carried out to test the hypothesis have indeed demonstrated that *transplantable* tumors can immunize recipients against graft acceptance of subsequent tumor inocula. Artifically immunosuppressed hosts have also been extensively used, and the increase in incidence of spontaneous tumors in them has been attributed to lowered surveillance against tumor antigens. However, several difficulties with the hypothesis have arisen. One is that *primary* spontaneous tumors, as opposed to retransplanted ones that may have undergone selection, display little or no ability to elicit a host immune reaction capable of interfering with their growth. Another significant point is that athymic *(nude)* mice, congenitally deficient in T-lymphocytes, do not show an excess of neoplasms (other than lymphoreticular ones). It appears that the elevated incidence of other kinds of tumors, previously reported in experimentally immunosuppressed animals, may therefore have been due to lowered defenses against tumor viruses rather than to lack of an immune surveillance response to tumor cell antigens. Whether immune surveillance in fact exercises any appreciable restraint on malignancies, as those diseases ordinarily occur, must therefore be considered an open question at the present time.

If the growth of strain-specific neoplasms in mice were due appreciably to an insufficiency of such a surveillance mechanism, this should result in a correlation, in allophenic animals, between the presence of lymphocytes of the genetically susceptible strain and the formation of a tumor in the target organ. In addition, those allophenic animals remaining free of tumors should have been "protected" by lymphoid cells predominantly of the nonsusceptible cell strain.

These questions have been examined in allophenic mice, in relation to formation of spontaneous mammary tumors (Mintz 1970a), liver tumors (Mintz 1970a, Condamine *et al.* 1971), and lung tumors (Mintz *et al.*1971). Mammary tumors occur in virtually all control females of the C3H strain and are attributable to a highly virulent mammary tumor virus transmitted in the milk. Mammary tumors are less frequent (43%) in C3Hf controls, which carry a less virulent form of the agent, are rare in C57BL/6 strain females, and are moderately frequent (40%) if C57BL/6 females are foster nursed on C3H mothers. Thus, allophenic females of the C3H ⟷ C57BL/6 and C3Hf ⟷ C57BL/6 strain combinations provide material for studies of autonomy vs. nonautonomy of expression of genetic susceptibility in mammary tumor formation. Allophenic males of the same two paired strain combinations may be used to study hepatoma formation, which is high (69%) in C3H control males, less frequent (33%) in C3Hf controls, and virtually absent in C57BL/6 males. Lung tumor development has been investigated in males as well as females of another allophenic strain combination, BALB/c ⟷ C57BL/6, because of the fairly high (56%) spontaneous incidence of these tumors in BALB/c controls and their rarity or absence in the C57BL/6 control strain.

In each of these studies, the animals were kept alive until tumors or other serious ailments developed. At autopsy, tumor and other tissue samples were fixed for histopathological inspection, and samples were genotypically characterized by means of strain-specific allelic differences. The marker loci employed for genetic typing coded for electrophoretic differences in enzymes, for histocompatibility (*H-2* locus) differences, detectable by serology or by graft tests, for immunoglobulin allotypes, or for other phenotypic differences.

The results indicated that lymphoid cells of the low-tumor-strain genotype afforded no protection whatever against formation of mammary, liver, or lung tumors (Mintz 1970a, Mintz *et al.* 1971). For example, in the mammary tumor experimental series, the sampled lymphoid tissues of different allophenic individuals contained between 10 and 100% of the *nonsusceptible* (C57BL/6) cell strain; 7S G2a immunoglobulins also comprised a majority of the C57BL/6 genotypic class. Nevertheless, mammary tumors developed in these same animals and were usually of the susceptible (C3H or C3Hf) genotype. As already pointed out, the animals showed normal competence to reject allogeneic skin grafts. Therefore, a deficiency in immunological surveillance against "tumor antigens" appears not to be at issue in development of these types of spontaneous neoplasms. (Immune response to oncogenic viruses may be important, although presence of the viruses at a young age, in these experiments, would be expected to lead to tolerance of them.)

Genotypic analyses of other tissues in the allophenic animals also revealed that susceptibility was largely localized to the cells of the potentially malignant tissue. This was concluded from the fact that the genotypes of the tumors themselves and of extrinsic tissues often differed within an individual. In each animal, target-tissue cells of each genotype were simultaneously exposed to the same systemic factors, yet it was generally only the susceptible cell strain that became tumorous.

Cell-localized genetic specificity within the target organ was also striking. In liver or mammary gland, for example, the susceptible and nonsusceptible cell strains were sometimes intimately admixed. This was especially striking

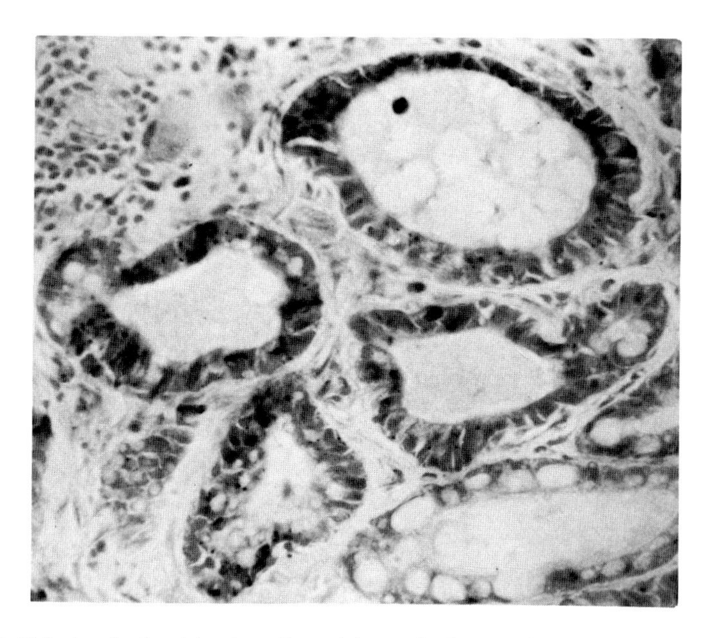

FIG. 2 A histochemically stained section of hyperplastic mosaic mammary gland from a C3H⟷C57BL/6 allophenic female. The two strains have low and high activity, respectively, of the enzyme β-glucuronidase. The pale alveolar cells are tumor susceptible (C3H strain) and are closely intermingled with darkly stained, low-susceptibility cells (C57BL/6 strain). (Reprinted, with permission of Academic Press, Inc., from Mintz 1976.)

in mammary tumor formation. When hyperplastic (premalignant) mammary gland sections were histochemically stained to reveal genotypic cellular identities, through quantitative strain differences in β-glucuronidase activity, fine-grained interspersions of the two cell strains were often seen, even within single alveoli (Figure 2). Yet it was the high-tumor strain of cells that became malignant, as in controls of that pure strain.

The morphology of such mosaic, early hyperplastic alveolar nodules is indistinguishable from that of C3H pure-strain nodules (Figure 3), but graft tests proved that this was not due to transformation of the low-tumor-strain C57BL/6 cells. When halves of single mosaic nodules were grafted separately to pure-strain recipients, each recipient selectively rejected only the allogeneic cells. The surviving cells in each graft then grew out, forming, in the C57BL/6 hosts, a normal mammary tree and, in the C3H hosts, another hyperplastic nodule (Mintz and Slemmer 1969). Eventually, such C3H nodules became frankly malignant. Thus, normal (C57BL/6 strain) cells had been entrapped in nodules whose morphology and malignant progression were due primarily to their C3H cells. From these and further transplant experiments with genetically mosaic nodules, a hypothesis of early mammary tumorigenesis in ordinary (single genotype) animals has been

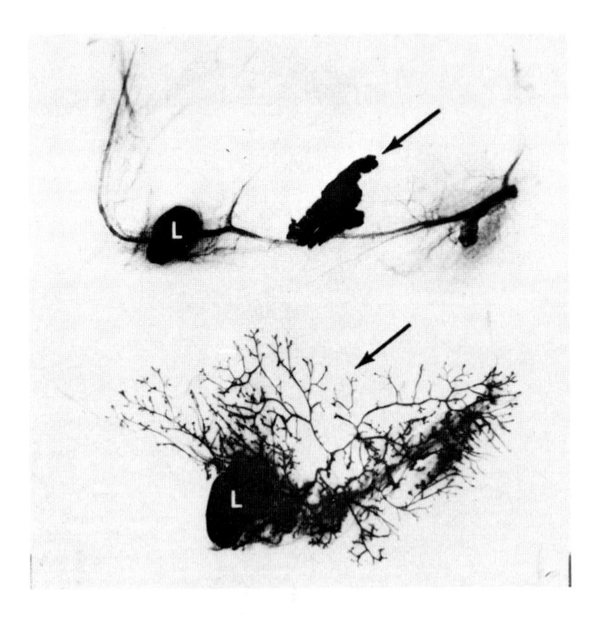

FIG. 3. Halves of a single hyperplastic alveolar nodule from a C3H ⟷ C57BL/6 allophenic mouse, following grafting to pure-strain hosts whose histocompatibility alleles differ. In a C3H host (top), only the C3H or high-tumor cells survive, and again produce a premalignant nodule (arrow). In a C57BL/6 host (bottom), the low-tumor cells are liberated and form a normal outgrowth. (From Mintz and Slemmer 1969.)

advanced: According to the hypothesis, normal cells may usually be closely associated with transformed ones at first and may, in fact, be indispensable for the ultimate capacity of the latter to grow and proliferate independently. Perhaps a similar supportive role of associated normal cells may also characterize the early stages of some other malignancies.

MONOCLONAL VS. MULTICLONAL ORIGIN OF TUMORS

The question of whether a tumor arises from one cell, as a single clone, or from more than one cell is related to the possibility of somatic cell mutation as the basis for tumorigenesis. Early in this century, since the time of Boveri, chromosomal or genetic changes have been considered as candidates for this role. More recently, it has been pointed out that if a tumor usually arises as the result of a mutation (whether spontaneous or induced) in a somatic cell, the relative rarity of mutations should generally result in single-clone tumors (Burnet 1974).

What has usually been overlooked in investigations of the problem is the fact that when a tumor is macroscopically detectable, it has already passed through many cell generations. During that time, selection may have amplified some clones and eliminated others. Therefore, a tumor that is monoclonal in *composition* may have been multiclonal in *origin.*

Most tumors at the time of sampling are in fact aneuploid and are individually differentiated chromosomally. The "stemline" of the tumor is believed to have been derived from a single variant cell that made its appearance some time *after* tumor initiation (Makino 1956, Hauschka and Levan 1958). According to the stemline concept, the variant cell selectively survives because of superior proliferative or other characteristics, and it may generate subpopulations of cells with still further genetic changes (Nowell 1974). In this way, new characteristics make their appearance in a stepwise fashion during tumor progression (Foulds 1954).

Indeed, the well-documented picture of tumor progression is consistent with many reports (reviewed by Friedman and Fialkow 1976) that most human tumors are monoclonal in composition. The markers employed in these studies are chromosomal ones in some cases. Many others have relied on heterozygosity for certain X-linked genes. Activation of only one such gene of a pair per somatic cell in heterozygous females results in cellular mosaicism with respect to gene function (rather than to gene structure, as in allophenic mice). The presence of only one phenotype in a tumor has often been assumed to signify not only monoclonal composition but also monoclonal origin. Human tumor studies have relied most heavily on variants of the X-linked enzyme glucose-6-phosphate dehydrogenase (G-6-PD), which are electrophoretically separable in tissue homogenates. Among the most extensively sampled, apparently monoclonal tumors are uterine leiomyomas, chronic lymphatic leukemia, and Burkitt's lymphoma.

Apart from the difficulty of employing a postselection situation to discern obscured origins, there is the further complication that *if* a marker is sensitive enough to reveal a minor clone in a tissue homogenate, it should also reveal, confusingly, admixed nontumorous cells, including normal tissue parenchyma, stroma, blood vessels, and blood cells; all of the latter would comprise two enzyme phenotypes in an X-linked heterozygote. An example of the difficulty is seen in human hereditary trichoepitheliomas, which are mosaic for G-6-PD. In these tumors, admixed nontumorous cells are evident in histological sections (Friedman and Fialkow 1976). Hereditary neurofibromas may also show two G-6-PD types, yet they have been interpreted as truly multiclonal, due to an elevated tendency for all neural cells of that tumor-prone genotype to become tumorous. The possibility nevertheless remains that such apparently multiclonal tumors may have undergone less progression and selection since their inception than have presumed single-clone tumors.

Allophenic mice have also generated tumors in which cell lineages have been examined (Mintz 1970a). Here, the different genotypic populations may bear many biochemical, chromosomal, or histocompatibility markers, either on autosomes or on sex chromosomes. Isozymic markers not detectable in blood and stroma are available (e.g., malate dehydrogenase electrophoretic variants), but they are relatively insensitive, as they would fail to detect a minor population of approximately 5%. The most sensitive marker, glucosephosphate isomerase, has the disadvantage that it is ubiquitous. Tumor homogenates generally show both electrophoretic variants of the enzyme, if that animal also has blood mosaicism; but in poorly vascularized tumors, for example in the lung, it is clear that the tumor itself consists entirely or largely of one cell-strain (Mintz *et al.* 1971).

These handicaps and ambiguities would be overcome if histochemical visualization of each strain were possible. The normal vs. tumorous status of the cells could then be directly evaluated at the same time as their genetic identity was ascertained. At present, no appropriate histochemical markers are known for human tissues. However, two such markers are available for tumor analyses in allophenic mice (Condamine *et al.* 1971, Dewey *et al.* 1976). Activity of the enzymes β-glucuronidase and β-galactosidase varies in different inbred strains of mice, depending in each case on alleles at a single locus. Tumors may be analyzed histochemically for their strain composition because the malignant state does not markedly alter the relative enzyme activities in tumors of pure-strain controls.

Twelve spontaneous hepatomas were examined from allophenic mice of strain combinations comprising cells with low vs. high activities of β-glucuronidase. In nine of the tumors, the malignant cells were of only the low-activity strain; one hepatoma included only high-activity-strain tumor cells. (Some of the tumors contained normal reticuloendothelial cells of the opposite pure strain or were surrounded by normal hepatocytes of both strains.) However, two of the twelve tumors were clearly genetically mosaic, with malignant hepatocytes of both genotypes (Figure 4) (Condamine *et al.* 1971). Of special relevance for the present discussion is the fact that the mosaicism was not diagnosed when homogenates of samples from the same two tumors were typed electrophoretically for strain-specific isozymes. Occurrence of two multiclonal hepatomas in so small a sample examined histochemically is all the more noteworthy because in only one of the input strains (C3H) do the controls have a high hepatoma incidence; in the other (C57BL/6 or BALB/c), the control incidence is very low. Thus, the odds were heavily weighted against finding tumors of both genotypes. Therefore, multiclonality in hepatomas, and possibly in some other kinds of tumors, may be more common than analyses of tumors from allophenic mice have shown. In many, multiclonal composition may go undetected because all clones are of the same (i.e., the more susceptible) genotype.

A mosaic tumor may truly signify origin from two (or more) transformed cells by nonmutational means, or it may result from a chance juxtaposition of two independently originating tumors. Intercellular spread from a single transformed clone to neighboring cells by transmission of some form of tumor agent is still another possibility. In order to help distinguish among the alternatives by increasing the chances of obtaining mosaic tumors, we are extending these studies in allophenic mice to strain combinations in which both strains, rather than only one, are of high-susceptibility genotypes and in which a histochemical marker is also present. At the same time, histocompatibility-strain differences are included, as in earlier experiments (Mintz and Slemmer 1969), so that if

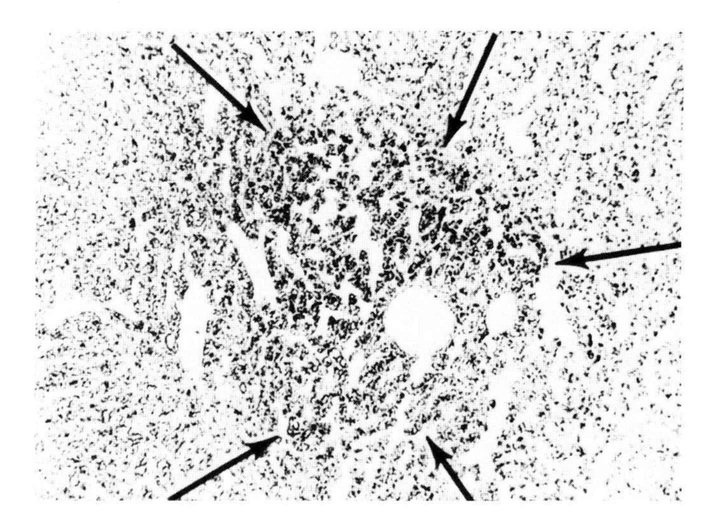

FIG. 4. A hepatoma from a C3H ⟷ C57BL/6 allophenic mouse, after staining for β-glucuronidase activity. Multiclonal origin is indicated by the presence of both C57BL/6 cells (dark, arrows) and C3H (light) cells. (From Condamine *et al.* 1971.)

both cell strains are actually malignant, each will be capable of surviving and producing a tumor in the histocompatible strain after pieces are grafted to pure-strain recipients.

NORMALIZATION OF TERATOCARCINOMA CELLS IN MOSAIC EMBRYOS

It is admittedly difficult to obtain direct and conclusive proof of either the mutational or the nonmutational origin of a tumor. If a particular malignancy had in fact come about without any material change in the genes, it might be possible to demonstrate that its genome was intact by restoring the generative cells of the tumor to complete normalcy—provided no substantial chromosomal or genetic changes had occurred secondarily during the period of malignancy. Such a "normalization" would, in effect, signify that *altered control of gene function,* rather than alteration of gene structure, had been at issue in the initial neoplastic conversion.

The full spectrum of gene expression occurs only if all specialized cell types are produced. This would require the differentiation of virtually an entire individual from stem cells of the tumor. A malignancy of such totipotent, developmentally primitive cells would therefore provide the most promising material for tests of nonmutational transformation.

Teratomas have been suspected, but not known with certainty, of being tumors derived from developmentally primitive cells. They have been described as "neoplasms composed of multiple tissues of kinds foreign to the part in which they arise" (Willis 1967). The most malignant of these tumors are comprised entirely or mainly of relatively undifferentiated, embryonal carcinoma (or teratocarcinoma) cells and few, if any, differentiated tissues. The most interesting teratomas are those in which there are substantial numbers of embryonal carcinoma stem cells along with various differentiated tissues, chaotically arranged. These teratomas have been especially well studied in the mouse (Stevens 1967a, Pierce 1967). Maintenance by serial transplantation (in the strain of origin) is possible as long as the proliferative stem cells are retained. An experiment by Kleinsmith and Pierce (1964) clearly proved that the differentiated cell types in the tumors arise from the embryonal carcinoma cells: When single carcinoma cells were transplanted subcutaneously, a variety of tissues was obtained.

While the embryonal carcinoma cells were thus shown to be developmentally multipotential, evidence for their totipotency was still lacking, as mouse teratomas are invariably devoid of certain tissues, e.g., liver, kidney, lung, thymus, and immunocompetent cells (Stevens and Hummel 1957, Stevens 1970). Moreover, many other tissues are incompletely differentiated. The possibility thus remained that genes required for differentiation of the deficient tissues may have been mutated or deleted when the neoplastic transformation occurred.

It should be pointed out that an argument against a mutational etiology of mouse teratomas is the high frequency with which these tumors may be experimentally induced. While the tumors tend to occur spontaneously in only two inbred strains (in the testes of 129-strain males and the ovaries of LT-strain females), they are readily obtained in many strains by simply transplanting either early-stage embryos or fetal germinal ridges containing germ cells to an ectopic site, such as the testis capsule or kidney capsule (Stevens 1967a, 1970). The transplantable tumors are malignant, according to conventional standards of pathology: they may invade and metastasize. The stem cells divide rapidly and cause the death of the host within about a month.

Malignant teratocarcinoma cells would scarcely be expected to realize any potential for full differentiation in an environment other than that of an early embryo, inasmuch as normal, totipotent embryo cells do not fully differentiate in a transplant site or in culture. Thus, a counterpart of the mosaic or allophenic mouse experiment seemed appropriate. By associating teratocarcinoma cells with genetically marked normal embryo cells, instead of associating two genetic types of embryo cells, as had been done previously (Mintz 1971a), the gene control mechanisms of the malignant cells might be induced to return to normal.

The tumor chosen for the initial normalization tests (Mintz *et al.* 1975, Mintz and Illmensee 1975) was one grown only in vivo by transplantation, because it seemed less likely that subtle genetic changes might have occurred than in the then available teratocarcinoma lines adapted to long-term tissue culture. This transplantable teratoma had been experimentally produced by Stevens in 1967 by transfer of a six-day mouse embryo beneath the testis capsule of an adult recipient. A solid tumor formed there, and when minced pieces were introduced into the body cavity, they became a modified ascites. The ascites contained multicellular "embryoid bodies" with some limited resemblance to early embryos. The young, small-size embryoid bodies comprised only a "core" of embryonal carcinoma cells surrounded by a "rind" of yolk sac-like epithelial cells. The ascites tumor (OTT 6050) had been carried as a highly malignant one for over 200 transplant generations spanning eight years (Figure 5). Embryonal carcinoma cells from cores indeed proved to be euploid (Mintz *et al.* 1975). Very little differentiation occurs in the embryoid body state; this may have had the effect of reducing selection for rapidly growing aneuploid variants of the undifferentiated cells.

Cleavage-stage blastomeres and embryonal carcinoma cells were found not to adhere well (Mintz *et al.* 1975). Therefore, the availability of a method (Lin 1966) for microinjecting cells into the cavity of slightly older embryos, in the blastocyst stage, made it possible to entrap the tumor cells where they might become integrated into the embryo (see Figure 5) (Mintz and Illmensee 1975). While the blastocyst at this time has some 64 to 128 cells, only about three cells, in the inner cell mass region, are believed to be the progenitors of the entire embryo (Mintz 1970b), as distinct from precursors of extraembryonic tissues; therefore, only small numbers (one to five) of tumor cells were introduced. Many genetic markers distinguishing the two input cell strains were used, as indicators of blastocyst- and tumor-derived cells in any tissues that might form

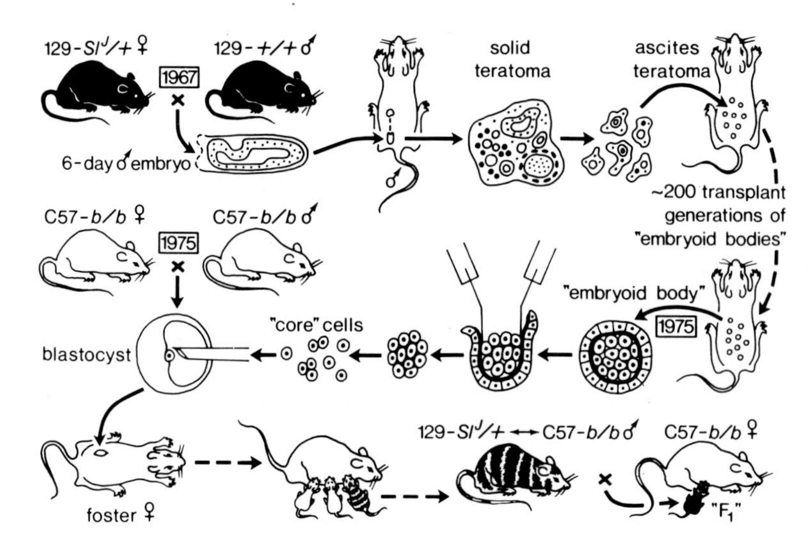

FIG. 5. Diagram of the experiment leading to complete reversal of malignant mouse teratocarcinoma cells to normalcy, after injection into a normal blastocyst. Normal somatic tissues form and include functional tumor-derived cells. Functional gametes are also produced, yielding progeny. See text for details. (Modified from Mintz and Illmensee 1975.)

and as evidence that tumor-derived cells were able to synthesize tissue-specific normal products. Since the OTT 6050 teratoma arose in the inbred 129 strain, the alleles of that strain were tentatively assumed to be present, even though most of those genes had not been expressed in the previous eight-year history of the tumor.

The results of this experiment were strikingly decisive. Healthy mosaic mice were obtained, in which functional normal cells derived from the tumor coexisted along with normal embryo-derived cells (Mintz and Illmensee 1975, Illmensee

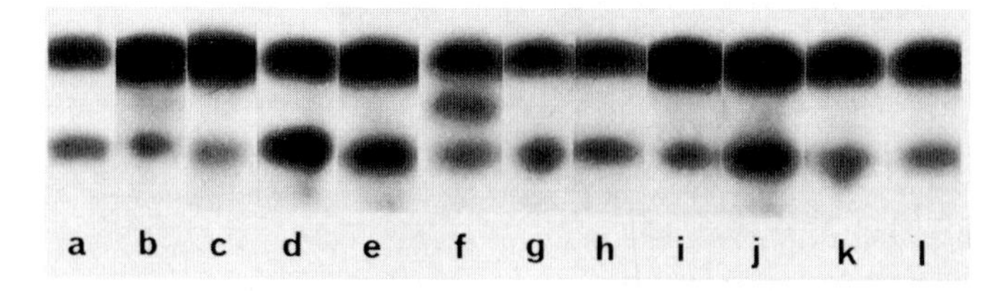

FIG. 6. Starch gel electrophoresis of glucosephosphate isomerase strain variants from tissues of a mosaic mouse produced by injecting a single teratocarcinoma cell into a blastocyst, as in Figure 5. The tumor-strain variant is slow migrating. After the control mixture in slot (a), the mosaic tissues (b-l), all with tumor-derived normal components, include blood, brain, spleen, heart, skeletal muscle (with a hybrid enzyme band due to heterokaryon formation), kidney, reproductive tract, liver, gut, thymus, and lung. (From Illmensee and Mintz 1976.)

and Mintz 1976). In cases of single teratocarcinoma cell injections into blastocysts, it was clear that the full gamut of somatic tissues could develop clonally from one totipotent cell (Figure 6). Tissue-specific proteins that are also genetically specific for the 129 (tumor) strain of origin provided evidence of normal function. Examples of tissue products coded for by 129-strain alleles in tumor-derived cells included eumelanin and phaeomelanin pigments in the coat, adult hemoglobin, 7S G1 and G2a classes of immunoglobulins (Figure 7), and liver proteins produced by hepatocytes and detected after excretion in the urine. Differentiation of mature thymus, lung, kidney, and plasma cells was particularly significant because of the complete absence of these tissues from the teratomas themselves. Possibly the development of such tissues may depend upon inductive interactions between embryonic components that either fail to form in the tumors or that fail to be brought into the appropriate associations.

Differentiation also revealed, by means of coat color, the presence of a tumor-contributed gene, *steel* (Sl^J) (Mintz and Illmensee 1975), not previously known to be present in the teratocarcinoma cells. From a search of the records (L. C. Stevens, personal communication), this gene was apparently transmitted from the mother of the six-day embryo used to produce the tumor in 1967; she had been genotypically $Sl^J/+$, the Sl^J gene having been maintained in some animals of the strain by forced heterozygosity. Thus, the restoration of orderly gene expression in differentiation had uncovered not only hitherto silent genes, whose presence was expected in 129-strain cells, but also an unsuspected gene.

Two other reports of teratocarcinoma cell injections into blastocysts have offered some evidence in support of donor cell participation in embryogenesis, although neither has shown developmental totipotency of the cells nor their complete reversibility to normalcy. In one study (Brinster 1974), only one genetic marker (pigmentation) was used, so that it was only possible to test the differentiation of a single cell type; one animal with a few pigmented stripes of the tumor-strain type was obtained. In the other study (Papaioannou *et al.* 1975), the donor cells were from aneuploid cell culture lines; only two genetic markers (pigmentation and glucosephosphate isomerase isozymes) were used. Because large numbers (20 to 40) of embryonal carcinoma cells were injected into each blastocyst, the developmental possibilities of any given tumor cell could not

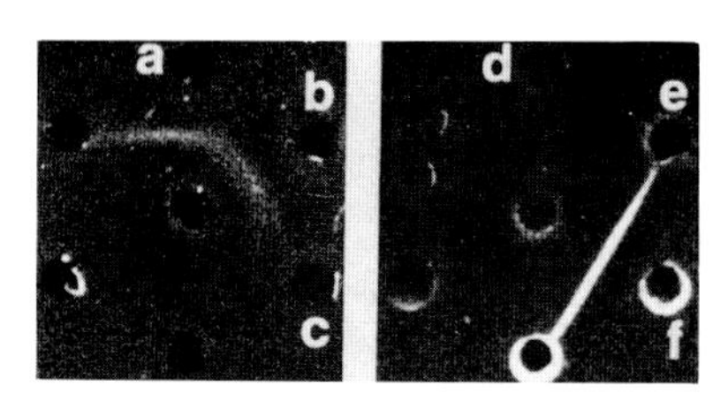

FIG. 7. Tumor-strain allotypes of 7S G1 and G2a immunoglobulins from a mosaic mouse produced as in Figure 5. Anti-129-type serum (left, center well) reacts with 129 control serum (a) and serum of the experimental mouse (b), but not with a C57BL/6 (blastocyst-strain) control (c). Anti-C57-type serum (right, center) reacts with a C57 control (f), but with neither a 129 control (d) nor the experimental mouse serum (e). (From Mintz and Illmensee 1975.)

be evaluated. The introduction of many malignant cells may also have prevented their integration, as this could account for the fact that tumors were found in most of these experimental animals.

It should be emphasized that, in our experiments, normalization of teratocarcinoma cells after injection into blastocysts is a stable phenomenon. Normal, tumor-derived tissues from the mosaic mice did not revert to the tumorous state when grafted to other hosts in ectopic sites. From our single-cell blastocyst injections, it appears that if the teratocarcinoma cells were successfully integrated, they were effectively normalized and remained normal in the allophenic animals. In only a small minority, in which absence of 129-strain cells in the animal's normal tissues showed failure of donor cell integration, did the carcinoma cells continue their malignant habit without contributing to the embryo, as if in an ordinary subcutaneous transplant to an adult (R. P. Custer, K. Illmensee, and B. Mintz, unpublished data).

One would like to know the mechanisms whereby integration of the carcinoma cells into a normal early embryo results in their normalization. It is tempting to hypothesize that cell surface compatibilities are involved. While this may be the case, there is as yet no definitive evidence to support such a view. One indication to the contrary is the fact that early embryo cells do not adhere well to embryonal carcinoma cells, although each cell type adheres strongly to its own kind (Mintz *et al.* 1975). In another approach to the question, we prepared an antiserum in syngeneic mice against our known totipotent embryonal carcinoma cells (from embryoid bodies) and while the serum reacted against preimplantation embryo cells, it also cross-reacted with postimplantation embryos and with various adult tissues, from which nonspecificity may be inferred (Dewey *et al.* 1977). The picture is far from clear at present, inasmuch as similar results, indicating nonspecificity, have been described by another laboratory (Stern *et al.* 1975), but other workers, using nullipotent or nondifferentiating embryonal carcinoma cells as the immunogen, have reported evidence for a major embryo-specific antigen shared only by the carcinoma cells and germ cells (Artzt *et al.* 1973, Gooding and Edidin 1974).

The point must be stressed that the cells used to inject blastocysts in these experiments show every indication of being a malignant population, not a mixture of malignant and normal cells: Their sustained transplantability far exceeds the number of cell generations that tissue grafts or untransformed cells are able to survive. And the injection of single cells, rather than groups of cells, fails to reduce the number of "takes," as would have occurred if the population were heterogeneous and only initially normal cells were capable of contributing to normal tissue development (Mintz and Illmensee 1975, Illmensee and Mintz 1976).

Even more striking than the full range of somatic differentiation seen in the best of our blastocyst injections with embryonal carcinoma cells was the formation in some individuals of fully functional sperms derived from the carcinoma cell lineage (Mintz and Illmensee 1975), which is of the X/Y or male sex chromo-

FIG. 8. Four normal progeny produced from a mosaic male (left, rear) and his normal mate. The male was obtained as in Figure 5. The presence of teratocarcinoma-lineage sperms in this male is shown by the two agouti offspring (left, front), one of which (far left) has a paler coat due to transmission of the *steel* gene from the tumor cells. The two black progeny (right, front) are from blastocyst-derived sperms. (Reprinted, with permission of Academic Press, Inc., from Mintz 1976, based on data of Mintz and Illmensee 1975.)

some type (Dunn and Stevens 1970). (Earlier studies [reviewed in Mintz 1974] have shown that functional gametes do generally develop in allophenic mice, even including X/X ⟷ X/Y mosaics, but only from the germ cells whose sex chromosome type conforms to the morphological sex type of the adult host.) The fertile males with germ cells derived from the X/Y teratocarcinoma were mated to normal females of the original blastocyst strain; their F_1 progeny (Figure 8) were entirely normal and exhibited dominant and codominant 129-strain alleles and the *steel* gene. Thus, final proof of genomic integrity was provided by transmission of the genes in the male germ line and their re-expression in the development of the "tumor progeny."

Similar results have since been obtained with a spontaneous, chromosomally female (X/X) transplantable mouse teratocarcinoma of the LT strain. Blastocyst injections with cells from an in vivo tumor have yielded the full range of normal somatic cell differentiation, as well as successful formation of functional oocytes and transmission of tumor-strain genes to F_1 progeny (K. Illmensee and B. Mintz, unpublished data).

The experimental use of genotypic cellular mosaicism in early embryos has therefore furnished the first unequivocal case of complete and stable reversal of an animal malignancy to normalcy.

NONMUTATIONAL AND MUTATIONAL TUMORS

The results offer strong support for the interpretation that the initial conversion to malignancy in these tumors came about through induced changes in gene function, rather than gene structure, due to an anomalous environment. Normal gene function was restored when the malignant stem cells were returned to the proper environment. One cannot rule out the possibility that the tumor cells might have a genetic deficiency specifically preventing their autonomous development into embryos, i.e., embryonal carcinoma cells are dependent on embryo associations to become normalized and do not themselves form embryos when introduced in utero (B. Mintz, unpublished data). Nevertheless, the evidence favoring a *nonmutational* etiology of these tumors is not only compelling but is in fact stronger and more direct—testing, as it does, the presence and function of so many genes—than the data indicating a possible *mutational* basis for some other tumors.

Imperfect as our present knowledge may be, it now appears likely that mutational and nonmutational classes of cancers exist. Many tumors may be caused by somatic mutation, as is widely believed. The evidence includes the fact that many known carcinogens have mutagenic properties (McCann and Ames 1976). Statistical analyses of the incidence of certain human childhood malignancies, e.g., retinoblastoma, Wilm's tumor, and neuroblastoma, suggest that two mutational events (one in the germ line and one in the soma, or both in the soma) are involved (Knudson 1973). Chromosome instability syndromes provide another class of possible mutational tumors, as in Bloom's syndrome (German 1974); here the chromosome changes antedate the cancers and hence may be causal to them. The aneuploidy common in nonhereditary tumors may, on the other hand, occur as a consequence of malignant progression rather than a cause of the cancer. Experimental tumors are in fact often euploid at first and generate variant subpopulations (Nowell 1974). The presence of a specific chromosomal change in a malignancy, first observed in human chronic myelogenous leukemia (Nowell and Hungerford 1960), has led other malignancies to come under scrutiny for specific chromosomal deviations (Rowley 1974). However, it remains difficult to rule out the possibility that they are merely consistent consequences of earlier causal events. A consistent sequence of karyotypic changes has actually been observed in patients with chronic myelogenous leukemia (Prigogina and Fleischman 1975).

Even in a probable nonmutational cancer such as the mouse teratoma, genetic "susceptibility" factors may play various roles. For example, teratomas are not experimentally inducible in all mouse strains, and they occur spontaneously in only 129-strain males and LT-strain females, as gonadal tumors from parthenogenetically developing germ cells (Stevens 1970, Stevens and Varnum 1974). In the latter cases, heritable factors may have no direct influence at all on carcinogenesis. They may instead promote parthenogenetic activation of germ cells in situ (e.g., due to instability of cortical granules), with the malignancy ultimately caused by the inappropriate local environment. Teratomas in the human population may also show hereditary components, as in the six kindreds described by Ashcraft and Holder (1974).

Nonmutational cancers may result from physiological changes in the cell or tissue milieu. Conditions conducive to cancers of this sort might include an anomalous environment, local trauma, or hormonal imbalance. In fact, the ease of induction of certain tumors via hormonal changes suggests that carcinogenesis may occur nonmutationally in many of those instances. Agents apparently capable of causing *mutational* cancers (e.g., radiation, chemical mutagens, and some viruses) may conceivably also sometimes trigger nonmutational changes leading to neoplasia by disrupting the cellular environment. Or some "carcinogenic" compounds might act nonmutationally by blocking a metabolic pathway or by binding to the cell's DNA, RNA, or other macromolecules.

Complete phenotypic reversal to normalcy after transfer to a normal environment has clearly been obtained in plant teratomas (Braun 1959). This work has significantly underscored the possibility of an epigenetic etiology for some tumors. Nevertheless, different mechanisms probably mediate reversal of plant as compared to animal teratomas. The plant reversals may thus not offer a model of tumor behavior generalizable to animal tumors, as has been suggested (Braun 1969). Major differences in the normalization process are evident: In the plant teratomas, a series of graftings to the growing tips of healthy host plants is required, and reversal is gradual; it is compatible with marked departure from euploidy. In the mouse teratomas, a single transfer to the blastocyst cavity of a normal host embryo rapidly leads to normal participation in differentiation (Mintz and Illmensee 1975), which is stable. Complete differentiation is perhaps less surprising in plant teratoma cells than in animal teratomas, given the much greater developmental potentialities of ordinary somatic cells of plants than those of animals (Steward *et al.* 1970).

Another animal tumor frequently, and perhaps prematurely, described as successfully normalized and made to differentiate is the Lucké renal adenocarcinoma of viral origin in the frog. When several nuclei from tumors in triploid frogs were injected into individual, enucleated eggs of diploid provenance, some triploid larvae that developed as far as the swimming stage were obtained (McKinnell *et al.* 1969). Despite the great interest of these results, which show the multipotentiality of kidney nuclei, it is not yet possible to conclude that reversal of *malignant*-cell nuclei had taken place, as the tumors of origin contain nontumorous blood vessels and connective tissue and possibly admixed normal kidney parenchymal cells, all triploid in this experiment.

Partial tumor cell normalization is a well-documented phenomenon in Friend virus-induced murine erythroleukemia cells. After their exposure to dimethyl sulfoxide, most of the erythroblasts undergo some maturation to normoblasts and show increased hemoglobin synthesis (Friend *et al.* 1971). Erythroleukemic cells also seem to respond to a normal environment, such as the spleen of an irradiated nonleukemic host, where they have been reported to undergo normal self-renewal and differentiation, effectively "rescuing" the host (Matioli 1973). It remains to be determined whether the reduction of malignancy entails some reversible change in properties of the causal viral agent and whether heterogeneous cell populations might be involved.

The possibility of reversal in transformed cultured cells has also been examined in many other studies that will not be reviewed here. These show that cells may lose their transformed phenotypes if an oncogenic virus is lost or if the cellular chromosome constitution changes in specific ways.

Although the foregoing examples are excerpted from an extensive literature (see references in Willis 1967, Braun 1969, Friend *et al.* 1971, Pierce 1974), it appears that no instance of reversal of an animal tumor as conclusive and extensive as that of the mouse teratocarcinoma has been described. The mode of its experimental normalization may provide a paradigm for some malignancies of more specialized mammalian tissues: If a specialized tissue tumor had arisen without mutation, its malignant cells might be induced to differentiate normally if placed in the normal stem cell developmental environment of that particular tissue. In most cases, the appropriate environment would probably not be the blastocyst. With this aim in view, we are now screening some primary and transplantable mouse tumors for "normalization" candidates. The more promising ones would be expected to be still fairly euploid and well differentiated. New microsurgical techniques will undoubtedly have to be devised for introduction of malignant cells in appropriate sites during relatively later stages of embryogenesis.

DEVELOPMENTAL ORIGINS OF MALIGNANCY

In recent years, there has been an increasing appreciation of the possibility that neoplasia may occur fundamentally as an aberration of cell differentiation (e.g., Pierce 1967, Markert 1968). Evidence taken as support for this view has included the following: production by tumors of substances ordinarily found only in fetal stages of the corresponding normal tissue, e.g., fetal liver isozymes (Schapira *et al.* 1975) or alpha-fetoprotein (Abelev 1974) in hepatomas, formation of products usually found in other tissues, e.g., hormone secretion by lung tumors (Lipsett 1965), and shifts in histological composition of a tumor toward a more differentiated or benign state, e.g., the transition of a neuroblastoma toward a ganglioneuroma (Cushing and Wolbach 1927). However, some of these features, such as the synthesis of fetal or atypical products, may be epiphenomena rather than primary ones.

Perhaps the most surprising hiatus in our knowledge of tumor etiology is the fact that we know so little about the identity of the target cells. The view, held by many, that relatively specialized cells become "dedifferentiated" is difficult to reconcile with all of the data. Among other difficulties with the hypothesis is the fact that terminally differentiated cells are generally nonmitotic, whereas neoplasms are characterized by a population of proliferating cells. It is the normal stem cells, which are responsible for the growth and differentiation of tissues, that are, instead, the logical candidates for the role of target cells in neoplasia. This point of view, as Pierce (1967) has emphasized, would account not only for the proliferative requirements of the malignancy but also for the frequent presence in tumors of a variety of cell types or stages.

In only a few tissues has the progression from stem cells been worked out, e.g., in the hematopoietic tissues, germinal layers of the skin, epithelium of the intestinal crypts, and spermatogonial cells. These are "stem cell renewal" systems (Leblond and Walker 1956) in which cell production continues to be balanced by cell loss. It seems likely that analogous systems, without continuous renewal, occur in all tissues at some time in their developmental history. Our understanding of normal differentiation would undoubtedly be greatly increased if stem cells could be recognized and characterized in all tissues.

A basic stem cell scheme of normal differentiation is diagrammed in the left-hand panel of Figure 9. As shown, there is a hierarchy of stem cells of increasing degrees of differentiation, starting with the most primitive (i.e., totipotent) ones and proceeding through progressive levels of specialization to form the stem cells of a particular tissue type. At any level, the mitotic progeny of a stem cell may remain like the parent cell (proliferative option) or differentiate further (differentiative option). According to the hypothesis shown in the right-hand panel of Figure 9, any normal stem cell (e.g., "specialized stem cell I") in the developmental hierarchy may undergo transformation—whether as the result of mutational or nonmutational triggering events—and become a malignant stem cell. In effect, the malignant stem cells' progeny retain their proliferative option but suffer *a diminution or loss of their differentiative option.*

Tumors are roughly characterizable histologically as undifferentiated (anaplastic), poorly differentiated, or well differentiated. As indicated in Figure 9, these alternatives could arise in either of two ways: as the result of the hierarchical

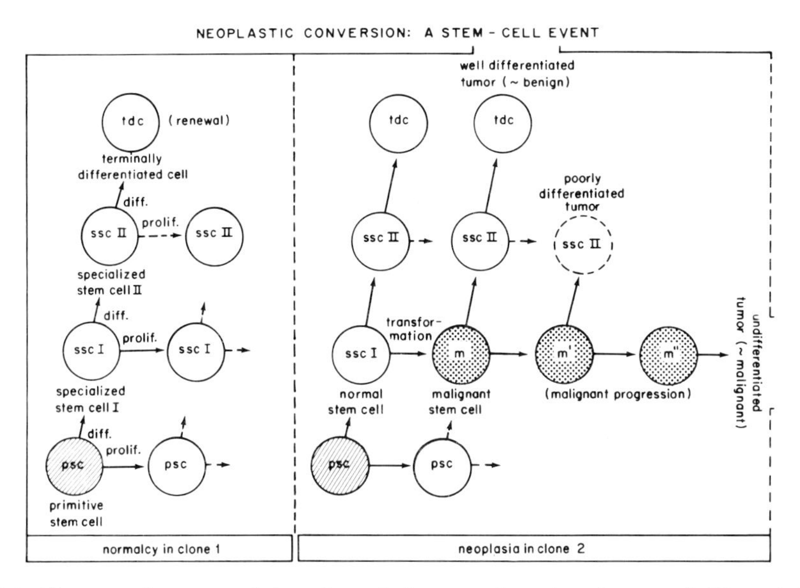

FIG. 9. Diagrammatic representation of neoplasia as an aberration of stem-cell differentiation. See text for details. (Reprinted, with permission of Academic Press, Inc., from Mintz 1976.)

level of the target normal stem cells or of retention of varying degrees of differentiation by the malignant stem cells. If the latter (m) continue to undergo malignant progression involving natural selection for more rapidly dividing variants (m′ and m″), the ultimate, highly modified neoplastic stem cell might have virtually no capacity for differentiation. Thus, the composition of the tumor could shift in time toward diminished differentiation, perhaps along with chromosomal changes resulting in aneuploidy. At the other extreme, a relatively well-differentiated tumor implies substantial retention of stem cell differentiation and is more likely to remain close to euploidy. If the stem cells are lost or retain a modest degree of proliferation, the tumor may become benign.

Pierce (1967, 1974) has aptly termed tumorigenesis a "caricature of stem cell renewal" and has obtained in vivo experimental data in support of this interpretation. For example, when transplantable squamous cell carcinomas of rats were labeled with ^{3}H-thymidine, label appeared first in the undifferentiated cells and only substantially later in the well-differentiated squamous cell "pearls" (Pierce and Wallace 1971). Proliferative stem cells that migrated into the pearls had apparently become differentiated. They subsequently ceased mitosis and became benign, i.e., incapable of forming a tumor upon retransplantation.

Many kinds of stem cells have complex life histories, e.g., the cells in spermatogonial renewal. A hypothesis that suggests itself is that conversion to malignancy may be possible in only a limited part of that history. Neoplastic conversion of totipotent mammalian cells may prove to be a case in point. Totipotency characterizes not only germ cells but also zygotes, embryo cells in preimplantation stages (Mintz 1974), and some embryo cells of the postimplantation egg cylinder (Stevens 1970, Levak-Svajger and Svajger 1974). Is malignant conversion in fact possible in only one stage of the life history, as proposed here, or in many or all stages? The evidence seems at first to imply that teratocarcinogenesis may occur in virtually any stage. Primordial germ cells, or embryos of any stage through the early egg cylinder, may form teratocarcinomas in an ectopic site (Stevens 1967, 1970); spontaneous testicular or ovarian tumors may develop from germ cells in situ (Linder 1969, Stevens and Varnum 1974); and spontaneous teratomas may also occasionally be found in other sites, after having possibly arisen from "lost" germ cells or embryo cells (Willis 1967). However, the actual transformational event, as distinct from cell proliferation, may be occurring in only one developmental stage, shared ultimately by all.

According to one long-standing opinion, all teratomas come from germ cells; according to another, from somatic embryo cells (see the reviews by Stevens 1967a, 1970, Pierce 1967, Damjanov and Solter 1974). As Stevens (1970) has pointed out, the germinal and somatic viewpoints converge, in that parthenogenetically developing germ cells form early embryos and embryos that give rise to teratomas may first have to form primordial germ cells. In an effort to break into this cycle, he ectopically transplanted fetal mouse genital ridges from matings expected to yield 25% of embryos homozygous for a genotype *(Sl/Sl)* that causes virtual absence of primordial germ cells (Stevens 1967b). Teratomas failed to form in approximately that percent of the embryos; this pointed to the germ cells as the target cells in teratocarcinogenesis. But absence of germ cells in transferred fetal germinal ridges of the *Sl/Sl* genotype leaves no source from which an early embryo might develop in the grafts. We have, in my laboratory, chosen another place to break into the cycle of events. Genetically sterile, early (six-day) embryos (rather than later-stage germinal ridges) and their presumptively fertile littermates are being transplanted to an ectopic site *before* the time of primordial germ cell formation in controls: It remains to be seen whether teratocarcinogenesis occurs in the presumptively sterile embryos, in which case the target cell would have to be a somatic not a germ cell. The significance of this experiment is that it attempts to distinguish between two events: a *potentiating circumstance* and actual *malignant conversion.* An example of the former would be any condition under which cells are inappropriately stimulated to divide, e.g., germ cell parthenogenesis in the gonad. But this need not in itself constitute neoplastic conversion. A further, still unknown "transformational" event would be required.

Another possible clue to the stage of inception of teratocarcinomas might be obtained by identifying the normal cell stage to which the malignant stem cells most closely correspond. We have, therefore, been comparing the soluble proteins of teratocarcinoma cells with those of normal early embryo cells of various stages. From preliminary results of acrylamide gel electrophoretic separations, there are many differences between teratocarcinoma cells and normal embryo cells, although both are developmentally totipotent (Mintz *et al.* 1975).

From the fact that transplanted embryos older than six and one-half days can no longer form teratocarcinomas (Stevens 1970, Damjanov and Solter 1974), and the observation that isolated and transplanted "ectoderm" of egg cylinder stages can form teratocarcinomas (Diwan and Stevens 1976), it appears that the day 6 "ectoderm" may still be totipotent and may be the stage appropriate for neoplastic conversion. Germ cells or embryo cells finding themselves in "potentiating circumstances" at earlier stages may have to reach the equivalent period to be susceptible to malignant conversion.

MOSAIC MICE FROM MUTAGENIZED TERATOCARCINOMA CELLS

In confronting the question whether malignant teratocarcinoma cells are developmentally totipotent, I had in view two objectives. One, already discussed, concerned the problem of malignancy: Successful normalization of this particular tumor would identify it as the result of a nonmutational developmental disturbance, in this case, of the most primitive stem cells, hence as a possible model for some kinds of tumors of more specialized stem cells. The other objective concerned the problem of differentiation: If these malignant cells did indeed prove to be totipotent, they could in effect serve as "surrogate eggs." That is, their totipotency would make available to the experimenter a ready source of cells far more numerous than are laboratory mouse eggs and able, unlike eggs, to proliferate their own cell type in culture. Therefore, they could serve as the objects of novel, large-scale mutagenesis experiments in vitro, where they could provide specifically selected mutations in a cell type capable of being returned to an in vivo environment for the full range of differentiation. Differentiation of somatic tissues would test gene expressions and their developmental consequences. Differentiation of germ cells would permit the mutation to be mapped by recombination in meiosis and would lead to production of all-mutant progeny. For all of the above objectives, the indispensable first step is the association of the malignant cells with genetically marked normal early embryo cells, to obtain viable mice with cellular genetic mosaicism.

In conventional mammalian mutagenesis experiments, e.g., in mice, germ cells (or embryos) are exposed to irradiation or mutagenic agents and allowed to give rise to progeny without prior selection, except for negatively self-selected lethals. The mutant offspring are usually recognized by gross defects (e.g., behavioral or color changes), which are often parts of complex syndromes. When the molecular basis for the change has been sought, it has usually remained unidentified. Somatic cell geneticists have attempted to obtain defined biochemical mutations in mammalian cells by utilizing some of the approaches of microbial geneticists, e.g., by employing selective systems in mutagenized cell cultures (e.g., Puck 1971). Nevertheless, the kinds of cells employed in such experiments undergo little or no differentiation. Moreover, the genetic status of variations presumed to be due to mutations in cultured cells is difficult to establish definitively in a purely mitotic lineage, even with cell hybridization techniques. Only by transmission and orderly recombination during meiosis in the germ line can mutant genes be identified and mapped in detail.

The new experimental system shown here (Figure 10) and outlined previously (Mintz *et al.* 1975, Mintz and Illmensee 1975, Mintz 1976), combines the advantages of the in vitro and in vivo approaches. First, cultures of totipotent malignant teratocarcinoma stem cells are exposed to a mutagenic agent. Special media or conditions are then used to select for desired phenotypes, and the mutant clones are expanded. Next, a mutant cell is introduced into a blastocyst of another strain, and the blastocyst is surgically transferred to a foster mother. Some mosaic mice with both mutant (tumor-derived) and wild-type (blastocyst-derived) cells would be obtained. In such individuals, cells with mutations that might otherwise be lost through lethality to the organism might be "rescued" by coexistence with normal cells, as in viable allophenic mice with both lethally anemic *(W/W)* and wild-type cells (Mintz 1970b). The mosaic animals from mutagenized teratocarcinoma cell injections into blastocysts would undoubtedly vary considerably in the tissue distributions of their mutant cells, as was true in earlier allophenic populations; this would be an important tool for identifying the primary tissue in which mutant gene action is critical for the organismic phenotype. Animals with teratocarcinoma-derived germ cells would, in matings to mice of an appropriate (e.g., blastocyst) strain, yield F_1 progeny heterozygous for the mutation in all their diploid cells. Dominant or semidominant phenotypes

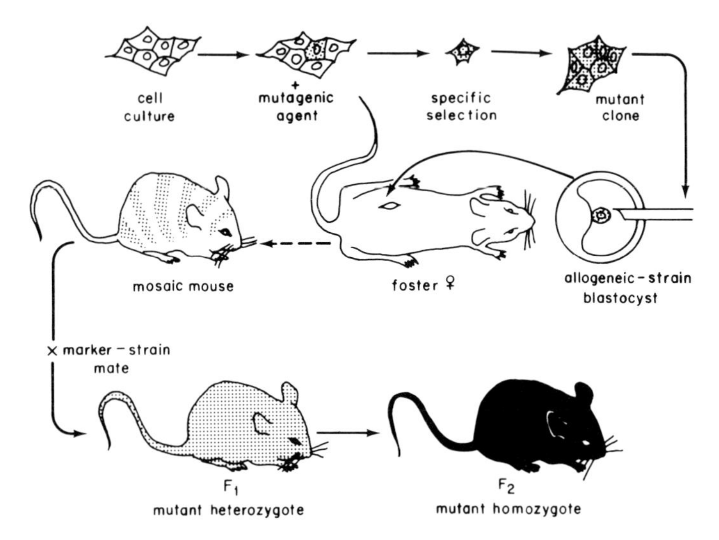

FIG. 10. The diagram summarizes a new scheme for generating specific genetic and biochemical probes of mammalian differentiation and for obtaining mouse models of human genetic diseases. Developmentally totipotent mouse teratocarcinoma cells are first mutagenized (shaded cells) during an in vitro culture period, and mutant clones bearing a specific biochemical (e.g., enzymatic) lesion are selected by manipulating the culture medium or conditions. A cell from such a clone is then injected into a mouse blastocyst of a genetically different strain, and the injected blastocyst is surgically transferred to the uterus of a pseudopregnant foster mother. In the resultant allophenic mouse, with differentiated tumor-derived mutant cells in some or all of its somatic tissues, some effects of the mutation might be analyzable. Gametes produced from the tumor cell lineage would yield F_1-mutant heterozygotes, for study of developmental effects of dominant or codominant mutations. For recessive mutations, homozygous F_2 progeny would be obtained. (Reprinted, with permission of Academic Press, Inc., from Mintz 1976.)

could be studied in differentiation of the F_1; F_2 homozygous segregants would be produced for expression of recessive mutant genes. Mapping of the mutation in relation to known genes would be done through the transmission studies.

This scheme would lend itself to analyses of numerous problems in mammalian differentiation, genetics, and metabolism. One of the most promising possibilities is that the system could be used to create animal models of human genetic diseases. Many serious human diseases, for which no animal model is available, are known to be caused by specific mutant genes that cause known enzymatic lesions. The diseases often have a complex clinical picture, and the chain of events has remained obscure, at least partly because of prenatal cascades of effects not readily observable in human subjects. In mosaic mice produced from specifically selected mutagenized teratocarcinoma cells, the fortuitous presence of mutant cells in only one or a few tissues of some individuals could enable the primary focus of the disorder to be identified. In the F_1 or F_2 progeny, the entire animal would serve as a model in which the developmental progression of the disease could be analyzed, and experimental cures attempted.

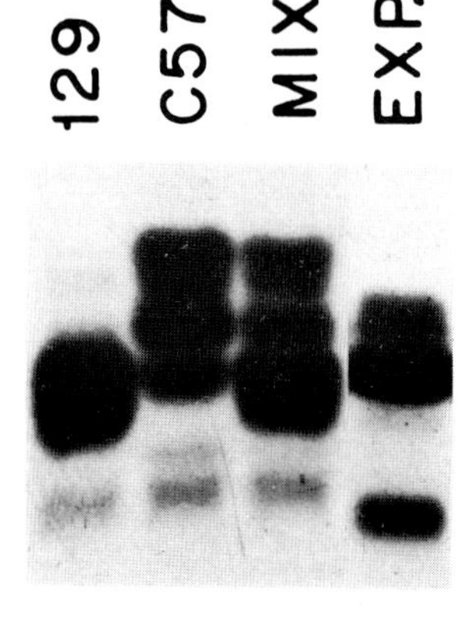

FIG. 11. Acrylamide gel electrophoresis of strain variants of major urinary proteins, including 129 (teratocarcinoma-strain) and C57 (blastocyst-strain) controls and a 1 : 1 control mixture. An experimental mouse, produced as shown in Figure 10, contains some 129-strain type and therefore has functional hepatocytes derived from the mutagenized teratocarcinoma cells, which were in this case specifically selected for deficiency of the X-linked enzyme HGPRT. (From unpublished data of Dewey, D. Martin, Jr., G. Martin, and Mintz.)

An example of a human hereditary disorder that would lend itself to such an analysis is the Lesch-Nyhan syndrome (Lesch and Nyhan 1964), a metabolic derangement showing excess uric acid synthesis and bizarre behavioral manifestations, due to deficiency of the X-linked enzyme, hypoxanthine-guanine phosphoribosyl-tranferase (HGPRT) (Seegmiller *et al.* 1967). In collaboration with Dr. Michael J. Dewey in my laboratory, we have injected mouse blastocysts with teratocarcinoma cells from a mutagenized line selected for survival in 6-thioguanine. The mutant cells, previously grown and selected by Drs. David Martin, Jr., and Gail Martin (at the University of California Medical School at San Francisco), are grossly deficient in HGPRT activity. We have recently obtained several successful cases of allophenic mice in which one or more tissues clearly contain cells of the mutant strain. An example of evidence for their presence in the liver, as well as their differentiation into functional hepatocytes, is seen in Figure 11. The analysis is conducted in living animals, indirectly, by means of electrophoretic strain differences in major urinary proteins produced in the liver. This marks the first instance of mammalian organismic differentiation from preselected mutant cells by a "parasexual" route. Further experiments on this and other mutational defects are in progress in our laboratory and promise to yield an exciting new period in the experimental study of mammalian development and disease.

CONCLUSION

As was true in microbial physiology, genetic mutants are also the most promising tool for analyzing control mechanisms in higher organisms and the basis for differentiation and its aberrations, including neoplastic disease. But two options are needed: First, it is indispensible to have a means of examining these processes in the organism itself. This has been accelerated by constructing whole animals that are, from earliest embryonic life onward, mosaics of genetically marked cells. The remaining requirement is some substantial measure of choice of appropriate mutations. The newer work described here may help to provide that choice.

ACKNOWLEDGMENTS

This program has been supported by U.S. Public Health Service Grants HD-01646, CA-06927, and RR-05539, and by an appropriation from the Commonwealth of Pennsylvania. I am grateful to many colleagues and to the Institute for Cancer Research for their participation, stimulus, and encouragement.

REFERENCES

Abelev, G. I. 1974. α-fetoprotein as a marker of embryo-specific differentiations in normal and tumor tissues. Transplant. Rev. 20:3–37.

Ashcraft, K. W., and T. M. Holder. 1974. Hereditary presacral teratoma. J. Pediatr. Surg. 9:691–697.

Artzt, K., P. Dubois, D. Bennett, H. Condamine, C. Babinet, and F. Jacob. 1973. Surface antigens common to mouse cleavage embryos and primitive teratocarcinoma cells in culture. Proc. Natl. Acad. Sci. USA 70:2988–2992.

Braun, A. C. 1959. A demonstration of the recovery of the crown-gall tumor cell with the use of complex tumors of single-cell origin. Proc. Natl. Acad. Sci. USA 45:932–938.

Braun, A. C. 1969. The Cancer Problem: A Critical Analysis and Modern Synthesis. Columbia University Press, New York.

Brinster, R. L. 1974. The effect of cells transferred into the mouse blastocyst on subsequent development. J. Exp. Med. 140:1049–1056.

Burnet, M. 1974. The biology of cancer, *in* Chromosomes and Cancer, J. German, ed., John Wiley and Sons, New York, pp. 21–38.

Condamine, H., R. P. Custer, and B. Mintz. 1971. Pure-strain and genetically mosaic liver tumors histochemically identified with the β-glucuronidase marker in allophenic mice. Proc. Natl. Acad. Sci. USA 68:2032–2036.

Cushing, H., and S. B. Wolbach. 1927. The transformation of a malignant paravertebral sympathicoblastoma into a benign ganglioneuroma. Am. J. Pathol. 3:203–220.

Damjanov, I., and D. Solter. 1974. Experimental teratoma. Curr. Top. Pathol. 59:69–130.

Dewey, M. J., A. G. Gervais, and B. Mintz. 1976. Brain and ganglion development from two genotypic classes of cells in allophenic mice. Dev. Biol. 50:68–81.

Dewey, M. J., J. D. Gearhart, and B. Mintz. 1977. Cell surface antigens of totipotent mouse teratocarcinoma cells grown *in vivo:* Their relation to embryo, adult, and tumor antigens. Dev. Biol. 55:359–374.

Diwan, S. B., and L. C. Stevens. 1976. Development of teratomas from the ectoderm of mouse egg cylinders. J. Natl. Cancer Inst. 57:937–942.

Dunn, G. R., and L. C. Stevens. 1970. Determination of sex of teratomas derived from early mouse embryos. J. Natl. Cancer Inst. 44:99–105.

Foulds, L. 1954. The experimental study of tumor progression: A review. Cancer Res. 14:327–339.

Friedman, J. M., and P. J. Fialkow. 1976. Cell marker studies of human tumorigenesis. Transplant. Rev. 28:2–33.

Friend, C., W. Scher, J. G. Holland, and T. Sato. 1971. Hemoglobin synthesis in murine virus-induced leukemic cells *in vitro:* Stimulation of erythroid differentiation by dimethyl sulfoxide. Proc. Natl. Acad. Sci. USA 68:378–382.

German, J. 1974. Bloom's syndrome. II. The prototype of human genetic disorders predisposing to chromosome instability and cancer, *in* Chromosomes and Cancer, J. German, ed., John Wiley and Sons, New York, pp. 601–617.

Gooding, L. R., and M. Edidin. 1974. Cell surface antigens of a mouse testicular teratoma. Identification of an antigen physically associated with H-2 antigens on tumor cells. J. Exp. Med. 140:61–78.
Hauschka, T. S., and A. Levan. 1958. Cytological and functional characterization of single cell clones isolated from the Krebs-2 and Ehrlich ascites tumors. J. Natl. Cancer Inst. 21:77–111.
Illmensee, K., and B. Mintz. 1976. Totipotency and normal differentiation of single teratocarcinoma cells cloned by injection into blastocysts. Proc. Natl. Acad. Sci. USA 73:549–553.
Kleinsmith, L. J., and G. B. Pierce, Jr. 1964. Multipotentiality of single embryonal carcinoma cells. Cancer Res. 24:1544–1551.
Knudson, A. G., Jr. 1973. Mutation and human cancer. Adv. Cancer Res. 17:317–352.
Leblond, C. P., and B. E. Walker. 1956. Renewal of cell populations. Physiol. Rev. 36:255–276.
Lesch, M., and W. L. Nyhan. 1964. A familial disorder of uric acid metabolism and central nervous system function. Am. J. Med. 36:561–570.
Levak-Svajger, B., and A. Svajger. 1974. Investigation on the origin of the definitive endoderm in the rat embryo. J. Embryol. Exp. Morphol. 32:445–459.
Lin, T. P. 1966. Microinjection of mouse eggs. Science 151:333–337.
Linder, D. 1969. Gene loss in human teratomas. Proc. Natl. Acad. Sci. USA 63:699–704.
Lipsett, M. B. 1965. Humoral syndromes associated with cancer. Cancer Res. 25:1068–1073.
Makino, S. 1956. Further evidence favoring the concept of the stem cell in ascites tumors of rats. Ann. NY Acad. Sci. 63:818–830.
Markert, C. L. 1968. Neoplasia: A disease of cell differentiation. Cancer Res. 28:1908–1914.
Matioli, G. 1973. Friend leukemic mouse stem cell reversion to normal growth in irradiated hosts. J. Reticuloendothel. Soc. 14:380–386.
McCann, J., and B. N. Ames. 1976. Detection of carcinogens as mutagens in the *Salmonella/* microsome test: Assay of 300 chemicals. Proc. Natl. Acad. Sci. USA 73:950–954.
McKinnell, R. G., B. A. Deggins, and D. D. Labat. 1969. Transplantation of pluripotential nuclei from triploid frog tumors. Science 165:394–396.
Mintz, B. 1962. Formation of genotypically mosaic mouse embryos. Am. Zool. 2:432.
Mintz, B. 1965. Genetic mosaicism in adult mice of quadriparental lineage. Science 148:1232–1233.
Mintz, B. 1967. Gene control of mammalian pigmentary differentiation. I. Clonal origin of melanocytes. Proc. Natl. Acad. Sci. USA 58:344–351.
Mintz, B. 1969. Developmental mechanisms found in allophenic mice with sex chromosomal and pigmentary mosaicism, *in* The First Conference on the Clinical Delineation of Birth Defects (Original Articles Series Vol. 5), D. Bergsma and V. McKusick, eds., National Foundation, New York, pp. 11–22.
Mintz, B. 1970a. Neoplasia and gene activity in allophenic mice, *in* Genetic Concepts and Neoplasia (The University of Texas System Cancer Center 23rd Annual Symposium on Fundamental Cancer Research), Williams and Wilkins Co., Baltimore, pp. 477–517.
Mintz, B. 1970b. Gene expression in allophenic mice, *in* Control Mechanisms in Expression of Cellular Phenotypes (Symposium of the International Society for Cell Biology) Vol. 9., H. Padykula, ed., Academic Press, New York, pp. 15–42.
Mintz, B. 1971a. Allophenic mice of multi-embryo origin, *in* Methods in Mammalian Embryology, J. Daniel, Jr., ed. Freeman, San Francisco, pp. 186–214.
Mintz, B. 1971b. Clonal basis of mammalian differentiation, *in* Control Mechanisms of Growth and Differentiation (25th Symposium of the Society for Experimental Biology), D. D. Davies and M. Balls, eds., Cambridge University Press, pp. 345–370.
Mintz, B. 1971c. Genetic mosaicism *in vivo:* Development and disease in allophenic mice. Fed. Proc. 30:935–943.
Mintz, B. 1974. Gene control of mammalian differentiation. Annu. Rev. Genet. 8:411–470.
Mintz, B. 1976. Gene expression in neoplasia and differentiation. Harvey Society Lectures (In press).
Mintz, B., and W. W. Baker. 1967. Normal mammalian muscle differentiation and gene control of isocitrate dehydrogenase synthesis. Proc. Natl. Acad. Sci. USA 58:592–598.
Mintz, B., R. P. Custer, and A. J. Donnelly. 1971. Genetic diseases and developmental defects analyzed in allophenic mice. Int. Rev. Exp. Pathol. 10:143–179.
Mintz, B., and K. Illmensee. 1975. Normal genetically mosaic mice produced from malignant teratocarcinoma cells. Proc. Natl. Acad. Sci. USA 72:3585–3589.

Mintz, B., K. Illmensee, and J. D. Gearhart. 1975. Developmental and experimental potentialities of mouse teratocarcinoma cells from embryoid body cores, *in* Teratomas and Differentiation, M. Sherman and D. Solter, eds., Academic Press, New York, pp. 59–82.

Mintz, B., and J. Palm. 1969. Gene control of hematopoiesis. I. Erythrocyte mosaicism and permanent immunological tolerance in allophenic mice. J. Exp. Med. 129:1013–1027.

Mintz, B., and S. Sanyal. 1970. Clonal origin of the mouse visual retina mapped from genetically mosaic eyes. Genetics 64:43–44.

Mintz, B., and W. K. Silvers. 1967. "Intrinsic" immunological tolerance in allophenic mice. Science 158:1484–1487.

Mintz, B., and G. Slemmer. 1969. Gene control of neoplasia. I. Genotypic mosaicism in normal and preneoplastic mammary glands of allophenic mice. J. Natl. Cancer Inst. 43:87–95.

Moore, W. J., and B. Mintz. 1972. Clonal model of vertebral column and skull development derived from genetically mosaic skeletons in allophenic mice. Dev. Biol. 27:55–70.

Nowell, P. C. 1974. Chromosome changes and the clonal evolution of cancer, *in* Chromosomes and Cancer, J. German, ed., John Wiley and Sons, New York, pp. 267–285.

Nowell, P. C., and D. A. Hungerford. 1960. A minute chromosome in human chronic granulocytic leukemia. Science 132:1497.

Papaionnou, V. E., M. W. McBurney, R. L. Gardner, and M. J. Evans. 1975. Fate of teratocarcinoma cells injected into early mouse embryos. Nature 258:70–73.

Pierce, G. B. 1967. Teratocarcinoma: Model for a developmental concept of cancer. Curr. Top. Dev. Biol. 2:223–246.

Pierce, G. B. 1974. The benign cells of malignant tumors, *in* Developmental Aspects of Carcinogenesis and Immunity, T. J. King, ed., Academic Press, New York, pp. 3–22.

Pierce, G. B., Jr., and T. F. Beals. 1964. The ultrastructure of primordial germinal cells of the fetal testes and of embryonal carcinoma cells of mice. Cancer Res. 24:1553–1567.

Pierce, G. B., L. C. Stevens, and P. K. Nakane. 1967. Ultrastructural analysis of the early development of teratocarcinomas. J. Natl. Cancer Inst. 39:755–773.

Pierce, G. B., and C. Wallace. 1971. Differentiation of malignant to benign cells. Cancer Res. 31:127–134.

Prehn, R. T. 1976. Do tumors grow because of the immune response of the host? Transplant. Rev. 28:34–42.

Prigogina, E. L., and E. W. Fleischman. 1975. Certain patterns of karyotype evolution in chronic myelogenous leukemia. Chromosome abnormalities in CML. Humangenetik 30:113–119.

Puck, T. T. 1971. Biochemical and genetic studies on mammalian cells. In Vitro 7:115–119.

Rowley, J. D. 1974. Do human tumors show a chromosome pattern specific for each etiologic agent? J. Natl. Cancer Inst. 52:315–320.

Schapira, F., A. Hatzfeld, and A. Weber. 1975. Resurgence of some fetal isozymes in hepatoma, *in* Isozymes III. Developmental Biology, C. L. Markert, ed., Academic Press, New York, pp. 987–1003.

Seegmiller, J. E., F. M. Rosenbloom, and W. N. Kelly. 1967. Enzyme defect associated with a sex-linked human neurological disorder and excessive purine synthesis. Science 155:1682–1684.

Stern, P. L., G. R. Martin, and M. J. Evans. 1975. Cell surface antigens of clonal teratocarcinoma cells at various stages of differentiation. Cell 6:455–465.

Stevens, L. C. 1967a. The biology of teratomas. Adv. Morphog. 6:1–31.

Stevens, L. C. 1967b. Origin of testicular teratomas from primordial germ cells in mice. J. Natl. Cancer Inst. 38:549–552.

Stevens, L. C. 1970. The development of transplantable teratocarcinomas from intratesticular grafts of pre- and postimplantation mouse embryos. Dev. Biol. 21:364–382.

Stevens, L. C., and K. P. Hummel. 1957. A description of spontaneous congenital testicular teratomas in strain 129 mice. J. Natl. Cancer Inst. 18:719–747.

Stevens, L. C., and D. Varnum. 1974. The development of teratomas from parthenogenetically activated ovarian mouse eggs. Dev. Biol. 37:369–380.

Steward, F. C., P. V. Ammirato, and M. O. Mapes. 1970. Growth and development of totipotent cells. Some problems, procedures and perspectives. Annals of Botany 34:761–787.

Willis, R. A. 1967. Pathology of Tumors. Butterworth and Co., London.

Stage of Cellular Differentiation in Tumor Cell Origin—Germ Cells

Cell Differentiation and Neoplasia, edited by Grady F. Saunders. Raven Press, New York © 1978.

Neoplasms as Caricatures of Tissue Renewal

G. B. Pierce and W. F. Cox, Jr.

Department of Pathology, University of Colorado Medical Center, Denver, Colorado 80262

Most investigators believe that cancer cells are derived by somatic mutation from normal differentiated cells. Dedifferentiation is believed to result with the loss of phenotypic traits and evolution of malignant cells. This is a simple explanation of the stability and heritability of neoplastic change. Stability and heritability of phenotype is also characteristic of differentiation, and since all other tissues, of whatever species, are derived by the process of cell division, differentiation, and organization, our laboratory has focused on the developmental aspects of tumors (Pierce 1967, Pierce and Dixon 1959, Pierce and Wallace 1971, Pierce *et al.* 1960, Wylie *et al.* 1973).

As the title of this paper implies, these researches have led to the concept that tumors are caricatures of the process of renewal of normal tissues. Eventually, the differentiation that occurs in tumors may be enhanced to serve as an alternative to cytotoxic therapy. In the interim, it is surprising what "sense" tumors make when viewed developmentally.

The first direct demonstration that differentiation occurred in tumors was obtained in studies of teratocarcinomas (Pierce and Dixon 1959, Pierce *et al.* 1960). Teratocarcinomas are so named because they contain elements representing each of the three embryonic germ layers, and they are malignant because they contain an undifferentiated, highly malignant tissue that has been called embryonal carcinoma, because of its resemblance to normal embryonic epithelium (Dixon and Moore 1953). These tumors occur in human beings, mice (Stevens and Little 1954), plants (Braun 1972), and other species. The experiments to be reviewed here were performed on transplantable teratocarcinomas first described by Stevens and Little (1954) of strain 129 mice. Many of the differentiated tissues in teratocarcinomas of strain 129 mice show organization. For example, a mass of cartilage may have endochondral ossification on one side with marrow cells occupying spaces between spicules of bone. Striated muscle may insert into the bone. The most extreme examples of organization have been called embryoid bodies because of their resemblance to mouse embryos. The most developed example of an embryoid body visualized to date closely resembles an eight and one-half day mouse embryo with neural folds, notochord, mesenchyme, somites, and an inverted layer of endoderm.

In attempting to convert the teratocarcinoma to the ascites, we inadvertently discovered a method of mass producing embryoid bodies (Pierce and Dixon

1959). Some of these bodies were large and cystic, measuring 5 or 6 mm in diameter, but the majority were small and visible only with the microscope. The large ones were encased in a layer of endoderm beneath which was a loose mesenchyme sometimes forming sinusoids with hematopoietic cells and containing masses of embryonal carcinoma. When such embryoid bodies were transplanted subcutaneously in strain 129 mice, teratocarcinomas developed (Pierce and Dixon 1959). These contained the 12 or 15 somatic tissues characteristic of the tumors. This was the first direct demonstration that differentiation occurred in neoplastic cells.

In an effort to determine which of the three embryonic germ layers was responsible for teratocarcinogenesis, a classical embryologic experiment was performed (Pierce *et al.* 1960). A large number of small embryoid bodies was transplanted singly and subcutaneously in animals, and six transplantation sites were examined daily for the first 12 days of the experiment by serial, histologic section. It turned out that the endoderm and mesoderm had limited capacity for proliferation and rapidly differentiated into mature derivatives of their respective germ layers. The embryonal carcinoma cells invaded through the wall of the embryoid body and proliferated into a small nodule of undifferentiated cells. These nodules of embryonal carcinoma cells eventuated the new teratocarcinoma with its multiplicity of tissues (Pierce *et al.* 1960).

At this point, it was desirable to know the biological nature of the tissues derived from embryonal carcinoma cells. Were they benign or malignant? To answer this question, it was necessary to separate the differentiated tissues from the embryonal carcinoma and observe their behavior on a host. To this end, extremely large embryoid bodies were grafted subcutaneously in animals. In the largest of these, the embryonal carcinoma had either died or completely differentiated, leaving elements representing each of the germ layers. These differentiated tissues persisted in the subcutaneous space of the animals for as long as six months. On the basis of their inability to cause destruction of their host, we concluded that they were benign (Pierce *et al.* 1960).

For confirmation of these studies, a cloning experiment was done in which small embryoid bodies composed of nodules of embryonal carcinoma overlaid by a single layer of proximal endodermal cells were dissociated with trypsin, and single cells were aspirated into microcapillary pipettes. The cells were transferred to the intraperitoneum of animals. Of about 372 such transfers, 42 developed into teratocarcinomas containing brain, bone, muscle, and glands characteristic of the tumors. It was concluded that embryonal carcinoma cells were the multipotential stem cells of teratocarcinomas capable of differentiating into functional and benign cells (Kleinsmith and Pierce 1964).

As was described by Dr. Mintz (see pages 27 to 53, this volume), three laboratories have now extended these studies to show that the cells which we characterized as multipotential and benign derivatives of embryonal carcinoma are in fact capable of participating in normal development evolving normal animals (Brinster 1975, Mintz *et al.* 1975, Papaioannou *et al.* 1975). This is an elegant confirmation of the experiments of Braun, who demonstrated that the tissues derived by differentiation of malignant cells in the plant teratoma system were, in fact, normal (Braun 1972).

In an effort to determine whether or not the lessons learned from study of teratocarcinomas applied to other tumors, autoradiographic studies with light and electron microscopes were performed on squamous cell carcinomas (Pierce and Wallace 1971) and adenocarcinomas of the breast (Wylie *et al.* 1973) at intervals following pulses of tritiated thymidine. In each situation, fully differentiated and functional cells were observed that were incapable of incorporating tritiated thymidine into their DNA. Undifferentiated cells picked up the label and subsequently differentiated into either keratinizing squamous cells or lactating breast cells. These cells were incapable of further synthesis of DNA, and when the squamous cells were selectively transplanted, they were incapable of giving rise to a tumor (Pierce and Wallace 1971). Whereas the teratocarcinoma had been considered a caricature of embryogenesis, these tumors were considered to be caricatures of tissue genesis.

The realization that highly malignant tumors could contain significant numbers of senescent or normal cells has interesting implications for immunology, electron microscopy, and biochemistry of tumors. Epithelial cells such as those of skin, bronchus, germinal epithelium, endometrium, cells lining the gastrointestinal tract, or mammary epithelium, when senescent, are shed from the body. Tumors lose this mechanism for getting rid of senescent progeny, and the breakdown products of these cells are presented to the reticuloendothelial system. If the antigens elicit an immune response, the tumor may become infiltrated with immunocytes, and in such tumors the prognosis is invariably better than in tumors lacking the immune response. The question to be answered is whether the prognosis is good because of the immune response or because the tumor is differentiating senescent cells. Similarly, biochemical analysis of tumors must take into account the fact that the malignant properties are confined to the stem cell population and possibly to some of their partially differentiated progeny. Even tumors that appear relatively homogeneous by light microscopy are extremely heterogeneous when viewed by the electron microscope. The percentage of cells in a tumor that have differentiated to a benign state varies widely from tumor to tumor, resulting in more biochemical variation between tumors than between tumors and normal tissue. The heterogeneity of cells in tumors must also be taken into account by electron microscopists interested in comparing membrane differentiation of normal and malignant tissues. It would be necessary to compare these differentiations between normal stem cells and malignant stem cells, if the results are to be valid. Unfortunately, understanding of the cellular heterogeneity of tumors imposed by differentiation, a heterogeneity that cannot be eliminated by cloning, creates more problems than it solves, at this time. When it is taken into account, the interpretation of immunological, biochemical, and ultrastructural studies will be much more meaningful.

The next important area of interest is the in vivo cell of origin of carcinomas and the relationship of benign and malignant tumors (Pierce 1974). The in vivo cell of origin of only one type of tumor has been proven conclusively. Stevens has shown that the primordial germ cell is the normal counterpart of embryonal carcinoma cells of teratocarcinoma (Stevens 1967, Stevens and Little 1954). In an electron microscopic study, it was observed that the state of differentiation of embryonal carcinoma cells closely approximated that of primordial germ cells (Pierce *et al.* 1967). This raised the possibility that rather than developing from well-differentiated cells, malignant stem cells might take origin from undifferentiated cells. This thought was reinforced by the observation of Franks and Wilson (1970) who observed comparable degrees of differentiation of normal fibroblasts and fibroblasts transformed by oncogenic viruses.

An ultrastructural comparison of the state of differentiation of stem cells of normal and malignant breast tumors has been made (Pierce *et al.* 1977). The stem cells were identified by autoradiography with light and electron microscopes, after pulses of tritiated thymidine. It turned out that the normal stem cells of the breast are as undifferentiated as are the malignant ones, confirming the idea that the target in carcinogenesis was probably an undifferentiated cell. If the target in carcinogenesis is an undifferentiated cell, then the whole concept of dedifferentiation as an essential mechanism in carcinogenesis can be effectively bypassed.

Confirmatory studies for the idea that normal and malignant stem cells should closely resemble each other was sought in studies comparing the ultrastructure of normal and malignant colonic stem cells (Pierce *et al.* 1977). Colonic tissue was chosen because it contains a mucinous series of cells, a vacuolated series that gives rise to the columnar epithelium and granular endocrine cells (argentaffin?), which are believed to be of neural crest origin. When identified with the electron microscope, the normal and malignant counterparts, from the standpoint of differentiation, were indistinguishable from each other. During these studies, an exceedingly undifferentiated cell was observed in the adenocarcinomas of the colon (Pierce *et al.* 1977). The nucleus had marginated chromatin, a light-staining cytoplasm containing numerous polysomes unattached to membranes, and a markedly infolded plasma membrane (see Figure 5). These membranes were connected to those of adjacent cells by desmosomes. Similar cells were found in the normal colon. Apparently they had been confused with lymphocytes previously, but the presence of desmosomes establishes their epithelial nature. It was postulated that they were undifferentiated colonic cells, possibly a tissue cytoblast, comparable to the hemocytoblast of the marrow. The hemocytoblast can differentiate into each of the hematopoietic stem lines, and it was postulated that the undifferentiated colonic cell might differentiate into mucous, vacuolated, and argentaffin cells. There was some evidence from the work of Chang and Leblond (1971) for the origin of argentaffin cells from an agranular precursor, but the idea that the argentaffin cells and the mucous and vacuolated cells should have a common origin was incompatible with the neural crest origin of the argentaffin cells. To examine this hypothesis, suspensions of adenocarcinoma cells were made, and single cells were transplanted into rats. To date, four tumors have developed from these transplants (Figures 1–5). As indicated in Table 1, there are marked variations in the cellular composition of the resulting cloned tumors, but each contains vacuolated, mucinous, and argentaffin cells. Thus, it would appear that the undifferentiated colonic cell is a multipotential stem cell capable of differentiating into each of the three differentiated cells of the colon and that the argentaffin cells are of endodermal origin (Cox and Pierce, unpublished observations).

The observation that normal and neoplastic stem cells are similar in their degree of differentiation sheds new light on the relationship of benign and malignant tumors. When chemical carcinogens are applied to the skin of an animal, benign tumors appear first, followed much later by the appearance of malignant ones. This has led to the concept that benign tumors are a stage in the development of malignant ones. An alternative explanation would be that the state of differentiation of the responding cell could determine the state of differentiation of the neoplastic stem cells derived from it and that the order of appearance of tumors would depend on how well the neoplastic stem cell could respond to the environment in which it found itself. For example, if an almost completely differentiated cell still capable of a round of mitosis responded to a carcinogenic stimulus by becoming neoplastic, it would become a neoplastic stem cell closely resembling the normal, and its potential would be such that it would form a benign tumor. The benign tumor would find all of its requirements in the normal

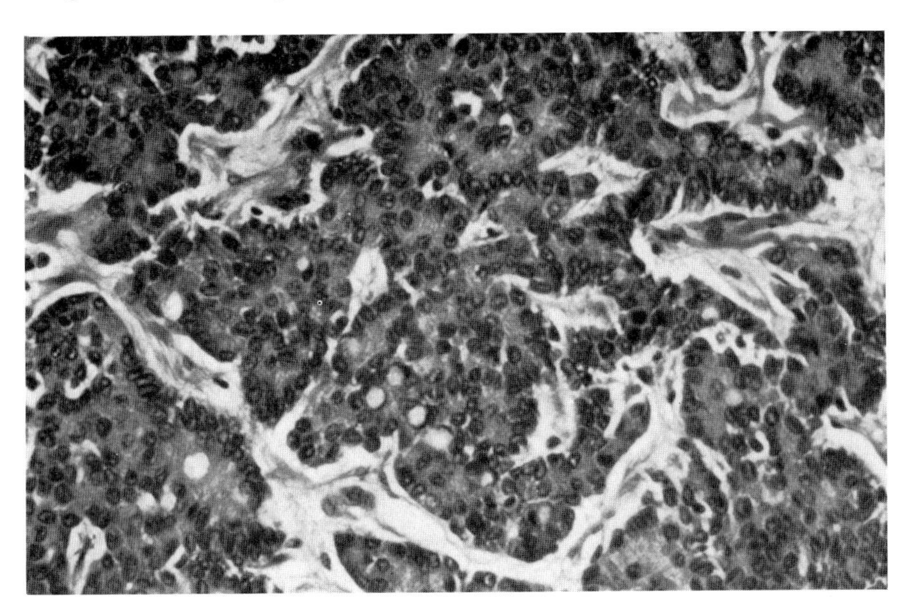

FIG. 1. Stock tumor illustrating the typical features of an adenocarcinoma composed of signet-ring and columnar cells. Granular endocrine cells are not easily recognized in tissues stained by hematoxylin and eosin (H&E). (X 1000)

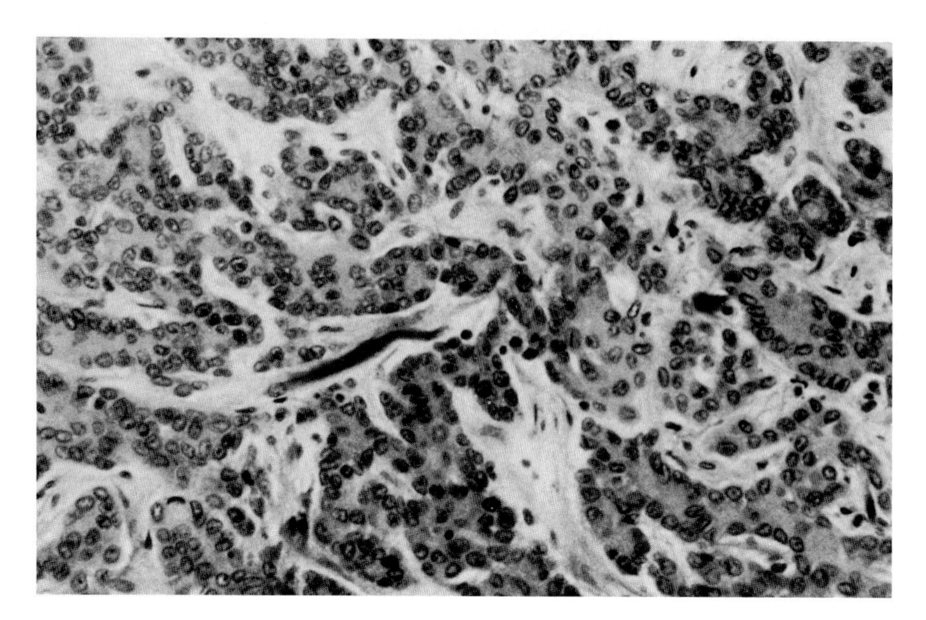

FIG. 2. Tumor clone 3232 (see Table 1). Note the adenocarcinomatous pattern of this tumor, composed of 96% granular endocrine cells and occasional goblet and columnar cells. It is of importance to note that this tumor, composed primarily of cells thought to be of neuroendocrine origin, has the same pattern as tumors which are of endodermal origin (see Figures 3 and 4). (X 1000, H & E)

tissue, and it would express its phenotype maximally and grow quickly. On the other hand, the least-differentiated normal stem cell would give rise to an equally undifferentiated malignant stem cell, which would be unable to express its phenotype because it would be so altered that it would not find its optimal growth requirements in the tissue of origin (Pierce 1974). There is compelling evidence from the experiments of Grobstein and Zwilling (1953), Fisher and Fisher (1959), and in the clinical situation of dormancy to indicate that a threshold number of cells is required for a normal differentiation to take place and for cells to express the malignant phenotype. The long latent period in carcinogenesis can be attributed to the growth-controlling effect of the normal environment on undifferentiated malignant stem cells.

This postulate raises another interesting postulate—Why should an almost terminally differentiated cell respond differently to a carcinogenic insult than its undifferentiated counterpart? It is our postulate that the explanation lies in the structure of the chromosomes.

In conclusion, we have attempted to bring you up to date on the research in our laboratory, which supports the view that tumors, like other tissues, are composed of proliferation compartments and differentiated compartments. The result is a caricature of the renewal process of normal tissue and suggests attempting to direct differentiation as an alternative to cytotoxic therapy of tumors.

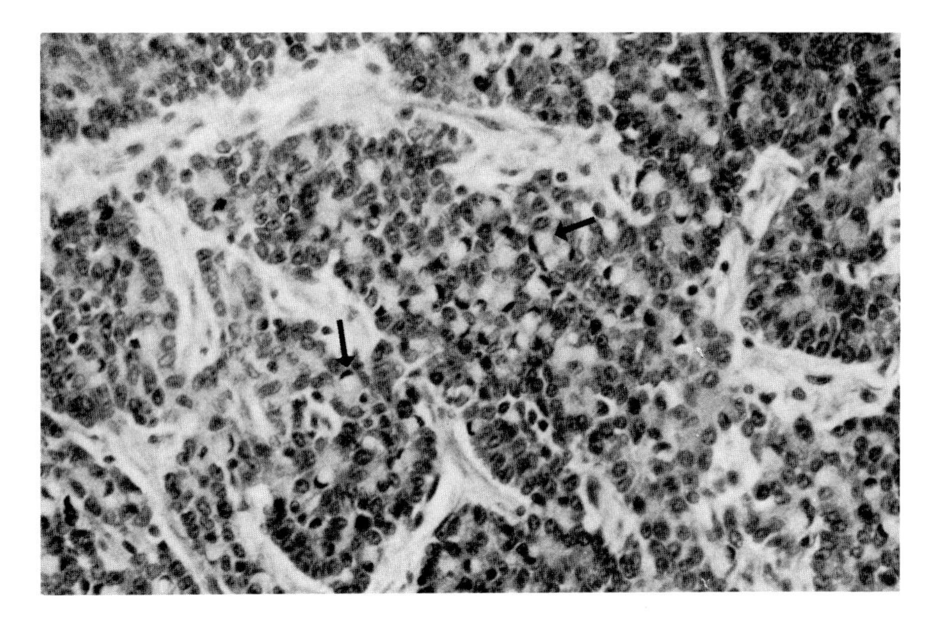

FIG. 3. Tumor clone 3236 (see Table 1). Note the adenocarcinomatous pattern of this tumor composed of equal numbers of vacuolated and mucinous cells (arrows). (X 1000, H & E)

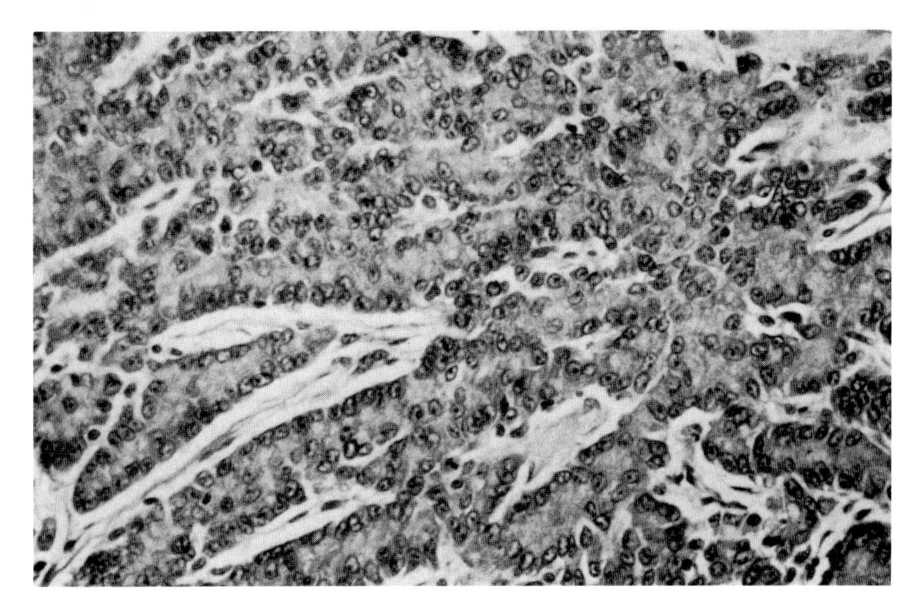

FIG. 4. Tumor clone 3318 (see Table 1). Note the adenocarcinomatous pattern of this tumor composed almost exclusively of columnar cells. (X 1000, H & E)

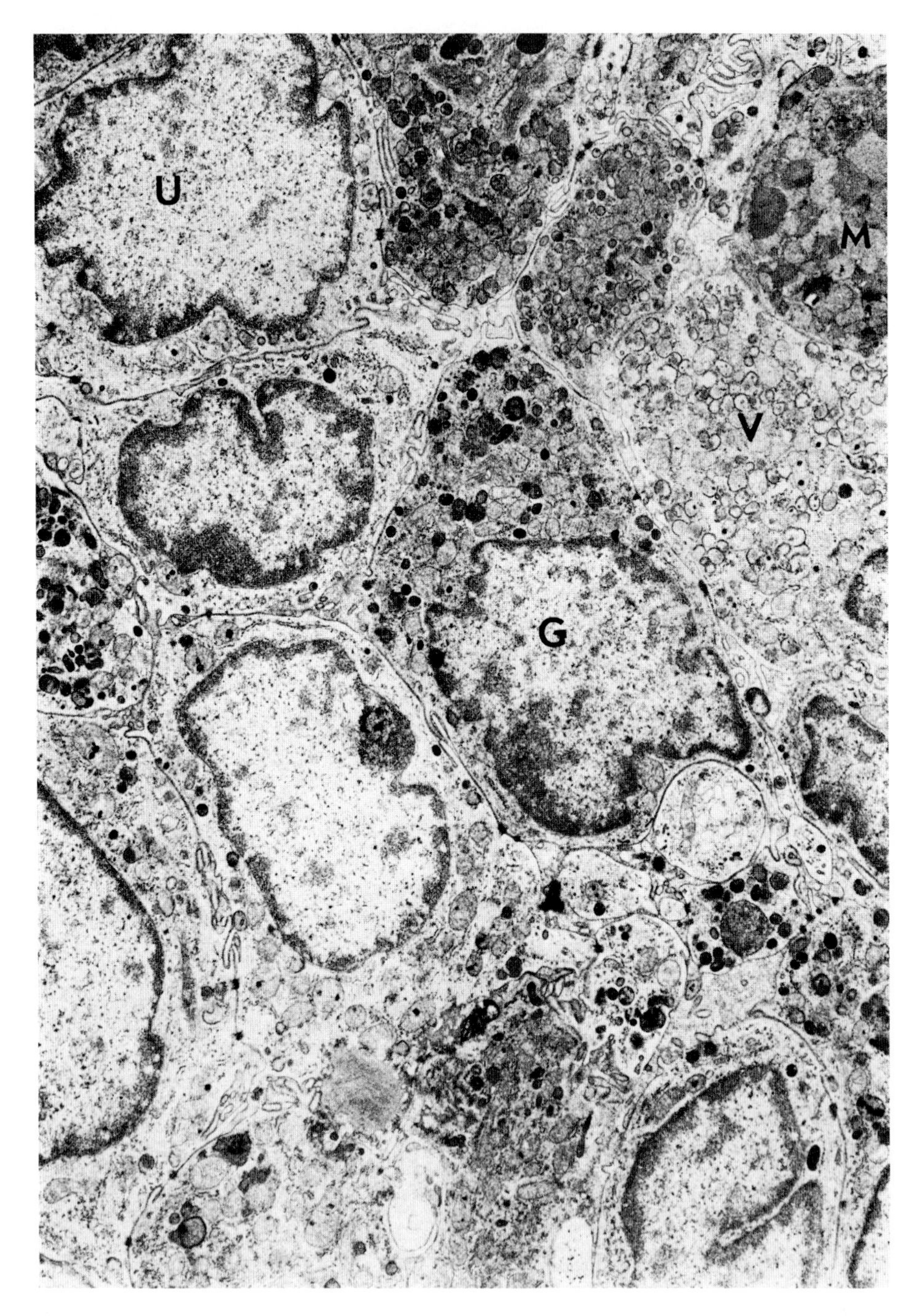

FIG. 5. Electron micrograph of clone 3232 illustrating the undifferentiated cells (U), granular endocrine cells (G), vacuolated cells (V), and mucinous cells (M). Note also that there are small granules in the undifferentiated cells. (OsO_4 fixative, X 6300)

TABLE 1. *Cell population of cloned rat colon adenocarcinoma*

Tumor number	Total count*	Mucinous cells	Vacuolated columnar cells	Granular endocrine cells
3232	1002	36 (3.6%)	8 (0.8%)	958 (95.6%)
3236	1018	444 (43.7%)	569 (55.8%)	5 (0.5%)
3317	1210	628 (52%)	542 (44%)	40 (4.0%)
3318	1038	64 (6.2%)	944 (90.9%)	30 (2.9%)

* Based on 10 HPF per tumor.

ACKNOWLEDGMENTS

This investigation was supported by Grant CA-15823 awarded by the National Cancer Institute, Grant AM-15663 awarded by the National Institute of Arthritis, Metabolic and Digestive Diseases, Department of Health, Education and Welfare, and by Grant PDT-23R from the American Cancer Society.

REFERENCES

Braun, A. C. 1972. The usefulness of the plant tumor system for studying the basic cellular mechanisms that underlie neoplastic growth generally, in Cell Differentiation, R. Harris, P. Allin, and D. Viza, eds., Munksgaard, Copenhagen, pp. 115–118.

Brinster, R. L. 1975. Can teratocarcinoma cells colonize the mouse embryo? *in* Teratomas and Differentiation, M. I. Sherman, and D. Solter, eds., Academic Press, Inc., New York, pp. 51–58.

Chang, G., and C. P. Leblond. 1971. Renewal of the epithelium in the descending colon of the mouse. Am. J. Anat. 131:73–99.

Dixon, F. J., Jr., and R. A. Moore. 1953. Testicular tumors: Clinicopathologic study. Cancer 6:427–454.

Franks, L. M., and P. D. Wilson. 1970. "Spontaneous" neoplastic transformation in vitro: The ultrastructure of the tissue culture cell. Eur. J. Cancer 6:517–523.

Fisher, B., and E. Fisher. 1959. Experimental evidence in support of the dormant tumor cell. Science 130:918–919.

Grobstein, C., and E. Zwilling. 1953. Modification of growth and differentiation of chorioallantoic grafts of chick blastoderm pieces after cultivation at a glass-clot interface. J. Exp. Zool. 122:259–284.

Kleinsmith, L. J., and G. B. Pierce. 1964. Multipotentiality of single embryonal carcinoma cells. Cancer Res. 24:1544–1551.

Mintz, B., K. Illmensee, and J. D. Gearhart. 1975. Developmental and experimental potentialities of mouse teratocarcinoma cells from embryoid body cores, *in* Teratomas and Differentiation, M. I. Sherman, and D. Solter, eds., Academic Press, Inc., New York, pp. 59–82.

Mintz, B. 1978. Genetic mosaicism and in vivo analyses of neoplasia and differentiation, *in* Cell Differentiation and Neoplasia (The University of Texas System Cancer Center 30th Annual Symposium on Fundamental Cancer Research, 1977), G. F. Saunders, ed., Raven Press, New York, pp. 27–53.

Papaioannou, V. E., M. W. McBurney, and R. L. Gardner. 1975. Fate of teratocarcinoma cells injected into early mouse embryos. Nature 258:70–73.

Pierce, G. B. 1967. Teratocarcinoma: Model for a developmental concept of cancer, *in* Current Topics in Developmental Biology, A. A. Moscona and A. Monroy, eds., Vol. 2. Academic Press, Inc., New York, pp. 223–246.

Pierce, G. B. 1974. Neoplasms, differentiation and mutations. Am. J. Pathol. 77:103–118.

Pierce, G. B., Jr., and F. J. Dixon, Jr. 1959. Testicular teratomas. I. Demonstration of teratogenesis by metamorphosis of multipotential cells. Cancer 12:573.

Pierce, G. B., and C. Wallace. 1971. Differentiation of malignant to benign cells. Cancer Res. 31:127–134.

Pierce, G. B., F. J. Dixon, and E. L. Verney. 1960. Teratocarcinogenic and tissue forming potentials of the cell types comprising neoplastic embryoid bodies. Lab. Invest. 9:583–602.
Pierce, G. B., L. C. Stevens, and P. K. Nakane. 1967. Ultrastructural analysis of the early development of teratocarcinomas. J. Natl. Cancer Inst. 39:755–773.
Pierce, G. B., P. K. Nakane, A. Martinez-Hernandez, and J. M. Ward. 1977. Ultrastructural comparison of differentiation of stem cells of colon and breast with their normal counterparts. J. Natl. Cancer Inst. 58:1329–1345.
Stevens, L. C. 1962. Testicular teratomas in fetal mice. J. Natl. Cancer Inst. 28:247–256.
Stevens, L. C. 1967. Origin of testicular teratomas from primordial germ cells in mice. J. Natl. Cancer Inst. 38:549–552.
Stevens, L. C., and C. C. Little. 1954. Spontaneous testicular teratomas in an inbred strain of mice. Proc. Natl. Acad. Sci. USA 40:1080–1087.
Wylie, C. V., P. K. Nakane, and G. B. Pierce. 1973. Degrees of differentiation in nonproliferating cells of mammary carcinoma. Differentiation 1:11–20.

Cell Differentiation and Neoplasia, edited by Grady F. Saunders. Raven Press, New York

Teratocarcinoma Cells and Cell Surface Differentiation

Michael Edidin, Suzanne Ostrand-Rosenberg, and Perry F. Bartlett

Biology Department, The Johns Hopkins University, Baltimore, Maryland 21218

A cell's position in the small aggregate constituting an early embryo and its contacts with other cells in the aggregate determine its fate in development (Gardner and Papaioannou 1975, Gardner and Rossant 1976). Logically, the factors affecting cell positioning must include elements of the cell surface, and a current opinion is that such elements (membrane-associated proteins and glycoproteins) may be detected and, to a first approximation, defined as antigens reacting either with antibodies or with elements of cellular immune responses. Such definition serves to map out differences between the surface structures of cells in early mammalian embryos and those of cells in adults. Thus, it indicates cells or molecules that may be implicated in normal development. The definition does not, however, indicate the function of the structures detected or even, indeed, if they have a normal function at all. Cell interactions are affected by many factors other than display of particular surface molecules, and it may be that association of surface structures in complexes, the mobility of single membrane elements (or of complexes) or the degree of cell metabolic activity are all as important as the species of structures displayed in determining the degree of interaction between neighboring cells or developmental fate of any one cell.

In this paper, I wish to deal with the least subtle approach to an analysis of cell surface differentiation in development, i.e., detection of embryo-associated surface antigens. The discussion will be further focused on a tumor model, teratocarcinoma, which has largely, though not exclusively (Baranska *et al.* 1970, Wiley and Calarco 1975), been used as an immunogen instead of actual embryos. I will briefly review the properties of the tumor model and the reactivities of some of the antisera prepared against it. The remainder of the paper will deal with work from my own laboratory, particularly recent work in which we attempt to show that different species share teratocarcinoma-defined antigens and work dealing with cellular immunity to the tumor. Four topics will be covered:

1) antigens of teratocarcinoma cells,
2) antigens shared by the tumor cells and normal embryos within a species,
3) cross-reactions between mouse and human teratocarcinoma,
4) cellular and humoral immunity to teratocarcinoma, especially in pregnancy.

TERATOMA, TERATOCARCINOMA, AND CELL SURFACE IMMUNOLOGY

Benign teratomas, disorganized mixtures of differentiated tissues, have been known at least since the 17th century and were recognized then as being of germinal origin (R. Plot, cited in Needham 1959). In our time, the analysis of development using these tumors has been greatly facilitated by the appearance in some mouse strains of testicular tumors, teratomas, and teratocarcinomas; the latter, being malignant, are propagable, either by animal passage or in cell culture (Stevens 1967, Pierce 1967, Sherman and Solter 1975). Lines established from primary teratocarcinomas grow and differentiate to characteristic tissue types; perhaps the most interesting grow in ascites as aggregates of cells differentiated to several types, at least superficially mimicking early mouse embryos. This range of differentiation may be obtained with both animal-grown and cultured teratocarcinoma lines, and the potential for differentiation of cultured teratocarcinoma cells may be tested in vivo. A further set of lines has been produced, which, in contrast to the pluripotent lines, fails to show any sign of differentiation at all, either in vivo or in vitro. These nullipotent lines provide the most uniform cell populations for use as immunogens and appear to consist solely of primitive stem cells, which, unlike the stem cells of pluripotent lines, have lost all ability to differentiate (Martin 1975).

Several sorts of antisera have been prepared to a variety of teratocarcinoma cells. Not surprisingly, the details of the reactions of these sera, especially with normal tissues including gametes (and with tumor cells other than teratocarcinoma), vary considerably. Also, all sera react with early mouse embryos and with a range of teratocarcinoma cells. The properties of three extensively described sera, including those from my laboratory, are described in a recent review by Erickson (1977), who comments usefully on the divergences in technique as well as in results between the laboratories whose results are summarized.

The most elegantly analyzed of all the sera produced has been that against the nullipotent teratocarcinoma cell line, F9 (Jacob 1977). This serum, produced in inbred mice of the tumor donor strain 129, reacts with sperm, with teratocarcinoma stem cells, and with early embryos, through around day 8 of development (Kemmler *et al.* 1976). On the basis of its reactions with sperm from wild-type and t-allele heterozygotes, it was suggested that the antiserum detected the wild-type allele of one t-complementation group, t^{w32}. The t-complex has been known for years to consist of a series of linked genes, many of which are lethal recessives that affected morphogenesis, especially of the tail, in heterozygotes (Dunn and Bennett 1964, Bennett 1975). In wild mouse populations, t-alleles are common and it has been shown that this prevalence is due to the high frequency of fertilization by t-bearing sperm, as opposed to wild-type sperm in matings by t/+ heterozygous males. The effects of t mutants on development and evidence suggesting that t-sperm had an advantage over wild-type sperm in actual fusion with and penetration of ova led to a view that genes of the complex specified cell surface structures that are important in cell-to-cell adhesions (Gluecksohn-Waelsch and Erickson 1971, Bennett *et al.* 1972). In the context of this model it was exciting to find that anti-F9 serum appeared to react less well with sperm from a particular t/+ heterozygote, t^{w32}/+, than with sperm from other heterozygotes or from +/+ animals. It further appeared that the F9 antigen was present only at stages at which t^{w32} appeared to act in homozygotes. As is often the case, this relatively simple model is not fully consistent with the biological complexity that it represents. It now appears that the F9 antigen is expressed much later in development than would be expected if it represented a transient stage of surface differentiation and, furthermore, that it cannot be said to be allelic to any one t-complementation group because it is absent from embryos homozygous for either one of two different t-alleles, members of different complementation groups (Kemmler *et al.* 1976, Jacob 1977). This contradiction reflects as much our lack of understanding of the genetic complexity of the t-complex as it does our lack of understanding of F9 antigens and their function in normal development. It is also unclear if there is an homologous genetic and surface antigen system in other species—it has been predicted that surface antigens involved in early development would be conserved in evolution, so that there ought to be a high degree of cross-reaction between species. There is some evidence that anti-F9 serum reacts with human spermatozoa (Buc-Caron *et al.* 1974). Also, it appears that anti-F9 serum reacts with human germinal cells, as well as with those of the mouse (Gachelin *et al.* 1976). On the other hand, Bobrow *et al.* (1975) could find no evidence for t-like mutants in man. Here, as in the examination of embryos with anti-F9, we are left with the impression that the evidence is far from complete. The prediction that there is extensive cross-reaction between embryo determinants in different species is important enough to require further experiment.

SURFACE ANTIGENS OF A PLURIPOTENTIAL TERATOCARCINOMA

In my laboratory, we have been chiefly concerned with the analysis of cells of a tumor line, 402AX, originally established as an embryoid body ascites by Leroy Stevens (1958). Although in the course of 10 years' passage the tumor has lost its ability to differentiate into masses of apparently histologically normal tissues, it still may produce a variety of cell types in culture, including cells apparently secreting blastocoelic fluid and mimicking the appearance of cultured isolated trophectoderm cells. The tumor was adapted to culture and rendered 8-azaguanine resistant, and it is now grown under a variety of regimens: in suspension, attached to plastic and passed every three days, or attached to plastic and passed every 12 days. The cells grown in suspension, TerCS, form small aggregates; cells grown on plastic appear small and undifferentiated when grown with passage every three days, while cells passaged every 12 days, TerC12, undergo some morphological diversification (D. Searles, unpublished observations). The xenoantiserum, rabbit antitumor, originally prepared to this tumor in 1970 (Edidin *et al.* 1971) was made against animal-grown tumor, a mixture of cell types. Our current xenoserum was made against cultured cells. It reacts with all of the teratocarcinoma sublines as well as with many other tumor cells, though not at all with fetal mouse fibroblasts.

Reactivity of the serum may be analyzed by absorption. When this is done, it is found that three separate families of antigens are detected by our reagent. The first, antigen I, is present on many tumor cells and on teratoma. It is not detectable by immunofluorescence on normal lymphocytes or on 3T3 cells. The latter, however, do react for this antigen after treatment with trypsin (Gooding and Edidin 1974). On cells bearing antigen(s) I, the antigen(s) appears to be physically associated with major histocompatibility antigens, H-2 antigens, since the two groups of antigens co-cap when treated with antibody and a fluorescent antiglobulin (Gooding and Edidin 1974). The pattern of reactivities is probably not due to antitumor virus antibodies. TerC itself is gp70 negative by immunofluorescence. Antigen I is found on cells of a blastocyst-derived MB line (Ostrand-Rosenberg *et al.* 1977), which appears to be virus negative by several criteria (M. Sherman, personal communication), as well as on the surface of EL-4 G^- cells that fail to express the G_{IX} surface antigen determined by the viral genome and apparently part of the major viral glycoprotein gp70 (Herberman *et al.* 1976); these cells are also negative for gp70 when tested with a high-titered antiserum. Finally, the antiserum does not react with cells of an ovarian teratocarcinoma line, which are gp70 positive when tested with an antigroup-specific antibody. A preliminary characterization of antigen I by gel electrophoresis of the materials precipitated from extracts of labeled antigen I-positive cells contains multiple peaks in the molecular weight range 40,000–100,000 (Gooding 1976).

If the anti-teratoma serum is absorbed on tumor cells positive for antigen, there remain reactivities against two other antigens, or families of antigens, II and III. Antigen II is found in some tumor cells, notably hepatomas, as well as in the teratoma, while antigen III appears to be unique to the teratoma cells. The presence of antigen II in hepatoma cells led us to suggest that it might be an alpha-fetoprotein (AFP). However, purified mouse AFP does not react with anti-teratocarcinoma serum in immunodiffusion of assays and does not block binding of this serum to 402AX cells.

H-2 ANTIGENS ON CELLS OF TERATOCARCINOMA 402AX

Teratocarcinomas resemble normal embryos in the morphology of their aggregates and in their differentiation. If this resemblance extends to surface antigen display, then, besides finding antigenic cross-reactions between tumor and embryo surfaces, we expect that the expression of histocompatibility antigens on teratocarcinoma cells approximates that on embryos. From a series of studies, it appears that preimplantation mouse embryos express only weak histocompatibility antigens (Edidin 1972, see also Muggleton-Harris and Johnson 1976).

Although there may be transient appearance of H-2 antigen just before implantation (Searle *et al.* 1976), embryos do not continually express H-2 antigens until around day 6 after implantation. It might be expected that teratocarcinoma cells, or at least the undifferentiated embryonal carcinoma cells, do not express these antigens. Artzt and Jacob (1974) showed that F9 cells, the nullipotential cells described in a previous section, failed to absorb anti-H-2 antibodies against the H-2^b antigens expected on cells derived from strain 129 mice. Other work by Jacob and his colleagues showed that a teratocarcinoma line capable of differentiating in culture, whose stem cells bore the F9 antigen, gradually lost F9 antigen and acquired H-2 reactivity as the cells differentiated (Jakob *et al.* 1973).

Our work with teratocarcinoma 402AX indicates that H-2 antigens cannot be detected at the surface of the tumor cells, either by fluorescence or in complement-dependent cytotoxicity assays, using both allo- and xeno-anti-H-2. Also, some antigens of the H-2 complex, or linked to it, are detectable by a bioassay system that greatly amplifies small degrees of antigenic stimulation. The H-2 antigens detected may be resident on a minority population of cells. However, recent work in our laboratory, using another highly amplifying assay, suggests that small amounts of some H-2 antigens are present on most cells of a tumor population.

The bioassay used was sensitization of adult animals to accelerated rejection of skin grafts. As few as 20,000 H-2^b lymphocytes administered intraperitoneally three days before grafting is sufficient to accelerate the rejection of a skin graft of a C57BL/10 (H-2^b) donor to a B10.Br (H-2^k) recipient. C57BL/10 and strain 129, the tumor strain of origin, are both H-2^b. Although 129 is a variant of this haplotype, H-2^{bc}, the two cross-sensitize. If teratocarcinoma cells, either cultured lines of 402AX or another line OTT6050 passaged in vivo, are used as immunogens (instead of lymphocytes), it is found that 6×10^6 teratocarcinoma cells contain H-2 antigens at a level approximating that found in 40,000–60,000 strain 129 lymphocytes (Table 1). This result may be explained either in terms of the differentiation of 1 to 2% of the tumor population or to a low average

TABLE 1. *Median survival times (MST) of test skin grafts from strain C57BL/10 on strain B10.Br recipients variously immunized before grafting*

Pretreatment*	MST (days)
Saline	10.0
2×10^4 129 lymphocytes	8.6
4×10^4 129 lymphocytes	8.0
6×10^6 teratocarcinoma, OTT6050	8.6
6×10^6 teratocarcinoma, 402AX	7.5

* Animals were injected intraperitoneally with saline or cell suspensions three days before test grafting. MST are calculated from data given in Edidin and Gooding (1975).

TABLE 2. *Damage to teratocarcinoma and other tumor cells by allo-anti-H-2 sera, measured in terms of uptake of ¹²⁵IUdR*

Target cell	% Inhibition of IUdR uptake by C3H anti-C.SW (anti-H-2^b) serum				% Inhibition of IUdR uptake by C.SW anti-C3H (anti-H-2^k) serum			
	1/160	640	2,560	10,240	1/160	640	2,560	10,240
TerC12	42	10	8	3	20	6	6	—
TerCS	68	40	25	5	5	3	2	—
C57BL/6 melanoma (H-2^b)	64	45	10	0	2	0	2	—
LM (H-2^k)	6	0	4	—	98	98	94	67

level of H-2 on all of the tumor cells. The first interpretation appears more plausible than the second. However, recently Mr. David Searles, in my laboratory, has re-examined H-2 antigens detected with a high-titered anti-H-2 serum, C3H anti-C3H.SW (made in a congeneic combination and hence able to react only with antigens of the H-2 complex or with those determined by closely linked genes on chromosome 17). With this serum, using an assay that measures antibody + complement-mediated damage to cells in terms of the cells' ability to incorporate ^{125}iododeoxyuridine (IUdR) into DNA, Searles finds expression of some elements of H-2 on undifferentiated teratocarcinoma cells. The magnitude of the effect is such that it is difficult to explain in terms of antibody binding to a minority tumor population.

The assay is performed in microtiter wells, using 10^4 target cells per well. Cells are exposed to the test antiserum for 15 minutes and then to selected rabbit complement for 30 minutes. Following treatment, the cells are incubated with 0.10 μCi ^{125}IUdR for 16 hours at 37°C. After harvesting and washing, the radioactivity of the cells is counted, and growth is compared to a standard suspension exposed to complement only. It is not clear what factors affect the cells' incorporation of radiolabel, and we must control carefully for the specificity of cell damage by antisera; but, as can be seen from Table 2, the assay appears to be sensitive and specific. H-2^k anti-H-2^b serum inhibits IUdR uptake by TerC12 and TerCs, cells that are genetically H-2^b, as well as by cells of a C57BL/6 melanoma, which express H-2^b antigens in more conventional assays. On the other hand, LM cells, H-2^k positive in conventional assays, are not damaged by the anti-H-2^b antibody. The reciprocal serum, H-2^b anti-H-2^k, may slightly damage TerC12 cells, does not appear to harm either TerCs or melanoma cells, and severely damages the positive controls, LM cells. Hence, it would appear that teratocarcinoma cells do express antigens determined by some elements of the H-2 complex. Whether or not similar expression of H-2 is seen in normal embryos by this method has not yet been determined.

ANTIGENS SHARED BETWEEN TERATOCARCINOMA 402AX AND NORMAL MOUSE EMBRYOS

Anti-402AX serum reacts well by immunofluorescence with all early stages of mouse development, from unfertilized ova through 6 to 7 day egg cylinder and somite embryos (Gooding *et al.* 1976). Closer analysis of the reactions of absorbed sera with embryos shows that antigen I, the antigen expressed on many tumor cells in physical association with H-2 antigens, is the predominant teratocarcinoma-defined antigen on embryos. Antigen I is detected in all stages of development through implantation but gradually is restricted to inner cell mass (ICM) cells, which are destined to form all of the embryo proper. Antigen I is detectable on trophectoderm of the early mouse blastocyst but cannot be found on the trophoblast cells that grow from a blastocyst "implanted" in vitro. Antigen II is transiently present on mural trophectoderm, that portion of the trophectoderm over the blastocoelic cavity, but not on the polar trophectoderm over the ICM of the maturing blastocyst. This antigen does not appear in more mature trophoblasts from cultured embryos. Antigen III has not been found in embryos, or indeed on tumor cells other than teratocarcinoma 402AX. However, we have not tested an "anti-III" serum against other teratocarcinoma and embryonal carcinoma cells.

We have not thoroughly studied later embryos for teratocarcinoma-antigen expression. It does appear that few cells of the 8-day embryo and no cells of the 9-day embryo react with the serum. "No cells" must be qualified, since we would not detect a few percent of cells that were positive in a whole embryo cell suspension. Indeed, preliminary results, examining primary germ cells of 12-day embryos, indicate that these precursors of gametogenic cells and gametes do bear some antigens absent from adult tissues and shared with 402AX (M. H. Johnson and M. Edidin, unpublished results). The embryo donors of the germ cells were not sexed, and we do not know if germ cells of both sexes are teratocarcinoma antigen positive. Our antiserum, unlike anti-F9 serum, does not react with male gonadal cells; since the 402AX antigens are expressed on unfertilized ova, we might expect that female, but not male, germ cells would be positive for them.

CROSS-REACTIONS BETWEEN MOUSE AND HUMAN TERATOCARCINOMA CELLS

It has been suggested (Buc-Caron *et al.* 1974) that surface features which are important in development ought to be conserved between mammalian species. If such features are detectable with antisera, then we might expect to find wide cross-reactions between teratocarcinomas and, indeed, embryos of different species. In collaboration with Dr. Michael Jewett, we have been able to test cultured human teratocarcinoma cells and sera from teratocarcinoma patients for cross-reactions with their mouse homologues. The cells are available as two lines, Tera-1 and Tera-2, put into culture some years ago and described by Jorgen Fogh (Fogh and Trimpe 1975). These cells do appear to bear antigens reacting with our anti-teratoma serum, although the expression of the antigens is dependent upon the culture density to which the cells have been grown (Table 3). It further appears that the antigenic-cross-reacting determinants lie hidden by charged, sialic acid-bearing molecules, since digestion of cells with neuramini-

TABLE 3. *Reaction of rabbit anti-mouse teratocarcinoma serum with human teratocarcinoma cells and fibroblasts*

Cell	Culture state	Neuraminidase digestion	% Fluorescence positive*
Tera-2	subconfluent	—	3
Tera-2	subconfluent	+	75
Tera-2	confluent	—	0
Tera-2	confluent	+	4
Flow 2000	subconfluent	—	0
Flow 2000	subconfluent	+	7

* Cells were reacted with anti-402AX serum or with normal rabbit serum or with 402AX-absorbed anti-402AX. Binding of antibody was visualized with fluorescent goat anti-rabbit Ig. Normal rabbit serum or 402AX absorbed antiserum never stained more than 8% of Tera-2 cells, regardless of culture density or enzyme treatment.

dase is usually required to show reactivity with anti-402AX. Cross-reaction is dependent upon cell density, since even digested cells do not react if harvested from confluent cultures. Of course, the expression of shared antigens and its dependence upon digestion of the cells' surface may involve more subtle variations in antigen display than simple masking by charged groups. We have not yet elaborated this point.

Human patient serum, from teratocarcinoma patients or from age-matched control patients with other malignancies, reacts with many mouse cells, both normal and malignant. After absorption on normal mouse cells the sera of some teratocarcinoma patients, but no sera from other patients, do react with the mouse teratocarcinoma by immunofluorescence; we have not looked at sufficient sera to estimate the frequency of positive, apparently specific, reactions or to correlate the positive reactions with the state of a patient's disease.

At present then we can say only that there is a strong cross-reaction between mouse and human teratocarcinomas, although the mouse-defined antigens are usually masked on the human tumor. We might speculate that they have been modified in the evolution of the surface structures controlling cell position in early human embryos, but we are unable to test this.

CELLULAR IMMUNITY TO TERATOCARCINOMA 402AX

Grafts of tumor 402AX to the kidney capsule of allogeneic or syngeneic hosts provoke a lymphoid cell infiltrate similar to that seen when allogeneic embryo grafts are implanted under the capsule (Edidin *et al.* 1974). The intensity of the reaction is dependent upon the degree of prior immunization of the graft recipient, and the size of the tumor is generally inversely proportional to the intensity of the infiltrate. Spleen and lymph node cells from syngeneic animals either carrying the tumor or systemically immunized with sonicated tumor cells can inhibit the growth of 402AX colonies in vitro in a long-term cell growth assay. A body of experiments indicates that cells bearing foreign antigenic determinants, for example, viruses, cannot be lysed by cytotoxic T-cells unless the targets are syngeneic at the H-2 locus with the attackers (Blanden *et al.* 1975). Because teratocarcinoma cells appear by most assays to lack H-2 antigens, we would expect these cells to resist lysis by lymphocytes directed against their tumor-associated antigens. Indeed, 402AX, like F9 (Goldstein *et al.* 1976), is not lysed by cytotoxic T-cells when lysis is assayed for one to three hours using release of ^{51}Cr as a measure of cell damage. Not only cells from tumor-immunized animals but also syngeneic lymphocytes activated by concanavalin A, which are generalized killers, fail to lyse 402AX. However, it is not clear if the cellular reaction to grafts of teratocarcinoma 402AX is best modeled in vitro by an acute assay for cell lysis; and, as mentioned above, in a long-term biological assay, colony inhibition, it did appear that 402AX cells were damaged by some cells in immune spleens.

We have further developed a long-term assay of cell-mediated immunity which indicates that 402AX cells can be destroyed by syngeneic strain 129 T-cells (P. F. Bartlett, B. Fenderson, and M. Edidin, in preparation). The assay of cytostasis is based on that described by Chia and Festenstein (1973), as modified by Greenberg *et al.* (1975). In this assay, target cells are plated into microwells with varying numbers of attacker cells. After 48 hours of co-incubation, $^{125}IUdR$ is added to each well; and 24 hours later, cells are washed, collected onto glass fiber filters, and counted. In our hands, this assay gives good specificity of cell damage and, combined with the use of attacker cells sensitized in a purely in vitro culture system, appears to assay long-term cell damage by T-lymphocytes. The controls for nonspecific activation of attackers are given in Table 4. While culture with any lymphocytes slightly inhibits IUdR uptake, only specifically immune cells are strongly inhibitory. In Table 5 we summarize results of this assay using attacker cells sensitized to irradiated standard 402AX cells and a variety of tumor and normal cultured cell targets, including the human teratocarcinoma line, Tera-2. It will be seen that the 402AX-sensitized cells are capable of damaging a wide range of tumor cells, all of which have previously been shown to share antigen I with teratocarcinoma cells. The sensitized attackers are effective at low attacker:target cell ratios and do not appear

TABLE 4. *Specificity of sensitization in vitro of lymphocytes cultured with various stimulating cells*

Stimulating antigen	IUdR uptake as % of controls*
TerCS	43
129 fibroblasts	78
Medium only	86
None. Freshly prepared lymphocytes	78

* Control cultures received no lymphocytes at all. They incorporated an average of 14,000 CPM from $^{125}IUdR$. Lymphocyte:target was 50:1.

TABLE 5. *Inhibition of cell uptake of IUdR by lymphocytes sensitized to TerC*

Target	% Inhibition relative to controls* at lymphocyte : target (ratios) 12 : 1	25 : 1	50 : 1	100 : 1	200 : 1
TerCS	10	30	40	50	58
129 embryo fibroblasts	−10†	−5	−20	−20	−25
3T12 (BALB/c)	0	40	50	55	—
3T3 (BALB/c)	0	0	5	8	—
EL-4 (C57BL/6) G^+	0	20	38	45	—
EL-4 (C57BL/6) G^-	0	12	35	38	—
Ter2 (human teratocarcinoma)	0	20	25	30	40
Flow 2000 fibroblasts (human embryo)	0	−15	−20	−20	−22

* Controls always received numbers of freshly prepared splenic lymphocytes equal to the number of sensitized attacker cells in experimental wells.

† Negative numbers indicate IUdR incorporation greater than controls—growth stimulation.

to react to RNA tumor virus determinants, because the teratocarcinoma itself, as well as EL4 G^- cells, does not react with a strong antiserum to the membrane-associated viral glycoprotein, gp70. The cross-reaction extends to human tumor cells, although we have not tested a wide range of target cells. These data reinforce our belief that a major surface antigen of 402AX cells is expressed over a range of tumors. It is not clear if this antigen is identical to or part of sero-antigen I, but this may be tested, preferably by immunizing with the purified antigen.

At least a majority, if not all, of the effector cells in our assay appear to be T-lymphocytes, since anti-Thy-1 serum + complement depletes the population of sensitized effector cells. Passage of the lymphoid population over a nylon wool column, or plating in the presence of serum to allow adherence of macrophages, does not diminish the population's activity after sensitization by irradiated TerC.

Cells sensitized to TerC in culture are also reactive in vitro against cultured embryos (B. Fenderson, unpublished observations). The details of this method have not been elaborated, so our assay may be less than optimum, but it appears that 10^5 lymphocytes will inhibit the growth, measured in terms of ^{125}IUdR uptake, of four cultured blastocysts (Table 6). We hope to use this system to

TABLE 6. *Effect of TerC-sensitized lymphocytes on the growth of blastocysts in culture*

Culture conditions	CPM* uptake from ^{125}IUdR
4 blastocysts alone	6711
4 blastocysts + lymphocytes previously cultured with medium alone	6228
4 blastocysts + lymphocytes sensitized to TerCS	4750

* CPM: counts per minute

test both embryo-tumor cross-reactions and to measure maternal immunity to the antigens shared between tumor and TerC. For the present, we have only been able to assay peritoneal exudate cells (plated for 24 hours to remove macrophages and other adherent cells) against TerC targets. The cellular basis of the response in this assay is obscure, and it may be that more than one lymphoid population is acting. The result, however, is clear—culture of TerC targets with peritoneal exudate populations results in significant stimulation of cell growth, as measured by IUdR uptake, when compared to the effect of peritoneal cells from age-matched virgin mice. Increased uptake is also found if embryo-derived, MB cells (Sherman 1975) are used as targets. If this effect in fact represents a physiological growth stimulation rather than a pathological aberration in transport or repair processes due to the attack, we may be detecting an immune process that acts to promote rather than inhibit growth of the embryo and fetus in utero.

CONCLUSION

This sketch describes the beginning of a search for function of surface antigens defined by anti-teratoma sera. We now have an inventory of antigens shared by teratocarcinoma and normal embryos; some of these antigens apparently are shared by different species as well. It remains to be seen how immunity to these antigens might interfere with or perhaps even abet embryonic development. Once this second step is taken, we can then press further to learn if appearance of surface antigens, or the manner in which they are arranged, or both, serve to determine the multitude of transactions between cells of the early mammalian embryo.

ACKNOWLEDGMENTS

Original work described here was supported by Grant AM11202, National Institutes of Health, to M.E., and by Contract N01-CB-43922 with the National Cancer Institute. S.O.R. is a postdoctoral fellow of the NCI. This is contribution number 893 from the Department of Biology, The Johns Hopkins University, Baltimore, Maryland.

REFERENCES

Artzt, K., and F. Jacob. 1974. Absence of serologically detectable H-2 on primitive teratocarcinoma cells in culture. Transplantation 17:632–634.

Baranska, W., P. Koldovsky, and H. Koprowski. 1970. Antigenic study of unfertilized mouse eggs: Cross reactivity with SV40-induced antigens. Proc. Natl. Acad. Sci. USA 67:193–199.

Bennett, D. 1975. The T-locus in the mouse: A review. Cell 6:441–454.

Bennett, D., E. A. Boyse, and L. J. Old. 1972. Cell surface immunogenetics in the study of morphogenesis, *in* Cell Interactions (3rd Lepetit Colloquium), L. G. Silvestri, ed., North-Holland Publishing Co., Amsterdam, pp. 247–262.

Blanden, R. V., P. C. Doherty, M. B. C. Dunlop, I. D. Gardner, and R. M. Zinkernagel. 1975.

Genes required for cytotoxicity against virus-infected target cells in K & D regions of H-2 complex. Nature 254:269–270.

Bobrow, M., J. G. Bodmer, W. F. Bodmer, H. O. McDevitt, J. Lorber, and P. Swift. 1975. The search for a human equivalent of the mouse T-locus—Negative results from a study of HL-A types in spina bifida. Tissue Antigens 5:234–237.

Buc-Caron, M. H., G. Gachelin, M. Hofnung, and F. Jacob. 1974. Presence of a mouse embryonic antigen on human spermatozoa. Proc. Natl. Acad. Sci. USA 71:1730–1733.

Chia, E., and H. Festenstein. 1973. Specific cytostatic effect of lymph node cells from normal and T-cell-deficient mice on syngeneic tumor target cells in vitro and its specific abrogation by body fluids from syngeneic tumor-bearing mice. Eur. J. Immunol. 3:483–487.

Dunn, L. C., and D. Bennett. 1964. Abnormalities associated with a chromosome region in the mouse. Science 144:260–267.

Edidin, M. 1972. Histocompatibility genes, transplantation antigens and pregnancy, *in* Transplantation Antigens, B. D. Kahan and R. Reisfeld, eds., Academic Press, New York, pp. 75–114.

Edidin, M., and L. R. Gooding. 1975. Teratoma-defined and transplantation antigens in early mouse embryos, *in* Teratomas and Differentiation, M. I. Sherman and D. Solter, eds., Academic Press, New York, pp. 109–121.

Edidin, M., H. L. Patthey, E. J. McGuire, and W. D. Sheffield. 1971. An antiserum to "embryoid body" tumor cells that reacts with normal mouse embryos, *in* Embryonic and Fetal Antigens in Cancer, N. G. Anderson and J. H. Coggin, Jr., eds., National Technical Information Service, Springfield, Va., pp. 239–248.

Edidin, M., L. R. Gooding, and M. H. Johnson. 1974. Surface antigens of normal early embryos and a tumor model system useful for their further study, *in* Immunological Approaches to Fertility Control, E. Diczfalusy, ed., Karolinska Institutet, Stockholm, pp. 336–356.

Erickson, R. P. 1977. Differentiation and other alloantigens of spermatozoa, *in* Immunobiology of Gametes, M. Edidin and M. H. Johnson, eds., Cambridge University Press, Cambridge, pp. 85–107.

Fogh, J., and G. Trimpe. 1975. New human tumor cell lines, *in* Human Tumor Cell Lines *In Vitro,* J. Fogh ed., Plenum Press, New York, pp. 115–159.

Gachelin, G., M. Fellous, J. L. Guenet, and F. Jacob. 1976. Developmental expression of an early embryonic antigen common to mouse spermatozoa and cleavage embryos and to human spermatozoa: Its expression during spermatogenesis. Dev. Biol. 50:310–320.

Gardner, R. L., and V. E. Papaioannou. 1975. Differentiation in the trophectoderm and inner cell mass, *in* The Early Development of Mammals, M. Balls and A. E. Wild, eds., Cambridge University Press, London, pp. 107–132.

Gardner, R. L., and J. Rossant. 1976. Determination during embryogenesis, *in* Embryogenesis in Mammals, Ciba Foundation Symposium 40, Elsevier, Amsterdam, pp. 5–18.

Gluecksohn-Waelsch, S., and R. P. Erickson. 1971. Cellular membranes. A possible link between H-2 and T-locus effects, *in* Immunogenetics of the H-2 System, A. Lengerová and M. Vojtišková, eds., S. Karger, Basel, pp. 120–122.

Goldstein, P., F. Kelley, P. Auner, and G. Gachelin. 1976. Role of H-2 in T-cell-mediated cytolysis: Sensitivity of H-2-less target cells. Nature 262:693–695.

Gooding, L. R. 1976. Expression of early fetal antigens on transformed mouse cells. Cancer Res. 36:3499–3502.

Gooding, L. R., and M. Edidin. 1974. Cell surface antigens of a mouse testicular teratoma. Identification of an antigen physically associated with H-2 antigens on tumor cells. J. Exp. Med. 140:61–78.

Gooding, L. R., Y. C. Hsu, and M. Edidin. 1976. Expression of tumor-associated antigens on murine ova and early embryos. Dev. Biol. 49:479–486.

Greenberg, A. H., L. Shen, and A. Medley. 1975. Characteristics of the effector cells mediating cytotoxicity against antibody-coated target cells. Immunology 29:719–729.

Herberman, R. B., H. Kirchner, H. T. Holden, M. Glaser, S. Haskill, and G. D. Bonnard. 1976. Cell mediated immunity in murine virus tumor systems, *in* Tumor Virus Infections and Immunity, R. Crowell, H. Friedman, and J. Prier, eds., University Park Press, Baltimore, pp. 147–164.

Jacob, F. 1977. Mouse teratocarcinoma and embryonic antigens. Immunological Reviews 33:1–32.

Jakob, H., T. Boon, J. Gaillard, J. F. Nicolas, and F. Jacob. 1973. Teratocarcinoma de la souris:

Isolement, culture et proprietes des cellules a potentialities multiple. Ann. Microbiol. (Instr. Pasteur) 124B:269–282.

Kemmler, R., C. Babinet, H. Condamine, G. Gachelin, J. L. Guenet, and F. Jacob. 1976. Embryonal carcinoma antigen and the T/t locus of the mouse. Proc. Natl. Acad. Sci. USA 73:4080–4084.

Martin, G. R. 1975. Teratocarcinomas as a model system for the study of embryogenesis and neoplasia: A review. Cell 5:229–243.

Muggleton-Harris, A. L., and M. H. Johnson. 1976. The nature and distribution of serologically detectable alloantigens on the preimplantation mouse embryo. J. Embryol. Exp. Morphol. 35:59–72.

Needham, J. 1959. A History of Embryology. 2nd ed. Abelard-Schuman, New York.

Ostrand-Rosenberg, S., C. Hammerberg, M. Edidin, and M. I. Sherman. 1977. Expression of histocompatibility-2 antigens on cultured cell lines derived from mouse blastocysts. Immunogenetics 4:127–136.

Pierce, G. B. 1967. Teratocarcinoma: Model for a developmental concept of cancer. Curr. Top. Dev. Biol. 2:223–246.

Searle, R. F., M. H. Sellens, J. Elson, E. J. Jenkinson, and W. D. Billington. 1976. Detection of alloantigens during preimplantation development and early trophoblast differentiation in the mouse by immunophoxidase labeling. J. Exp. Med. 143:349–359.

Sherman, M. I. 1975. Long term culture of cells derived from mouse blastocysts. Differentiation 3:51–67.

Sherman, M. I., and D. Solter, eds. 1975. Teratomas and Differentiation. Academic Press, New York.

Stevens, L. C. 1958. Studies on transplantable testicular teratomas of strain 129 mice. J. Natl. Cancer Inst. 20:1257–1276.

Stevens, L. C. 1967. The biology of teratomas. Adv. Morphog. 6:1–31.

Wiley, L. M., and P. G. Calarco. 1975. The effects of anti-embryo sera and their localization on the cell surface during mouse preimplantation development. Dev. Biol. 47:407–418.

Stage of Cellular Differentiation in Tumor Cell Origin—Tissue Differentiation and Oncogenesis

Cell Differentiation and Neoplasia, edited by Grady F. Saunders. Raven Press, New York

Developmental Genetics of Neural Tumors in Man

Alfred G. Knudson, Jr. and Anna T. Meadows

Institute for Cancer Research, Fox Chase Cancer Center, Philadelphia, Pennsylvania 19111; and The Children's Hospital of Philadelphia, Philadelphia, Pennsylvania 19104

There is, in man, a set of dominant genes that is characterized by a strong predisposition to neoplasia (Knudson *et al.* 1973). Each gene in the set has its own typical tissue specificity or specificities, suggesting that the corresponding set of normal alleles is concerned with differentiation. When more than one tumor gene affects the same tissue, an opportunity to investigate, in more detail, the features of developmental specificity is presented. An example is the group of genes predisposing to tumors of the sympathetic nervous system (Knudson and Meadows 1976). A still greater opportunity to study tissue-specific developmental genes is presented by genetic tumors of the nervous system as a whole, since that system is the most complex and since more tumor mutations are known for it than for any other. Our purpose here is to present such an analysis.

Tumors arise in all parts of the nervous system, but particularly in the brain, retina, sympathetic nervous system, and peripheral nerves. Histologic types of neural tumors include retinoblastoma, neuroblastoma, pheochromocytoma, medulloblastoma, and the various gliomas. Familial cases of all of these tumors have been reported, and it is from an analysis of these familial reports that we can learn about the developmental specificities of the genes responsible for these familial cases.

RETINOBLASTOMA

The paradigm of the hereditary neural tumor is retinoblastoma, because it has been known for many years and because a significant cure rate has permitted survivors to reproduce. It is now clear that about 40% of cases are heritable in dominant fashion and that the remaining 60% are not hereditary (Knudson 1971). Even the heritable cases often have no family history of the disease, probably because they are the result of new mutations occurring in parental germ cells. Individuals bearing a germinal mutation develop a mean number of three to four tumors, whereas those not bearing such a mutation are at a risk of approximately one per 30,000 of developing one tumor. The germinal mutation increases tumor risk about 100,000 times (Knudson 1971, Knudson *et al.* 1975).

Not only do genetically predisposed individuals usually develop more than one tumor, but they also develop their first tumors at an earlier age than do individuals with the nonhereditary form. But for both forms, there is a limitation upon the age at which the tumor may develop. Only rarely does retinoblastoma develop beyond the age of six years. This is almost certainly a direct consequence of the fact that the target cell for transformation, the retinoblast, disappears by differentiation to a postmitotic state incompatible with tumor expression.

The gene for retinoblastoma imparts a predisposition to some other tumors as well, particularly to osteogenic sarcoma, which affects at least 1% of gene carriers (Jensen and Miller 1971, Kitchin and Ellsworth 1974, Abramson *et al.* 1976). This incidence of osteogenic sarcoma is considerably increased by X-irradiation (Sagerman *et al.* 1969). In spite of these other tumors, the retinoblastoma gene is highly specific, in that all other spontaneously occurring tumors have a far lower incidence than retinoblastoma in gene carriers.

Even so, the retinoblastoma gene is not sufficient to produce retinoblastoma. Every cell in the retina carries this mutation, yet only three to four cells on average develop into tumor cells; at the level of the cell, oncogenesis is a rare event. This suggests to us that at least one other event is necessary for neoplastic transformation (Knudson 1971). Since the frequency of such an event would be of the order of magnitude of one per million cells, it is possible that it is a somatic mutation (Knudson 1971). This would be compatible also with the observation that neoplastic transformation in bone is greatly enhanced by a known mutagen, the X-ray (Strong and Knudson 1973). Accordingly, oncogenesis, in the case of retinoblastoma, has been viewed as a two-step (mutation) process. In the hereditary form, one step is inherited, the other somatic; in the nonhereditary form, both steps are somatic (Knudson 1971). The process is essentially the same for both forms.

Rarely, a prezygotic event is manifest in all cells of the host in the form of a deletion in the long arm of chromosome 13. In such instances, the age-specific incidence of retinoblastoma and the probability of bilateral tumor are not significantly different from the usual prezygotic cases, in which no chromosomal abnormality is found, suggesting that the gene for retinoblastoma is located in the long arm of this chromosome (Knudson *et al.* 1976). In some instances, therefore, genetic predisposition is imparted by a visible deletion of the site, in others by a submicroscopic change of unknown nature. The deletion cases indicate that predisposition can be associated with the elimination of the product of the gene on one chromosome 13. The location of a second event (mutation) is, of course, unknown. One possibility is that it occurs at the homologous site on the other chromosome 13, thus rendering the affected cell homozygously defective for the gene. Retinoblastoma would thereby be a single (recessive) gene disorder at the cellular level and, consequently, a single molecular event.

TUMORS OF THE SYMPATHETIC NERVOUS SYSTEM

Familial cases of three tumors of the sympathetic nervous system have been reported. Two of these, neuroblastoma and ganglioneuroma, are related developmentally and have been found in the same families, even in the same patient. With rare exception, neither of these occurs in the same families that develop the third tumor, pheochromocytoma.

Neuroblastoma and Ganglioneuroma

Neuroblastoma, like retinoblastoma, is typically a childhood neoplasm, with a peak in age-specific incidence in the first five years of life. Fewer data on the dominant inheritance of some cases are available than for retinoblastoma because there have been far fewer survivors, but enough familial cases have been reported to permit the conclusion that some cases are attributable to a dominant gene (Knudson and Strong 1972, Knudson and Meadows 1976). These hereditary cases may have more than one tumor, and the tumors appear at an earlier median age than in nonhereditary cases. As is true of retinoblastoma, most of these so-called hereditary cases do not have a positive family history of neuroblastoma but rather are new germinal mutations. It is estimated that these hereditary, or prezygotic, cases comprise 22% of all neuroblastoma cases. The remainder are nonhereditary, or postzygotic, cases (Knudson and Strong 1972). The generalizations regarding hereditary and nonhereditary forms of the same tumor seem to be applicable to neuroblastoma. The probability that a normal individual will develop neuroblastoma is of the order of magnitude of 10^{-4}, whereas the gene carrier will develop a mean number of approximately one tumor, a relative risk of 10,000.

A congenital anomaly has been reported in a few neuroblastoma patients or their relatives, which may have pathogenetic significance. This condition is aganglionosis, or Hirschsprung's disease (Knudson and Meadows 1976). The particular form of this condition associated with neuroblastoma involves a long segment of colon, the form that is usually found when Hirschsprung's disease itself is hereditary (Passarge 1967). Neuroblastoma is not found in the myenteric ganglia, but that may simply be a result of the fact that these ganglia are populated by the migration of cells which have differentiated beyond the neuroblastic stage. The same mutation predisposes to one developmental disorder (neoplasm) in the sympathetic branch of the autonomic nervous system and to another (failure of formation or of migration) in the parasympathetic branch.

The neuroblastoma gene has a specificity that includes the more mature benign tumor, ganglioneuroma. The two tumors have been found together in the same family, in the same patient, and even in the same tumor mass. A phenomenon of great clinical importance is the occasional spontaneous or treatment-induced maturation of neuroblastoma into ganglioneuroma. This phenomenon has also been observed in vitro in that neuroblasts from neuroblastoma often differentiate into cells with many cytological and biochemical properties of ganglion cells.

Another occasional feature of neuroblastoma, that of spontaneous regression, may be related to spontaneous maturation. The problem of regression must be related to normal development of the fetal adrenal medulla from nests of neuroblasts with actively dividing centers (Turkel and Itabashi 1974). These nests are found even in small numbers in newborn infants but disappear, apparently by differentiation. Occasionally, large cell clusters are found and are called neuroblastoma in situ (Beckwith and Perrin 1963). These often regress spontaneously and may simply be unusually large clusters of normally differentiating cells. In other instances, particularly in those occurring more than three months after birth, they may represent small neuroblastomas, especially since these have been found in the adrenals of patients with unquestionable neuroblastoma. The problem of regression of neuroblastoma is obviously confounded by the normal pattern of development in the adrenal medulla, but it is strongly suggestive that the action of a neuroblastoma mutation is to interfere with this normal developmental process.

Pheochromocytoma

Pheochromocytoma is a benign tumor of the sympathetic nervous system that actively secretes norepinephrine and/or epinephrine in large amounts. Over 20% of cases are apparently associated with a dominant gene specific for this tumor (Knudson and Strong 1972). Multiple tumors are by no means rare in the hereditary form; in fact, the mean number of tumors per gene carrier has been estimated to be at least two, for reported cases (Knudson and Strong 1972). These hereditary cases occur at a median age of approximately 20 years, whereas nonhereditary cases show a median age of 35 years or so. Thus, the time scale for tumor appearance is greatly extended relative to that for neuroblastoma. Pheochromocytoma is rarely observed before the age of five years, a phenomenon that has been explained by a growth phase (from a single tumor cell) of 5 to 10 years for pheochromocytoma generally (Knudson and Meadows 1976). However, it may be observed even after the age of 50 years.

Although this dominant form of pheochromocytoma may occur in any segment of the sympathetic nervous system, its specificity is great, in that it is not associated with either neuroblastoma or ganglioneuroma. A second form of pheochromocytoma mutation has, on the one hand, greater specificity, in that it produces tumors only in the adrenal medulla, and, on the other hand, less specificity, in that it also predisposes to medullary carcinoma of the thyroid and parathyroid adenomas (Sipple 1961). Sipple's disease accounts for approximately 2% of all pheochromocytomas. As with the more common, dominant pheochromocytoma gene the mean number of tumors steadily increases with age, reaching a number of four or so by age 50 years for both pheochromocytoma and thyroid carcinoma (Knudson and Strong 1972). How a mutant gene can predispose equally to tumors in two different tissues is unknown, but it is surely relevant that the epinephrine-secreting cells of the adrenal medulla and the thyrocalcitonin-secreting cells of the thyroid medulla have a common embryologic origin in the neural crest (Bolande 1974). This is not a full explanation, however, because many other neural crest derivatives are not affected.

Perhaps a clue to the nature of the shared genetic function will be provided by a third pheochromocytoma mutation, one closely resembling Sipple's disease but one that additionally causes plexiform neuromas of the lips, tongue, buccal mucosa, and eyelids—the so-called mucosal neuroma syndrome (Gorlin *et al.* 1968). These neuromas are functionally related to the thyroid medulla in that they too can form thyrocalcitonin (Brown *et al.* 1975).

BRAIN TUMORS

The subject of brain tumors is vastly more complex than those already considered, owing to the wide diversity of histologic types and to the lack of differentiation that is seen in some brain tumors. The brain tumors considered here are only those that arise from neural elements; tumors such as lymphoma of the brain are excluded. The most characteristic and also most common tumors are those that demonstrate some degree of neuroglial differentiation. These are the gliomas, a group that includes astrocytoma, oligodendroglioma, and ependymoma. Gliomas are found at all ages and are the principal brain tumors of adults. A second characteristic tumor is the medulloblastoma, a special tumor that arises only in the cerebellum. Finally, there are some tumors of early childhood that are so undifferentiated that they are difficult to classify; these include the medulloepithelioma and the cerebral neuroblastoma, both of which are thought to arise from early embryonic stages (Rubinstein 1972). Both of these tumors are so rare and so controversial that they are not discussed here. There are also no reported familial cases of either.

Gliomas

Gliomas may be found throughout the central nervous system, including the spinal cord. Although they occur at all ages, those arising in the posterior fossa are more characteristic of children. The commonest tumor in this group is the astrocytoma, which may be well differentiated and slowly growing (grades I and II) or more malignant (grades III and IV). The latter were formerly classified as glioblastoma multiforme. Less common is the oligodendroglioma, which most often originates in the cerebrum. Ependymomas arise from cells lining the ventricular system, the most common site of origin being the floor of the fourth ventricle. The most undifferentiated tumor in the glioma category is the polar spongioblastoma, again a tumor arising near the ventricular system; it occurs chiefly in the first two decades of life.

The occurrence of gliomas at all ages contrasts sharply with the childhood occurrence of medulloblastoma, medulloepithelioma, and central neuroblastoma. This phenomenon is evidently explainable on the basis of continued mitotic activity of the normal cells that give rise to gliomas. Neuronal elements, except those that produce medulloblastoma, are generated early from the primitive neuroepithelium and migrate postmitotically to their ultimate sites in the brain. On the other hand, neuroglial elements, generated from primitive bipolar spongioblasts, continue to show mitotic activity after migration (Rubinstein 1972).

The familial occurrence of gliomas has been reported numerous times (Schoenberg *et al.* 1975, Horton 1976). Most often, these reports have concerned sibs (including twins) and half-sibs, but there have also been several reports involving transmission from parents to one or more child and several involving close relatives spanning two or three generations. These familial patterns have been very similar to those observed for retinoblastoma and neuroblastoma under conditions of high mortality. It is a reasonable conclusion that some fraction of gliomas, as yet undetermined quantitatively, can be attributed to predisposition by an incompletely penetrant dominant gene (Knudson *et al.* 1973, Knudson, in press).

The genetic specificity of such a gene within the category of glioma is not great in general, although evaluation is hampered by the fact that in many reports the diagnosis is simply "glioma." Astrocytomas of all grades are the most common and often the only specified tumors found in a family. Astrocytoma has been associated in the same family with oligodendroglioma or ependymoma as well. It is noteworthy that more than one glial element may be present in the same tumor. A remarkable example is the one reported by Fairburn and Urich (1971) of identical twins, each of whom expired from a glioma that consisted of two distinct parts, one oligodendroglial and one astrocytic. In one twin the tumor was cerebral, in the other cerebellar.

One conclusion that can be reached at this time is that there is at least one dominant gene that predisposes to glioma, but the specificity does not extend within this category with respect to histologic type or location. The common denominator for the target cell for transformation appears to be embryologic descent from the bipolar spongioblast. The polar spongioblastoma is so rare a tumor that it cannot be decided whether the specificity of the dominant "glioma" gene extends to it.

Medulloblastoma

Medulloblastoma is a unique tumor, typically occurring in childhood, very malignant, and arising in the cerebellum. The cell of origin is a special neural element that arises from the neuroepithelial roof of the fourth ventricle and migrates, while still mitotic, to the external granular layer of the cerebellar cortex. It is, then, an exception to the rule that neuronal, as opposed to neuroglial, cells may migrate while still mitotic. However, these cells do not long continue mitosis, a fact that probably accounts for the typical occurrence of this tumor in children. This purely neural nature has been brought into question by the observation that while most of these tumors remain undifferentiated and a few form distinct ganglion cells, rare examples show spongioblastic differentiation, suggesting a neuroglial potential (Rubinstein 1972).

Medulloblastoma has been reported seven times in sibs (including twins) or half-sibs (Yamashita *et al.* 1975). It has not been reported in successive generations. The median age of these familial cases is younger than is the median age for medulloblastoma generally. In one instance, medulloblastoma and a glioma (ependymoma) were found in sibs, although the child with medulloblastoma was abnormal in that she had Turner's syndrome (Pendergrass *et al.* 1974).

Although the data are too few to draw any conclusion regarding the nature of a hereditary fraction of medulloblastoma cases, it is of great interest that this tumor, but not gliomas, is found in some patients with a known dominantly inherited syndrome, the nevoid basal cell carcinoma syndrome (Herzberg and Wiskemann 1963, Horton 1976). These patients develop the tumor at an earlier than usual age, and, in fact, some patients are treated for the tumor before the syndrome is apparent. The radiotherapy administered for the tumor can induce large numbers of basal cell carcinomas with a short latent period in the overlying skin (Strong 1977).

There is, therefore, at least one dominant gene that predisposes to medulloblastoma specifically. There seems to be no overlap in the specificity of this gene and the gene(s) that predisposes to glioma. It is probable that this separation reflects the distinctively separate embryonic origins of the two cell lineages.

NEUROFIBROMATOSIS

The genetic predispositions that have been considered so far are specific for one tumor or a related group of tumors. But the dominant gene for neurofibromatosis, or von Recklinghausen's disease, demonstrates less specificity, predisposing to an array of neural tumors. The usual tumor is the neurofibroma, a neural sheath tumor classified by some as a harmartoma rather than a true neoplasm. This suggestion is supported by the finding that they are of multicellular origin (Fialkow *et al.* 1971).

Pheochromocytoma occurs in approximately 10% of neurofibromatosis patients who attain the age of 60 years (Knudson and Strong 1972). This incidence, and the incidence of bilateral tumor, is much lower than for the two dominant pheochromocytoma genes already discussed. Ganglioneuroma is occasionally found in neurofibromatosis, and there are even a few case reports of neuroblastoma (Knudson and Meadows 1976).

Gliomas are among the more common tumors seen in this disease, especially in its central form characterized by cranial nerve neurofibromas and meningiomas. The gliomas encountered include especially astrocytoma, spongioblastoma, and ependymoma. The single most common tumor involves the optic nerve (Horton 1976).

There are two reports of medulloblastoma in patients with neurofibromatosis (Corkill and Ross 1969, Meadows *et al.*, in press), suggesting that this dominant gene has broad specificity within the nervous system. The basis for this phenomenon is, of course, unknown, but the findings do suggest that neural tumors have some genetic feature in common.

DISCUSSION

There is a set of dominant genes that demonstrate considerable specificity for at least five of the major neural tumors of man. Another gene also predisposes to tumors, primarily of the nervous system, with much less specificity.

A model for such dominant tumors is retinoblastoma. According to this model, tumors exist in both hereditary and nonhereditary form, each involving two mutations in the transformation of the normal precursor cell to the first tumor cell. The possibility has been noted that neoplasia results from homozygosity for these mutations at specific chromosomal loci, different for each tumor. A consequence of homozygosity is that each tumor results from a single molecular defect.

The occurrence of so many tumor genes expressed within one highly differentiated system provides clues to the function of cancer genes in general. The requirement that these genes act upon cell lines which retain mitotic potential limits their expression, in some instances, to the first few years of life (retinoblastoma, neuroblastoma, spongioblastoma, and medulloblastoma). In each case, further differentiation of the affected cell line is arrested or impaired. More differentiated cell lines that retain mitotic activity for longer periods can develop into neoplasms in adulthood (pheochromocytoma and most gliomas).

We conclude that the normal alleles of neural tumor genes perform crucial functions in the development of the nervous system. Explication of the fundamental defects in the tumor cells would thereby characterize a major class of genes active in morphogenesis.

There is no direct evidence bearing on the nature of these molecular events. It has been suggested elsewhere that in tumors of the sympathetic nervous system the presumptive single molecular defect involves abnormal cellular membranes, both external and internal (Knudson and Meadows 1976). Such an explanation, on one hand, suggests that loss of cell:cell interaction is the basis for the loss of control of cell division so characteristic of cancer and, on the other hand, may explain such intracellular phenomena as the loss of binding capacity for catecholamines by the membranous granules of neuroblastoma cells (Page and Jacoby 1964).

ACKNOWLEDGMENTS

This work was supported in part by an appropriation from the Commonwealth of Pennsylvania and by Contract N01-23286 from the Division of Cancer Cause and Prevention, National Cancer Institute.

REFERENCES

Abramson, D. H., R. M. Ellsworth, and L. E. Zimmerman. 1976. Nonocular cancer in retinoblastoma survivors. Trans. Am. Acad. Ophthalmol. Otolaryngol. 81:454–457.

Beckwith, J. B., and E. V. Perrin. 1963. In situ neuroblastoma: A contribution to the natural history of neural crest tumors. Am. J. Pathol. 43:1089–1104.

Bolande, R. P. 1974. The neurocristopathies: A unifying concept of disease arising in neural crest maldevelopment. Hum. Pathol. 5:410–429.

Brown, R. S., E. Colle, and A. H. Tashjian. 1975. The syndrome of multiple mucosal neuromas and medullary thyroid carcinoma in childhood. J. Pediatr. 86:77–83.

Corkill, A. G. L., and C. F. Ross. 1969. A case of neurofibromatosis complicated by medulloblastoma, neurogenic sarcoma, and radiation-induced carcinoma of the thyroid. J. Neurol. Neurosurg. Psycht. 32:43–47.

Fairburn, B., and H. Urich. 1971. Malignant gliomas occurring in identical twins. J. Neurol. Neurosurg. Psychiatry 34:718–722.

Fialkow, P. J., R. W. Sagebiel, S. M. Gartler, and D. L. Rimoin. 1971. Multiple cell origin of hereditary neurofibromas. N. Engl. J. Med. 284:298–300.

Gorlin, R. J., H. O. Sedano, R. A. Vickers, and J. Crevenka. 1968. Multiple mucosal neuromas, pheochromocytoma, and medullary carcinoma of the thyroid—A syndrome. Cancer 22:293–300.

Herzberg, J. J., and A. Wiskemann. 1963. Die fünfte Phakormatose, Basalzellnaevus mit familiärer Belastung und Medulloblastom. Dermatologica 126:106–173.

Horton, W. A. 1976. Genetics of central nervous system tumors, *in* Cancer and Genetics (Birth Defects Original Article Series, Vol. 12, No. 1), D. Bergsma, ed., The National Foundation-March of Dimes, Alan R. Liss, New York, pp. 91–97.

Jensen, R. D., and R. W. Miller. 1971. Retinoblastoma—Epidemiologic characteristics. N. Engl. J. Med. 285:307–311.

Kitchin, F. D., and R. M. Ellsworth. 1974. Pleiotropic effects of the gene for retinoblastoma. J. Med. Genet. 11:244–246.

Knudson, A. G. 1971. Mutation and cancer: Statistical study of retinoblastoma. Proc. Natl. Acad. Sci. USA 68:820–823.

Knudson, A. G. 1977. Genetics and etiology of human cancer, *in* Advances in Human Genetics, H. Harris and K. Hirschhorn, eds., Vol. 8. Plenum Press, New York, pp. 1–66.

Knudson, A. G., and A. T. Meadows. 1976. Developmental genetics of neuroblastoma. J. Natl. Cancer Inst. 57:675–682.

Knudson, A. G., and L. C. Strong. 1972. Mutation and cancer: Neuroblastoma and pheochromocytoma. Am. J. Hum. Genet. 24:514–532.

Knudson, A. G., H. W. Hethcote, and B. W. Brown. 1975. Mutation and childhood cancer: A probabilistic model for the incidence of retinoblastoma. Proc. Natl. Acad. Sci. USA 72:5116–5120.

Knudson, A. G., L. C. Strong, and D. E. Anderson. 1973. Heredity and cancer in man. Prog. Med. Genet. 9:113–158.

Knudson, A. G., A. T. Meadows, W. W. Nichols, and R. Hill. 1976. Chromosomal deletion and retinoblastoma. N. Engl. J. Med. 295:1120–1123.

Meadows, A. T., G. J. D'Angio, V. Mike, A. Banfi, C. Harris, R. D. T. Jenkin, and A. Schwartz. 1977. Patterns of second malignant neoplasms in children. Cancer (In press).

Page, L. B., and G. A. Jacoby. 1964. Catecholamine metabolism and storage granules in pheochromocytoma and neuroblastoma. Medicine 42:379–386.

Passarge, E. 1967. Genetics of Hirschsprung's disease: Evidence for heterogeneous etiology and a study of sixty-three families. N. Engl. J. Med. 276:138–143.

Pendergrass, T. W., J. F. Fraumeni, and E. L. Fagan. 1974. Brain tumors in sibs, one with the Turner syndrome. J. Pediatr. 85:875.

Rubinstein, L. J. 1972. Cytogenesis and differentiation of primitive central neuroepithelial tumors. J. Neuropathol. Exp. Neurol. 31:7–26.

Sagerman, R. H., J. R. Cassady, P. Tretter, and R. M. Ellsworth. 1969. Radiation induced neoplasia following external beam therapy for children with retinoblastoma. Am. J. Roentgenol. Radium Ther. Nucl. Med. 105:529–535.

Schoenberg, B. S., G. G. Glista, and T. J. Reagen. 1975. The familial occurrence of glioma. Surg. Neurol. 3:139–145.

Sipple, J. H. 1961. The association of pheochromocytoma with carcinoma of the thyroid gland. Am. J. Med. 31:163–166.

Strong, L. C., and A. G. Knudson. 1973. Second cancers in retinoblastoma. Lancet 2:1086.

Strong, L. C. 1977. Theories of pathogenesis: Mutation and cancer, *in* The Genetics of Human Cancer, J. J. Mulvihill, R. W. Miller, and J. F. Fraumeni, eds., Raven Press, New York, pp. 401–415.

Turkel, S. B., and H. H. Itabashi. 1974. The natural history of neuroblastic cells in the fetal adrenal gland. Am. J. Pathol. 76:225–243.

Yamashita, J., H. Handa, and M. Toyama. 1975. Medulloblastoma in two brothers. Surg. Neurol. 4:225–227.

Cell Differentiation and Neoplasia, edited by Grady F. Saunders. Raven Press, New York

Neoplastic Neuroepithelial Differentiation in an Experimental Transplantable Teratoma

Mary M. Herman and Scott R. VandenBerg

Department of Pathology (Neuropathology), Stanford University School of Medicine, Stanford, California 94305

The nervous system is the second most frequent site of primary tumors in children. These include an appreciable proportion of embryonal neoplasms. Embryonal neoplasms that originate within the central nervous system (CNS) are particularly complex in their identification and classification, although to some extent they can be equated to distinct stages of CNS cytogenesis (Rubinstein 1972). Moreover, they appear to possess some of the capabilities for divergent differentiation that are normally found in neural cytomorphogenesis (Artzt and Bennett 1972, Rubinstein 1976). For this reason, the mouse transplantable teratoma was considered as a possible model for the study of neoplastic differentiation in embryonal tumors originating within the CNS.

Most experimental CNS tumors that have been produced by chemical carcinogens, such as the N-nitroso alkylating agents (Swenberg 1976), or by oncogenic viruses (Bigner and Pegram 1976) are, on the whole, comparable to human gliomas composed of adult cell types. A small number of animal tumors, induced by the cerebral inoculation of human adenovirus type 12 or the JC strain of human papovavirus, which causes progressive multifocal leukoencephalopathy, are of embryonal character, equivalent to the human neuroblastoma (Mukai and Kobayashi 1973), retinoblastoma (Kobayashi and Mukai 1974), or medulloblastoma (ZuRhein and Varakis, in press). An alternative system for the study of neuro-oncogenesis in which further differentiation might be expressed is, as we have proposed, the neuroepithelial component of a differentiating transplantable mouse testicular teratoma. Malignant testicular teratomas that are both multipotential and transplantable arise spontaneously in specific inbred strains of mice (Stevens 1967), but their incidence can be increased by genetic and environmental factors (Stevens 1973). They can also be produced by grafting embryos into the testes or beneath the renal capsules of adult mice of syngeneic strains before the critical stages of mesoderm formation (Stevens 1970, Damjanov *et al.* 1971a). These various teratomas display divergent differentiation, including partial neural differentiation, from primitive stem cells into derivatives of the three classic germ layers.

Our laboratory has been studying the OTT-6050 line of transplantable mouse teratoma for the past five years. We have made detailed morphologic observations of both the ascitic and solid forms of this line (Stevens 1970) to determine the range of differentiation of tissues derived from the classic germ layers, and we have correlated the morphologic with the biochemical studies. We have also characterized two different alkaline phosphatases in the solid tumor and in the embryoid bodies and have correlated the data with histochemical observations (Wada *et al.* 1976). Work on the characterization of nonhistone chromosomal proteins (NHCP) in the solid tumor and their comparison with NHCP in the brains of newborn and adult strain-related mice is in progress (Choie *et al.*, in press). We have also studied the in vitro behavior of the ascitic and solid forms of the teratoma and have been particularly interested in obtaining neuroepithelial differentiation from explanted embryoid bodies. Our eventual goal is to develop a spectrum of neuroepithelial differentiation in vitro from the embryoid bodies, and a transplantable homogeneous neuroepithelial tumor, both of which can be used to determine the factors influencing neuroepithelial differentiation and maturation.

The OTT-6050 line of transplantable mouse teratoma derived from an intratesticular implant of a 6-day-old 129/Sv embryo into the testis of an adult F_1 (A/He $\times$ 129) host in February 1967 (Stevens 1970) was kindly provided by Dr. Leroy C. Stevens of the Jackson Laboratory, Bar Harbor, Maine, and has been maintained subsequently at the Jackson Laboratory in ascitic form. The tumor in both ascitic and solid subcutaneous forms has been carried in our laboratory for approximately 60 transplant generations. Syngeneic hosts are 5- to 8-week-old 129/J female mice. Methods for transplantation have been detailed elsewhere (VandenBerg *et al.* 1975). After several subcutaneous passages, a solid tumor containing a high percentage (80 to 90%) of neuroepithelial cells was obtained, as confirmed by serial sections of one entire tumor at 5 μm intervals.

LIGHT AND ELECTRON MICROSCOPIC STUDIES

In morphologic studies of the solid tumor (VandenBerg *et al.* 1975, Herman *et al.* 1975) the following observations were made. The most primitive form of neuroepithelial differentiation corresponded to compact collections, or tubules, of primitive neuroepithelial cells, which, when in tubules, demonstrated a well-defined luminal limiting membrane and an external limiting membrane *(primitive medullary rosettes);* mitotic figures were frequent and were always located adjacent to the internal (luminal) limiting membrane. These areas closely resemble the developing mammalian neural tube or its human neoplastic equivalent, the rare medulloepithelioma. By electron microscopy, the primitive neuroepithelial cells had a scanty granular endoplasmic reticulum, which is one of the hallmarks of germinal matrix cells (Fujita and Fujita 1963, Wechsler 1966, Herman and Kaufman 1966, Wechsler and Mehler 1967), and nuclear characteristics that, together with other features, clearly distinguished them from stem cells. The apices of the cells, joined by short zonulae adherentes, often formed discrete bulges that contained numerous polyribosomes, centrioles, and rare mature cilia with a 9 + 0 pattern of microtubules. Microvilli were absent.

Frequent transitions between the primitive medullary rosettes and the more differentiated ependymoblastomatous or ependymal rosettes were seen. Compared with the primitive medullary rosettes, cells of these transitional forms showed increased development of granular endoplasmic reticulum and the Golgi apparatus. The cell polarity became accentuated. In the early transitional stages, microvilli formation was evident. Later the microvilli were more numerous and better developed, and the apices of the rosette-forming cells had more elaborate and lengthier intercellular junctions with the features of zonulae adherentes.

Compact, multilayered columnar cells, often with one or more mitotic figures in a juxtaluminal position, frequently formed rosettes and tubules and were designated *ependymoblastomatous rosettes.* They are similar to those found in the rare human ependymoblastoma. By electron microscopy the apices of these cells were joined by numerous, lengthy intercellular junctions of the zonula adherens type and often contained cilia which were usually of the 9 + 2 microtubular pattern but were occasionally 9 + 3, 9 + 4, 10 + 2, or 10 + 3. Numerous surface microvilli projected into the lumen.

Mature *ependymal rosettes* were also found in some fields. They were composed of one or occasionally two layers of cells and were devoid of mitotic figures. Their lumen contained numerous cilia with the typical 9 + 2 microtubular pattern, and the apices of the cells bordering the lumen were joined by long elaborate junctions of the zonula adherens type. The mature ependymal areas correspond to the human ependymoma.

Areas of *astrocytic* and *oligodendroglial differentiation* were found, and these correspond to the human astrocytoma and oligodendroglioma. In the astrocytic areas, the cells contained fibrillated phosphotungstic acid hematoxylin (PTAH)-positive processes. Astrocytic differentiation was confirmed by immunofluorescence (VandenBerg *et al.* 1975), immunohistochemical (horseradish peroxidase-labeled antibody method), and immunoradiometric methods (Ludwin *et al.* 1976) for the demonstration of glial fibrillary acidic (GFA) protein, a protein specific for astrocytes (Eng *et al.* 1971, Bignami *et al.* 1972, Bignami and Dahl 1974). However, astrocytic filaments were sparse at the fine structural level.

Various stages of *ganglionic differentiation* were present in some regions. Both neuroblasts, occasionally arranged in well-defined rosettes without a lumen but with a distinct central eosinophilic fibrillary matrix *(neuroblastic rosettes),* and more mature neurons were noted. The Bielschowsky silver impregnation technique showed numerous delicate axons coursing through the ganglionic areas, and occasionally unipolar or bipolar processes were demonstrated to originate from the perinuclear cytoplasm of a neuroblast. Occasional fields of more mature neurons were shown by electron microscopy to contain stacks of granular endoplasmic reticulum in the perikaryon, numerous compact neurites filled with microtubules, and junctional contacts and synapses. A few cell processes contained dense core vesicles measuring 850–1150 Å in outer diameter, with the central core of variable density and measuring 570–715 Å; occasional dense core vesicles were also noted in presynaptic terminals. Myelin was never found, either by light or electron microscopy or by two-site immunoradiometric assay for myelin basic protein (Ludwin *et al.* 1976). The neuroblastic areas correspond to the neuroblastoma and the more differentiated areas to the human ganglioneuroma or ganglioglioma. Some of these morphologic features have been observed in other experimental mouse teratomas in previous studies by Damjanov *et al.* (1973) and Tresman and Evans (1975); myelin was found in the first study but not in the second, and synapses were described in both studies.

The demonstration of ependymal, astrocytic, and neuronal differentiation in areas in continuity with primitive neuroepithelial cells confirms the utility of the OTT-6050 teratoma for the study of neoplastic neurocytogenesis. Cells either were undifferentiated or showed cytologic features of maturation along a particular line, thus following the general rule that selective utilization of genetic material for differentiation is expressed by a process in which the selection of one effective genome usually automatically suppresses the utilization of others.

Intracisternal A-particles were found in stem cells of the OTT-2466 and OTT-6050 teratomas and in more differentiated cell types in the OTT-6050 line, including primitive endodermal cells, ciliated respiratory epithelium, cells of primitive neuroepithelial rosettes, neuroblasts, and primitive glial cells (Spence *et al.* 1975). Both parent strains, A/He and 129/Sv, also contained the virions. The particles have been noted previously in the embryoid bodies of the OTT-6050 teratoma (Teresky *et al.* 1974). Their biological role has not yet been clarified, despite the fact that numerous structural, immunologic, and biochemical studies, and tests for oncogenicity have been performed.

NEUROCHEMICAL STUDIES

Concurrent with the morphological studies, we undertook neurochemical characterization of solid tumors of the OTT-6050 teratoma (Orenberg *et al.* 1976). We compared tumors that had neuroepithelial differentiation as either the major or the minor component with the brains of hybrid newborn and adult mice [129/J $\times$ F_1 (Balb/c $\times$ 129)]. We found that the tumors contained cyclic AMP (cAMP), serotonin (5-HT), and enzymes of the serotonergic, adrenergic, and cholinergic systems; specifically, the enzymes included tryptophan hydroxylase (TPH), aromatic amino acid decarboxylase (AADC), monoamine oxidase (MAO), tyrosine hydroxylase (TH), dopamine-β-hydroxylase (DBH), choline acetyltransferase (ChAc), and acetylcholinesterase (AChE). Compared with pooled adult brains, the levels of cAMP and the above enzymes in the tumors with increased proportions of neuroepithelial cells ranged approximately as follows: cAMP, 50–75%; TPH, 10–200%; AADC, 60–120%; MAO, 10–25%;

TH, 25–130%; DBH, 20–25%; ChAc, 10–40%; and AChE, 10–35%. 5-HT, which shows a predominantly postnatal development in rodent brains (Loizou 1972), was found in unusually high concentrations in one tumor in this group, with levels 300% greater than in adult brains. Levels of cAMP, 5-HT, and the above enzymes were lower in the tumors composed predominantly of stem cells and nonneuroepithelial cells as compared with tumors composed predominantly of neuroepithelial cells. Collectively, the trends therefore suggested an increase of cAMP and an increased activity of the synthesizing enzymes TPH, AADC, TH, and DBH in tumors with increased proportions of neuroepithelial cells.

In previous neurochemical studies by others, 5-HT and 5-hydroxyindoleacetic acid were found in a transplantable teratoma in C_3H mice in which exclusive neuroepithelial differentiation was present (Jakupčević *et al.* 1974). Others have noted AChE in cultured OTT-6050 cells (Gearhart and Mintz 1974, 1975, Levine *et al.* 1974). The presence of biogenic amine and cholinergic transmitters with their associated enzymes and of transmitter-sensitive adenylate cyclase has been demonstrated in both cloned and uncloned C-1300 neuroblastoma cell lines (see Orenberg *et al.* 1976, Prasad and Sinha 1978, pages 111 to 141, this volume, for review), despite the fact that morphologic differentiation to mature neuronal elements does not occur (Ruffner and Grieshaber 1974). A cell line derived from a murine ganglioneuroblastoma also failed to demonstrate a correlation between morphologic characteristics and biochemical differentiation (Ruffner and Smith 1974). Thus, obvious limitations exist in the use of these systems as models for the spontaneous differentiation that occasionally occurs in human neuroblastomas and ganglioneuroblastomas.

The neurochemical findings in our studies of the OTT-6050 teratoma therefore confirm the neuroepithelial differentiation and neuronal maturation noted morphologically. They also suggest that the neuroepithelial component of this tumor may be a more useful system for the structural and biochemical characterization of neuronal differentiation in primitive neural neoplasms than that of the C-1300 neuroblastoma or murine ganglioneuroblastoma.

TISSUE CULTURE STUDIES

In vitro studies on neuroepithelial differentiation from the embryoid bodies derived from ascitic fluid of the OTT-6050 teratoma are in progress in our laboratory. Before culturing, the embryoid bodies were filtered by gravity sedimentation through a double-layered nylon screen with 74 μm pores so as to exclude the larger bodies in which internal differentiation along a number of different lines is known to occur (Gearhart and Mintz 1974, Martin and Evans 1975). The filtered bodies were then tested by both immunofluorescence (VandenBerg *et al.* 1976a) and two-site immunoradiometric assay (courtesy of Dr. L. F. Eng) for the presence of GFA protein, and the protein was found to be absent. Immediately before explanting, the glass coverslips to be used as substrates were conditioned with sequential treatments with argon gas plasma in a Plasmod generator (Tegal Corporation, Richmond, California) to facilitate attachment of the embryoid bodies. Further description of the tissue culture and GFA protein immunofluorescence methods, and full details of the medium used in these studies are given elsewhere (VandenBerg *et al.* 1976a).

The results of the tissue culture studies on neuroepithelial differentiation are as follows (VandenBerg *et al.* 1976a). Within 24 hours the embryoid bodies had attached, and outgrowths of primitive, multilayered neuroepithelial cells were present. Within 48 hours some of the primitive neuroepithelial cells demonstrated faint, hair-like positivity in their cytoplasm when tested by GFA protein immunofluorescence. Within 72 hours there was a divergence of two cell types, both occurring in mitotically active populations. One population was *astrocytic,* as confirmed in later cultures by positive immunofluorescence to GFA protein, which was more intense and more typically condensed than the early hair-like positivity noted within 48 hours. After one week in culture, the tapering astrocytic cell processes formed complex meshes. The other cell population was *neuroblastic,* as confirmed in later cultures by specific argyrophilia and by a modified Falck-Hillarp histochemical technique (Falck *et al.* 1962, VandenBerg *et al.* 1976b) that involved an initial immersion fixation of the sample in 4% (w/v) paraformaldehyde in O.1 M phosphate buffer (pH 6.8 at 14–16° C) and a final exposure of the sample to formaldehyde gas (generated from paraformaldehyde equilibrated to 80% humidity) at 80° C for 45 minutes. Samples for negative controls were immersed in saline and heated to 80° C in room air. To confirm the specificity of the histochemical fluorescence, positive samples were treated with alcoholic sodium borohydride as previously described (Corrodi *et al.* 1964), and the fluorescence was regenerated by exposing the samples to formaldehyde gas at 80° C for one hour.

Histochemical fluorescence could be demonstrated only following pretreatment of the cultures for 24 hours with a monoamine oxidase inhibitor (pargyline hydrochloride, 0.2 mM final concentration). The fluorescence was considerably enhanced by a 24-hour pretreatment with the following growth medium supplements: 1.0 mM dibutyryl cAMP, 0.17 mM sodium ascorbate, 2.0 mM tryptophan, 0.15 mM nialamide, and 0.15 mM pargyline hydrochloride. We suspect that the positive fluorescence obtained under this set of conditions represents 5-HT because our previous neurochemical studies of the solid tumors showed that the serotonergic pathway is the best developed of all the biogenic amine or cholinergic pathways and that loading the nutrient medium with large amounts of tryptophan seemed to intensify the histochemical fluorescence. However, this assumption needs to be confirmed by biosynthetic experiments that are presently underway in our laboratory.

In summary, divergent neuroepithelial differentiation analogous to embryonic neurocytogenesis occurred in vitro within mitotically active cell populations as an early event and proceeded without apparent tissue relationships to other germ layer derivatives. The divergent diferentiation was characterized by astrocytes and neuroblasts. As indicated by GFA protein expression, the appearance of astrocytes of typical morphology was preceded by biochemical differentiation. The second cell type, the neuroblasts, failed to demonstrate GFA protein and had a small perikaryon with slender bipolar processes that in later cultures were positive for biogenic amines by histofluorescence and were argyrophilic by the Bodian Protargol method.

Others who have studied the spectrum of neuroepithelial differentiation from the embryoid bodies of experimental mouse teratomas using various cell or tissue culture systems have described primitive, neural, tube-like epithelium and rosettes (Pierce and Verney 1961, Kahan and Ephrussi 1970, Rosenthal *et al.* 1970, Martin and Evans 1975), and cells interpreted as glia (Lehman *et al.* 1974) and neurons (Levine *et al.* 1974, Teresky *et al.* 1974, Martin and Evans 1975). In those studies, arguments for glial differentiation were based on the cells' affinity for silver impregnation, the demonstration of glial morphologic features by the silver technique, the demonstration of microfilaments by electron microscopy, and the absence of neuronal markers such as choline acetyltransferase and tyrosine hydroxylase. Presumptions of neuronal differentiation were based on cell morphologic features demonstrable with hematoxylin and eosin and on the appearance of the brain- and fetal muscle-specific creatine phosphokinase and of acetylcholinesterase, an enzyme that is specific for muscle cells as well as for neurons.

In our studies, the early appearance of GFA protein in the embryoid body cultures suggests that in neoplastic neuroglia the expression of specific cellular products may precede structural differentiation. Furthermore, the early appearance of a morphologically distinguishable neuroepithelial cell population, the positive immunofluorescence for GFA protein demonstrable in only one neuroepithelial cell type, and the selective argyrophilia of the other, analogous to human fetal cerebrum in culture (Antanitus *et al.* 1975), are all consistent with the view that divergent differentiation occurred in vitro from primitive neuroepithelial cells. This possibility is given further credence by the following: (1) the use of a filtration procedure before explanting in order to eliminate large embryoid bodies that may have contained differentiated cells in the central core, (2) the failure to detect specific immunofluorescence for GFA protein in either the attached bodies or the filtrates of ascitic fluid processed for culture, confirmed in the latter case by assays for GFA protein, and (3) the typical blastocyst-like morphology of the attached embryoid bodies in culture. The conclusion is, moreover, in accord with two other lines of experimental evidence: (1) the stem cells in the embryoid bodies are multipotential (Kleinsmith and Pierce 1964, Kahan and Ephrussi 1970, Rosenthal *et al.* 1970, Martin and Evans 1975, Mintz and Illmensee 1975, Illmensee and Mintz 1976), and (2) morphologic and biochemical analyses of OTT-6050 embryoid bodies using other tissue-spe-

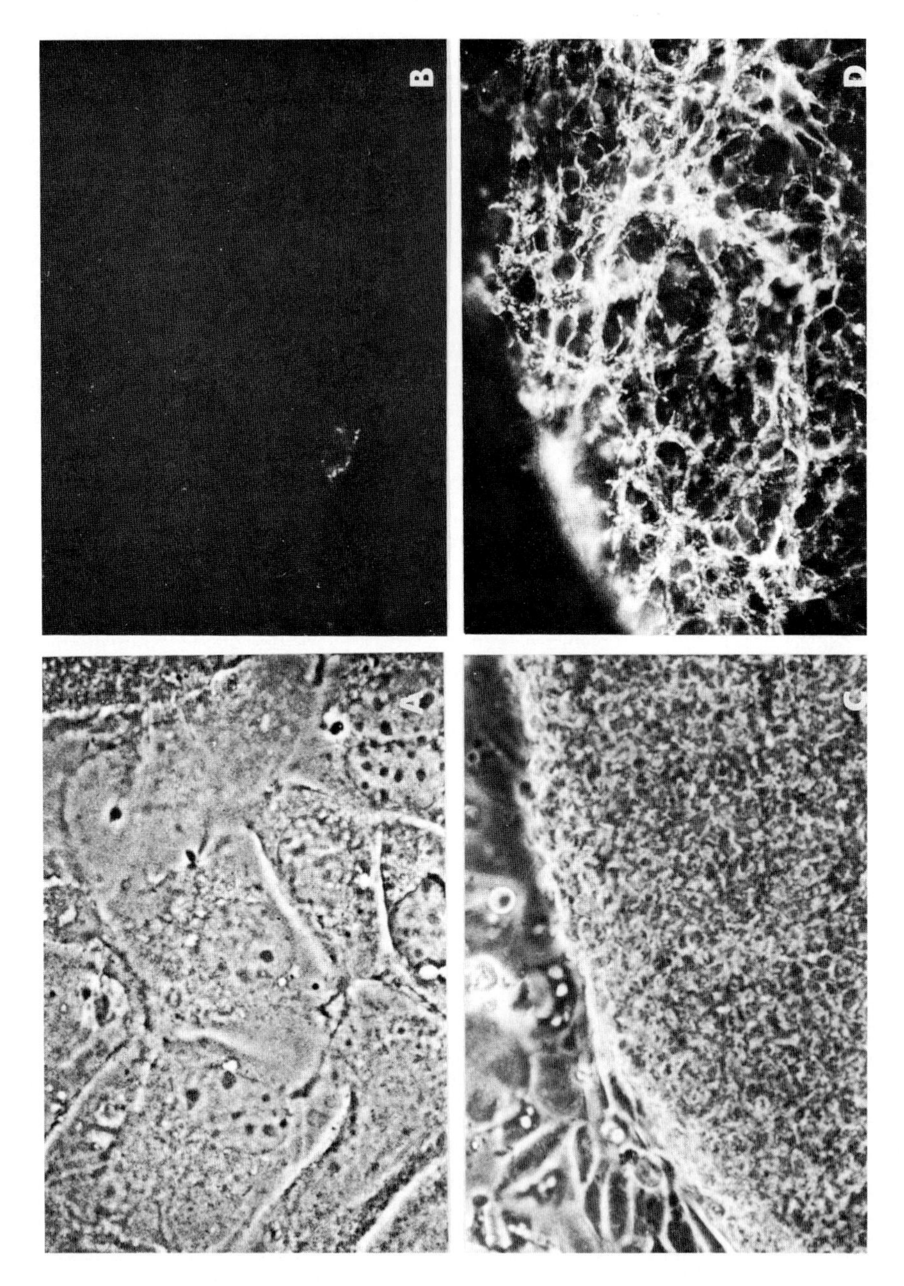

FIG. 1. Indirect immunofluorescence reaction of liver-spleen absorbed, brain-associated cell surface antiserum with living cultures of ascitic embryoid bodies. Specimens were fixed with 2% (w/v) paraformaldehyde in 0.1 M phosphate buffer following the immunofluorescence technique. **A,** phase contrast of the large epithelial-like stem cells derived from the attached embryoid bodies after 12 days in vitro (original X 690); **B,** negative immunofluorescence reaction

cific markers (Bernstine *et al.* 1973, Levine *et al.* 1974, Gearhart and Mintz 1974, 1975) have demonstrated that the progressive differentiation that subsequently occurred in vitro was not present prior to culture.

ORGAN CULTURE SYSTEM STUDIES

Neuroepithelial differentiation from embryoid bodies grown on spongefoam matrices (Spongostan, Ferrosan, Denmark) in a three-dimensional organ culture system is being studied in vitro in our laboratory. Before culturing, the embryoid bodies were filtered, and the spongefoam was pretreated with an argon plasma as in the tissue culture studies. Our initial, in vitro organ culture studies were for 90 days. We explanted the gravity-sedimented bodies onto spongefoam saturated with nutrient medium, similar to that used in the tissue culture studies (VandenBerg *et al.* 1976a), and then submerged the foam in a well of a microtiter dish (Multiwell tissue culture plate, Falcon plastics, Oxnard, California) filled with medium. After two weeks in the well, the spongefoam and attached embryoid bodies were transferred onto stainless steel grids in an organ culture dish (Falcon Plastics), as used for culture of human and experimental CNS tumors in our laboratory.

To date, we have achieved limited neuroepithelial differentiation in the cultures after 39 days in vitro. We have found fields of primitive neuroepithelial cells in continuity with collections of stem cells, primitive neuroepithelial rosettes that resemble a transitional form between medullary epithelium and the ependymoblastomatous rosettes, and ependymal rosettes. Very limited differentiation occurred in spongefoam cultures maintained for 39 days in the wells. After 39 days in vitro, all cultures became increasingly necrotic. We are currently modifying our techniques in an attempt to achieve long-term in vitro cultures with the full spectrum of neuroepithelial differentiation that is found in the solid tumors in the host.

STUDIES WITH BRAIN-ASSOCIATED CELL SURFACE ANTISERUM

Recently we have developed a heterologous immune serum, which was prepared against a neonatal F_1 (129J $\times$ C_3H/HeJ) brain fraction enriched with surface membranes (VandenBerg *et al.,* in press). Repeated absorptions with liver-spleen membrane fractions to remove nonneural, antimurine activity showed antibrain activity, assayed by microcomplement fixation, which could be completely absorbed with solid teratoma tissue. Indirect immunofluorescence

---

of the same cells (original X 690); **C,** phase contrast of a multilayered cluster of primitive neuroepithelial cells (bottom) adjacent to a nonneural monolayer (top) after 20 days in vitro (original X 270); **D,** edge of the same cluster showing positive immunofluorescence of the neuroepithelial cells with a highly cellular meshwork of variably elongated processes. Note the absence of fluorescence in the adjacent nonneural monolayer (original X 340).

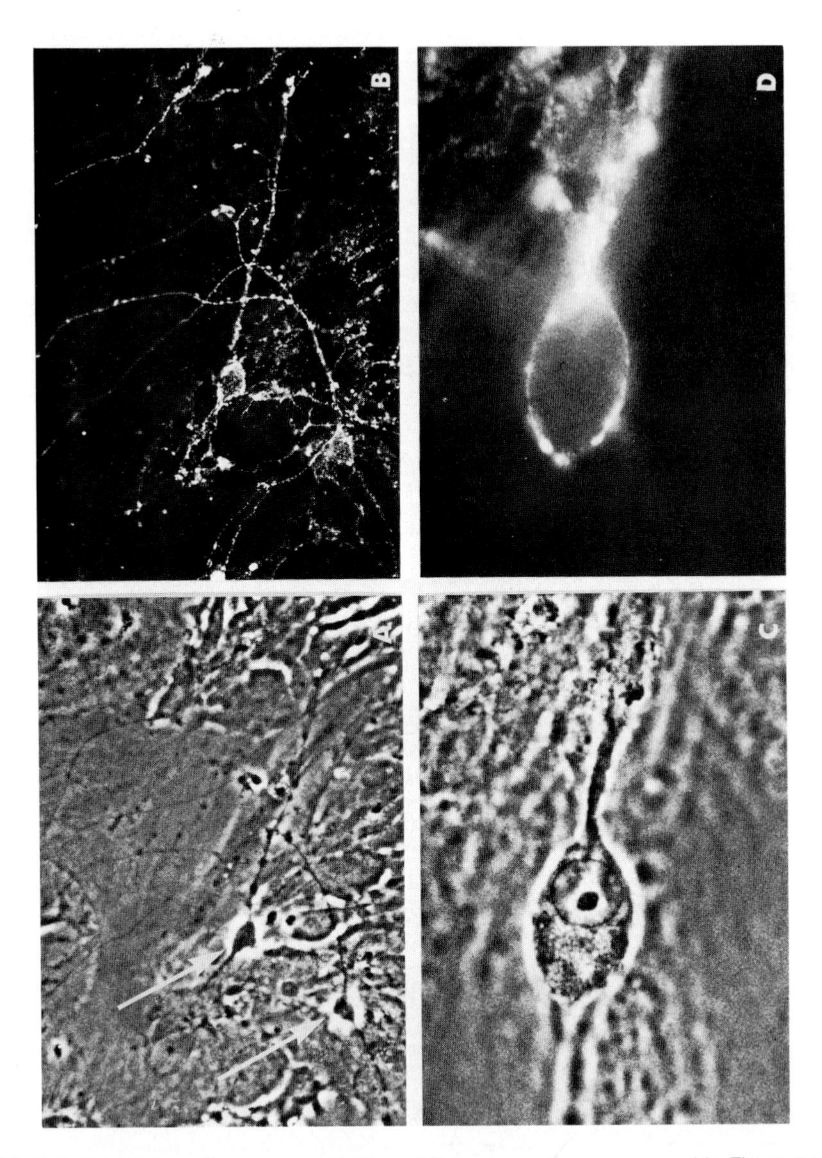

FIG. 2. Indirect immunofluorecence reaction of the same antiserum as used in Figure 1 with living embryoid body cultures after 12 days in vitro and fixed as in Figure 1. **A,** phase contrast of two neuroblasts (arrows) with delicate elongated processes extending over a cell monolayer (original X 265); **B,** positive immunofluorescence limited to the same neuroblasts showing activity on both the cell bodies and the extensive webs of processes (original X 280); **C,** phase contrast of a typical neuroblast with a neurite and a cell body demonstrating a prominent nucleolus (original X 830); **D,** positive immunofluorescence of the same neuroblast showing the cell surface distribution of the reactive antigens (original X 830).

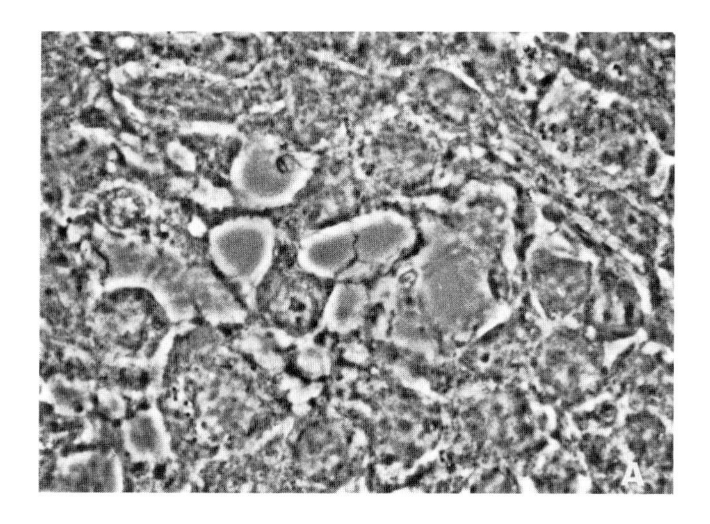

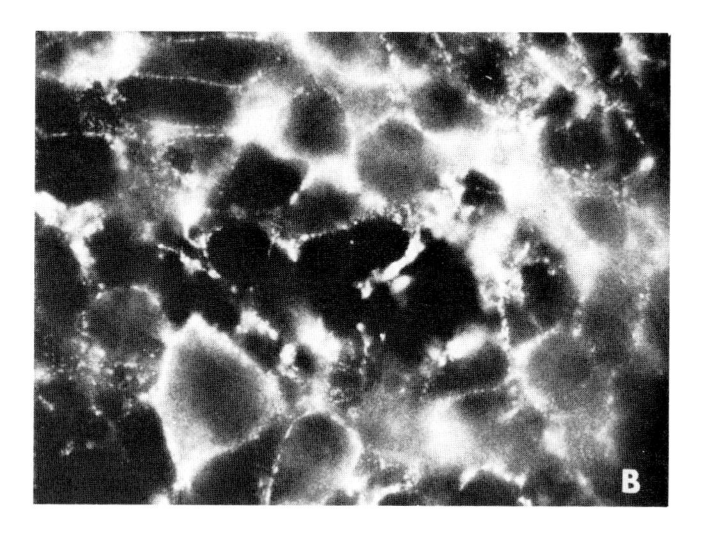

FIG. 3. Indirect immunofluorescence reaction of the same antiserum as used in Figure 1, with differentiating embryoid body cultures as described in Figure 2. **A,** phase contrast of a highly cellular group of differentiating glia with arcades of stout cytoplasmic processes (original X 1035); **B,** positive surface immunofluorescence of the same cells (original X 1035).

using the absorbed immune serum applied to living cultures of differentiating embryoid bodies on glass coverslips demonstrated fluorescence confined specifically to neuroepithelial cell surfaces (Figures 1–3). The reactive neural cells, identified by previously described techniques (VandenBerg *et al.* 1976a), consisted of neuroblasts (Figure 2) and glial cells (Figure 3), in addition to the more primitive neuroepithelial cells (Figure 1). In sections of frozen tissue, the absorbed serum recognized antigenic sites in all examined areas of both the neonatal and adult mouse brain (except for myelinated fiber tracts in the basal ganglia), and only within neuroepithelial cell populations of the solid transplants of the teratoma. The removal of all antibrain activity by absorption with the solid teratomas suggests that all the brain antigenic sites recognized by the immune serum were present also in the tumor. The lack of recognition of myelinated tracts may reflect the low level of myelin immunogens present in the neonatal brain tissue used in our experiments.

The demonstration of brain-associated surface antigenic sites on the differentiating neuroepithelial tissues of OTT-6050 extends the findings by others that totipotential stem and/or endodermal cells of several lines of mouse teratoma share certain surface antigenic sites with normal embryos (Artzt *et al.* 1973, Babinet *et al.* 1975, Stern *et al.* 1975, Gooding *et al.* 1976). Recently, Dewey *et al.* (1977) using syngeneic immunosera against small embryoid bodies of OTT-6050 have found surface antigenic sites that are shared not only with normal embryos of both early and later stages but also with certain normal adult tissues including brain tissue. The need for further extensive testing of selective cross-reactivity and tissue specificity, however, precludes any speculation about the organ-specific nature of these antigens on the totipotential stem cells. The presence of embryonic cell surface antigens on the teratoma stem cells reported by other workers (Artzt *et al.* 1973, Babinet *et al.* 1975, Stern *et al.* 1975, Gooding *et al.* 1976, Dewey *et al.* 1977), and the expression of brain-associated antigenic sites on teratoma neural cells, which are not found on totipotential stem cells, may provide a means of identifying early neuroepithelial differentiation in living cell populations.

ALKALINE PHOSPHATASE AND NONHISTONE CHROMOSOMAL PROTEIN STUDIES

Two other biochemical features of the OTT-6050 teratoma are being pursued in collaboration with other laboratories in order to provide possible additional markers for the study of the neural differentiation in this tumor. Earlier studies by others demonstrated alkaline phosphatase as a biochemical marker for the stem cells of the mouse teratoma (Damjanov *et al.* 1971b, Bernstine *et al.* 1973) and compared certain functional characteristics of crude enzyme extracts from the tumor with murine intestine, liver, kidney, and placental isoenzymes (Bernstine *et al.* 1973, Bernstine and Ephrussi 1975). We have partially purified alkaline phosphatase from the embryoid bodies and solid tumors of the OTT-6050 line and, using the technique of electrophoresis in polyacrylamide gels, have compared the teratoma enzyme with the isoenzymes in the kidney and placenta of strain-related mice (Wada *et al.* 1976). Covalent $^{32}PO_4$-labeling of the alkaline phosphatases and electrophoresis in polyacrylamide gels in sodium dodecyl sulfate (SDS) were used to compare the subunit molecular weights of the enzymes. The results indicate that the mouse teratoma enzyme is distinct from the kidney and placental isoenzymes. Concurrent histochemical studies localized the enzyme to the stem cells of the tumor, thus implying that the alkaline phosphatase in these cells is a distinct isoenzyme. We also found that the embryoid bodies contained a second alkaline phosphatase, which may correspond to the placental isoenzyme. Since the outer cell layer of the embryoid bodies demonstrated alkaline phosphatase activity with histochemical methods, the latter enzyme may be attributable to this layer.

Data on the chromosomal proteins of mouse teratoma have been reported from other laboratories (Jami and Loeb 1975, Loeb *et al.* 1976, Paulin *et al.* 1976), and both similarities and distinctions were noted between the nonhistone chromosomal proteins (NHCP) of adult 129/Sv mouse brain and a cloned, undifferentiated aneuploid cell line derived from OTT-6050 (Jami and Loeb 1975). However, no studies have investigated the NHCP of the differentiating solid teratomas in vivo for a comparison with NHCP of the mouse brain at different stages of development. To determine whether specific nonhistone chromosomal proteins are related to neuroepithelial differentiation or to stages of this differentiation, we have compared these proteins in fetal, newborn, and adult brains of strain-related mice with those in adult livers and kidneys of the same strain using SDS polyacrylamide gel electrophoresis and, in turn, with those in the OTT-6050 teratoma (Choie *et al.*, in press). One-dimensional electrophoresis demonstrated numerous qualitative similarities in the fetal, neonatal, and adult brain NHCP profiles, but quantitative differences were also evident. The NHCP composition of the adult brain was clearly distinct from that of the liver and kidney and dissimilar from that of the teratoma. A number of possibilities may account for the differences between brain and teratoma NHCP: (1) all the brain tissues analyzed represent completed stages of maturation, whereas the teratoma is composed of a wide range of cell types varying in proportion from tumor to tumor and including stem cells, a fraction of nonneuroepithelial elements, and neuroepithelial cells at various stages of differentiation, and (2) the NHCP profile of the teratoma may reflect its neoplastic state. Workers have shown previously that neoplastically transformed cells demonstrate NHCP that are distinguishable from controls of the same origin (Forger *et al.* 1976, Krause and Stein 1974). To determine whether there are NHCP specific for mouse neuroepithelial elements, each NHCP fraction must be analyzed in greater detail; to this end, studies utilizing a two-dimensional separation of the chromatin proteins are underway. If proteins common to both the brain and tumor can be found, the next step will be their isolation and purification.

CONCLUSIONS

In summary, we have characterized the neuroepithelial differentiation in the mouse teratoma OTT-6050, using morphological, biochemical, immunological, and in vitro methods. Our long-range goals are to obtain a spectrum of divergent neuroepithelial differentiation in vitro from totipotential stem cells so that we may have a system for analyzing the factors operative in neural differentiation. Using biochemical or immunological markers such as the GFA protein, non-histone chromosomal proteins that could be specific for neuroepithelial cells, alkaline phosphatase specific for stem cells of the teratoma, and the brain-associated neuroepithelial antiserum, we shall attempt to monitor more precisely the critical time period during which a multipotential stem cell becomes a determined neuroepithelial cell. We will also attempt to develop a transplantable homogeneous neuroepithelial tumor line composed of a complete spectrum of divergent neuroepithelial elements including, possibly, mature neurons.

ACKNOWLEDGMENTS

This investigation was supported by Grant No. 11689, awarded by the National Cancer Institute, Department of Health, Education and Welfare. The advice and encouragement of Dr. L. J. Rubinstein are gratefully acknowledged.

REFERENCES

Antanitus, D. S., B. H. Choi, and L. W. Lapham. 1975. Immunofluorescence staining of astrocytes in vitro using antiserum to glial fibrillary acidic protein. Brain Res. 89:363–367.

Artzt, K., and D. Bennett. 1972. A genetically caused embryonal ectodermal tumor in the mouse. J. Natl. Cancer Inst. 48:141–158.

Artzt, K., P. Dubois, D. Bennett, H. Condamine, C. Babinet, and F. Jacob. 1973. Surface antigens common to mouse cleavage embryos and primitive teratocarcinoma cells in culture. Proc. Natl. Acad. Sci. USA 70:2988–2992.

Babinet, C., H. Condamine, M. Fellous, G. Gachelin, R. Kemler, and F. Jacob. 1975. Expression of a cell surface antigen common to primitive mouse teratocarcinoma cells and cleavage embryos during embryogenesis and spermatogenesis, *in* Teratomas and Differentiation, M. I. Sherman and D. Solter, eds. Academic Press, New York, pp. 101–107.

Bernstine, E. G., and B. Ephrussi. 1975. Alkaline phosphatase activity in embryonal carcinoma and its hybrids with neuroblastoma, *in* Teratomas and Differentiation, M. I. Sherman and D. Solter, eds., Academic Press, New York, pp. 271–287.

Bernstine, E. G., M. L. Hooper, S. Grandchamp, and B. Ephrussi. 1973. Alkaline phosphatase activity in mouse teratoma. Proc. Natl. Acad. Sci. USA 70:3899–3903.

Bignami, A., and D. Dahl. 1974. Astrocyte-specific protein and neuroglial differentiation: An immunofluorescence study with antibodies to the glial fibrillary acidic protein. J. Comp. Neurol. 153:27–38.

Bignami, A., L. F. Eng, D. Dahl, and C. T. Uyeda. 1972. Localization of the glial fibrillary acidic protein in astrocytes by immunofluorescence. Brain Res. 43:429–435.

Bigner, D. D., and C. N. Pegram. 1976. Virus-induced experimental brain tumors and putative associations of viruses with human brain tumors: A review, *in* Advances in Neurology, Vol. 15, R. A. Thompson and J. R. Green, eds., Raven Press, New York, pp. 57–83.

Choie, D. D., E. C. Friedberg, S. R. VandenBerg, and M. M. Herman. 1977. Non-histone chromosomal proteins in mouse brain at different stages of development and in a transplantable mouse teratoma. J. Neurochem. (In press).

Corrodi, H., N-Å. Hillarp, and G. Jonsson, 1964. Fluorescence methods for the histochemical demonstration of monoamines. 3. Sodium borohydride reduction of the fluorescent compounds as a specificity test. J. Histochem. Cytochem. 12:582–586.

Damjanov, I., D. Solter, M. Belicza, and N. Skreb. 1971a. Teratomas obtained through extrauterine growth of seven-day mouse embryos. J. Natl. Cancer Inst. 46:471–480.

Damjanov, I., D. Solter, and N. Skreb. 1971b. Enzyme histochemistry of experimental embryo-derived teratocarcinoma. Z. Krebsforsch. 76:249–256.

Damjanov, I., D. Solter, and D. Šerman. 1973. Teratocarcinoma with the capacity for differentiation restricted to neuro-ectodermal tissue. Virchows Arch. [Zellpathol.] 13:179–195.

Dewey, J. J., J. D. Gearhart, and B. Mintz. 1977. Cell surface antigens of totipotent mouse teratocarcinoma cells grown in vivo: Their relation to embryo, adult and tumor antigens. Dev. Biol. 55:359–374.

Eng, L. F., J. J. Vanderhaeghen, A. Bignami, and B. Gerstl. 1971. An acidic protein isolated from fibrous astrocytes. Brain Res. 28:351–354.

Falck, B., N.-Å, Hillarp, G. Thieme, and A. Torp. 1962. Fluorescence of catechol amines and related compounds condensed with formaldehyde. J. Histochem. Cytochem. 10:348–354.

Forger, J. M., D. D. Choie, and E. C. Friedberg. 1976. Non-histone chromosomal proteins of chemically transformed neoplastic cells in tissue culture. Cancer Res. 36:258–262.

Fujita, H., and S. Fujita. 1963. Electron microscopic studies on the neuroblast differentiation in the central nervous system of domestic fowl. Z. Zellforsch. Mikrosk. Anat. 60:463–478.

Gearhart, J. D., and B. Mintz. 1974. Contact-mediated myogenesis and increased acetylcholinesterase activity in primary cultures of mouse teratocarcinoma cells. Proc. Natl. Acad. Sci. USA 71:1734–1738.

Gearhart, J. D., and B. Mintz. 1975. Creatine kinase, myokinase, and acetylcholinesterase activities in muscle-forming primary cultures of mouse teratocarcinoma cells. Cell 6:61–66.

Gooding, L. R., Y-C Hsu, and M. Edidin. 1976. Expression of teratoma-associated antigens on murine ova and early embryos. Dev. Biol. 49:479–486.

Herman, L., and S. L. Kaufman. 1966. The fine structure of the embryonic mouse neural tube with special reference to cytoplasmic microtubules. Dev. Biol. 13:145–162.

Herman, M. M., J. C. Sipe, L. J. Rubinstein, S. R. VandenBerg, A. M. Spence, and J. Vraa-Jensen. 1975. An experimental mouse testicular teratoma as a model for neuroepithelial neoplasia and differentiation. II. Electron microscopy. Am. J. Pathol. 81:421–444.

Illmensee, K., and B. Mintz. 1976. Totipotency and normal differentiation of single teratocarcinoma cells cloned by injection into blastocysts. Proc. Natl. Acad. Sci. USA 73:549–553.

Jakupčević, M., Z. Lacković, I. Damjanov, and M. Bulat. 1974. Biogenic amines in a retrotransplantable neurogenic teratocarcinoma. Experientia 30:652–653.

Jami, J., and J. E. Loeb. 1975. Chromatin proteins of mouse primitive teratocarcinoma cells: Electrophoretic comparison with somatic tissues of the mouse, *in* Teratomas and Differentiation, M. I. Sherman and D. Solter, eds., Academic Press, New York, pp. 221–233.

Kahan, B. W., and B. Ephrussi. 1970. Developmental potentialities of clonal in vitro cultures of mouse testicular teratoma. J. Natl. Cancer Inst. 44:1015–1036.

Kleinsmith, L. J., and G. B. Pierce, Jr. 1964. Multipotentiality of single embryonal carcinoma cells. Cancer Res. 24:1544–1551.

Kobayashi, S., and N. Mukai. 1974. Retinoblastoma-like tumors induced by human adenovirus type 12 in rats. Cancer Res. 34:1646–1651.

Krause, M. O., and G. S. Stein. 1974. Modifications in the chromosomal proteins of SV-40 transformed WI-38 human diploid fibroblasts. Biochem. Biophys. Res. Commun. 59:796–803.

Lehman, J. M., W. C. Speers, D. E. Swartzendruber, and G. B. Pierce. 1974. Neoplastic differentiation: Characteristics of cell lines derived from a murine teratocarcinoma. J. Cell Physiol. 84:13–28.

Levine, A. J., M. Torosian, A. J. Sarokhan, and A. K. Teresky. 1974. Biochemical criteria for the in vitro differentiation of embryoid bodies produced by a transplantable teratoma of mice. The production of acetylcholine esterase and creatine phosphokinase by teratoma cells. J. Cell Physiol. 84:311–318.

Loeb, J. E., E. Ritz, C. Creuzet, and J. Jami. 1976. Comparison of chromosomal proteins of mouse primitive teratocarcinoma, liver and L cells. Exp. Cell Res. 103:450–454.

Loizou, L. A. 1972. The postnatal ontogeny of monoamine-containing neurones in the central nervous system of the albino rat. Brain Res. 40:395–418.

Ludwin, S. K., L. F. Eng, S. R. VandenBerg, J. C. Kosek, and M. M. Herman. 1976. Glial differentiation in an experimental mouse teratoma. (Abstract) J. Neuropath. Exp. Neurol. 35:102.

Martin, G. R., and M. J. Evans. 1975. Multiple differentiation of clonal teratocarcinoma stem cells following embryoid body formation in vitro. Cell 6:467–474.

Mintz, B., and K. Illmensee. 1975. Normal genetically mosaic mice produced from malignant teratocarcinoma cells. Proc. Natl. Acad. Sci. USA 72:3585–3589.

Mukai, N., and S. Kobayashi. 1973. Human adenovirus-induced medulloepitheliomatous neoplasms in Sprague-Dawley rats. Am. J. Pathol. 73:671–690.

Orenberg, E. K., S. R. VandenBerg, J. D. Barchas, and M. M. Herman. 1976. Neurochemical studies in a mouse teratoma with neuroepithelial differentiation. Presence of cyclic AMP, serotonin and enzymes of the serotonergic, adrenergic and cholinergic systems. Brain Res. 101:273–281.

Paulin, D., J.-F. Nicolas, M. Jacquet, H. Jakob, F. Gros, and F. Jacob. 1976. Comparative protein patterns in chromatins from mouse teratocarcinoma cells. Exp. Cell Res. 102:169–178.

Pierce, G. B., Jr., and E. L. Verney. 1961. An in vitro and in vivo study of differentiation in teratocarcinomas. Cancer 14:1017–1029.

Prasad, K. N., and P. K. Sinha. 1978. Regulation of differentiated functions and malignancy in neuroblastoma cells in culture, *in* Cell Differentiation and Neoplasia (The University of Texas System Cancer Center 30th Annual Symposium on Fundamental Cancer Research, 1977) G. F. Saunders, ed., Raven Press, New York, pp. 111–141.

Rosenthal, M. D., R. M. Wishnow, and G. H. Sato. 1970. In vitro growth and differentiation of clonal populations of multipotential mouse cells derived from a transplantable testicular teratocarcinoma. J. Natl. Cancer Inst. 44:1001–1014.

Rubinstein, L. J. 1972. Cytogenesis and differentiation of primitive central neuroepithelial tumors. J. Neuropathol. Exp. Neurol. 31:7–26.

Rubinstein, L. J. 1976. Current concepts in neuro-oncology, *in* Advances in Neurology, Vol. 15, R. A. Thompson and J. R. Green, eds., Raven Press, New York, pp. 1–25.

Ruffner, B. W., Jr., and D. M. Grieshaber. 1974. Biochemical differentiation of a murine neuroblastoma in vitro and in vivo. Cancer Res. 34:551–558.

Ruffner, B. W., Jr., and M. Smith. 1974. Biochemical differentiation of a murine ganglioneuroblastoma in tissue culture. Exp. Cell Res. 89:442–447.

Spence, A. M., S. R. VandenBerg, and M. M. Herman. 1975. Intracisternal A-particles in transplantable murine teratomas. Beitr. Pathol. 155:428–434.

Stern, P. L., G. R. Martin, and M. J. Evans. 1975. Cell surface antigens of clonal teratocarcinoma cells at various stages of differentiation. Cell 6:455–465.

Stevens, L. C. 1967. The biology of teratomas, *in* Advances in Morphogenesis, Vol. 6, M. Abercrombie and J. Brachet, eds. Academic Press, New York, pp. 1–31.

Stevens, L. C. 1970. The development of transplantable teratocarcinomas from intratesticular grafts of pre- and post-implantation mouse embryos. Dev. Biol. 21:364–382.

Stevens, L. C. 1973. A new inbred subline of mice (129/terSv) with a high incidence of spontaneous congenital testicular teratomas. J. Natl. Cancer Inst. 50:235–242.

Swenberg, J. A. 1976. Chemical induction of brain tumors, *in* Advances in Neurology, Vol. 15, R. A. Thompson and J. R. Green, eds. Raven Press, New York, pp. 85–99.

Teresky, A. K., M. Marsden, E. L. Kuff, and A. J. Levine. 1974. Morphological criteria for the in vitro differentiation of embryoid bodies produced by a transplantable teratoma of mice. J. Cell Physiol. 84:319–332.

Tresman, R. L., and M. J. Evans. 1975. A light and electron microscopical study of the nervous tissue of mouse teratomas. J. Neurocytol. 4:301–314.

VandenBerg, S. R., M. M. Herman, S. K. Ludwin, and A. Bignami. 1975. An experimental mouse testicular teratoma as a model for neuroepithelial neoplasia and differentiation. I. Light microscopic and tissue and organ culture observations. Am. J. Pathol. 79:147–168.

VandenBerg, S. R., J. E. Hickey, and M. M. Herman. 1977. Brain-associated surface antigens on neuroepithelial cells in a transplantable mouse teratoma. Acta Neuropathol. (In press).

VandenBerg, S. R., S. K. Ludwin, M. M. Herman, and A. Bignami. 1976a. In vitro astrocytic differentiation from embryoid bodies of an experimental mouse testicular teratoma. Am. J. Pathol. 83:197–212.

VandenBerg, S. R., S. K. Ludwin, M. M. Herman, and A. Bignami. 1976b. In vitro astrocytic differentiation from embryoid bodies of an experimental mouse testicular teratoma. (Abstract) J. Neuropathol. Exp. Neurol. 35:354.

Wada, H. G., S. R. VandenBerg, H. H. Sussman, W. E. Grove, and M. M. Herman. 1976. Characterization of two different alkaline phosphastases in mouse teratoma: Partial purification, electrophoretic and histochemical studies. Cell 9:37–44.

Wechsler, W. 1966. Die Feinstruktur des Neuralrohres und der neuroektodermalen Matrixzellen am Zentralnervensystem von Hühnerembryonen. Z. Zellforsch. Mikrosk. Anat. 70:240–268.

Wechsler, W., and K. Mehler. 1967. Electron microscopy of neuronal and glial differentiation in the developing brain of the chick. Prog. Brain Res. 26:93–144.

ZuRhein, G. M., and J. N. Varakis. 1977. Perinatal induction of medulloblastomas in Syrian hamsters by JC virus, a human papovirus, *in* Conference on Perinatal Carcinogenesis, M. Rice, ed. Natl. Cancer Inst. Monogr. (In press).

Cell Differentiation and Neoplasia, edited by Grady F. Saunders. Raven Press, New York

Regulation of Differentiated Functions and Malignancy in Neuroblastoma Cells in Culture

Kedar N. Prasad and Pramod K. Sinha

Department of Radiology, University of Colorado Medical Center, Denver, Colorado 80262

The relationship between differentiation and malignancy is not well understood. Based on several studies, the following general assumptions can be made. (1) The malignant properties acquired by the cells are probably due to mutational events, and these properties are dominant and heritable from one cell generation to another. (2) Some properties of the cancer cells are unique to malignancy; some are similar to those of embryonic cells, and some are characteristic of differentiated cells. Some genomes of embryonic features re-express in the cancer cells derived from "differentiated" cells, and many features of "differentiated" cells continue to express at varying degrees in spite of malignant transformation. Some genomes of embryonic features continue to express in the cancer cells derived from embryonic cells, but some genomes of differentiated functions express at varying degrees in spite of malignant transformation of embryonic cells. (3) A malignant cell may be transformed to a "differentiated" and non-malignant state but may retain some characteristic of tumor cells (e.g., tumor-specific antigen). Conversely, a cell may express many differentiated functions at varying degrees, mostly at low levels, but may still be malignant. From these assumptions, it appears that the relationship between differentiation and malignancy could be very complex. Therefore, the identification of endogenous and exogenous factors that regulate the expression of differentiation and/or malignancy in mammalian cells in culture may provide useful biological tools for probing the link between differentiation and malignancy and may help in exploring the mechanisms of expression of malignancy as well as new treatment modalities. Indeed, using neuroblastoma cells in culture as a model system, many exogenous and endogenous factors that increase the expression of one or more differentiated functions have been identified. The tumorigenicity of cells treated with some of these agents is reduced.

In order to improve the therapy of tumors, the discovery of agents that cause differentiation is as important as the discovery of agents that kill tumor cells without significantly affecting the host's immune system. Indeed, we have discovered both types of agents for neuroblastoma cells in culture. For example, cyclic

AMP-stimulating agents cause morphological and biochemical differentiation. Sodium butyrate kills human neuroblastoma cells in a highly selective manner. Sodium ascorbate (vitamin C) markedly potentiates the effect (i.e., reduces cell number) of X-irradiation and some chemotherapeutic agents in mouse neuroblastoma cells, but it produces no such effect in mouse fibroblasts or rat glioma cells.

In order to establish the possible link between the expression of differentiated functions and malignancy, it is essential to examine the regulation of expression of each individual differentiated function separately.

REGULATION OF DIFFERENTIATION AND MALIGNANCY

Regulation of Neurite Formation

A mature and nonmalignant neuron contains well-defined axons and dendritic processes. The neurites that are extended in neuroblastoma cells appear similar in appearance to the axons and dendrites of normal neurons (Ross *et al.* 1975). Although these processes are electrically excitable and are capable of generating action potentials (Nelson *et al.* 1969, Ross *et al.* 1975), cells having such processes may not express many of the biochemical differentiated functions (Prasad 1973, Prasad and Kumar 1974). Agents such as dibutyryl cyclic AMP, PGE_1 inhibitors of phosphodiesterase, serum-free medium, X-irradiation, 5-bromodeoxyuridine, tri-fluoro-methyl-2-deoxyuridine, 6-thioguanine, cytosine arabinoside, methotrexate, glial extract, hypertonic medium, and nerve growth factor increase the extent of neurite formation (see review by Prasad 1975). The neurites in mouse (Figure 1A-C) and human neuroblastoma cells (Figure 2A-E) are formed after treatment of cells with various agents. However, the rate of expression of differentiated phenotype in human neuroblastoma cells is much slower than that in mouse cells. Figure 3 shows the changes in the levels of cyclic AMP as a function of time after treatment of mouse neuroblastoma cell culture with prostaglandin E_1 (PGE_1) and 4-(3-butoxy-4-methoxybenzyl)-2-imidazolidinone (R020-1724) (Prasad 1973). It should be noted that the level of cyclic AMP remained high for a period of four days. This is in contrast to nonneural cells in culture in which the increase in the level of cyclic AMP after treatment with cyclic AMP-stimulating agents is a transient one. Some agents that do not increase the cyclic AMP level also induce neurites, indicating that the expression of neurites in neuroblastoma cells involves at least two modes of regulation, one of which appears to be cyclic AMP.

Various agents induce neurites by promoting the organization of microtubules and microfilaments (Prasad and Hsie 1971, Prasad 1973), since the expression of differentiated phenotype is blocked by vinblastine sulfate (which interferes with the assembly of microtubules) and cytochalasin B (which interferes with the assembly of microfilaments). Cytochalasin B and vinblastine sulfate block R020-1724-induced neurites in mouse neuroblastoma cells; however, these agents do not decrease the R020-1724-stimulated increase in the cyclic AMP level (Prasad 1973), indicating further the requirement of the organization of microtubules and microfilaments for the expression of neurites. The expression of differentiated phenotype does not require the synthesis of new RNA (Prasad and Hsie 1971), since cyclic AMP-stimulating agents induce neurites in the presence of actinomycin D (which inhibits RNA synthesis) and in enucleated neuroblastoma cells (Miller and Ruddle 1974). However, there is some controversy regarding the requirement of new protein synthesis. Neurites induced in serum-free medium do not require the synthesis of new protein (Seeds *et al.* 1970); however, 5-bromodeoxyuridine- and cyclic AMP-induced neurites require the synthesis of new protein (Schubert and Jacob 1970, Prasad and Hsie 1971). This difference may be due to differences in the experimental condition. It has been suggested that the neurite formation can be initiated in the absence of protein synthesis but soon reaches a point beyond which de novo synthesis of protein is required (Schubert *et al.* 1971). The neurite formation in some mouse neuroblastoma cells requires strong interaction between the cell and the surface of the culture dish (Schubert and Jacob 1970), although this may not be equally true in another mouse neuroblastoma clone (Prasad 1973). Agents such as X-irradiation and 6-thioguanine induce neurites in mouse cells and 5-bromodeoxyuridine in human cells without changing the intracellular level of cyclic AMP (Prasad 1973, Prasad *et al.* 1972).

The neurites that are formed after treatment of neuroblastoma cells with various agents may be reversible or irreversible after the removal of the drug. In a sensitive clone, cyclic AMP-stimulating agents produce irreversible effect on neurite formation. The irreversibility of neurite formation after treatment with agents that elevate the intracellular level of cyclic AMP is confirmed in some clones by the following observations: When the "differentiated" cells, four days after treatment with PGE_1 or R020-1724, are removed from dishes by means of Viokase solution and replated in separate dishes, cells attach and form long neurites within 24 hours, even though no drug is present in the medium (Prasad and Kumar 1974). The number of morphologically differentiated cells in the newly plated dishes is similar to that in dishes from which the drug is not removed. This indicates that the cellular factors which control the expression of differentiated phenotype remain functional after subculture. The culture must be maintained in the presence of drugs (PGE_1 and R020-1724) for at least four days before the irreversibility of neurites can be observed. From the analysis of many studies it is now apparent that the difference in the cyclic AMP effects observed in various laboratories may be due to one of the following: (1) A difference in clone may account for the difference in results. Most clones are sensitive to PGE_1 and R020-1724 in causing neurite formation. However, some clones, irrespective of their neuronal cell type, are sensitive to PGE_1 but not to phosphodiesterase inhibitor and vice versa (Prasad *et al.* 1974). The clone that is insensitive to inhibitors of cyclic AMP phosphodiesterase is also unresponsive to dibutyryl cyclic AMP. (2) Our preliminary data show

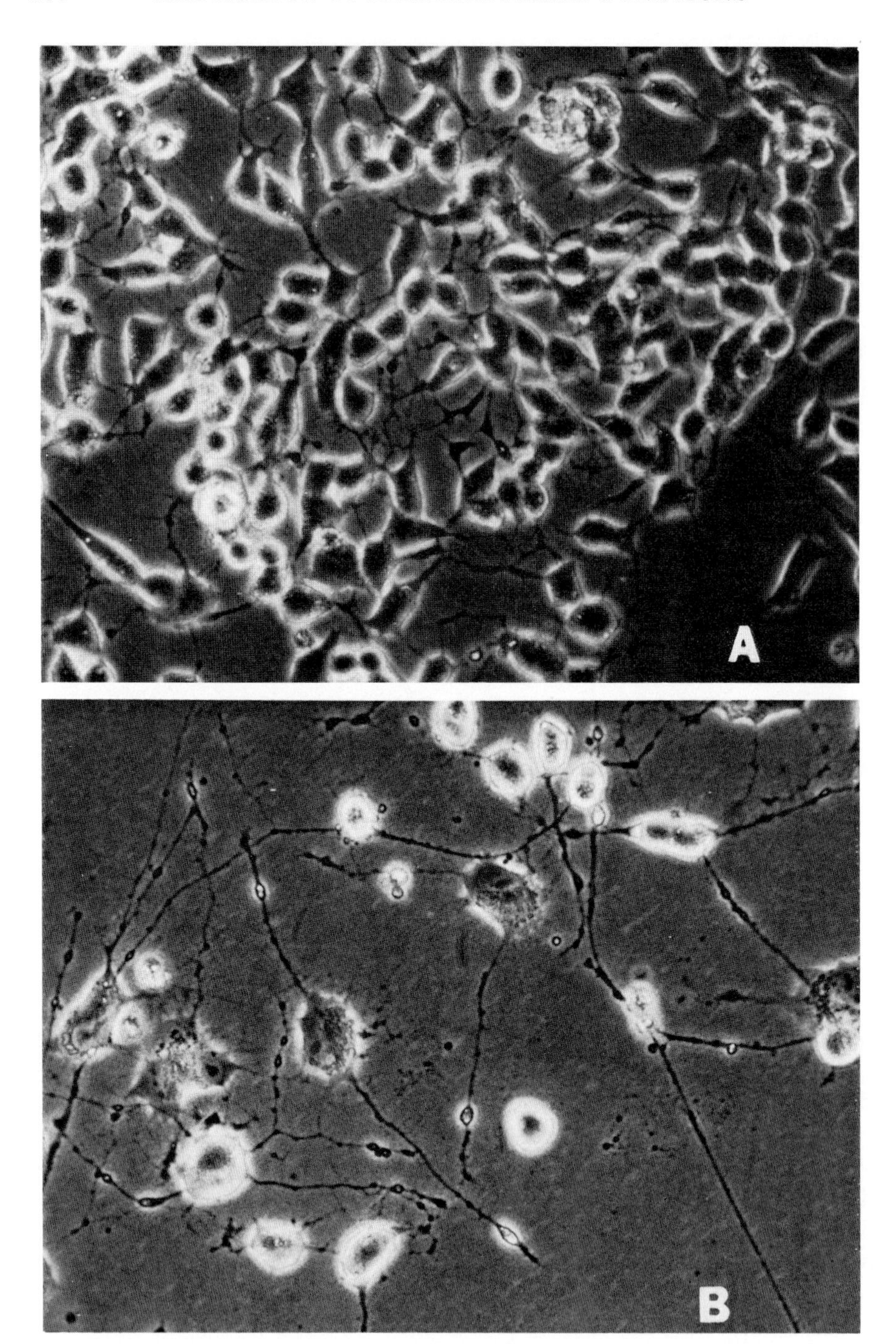
A
B

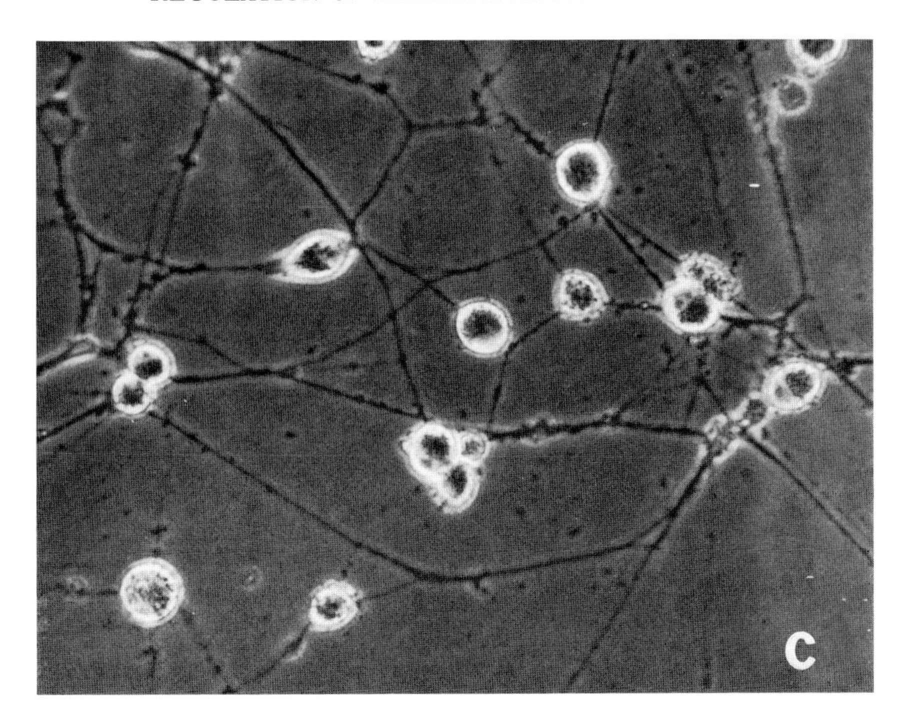

FIG. 1. Phase contrast micrographs of mouse neuroblastoma cells (NBA_2[1] clone) in culture. Cells (50,000) were plated in Falcon plastic dishes (60 mm), and prostaglandin E_1 (PGE_1) (10 μg/ml) and 4-(3-butoxy-4-methoxybenzyl)-2-imidazolidinone (R020–1724) (200 μg/ml) were added separately 24 hours later. The medium and drug were changed at 3, 5, 8, and 11 days after treatment. The control culture (A) shows that cells grow in clumps and that some of them have short cytoplasmic processes. PGE_1-treated culture four days after treatment (B) shows formation of long neurites. PGE_1-treated culture 14 days after treatment (C) shows that the remaining cells maintain their differentiated phenotype and show an extensive network of thickened fibers. (Reproduced, with permission of Cold Spring Harbor Laboratory, from Prasad and Kumar 1974.) (original X 131)

that the expression of differentiated phenotype is markedly altered in different types of serum (Prasad *et al.* 1977). Table 1 shows that the expression of neurites in control cells (NBP_2 clone) varies from about 2% in newborn calf serum to 77% in agamma globulin calf serum. In addition, the effects of PGE_1 and R020-1724 on morphological differentiation markedly vary depending on the type of serum. PGE_1 is least effective in fetal calf serum (14%), and it is most effective in heat-inactivated fetal calf serum (81%). Similarly, R020-1724 is least effective in fetal calf serum (26%), and it is most effective in agamma globulin newborn calf serum (75%). It should be noted that the concentration requirement of each drug for a maximal effect on morphological differentiation also varies as a function of the type of serum. It is unknown if the expression of biochemical differentiated functions and the growth rate are modified in different types of serum. It is also unknown if the alterations in the expression of

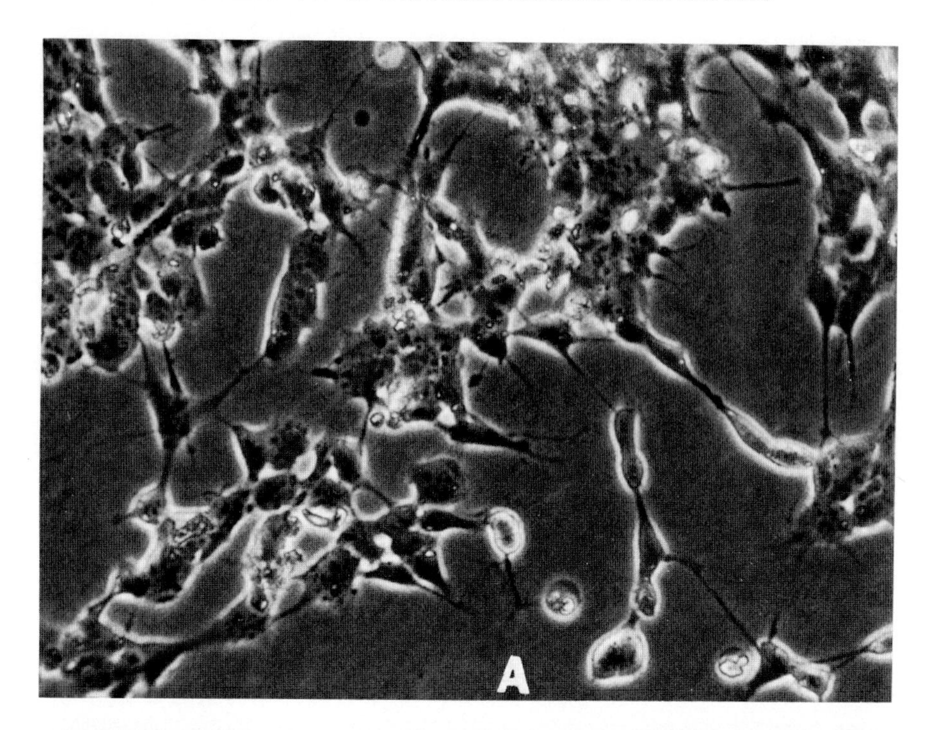

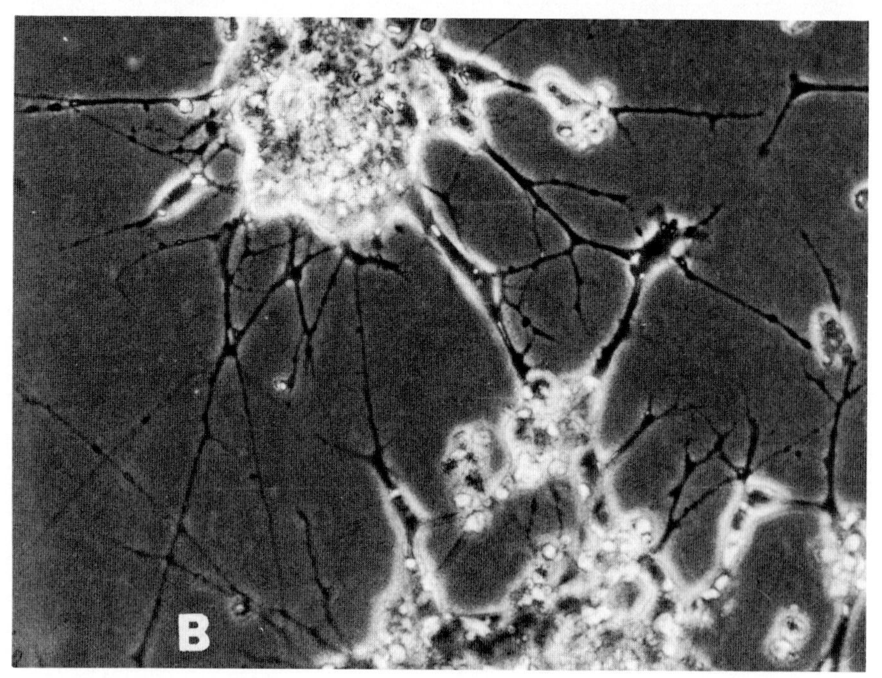

FIG. 2A, 2B.

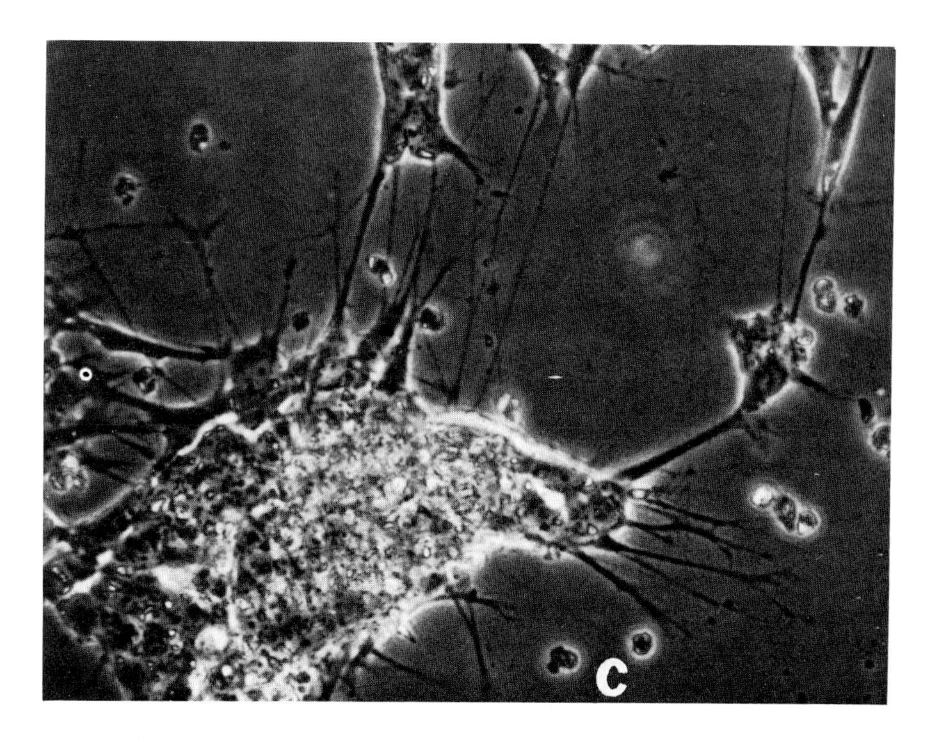

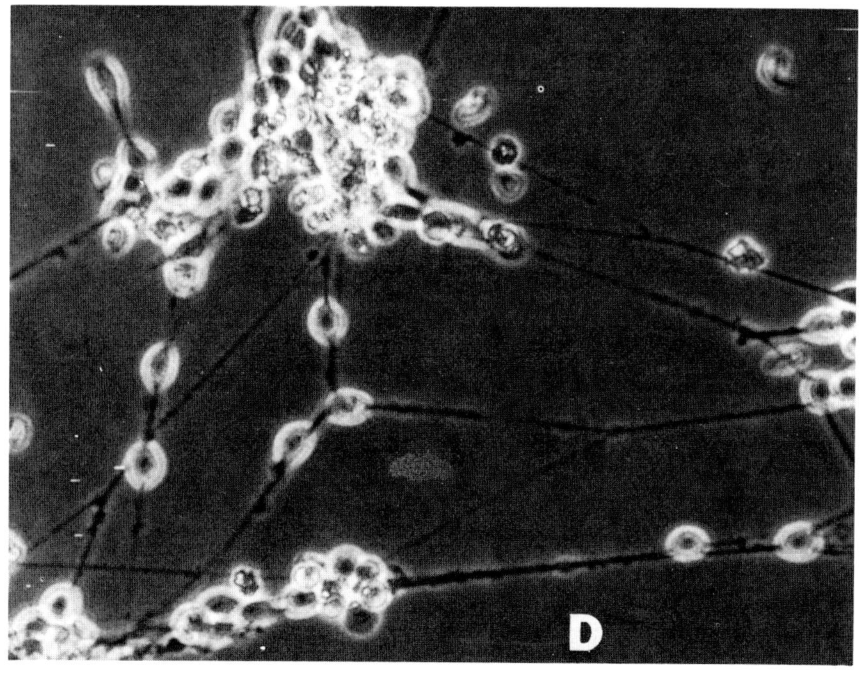

FIG. 2C, 2D. (Legend on page 118.)

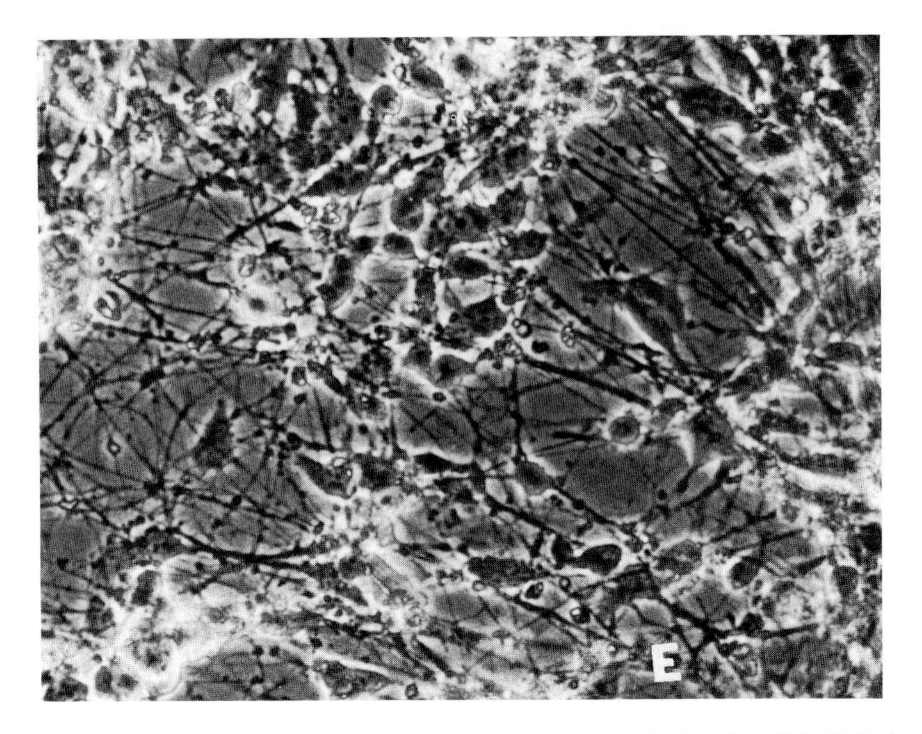

FIG. 2. Phase contrast micrographs of human neuroblastoma cells in culture (IMR-32 line). Cells were plated in Falcon plastic dishes (60 mm), and papaverine (2.5 μg/ml), sodium butyrate (0.5 mM), serum-free medium (SFM), and 5-bromodeoxyuridine (5-BrdU) (2.5 μM) were added individually four days after plating. The drug and medium were changed every two to three days, and the cultures were maintained for 10 to 13 days. The control culture (A) shows that cells grow in clumps and exhibit no spontaneous morphological differentiation (cytoplasmic processes greater than 50 μm in length). Papaverine-treated culture 10 days after treatment (B) shows the formation of extensive neurites. Many cell deaths occurred during this period. Sodium butyrate-treated culture 10 days after treatment (C) also shows the formation of extensive neurites. Considerable cell death occurred during this period. SFM-treated culture three days after treatment (D) and 5-BrdU-treated cultures 10 days after treatment (E) show extensive neurite formation. (Reproduced, with the permission of Cold Spring Harbor Laboratory, from Prasad and Kumar 1974, and of Cancer, from Prasad and Kumar 1975). (original X 131)

differentiated functions in various types of serum are linked with the changes in the cyclic AMP system.

The inhibition of cell division is not a prerequisite for the expression of neurites. On the contrary, the neurites are formed in the dividing mouse neuroblastoma cells. Cyclic AMP (Prasad and Hsie 1971) induces neurites prior to inhibition of cell division. 5-bromodeoxyuridine induces neurites at a concentration that does not inhibit DNA synthesis (Schubert and Jacob 1970). Several other differentiated functions such as neural specific enzymes (Augusti-Tocco *et al.* 1970), synthesis of catecholamines (Helson and Biedler 1973), nervous system-specific protein (Herschmann and Lerner 1973), increase in the intracellular level of cyclic AMP (Prasad 1973), level of cyclic AMP phosphodiesterase activity (Sinha and Prasad, in press), and the stimulation of adenylate cyclase activity by neurotransmitters (Prasad and Gilmer 1974, Prasad *et al.* 1975c) can be expressed in the dividing nerve cells, although some of these functions may be present at extremely low levels. They may, however, fully express after cessation of cell division, at which time other new differentiated functions may also appear.

The expression of neurites appears to be independent of malignancy, since the neurites of varying sizes may be formed in the dividing neuroblastoma cells. A neuroblastoma cell may permanently stop cell division without the expression of long neurites. The neurites are also expressed in the absence of any increase in the expression of many biochemical differentiated functions.

Membrane Change

Mouse neuroblastoma cells show strong agglutination in the presence of concanavalin A (Con A) or wheat germ agglutinin (WGA), whereas "differentiated" cells induced by inhibitor of cyclic AMP phosphodiesterase (R020-1724) show very little (Prasad 1973). However, the "differentiated" cells induced by dibutyryl cyclic AMP or PGE_1 show agglutination in the presence of Con A and WGA

TABLE 1. *Effect of different types of sera on the expression of differentiated phenotype in mouse neuroblastoma cells in culture**

Types of sera	% of morphologically differentiated cells		
	Control	PGE_1	R020–1724
Fetal calf (FC)	11 ± 2†	14 ± 2 (5 μg/ml)	26 ± 3 (100 μg/ml)
Heat-inactivated FC	~2 ± 0	81 ± 5 (15 μg/ml)	51 ± 4 (200 μg/ml)
Dialyzed FC	6 ± 1	72 ± 3 (2 μg/ml)	63 ± 4 (100 μg/ml)
Agamma globulin FC	Highly toxic		
Newborn calf (NBC)	~2 ± 0	20 ± 2 (10 μg/ml)	60 ± 4 (100 μg/ml)
Heat-inactivated NBC	~5 ± 0	34 ± 3 (10 μg/ml)	40 ± 3 (100 μg/ml)
Agamma globulin NBC	~2 ± 0	41 ± 3 (10 μg/ml)	75 ± 4 (200 μg/ml)
Calf	Highly toxic		
Heat-inactivated calf	Highly toxic		
Dialyzed calf	28 ± 3	56 ± 4 (10 μg/ml)	93 ± 3 (100 μg/ml)
Agamma globulin calf	77 ± 3	92 ± 2 (5 μg/ml)	89 ± 3 (100 μg/ml)

* Cells (25,000) were plated in Falcon plastic dishes (60 mm) and prostaglandin E_1 (PGE_1) and 4-(3-butoxy-4-methoxybenzyl)-2-imidazolidinone (R020–1724) were added separately, 24 hours after plating. The drug and medium were changed at days 2 and 3, and the percent of morphologically differentiated cells (processes > 50 μm in length) were determined at four days after treatment. The value in parentheses represents an optimal concentration. Each value represents an average of nine samples.

† Standard deviation.

to an extent similar to that observed in malignant neuroblastoma cells. Thus, changes in the agglutination sites are not necessarily linked with the "differentiation" of neuroblastoma cells. Therefore, it is unlikely that they can be linked with the malignant transformation of nerve cells. The lack of agglutination in R020-1724-treated cells may be due to the unique effect of this agent on membrane. The transport of glucose does not significantly change in cyclic AMP-induced differentiated cells (Prasad *et al.* 1974).

Regulation of Neurotransmitter Metabolizing Enzymes

Tyrosine Hydroxylase

A mature and nonmalignant neuron expresses the activities of neural specific enzymes at high levels, but in neuroblastoma cells they are expressed at low levels. The activity of tyrosine hydroxylase (a rate-limiting enzyme in the biosynthesis of catecholamines) markedly increases in cyclic AMP-induced "differentiated" mouse neuroblastoma cells (Waymire *et al.* 1972, Richelson 1973). This increase in enzyme activity can be blocked by cycloheximide but not by actinomycin D, indicating that cyclic AMP effect on tyrosine hydroxylase (TH) activity is primarily at the translational level. Certain agents (phosphodiesterase inhibitors, PGE_1, analogs of cyclic AMP, sodium butyrate, and 5-bromodeoxyuridine), which increase the cyclic AMP level, also elevate tyrosine hydroxylase activity in mouse neuroblastoma cells (Prasad 1975). Cyclic AMP-stimulating agents also increase TH activity in human neuroblastoma cells (Prasad and Kumar 1974). Some agents (X-irradiation, 6-thioguanine, 5′-AMP, and cytosine arabinoside), which do not elevate the cyclic AMP level, do not increase tyrosine hydroxylase activity. On the other hand, serum-free medium increases the cyclic AMP level in mouse neuroblastoma cells (Prasad 1973) but does not increase tyrosine hydroxylase activity (Kates *et al.* 1971, Waymire *et al.* 1972), indicating

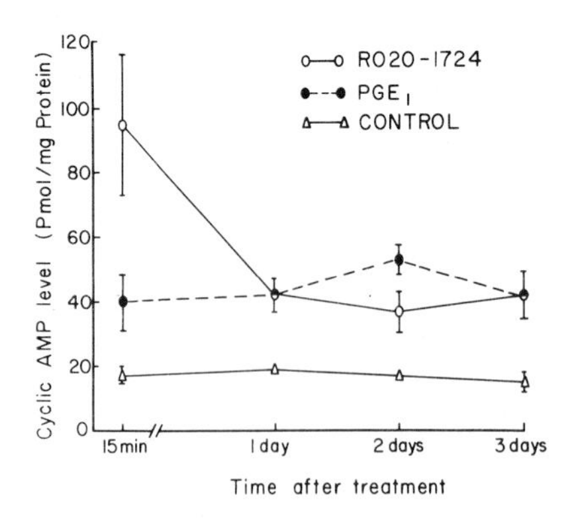

FIG. 3. Changes in the cyclic AMP level during morphological differentiation. Cells (0.5×10^6) of clone $NBA_2(1)$ were plated in large Falcon plastic flasks (75 cm^2) and prostaglandin E_1 (PGE_1) (10 $\mu g/ml$) and 4-(3-butoxy-4-methoxybenzyl)-2-imidazolidinone (R020–1724) (200 $\mu g/ml$) were added separately 24 hours later. The control cultures received an equivalent volume of alcohol. The drug and medium were changed two days after treatment, and the level of cyclic AMP was determined three days after treatment. Each value represents an average of six to eight samples. The bar at each point is standard deviation. (Reproduced, with permission of Cold Spring Harbor Laboratory, from Prasad and Kumar 1974.)

that cyclic AMP probably requires a second factor, which is present in the serum, for increasing tyrosine hydroxylase activity. The tyrosine hydroxylase activity increases after treatment of human neuroblastoma cells with 5-bromodeoxyuridine (Prasad *et al.* 1973a), and after treatment of mouse cells with 5-(3,3-dimethyl-1-triazeno)-imidazole-4-carboxamide (DTIC) (Culver *et al.* 1977). These agents do not increase the cyclic AMP levels. Thus, the activity of tyrosine hydroxylase is regulated by at least two modes, one of which involves the changes in the level of cyclic AMP. 6-thioguanine and X-irradiation induce neurites without increasing TH activity, indicating that morphological differentiation and TH activity are independently regulated (Prasad 1975).

Choline Acetyltransferase and Acetylcholinesterase

The activities of choline acetyltransferase (CAT) and acetylcholinesterase (AchE) are markedly increased in cyclic AMP-induced "differentiated" cells, as well as in cells treated with X-irradiation, 6-thioguanine, and 5′-AMP, which do not increase the cyclic AMP level (Prasad 1973). The increase in enzyme activity is blocked by cycloheximide but not by actinomycin D, indicating the cyclic AMP effects on CAT and AchE are primarily at the translational level. These data show that the activities of CAT and AchE are regulated by at least two modes, one of which involves changes in the cyclic AMP level. The activities of CAT and AchE are inversely related to the growth rate.

The activities of tyrosine hydroxylase, choline acetyltransferase, and acetylcholinesterase are expressed more fully in neuroblastoma cells when such cells are induced to differentiate by raising the intracellular level of cyclic AMP. Since the "differentiated" cells lose the tumorigenicity, it is possible that a defect in the cyclic AMP system may be responsible for malignant transformation as well as for expression of low activities of neural-specific enzymes, but the expression of malignancy and low enzyme activity may be only casually related. For example, a mutant clone of neuroblastoma cells isolated from tyrosine-free medium contains tyrosine hydroxylase activity that is similar to that observed in brain (Breakefield and Nirenberg 1974). The activities of acetylcholinesterase and choline acetyltransferase can be increased in cells that maintain tumorigenicity (Prasad 1973).

Catechol-O-Methyltransferase

Catechol-o-methyltransferase (COMT) degrades catecholamines in mammalian cells. The enzyme activity is not increased in cyclic AMP-induced "differentiated" mouse neuroblastoma cells, suggesting that cyclic AMP is not involved in the regulation of COMT activity; however, the COMT activity is increased after treatment with X-irradiation, sodium butyrate, and 5-bromodeoxyuridine (Prasad 1973). In human neuroblastoma cells, the COMT activity is not detectable until the cells are treated with 5-bromodeoxyuridine (Prasad *et al.* 1973a).

The effect of 5-bromodeoxyuridine on COMT activity is similar in both mouse and human neuroblastoma cells in culture. Again, the expression of malignancy is not related to a change in the activity of COMT.

Synthesis and Phosphorylation of Chromosomal and Nonchromosomal Proteins

An elevation of the intracellular level of cyclic AMP in neuroblastoma cells markedly reduced the synthesis of histone (Figure 4) and phosphorylation of H_1-histone (Figure 5) (Lazo *et al.* 1976). Since the above changes occur at the time of inhibition of cell division and DNA synthesis (Prasad *et al.* 1972), they may be important biological signals for the dividing neuroblasts to "turn off" cell division. The occurrence of histone synthesis and phosphorylation of H_1-histone may be necessary biological events for the proliferating system (differentiated or embryonic); but for the embryonic nerve cells, which must eventually stop cell division, a continuation of these events after a specified time during development may be indicative of malignant change.

We have reported that there was no prominent change in the synthesis or phosphorylation of nonhistone chromosomal proteins in cyclic AMP-induced "differentiated" cells (Lazo *et al.* 1976). Other investigators (Burdman 1972, Fujitani and Holoubek 1974) have been unable to demonstrate any significant changes in nonhistone chromosomal proteins during development of rat brains.

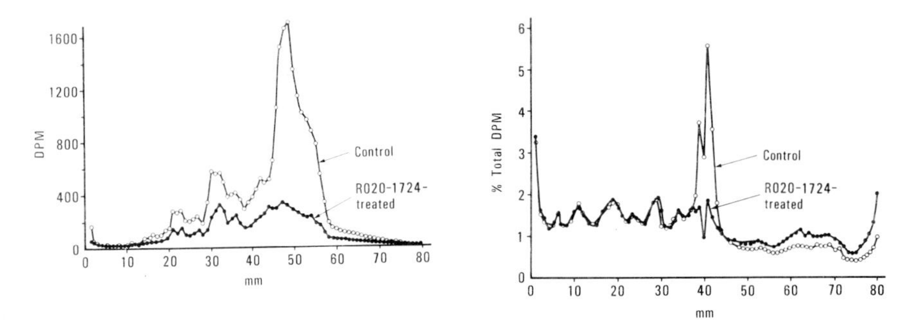

FIG. 4, left. Distribution of radiolabeled amino acids in SDS polyacrylamide gels of various histone fractions (0.25 N HCl fraction) from control and cyclic AMP-induced "differentiated" mouse neuroblastoma cells. Cells were treated with PGE_1 or R020–1724 for three days. Fresh growth medium was added one hour before the addition of radioactive amino acids. Control and treated cells were labeled with ^{14}C- or ^{3}H-amino acids for 2 hours. Total DPM $^{14}C = 25{,}133$; $^{3}H = 8{,}975$. Gels were run for 17.5 hours at 6 ma/gel. (Reproduced, with permission of Exp. Cell Res., from Lazo *et al.* 1976.)

FIG. 5, right. Distribution of radiolabeled phosphate in SDS polyacrylamide gels of various histone fractions (0.25 N HCl fraction) from control and cyclic AMP-induced "differentiated" mouse neuroblastoma cells. Cells were treated with PGE_1 or R020–1724 for three days. Fresh growth medium was added one hour before the addition of radioactive phosphate. Control and treated cells were labeled with ^{32}P or ^{33}P for 2 hours. Total DPM $^{32}P = 298{,}079$; $^{33}P = 108{,}670$. Gels were run for 21 hours at 6 ma/gel. (Reproduced, with permission of Exp. Cell Res., from Lazo *et al.* 1976.)

However, we have observed a small decrease in the synthesis and a small increase in phosphorylation of 40,000 dalton peptides in cyclic AMP-induced "differentiated" cells. The significance of this change in nonhistone protein is unknown. It has been suggested (Elgin *et al.* 1973) that a nonhistone protein of the 40,000–50,000 dalton range may be involved in DNA replication in mammalian cells. If this is true, the changes in the synthesis and phosphorylation of 40,000 dalton peptides in the "differentiated" cells may be a reflection of the inhibition of DNA synthesis that occurs in these cells (Prasad *et al.* 1972).

Reduction in Tumorigenicity of "Differentiated" Cells

The tumorigenicity of differentiated cells is markedly reduced or completely abolished, depending on the experimental condition. For example, in an uncloned cell line, analogs of cyclic AMP, R020-1724 (phosphodiesterase inhibitor), and PGE_1 only partially reduced the tumorigenicity of differentiated cells, whereas the combination of PGE_1 with either analogs of cyclic AMP or a phosphodiesterase inhibitor completely abolished the tumorigenicity of "differentiated" cells (Table 2). This is consistent with the observation that the tumor contains variant cells, some of which are sensitive to PGE_1 but not to R020-1724, and vice versa (Prasad 1972). We have repeated the above experiment using a clonal line (NBP_2). This clone is relatively more sensitive to R020-1724 than to PGE_1. In order to suppress the tumorigenicity of differentiated cells, we found it essential to treat the cells with both PGE_1 and R020-1724 for four days. Of the mice that failed to develop tumors after subcutaneous injection of "differentiated" cells, 30% rejected the malignant cells (0.25×10^6) administered subcutaneously

TABLE 2. *Incidence of tumors after subcutaneous injection of control and differentiated neuroblastoma cells**

Treatment	Number of animals	Incidence of tumors (% of total)
Control cells treated with or without solvent	30	100
dbcAMP	15	50
R020–1724	15	40
8-benzylthio cAMP	15	60
PGE_1	15	20
PGE_1 + dbcAMP	15	0
PGE_1 + 8-benzylthio cAMP	15	0
PGE_1 + R020–1724	16	0

* Tumorigenicity of differentiated cells. Cells (10^5) were plated in 60 mm Falcon plastic dishes and treated with drugs 24 hours later. dbcAMP (0.5 mM), R020–1724 (200 μg/ml), 8-benzylthio cAMP (400 μg/ml), or PGE_1 (10 μg/ml) were added individually or in combination with PGE_1 (10 μg/ml). After four days of incubation, control and differentiated cells (0.25×10^6) were injected subcutaneously into male A/J mice (6–8 weeks old). Cell viability in the control and drug-treated cultures was 90–95%. Animals were observed for 60 days. (These data are taken from Prasad 1972.)

(Prasad *et al.* 1976b). This suggested, for the first time, that the differentiated neuroblastoma cells may act as a strong antigen against malignant neuroblastoma cells.

Regulation of the Intracellular Level of Cyclic AMP

Since an elevation of the intracellular level of cyclic AMP in neuroblastoma cells induces many differentiated functions that are characteristic of mature neurons and reduces the tumorigenicity, it would be reasonable to assume that there is a defect(s) in the cyclic AMP system which causes low levels of cyclic AMP and thereby causes malignancy and prevents the cells from expressing differentiation at a maximal level. The identification of such a defect(s) may provide new insight into the mechanism of regulation of differentiation and may clarify the relationship between differentiation and malignancy. Many exogenous and endogenous factors affect the intracellular level of cyclic AMP. These include relative activity of adenylate cyclase and cyclic AMP phosphodiesterase, response of adenylate cyclase to neurotransmitters, and levels of cyclic AMP-binding proteins. Therefore, the study on the regulation of cyclic AMP level in neuroblastoma cells is essential.

Adenylate Cyclase Activity

The basal level of adenylate cyclase activity and the sensitivity of adenylate cyclase to neurotransmitters in homogenates vary from one clone of mouse neuroblastoma cells to another (Prasad and Gilmer 1974, Prasad and Kumar 1974, Prasad *et al.* 1975d). In homogenates of cyclic AMP-induced "differentiated" neuroblastoma cells, we have made the following observations: (1) The sensitivity of adenylate cyclase to dopamine, norepinephrine (Figure 6), and

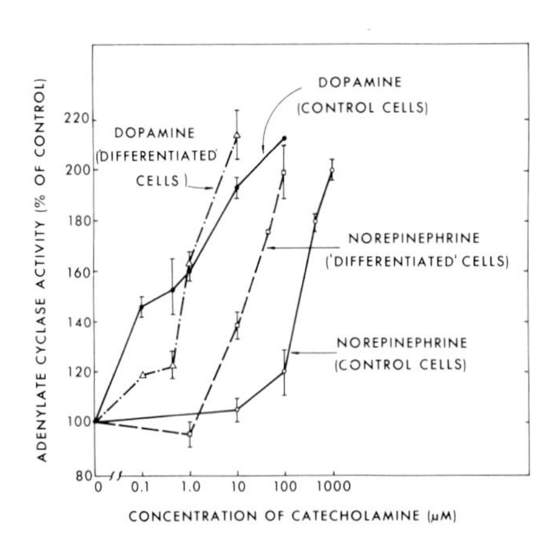

FIG. 6. Changes in adenylate cyclase activity in homogenates of control and "differentiated" mouse neuroblastoma cells $NBA_2(1)$ after treatment with dopamine and norepinephrine. 4-(3-butoxy-4-methoxybenzyl)-2-imidazolidinone (R020–1724), a specific inhibitor of cyclic AMP phosphodiesterase, was used to induce "differentiation." The basal activities of adenylate cyclase in control (15 ± 1.4 pmol/mg of protein/minute) and "differentiated" (21 ± 1 pmol/mg of protein/minute) cells were considered 100% control values; the adenylate cyclase values of treated cells were expressed as percentage of control. Each value represents an average of 8 to 12 samples. The bar at each point is standard deviation. (Reproduced, with permission of Proc. Natl. Acad. Sci. USA, from Prasad and Gilmer 1974.)

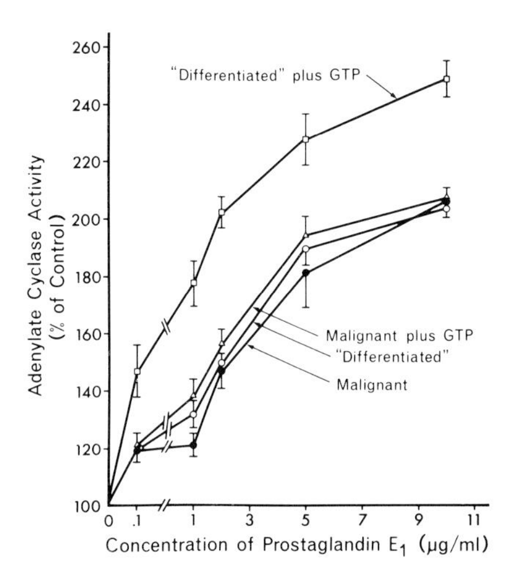

FIG. 7. Changes in adenylate cyclase activity in homogenates of malignant and "differentiated" mouse neuroblastoma cells (NBP_2) after treatment with prostaglandin E_1 and guanosine triphosphate (10 μM). 4-(3-butoxy-4-methoxybenzyl)-2-imidazolidinone (R020–1724), a specific inhibitor of cyclic AMP phosphodiesterase, was used to induce differentiation. The basal activities of adenylate cyclase in malignant (23 ± 1 pmol/mg of protein/minute) and differentiated (25 ± 1 pmol/mg of protein/minute) cells were considered 100% control values; the adenylate cyclase values of treated cells were expressed as percentage of control. Each value represents an average of six samples. The bar at each point is standard deviation. (Reproduced, with permission of Cancer Res., from Prasad *et al.* 1975c.)

acetylcholine increases; (2) the sensitivity of adenylate cyclase to PGE_1 and guanosine triphosphate (GTP) does not change; (3) GTP potentiates the PGE_1-stimulated adenylate cyclase activity (Figure 7); and (4) high concentrations of calcium, magnesium, and manganese inhibit adenylate cyclase activity. Thus, the decrease in the sensitivity of adenylate cyclase to neurotransmitters, ions, and GTP appears to be associated with malignancy; however, these may be secondary lesions, since the stimulatory effect of dopamine and norepinephrine on the cyclic AMP level in the malignant cells cannot be measured (Table 3) until the activity of cyclic AMP phosphodiesterase is inhibited (Prasad *et al.* 1974, Sahu and Prasad 1975). PGE_1 stimulates the intracellular level of cyclic AMP in varying degrees in most of the mouse clones (Gilman and Nirenberg 1971, Prasad 1973, Hamprecht and Schultz 1973, Blume *et al.* 1973, Sahu and Prasad 1975), but in one clone (NBE^-), it does not increase the cyclic AMP level significantly until the phosphodiesterase activity is inhibited (Table 3). PGE_1 in combination with a phosphodiesterase inhibitor increases the cyclic AMP level more than that produced by the individual agent alone. A similar observation was made for adenosine (Prasad 1975). Thus, the rate-limiting factor in the accumulation of the cyclic AMP following the treatment of neuroblastoma cells with neurotransmitters, adenosine, and PGE_1 appears to be the level of cyclic AMP phosphodiesterase activity. Therefore, we have suggested that an increase in cyclic AMP phosphodiesterase activity in the dividing neuroblasts may be one of the early lesions of malignancy of nerve cells (Prasad *et al.* 1974).

Cyclic AMP Phosphodiesterase Activity

The cyclic AMP phosphodiesterase activity is markedly increased (Prasad 1973, Kumar *et al.* 1975) in cyclic AMP-induced "differentiated" mouse neuro-

TABLE 3. *Effect of neurotransmitters and PGE_1 on cyclic AMP level in neuroblastoma clones**

Treatment	cyclic AMP levels in various clones (pmoles/mg protein)				
	NBA_2 (1)++	NBA_2 (1)	NBE^- (A)	$NBDB^-$	NBP_2
Control, dopamine, or norepinephrine	25 ± 4†	14 ± 2	19 ± 2	19 ± 4	11 ± 2
R020–1724	41 ± 2	48 ± 7	107 ± 5	26 ± 4	65 ± 8
PGE_1	31 ± 4	20 ± 2	22 ± 2	56 ± 9	19 ± 2
R020–1724 + dopamine	117 ± 16	233 ± 14	139 ± 4	162 ± 49	66 ± 5
R020–1724 + norepinephrine	132 ± 6	191 ± 19	120 ± 11	122 ± 36	67 ± 4
R020–1724 + PGE_1	$2,563 \pm 121$	$2,803 \pm 173$	$1,953 \pm 118$	$2,426 \pm 124$	$2,404 \pm 212$

* $NBA_2(1)^{++}$ contains tyrosine hydroxylase (TH) but no choline acetyltransferase (ChA) and has been in culture less time than $NBA_2(1)$; $NBA_2(1)$ contains TH but no ChA; $NBE^-(A)$ contains ChA but no TH; $NBDB^-$ contains neither TH nor ChA; NBP_2 contains both ChA and TH. (These data are taken from Sahu and Prasad 1975.)

† Standard deviation.

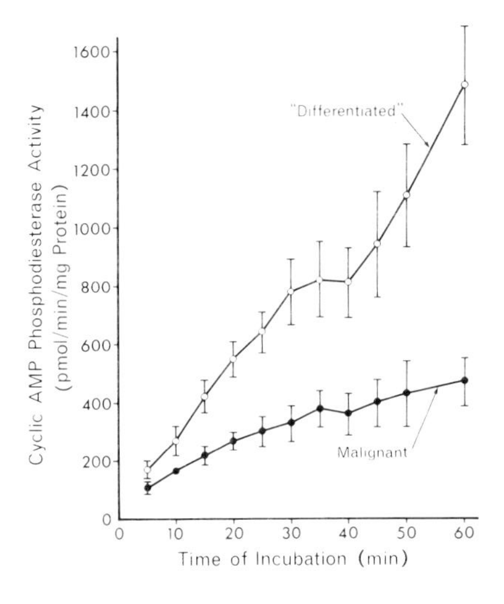

FIG. 8. Effect of magnesium on phosphodiesterase activity in homogenates of malignant and differentiated mouse neuroblastoma cells. 4-(3-butoxy-4-methoxybenzyl)-2-imidazolidinone, an inhibitor of cyclic AMP phosphodiesterase, was used to induce differentiation. Each value represents an average of 8 to 12 samples. (Reproduced, with permission of Cancer Res., from Kumar and Prasad 1975.)

blastoma cells in culture (Figure 8). This increase in enzyme activity was blocked by cycloheximide but not by actinomycin D. The activity of cyclic AMP phosphodiesterase also increases during exponential growth and reaches a maximal value at confluency (Sinha and Prasad, in press). The increase in enzyme activity during growth period occurs without any change in the intracellular level of cyclic AMP and is blocked by both cycloheximide and actinomycin D. Thus, the activity of cyclic AMP phosphodiesterase in neuroblastoma cells is regulated by at least two modes, namely, cyclic AMP and growth.

Binding of Cyclic AMP with Proteins

Since the intracellular levels of cyclic AMP and cyclic AMP phosphodiesterase activity increase in "differentiated" neuroblastoma cells, the cells must develop the mechanism of protecting the formed cyclic AMP from the enzymatic hydrolysis. The protein-bound cyclic AMP is not hydrolyzed by phosphodiesterase; therefore, we have measured the binding of cyclic AMP with proteins of malignant and "differentiated" neuroblastoma cells. Indeed, the binding of cyclic AMP with soluble (100,000 X g supernatant) and pellet proteins from cyclic AMP-induced "differentiated" neuroblastoma cells increased by about twofold (Prasad *et al.* 1975e). In malignant and "differentiated" mouse neuroblastoma cells, two peaks of binding proteins are observed, and the magnitude of each of these peaks increases in "differentiated" cells (Figure 9). An increase in cyclic AMP binding occurs 24 hours after treatment of neuroblastoma cells with PGE_1 (Figure 10) and is completely blocked by cycloheximide but not by actinomycin D (Prasad *et al.* 1976a). The cyclic AMP binding is heat labile and is sensitive to the action of protease. These data suggest that an increase in cyclic AMP binding in "differentiated" cells is due to an increase in the levels of binding proteins. X-irradiation and 6-thioguanine, which increase total protein, inhibit

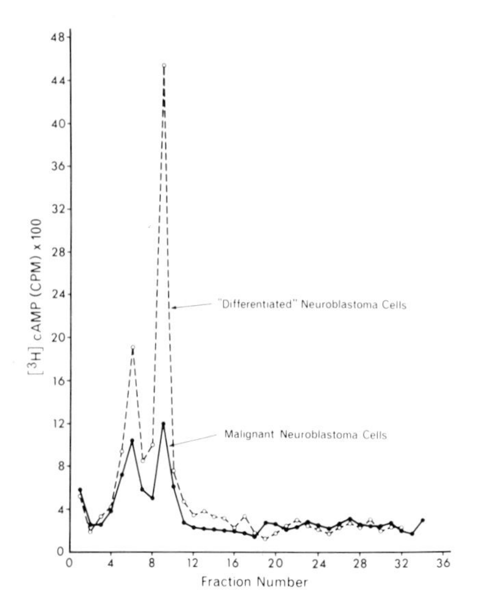

FIG. 9. Binding of cyclic(^{3}H)AMP with protein fractions obtained after polyacrylamide gel electrophoresis. 4-(3-butoxy-4-methoxybenzyl)-2-imidazolidinone (R020–1724), a specific inhibitor of cyclic AMP phosphodiesterase, was used to induce differentiation. The soluble proteins of malignant and "differentiated" neuroblastoma cells were applied on the gel. The experiments were repeated twice. (Reproduced, with permission of Biochem. Biophys. Res. Commun., from Prasad *et al.* 1975e.)

cell division and induce some differentiated functions (Prasad 1973) in neuroblastoma cells without changing the cellular cyclic AMP; however, they do not elevate the level of cyclic AMP-binding proteins. This indicates that neither the inhibition of cell division nor the increase in total protein is sufficient to increase the levels of binding proteins.

The increased amount of binding proteins in "differentiated" cells may in part account for the existence of high intracellular levels of cyclic AMP, even in the presence of elevated cyclic AMP phosphodiesterase activity. Thus, one of the mechanisms by which neuroblastoma cells could maintain a high intracellular level of cyclic AMP is to increase the amount of binding proteins during differentiation. Glial and L-cells, after treatment with PGE_1 or R020-1724,

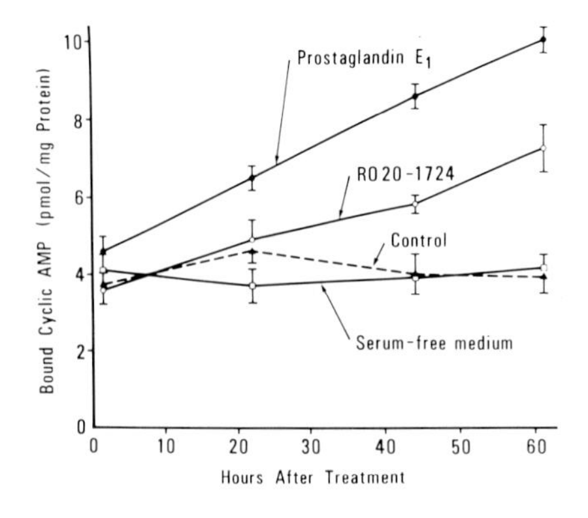

FIG. 10. Binding of cyclic AMP with soluble proteins of neuroblastoma cells as a function of treatment time. PGE_1, R020–1724, and serum-free medium were added separately to culture, and the binding assay was performed at 3 hours, 1 day, 2 days, and 3 days after treatment. The incubation mixture contained 100 μg protein. Each value represents an average of six samples. (Reproduced, with permission of Cancer Res., from Prasad *et al.* 1976c.)

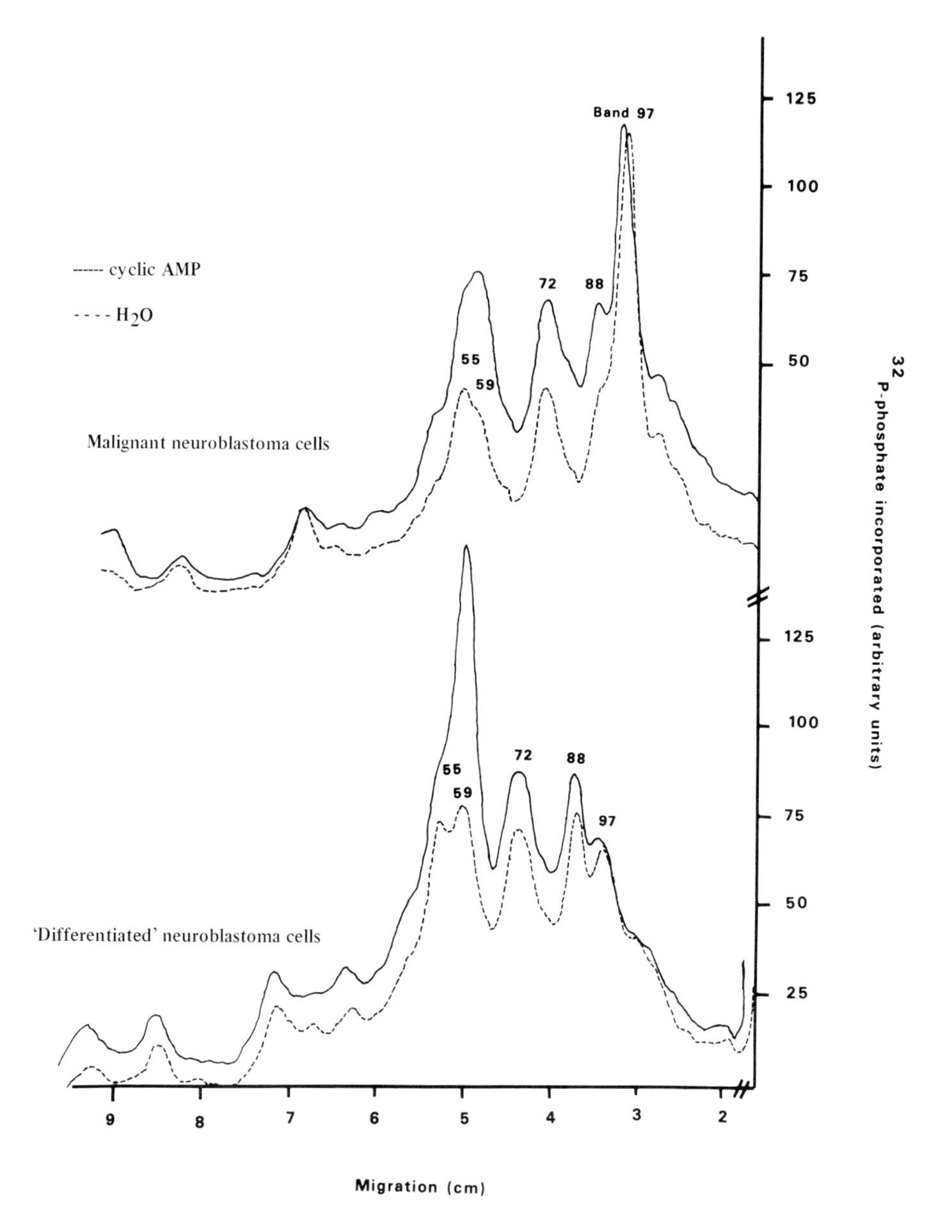

FIG. 11. Autoradiogram tracing of phosphorylated proteins in the cytosol of malignant and cyclic AMP-induced "differentiated" neuroblastoma cells in culture. It should be noted that the incorporation of ^{32}P into proteins from gamma ^{32}P-ATP is a net result of the action of protein kinase(s) and phosphoprotein phosphatase(s) activities. Therefore, the term phosphorylation refers to changes in the amount of radioactive phosphate incorporation into specific bands under standard assay conditions and does not refer to the capacity of specific enzymes. (Reproduced, with permission of Nature, from Ehrlich *et al.* 1977.)

fail to develop such a mechanism. This, in part, may account for the reversibility of cyclic AMP effects on nonneural tumor cells soon after the removal of the drug. Thus, there is a defect in the regulation of cyclic AMP-binding proteins in rat glioma cells in culture, or the level of binding proteins in rat glioma cells is not regulated by cyclic AMP. The increase in the levels of binding proteins in "differentiated" neuroblastoma cells is associated with the irreversible effect of cyclic AMP, but this is not the cause of irreversibility, since the increase in binding proteins occurs in dividing nerve cells (24 hours after treatment with PGE_1). However, this may be one of the important events that is essential for maintaining the irreversible neural differentiation.

If the results on cyclic AMP-binding proteins can be applicable to the in vivo condition, the rationale for developing a new therapeutic approach for the treatment of neuroblastoma would differ from that for nonneural tumors; e.g., elevations of the intracellular levels of cyclic AMP and cyclic AMP-binding proteins may be of therapeutic value for neuroblastoma and nonneural tumors, respectively.

Cyclic AMP-Dependent Phosphorylation

The binding of cyclic AMP with the regulatory subunits results in dissociation of the regulatory proteins from the catalytic units, with subsequent activation of the latter (Gill and Garren 1970, Tao *et al.* 1970, Reimann *et al.* 1971), which then phosphorylate various proteins. If this is true during differentiation of neuroblastoma cells, the cyclic AMP-dependent phosphorylation activity should increase in "differentiated" cells. We have first investigated this by measuring the phosphorylation activity of 100,000 × g supernatant and pellet fraction in the presence of added histone. We obtained the following results: (1) Cyclic AMP-dependent phosphorylation of H_1-histone did not significantly change in the supernatant and pellet fractions of "differentiated" neuroblastoma cells (Prasad *et al.* 1976a). (2) Cyclic AMP-dependent phosphorylation of H_2b histone decreased by about twofold in the supernatant fraction of "differentiated" neuroblastoma cells (Prasad, unpublished observations). These data suggest that the level of cyclic AMP-dependent phosphorylation activity in the supernatant fraction either decreases or shows no change at the time when the level of cyclic AMP-binding proteins increases. However, the phosphorylation activity of exogenous histones may not be a true reflection of endogenous cyclic AMP-dependent phosphorylation activity of specific cellular proteins. Indeed, we have observed (Ehrlich *et al.* 1977) that the cyclic AMP-dependent phosphorylation activity of cytosol proteins markedly increases, but cyclic AMP-independent phosphorylation of another protein decreases by about twofold in "differentiated" cells (Figure 11). There is also an increase in cyclic AMP-independent phosphorylation activity of two specific proteins in the crude nuclear fractions of "differentiated" cells (Figure 12).

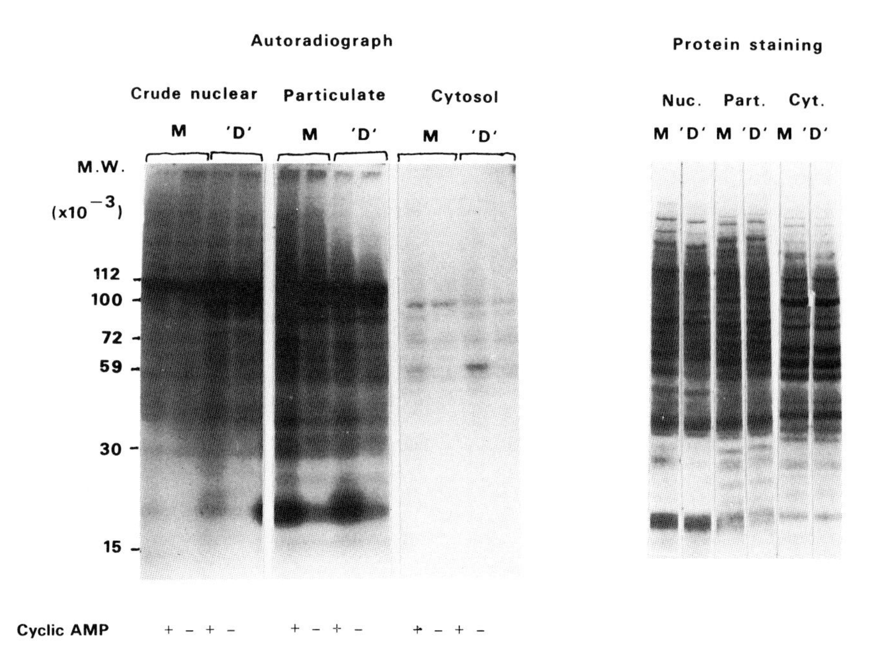

FIG. 12. Autoradiogram of phosphorylated proteins of various subcellular fractions obtained from malignant and cyclic AMP-induced "differentiated" neuroblastoma cells in culture. It should be noted that the incorporation of ^{32}P into proteins from gamma ^{32}P-ATP is a net result of the action of protein kinase(s) and phosphoprotein phosphatase(s) activities. Therefore, the term phosphorylation refers to changes in the amount of radioactive phosphate incorporation into specific bands under standard assay conditions and does not refer to the capacity of specific enzymes. (Reproduced, with permission of Nature, from Ehrlich *et al.* 1977.)

POLY A-CONTAINING CYTOPLASMIC RNA

The cytoplasms of cyclic AMP-induced "differentiated" neuroblastoma cells had more poly A-containing RNA (Bondy *et al.* 1974), presumed to be messenger RNA, than that of the malignant cells (Figure 13). This increase in poly A-containing cytoplasmic RNA was in part due to an increase in the rate of transport of mRNA from nucleus to cytoplasm (Prasad *et al.* 1975b). The increase in mRNA contents in "differentiated" cells was consistent with the observation that the activities of many enzymes increase in differentiated cells. However, the quantitative amount of genetic materials available for the transcription might remain relatively constant during differentiation. This was substantiated by the similarity in binding of dactinomycin with nuclei of malignant and "differentiated" cells. The binding of actinomycin D has been used to estimate the amount of DNA within nuclei available for the template activity (Kernell *et al.* 1971, Watson 1974). The lack of any quantitative change in template activity during cyclic AMP-induced "differentiation" of neuroblastoma cells could be due to the fact that some genetic information is activated while some is suppressed

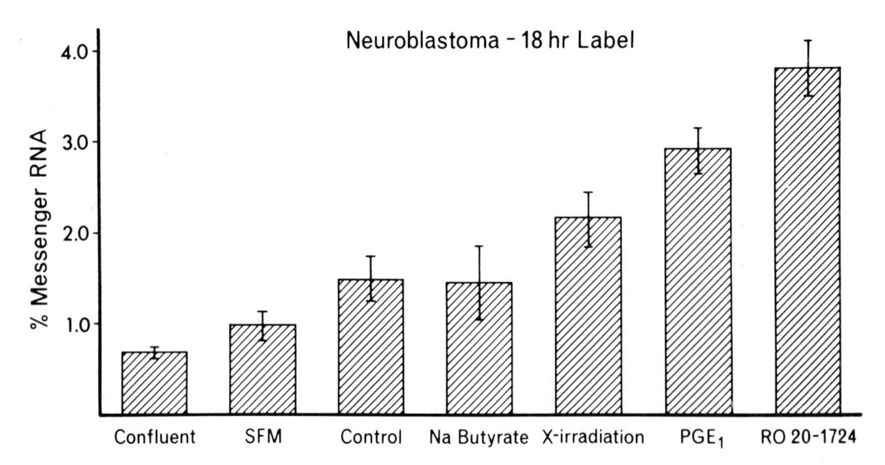

FIG. 13. Proportion of labeled cytoplasmic RNA that is messenger in mouse neuroblastoma cells after various treatments. Neuroblastoma cells were plated in Falcon plastic flasks (75 cm^2) and serum-free medium (SFM), sodium butyrate (Na butyrate), prostaglandin E_1 (PGE_1), 4-(3-butoxy-4-methoxybenzyl)-2-imidazolidinone (R020–1724), or X-irradiation was given 24 hours later. The fresh growth medium and drug were changed every day. ^{3}H-adenosine (2 μCi/ml) and fresh drug were added 54 hours after treatment, and cells were further incubated for 18 hours. The polyadenylic acid containing cytoplasmic RNA in control and treated cells was determined. Each value represents an average of at least six samples ± standard error of mean. (Reproduced, with permission of Exp. Cell Res., from Prasad *et al.* 1975b.)

during differentiation. Indeed, the activities of several enzymes in "differentiated" neuroblastoma cells were elevated, but the synthesis of histone, phosphorylation of H_1-histone, and cyclic AMP-independent phosphorylation activity of cytosol proteins were considerably reduced (Prasad 1975, Lazo *et al.* 1976, Ehrlich *et al.* 1977).

REGULATION OF THE INTRACELLULAR LEVEL OF CYCLIC GMP AND ITS RELATIONSHIP TO MALIGNANCY

Mouse neuroblastoma cells also have cyclic GMP (Prasad *et al.* 1976b), the concentration of which is 100-fold less than cyclic AMP (Table 4). In addition to the presence of cyclic GMP phosphodiesterase, which is distinct from cyclic AMP phosphodiesterase (Prasad *et al.* 1975a), the extremely low binding affinity of cyclic GMP with proteins (10 times less than cyclic AMP) may in part account for the low intracellular level of cyclic GMP. Preliminary study indicates that PGE_1 or R020-1724 does not increase the intracellular level of cyclic GMP 15 minutes after treatment; these agents under a similar experimental condition are known to increase the cyclic AMP level. Acetylcholine increases the intracellular level of cyclic GMP by about twofold; however, it does not significantly change the level of cyclic AMP. Acetylcholine does not antagonize the effect of PGE_1 or R020-1724 on differentiation and does not affect growth rate or

TABLE 4. *Effect of various agents on the intracellular level of cyclic nucleotides in neuroblastoma cells*

Treatment	Dose (μg/ml)	Cyclic AMP level (pmoles/g protein)	Cyclic GMP level (pmoles/g protein)
Control		5,000 ± 400*	50 ± 7
Acetylcholine	100	5,300 ± 450	109 ± 12
R020–1724	200	29,000 ± 400	39 ± 6
Papaverine	25	32,000 ± 600	84 ± 9
PGE_1	10	8,500 ± 500	41 ± 8

* Standard deviation
(These values are taken from Prasad *et al.* 1976b.)

differentiation in mouse neuroblastoma cells in culture (Prasad *et al.* 1975d). The addition of either exogenous cyclic GMP or N^2-2′0-dibutyryl cyclic GMP into cultures of mouse or human neuroblastoma cells inhibits cell division without causing differentiation and when added with PGE_1 or R020-1724 fails to antagonize the expression of differentiation (Prasad 1975). These data suggest that cyclic GMP neither has any role in regulating the expression of differentiated functions in neuroblastoma cells nor antagonizes the effect of cyclic AMP, and possibly has no role in the expression or maintenance of malignancy. Further work is needed to substantiate this.

A WORKING HYPOTHESIS FOR THE MALIGNANCY OF NERVE CELLS

Figure 14 shows a diagrammatic model to explain the malignancy of nerve cells (Prasad 1974). This model suggests that a mutation in the regulatory gene for cyclic AMP phosphodiesterase within a single and/or a group of dividing nerve cells may result from viruses, chemical carcinogens, ionizing radiation, or any combination of these agents; and this mutational change may increase phosphodiesterase activity in mutated cells. The high cyclic AMP phosphodiesterase activity could lead to low levels of cyclic AMP, which in turn could cause low levels of adenylate cyclase activity and insensitivity of adenylate cyclase to neurotransmitters. These changes might then prevent the expression of neuronal differentiated functions in the mutated nerve cell, causing it to become a cancer cell. The regulatory genes for other differentiated functions in the daughter cancer cells may consequently become more susceptible to mutagenic changes, which then may produce further molecular lesions. This may account for the fact that the neuroblastoma cells obtained from a tumor differ quantitatively and qualitatively from one another with respect to expression of cellular properties and to sensitivity to different drugs. The mutated nerve cell appears to maintain the capacity to differentiate into various forms of nerve cells. This is supported by the fact that the neuroblastoma contains four major types of nerve cells (Amano *et al.* 1972, Prasad *et al.* 1973b, Knapp and Mandell

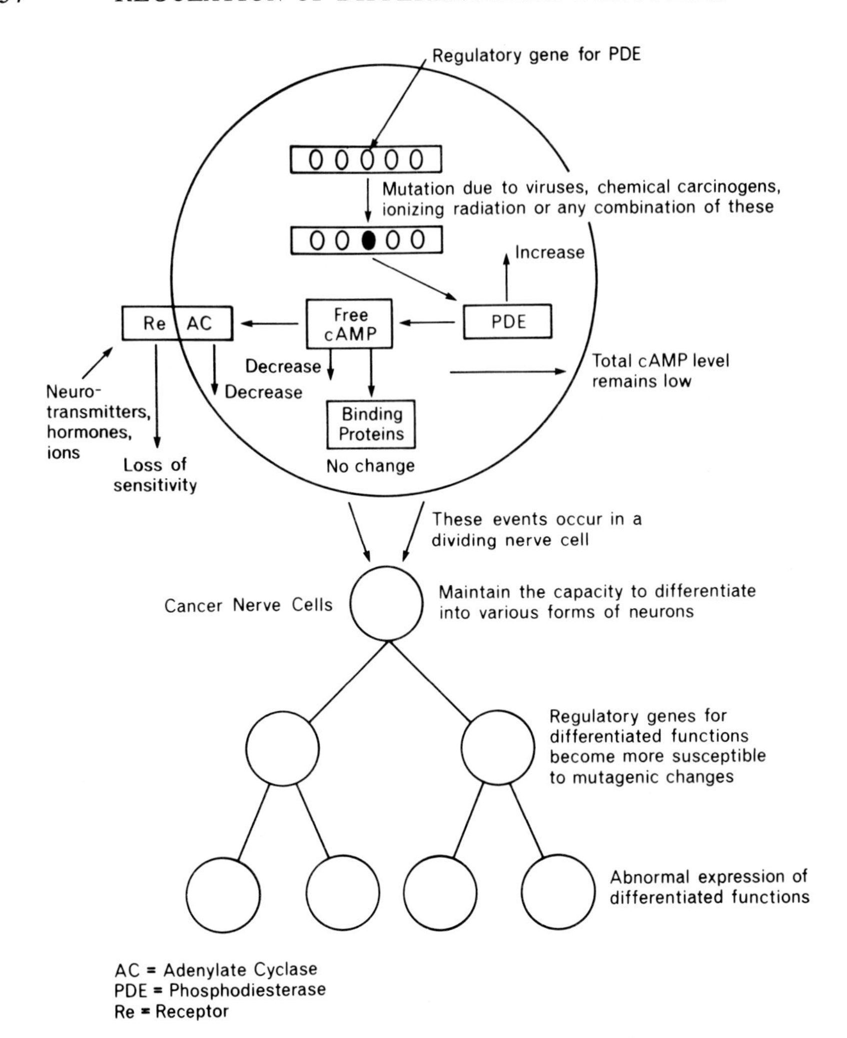

FIG. 14. A diagrammatic model to explain the postulated mechanism for the development of cancer of nerve cells. Adenosine 3′,5′-cyclic monophosphate = cAMP; cAMP phosphodiesterase = PDE; adenylate cyclase = AC; receptor = Re; dopamine = DA; norepinephrine = NE. (Reproduced, with permission of Differentiation, from Prasad 1974.)

1974)—(1) adrenergic cells, (2) cholinergic cells, (3) sensory-like cells, and (4) serotonergic cells. However, the differentiated functions of these nerve cells are not adequately expressed, and, therefore, they continue to divide. Whether or not the first cancer nerve cell will lead to the formation of a detectable neoplasm depends on the host's immunological environment. If the host's immunological environment is normal, the mutated nerve cell may be rejected, and no malignant lesion will ever appear. On the other hand, if the immunological environment of the host is abnormal, the malignant neoplasm may become detectable in a few months or years after the appearance of the first cancer cell. The following experimental data support the proposed hypothesis: (1) a raised intracellular level of cyclic AMP in neuroblastoma cells irreversibly induces several differentiated functions that are characteristic of mature neurons (Prasad 1975); (2) the inhibition of cyclic AMP phosphodiesterase activity reduces the oncogenicity of "differentiated" cells (Prasad 1972); (3) although dopamine and norepinephrine-sensitive adenylate cyclases are demonstrable in vitro (Prasad and Gilmer 1974), these agents do not increase the cyclic AMP level until the phosphodiesterase activity is inhibited (Prasad *et al.* 1974, Sahu and Prasad 1975); (4) a relatively high phosphodiesterase activity in neuroblastoma cells is associated with low activity of adenylate cyclase and a low level of cyclic AMP (Prasad *et al.* 1975d). Although the basal level of cyclic AMP phosphodiesterase activity in "differentiated" mouse neuroblastoma cells increases (Prasad 1973, Kumar *et al.* 1975), the intracellular level of cyclic AMP also increases (Prasad and Kumar 1974). The association of a high level of cyclic AMP with the high level of cyclic AMP phosphodiesterase activity may be due to an increased level of cyclic AMP-binding proteins (Prasad *et al.* 1975e). The protein-bound cyclic AMP may be relatively less susceptible to enzymatic hydrolysis.

Our evidence does not suggest that a defect in the adenylate cyclase regulation is one of the early lesions of the malignancy of nerve cells. For example, the adenylate cyclase activity in neuroblastoma homogenates can be stimulated by dopamine, norepinephrine, PGE_1, and acetylcholine (Prasad 1975); however, only PGE_1 increases the intracellular level of cyclic AMP without the inhibition of cyclic AMP phosphodiesterase activity (Prasad 1973). In one clone of neuroblastoma cells, even PGE_1 does not increase the cyclic AMP level until the phosphodiesterase activity is inhibited (Sahu and Prasad 1975). Thus, the rate-limiting factor in the measurement of neurotransmitters' response is the high activity of phosphodiesterase.

The present model assumes that the malignant transformation occurs in either a single and/or in a group of dividing nerve cells. There is some indirect evidence to support this. For example, these tumors are generally detectable in children under five years of age. In addition, the neuroblastomas contain more than one type of nerve cell, indicating that the malignant transformation occurs in nerve cells that are not yet committed to form any one neural cell type. There is no evidence that the mature neurons ever undergo malignant transformation, although the supporting elements of adult nervous tissue do so.

EFFECT OF VITAMIN C

Mouse neuroblastoma cells are more sensitive (i.e., reduction in cell number) to sodium ascorbate (vitamin C) than mouse fibroblasts (L-cells) and rat glioma (Prasad and Sinha 1977). Sodium ascorbate markedly potentiates (Table 5) the effect of several therapeutic agents such as X-irradiation (30-fold), vinblastine sulfate (17-fold), 6-thioguanine (2-fold), 5-fluorouracil (11-fold), Cytoxan

TABLE 5. *Effect of combined treatment with sodium ascorbate and therapeutic agents on mouse neuroblastoma cells in culture**

Treatment	Number of cells X 10^4
Control	259 ± 2.7†
Na ascorbate (200 μg/ml)	256 ± 22
X-irradiation (600 rads)	15.1 ± 2.0
Na ascorbate + X-irradiation	0.5 ± 0.1
Vinblastine (0.01 μg/ml)	6.0 ± 2
Na ascorbate + vinblastine	0.35 ± 0.1
6-thioguanine (0.5 μM)	13.0 ± 6
Na ascorbate + 6-thioguanine	6 ± 1.0
5-fluorouracil (0.05 μg/ml)	43 ± 6.0
Na ascorbate + 5-fluorouracil	4.1 ± 1.4
Cytoxan (800 μg/ml)	110 ± 13
Na ascorbate + Cytoxan	6.1 ± 2.1
Adriamycin (.04 μg/ml)	2.6 ± 0.4
Na ascorbate + Adriamycin	.04 ± .01
CCNU (40 μg/ml)	12.9 ± 3.0
Na ascorbate + CCNU	4.3 ± 1.6
R020–1724 (200 μg/ml)	15 ± 4
Na ascorbate + R020–1724	0.20 ± 0.1
Sodium butyrate (0.5 mM)	33 ± 7
Na ascorbate + Na butyrate	12 ± 3

* Neuroblastoma cells (50,000) were plated in Falcon tissue culture dishes (60 mm). Sodium ascorbate was added immediately after the addition of drug or immediately after X-irradiation 24 hours after plating. The medium and drug were changed every day, and the cell number was counted by a hematocytometer at four days after treatment. Each value represents an average of at least eight samples.

† Standard deviation.

(18-fold), Adriamycin (60-fold), 1-(2-chloroethyl)-3-cyclohexyl-1-nitrosourea (CCNU) (3-fold), R020-1724 (75-fold), and sodium butyrate (3-fold) on mouse neuroblastoma cells; however, it produces no similar effects on mouse L-cells or rat glioma cells. Although sodium ascorbate potentiates the effect of bleomycin sulfate by twofold, it produces a similar effect on fibroblasts and rat glioma cells. On the other hand, the cytotoxic effect of 5-(3,3-dimethyl-1-triazeno)-imidazole-4-carboxamide (DTIC) is completely prevented by sodium ascorbate; however, the effect of methotrexate is only partially prevented by sodium ascorbate. If the present results could be applicable to human tumor in vivo, the addition of sodium ascorbate with certain chemotherapeutic agents may markedly increase their cytotoxic effect on tumor cells in a highly selective manner without affecting any further the host's immune system. The fact that the addition of catalase (200 μg/ml) completely prevents the cytotoxic effect of high concentrations of sodium ascorbate, as well as the potentiating effect of sodium ascorbate in combination with a chemotherapeutic agent, suggests that sodium ascorbate inhibits catalase activity in the cell, which then allows the accumulation of hydrogen peroxide (H_2O_2) resulting in cell death. This indirectly indicates that catalase activity in neuroblastoma cells may be a predominant mode of destroying cellular

H_2O_2. We, therefore, speculate that the tumor cells in which catalase is the predominant mode of destroying H_2O_2 would respond to the potentiating effect of sodium ascorbate in a manner observed for neuroblastoma cells in culture.

CAN "DIFFERENTIATED" CELLS ACT AS A STRONG ANTIGEN?

Human neuroblastoma cells in culture possess tumor-specific transplantation antigens against which an immune reaction can be demonstrated by using the colony inhibition assay (Helström *et al.* 1968). We have obtained preliminary evidence which indicates that the "differentiated" cells may act as a strong antigen. Uncloned cells treated with PGE_1 and R020-1724 for four days do not produce tumors when injected into mice (Prasad 1972), whereas untreated cells produce tumors in all cases. Many of the mice that fail to develop tumors after subcutaneous injection of "differentiated" cells reject the subcutaneously administered malignant neuroblastoma cells (0.25×10^6) (Prasad *et al.* 1976b). The "differentiated" cells may retain the tumor-specific antigens and, therefore, may eventually reject themselves. The above phenomenon remains to be fully established. However, it has been observed clinically that the spontaneous regression of neuroblastoma tumors begins with the transformation to ganglioneuromas, which also eventually disappear. If these assumptions are correct, the agent that causes differentiation of neuroblastoma cells may be useful in the management of human neuroblastoma tumors in a highly selective manner.

SUGGESTION OF A NEW APPROACH FOR THE TREATMENT OF NEUROBLASTOMAS

Based on the current data on neuroblastoma cells in culture, our previous suggestion for the therapy of neuroblastoma tumor (Prasad *et al.* 1974) should be modified as follows: (1) Sodium ascorbate and sodium butyrate are given daily for the entire period of treatment. Sodium ascorbate is a relatively nontoxic compound. The oral administration of sodium butyrate (6–10 g/day) for four months produced no detactable toxic effects in a child with neuroblastoma (L. Furman, personal communication). (2) After five days of the above treatment, cyclophosphamide, vincristine, Adriamycin, and 5-fluorouracil are administered in sequence and in the dosages and at time intervals that are currently in use. It is hoped that after completion of the second phase, most of the tumor cells will be killed. (3) After the second phase, the differentiating agents such as cyclic AMP-stimulating agents (inhibitors of phosphodiesterase like papaverine and R020-1724 and stimulators of adenylate cyclase like prostaglandin E_1) are continuously infused for at least four days. The administration of PGE_1 in the presence of an inhibitor of cyclic AMP phosphodiesterase is suggested because the neuroblastoma tumor contains cells whose sensitivity to PGE_1 and phosphodiesterase inhibitor markedly varies. However, the combination of PGE_1 and phosphodiesterase inhibitor increases the intracellular level of cyclic AMP in all clones (Sahu and Prasad 1975). It is hoped that the administration of differentiating agents would produce differentiation in many of the remaining cells. The differentiated cells may evoke an immune response that would kill many of the drug-resistant tumor cells. (4) After the administration of the differentiating agents, the nonspecific immune stimulants such as Bacillus Calmette-Guerin (BCG) should be administered. It is hoped that the further stimulation of the host's immune system would help the body to reject the residual drug-resistant tumor cells. Phases 1, 3, and 4 should be repeated at least twice after the clinical tumor-free state has been achieved. Because of the carcinogenic effect of ionizing radiation, the use of radiation for the treatment of neuroblastoma is not recommended until an extremely emergency condition exists. This proposed concept is being tested in other neural crest tumor cell lines such as melanoma and Wilm's tumor.

CLINICAL TRIAL OF DIFFERENTIATING AGENTS

Dr. L. Helson of Memorial Hospital, New York, has been using combinations of cytotoxic drugs, such as cyclophosphamide and vincristine, with differentiating agents, such as papaverine and tri-fluoro-methyl-2-deoxyuridine, in metastatic neuroblastoma patients, some of whom were unresponsive to previous treatments and others who were previously untreated. Although a marked regression of tumor was observed in all patients, and the conversion from neuroblastoma to ganglioneuroma was observed in cases where biopsies were taken and examined (Helson 1975), the response of the previously untreated patients appears to be the best. Dr. Helson has indicated (personal communication) that 17 of 19 previously untreated patients remain free of disease between 12–24 months.

CONCLUSIONS

The following conclusions can be made: (1) Cyclic AMP appears to be one of the important factors in induction as well as in regulation of several differentiated functions in mammalian nerve cells. A low level of cyclic AMP in dividing nerve cells, which results from an increase in cyclic AMP phosphodiesterase activity due to a mutation on the regulatory gene of this enzyme, is responsible for the expression of malignancy and "abnormal differentiation." (2) The exact relationship between differentiation and malignancy remains to be clarified. However, it appears that no one individual differentiated function is linked with malignancy. This is not surprising because the expression of many of these functions is independently regulated. However, when several of the differentiated functions express at maximal levels, the tumorigenicity of such cells is abolished. Thus, the expression of malignancy and abnormal differentiation appeared to be linked in neuroblastoma cells in culture. (3) The increased level of cyclic AMP-binding proteins provides one of the important intracellular mechanisms for protecting the formed cyclic AMP from enzymatic hydrolysis during differentiation of neuroblastoma cells. (4) Reduction in histone synthesis and in H_1-

histone phosphorylation may be an important biological signal for the dividing neuroblasts to "turn off" cell division. If these events do not occur, it might be indicative of malignant change.

REFERENCES

Augusti-Tocco, G., G. Sato, P. Calude, and D. Potter. 1970. Clonal cell lines of neurons, *in* Control Mechanisms in the Expression of Cellular Phenotypes, H. A. Padykula, ed., Academic Press, Inc., New York, pp. 109–120.

Amano, T., E. Richelson, and M. Nirenberg. 1972. Neurotransmitter synthesis by neuroblastoma clones. Proc. Natl. Acad. Sci. USA 69:258–263.

Blume, A., C. Dalton, and H. Sheppard. 1973. Adenosine mediated elevation of cyclic 3′,5′-adenosine monophosphate concentrations in cultured mouse neuroblastoma cells. Proc. Natl. Acad. Sci. USA 70:3099–3102.

Bondy, S. C., K. N. Prasad, and J. L. Purdy. 1974. Neuroblastoma: Drug-induced differentiation increases proportion of cytoplasmic RNA that contains polyadenylic acid. Science 186:359–361.

Breakefield, X. O., and M. W. Nirenberg. 1974. Selection for neuroblastoma cells that synthesize certain transmitters. Proc. Natl. Acad. Sci. USA 71:2530–2533.

Burdman, J. A. 1972. The relationship between DNA synthesis and the synthesis of nuclear proteins in rat brain during development. J. Neurochem. 19:1459–1469.

Culver, B., S. K. Sahu, A. Vernadakis, and K. N. Prasad. 1977. Effects of 5-(3,3,-dimethyl-1-triazeno)imidazole-4-carboxamide (NSC 45388, DTIC) on neuroblastoma cells in culture. Biochem. Biophys. Res. Commun. 76:778–783.

Ehrlich, Y. H., E. G. Brunngraber, P. K. Sinha, and K. N. Prasad. 1977. Specific alterations in phosphorylation of cytosol proteins from differentiating neuroblastoma cells grown in culture. Nature 265:238–241.

Elgin, S. C. R., J. B. Boyd, L. E. Hood, W. Wray, and F. C. Wu. 1973. A prologue to the study of the nonhistone chromosomal proteins. Cold Spring Harbor Symp. Quant. Biol. 38:821–833.

Fujitani, H., and V. Holoubek. 1974. Nonhistone nuclear proteins of rat brain. J. Neurochem. 23:1215–1224.

Gill, G. N., and L. D. Garren. 1970. A cyclic-3′,5′-adenosine monophosphate dependent protein kinase from the adrenal cortex: Comparison with a cyclic AMP binding protein. Biochem. Biophys. Res. Commun. 39:335–343.

Gilman, A., and M. W. Nirenberg. 1971. Regulation of adenosine 3′,5′-cyclic monophosphate metabolism in cultured neuroblastoma cells. Nature 234:356–357.

Hamprecht, B., and J. Schultz. 1973. Stimulation by prostaglandin E_1 of adenosine 3′,5′-cyclic monophosphate formation in neuroblastoma cells in the presence of phosphodiesterase inhibitors. FEBS Lett. 34:85–89.

Helström, J. G., K. E. Helström, G. E. Pierce, and A. H. Bill. 1968. Demonstration of cell bound and humoral immunity against neuroblastoma cells. Proc. Natl. Acad. Sci. USA 60:1231–1238.

Helson, L. 1975. Management of disseminated neuroblastoma. CA 25:264–268.

Helson, L. and J. L. Biedler. 1973. Catecholamines in neuroblastoma cells from human bone marrow, tissue culture and murine C-1300 tumor. Cancer 31:1087–1091.

Herschman, H. R., and M. P. Lerner. 1973. Production of a nervous system specific protein (14-3-2) by human neuroblastoma cells in culture. Nature New Biol. 241:242–244.

Kates, J. R., R. Winterton, and K. Schlesinger. 1971. Induction of acetylcholinesterase activity in mouse neuroblastoma tissue culture cells. Nature 229:345–346.

Kernell, A. M., L. Bolound, and N. R. Ringertz. 1971. Chromatin changes during erythropoiesis. Exp. Cell Res. 65:1–6.

Knapp, S., and A. J. Mandell. 1974. Serotonin biosynthetic capacity of mouse C-1300 neuroblastoma cells in culture. Brain Res. 66:547–551.

Kumar, S., G. Becker, and K. N. Prasad. 1975. Cyclic AMP phosphodiesterase activity in malignant and cyclic AMP-induced "differentiated" neuroblastoma cells. Cancer Res. 35:82–87.

Lazo, J. S., K. N. Prasad, and R. W. Ruddon. 1976. Synthesis and phosphorylation of chromatin associated proteins in cAMP-induced "differentiated" neuroblastoma cells in culture. Exp. Cell Res. 100:41–46.

Miller, R. A., and F. H. Ruddle. 1974. Enucleated neuroblastoma cells form neurites when treated with dibutyryl cyclic AMP. J. Cell Biol. 63:295–299.
Nelson, P., W. Ruffner, and M. Nirenberg. 1969. Neuronal tumor cell with excitable membrane grown *in vitro.* Proc. Natl. Acad. Sci. USA 64:1004–1010.
Prasad, K. N. 1972. Cyclic AMP-induced differentiated mouse neuroblastoma cells lose tumorgenic characteristics. Cytobios 6:163–166.
Prasad, K. N. 1973. Role of cyclic AMP in the differentiation of neuroblastoma cell culture, *in* The Role of Cyclic Nucleotides in Carcinogenesis, J. Schultz and H. G. Gratzner, eds., Academic Press, Inc., New York, pp. 207–237.
Prasad, K. N. 1974. Abnormal regulation of cyclic AMP phosphodiesterase: A hypothesis for the development of cancer of nerve cells. Differentiation 2:367–369.
Prasad, K. N. 1975. Differentiation of neuroblastoma cells in culture. Biol. Rev. 50:129–165.
Prasad, K. N. 1977. Differentiation and growth of neuroblastoma cells and serum types. Trans. Am. Soc. Neurochem. 8:87A.
Prasad, K. N., G. Becker, and K. Tirpathy. 1975a. Differences and similarities between guanosine 3′,5′-cyclic monophosphate phosphodiesterase and adenosine 3′,5′-cyclic monophosphate phosphodiesterase activities in neuroblastoma cells in culture. Proc. Soc. Exp. Biol. Med. 149:757–762.
Prasad, K. N., S. C. Bondy, and J. L. Purdy. 1975b. Polyadenylic acid containing cytoplasmic RNA increases in adenosine 3′ : 5′-cyclic monophosphate-induced "differentiated" neuroblastoma cells in culture. Exp. Cell Res. 94:88–94.
Prasad, K. N., D. Fogleman, M. Gaschler, P. K. Sinha, and J. L. Brown. 1976a. Cyclic nucleotide-dependent protein kinase activity in malignant and cyclic AMP-induced differentiated neuroblastoma cells in culture. Biochem. Biophys. Res. Commun. 68:1248–1255.
Prasad, K. N., and K. N. Gilmer. 1974. Demonstration of dopamine-sensitive adenylate cyclase in malignant neuroblastoma cells and change in sensitivity of adenylate cyclase to catecholamines in "differentiated" cells. Proc. Natl. Acad. Sci. USA 71:2525–2529.
Prasad, K. N., K. N. Gilmer, S. K. Sahu, and G. Becker. 1975c. Effect of neurotransmitters, guanosine triphosphate, and divalent ions on the regulation of adenylate cyclase activity in malignant and adenosine cyclic 3′,5′-monophosphate-induced "differentiated" neuroblastoma cells. Cancer Res. 35:77–88.
Prasad, K. N., and A. W. Hsie. 1971. Morphological differentiation of mouse neuroblastoma cells induced in vitro by dibutyryl adenosine 3′ : 5′-cyclic monophosphate. Nature New Biol. 233:141–142.
Prasad, K. N., and S. Kumar. 1974. Cyclic AMP and the differentiation of neuroblastoma cells, *in* Control of Proliferation in Animal Cells, B. Clarkson and R. Baserga, eds., Cold Spring Harbor Laboratory, Cold Spring Harbor, New York, pp. 581–594.
Prasad, K. N., and S. Kumar. 1975. Role of cyclic AMP in differentiation of human neuroblastoma cells in culture. Cancer 36:1338–1343.
Prasad, K. N., S. Kumar, G. Becker, and S. K. Sahu. 1975d. Role of cyclic nucleotides in differentiation of neuroblastoma cells in culture, *in* Cyclic Nucleotides in Diseases, B. Weiss, ed., University Park Press, Baltimore, pp. 45–66.
Prasad, K. N., B. Mandal, and S. Kumar. 1973a. Human neuroblastoma cell culture: Effect of 5-bromodeoxyuridine on morphological differentiation and levels of neural enzymes. Proc. Soc. Exp. Biol. Med. 144:38–42.
Prasad, K. N., B. Mandal, J. C. Waymire, G. J. Less, A. Vernadakis, and N. Weiner. 1973b. Basal levels of neurotransmitter synthesizing enzymes and effect of cyclic AMP agents on the morphological differentiation of isolated neuroblastoma clones. Nature New Biol. 241:117–119.
Prasad, K. N., S. K. Sahu, and S. Kumar. 1974. Relationship between cyclic AMP and differentiation of neuroblastoma cells in culture, *in* Differentiation and Control of Malignancy of Tumor Cells, W. Nakahara, T. Ono, T. Sugimura, and H. Sugano, eds., University of Tokyo Press, Tokyo, pp. 287–309.
Prasad, K. N., S. K. Sahu, and P. K. Sinha. 1976b. Cyclic nucleotides in the regulation of expression of differentiated functions in neuroblastoma cells. J. Natl. Cancer Inst. 57:619–631.
Prasad, K. N., and P. K. Sinha. 1977. Vitamin C increases the lethal effect of therapeutic agents on neural tumor cells in culture. Proc. Am. Assoc. Cancer Res. 18:95a.
Prasad, K. N., P. K. Sinha, S. K. Sahu, and J. L. Brown. 1975e. Binding of cyclic nucleotides with soluble proteins increases in "differentiated" neuroblastoma cells in culture. Biochem. Biophys. Res. Commun. 66:131–138.

Prasad, K. N., P. K. Sinha, S. K. Sahu, and J. L. Brown. 1976c. Binding of cyclic nucleotides with proteins in malignant and cyclic AMP-induced "differentiated" neuroblastoma cells in culture. Cancer Res. 36:2290–2296.

Prasad, K. N., J. C. Waymire, and N. Weiner. 1972. A further study on the morphology and biochemistry of X-ray and dibutyryl cyclic AMP-induced differentiated neuroblastoma cells in culture. Exp. Cell Res. 74:110–114.

Reimann, E. M., C. O. Brostrom, J. D. Corbin, C. A. King, and E. G. Krebs. 1971. Separation of regulatory and catalytic subunits of the cyclic 3′,5′-adenosine monophosphate-dependent protein kinase(s) of rabbit skeletal muscle. Biochem. Biophys. Res. Commun. 42:187–194.

Richelson, E. 1973. Stimulation of tyrosine hydroxylase activity in an adrenergic clone of mouse neuroblastoma by dibutyryl cyclic AMP. Nature New Biol. 242:175–177.

Ross, J. R., J. B. Olmsted, and J. L. Rosenbaum. 1975. The ultrastructure of mouse neuroblastoma cells in tissue culture. Tissue Cell 1:106–136. .

Sahu, S. K., and K. N. Prasad. 1975. Effect of neurotransmitters and prostaglandin E_1 on cyclic AMP levels in various clones of neuroblastoma cells in culture. J. Neurochem. 24:1267–1269.

Schubert, D., and F. Jacob. 1970. 5-bromodeoxyuridine-induced differentiation of a neuroblastoma. Proc. Natl. Acad. Sci. USA 67:247–254.

Schubert, D., S. Humphreys, F. Vitry, and F. Jacob. 1971. Induced differentiation of a neuroblastoma. Dev. Biol. 52:514–546.

Seeds, N. W., A. G. Gilman, T. Amano, and M. W. Nirenberg. 1970. Regulation of axon formation by clonal lines of neural tumor. Proc. Natl. Acad. Sci. USA 66:160–167.

Sinha, P. K., and K. N. Prasad. 1977. A further study on the regulation of cyclic nucleotide phosphodiesterase activity in neuroblastoma cells. Effect of growth. In Vitro (In press).

Tao, M., M. L. Salas, and F. Lipmann. 1970. Mechanism of activation by adenosine 3′ : 5′-cyclic monophosphate of a protein phosphokinase from rabbit reticulocytes. Proc. Natl. Acad. Sci. USA 67:408–414.

Watson, W. E. 1974. The binding of actinomycin D to the nuclei of exotomized neurons. Brain Res. 65:317–322.

Waymire, J. C., N. Weiner, and K. N. Prasad. 1972. Regulation of tyrosine hydroxylase activity in cultured mouse neuroblastoma cells. Elevations induced by analogs of adenosine 3′ : 5′-cyclic monophosphate. Proc. Natl. Acad. Sci. USA 69:2241–2245.

Stage of Cellular Differentiation in Tumor Cell Origin—Immunogenesis and Neoplasia

Cell Differentiation and Neoplasia, edited by Grady F. Saunders. Raven Press, New York

Plasmacytomagenesis and the Differentiation of Immunoglobulin-Producing Cells

Michael Potter and Michael Cancro

National Cancer Institute, Laboratory of Cell Biology, Bethesda, Maryland 20014; and Department of Pathology, University of Pennsylvania, Philadelphia, Pennsylvania 19104

The development of the immunoglobulin(Ig)-producing cell in the mammalian organism proceeds in several discrete stages (Figure 1). Although these stages may take place at different tissue sites or can be temporally separated, they nonetheless form a sequential program. The final maturation step is the formation of an Ig-secreting cell that has limited ability to divide. This multistage, vertical form of differentiation provides a background for the study of neoplastic development in a highly specialized cell. A brief summary of the major steps in this differentiation are given.

In mammals, stem cells of mesenchymal origin become differentiated for immunoglobulin production in the liver during fetal life and in the bone marrow during adult life (Owen *et al.* 1975, Melchers *et al.* 1975, Osmond and Nossal 1974a,b, Osmond *et al.,* 1975). The immediate precursor of the lymphoid cell is not known, but there is evidence for a common stem cell that serves both the hematopoietic and lymphoid systems (see Miller and Phillips 1975). There is also evidence that the entire immunoglobulin-producing potential may be regenerated from a single (CFU-S) cell (Trentin *et al.* 1967).

The key events that signal the beginning of immunoglobulin production are associated with determinative changes that involve immunoglobulin structural genes. The process is only understood by its end results and has not been actively studied. Immunoglobulin synthesis depends on the formation of two covalent DNA templates in each cell, one for the heavy (H) chain and one for the light (L) chain. Each chain, however, is controlled by two structural genes: V_L and C_L for L and V_H and C_H for H. Since genomes contain information for many V-genes and a single species-specific C-gene, the differentiation process is thought to depend on a mechanism that joins one V-gene to a corresponding C-gene. This process, which takes place on an autosomal chromosome, occurs on only one of the two homologues. Most Ig-producing cells are restricted to making Ig with only one V_L and one V_H (there are some exceptions [see Morse *et al.* 1976]). It is assumed that this restriction may be controlled in part by V-C joining. A notable exception to the single L- and H-chain template per cell is the evidence of simultaneous production of more than one H-chain per

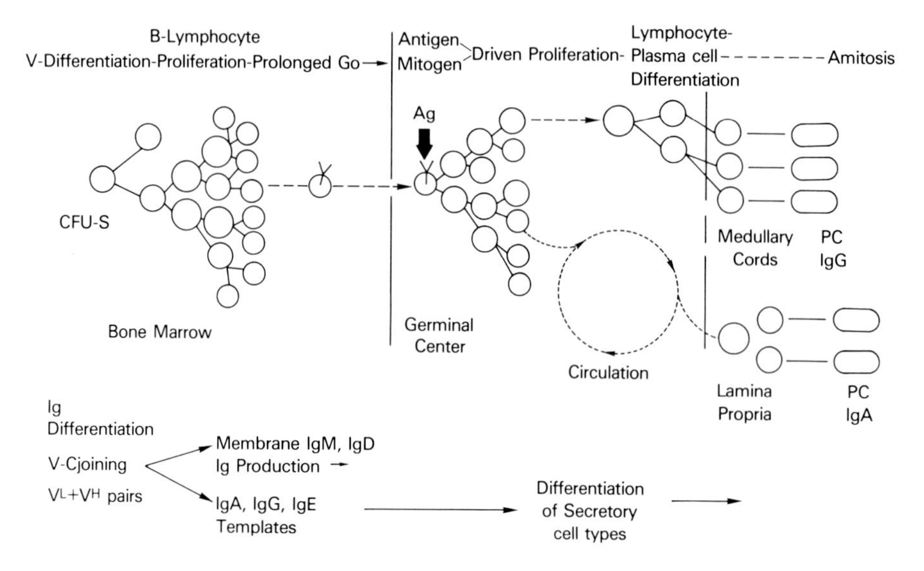

FIG. 1. Discrete stages in the development of the immunoglobulin-producing cell in mammals.

cell, e.g., the formation of $C\mu$-V_{Hx}, $C\gamma$-V_{Hx} heavy chains in one cell, is now well documented. It is thought that during differentiation of the H-chain genes, multiple templates are formed, and these are switched on and off by separate processes that modulate during maturation. In the early stages of Ig-producing cell development, membrane Ig (IgM, IgD) is produced, while in later secretory stages, IgA, IgG, and IgE are produced. During differentiation of Ig-forming cells, a variety of cell types are formed by activating a different V_L and V_H combination in each cell line. Little is known about this process, i.e., whether it is random or somehow restricted.

In the adult mouse, a large daily production of immunoglobulin-differentiated lymphocytes occurs (Osmond *et al.* 1975, Melchers *et al.* 1975), but it is not clear how much of this represents a continuous differentiation of the B-cell repertoire (i.e., clones expressing novel V_LV_H pairs), how much represents the renewal of established clones, and how much is contributed by memory cells (i.e., the progeny of cells stimulated by antigen). Regardless of the relative proportion of these components, bone marrow proliferation represents the generative phase in the development of the immunoglobulin-producing cell. Although there are changes in cell size and the density of immunoglobulin on the cell surface (Melchers *et al.* 1975), this phase is characterized by the differentiation of a cell that produces membrane-associated immunoglobulin (IgM or IgD) and culminates in a mature, virgin B-lymphocyte that is arrested in a prolonged G_0 stage of the cell cycle. Subsequently, B-lymphocytes leave the bone marrow, enter the circulation, and populate the lymphoreticular tissues.

The second stage of differentiation is associated with antigen interaction. Anatomically this takes place in the germinal centers of lymph nodes, spleen, and Peyer's patches. Interactions with antigenic and mitogenic factors trigger the transformation of the lymphocyte from a cell that produces membrane-associated IgM and/or IgD to a cell that secretes immunoglobulin of one of the major physiological classes (IgM, IgA, IgE, IgG). Antibody-secreting cells may also arise in the bone marrow after intravenous injection of antigen (Benner *et al.* 1974a,b, 1977), and lymphocytes from bone marrow can be stimulated in vitro to differentiate into Ig-secreting cells (Kearney and Lawton 1975a). Thus, under some conditions, B-lymphocytes can be directly differentiated into plasma cells in the generative site. The final phase of development is the maturation of the Ig-secreting cell to a presumably senescent, nondividing state. Ig-secreting cells are found in the medullary cords of lymph nodes and in the lamina propria of the gut. In the former, the mature cells presumably migrate from cortex to medulla, while in the latter location they probably contact antigen in Peyer's patches, enter the circulation, and "home" into the new final microenvironment (see Lamm 1976).

PLASMACYTOMAGENESIS

There are several systems in which the process of plasmacytomagenesis can be studied. Bazin *et al.* (1972, 1973) have described spontaneous ileocecal lymph node Ig-secreting tumors in the Lou/W inbred strain in the rat. A high proportion of these tumors secrete IgE. In man, multiple myeloma is a plasmacytomatous process that arises in the bone marrow. In the mouse, there are rare, spontaneous ileocecal plasmacytomas that arise in the lamina propria (Dunn 1957). Here we shall confine the discussion to a more abundant source—the induction of peritoneal plasmacytomas by the intraperitoneal injection of mineral oil and related substances in BALB/c and NZB mice (see Potter 1975). This model system was originally discovered by Merwin and Algire (1959), who found that BALB/c mice implanted intraperitoneally with Millipore diffusion chambers or other plastic materials (Merwin and Redmon 1963) developed fibrosarcomas and plasmacytomas. This led to the finding that other nonmetabolizable materials, such as mineral oils (Potter and Robertson 1960, Potter and Boyce 1962), also induced plasmacytomas in BALB/c mice. Most induction experiments have been done with mineral oils or the pure alkane, pristane (2,6,10,14-tetramethylpentadecane) (Anderson and Potter 1969), because of the ease with which they may be handled. Both the implantation of plastics and the injection of oils induces the formation of a chronic granulomatous tissue on the peritoneal surfaces (Merwin and Redmon 1963, Potter and MacCardle 1964, Cancro and Potter 1976), and the developing plasmacytoma is usually found in close association with this tissue. It was found that three 0.5 ml injections of oil given intraperitoneally at two-month intervals were more effective than a single dose.

Using the three i.p. dose procedure, several major biological features of the system were examined. A striking strain specificity of susceptibility to plasmacy-

TABLE 1. *Susceptibility of standard inbred strains and F_1 hybrids to peritoneal oil granuloma plasmacytomagenesis*

Strain	Total no. of PCT	%	Reference	F_1 hybrid	Total no. of PCT	%	Reference
BALB/c	219/373	58	a	C57BL x BALB/c	16/227	8	a
NZB	18/51	35	b	BALB/c x DBA/2	0/82	0	a
C57BL	3/103	3	a	BALB/c x NZB	79/163	49	b
DBA/2	0/36	0	a	BALB/c x SJL	2/53	4	c
SWR	0/30	0	a	BALB/c x NZC	0/56	0	b
A/He	0/26	0	a	BALB/c x C3H	1/28	3	b
AL/N	1/59	0	a				

References: a. Potter 1975 b. Warner 1975 c. McIntire and Princler 1969.

toma induction exists, and a summary of strains tested in different laboratories is given in Table 1. Two unrelated, inbred strains of mice are susceptible to plasmacytoma development—BALB/c and NZB (Potter *et al.* 1975, Warner 1975). Strain susceptibility suggests that specific genes, which act in the development of immunoglobulin-producing cells, are involved in plasmacytomagenesis. There are very few clues pointing to a similar inductive mechanism between BALB/c and NZB strains. While they have the same MHC haplotype (H-2^d), they have different Ig allotypes. They also differ greatly in expression of type C RNA viruses. The BALB/c strain carries several vertically transmissable C-type RNA viruses (Aaronson and Stephenson 1973), and these are usually activated when present in mice over six months of age. In contrast, NZB mice produce large quantities of a xenotropic virus throughout their lives (Levy and Pincus 1970) and also develop autoimmune disease (Mellors 1966). Several workers have noted that it is difficult to induce tolerance in these two strains of mice, for apparently different reasons (Staples and Talal 1969, Lukic *et al.* 1975). Despite the low incidence of plasmacytomas in most strains, all strains examined so far develop a peritoneal granuloma in response to the injection of the mineral oil.

In the following sections, evidence will be presented which suggests that the progenitors of plasmacytomas in the BALB/c system (1) are derived from a relatively specific subpopulation of B-lymphocytes that are generally in a comparatively late stage of differentiation, (2) probably originate as rarely occurring members of that subpopulation which lack or escape normal regulatory constraints with respect to their proliferative capacity, and (3) attain clonal sizes that emerge as detectable plasmacytomas through interactions with unusual, perhaps selective, microenvironmental influences enhancing the expansion and survival of these rarely occurring precursors.

RELATIONSHIP OF ANTIGENIC STIMULATION TO PLASMACYTOMAGENESIS

The evidence strongly suggests that the plasmacytoma cell arises sometime after the activation of the B-lymphocyte. This includes two major stages of development: (1) the interaction with antigen, and (2) the development of the secretory cell. The precise mechanism of B-lymphocyte activation is a subject currently under contention. The basic area of disagreement centers about the role of antigen or hapten itself and whether the interaction of antigen with immunoglobulin receptor plays a role in initiating the activation step. One school of thinking, championed by G. Möller, proposes that antigen by itself is not sufficient for activation but that mitogenic signals operating through nonimmunoglobulin receptors on the cell are responsible (Möller *et al.* 1976). The other view considers the interaction of antigen with membrane-bound immunoglobulin receptor to be the important, if not essential, factor in activation (Feldman *et al.* 1975). A key point relevant to this discussion is whether B-lymphocyte activation can occur without specific binding of antigen by cell surface immunoglobulin. Coutinho and Möller (1973) have presented considerable evidence that a number of antigens, in particular those that can directly activate B-lymphocytes without T-lymphocyte assistance, are in fact polyclonal, nonspecific, mitogenic activators for the B-lymphocyte. Opitz *et al.* (1976) have found that cocultivation of unstimulated spleen cells and high numbers of macrophages led to the development of antibody-secreting cells. Splenic B-lymphocytes can be activated in vitro with a high degree of efficiency by lipopolysaccharide (LPS) to become Ig-secreting cell types (Andersson *et al.* 1977). These cell types express different antigen-binding properties, indicating the polyclonal nature of the activation. Kearney and Lawton (1975a, b) have induced bone marrow lymphocytes isolated from adults with LPS to develop into IgG- and IgA-secreting cell types. Fetal liver and splenic lymphocytes have also been induced in vitro to become IgG- and IgA-secreting cell types, and IgA secretion could be induced to occur three weeks before its normal ontogenetic development. Although these studies were carried out in vitro, they do raise the possibility that B-cells in vivo may be nonspecifically triggered. If the entire B-lymphocyte population were vulnerable to transformation, then the isotypes of immunoglobulin expressed in plasmacytomas should reflect the immunoglobulin-producing potential of the B-lymphocyte pool before interaction with antigen, and the distribution of the heavy chain classes represented in the myeloma population and the antigen-binding properties of myeloma proteins should be stochastic. This, however, does not appear to be the case, and the evidence obtained with the system suggests that antigens have played a role in the natural history of the plasmacytoma cell, i.e., the B-lymphocyte precursors of plasmacytoma (PCT) have been antigen activated.

Preponderance of IgA Myeloma Proteins in BALB/c Mice

One indication of this characteristic in BALB/c myeloma proteins is the preponderance of the IgA isotype (Table 2). If one calculates the frequency of isotypes expressed in tumors that actually secrete a heavy chain or complete immunoglobulin (many plasmacytomas either do not secrete or make only light chains), the IgA heavy chain class in these tumors is over 60%. In six studies (see Morse *et al.* 1976), IgA is the predominant class. Further, among all of

TABLE 2. *Frequency of immunoglobulin classes expressed by plasmacytomas*

Immunoglobulin production	No. of Tumors Merwin (1963)	Cohn* (1967)	McIntire (1969)	Potter (1970b)	Grey (1971)	Current study†
γM	0	0	2	0	0	7
γA	20	53	29	66	120	338
γGs						
γ1(γF)	0	9	2	10	19	54
γ2a(γG)	3	3	4	7	10	34
γ2b(γH)	0	6	5	11	11	69
γ3(γJ606)	0	0	0	0	1	6
Free κ-chain	2	3	5	6	0	13
Free λ-chain	0	0	0	0	0	1
Polysecreting	0	2	0	0	0	54
N.P.D.‡	2	26	5	9	0	202
	27	118	52	111	161	778

* Several tumors in this series not typed.

† Among heavy chain classes listed. λ light chains associated with heavy chains in 19 tumors: γA(13), γ2a(1), γ2b(2), γ1-γA(2), and γ2b-γA(1). Four hundred fifty-seven of the tumors were studied retrospectively.

‡ No paraprotein detected.

(Reprinted, with permission of J. Immunol., from Morse *et al.* 1976.)

the myeloma systems, including those in man and rat, only in the BALB/c system is IgA the major heavy chain class expressed (Potter 1976).

IgA differentiation is associated with secretory immunoglobulin production in the lamina propria of the gastrointestinal, respiratory, mammary, and salivary glands (see Lamm 1976). From a developmental point of view, IgA-producing cells develop in a characteristic fashion. In the gastrointestinal tract, the area most frequently studied, the precursors of IgA-secreting cells are sensitized with antigen in the Peyer's patches or mesenteric nodes. The sensitized cells, however, leave the site, enter the circulation, and then migrate into the lamina propria tissues instead of developing directly in situ to IgA-secreting cells. This requires a homing mechanism. That such a process operates is clearly manifested in the distribution of IgA-secreting cells in the lamina propria tissues of the mouse and other species, where they comprise 80% of the normal plasma cell population (Crabbé *et al.* 1970).

It should be noted here that the peritoneum (and this should also include the oil granuloma) is not a known natural site for IgA-producing cells. Pierce and Gowans (1975), however, have implicated the peritoneal cavity in IgA antibody formation. In a study of the immunogenic properties of choleratoxoid given by different routes, they found that the most efficient means for establishing a high density of specific antibody-producing cells in the lamina propria of the gut was to sensitize the rats first with an i.p. injection of choleratoxoid in complete Freund's adjuvants (a mineral oil containing adjuvant) and secondarily boosting two weeks later with intraintestinal antigen. They also demonstrated that the cells which populated the lamina propria of the gut circulated in the thoracic duct lymph before entering the lamina propria. The implication of the work is that the peritoneum can indeed be involved in an immune response that produces IgA-differentiated cells. Prolonged oral immunization can also start this chain of events (Pierce and Gowans 1975).

Cebra *et al.* (1976) have demonstrated IgA-producing cells with the T15-idiotype in Peyer's patches. This is an idiotype found on: (1) a group of phosphorylcholine-binding myeloma proteins, (2) natural antibodies, and (3) induced antibodies with phosphorylcholine specificity (see Potter, in press). This observation suggests that activated B-lymphocytes in the Peyer's patches are already differentiated to make IgA. Possibly such cells enter the circulation and then the peritoneum, where they are stimulated to complete their maturation to plasma cells. It has been known for some time that a small population of large lymphocytes in the thoracic duct lymph can be readily labeled with thymidine. These same cells migrate into the lamina propria of the gastrointestinal tract, where they become plasma cells. Williams and Gowans (1975) have demonstrated that these cells contain cytoplasmic and large amounts of cell surface IgA. In the lamina propria of the gut, plasma cells producing IgA probably have a limited capacity to divide (Mattioli and Tomasi 1973). The usual half-life is short (four to seven days), but some cells can persist for many weeks.

Antigen-Binding Properties of Myeloma Proteins

A second source of information about the natural history of the plasmacytomas in mice is from antigen-binding activities of myeloma proteins. In the BALB/c mouse system, a relatively large number of antigen-binding myeloma proteins have been found by screening with relatively few antigens. Early screening studies (Cohn 1967, Potter 1970a, 1971) revealed that polyvalent antigens containing haptens such as phosphorylcholine, β1,6-linked D-galactose, α1,3- and α1,6-linked glucoses (dextrans), β2,1- and β2,6-linked fructoses (levans) were often precipitated by myeloma proteins (Table 3).

It is difficult to estimate with accuracy the total number of proteins from which these antigen-binding myeloma proteins have been screened, but a reasonable number is 1000. These proteins originated at either the Salk Institute or at the National Institutes of Health. The rough yield is about 4.6% (see Table 2). The immunochemistry of these proteins has recently been reviewed (Potter, in press). These antigens could usually be found in the immunological environment, i.e., the microbial flora, food, or bedding (Potter 1970a, 1971). For each of the above-listed antigens, more than one myeloma protein has been found. In addition, there are other myeloma proteins that bind antigens produced by other organisms in the BALB/c microbial flora (Potter 1970a, 1971). Many studies of both natural antibodies and antibodies induced by immunization with these antigens have revealed that they are structurally and antigenically related to myeloma proteins with the same binding properties (Potter *et al.* 1976, Potter, in press).

The conclusion that seems plausible at present is that the population which produced these antibodies is in part related to the plasmacytomas and that a relationship between the cells producing antibodies to these polyvalent antigens and the antigen-binding myeloma proteins (see Table 3) may exist. This is a complex comparison that requires a knowledge of the structure of the natural immunogen and the heavy chain class of the natural or induced antibodies. We have recently found that antibodies induced to gum ghatti, a β1,6 galactan containing antigen (Mushinski and Potter, in press) and *Salmonella milwaukee* flagellin (Smith *et al.*, in press) are IgA in class and share idiotypes with myeloma

TABLE 3. *Antigen-binding myeloma proteins of BALB/c mice**

Hapten	Natural antigen source	Number of Balb/c myeloma proteins with binding activity		
		IgM	IgA	IgG
Phosphorylcholine	Strep. pneumoniae teichoic acid Ascaris suum cuticle Lactobacillus Ag.† Trichoderma Ag.† Aspergillus Ag.† Proteus morganii lps.†	0	12	1
βl,6D-Galactan	Wood shavings† Larchwood arabinogalactan Gum ghatti Wheat germ antigen† Mammalian lung galactan Helix pomatia galactan	0	8	0
α 1,3 Dextran	Leuconostoc meserentoides dextran (Bl355S)	1	2	0
α 1,6 Dextran	Leuconostoc and other dextrans	0	2	0
β2,1 Fructan	Bacterial levan† Inulin	0	13	0
β2,6 Fructan	Bacterial levan† Grass levan	0	1	1
Unknown	Flagellin (Salmonella, Pasteurella†)	0	2	0
Unknown	Proteus mirabilis lps† Salm. telaviv lps Salm. tranoroa lps	0	2	0

* Data from the National Institutes of Health and the Salk Institute. See Potter 1977 for references.

† Isolated from mouse.

proteins of the corresponding binding specificity. This finding suggests that normal cells differentiated both for V-specificity and H-chain class exist in the BALB/c mouse and that these cells may be the precursors of the plasmacytoma cells.

THE ORIGIN AND EXPANSION OF THE PLASMACYTOMA PRECURSOR POPULATION

The foregoing discussion indicates that the progenitors of the plasmacytoma represent a subset of those B-lymphocytes which have been antigen stimulated and which are at a later stage of their differentiation. Additionally, those cells that express the IgA isotype appear to be particularly disposed to plasmacytomagenesis. A critical question then is this: Are all cells at this stage of maturation in the BALB/c mouse essentially preneoplastic cells, or does only a portion possess neoplastic potential? A key characteristic of plasmacytomagenesis relevant to this problem is the rate at which plasmacytomas appear after the first injection of oil. Plasmacytomas appear as early as 135 days after the first injection of oil, but the majority appear between 200 and 400 days and continue for as long as 600 days after the injection of oil. For reasons not understood, tumor appearance is accelerated in mice treated with pristane rather than mineral oil (Potter 1975). The protracted appearance of the tumors suggests that even though the cumulative index is quite high (~ 60%), the appearance of potentially neoplastic cells may be a relatively infrequent event. If this is the case, it might be expected that factors which amplify or enhance the growth and viability of rare, potentially neoplastic cells would be required for, or certainly potentiate, the development of overt neoplasia. Recent investigations strongly suggest that the oil-induced granuloma may play such a role in BALB/c plasmacytomagenesis.

The criticial role of the (oil) granuloma in plasmacytomagenesis was strongly suggested by the fact that plasmacytomas always arose in the peritoneum and remained localized there (Potter and MacCardle 1964). A more detailed concept on the role of the granulomatous tissue in plasmacytomagenesis has come from recent experiments. The background for this work began with transplantation studies on primary plasmacytomas. For many years, all plasmacytomas were transplanted as large chunks of tissue implanted subcutaneously using a 13-gauge trochar. The success rate with this method was good, but many tumors took a considerable time to appear (6 to 12 months). Since plasmacytomas arose in the peritoneum, and tumors could be diagnosed in the primary host by finding free tumor cells in the ascites, many attempts were made to transplant tumors by inoculating free tumor cells into the peritoneal cavities of normal mice. These attempts gave very poor results, and the method was not used by us. It was found, however, that if the peritoneum was conditioned by a single intraperitoneal injection of mineral oil or pristane 30 to 60 days before transplantation, the tumors grew in virtually all of the mice inoculated, even when very low cell doses were used (Potter *et al.* 1972) (Table 4). This suggested that the conditions in the peritoneal tissues stimulated by the injection of oil were conducive to plasmacytoma growth and that the early-developing plasmacytoma depended upon this type of environment for growth.

Prompted by the findings of Takakura *et al.* (1966), who found that chronic "cortisonization" inhibited plasmacytomagenesis, we studied the effect of hydrocortisone on the transplantability of primary plasmacytoma cells (Cancro and Potter 1976). Hydrocortisone has a profound action in suppressing the entry of blood monocytes into the peritoneal cavity (Thomson and Van Furth 1970). When continuous hydrocortisone treatment (0.5 mg/day) was begun simultaneously with the first injection of pristane, the peritoneal oil granuloma did not develop, nor was there an accumulation of free adherent cells in the peritoneum (Cancro and Potter 1976). If 10^5 plasmacytoma cells from a primary syngeneic host are transplanted into a mouse that has been given pristane and placed on hydrocortisone treatment three days before transplantation, the incidence of progressively growing tumors is reduced from 99% (in animals receiving pristane only) to 21% (Figure 2). By contrast, if hydrocortisone treatment is not begun until three days after the injection of pristane, then the tumors grow in 100% of the mice. Three days is sufficient time to allow the peritoneum to become populated with a large number of exudative cells and monocytes. Even in the subsequent presence of hydrocortisone, these cells are able to develop into macrophages, adhere to the peritoneum, and form a granuloma. Mice treated continuously with thioglycollate broth develop a large population of peritoneal exudate cells but no granuloma. Like normal mice, however, thioglycollate-treated mice are not able to support primary plasmacytoma growth (Cancro and Potter 1976). The presence of free macrophages in the peritoneal space is thus not sufficient to promote the growth of the primary plasmacytoma cells. These findings suggest that the presence of a granulomatous tissue composed largely of macrophages that have adhered to the peritoneal surface is an essential factor in the support of primary plasmacytoma cell growth. The dependence on the granuloma for growth does not persist long, and the tumors gain the

TABLE 4. *Transplantation of 10^5 plasmacytoma cells from a primary host to conditioned hosts*

Conditioning Day 0	Day 3	No. of tumors	No. of mice with progressively growing tumors	
None	None	23	5/212	3%
Thioglycollate*		6	0/60	0%
Pristane	None	11	227/228	99%
Pristane/Cortisone	None	11	32/150	21%
Pristane	Cortisone	11	150/150	100%

* Thioglycollate broth was given every three days for the duration of the experiment.
(Data of Cancro and Potter 1976)

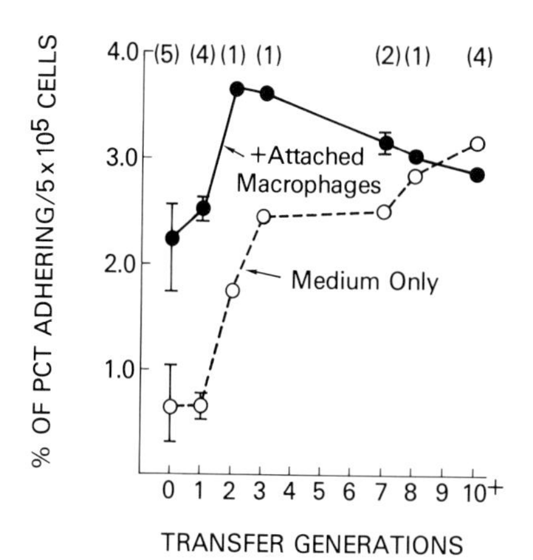

FIG. 2. In vitro adherence of plasmacytoma cells.

capacity to grow without a supporting oil granuloma after one or two transfer generations. It has been possible to show some aspects of this requirement for adherent cell layer in vitro. Two experiments are briefly described.

In the first type of experiment, the ability of primary plasmacytoma cells to adhere to a plastic surface was studied. Ascites containing both peritoneal exudate cells and tumor cells from a primary host were placed in plastic dishes, to remove adherent cells. Then 10^5 tumor cells were added to one of two types of plastic tissue culture dishes. Medium only was added to the control dishes, while the experimental dishes contained an adherent layer of peritoneal macrophages. After one hour of incubation, the free cells were decanted, and the plates were allowed to dry and were then stained. The tumor cells could be enumerated easily as intensely staining cells against the background of pale, flattened adherent cells. In primary and first transfer generation plasmacytomas, the percentage of cells that adhered in one hour ranged from 0.38 to 1.05%, while in the experimental dishes the range was 1.74 to 2.67% (a three to four times greater number of cells adhering) (see Figure 2). In contrast, later generation plasmacytomas (i.e., those that no longer required pristane conditioning for intraperitoneal growth) displayed equally efficient adherence in the absence of macrophages.

The second type of experiment explored a consequence of the process of interaction of plasmacytoma (PCT) cells with an adherent cell substratum. Here, 10^5 primary plasmacytoma cells were added to three different types of dishes (Figure 3, Table 5): (1) plastic tissue culture dishes containing medium only, (2) tissue culture dishes containing an adherent layer of irradiated macrophages or 3T3 cells, and (3) plastic bacteriological petri dishes (in which cells adhere inefficiently because the surfaces of these plates have not been treated appropriately) containing irradiated peritoneal macrophages or 3T3 cells. The plates were incubated for 16 hours after the addition of plasmacytoma cells, pulsed for four hours with ^{3}H-thymidine, and the incorporation of label into DNA

	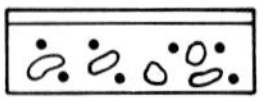 Petri Dish Suspended Adherent Cells	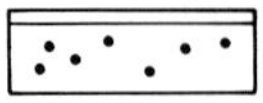 Tissue Culture Dish	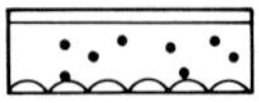 Tissue Culture Dish Attached Adherent Cells
	cpm x 10^3	cpm x 10^3	cpm x 10^3
PC73	33*	32*	71*
	39	30	82
PC65	27	26	58
	26*	25*	58
T830	18	20	92
PC105	18	29	88
Mean	26,800	30,200	74,800

* = 3T3; others = mϕ
Adherent cells irradiated 2000R

FIG. 3, TABLE 5. Effect of adherent cell layer on the incorporation of ^{3}H-thymidine into DNA of primary plasmacytoma cells.

was determined. As may be seen in Table 5, three times more incorporation was obtained when the cells were incubated with an adherent layer than in the other two types of dishes. These results indicate that the interaction of primary plasmacytoma cells with an adherent substratum apparently enhances DNA synthesis by the tumor cells, and the presence of macrophages or fibroblasts in an unattached state does not produce this enhancement. These experiments support the observations made in the in vivo transplantation studies and suggest that the presence of a substratum of attached macrophages permits the cells to divide more efficiently and may even sustain cells in mitosis. Additional experiments in this type of assay have shown that tumors whose in vivo growth is no longer dependent upon pristane conditioning exhibit equivalent levels of ^{3}H-thymidine incorporation in tissue culture dishes with or without macrophages.

In vitro studies of normal antibody formation using the Mishell-Dutton system have revealed a requirement for macrophages (Mishell and Dutton 1967, Mosier 1967) which can be replaced by the addition of mercaptoethanol (Click *et al.* 1972), an agent that in our studies also enhanced attachment to plastic tissue culture dishes and increased the incorporation of thymidine into DNA (though not to the degree observed with adherent cells) (Cancro and Potter, manuscript in preparation). Möller *et al.* (1976) have shown that adherent peritoneal exudate cells and fibroblasts (1) increase DNA synthesis in cultures of splenic B-lymphocytes from athymic nude mice, (2) are essential for the formation of antibody to the thymus-dependent antigen SRBC, and (3) stimulate polyclonal antibody synthesis by other B-lymphocytes in the culture. In their study they demonstrated that cell-free substances released by the macrophages could substitute for the macrophages or fibroblasts themselves. Desaymard and Ivanyi (1976) have found that macrophages in Marbrook-Diener culture systems stimulated polyclonal B-cell DNA synthesis and immunoglobulin production. These observations indicate normal B-lymphocytes that are being activated to become immunoglobulin-secreting cells also require the presence of adherent cells. The requirement for interaction with a suitable substratum of the developing plasmacytoma cell might reflect a normal mechanism associated with later stages of the activated B-lymphocyte. It is possible that the oil granuloma in vivo acts as a selecting factor by permitting cells with a requirement for adhesion to continue on in division. The peritoneal oil granuloma is an abnormal microenvironment. Probably more adherent macrophage sites are available than in a normal tissue. This situation may tend to detain cells in the adherent microenvironment when normally they would lose these associations. If these cells carry some sort of "mutation" associated with neoplastic properties (i.e., a "preplasmacytoma cell"), the permissive divisions on the granulomatous substratum could "promote" the development of the tumor. If during this process, C-type RNA viruses became activated or additional mutational events occurred, the regulation of growth could be further disturbed. Alternatively, it might be argued that the abnormal microenvironment encourages the cells to continue mitotic cycling when the normal process might be some form of programmed cell death. Cells with "mutations" that block the "turning-on" of the program leading to mitotic arrest could now selectively persist. There is suggestive evidence that some form of mutagenesis occurs in BALB/c plasmacytomagenesis. First, many plasmacytomas have abnormal karyotypes (Yosida *et al.* 1968, 1970). In a study of 14 primary plasmacytomas, Yosida *et al.* (1970) found that 11 had abnormal karyotypes, with median chromosome numbers in the near-tetraploid range; none were pure diploids. Of the three tumors with a preponderance of diploid cells, all developed abnormal karyotypes in early transfer generations. Moriwaki *et al.* (1971) transplanted one of these diploid tumors for a number of generations and found that it continuously produced new tetraploid lines.

It is imagined here that in plasmacytoma-susceptible strains of mice, germ-line genes cooperate in some way with intrinsic "mutational" events to raise the level of "mutants." The selecting ability of the oil granuloma now could permissively promote these "mutants" into fully neoplastic cell types. Thus, this model system resembles the two-stage model of epidermal carcinogenesis described by Berenblum (1975).

One of the predictions of the selection hypothesis is that the tumor cells might not only be differentiated to a similar degree but also have in common similar biochemical defects.

SUMMARY

1) The available evidence indicates that highly differentiated plasma cells late in maturation can be neoplastic. The time and site of origin of the neoplastic change cannot be identified with certainty but appears to involve late events

in the maturation of the Ig-producing cell. The inbred-strain susceptibility is a striking characteristic and implicates germ-line genes as playing a role in BALB/c plasmacytomagenesis. When these act or how they exert their effects has not been determined. At this stage, the genetic influence could be viewed as either *indirect,* i.e., not operating in the B-lymphocyte plasma cell lineage, or direct, i.e., actually governing a regulatory event in plasma cell development. The effect of the gene or genes may be only to increase the probability of neoplastic change and is not the sole determinant. Other mutational events (somatic mutations, C-type RNA virus production) may complement germ-line genes that govern susceptibility.

2) A powerful influence in plasmacytomagenesis is the promoting effect of the oil granuloma. Plasmacytomas in the original host appear to depend on this microenvironment for growth, in a manner reminiscent of the hormone-dependent tumors. The oil granuloma, then, may also act as a selecting factor that favors growth of potential or preplasmacytoma cells.

3) Plasma cells, like other cells of the immunohematopoietic tissues, develop in stages. The generative stages establish an available population for further maturation and development. In the generative stages of the immunoglobulin system of cells, immunoglobulin genes (V_L, V_H) that determine the binding specificity of an immunoglobulin are activated; usually only one V_L and one V_H gene from a large number of possible V-isotypes is selected in one cell. Thus, in the generative phase, the Ig-producing clonotypes are generated. The V_H gene can apparently be linked to several (possibly all) C_H genes (there are at least seven in the mouse). During the differentiation process, the type of C_H gene changes; early membrane Ig ($C\mu$, $C\gamma$) is produced. In secretory cells usually $C\gamma$, $C\gamma 2a$, $C\gamma 2b$, $C\gamma 3$, $C\alpha$, $C\epsilon$, or $C\mu$ is exclusively produced (some exceptions are known, e.g., the double producers).

The basic clues derived from plasmacytomas about their origin are provided by the myeloma proteins they produce—about 60% or more in BALB/c are IgA, and about 5% have thus far had antigen-binding activity for environmental (natural) antigens. These characteristics suggest plasmacytomagenesis occurs in a specific segment of the immunocyte population. The IgA-producing population is implicated here for the following reasons: (a) the IgA system of plasma cells is normally deployed in the lamina propria of gut and respiratory tracts; (b) antigen-sensitization is thought to occur in Peyer's patches, and the precursors of the lamina propria plasma cells later enter the circulation and then "home" to the lamina propria; (c) the peritoneum, although not a normal site for IgA production, could divert, detain and support circulating cells in transit from Peyer's patches to the lamina propria.

4) Despite the apparently advanced stage of maturation and differentiation in plasmacytoma cells, a basic question is still valid: Are these cells nonetheless arrested at a stage of development that is normally superceded by a mechanism that leads to cell death and elimination?

5) It is clear that the key to understanding neoplastic phenomenon of this

type is a thorough knowledge of the regulatory and developmental parameters which govern normal populations at each stage of differentiation. Through such knowledge, the aberrent elements active in neoplastic populations may be understood. It is, therefore, of utmost importance that the study of normal differentiation and regulation be intimately coupled with the examination of neoplasia.

REFERENCES

Aaronson, S. A., and J. R. Stephenson. 1973. Independent segregation of loci for activation of biologically distinguishable RNA C-type viruses in mouse cells. Proc. Natl. Acad. Sci. USA 70:2055–2058.

Anderson, P. N., and M. Potter. 1969. Induction of plasma cell tumors in BALB/c mice with 2,6,10,14-tetramethylpentadecane (pristane). Nature 222:994–995.

Andersson, J., A. Coutinho, W. Lernharft, and F. Melchers. 1977. Clonal growth and maturation to immunoglobulin secretion in vitro of every growth inducible B lymphocyte. Cell 10:27–34.

Bazin, H., A. Beckers, C. Deckers, and M. Moriamé. 1973. Transplantable immunoglobulin-secreting tumors in rats. V. Monoclonal immunoglobulins secreted by 250 ileocecal immunocytomas in Lou/WsI rats. J. Natl. Cancer Inst. 51:1359–1361.

Bazin, H., C. Deckers, A. Beckers, and and J. F. Heremans. 1972. Transplantable immunoglobulin-secreting tumours in rats. I. General features of Lou/WsI. Int. J. Cancer 10:568–580.

Benner, R., F. Meima, G. Van der Meulen, and W. B. Muiswinkel. 1974a. Antibody formation in mouse bone marrow. I. Evidence for the development of plaque forming cells in situ. Immunology 26:247–255.

Benner, R., F. Meima, and G. Van der Meulen. 1974b. Antibody formation in mouse bone marrow. II. Evidence for a memory dependent phenomenon. Cell. Immunol. 13:95–106.

Benner, R., A. Van Qudenaren, and H. de Ruiter. 1977. Antibody formation in mouse bone marrow. VII. Evidence against the migration of plaque-forming cells as the underlying cause for bone marrow plaque-forming cell activity. A study with parabiotic mice. Cell. Immunol. 29:28–36.

Berenblum, I. 1975. Sequential aspects of chemical carcinogenesis: Skin, *in* Cancer: A Comprehensive Treatise, F. F. Becker, ed., Vol. 1. Plenum Press, New York, pp. 323–351.

Cancro, M., and M. Potter. 1976. The requirement of an adherent substratum for the growth of developing plasmacytoma cells in vivo. J. Exp. Med. 144:1554–1566.

Cebra, J. J., P. J. Gearhart, R. Kamat, S. M. Robertson, and J. Tseng. 1976. Origin and differentiation of lymphocytes involved in the secretory IgA response. Cold Spring Harbor Symp. Quant. Biol. 41:201–215.

Click, R. E., L. Benck, and B. J. Alter. 1972. Immune response in vitro: Culture conditions for antibody synthesis. Cell. Immunol. 3:264–276.

Cohn, M. 1967. Natural history of the myeloma. Cold Spring Harbor Symp. Quant. Biol. 32:211–221.

Coutinho, A., and G. Möller. 1973. B cell mitogenic properties of thymus independent antigens. Nature New Biol. 245:11–15.

Crabbé, P. A., D. R. Nash, H. Bazin, H. Eyssen, and J. F. Heremans. 1970. Immunochemical observations on lymphoid tissues from conventional and germ free mice. Lab. Invest. 22:448–457.

Desaymard, C., and L. Ivanyi. 1976. Comparison of in vitro immunogenicity, tolerogenicity and mitogenicity of dinitrophenyl-levan conjugates with varying epitope density. Immunology 30:647–653.

Dunn, T. B. 1957. Plasma cell neoplasms beginning in the ileocecal area in strain C3H mice. J. Natl. Cancer Inst. 19:371–391.

Feldmann, M., J. G. Howard, and C. Desaymard. 1975. Role of antigen structure in the discrimination between tolerance and immunity by B-cells. Transplant. Rev. 23:78–97.

Grey, H. M., J. W. Hirst, and M. Cohn. 1971. A new mouse immunoglobulin IgG_3. J. Exp. Med. 133:289–304.

Kearney, J. F., and A. R. Lawton. 1975a. B lymphocyte differentiation induced by lipopolysaccharide. I. Generation of cells synthesizing four major immunoglobulin classes. J. Immunol. 115:671–676.

Kearney, J. F., and A. R. Lawton. 1975b. B lymphocyte differentiation induced by lipopolysaccharide. II. Response of fetal lymphocytes. J. Immunol. 115:677–687.
Lamm, M. E. 1976. Cellular aspects of immunoglobulin A. Adv. Immunol. 22:223–290.
Levy, J. A., and T. Pincus. 1970. Demonstration of biological activity of a murine leukemia virus of New Zealand black mice. Science 170:326–327.
Lukic, M. L., H. H. Wortis, and S. Leskowitz. 1975. A gene locus affecting tolerance to BCG in mice. Cell. Immunol. 15:457–463.
Mattioli, C. A., and T. B. Tomasi, Jr. 1973. The life span of IgA plasma cells from mouse intestine. J. Exp. Med. 138:452–460.
McIntire, K. R., and G. L. Princler. 1969. Prolonged adjuvant stimulation in germ free Balb/c mice: Development of plasma cell neoplasia. Immunology 17:481–487.
Melchers, F., R. E. Cone, H. von Boehmer, and J. Sprent. 1975. Immunoglobulin turnover in B-lymphocyte subpopulations. Eur. J. Immunol. 5:382–388.
Mellors, R. C. 1966. Autoimmune and immunoproliferative diseases of NZB/B1 mice and hybrids. Int. Rev. Exp. Pathol. 5:217–252.
Merwin, R. M., and G. H. Algire. 1959. Induction of plasma cell neoplasms and fibrosarcomas in BALB/c mice carrying diffusion chambers. Proc. Soc. Exp. Biol. Med. 101:437–439.
Merwin, R. M., and L. W. Redmon. 1963. Induction of plasma cell tumors and sarcomas in mice by diffusion chambers placed in the peritoneal cavity. J. Natl. Cancer Inst. 31:998–1007.
Metcalf, D. 1973. Colony formation in agar by murine plasmacytoma cells: Potentiation by hematopoietic cells and serum. J. Cell. Physiol. 81:397–410.
Metcalf, D. 1974. The serum factor stimulating colony formation in vitro by murine plasmacytoma cells: Response to antigens and mineral oil. J. Immunol. 113:235–243.
Miller, R. G., and D. Phillips. 1975. Development of B lymphocytes. Fed. Proc. 34:145–150.
Mishell, R., and R. W. Dutton. 1967. Immunization of dissociated spleen cultures from normal mice. J. Exp. Med. 126:423–442.
Mosier, D. E. 1967. A requirement for two cell types for antibody formation in vitro. Science 158:1573–1575.
Moriwaki, K., H. T. Imai, J. Yamashita, and T. H. Yosida. 1971. Ploidy fluctuations of mouse plasma-cell neoplasm MSPC-1 during serial transplantation. J. Natl. Cancer Inst. 47:623–637.
Möller, G., H. Lemke, and H. G. Opitz. 1976. The role of adherent cells in the immune response. Scand. J. Immunol. 5:269–280.
Morse, H. C., III, J. G. Pumphrey, M. Potter, and R. Asofsky. 1976. Murine plasma cells secreting more than one class of immunoglobulin heavy chain. I. Frequency of two or more M components in ascitic fluids from 788 primary plasmacytomas. J. Immunol. 117:541–547.
Mushinski, E. B., and M. Potter. 1977. Idiotypes on galactan binding myeloma proteins and anti-galactan antibodies in mice. J. Immunol. (In press).
Namba, Y., and M. Hanaoka. 1972. Immunocytology of cultured IgM forming cells of mouse. I. Requirement of phagocytic cell factor for the growth of IgM forming tumor cells in tissue culture. J. Immunol. 109:1193–1200.
Namba, Y., and M. Hanaoka. 1974. Immunocytology of IgM forming cells of mouse. II. Purification of phagocytic cell factor and its role in antibody formation. Cell Immunol. 12:74–84.
Opitz, H. G., U. Opitz, L. Lemke, R. Huget, and H. D. Flad. 1976. Polyclonal stimulation of lymphocytes by macrophages. Eur. J. Immunol. 6:457–461.
Osmond, D. G., R. G. Miller, and H. von Boehmer. 1975. Characterization of immunoglobulin-bearing and other small lymphocytes in mouse bone marrow by sedimentation and electrophoresis. J. Immunol. 114:1230–1236.
Osmond, D. G., and G. J. V. Nossal. 1974a. Differentiation of lymphocytes in mouse bone marrow. I. Quantitative radioautographic studies of antiglobulin binding by lymphocytes in bone marrow and lymphoid tissues. Cell. Immunol. 13:117–131.
Osmond, D. G., and G. J. V. Nossal. 1974b. Differentiation of lymphocytes in mouse bone marrow. II. Kinetics of maturation and renewal of antiglobulin-binding cells studied by double labeling. Cell. Immunol. 13:132–145.
Owen, J. J. T., M. C. Raff, and M. D. Cooper. 1975. Studies on the generation of B lymphocytes in the mouse. Eur. J. Immunol. 5:468–473.
Pierce, N. F., and J. L. Gowans. 1975. Cellular kinetics of the intestinal immune response to choleratoxoid in rats. J. Exp. Med. 142:1550–1563.
Potter, M. 1970a. Mouse IgA myeloma proteins that bind polysaccharide antigens of enterobacterial origin. Fed. Proc. 29:85–91.

Potter, M. 1970b. Myeloma proteins with antibody-like activity in mice, *in* Miami Winter Symposia. Vol. 2. North-Holland Publishing Co., Amsterdam, p. 397.
Potter, M. 1971. Antigen binding myeloma proteins in mice. Ann. N.Y. Acad. Sci. 190:306–321.
Potter, M. 1975. Pathogenesis of plasmacytomas in mice, *in* Cancer: A Comprehensive Treatise, F. F. Becker, ed., Vol. 1. Plenum Press, New York, pp. 161–182.
Potter, M. 1976. Tumors of immunoglobulin producing cells and thymus-derived lymphocytes, *in* 27th Mosbacher Colloquium, F. Melchers and K. Rajewsky, eds., Springer-Verlag, Berlin, pp. 141–172.
Potter, M. 1977. Antigen binding myeloma proteins of mice. Adv. Immunol. (In press).
Potter, M., and C. Boyce. 1962. Induction of plasma cell neoplasms in strain BALB/c mice with mineral oil adjuvants. Nature 193:1086–1087.
Potter, M., and R. C. MacCardle. 1964. Histology of developing plasma cell neoplasia induced by mineral oil in BALB/c mice. J. Natl. Cancer Inst. 33:497–515.
Potter, M., J. G. Pumphrey, and D. W. Bailey. 1975. Genetics of susceptibility to plasmacytoma induction. I. Balb/c AnN(C), C57 Bl/6N(B6), C57 Bl/Ka(BK), and $(C \times B6)F_1$, $(C \times BK)F_1$, and $C \times B$ recombinant-inbred strains. J. Natl. Cancer Inst. 54:1413–1417.
Potter, M., J. G. Pumphrey, and J. L. Walters. 1972. Growth of primary plasmacytomas in the mineral oil-conditioned peritoneal environment. J. Natl. Cancer Inst. 49:305–308.
Potter, M., and C. L. Robertson. 1960. Development of plasma cell neoplasms in BALB/c mice after intraperitoneal injection of paraffin oil adjuvant-heat killed staphylococcus mixtures. J. Natl. Cancer Inst. 25:847–861.
Potter, M., and J. L. Walters. 1973. Effects of intraperitoneal pristane on established immunity to the Adj-PC-5 plasmacytoma. J. Natl. Cancer Inst. 51:875–881.
Smith, A. M., J. Slack, and M. Potter. 1977. Restrictions in the immune response to flagellar proteins in inbred mice. Eur. J. Immunol. (In press).
Staples, P. J., and N. R. Talal. 1969. Relative inability to induce tolerance in adult NZB and NZB/NZW F_1 mice. J. Exp. Med. 129:123–139.
Takakura, K., B. Mason, and V. P. Hollander. 1966. Studies on the pathogenesis of plasma cell tumors. I. Effect of cortisol on the development of plasma cell tumors. Cancer Res. 26:596–599.
Thompson, J., and R. Van Furth. 1970. The effect of glucocorticosteroids on the kinetics of mononuclear phagocytes. J. Exp. Med. 131:429–442.
Trentin, J., N. Wolf, V. Cheng, W. Fahlberg, D. Weiss, and R. Bonhag. 1967. Antibody production by mice repopulated with limited numbers of clones of lymphoid cell precursors. J. Immunol. 98:1326–1337.
Warner, N. 1975. Autoimmunity and the pathogenesis of plasma cell tumor induction in NZB and hybrid mice. Immunogenetics 2:1–20.
Williams, A. F., and J. L. Gowans. 1975. The presence of IgA on the surface of the rat thoracic duct lymphocytes which contain internal IgA. J. Exp. Med. 141:335–345.
Yosida, T. H., H. J. Imai, and M. Potter. 1968. Chromosomal alteration and development of tumors. XIX. Chromosome constitution of tumor cells in 16 plasma cell neoplasms of BALB/c mice. J. Natl. Cancer Inst. 41:1083–1097.
Yosida, T. H., H. T. Imai, and K. Moriwaki. 1970. Chromosomal alteration and development of tumors. XXI. Cytogenetic studies of primary plasma-cell neoplasms induced in BALB/c mice. J. Natl. Cancer Inst. 45:411–418.

Cell Differentiation and Neoplasia, edited by Grady F. Saunders. Raven Press, New York
© 1978.

The Structure of H-2 Antigens and Their Role in Tumor Cell Recognition

B. A. Cunningham, R. J. Milner, J. W. Schrader, K. Reske, R. Henning, J. A. Ziffer, and G. M. Edelman

Department of Developmental and Molecular Biology, The Rockefeller University, New York, New York 10021

A variety of molecules are present on cell surfaces, and many of these may play direct roles in differentiation and the control of cell growth. Such molecules might serve to regulate critical metabolites, act as hormone receptors, direct specific cell-cell interactions, or control detection of cells by the immune system. Among the cell surface glycoproteins are a group of molecules characterized by their ability to influence the fate of tissue grafts between individuals of the same species. These glycoproteins are the major histocompatibility or transplantation antigens, designated HLA in man and H-2 in the mouse.

The rejection of tissue grafts in both man and mouse results in the generation of a special class of cytotoxic cells that are thymus-derived (T) lymphocytes (Eijsvoogel *et al.* 1972, Alter *et al.* 1973). These killer cells can destroy the cells of the graft and any cells similar to them. Although the grafted cells may differ from those of the recipient in many of their surface molecules, the key molecules recognized by the cytotoxic T-cells are the HLA or H-2 antigens.

Cytotoxic T-cells resembling those involved in allograft reaction are also involved in the destruction of neoplastic (Rouse *et al.* 1972) and virally infected cells (Doherty and Zinkernagel 1974). Moreover, recent studies indicate that these cytotoxic cells also recognize the major histocompatibility antigens (Zinkernagel and Doherty 1974). The apparent, special relationship between the major histocompatibility antigens and cytotoxic T-cells, therefore, could be of critical importance in providing a fundamental protective system against neoplastic cells, virally infected cells, or any abnormal cell. For this reason, we have begun to examine in detail the structure of the H-2 antigens of the mouse and their role in the lysis of tumor cells and virally infected cells by cytotoxic T-lymphocytes.

GENETICS OF THE H-2 SYSTEM

The H-2 antigens have been particularly well characterized in terms of their serology and genetics (Klein 1975), and these studies have raised a number of questions about the H-2 system and its relation to the immune system. The

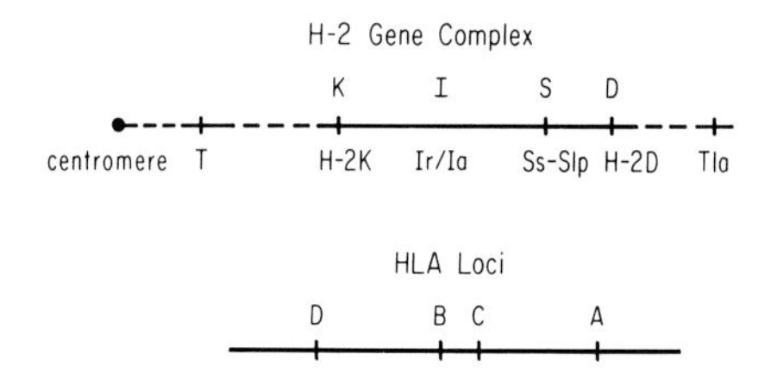

FIG. 1. The H-2 gene complex on chromosomes 17 of the mouse is bounded by the H-2K and H-2D loci and includes the I and S regions. The Tla and T loci are on the same chromosome but outside the H-2 complex. The H-2 complex resembles the HLA loci except that the HLA region (HLA-D) comparable to the H-2I region is outside the area bounded by the HLA-B and HLA-A loci.

genes for the H-2 antigens are located in two closely linked loci on the 17th chromosome of the mouse (Figure 1). Each of these loci, H-2K and H-2D, are probably single genes and are characterized by their extensive polymorphism. In addition, the products of these loci display two types of antigenic specificities: private specificities that are unique to each allele of each locus and public specificities that may be shared among alleles of either locus.

The H-2K and H-2D loci form the boundaries of a segment of the 17th chromosome called the H-2 gene complex. Between these loci are two other regions, I and S, which have also been examined in detail. The I region contains genes that determine the ability of an animal to respond to certain antigens. These genes are called immune response (Ir) genes. The products of the Ir genes are not known, but alloantisera have allowed the detection of certain glycoproteins on cell surfaces, the genes for which map in the I region. Current attention has been focused on establishing the relationship between the I region-associated (Ia) antigens and the Ir genes (see Katz and Benacerraf 1976).

The S region (Shreffler and Owen 1963) contains genes specifying a serum protein, Ss, and its sex-linked variant, Slp. Recent studies indicate that the Ss protein is the C4 component of complement (Lachman *et al.* 1975).

These data suggest that the H-2 complex contains a variety of genes that arose by duplication (Klein and Shreffler 1971). The products of these genes may have a special relationship to the immune system and may all serve a similar or related function. In addition, two regions, Tla and the t-complex (T) outside the H-2 complex but also on the 17th chromosome, may have similar properties. The Tla region codes for the thymus leukemia antigen (TL), which is found on the thymus cells of some strains of mice and on the leukemia cells of mice including those from strains that do not normally express it on their thymus cells (Boyse and Old 1969). The structure of the TL antigen resembles that of the H-2K and H-2D antigens (Ostberg *et al.* 1975, Vitetta *et al.* 1975), and its expression is reciprocally related to the expression of the H-2D antigen on cell surfaces.

The t-complex is also highly polymorphic (Bennett 1975). Genes in this locus appear to be important in embryonic development, and some mutants at this locus are lethal in homozygotes. Products of this locus are also expressed on teratocarcinoma cells.

STRUCTURE OF H-2 ANTIGENS

The genetics and serology of the H-2 gene complex and other loci on the 17th chromosome raise a number of questions about the products of these genes. For example: What is the basis of the polymorphism in the H-2 system and the nature of the public and private antigenic specificities? Are H-2K gene products evolutionarily related to H-2D gene products and to the products of other loci on the 17th chromosome? Are any of the molecules coded for by these genes related to antibodies? The answers to these questions and an understanding of the function of H-2 antigens depend on knowledge of the structure of H-2 antigens in solution and on the cell surface.

We have used a variety of techniques to devise a working model of H-2K and H-2D antigens (Henning *et al.* 1976a). This model is outlined in Figure 2. H-2 antigens on the cell contain two polypeptide chains, designated heavy chains and light chains. The two chains are held together only by noncovalent forces, and the H-2 molecule is firmly attached to the cell by interaction of the heavy chain with the plasma membrane. This interaction probably involves penetration of the lipid bilayer, and the H-2 heavy chain may extend into the interior of the cell.

H-2 molecules can be solubilized by disrupting the membranes with detergent or by treating the cell with the protease papain (Nathenson and Cullen 1974). When detergent is used, the polypeptide chains are left intact, but the molecule requires the continuous presence of detergent to stay in solution. In addition, on detergent extraction, at least one disulfide bond forms between two H-2 heavy chains giving a four-chain molecule (Henning *et al.* 1976a).

If cells or detergent-solubilized molecules are treated with papain, the enzyme cleaves peptide bonds in the heavy chain, giving a water soluble molecule (Fs) containing an intact light chain and a fragment (F_H) of the heavy chain. The F_H fragment appears to lack the portion of the molecule that interacts with the cell membrane and the half-cystinyl residues that form the disulfide bond in the detergent-solubilized protein.

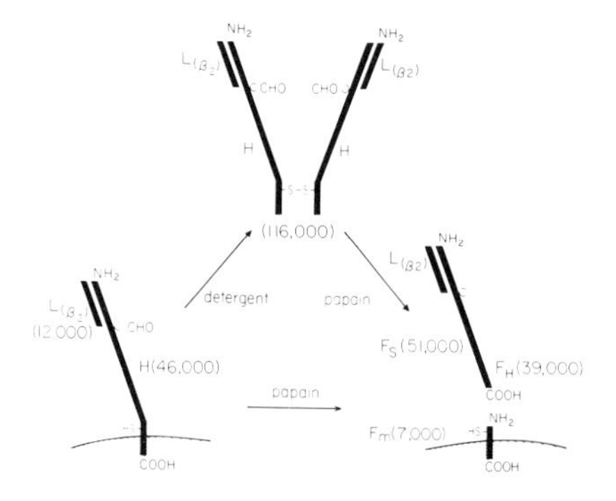

FIG. 2. Working model of the H-2 molecule on the cell surface and in solution. H-2 antigens were solubilized with nonionic detergents or the protease papain (Henning *et al.* 1976a). CHO designates carbohydrates.

	1				5					10					15					20
Human	Ile	Gln	Arg	Thr	Pro	Lys	Ile	Gln	Val	Tyr	Ser	Arg	His	Pro	Ala	Glu	Asn	Gly	Lys	Ser
Rabbit	Val	Gln	Arg	Ala	Pro	Asn	Val	Gln	Val	Tyr	Ser	Arg	His	Pro	Ala	Glu	Asn	Gly	Lys	Asp
Dog	Val	Gln	His	Pro	Pro	Lys	Ile	Gln	Val	Tyr	Ser	Arg	His	Pro	Ala	Glx	Asx	Gly	Lys	Pro
Mouse	Ile	Gln	Lys	Thr	Pro	Gln	Ile	Gln	Val	Tyr	Ser	Arg	His	Pro	Pro	Glu	Asn	Gly	Lys	Pro
G. Pig	Val	Leu	(His)	Ala	Pro	Glx	Val	Gln	Val	Tyr	—	—	His	(Pro)	Ala	Glu	Asn	(Gly)	—	—

	21				25					30					35
Human	Asn	Phe	Leu	Asn	Cys	Tyr	Val	Ser	Gly	Phe	His	Pro	Ser	Asp	Ile
Rabbit	Asn	Phe	Leu	Asn	Cys	Tyr	Val	Ser	Gly	Phe	His	Pro	Ser	Asp	Ile
Dog	Asx	Phe	Leu	Asx	Cys	Tyr	Val	Ser	Gly	Phe	His	Pro	—	Glx	Ile
Mouse	Asn	Ile	Leu	Asn	Cys	Tyr	Val	Thr	Glu	Phe	His	Pro	Pro	—	Ile

FIG. 3. Partial amino acid sequences of β_2-microglobulin from dog (Smithies and Poulik 1972), rabbit (Cunningham and Berggård 1975), mouse (Appella *et al.* 1976), and guinea pig (Berggård and Cunningham, in preparation) in comparison to the human protein (Cunningham *et al.* 1973). Underlined residues are those different from the human sequences.

The H-2 heavy chain is a glycoprotein with a molecular weight of about 45,000 (Nathenson and Cullen 1974). Comparison of the amino acid sequence of H-2 heavy chains and their F_H fragments (Henning *et al.* 1976a, Ewenstein *et al.* 1976) indicates that the amino terminus extends away from the cell, whereas the carboxyl terminal region of the polypeptide chain is more closely associated with the membrane. The heavy chains from different alleles of the H-2K and H-2D loci differ slightly in apparent molecular weight, but these differences are not correlated with the locus coding for the H-2 heavy chain.

The light chain (M.W. 12,000) is much smaller than the heavy chain. It contains no carbohydrate and is apparently identical in all H-2 molecules. This polypeptide chain has been shown to be identical to a urinary protein, β_2-microglobulin (Silver and Hood 1974, Rask *et al.* 1974), which we have characterized in a number of species (Cunningham 1976). The gene for β_2-microglobulin in man is not in the HLA-gene complex and is not even on the same chromosome (Goodfellow *et al.* 1975).

Many of the questions about H-2 antigens require detailed knowledge of the amino acid sequence of the polypeptide chains. The light chain has been well defined, and studies of this molecule have enhanced speculation that the H-2 antigens may be closely related to antibodies.

β_2-Microglobulin

Since its discovery in human urine in 1969 (Berggård and Bearn 1968), β_2-microglobulin has been identified in tissues and fluids from a number of species, including the mouse. Partial amino acid sequences have been determined for several of these proteins, and all are similar (Figure 3). Most of the variations

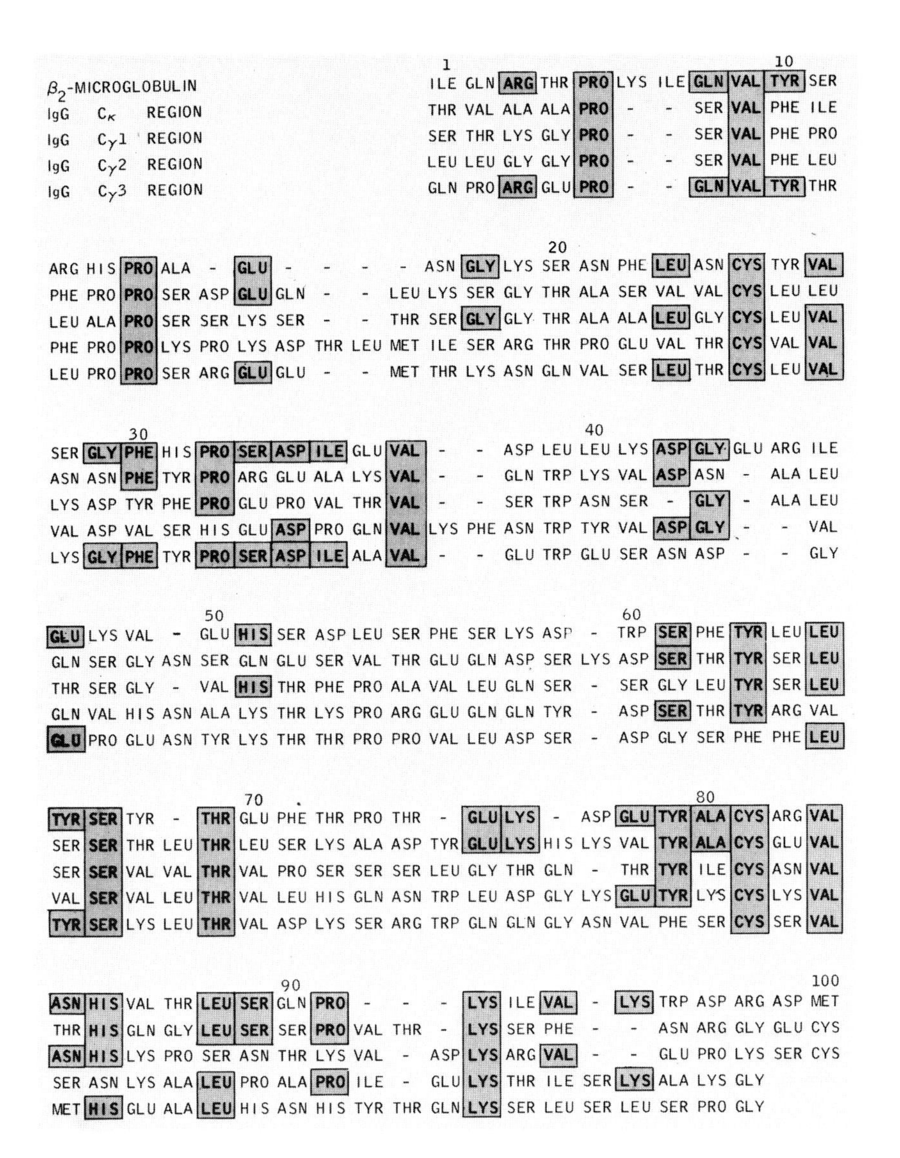

FIG. 4. Similarity of β_2-microglobulin to the constant homology regions (Edelman *et al.* 1969) of the light (Cκ) and heavy ($C\gamma_1$, $C\gamma_2$, and $C\gamma_3$) chains of the human immunoglobulin G_1 Eu (Peterson *et al.* 1972).

are clustered near the amino terminus where species-specific residues may appear. Otherwise, the sequences and amino acid compositions indicate that this is a highly conserved protein. More important, no variant has been detected within a species by serological techniques or by amino acid sequence analysis.

The human protein has been characterized in detail, including the determination of its complete amino acid sequence (Cunningham *et al.* 1973). Its most striking feature is its similarity with the constant homology regions of all immu-

		1				5					10					15
H-2K^b	NH_2	X	Pro	His	—	Leu	Arg	Tyr	Phe	Val	(Thr)	Ala	Val	—	Arg	Pro
K^d	NH_2	Met	X	His	—	X	Arg	Tyr	X	X	(Thr)	—	X	—	Arg	Pro
K^k	NH_2	Met	Pro	His	—	Leu	Arg	Tyr	Phe	His	—	Ala	Val	—	Ile	Pro
D^b	NH_2	—	Pro	—	—	—	Arg	Tyr	—	—	—	Ala	Val	—	Arg	Pro
D^d	NH_2	Met	X	His	—	Leu	Arg	Tyr	Phe	Val	(Thr)	Ala	Val	(Thr)	Arg	Pro
HLA-A2	NH_2	Gly	Ser	—	Ser	Met	Arg	Tyr	Phe	Phe	Thr	Ser	Val	Ser	Arg	Pro
A1,2	NH_2	—	Ser	—	Ser	Met	Arg	Tyr	Phe	Phe	Thr	Ser	Val	Ala	Arg	Pro
B7	NH_2	Gly	Ser	—	Ser	Met	Arg	Tyr	Phe	Tyr	Thr	Ser	Val	Ser	Arg	Pro
B8,13	NH_2	—	Ser	—	Ser	Met	Arg	Tyr	Tyr	Tyr	Ser	Ala	Val	Ser	Arg	Pro

	16				20					25		
H-2K^b	—	Leu	—	—	(Pro)	Arg	Tyr	—	—	—	(Leu)	Tyr
K^d	—	X	—	—	(Pro)	Arg	(Phe)	—	—	—	—	Tyr
K^k	—	Leu	—	Lys	Pro	Phe	Ala	—	—	—	—	Tyr
D^b	—	Leu	—	—	Pro	Arg	Tyr	—	—	—	—	Tyr
D^d	—	Phe	—	—	Pro	Arg	Tyr	—	—	—	—	Tyr
HLA-A2	Gly	—	Gly	Glu	—	—	Phe	Ile	Ala	Val	—	—
A1,2	Gly	—	—	—	—	—	—	—	—	—	—	—
B7	Gly	—	Gly	Glu	—	—	Phe	Ile	Ala	Val	—	—
B8,13	Gly	—	—	—	—	—	—	—	—	—	—	—

FIG. 5. Partial amino acid sequences of the heavy chains of H-2 antigens as determined by the radiochemical approach (Henning *et al.* 1976a, Cunningham *et al.,* in press, Ewenstein *et al.* 1976, Silver and Hood 1976, and Vitetta *et al.* 1976) and of HLA antigens as determined without radioactive labels (Terhorst *et al.* 1976) or with external radioactive label (Bridgen *et al.* 1976). An X denotes positions where residues seen in other H-2 sequences are not present; residues differing at positions within the H-2 and HLA sequences are underlined.

noglobulins (Figure 4) (Cunningham 1976). The homology between the amino acid sequence of β_2-microglobulin and these regions of immunoglobulin leaves little doubt that they share a common evolutionary origin. A key question has been whether this relationship extends to the H-2 heavy chain.

H-2 Heavy Chains

Analysis of the amino acid sequences of H-2 heavy chains has been limited because of the small amounts of material that can be isolated from cells. Recent development of microchemical techniques, however, has allowed sufficient sequence information to be obtained to provide some picture of these molecules.

The approach has been to grow cells in the presence of one or more radioactive amino acids and to isolate the labeled H-2 antigens by specific immune precipitation. The labeled H-2 heavy chain is then separated from the other components in the precipitate by preparative polyacrylamide gel electrophoresis in sodium dodecyl sulfate (SDS). The positions where labeled residues appear in the chain are then determined using the automatic sequencer. This approach has been used in a variety of laboratories (Henning *et al.* 1976a, Silver and Hood 1976, Vitetta *et al.* 1976, Ewenstein *et al.* 1976) and has given the sequences summarized in Figure 5.

All of the sequences are similar and products of H-2K alleles are not distinguished from products of H-2D alleles. These results support the hypothesis that at least these two regions arose by duplication of an ancestral gene. There are also distinct differences in the sequences that undoubtedly reflect the polymorphism but that cannot as yet be correlated with public or private specificities. The data do not answer the question of whether there is an evolutionary relationship between H-2 heavy chains and immunoglobulins. The sequences determined so far show no convincing homology; but they represent only a small percentage of the total structure, and the amino terminal portion of the chain could be less highly conserved than other regions.

The H-2 sequences are similar to sequences of HLA antigens determined by other techniques (Terhorst *et al.* 1976, Bridgen *et al.* 1976). This observation supports the previous conclusion that the major histocompatibility systems in the two species are similar and probably serve the same function.

Although the structural studies require considerable extension, they are beginning to set the framework and limitation for a consideration of H-2 function.

Role of H-2 Antigens in Cell Lysis

A variety of data indicate that H-2 antigens are intimately involved in the lysis of virally infected (Zinkernagel and Doherty 1974, Koszinowski and Ertl 1975) and chemically modified cells (Shearer *et al.* 1975) by cytotoxic T-cells. For example, cytotoxic T-cells generated against virally infected cells will lyse target cells infected with the same virus only if the target and stimulating cells share at least one H-2 antigen.

We (Schrader *et al.* 1975) and others (Germain *et al.* 1975) have found evidence for a similar role for H-2 antigens in the lysis of tumor cells by H-2 identical or syngeneic cytotoxic cells. The lysis of the target tumor cells is inhibited by antibodies to the H-2 antigens present on all of the cells (Figure 6). Although both the killer cells and the target cells have the same H-2 antigens, the H-2 antigen on the target cell appears to be particularly important (Schrader and Edelman 1976).

All of the data have led to a number of models for how H-2 antigens may be involved in cell-mediated lysis and suggest why a special relationship between the receptor on cytotoxic cells and H-2 antigens on target cells may be critical in allograft reactions and the lysis of virally infected cells and tumor cells. Some of these models are outlined in Figure 7.

Basically, the system is viewed in one of two ways: (1) either the cytotoxic

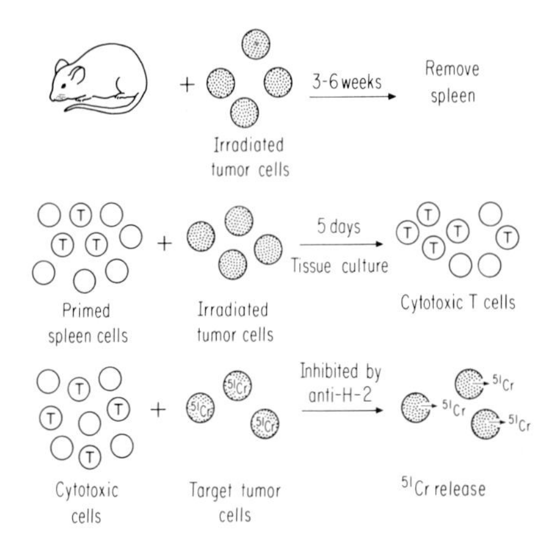

FIG. 6. Generation of cytotoxic T-cells against H-2 identical or syngeneic tumor cells includes primary stimulation in vivo and secondary stimulation in tissue culture. Assay is carried out on ^{51}Cr-labeled tumor cells. Cell lysis is blocked by antibodies to H-2 antigens present on both the cytotoxic cells and the target cells.

T-cell has a single receptor, and the H-2 molecule is part of or directly affects a single component on the target cell; or (2) the cytotoxic cell has two separate receptors that recognize the target antigen and the H-2 molecule independently. One-receptor models include consideration that H-2 and foreign (viral or tumor) antigens may form hybrid structures (Figure 7A) or that H-2 molecules may be modified by, or themselves modify, viral components (Figure 7B). A variant of the two-receptor model (Figure 7D) suggests that the H-2 and foreign antigens may be linked by a third molecule at or under the cell surface.

A variety of attempts are now underway to discriminate between these possibilities. One test, suggested to us by the notion of hybrid molecules (Figure 7A), was to look for close association of H-2 and viral antigens on the surfaces of tumor cells. Our first studies took advantage of the fact that molecules can be

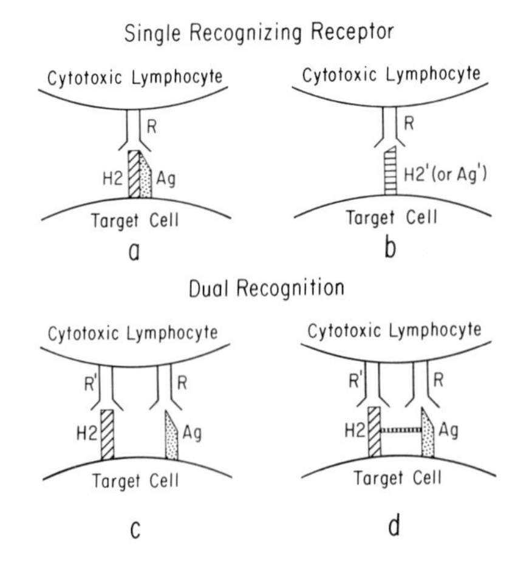

FIG. 7. Models for how cytotoxic T-cells might recognize both H-2 antigens and tumor (or viral) antigens on target cells. With one receptor, the T-cell could recognize a hybrid of the two antigens (a), or a modified antigen (b). Two-receptor or dual recognition could include totally independent target antigens (c) or antigens linked by a third molecule on or under the cell surface (d).

redistributed on the cell surface using divalent ligands such as antibodies to specific cell surface receptors.

Antibodies to molecules such as H-2 antigens cause redistribution of the receptors first into patches and then into caps, both of which can be seen by using fluorescent antibodies (Taylor *et al.* 1971). If one cell surface molecule co-caps with another, the two cell surface molecules are assumed to be associated on the cell. Such approaches have been used to show that H-2K antigens are not associated with H-2D antigens (Neauport-Sautes *et al.* 1973) but that β_2-microglobulin is associated with HLA antigens (Neauport-Sautes *et al.* 1974).

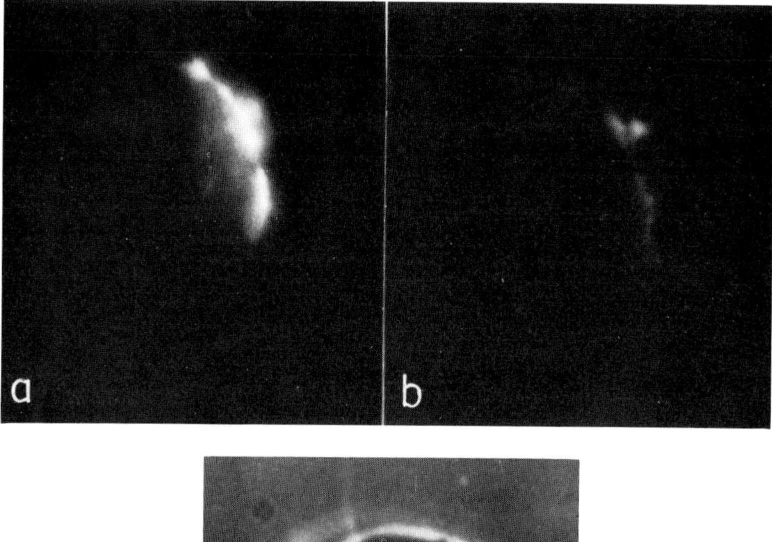

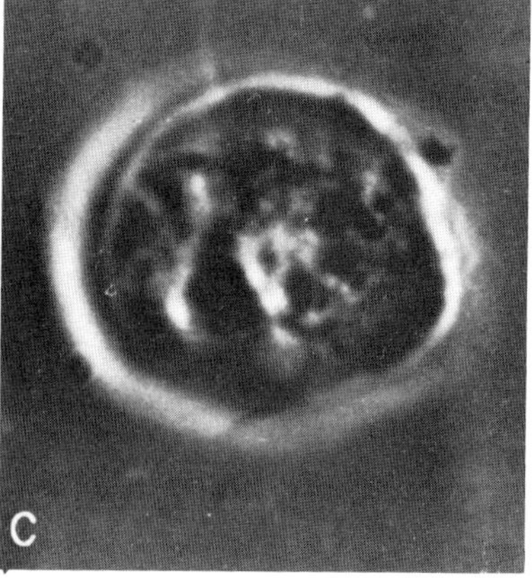

FIG. 8. Viral components on tumor cell surfaces are redistributed (co-capped) with H-2 antigens when cells are treated with antibodies to H-2 antigens (Schrader *et al.* 1975). (a) EL4 tumor cells showing capped H-2 antigens as detected by a fluorescein label; (b) the same cell showing that viral antigens, detected with a tetramethyl rhodamine label, are in the same position as the capped H-2 antigens; (c) the same cell under phase contrast.

We used antisera to H-2K antigens and antisera to Rauscher leukemia virus or its purified protein gp70 to look for co-capping of H-2 and viral antigens on the tumor cell EL4 (H-2^b) (Schrader *et al.* 1975). If H-2 antigens were first capped and detected with fluorescein-labeled antibodies and then fixed so that no further capping could take place, the viral components were detected with rhodamine-labeled antiviral antibodies in the same cap (Figure 8). These results suggest that H-2 and viral components are closely associated on the cell surface. We have recently carried out other experiments using these antibodies to block complement-mediated lysis (Henning *et al.* 1976b). These studies also suggest that H-2 and viral components are closely associated on the cell.

Definitive proof of such interactions would argue strongly for one-receptor models, although a two-receptor model in which the receptors are linked by a third molecule (Figure 7D) could not be excluded. Distinction between these models is difficult and if the distance between receptor molecules is small, the two models are essentially the same.

We have recently begun studies of the more general question of whether viral replication or the synthesis of new proteins is required for cell-mediated lysis to occur or for H-2 antigens to be restrictive (Schrader and Edelman 1977). In these experiments, we attempted to generate cytotoxic T-cells against cells coated with UV-inactivated Sendai virus. Such cytotoxic cells were produced, and both a primary response in vivo and a secondary response in vitro could be demonstrated. The cytotoxic activity was destroyed by anti-Thy 1.2 serum plus complement, indicating that the effector cells were T-cells. More

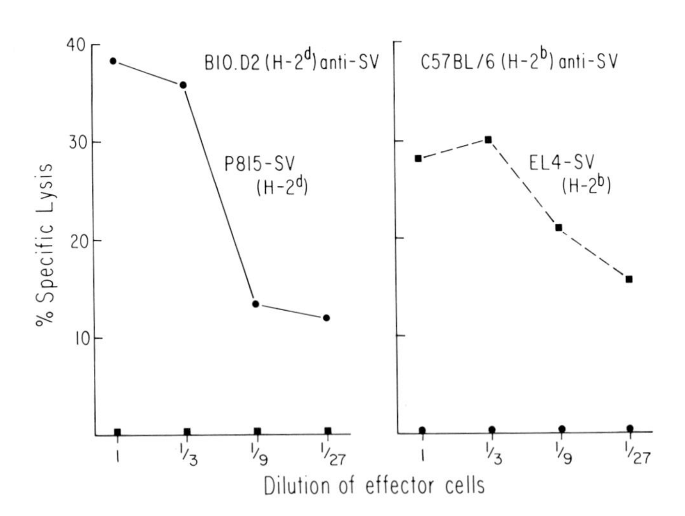

FIG. 9. The cell-mediated lysis of tumor cells coated with inactivated Sendai virus shows H-2 restriction (Schrader and Edelman 1977): B10.D2 (H-2^d) cytotoxic cells lyse virus-coated H-2^d (P815) (closed circles) but not virus-coated H-2^b (EL4) (closed squares) targets; conversely, C57BL/6 (H-2^b) cytotoxic cells lysed virus-coated H-2^b (EL4) but not virus-coated H-2^d (P815) targets.

important, the lysis of target cells showed H-2 restriction (Figure 9). H-2^b cytotoxic cells lysed H-2^b tumor cells coated with Sendai virus but did not lyse H-2^d cells coated with the same virus. Conversely, H-2^d cytotoxic cells lysed H-2^d but not H-2^b target cells.

To simplify interpretation of these experiments, it is important to consider whether or not synthesis of proteins coded for by the viral genome was needed. Kinetic studies indicated that a significant amount of lysis takes place within 30 minutes after exposing the target to inactivated virus. Sendai virus components are usually not produced even in cells infected with active virus until about 5 hours after infection (Lamb *et al.* 1976).

To test more rigorously for newly synthesized viral or cellular components, normal cells were labeled with radioactive amino acids and extracts of these cells compared with extracts of cells treated with inactive virus or active virus and similarly labeled (Figure 10). Cells treated with active virus gave four new components not seen in normal cells. The molecular weights of these components corresponded to the proteins known to be in Sendai virus (Lamb *et al.* 1976). The cells treated with inactive virus, however, did not synthesize these proteins to any significant extent, and the extracts from these cells did not differ from that of normal cells. These and other experiments suggest that the synthesis of new proteins is not required to generate the targets of the cytotoxic T-cell.

We currently favor the model (Figure 7A) suggesting that pre-existing H-2 and foreign antigens can form complexes on the cell surface and that these complexes are preferentially recognized by a single receptor on cytotoxic T-cells. Such complexes could present hybrid determinants to the T-cell, providing a basis for the requirement that both H-2 and foreign antigens be recognized. Animals heterozygous for H-2 antigens could be at an advantage, as they could provide more diversity for selection from the repertoire of available T-cell receptors. In an allogeneic response, the foreign antigens would simply be the H-2 antigens themselves.

This model requires more substantial evidence that complexes can be preferentially formed between H-2 and foreign antigens on the cell. In addition, we must provide evidence that such complexes generate hybrid determinants. If the allogenic system is comparable to the lysis of viral and tumor cells, such determinants might have an intimate relationship between the public and private specificities of the H-2 antigens. Alternatively, regions responsible for these interactions may provide the basis for some specificity in interacting with viruses of tumor antigens. Further consideration of this or any other model also depends on extending our knowledge of the T-cell receptor itself.

Regardless of which model is correct, current thinking suggests that the major histocompatibility antigens may function in normal immune mechanisms that protect the body against viral infection and tumorigenesis. A more precise description of these molecules and this mechanism could have far-reaching implications for our ability to understand and control a variety of diseases, particularly cancer.

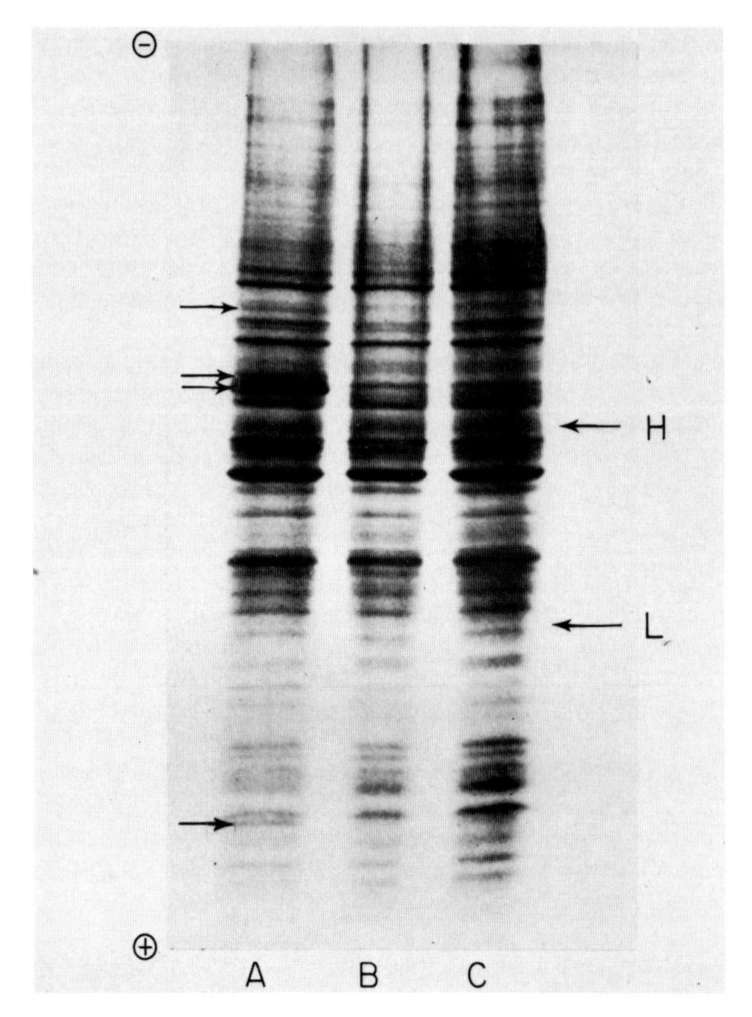

FIG. 10. Extracts of cells infected with inactive Sendai virus (A) and labeled with ^{35}S-methionine or ^{3}H-leucine show four proteins (arrows, left) not present in cells treated with inactive virus (B) as detected on SDS gel electrophoresis (Schrader and Edelman 1977). Extracts of cells treated with inactive virus are essentially identical to extracts of normal cells (C). H, L (arrows, right) denote positions of immunoglobulin light and heavy chain markers. Bands were detected by autoradiography.

ACKNOWLEDGMENTS

The work discussed here was supported by U.S. Public Health Service Grants AI-11378, AI-02973, and AM-04256 from the National Institutes of Health, and Grant RF-70095 from the Rockefeller Foundation. BAC is the recipient of an Irma T. Hirschl Career Scientist Award, and JWS was a Fellow of the Damon Runyon–Walter Winchell Cancer Fund.

REFERENCES

Alter, B. J., D. J. Schendel, M. L. Bach, F. H. Bach, J. Klein, and J. H. Stimpfling. 1973. Cell-mediated lympholysis. Importance of serologically defined H-2 regions. J. Exp. Med. 137:1303–1309.

Appella, E., N. Tanigaki, T. Natori, and D. Pressman. 1976. Partial amino acid sequence of mouse β_2-microglobulin. Biochem. Biophys. Res. Commun. 70:425–430.

Bennett, D. 1975. The T-locus of the mouse. Cell 6:441–454.

Berggård, I., and A. G. Bearn. 1968. Isolation and properties of a low molecular weight β_2-globulin occurring in human biological fluids. J. Biol. Chem. 243:4095–4103.

Boyse, E. A., and L. J. Old. 1969. Some aspects of normal and abnormal cell surface genetics. Ann. Rev. Genet. 3:269–290.

Bridgen, J., D. Snary, M. J. Crumpton, C. Barnstable, P. Goodfellow, and W. F. Bodmer. 1976. Isolation and N-terminal amino acid sequence of membrane-bound human HLA-A and HLA-B antigens. Nature 261:200–205.

Cunningham, B. A. 1976. Structure and significance of β_2-microglobulin. Fed. Proc. 35:1171–1176.

Cunningham, B. A., and I. Berggård. 1975. Partial amino acid sequence of rabbit β_2-microglobulin. Science 187:1079–1080.

Cunningham, B. A., R. Henning, R. J. Milner, K. Reske, J. A. Ziffer, and G. M. Edelman. 1976. Structure of murine histocompatibility antigens. Cold Spring Harbor Symp. Quant. Biol. 41:351–362.

Cunningham, B. A., J. L. Wang, I. Berggård, and P. A. Peterson. 1973. The complete amino acid sequence of β_2-microglobulin. Biochemistry 12:4811–4822.

Doherty, P. C., and R. M. Zinkernagel. 1974. T cell-mediated immunopathology in viral infections. Transplant. Rev. 19:89–120.

Edelman, G. M., B. A. Cunningham, W. E. Gall, P. D. Gottlieb, U. Rutishauser, and M. J. Waxdal. 1969. The covalent structure of an entire γG immunoglobulin molecule. Proc. Natl. Acad. Sci. USA 63:78–85.

Eijsvoogel, V. P., M. J. G. J. duBois, C. J. M. Melief, M. L. deGroot-Kooy, C. Koning, J. J. van Rood, A. van Leeuwen, E. deToit, and P. T. A. Schellekens. 1972. Position of a locus determining mixed lymphocyte reaction (MLR), distinct from the known HL-A loci, and its relation to cell-mediated lympholysis (CML), *in* Histocompatability Testing, 1972, J. Dausset and J. Colombani, eds., Munksgaard, Copenhagen, pp. 501–508.

Ewenstein, B. M., J. H. Freed, L. E. Mole, and S. G. Nathenson. 1976. Localization of the papain cleavage site of H-2 glycoproteins. Proc. Natl. Acad. Sci. USA 73:915–918.

Germain, R. N., M. E. Dorf, and B. Benacerraf. 1975. Inhibition of T lymphocyte-mediated tumor-specific lysis by alloantisera directed against the H-2 serological specificities of the tumor. J. Exp. Med. 142:1023–1028.

Goodfellow, P. N., E. A. Jones, V. van Heyningen, E. Solomon, M. Bobrow, V. Miggiano, and W. F. Bodmer. 1975. The β_2-microglobulin gene is on chromosome 15 and not in the HL-A region. Nature 254:267–269.

Henning, R., R. J. Milner, K. Reske, B. A. Cunningham, and G. M. Edelman. 1976a. Subunit structure, cell surface orientation, and partial amino acid sequences of murine histocompatability antigens. Proc. Natl. Acad. Sci. USA 73:118–122.

Henning, R., J. W. Schrader, and G. M. Edelman. 1976b. Anti-viral antibodies inhibit the lysis of tumor cells by anti-H-2 sera. Nature 263:689–691.

Katz, D. H., and B. Benacerraf, eds. 1976. The Role of Products of the Histocompatibility Gene Complex in Immune Responses. Academic Press, Inc., New York.

Klein, J. 1975. Biology of the Mouse Histocompatability-2 Complex. Springer-Verlag, New York.

Klein, J., and D. C. Shreffler. 1971. The H-2 model for major histocompatability systems. Transplant. Rev. 6:3–29.

Koszinowski, U., and H. Ertl. 1975. Lysis mediated by T cells and restricted by H-2 antigen of target cells infected with vaccinia virus. Nature 255:552–554.

Lachman, P. J., D. Grennan, A. Martin, and P. Demant. 1975. Identification of Ss protein as murine C4. Nature 258:242–244.

Lamb, R. A., B. W. J. Mahey, and P. W. Choppin. 1976. The synthesis of Sendai virus polypeptides in infected cells. Virology 69:116–131.

Nathenson, S. G., and S. E. Cullen. 1974. Biochemical properties and immunochemical-genetic relationships of mouse H-2 alloantigens. Biochim. Biophys. Acta 344:1–25.

Neauport-Sautes, C., F. Lilly, D. Silvestre, and F. M. Kourilsky. 1973. Independence of H-2K and H-2D antigenic determinants on the surface of mouse lymphocytes. J. Exp. Med. 137:511–526.

Neauport-Sautes, C., A. Bismuth, F. M. Kourilsky, and Y. Manuel. 1974. Relationship between HL-A antigens and β_2-microglobulin as studied by immunofluorescence on the lymphocyte membrane. J. Exp. Med. 139:957–968.

Ostberg, L., L. Rask, H. Wigzell, and P. A. Peterson. 1975. Thymus leukaemia antigen contains β_2-microglobulin. Nature 253:735–736.

Peterson, P. A., B. A. Cunningham, I. Berggård, and G. M. Edelman. 1972. β_2-microglobulin-A free immunoglobulin domain. Proc. Natl. Acad. Sci. USA 69:1697–1701.

Rask, L., J. B. Lindblom, and P. A. Peterson. 1974. Subunit structure of H-2 alloantigens. Nature 249:833–834.

Rouse, B. T., M. Rollinghoff, and N. L. Warner. 1972. Anti-θ serum-induced suppression of the cellular transfer of tumor-specific immunity to a syngeneic plasma cell tumor. Nature New Biol. 238:116–117.

Schrader, J. W., B. A. Cunningham, and G. M. Edelman. 1975. Functional interactions of viral and histocompatibility antigens at tumor cell surfaces. Proc. Natl. Acad. Sci. USA 72:5066–5070.

Schrader, J. W., and G. M. Edelman. 1976. Participation of the H-2 antigens of tumor cells in their lysis by syngeneic T cells. J. Exp. Med. 143:601–614.

Schrader, J. W., and G. M. Edelman. 1977. Joint recognition by cytotoxic T cells of inactivated Sendai virus and products of the major histocompatibility complex. J. Exp. Med. 145:523–539.

Shearer, G. M., T. G. Rehn, and C. A. Garbarino. 1975. Cell-mediated lympholysis of trinitrophenyl-modified autologous lymphocytes. Effector cell specificity to modified cell surface components controlled by the H-2K and H-2D serological regions of the murine histocompatibility complex. J. Exp. Med. 141:1348–1364.

Shreffler, D. C., and R. D. Owen. 1963. A serologically detected variant in mouse serum: Inheritance and association with the histocompatability-2 locus. Genetics 48:9–25.

Silver, J., and L. Hood. 1974. Detergent-solubilized H-2 alloantigen is associated with a small molecular weight polypeptide. Nature 249:764–765.

Silver, J., and L. Hood. 1976. Structure and evolution of transplantation antigens: Partial amino-acid sequences of H-2K and H-2D alloantigens. Proc. Natl. Acad. Sci. USA 73:599–603.

Smithies, O., and M. D. Poulik. 1972. Dog homologue of human β_2-microglobulin. Proc. Natl. Acad. Sci. USA 69:2914–2917.

Taylor, R. B., W. P. H. Duffus, M. C. Raff, and S. dePetris. 1971. Redistribution and pinocytosis of lymphocyte surface immunoglobulin molecules induced by anti-immunoglobulin antibody. Nature New Biol. 233:225–229.

Terhorst, C., P. Parham, D. L. Mann, and J. L. Strominger. 1976. Structure of HL-A antigens: Amino acid and carbohydrate compositions and NH_2-terminal sequences of four antigen preparations. Proc. Natl. Acad. Sci. USA 73:910–914.

Vitetta, E. S., J. D. Capra, D. G. Klapper, J. Klein, and J. W. Uhr. 1976. The partial amino-acid sequence of an H-2K molecule. Proc. Natl. Acad. Sci. USA 73:905–909.

Vitetta, E. S., J. W. Uhr, and E. A. Boyse. 1975. Association of a β_2-microglobulin-like subunit with H-2 and TL alloantigens on murine thymocytes. J. Immunol. 114:252–254.

Zinkernagel, R. M., and P. C. Doherty. 1974. Immunological surveillance against altered-self components by sensitized T lymphocytes in lymphocytic choriomeningitis. Nature 251:547–548.

Stage of Cellular Differentiation in Tumor Cell Origin—Hemopoietic Stem Cell Systems

Cell Differentiation and Neoplasia, edited by Grady F. Saunders. Raven Press, New York © 1978.

Interrelationship of Differentiation and Proliferation Control in Hemopoietic Stem Cells

Laszlo G. Lajtha, Brian I. Lord, T. Michael Dexter, Eric G. Wright, and Terence D. Allen

Paterson Laboratories, Christie Hospital and Holt Radium Institute, Manchester M20 9BX England

The hemopoietic stem cell is pluripotential in that even in the adult organism it is the ultimate ancestor of at least three main lines of hemic cells. The vast cell production in adult life of some 6×10^{15} red cells and about 2×10^{15} granulocytes (amounting to over 10 times the body weight) depends on two basic proliferative mechanisms: the maintenance proliferation of the stem cells and the amplifying proliferation in their descendant transit populations of more or less "committed" precursor cells. The variety of the cell classes or cell types produced depends on a number of differentiation steps, occurring in an hierarchical order, in each instance (apart from the so-called first step differentiation from the pluripotent stem cells) depending also on a certain stage of "maturity" having been reached in the respective precursor population (Lajtha and Schofield 1974).

The scheme of the "three-tier" interlinked cell populations is illustrated in Figure 1. As has been discussed elsewhere (Schofield and Lajtha 1976), the average amplification from the pluripotent stem cell to the final, nonproliferating end cells amounts to some 10 to 12 cell division cycles, i.e., an amplification of between 1,000–4,000 (assuming no cell loss during amplification).

What is perhaps not always realized is the number of cell divisions the stem cells must be able to undergo *as* stem cells to maintain their number throughout the life of the individual (because maintain this number they do). It is, of course, only in the mouse that the absolute numbers of the stem cells can be measured (Till and McCulloch 1961, Siminovitch *et al.* 1963), but in that animal, about 10^6 stem cells are being maintained throughout the animal's life, in spite of some 20% "removed" each day for differentiation. This means a linear production requiring some 160 cell division cycles on the average, in each stem cell, after the birth of the animal. In man, one is dealing with a 30 to 35 times longer lifespan of the individual and, taking into account the uncertainty of assumptions about absolute steady-state numbers, one might be dealing with an average self-maintaining capacity of stem cells of some 1,000–2,000 cell cycles.

Proliferation control has to operate, therefore, at two very different levels: control of the 10 to 12 amplifying divisions during the transient life of the

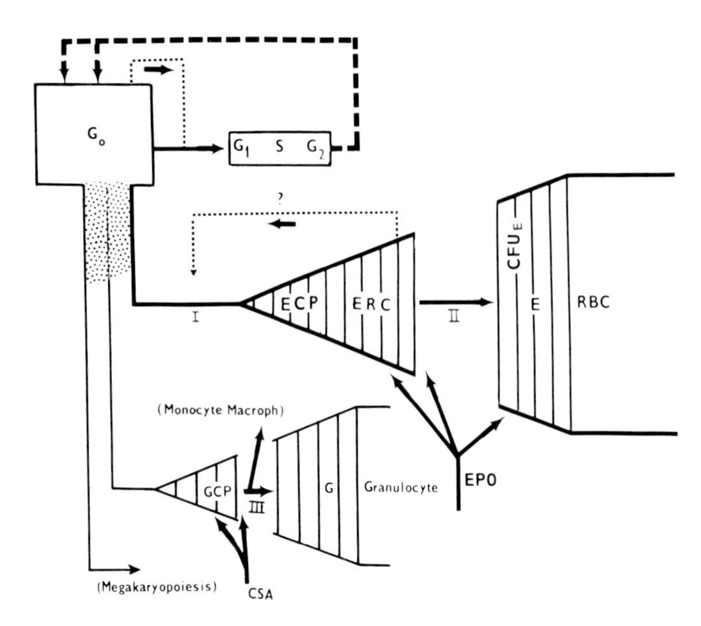

FIG. 1. Three-tier model of hemopoiesis. The pluripotent stem cell, either in the noncycling G_o state or in the cycle (G_1-S-G_2), is regulated by specific proliferation control mechanisms. The result of the "first step" differentiation (I) results in the production of "committed" precursor cells, e.g., the erythroid (ECP) or granulocytic (GCP), depending on the nature of the differentiation stimulus (not yet known). The committed cells undergo limited amplification before a second differentiation step (II) induces the second set transit populations, i.e., the pronormoblasts (E) or the promyelocytes and myelocytes (G). The second differentiation step is regulated by erythropoietin (EPo) or "colony-stimulating activity" (CSA) levels, respectively. Proliferation in the transit populations is controlled by cell cycle length modulation (see text), which determines the number of cell cycles (hence amplification) during the respective transit times.

differentiated and maturing "transit populations" and control of the steady-state maintenance of the stem cells.

Control of the transit population is perhaps the less critical of the two. Once a stem cell ceases to be a stem cell, by differentiating into one of the "committed" precursor cells, its lifespan is limited to probably not more than 10 to 12 days in man (until it reaches the nonproliferating end cell stage of mature erythrocyte or mature polymorphonuclear granulocyte). During that transit—which amounts to a "suicide" maturation—the number of amplifying divisions can modify very sensitively the number of cells produced; a single extra division would double, one division less would halve the cell production. Indeed the number of division cycles is controlled elastically, and on demand three to six extra cycles—amounting to 8 to 64 times output of cells—is quite feasible (Lajtha *et al.* 1971).

The maintenance control of the stem cells is more critical as a "long-term policy"; it is easier to compensate for increased transit amplification by decreasing stem cell differentiation than to try to compensate for stem cell proliferation by increasing "removal" for differentiation.

CONTROL OF THE TRANSIT POPULATIONS

Cell proliferation in the bone marrow spaces tends to be geared inversely to the specific cellularity in the space. Patt and Maloney (1972) accumulated data from 180 rabbit femora and showed that the lower the cellularity of the femur, the higher was the uptake of tritiated thymidine by the cells. Evidence for the cell line specificity of this apparent feedback proliferation control in the granulocytic, erythroid, and lymphocytic cells has been presented by Rytömaa and Kiviniemi (1968), Kivilaakso and Rytömaa (1971), Moorhead *et al.* (1969), and Houck *et al.* (1971). This has been confirmed (Table 1) using strict internal controls both in vitro (Lord *et al.* 1974a) and in vivo (Lord 1975, Lord *et al.* 1977a), and the possible physiological nature of the feedback inhibition was indicated by the easy reversibility of action of the specific inhibitors produced by the respective "end" cells (Lord *et al.* 1974b). An interesting aspect of these experiments is that the maximum inhibition achievable appears to result in a roughly 50 percent depression of proliferation of the cell populations in the erythroid and granulocytic cells respectively; in the lymphoid cells, over 90 percent depression is possible.

This raised the question of whether the erythrocyte and granulocyte control is affected by a cell cycle modulation (e.g., prolongation of the G_1 period) rather than possible complete inhibition. As has been pointed out earlier, prolongation of the cell cycle during a fixed maturation transit has a profound influence on the number of cells produced, by allowing a smaller number of cell division

TABLE 1. *Percentage change in SCM of hemopoietic cells when treated with mature blood cell extracts*

Test extract	Molecular weight range	Cell type tested: Normal lymphocytes	Cultured granulocytic cells	Fetal liver normoblasts
Lymph node extract (LNE)	500–1,000	+ 2	+ 4	+ 2
	1,000–10,000	+ 7	− 2	+ 5
	30,000–50,000	*+44*	+ 2	+ 2
Granulocyte extract (GCE)	*500–1,000*	+ 1	*+42*	− 3
	1,000–10,000	+ 1	*+13*	+ 4
	30,000–50,000	0	+ 4	+ 4
Erythrocyte extract (RCE)	*500–1,000*	− 1	− 2	*+51*
	1,000–10,000	− 1	− 2	*+48*
	30,000–50,000	− 1	− 4	*+46*

Changes in the structuredness of the cytoplasmic matrix (SCM) of cells treated with blood cell extracts, measured by the technique of fluorescence polarization (Cercek *et al.* 1973). Molecular weight fractions in italics indicate the reported active fractions (Houck *et al.* 1971, Rytömaa and Kiviniemi 1968). Percentage changes in italics illustrate significant changes in SCM, thus demonstrating cell line specificity of the extracts. (Reprinted, with permission of Br. J. Cancer, from Lord *et al.* 1974a.)

cycles to occur. This question is, however, not yet settled because a G_1 elongation in all cells, or a full inhibition of all the cells, could produce similar experimental results. Work is proceeding in our laboratories to attempt to distinguish between these possibilities.

Much less is known about proliferation control in the "second tier" of the hemopoietic populations, the "committed" erythroid or granuloid precursors. The experiments of Reissmann and his colleagues (1970, 1972) indicated that erythropoietin, the factor inducing the "second step differentiation" from the committed erythroid precursors to the early normoblasts, also stimulated amplification in the committed ERC (erythropoietin responsive cells) as well. This is unquestionably true for the experimental animals in which stem cell and ERC depletion has been achieved by Myleran. In the intact polycythemic mouse, in which endogenous erythropoietin levels are immeasurably low, there is no indication that the size of the ERC "compartment" decreases, even after 30 days of sustained polycythemia, but it appears that it can still be amplified further by large doses of erythropoietin (von Wangenheim and Schofield, unpublished data).

In the case of the committed granulocytic precursors (CFU-C) (as described by Bradley and Metcalf 1966, Pluznik and Sachs 1966) factors called colony-stimulating activity (CSA) are necessary for in vitro granulocytic colony development. It was a tempting comparison with erythropoietin to postulate that CSA is likewise essential for the second step differentiation from CFU-C to early granulocytes, as well as for the proliferation of CFU-C themselves. Lately, however, a long-term bone marrow culture system has been described (Dexter *et al.* 1973, Dexter and Lajtha 1974), and in its modified version (Dexter *et al.* 1977) both CFU-C proliferation and granulocytic differentiation can be maintained with no detectable CSA in the culture medium.

The role of these factors, and with it the nature of proliferation (amplification) control in the "committed" second tier cell population is not yet settled. It appears, however, that the factors extracted from end cells, and which modulate the cell cycle in the third tier, do not exercise any proliferation control at this level. ERC remain unaffected by red cell extracts (Lord *et al.* 1977a); granulocyte extracts affect neither the growth of the CFU-C population nor the production of colonies in agar culture (Lord *et al.* 1977b).

STEM CELL CONTROL

In the healthy (uninfected) steady-state mouse, most of the stem cells exist in the noncycling G_0 state, from which they can be triggered into cycle (Lajtha 1962a,b, 1963). From the difference in steam cell turnover states in normal spleen versus normal bone marrow (Guzman and Lajtha 1970), in phenylhydrazine-treated spleen versus bone marrow (Rencricca *et al.* 1970), and in shielded versus irradiated bone marrow (Croizat *et al.* 1970, Gidali and Lajtha 1972), it appears that the maintenance of the G_0 state of stem cells is under "local" control.

This "local" maintenance of the resting state can be overridden by a number of factors: direct stimulation of the effective pathways of cell proliferation (see review by Byron 1975) or by perhaps more "physiological" feedback information following depopulation of, for example, committed descendant cell populations. The latter can be achieved by a large depopulation of the ERC population following a large dose of erythropoietin (Guzman and Lajtha 1970) or by hydroxyurea (Vassort *et al.* 1973), which affect both ERC and CFU-C. Certainly the reverse situation holds. Under conditions of continuous irradiation, stem cell cycling is initially achieved by depopulation of all cell types in the bone marrow. The rate of cycling, however, remains the same whether the population is 20 percent of normal or 1 percent, and this probably arises because of the rapid return to normal of such populations as the committed erythroid precursor cells (Wu and Lajtha 1975). The results of partial body irradiation (which induces temporary cycling in the shielded femur) are difficult to analyze since both increased demand for differentiation and increased stem cell migration (Croizat *et al.* 1976) contribute to a decreased local concentration of stem cells—temporarily—in the unirradiated areas.

The question is this: What is the physiological mechanism by which the normal local proliferation control acts, and how is this overridden by physiological demands?

There are two reports in the literature concerning inhibitory factors acting on hemopoietic stem cell proliferation. Lord *et al.* (1976) found that a large molecular weight (50,000–100,000 daltons) saline extract from resting bone marrow inhibits stem cell cycling (reversing it to the normal resting value) but that such extracts cannot be prepared from bone marrow in which the stem cells are cycling (Table 2). They, furthermore, have shown that the active extract (fraction IV) is specific for the stem cells, having no effect on the cycling committed granulocytic precursor cells or other bone marrow cells. Frindel and Guigon (1977) found that the dialysate from fetal calf bone marrow, when injected into irradiated mice, decreased the cycling of the stem cells to normal (resting) values. Similarly prepared extracts from calf liver or extracts of lamb thymus, which inhibit lymphocyte proliferation (Kiger *et al.* 1972), had no effect on the stem cells.

These results (in their present state) are difficult to correlate. Fetal calf marrow is likely to contain cycling stem cells, a source which, in the case of mouse marrow, does not yield an inhibitory factor. Frindel and Guigon's extract (1977) appears to be a low molecular weight compound, as opposed to the fraction found by Lord *et al.* (1976). This latter is cell specific (only affects stem cells); the cell specificity of the fetal calf extract has yet to be investiaged.

Table 2 shows that conversely to a stem cell inhibitor (fraction IV from normal bone marrow), a stimulator (fraction III) of molecular weight 30,000–

TABLE 2. *Percentage of stem cells killed by tritiated thymidine*

Treatment		Resting stem cells	Proliferating stem cells
Control (no extract)		6	32
Fraction			
I.	NBME	—	37
	RBME	5	—
II.	NBME	—	27
	RBME	9	—
III.	NBME	8	25
	RBME	*37*	33
IV.	NBME	2	5
	RBME	12	*35*
V.	NBME	23*	31
	RBME	22*	29

Tritiated thymidine (^{3}HTdR) suicide measurements on normal bone marrow stem cells (resting) and regenerating bone marrow stem cells (proliferating) following treatment with fractionated extracts of normal bone marrow (NBME) and regenerating bone marrow (RBME). Fractionation is by molecular weight:

I	500–10,000 daltons
II	10,000–30,000 daltons
III	30,000–50,000 daltons
IV	50,000–100,000 daltons
V	$>$100,000 daltons

* These high values probably represent nonspecific stimulation of stem cells by macromolecules in the crude extract.

50,000 daltons can also be demonstrated but is only extractable from regenerating marrow. Since the committed precursor cells are cycling in the marrow, no evidence for cell specificity can be demonstrated for this stimulator. Recent unpublished experiments, however, have shown that fraction III of extracts from mature granulocytes or erythrocytes does not stimulate resting stem cells. The presence of this stimulator is, therefore, source specific and probably restricted to populations containing cycling stem cells. Frindel *et al.* (1976) described a similar effect—moderate stimulation of stem cell cycling by co-cultivation with hydroxyurea damaged bone marrow cells but not with undamaged bone marrow cells. These experiments, however, gave no indication of the size of the stimulatory molecules. In the absence of purified extracts from the cells—and until cell specificity can be demonstrated—the question of nonspecific proliferation effectors (as found by Byron 1975) cannot be excluded.

The two active extracts can, however, be used in combination to manipulate the cycling of the stem cells. For example, resting stem cells treated with stimulator increase their cycling but immediately become susceptible to inhibition by

TABLE 3. *Interaction of stem cell proliferation inhibitors and stimulators*

Test cells	Treatment	% Stem cells killed by ^{3}HTdR
Normal bone marrow	—	8.0 ± 4.5
Normal bone marrow	RBME-III	39.1 ± 6.1
Normal bone marrow	RBME-III* + NBME-IV	1.9 ± 3.3
Regenerating bone marrow	—	26.4 ± 7.2
Regenerating bone marrow	NBME-IV	5.8 ± 5.6
	NBME-IV* + RBME-III	38.1 ± 7.2
PHZ—bone marrow or regenerating bone marrow or fetal liver	—	43.8 ± 2.1
PHZ—bone marrow or regenerating bone marrow or fetal liver	PHZ—Spleen or normal bone marrow *	10.3 ± 4.3
PHZ—spleen or normal bone marrow	—	5.3 ± 1.5
PHZ—spleen or normal bone marrow	PHZ—bone marrow or regenerating bone marrow or fetal liver †	37.0 ± 2.1

Percentage of stem cells in DNA synthesis measured by the tritiated thymidine suicide test (Becker *et al.* 1965, Lord *et al.* 1974c.)

RBME-III Regenerating bone marrow extract—fraction III (30,000–50,000 daltons).

NBME-IV Normal bone marrow extract—fraction IV (50,000–100,000 daltons).

PHZ Phenylhydrazine-treated mice (Rencricca *et al.* 1970).

* RBME-III and NBME-IV added together or separated by one to two hours.

† Cells irradiated (800 rads X-rays) to eliminate viable stem cells.

the addition of fraction IV from normal bone marrow. The reverse situation also holds (Lord, Mori, and Wright, unpublished results). A similar effect can be demonstrated using cell populations containing cycling and noncycling stem cells. The addition of an irradiation "killed" cell population, which had contained cycling stem cells, to one containing noncycling stem cells very rapidly stimulates these stem cells into a proliferative mode. Conversely, a "killed" population containing noncycling stem cells is capable of inhibiting cycling stem cells (Wright and Lord, unpublished results). These experiments are summarized in Table 3. A combination of stem cell-specific proliferation inhibitors and even nonspecific stimulators may, therefore, have a promising role of manipulation of the drug sensitivity and recovery rate of the hemopoietic cells during regimes of cycle-active cytotoxic chemotherapy.

The mechanisms by which these changes in the proliferation status of stem cells is brought about is still not known. Nečas and Neuwirt (1976) suggested that the signal was determined by the population of cells actually in DNA synthesis. They found that it was only when the total marrow cellularity was reduced as a result of removing S phase cells only that stem cells were recruited into cycle, and they concluded that S phase stem cells may inhibit the entry of nonproliferating G_0 cells into the cell cycle. It is not clear under these circumstances, however, whether there was a relatively large depopulation of the second and early third tier cells that might trigger the change. Furthermore, it is difficult to explain exponential growth of the stem cell population in irradiated mice. This suggestion must, therefore, for the time being remain unproved.

STEM CELL MAINTENANCE IN VITRO

Evidence of the "locally" operating inhibitor of stem cell proliferation, described above, indicates that, in some way, it depends on local stem cell concentration. Whether this implies a direct feedback control, or control by some cell type other than the stem cells, is not known. Nor does the evidence available exclude the possibility that more than one inhibitory mechanism may operate under various in vivo conditions.

The study of stem cell proliferation and differentiation control has been hampered by the lack of a suitable "clean" in vitro system for stem cell maintenance. The study of the in vivo system is beset by the complexity of the possible reactions by the whole organism to the types of perturbations introduced (whole body irradiation, injections of cytotoxic drugs, or various impure cell extracts). While in vitro systems create their own problems—especially the question of how far removed they are from their normal physiological condition and their spatial distribution in the normal environment (Lord *et al.* 1975), notwithstanding their survival under the culture conditions—the parameters of in vitro systems are much easier to control.

Recently, a long-term bone marrow culture system has been developed (Dexter *et al.* 1973) in which stem cell maintenance and differentiation can be maintained for several weeks. In its present state, the system is as yet "monophyletic" in so far as only granulopoiesis has been satisfactorily demonstrated in it, including, however, the committed granulocytic precursor CFU-C. Once the culture is started, it represents a "chemostat" in which stem cells, committed precursor cells, and granulocytic cells are continuously produced. With the current "feeding" system, whereby one half of the medium and suspended cells are removed twice weekly, we are dealing with two population doublings per week for 12 to 16 weeks. Since the stem cells also double their number between "feedings," in spite of loss of stem cells for differentiation, they must undergo in excess of 20 to 30 cell cycles during the current culture periods (Table 4). The critical aspect for the long-term maintenance of stem cells, and therefore the bone

TABLE 4. *Co-culture of bone marrow + bone marrow cells (from normal mice)*

Weeks in culture	Cell count (x 10^{-6})	Pluripotent stem cells	Committed granulocyte precursor cells
2	0.8	140	1400
4	0.6	180	ND
5	0.7	160	ND
7	0.9	110	1600
9	0.9	240	ND
12	0.7	190	2500
15	1.1	0	350

Pluripotent and committed precursor cells in the culture suspension assayed by the spleen colony technique (Till and McCulloch 1961) and the agar culture technique (Bradley and Metcalf 1966, Pluznik and Sachs 1966), respectively.
ND—Not done.

marrow culture as such, is the establishment of a suitable adherent layer of cells *on* which stem cells and granulopoiesis can be maintained.

The source of this adherent layer is bone marrow, and by the time a suitable adherent layer is established, it contains three main cell types: elongated macrophage-type cells, flattened "epithelioid" cells, and monstrous giant fat-containing cells (Figure 2). When such adherent layers are then seeded with fresh bone marrow cells, the "successful" cells settle *on* the adherent epithelioid or macrophage-like cells (if they settle on the glass surface they appear to degenerate quickly into nonhemopoietic forms). It is noteworthy that the active cell production appears to occur in the vicinity of the giant fat cells; indeed, it appears that without them, no successful culture can be established, and their number appears to regulate the level of hemopoiesis (i.e., the number of stem cells and granulocytes maintained and produced).

The metabolic role of the cells in the adherent layer, particularly that of the giant fat cells, is not known. Whether they represent essential stimulators under the culture conditions, or whether they represent some other, essentially "permissive" condition necessary for stem cell survival, is yet to be settled.

Furthermore, the culture does not represent a steady state, as in a normal animal. On refeeding, the stem cell population grows rapidly and then settles as a pseudosteady state until the next refeeding. As the method is operated at present, this has been carried on for more than 20 population doublings.

Nevertheless, this is the first time that anything like this degree of self-maintenance could be kept for such lengths of time. Granulopoietic differentiation obviously does occur in the cultures (interestingly in spite of undetectable CSA levels in the medium). The pluripotent stem cells are maintained functionally intact (when tested by injecting them into irradiated recipient mice, they protect the animals and produce erythroid, granulocytic, and megakaryocytic colonies in their spleens); therefore, with appropriate modification of the culture condition, it should be possible to produce erythropoiesis as well.

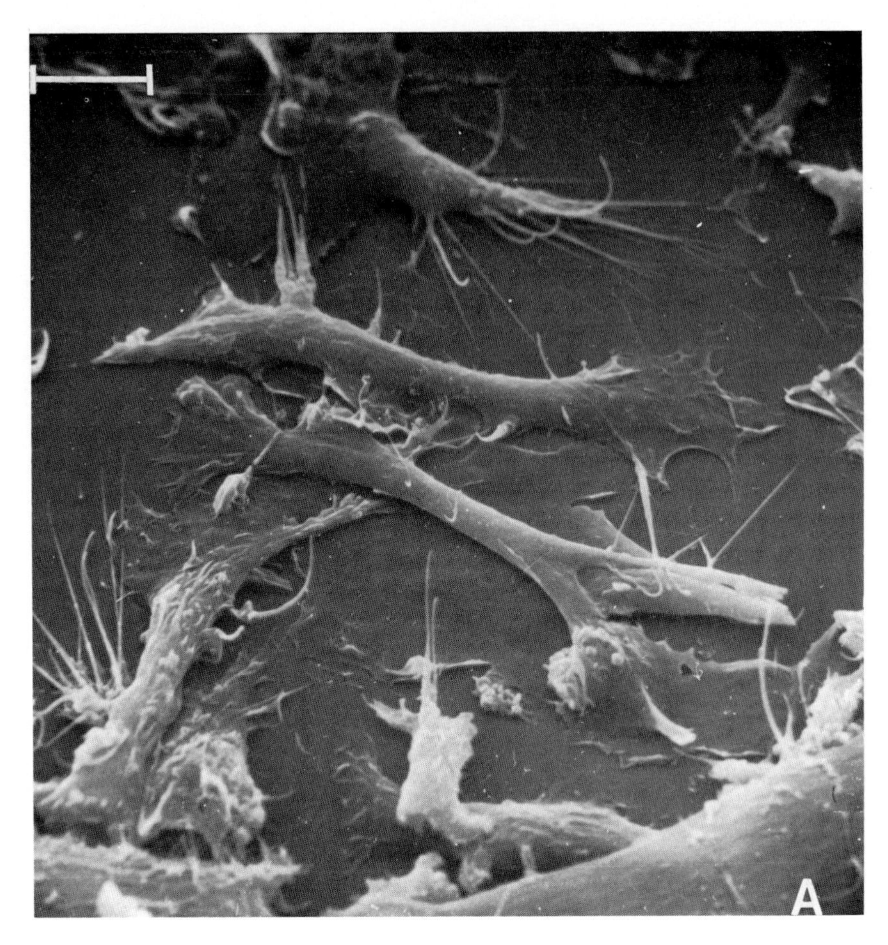

FIG. 2A. Scanning electron micrograph of cultured bone marrow. Attached mononuclear phagocytic cells that are extended along a single axis, with numerous cytoplasmic projections. Scale bar = 10 μm

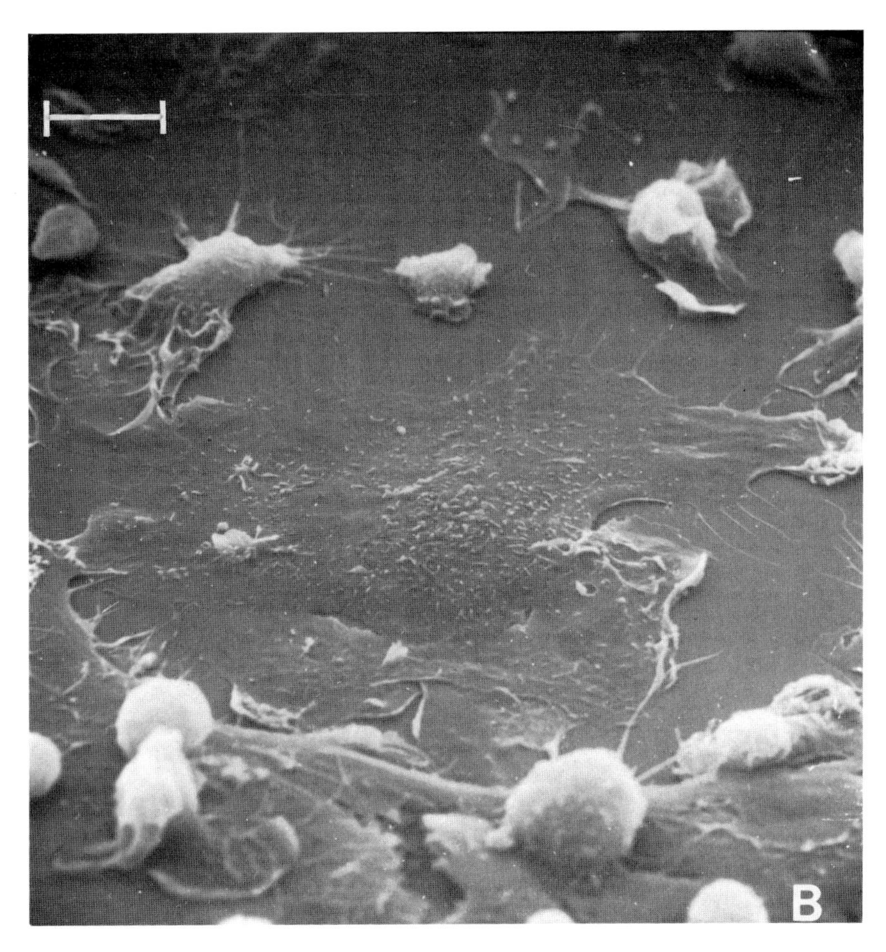

FIG. 2B. Scanning electron micrograph of cultured bone marrow. Attached epithelial cell showing its flattened morphology. Scale bar = 20μm

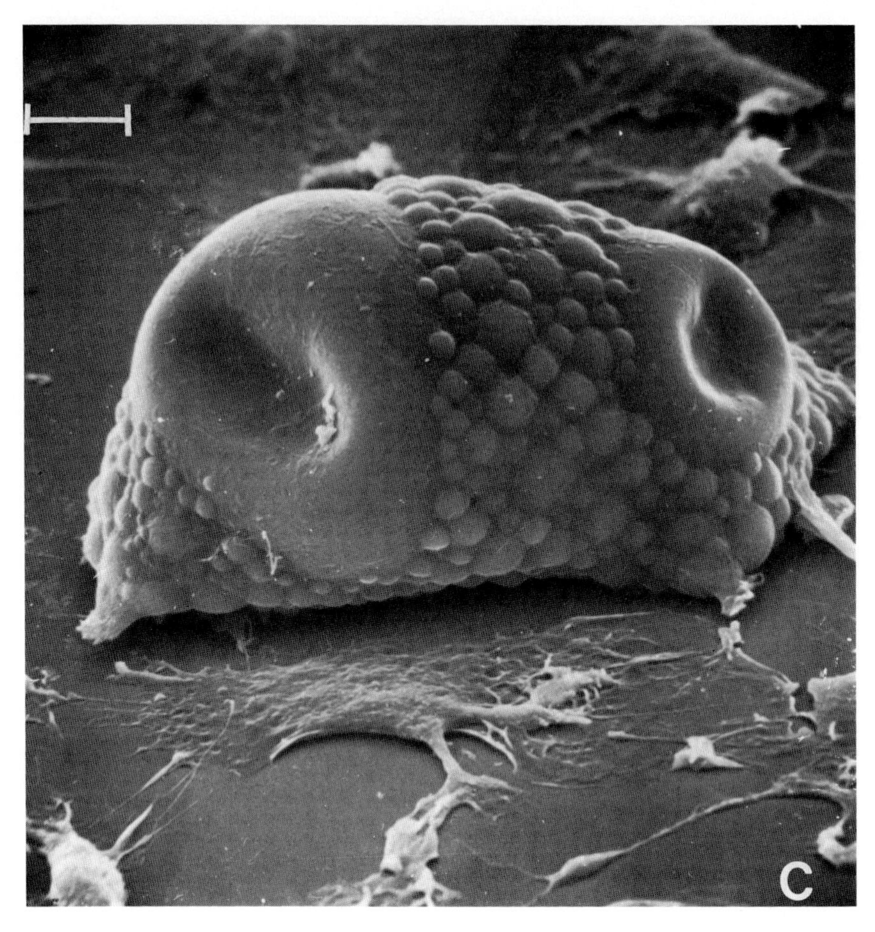

FIG. 2C. Scanning electron micrograph of cultured bone marrow. "Giant fat" cell in the early stages of lipid accumulation, showing on its surface two large vacuoles and numerous small ones. The vermiform ridges on the surface are the mitochondria in the very thin layer of cytoplasm that surrounds the lipid-containing vacuoles. Scale bar = 10 μm

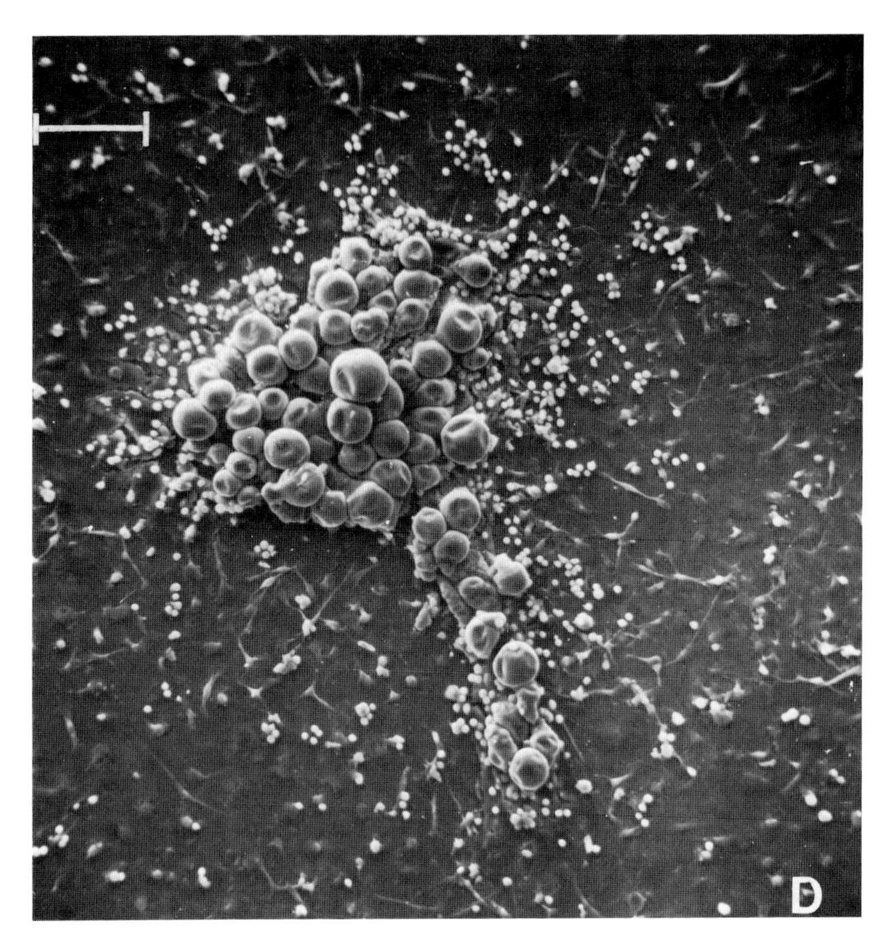

FIG. 2D. Scanning electron micrograph of cultured bone marrow. Characteristic accumulation of granulocytes clustered around the periphery of a giant fat cell. Scale bar = 100 μm (Figures 2A-D are reproduced, with the permission of the Press of the Wistar Institute of Anatomy and Biology, from Dexter *et al.* 1977, in press.)

This means that an assay system is available for the study of the differentiation control of stem cells (the so-called first step differentiation of Figure 1), at least in respect of the production of the committed second tier granulocytic precursors. The same method applied to human bone marrow may give—albeit by inference—a method for the study of human pluripotent stem cells.

CONCLUSION

The foregoing illustrates the state of flux in the study of hemopoietic proliferation and differentiation control. Some cell line-specific proliferation-controlling factors of likely physiological nature have been demonstrated. The methodology for assaying both proliferation and differentiation in a cell-specific sense is improving. Some of the factors available might be exploited in attempts to manipulate the hemopoietic system for clinical purposes. The field of study is still in a very early stage of its development, but not without reasonable promises for the future.

REFERENCES

Becker, A. J., E. A. McCulloch, L. Siminovitch, and J. E. Till. 1965. The effect of differing demands for blood cell production on DNA synthesis by hemopoietic colony-forming cells of mice. Blood 26:296–308.

Bradley, T. R., and D. Metcalf. 1966. The growth of mouse bone marrow cells *in vitro*. Aust. J. Exp. Biol. Med. Sci. 44:287–300.

Byron, J. W. 1975. Manipulation of the cell cycle of the hemopoietic stem cell. Exp. Hematol. 3:44–53.

Cercek, L., B. Cercek, and C. H. Ockey. 1973. Structuredness of the cytoplasmic matrix and Michaelis-Menten constants for the hydrolysis of FDA during the cell cycle in Chinese hamster ovary cells. Biophysik 10:187–194.

Croizat, H., E. Frindel, and M. Tubiana. 1970. Proliferative activity of the stem cells in the bone marrow of mice after single and multiple irradiations (total or partial body exposure). Int. J. Radiat. Biol. 18:347–358.

Croizat, H., E. Frindel, and M. Tubiana. 1976. Abscopal effect of irradiation on haemopoietic stem cells of shielded bone marrow—role of migration. Int. J. Radiat. Biol. 30:347–358.

Dexter, T. M., T. D. Allen, and L. G. Lajtha. 1977. Conditions controlling the proliferation of haemopoietic stem cells. J. Cell Physiol. 91:335–344.

Dexter, T. M., T. D. Allen, L. G. Lajtha, R. Schofield, and B. I. Lord. 1973. Stimulation of differentiation and proliferation of haemopoietic cells *in vitro*. J. Cell Physiol. 82:461–474.

Dexter, T. M., and L. G. Lajtha. 1974. Proliferation of haemopoietic stem cells in vitro. Br. J. Haematol. 28:525–530.

Frindel, E., H. Croizat, and F. Vassort. 1976. Stimulating factors liberated by treated bone marrow and *in vitro* effect on CFU kinetics. Exp. Hematol. 4:56–61.

Frindel, E., and M. Guigon. 1977. Inhibition of CFU entry into cycle by a bone marrow extract. Exp. Hematol. 5:74–76.

Gidali, J., and L. G. Lajtha. 1972. Regulation of haemopoietic stem cell turnover in partially irradiated mice. Cell Tissue Kinet. 5:147–157.

Guzman, E., and L. G. Lajtha. 1970. Some comparisons of the kinetic properties of femoral and splenic haemopoietic stem cells. Cell Tissue Kinet. 3:91–98.

Houck, J. C., H. Irausquin, and S. Leikin. 1971. Lymphocyte DNA synthesis inhibition. Science 173:1139–1141.

Kiger, N., I. Florentin, and G. Mathé. 1972. Some effects of a partially purified lymphocyte-inhibiting factor from calf thymus. Transplantation 14:448–454.
Kivilaakso, E., and T. Rytömaa. 1971. Erythrocyte chalone, a tissue specific inhibitor of cell proliferation in the erythron. Cell Tissue Kinet. 4:1–9.
Lajtha, L. G. 1962a. Problems of bone marrow cell kinetics. Postgrad. Med. J. 38:41–47.
Lajtha, L. G. 1962b. Stem cell kinetics and erythropoietin, *in* Erythropoiesis, L. O. Jacobson and M. Doyle, eds., Grune and Stratton, New York and London, pp. 140–150.
Lajtha, L. G. 1963. Macromolecular aspects of the cell cycle. J. Cell Comp. Physiol. (Suppl. 1) 62:143–145.
Lajtha, L. G., C. W. Gilbert, and E. Guzman. 1971. Kinetics of haemopoietic colony growth. Br. J. Haematol. 20:343–354.
Lajtha, L. G., and R. Schofield. 1974. On the problem of differentiation in haemopoiesis. Differentiation 2:313–320.
Lord, B. I. 1975. Modification of granulocytopoietic cell proliferation by granulocyte extracts. Boll. Ist. Sieroter. Milanese 54:187–194.
Lord, B. I., L. Cercek, B. Cercek, G. P. Shah, T. M. Dexter, and L. G. Lajtha. 1974a. Inhibitors of haemopoietic cell proliferation?: Specificity of action within the haemopoietic system. Br. J. Cancer 29:168–175.
Lord, B. I., L. Cercek, B. Cercek, G. P. Shah, and L. G. Lajtha. 1974b. Inhibitors of haemopoietic cell proliferation: Reversibility of action. Br. J. Cancer 29:407–409.
Lord, B. I., L. G. Lajtha, and J. Gidali. 1974c. Measurement of the kinetic status of bone marrow precursor cells: Three cautionary tales. Cell Tissue Kinet. 7:507–515.
Lord, B. I., K. J. Mori, E. G. Wright, and L. G. Lajtha. 1976. An inhibitor of stem cell proliferation in normal bone marrow. Br. J. Haematol. 34:441–445.
Lord, B. I., G. P. Shah, and L. G. Lajtha. 1977a. The effects of red blood cell extracts on the proliferation of erythrocyte precursor cells *in vivo*. Cell Tissue Kinet. 10:215–222.
Lord, B. I., N. G. Testa, and J. H. Hendry. 1975. The relative spatial distributions of CFU_s and CFU_c in normal mouse femur. Blood 46:65–72.
Lord, B. I., N. G. Testa, E. G. Wright, and R. K. Banerjee. 1977b. Lack of effect of a granulocyte proliferation inhibitor on their committed precursor cells. Biomedicine 26:163–169.
Moorhead, J. F., E. Paraskova-Tchernozemska, A. J. Pirrie, and C. Hayes. 1969. Lymphoid inhibitors of human lymphocyte DNA synthesis and mitosis *in vitro*. Nature 224:1207–1208.
Nečas, E., and J. Neuwirt. 1976. Control of haemopoietic stem cell proliferation by cells in DNA synthesis. Br. J. Haematol. 33:395–400.
Patt, H. M., and M. A. Maloney. 1972. Relationship of bone marrow cellularity and proliferative activity: A local regulatory mechanism. Cell Tissue Kinet. 5:303–309.
Pluznik, P. H., and L. Sachs. 1966. The induction of colonies of normal "mast" cells by a substance in conditioned medium. Exp. Cell Res. 43:553–563.
Reissmann, K. R., and S. Samorapoompichit. 1970. Effect of erythropoietin on proliferation of erythroid stem cells in the absence of transplantable colony forming units. Blood 36:287–296.
Reissmann, K. R., and K. B. Udupa. 1972. Effect of erythropoietin on proliferation of erythropoietin-responsive cells. Cell Tissue Kinet. 5:481–490.
Rencricca, N. J., V. Rizzoli, D. Howard, P. Duffy, and F. Stohlman, Jr. 1970. Stem cell migration and proliferation during severe anemia. Blood 36:764–771.
Rytömaa, T., and I. Kiviniemi. 1968. Control of granulocyte production. I. Chalone and antichalone, two specific humoral regulators. Cell Tissue Kinet. 1:329–340.
Schofield, R., and L. G. Lajtha. 1976. Cellular kinetics of erythropoiesis, *in* Congenital Disorders of Erythropoiesis (Ciba Foundation Symposium 37) (New Series) Elsevier/Excerpta Medica/North Holland Publishing Co., Amsterdam, pp. 3–24.
Siminovitch, L., E. A. McCulloch, and J. E. Till. 1963. The distribution of colony-forming cells among spleen colonies. J. Cell Comp. Physiol. 62:327–336.
Till, J. E., and E. A. McCulloch. 1961. A direct measurement of the radiation sensitivity of normal mouse bone marrow cells. Radiat. Res. 14:213–222.
Vassort, F., M. Winterholer, E. Frindel, and M. Tubiana. 1973. Kinetic parameters of bone marrow stem cells using *in vitro* suicide by tritiated thymidine or by hydroxyurea. Blood 41:789–796.
Wu, Chu-tse, and L. G. Lajtha. 1975. Haemopoietic stem cell kinetics during continuous irradiation. Int. J. Radiat. Biol. 27:41–50.

Cell Differentiation and Neoplasia, edited by Grady F. Saunders. Raven Press, New York

Regulation of Proliferation and Maturation at Early and Late Stages of Erythroid Differentiation

Norman N. Iscove

Basel Institute for Immunology, CH4058 Basel, Switzerland

Red cell formation is a multistage process. It begins when multipotential hemopoietic stem cells give rise to progeny committed to erythroid differentiation. These cells go on to proliferate and mature under the influence of the glycoprotein hormone erythropoietin. As maturation continues, the cells acquire morphological markers which permit their visual identification beginning at the proerythroblast stage. These cells mark the beginning of a terminal sequence, which will include the onset of hemoglobin synthesis and end with the formation of mature erythrocytes.

Because cells from the proerythroblast stage onward bear recognized morphological and biochemical markers, they are accessible to direct study. As detailed elsewhere in this volume, information on these later stages has been derived from studies on normal populations as well as on cells transformed by Friend and related viruses. Until very recently, however, direct assays were not available for cells at stages preceding the terminal program. The breakthrough occurred when it was observed that when erythroid precursors were immobilized in appropriate culture medium containing erythropoietin they would proliferate to form colonies (Stephenson *et al.* 1971, Axelrad *et al.* 1974). In this way it became possible to enumerate directly precursors that could not be identified using any known markers. It is my purpose to describe recent advances, based on colony methods, in charting the area between the multipotent stem cells and the onset of the terminal maturation program.

COLONY ASSAYS FOR ERYTHROID PRECURSORS IN CULTURE

CFU-E

When murine hemopoietic cells are cultivated in medium immobilized with either fibrin or methyl cellulose, colonies of hemoglobin-synthesizing cells become identifiable at 36 to 48 hours of culture (Figure 1). The colonies range in size from as many as 60 cells to an arbitrary, lower scoring limit of eight cells, and by 48 hours, the cells stain positively for hemoglobin. After this time there is no further proliferation within these colonies. Rather, the cells

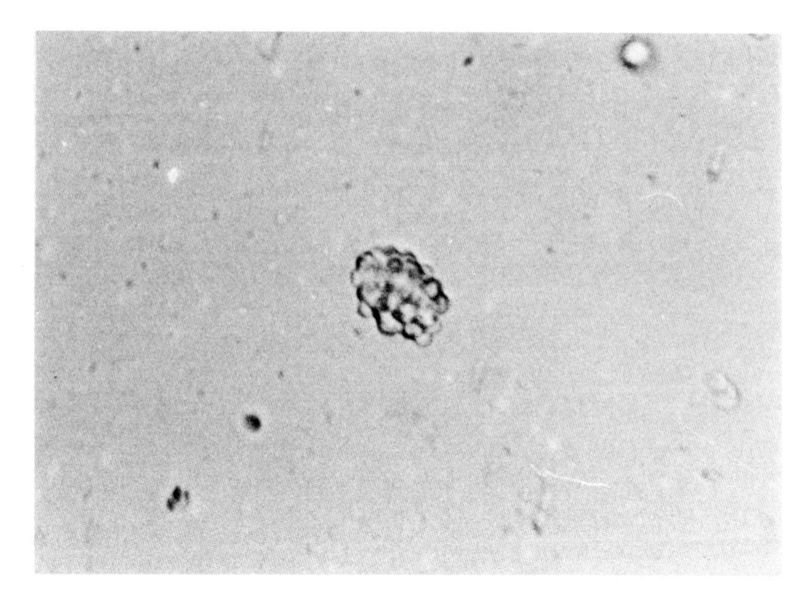

FIG. 1. An erythroid colony after 40 hours of incubation. Such colonies are easily identified in methyl cellulose cultures by their tightly agglutinated configuration and by the very small size of the individual cells.

undergo terminal maturation, extruding their nuclei to become mature erythrocytes by the third day of culture. Lysis soon follows, and the colonies disappear from the plates.

The originating cells for these colonies have been designated CFU-E, or erythroid colony-forming units (Stephenson *et al.* 1971). Colony formation by CFU-E is strictly dependent on the presence of erythropoietin. In methyl cellulose cultures, the number of colonies is a linear function of the number of cells plated over a very wide range and down to limiting dilutions (Iscove and Sieber 1975). The frequency of CFU-E in adult mouse bone marrow is about 600 per 10^5 nucleated bone marrow cells.

BFU-E

Although colony development by CFU-E ceases by 48 hours, other erythroid colonies continue to grow in the presence of erythropoietin. Because they are larger and more dispersed than those generated by CFU-E, these colonies have been called "bursts," and their cells of origin, BFU-E (Axelrad *et al.* 1974). Initially, bursts do not contain hemoglobin-positive cells and are therefore not recognizable as erythroid. Some become identifiable after as early as three days of culture, while others do not contain recognizable erythroblasts until eight or more days (N. Iscove, unpublished observations, Gregory 1976). Those bursts that can be identified early tend to cease growth early, so that the largest colonies are those that do not become identifiably erythroid before eight to nine days of culture. The largest ones contain 10^4 or more cells by the time they stop growing (Figure 2).

In methyl cellulose cultures, the number of large bursts scored at 10 to 12 days is also linearly related to the number of cells plated, over a wide range progressing to limiting dilution (Iscove and Sieber 1975). The frequency of 10- to 12-day BFU-E is about 25 per 10^5 nucleated adult mouse marrow cells.

REQUIREMENTS FOR PROLIFERATION AND MATURATION IN CULTURE

Thiols

The growth-promoting activity of sulfhydryl-containing compounds, such as β-mercaptoethanol or α-thioglycerol, was originally reported in cultures of lymphoid cells (Click *et al.* 1972, Broome and Jeng 1973). Later, these compounds were found to exert a very similar effect on growth of CFU-E and BFU-E, enhancing plating efficiency of CFU-E severalfold (Iscove and Sieber 1975). Their mechanism of action has not yet been clarified. They are active at very low concentration ($0.5–1 \times 10^{-4}$M), as well as under conditions in which they are known or likely to be in the oxidized disulfide form (Broome and Jeng 1973, Iscove and Sieber 1975). These observations tend to rule out any role for them based on simple alteration of the redox potential of the medium.

Serum

With rare exceptions, serum is universally employed for culture of mammalian cells in order to provide substances not present in the defined portion of culture media. Since serum contains many different active agents, it represents a set of variables that are not under complete experimental control. This lack of definition leads not only to problems of reproducibility but also to confusion when one wishes to assay and identify substances in biological fluids or conditioned media which are active in vitro.

The role of serum in cultures of murine hemopoietic cells has recently been substantially clarified. The support of colony formation by CFU-E appears mainly attributable to four normal serum constituents—albumin, transferrin, selenium, and phospholipid (Guilbert and Iscove 1976). Addition of these substances to the medium reduces by 100-fold the amount of serum required by CFU-E and totally replaces serum for growth of granulocyte/macrophage colonies (Guilbert and Iscove 1976, L. Guilbert and N. Iscove, work in progress). Although colony formation by BFU-E can be shown to depend on the same additions (Figure 3), it appears that the requirements for BFU-E exceed those

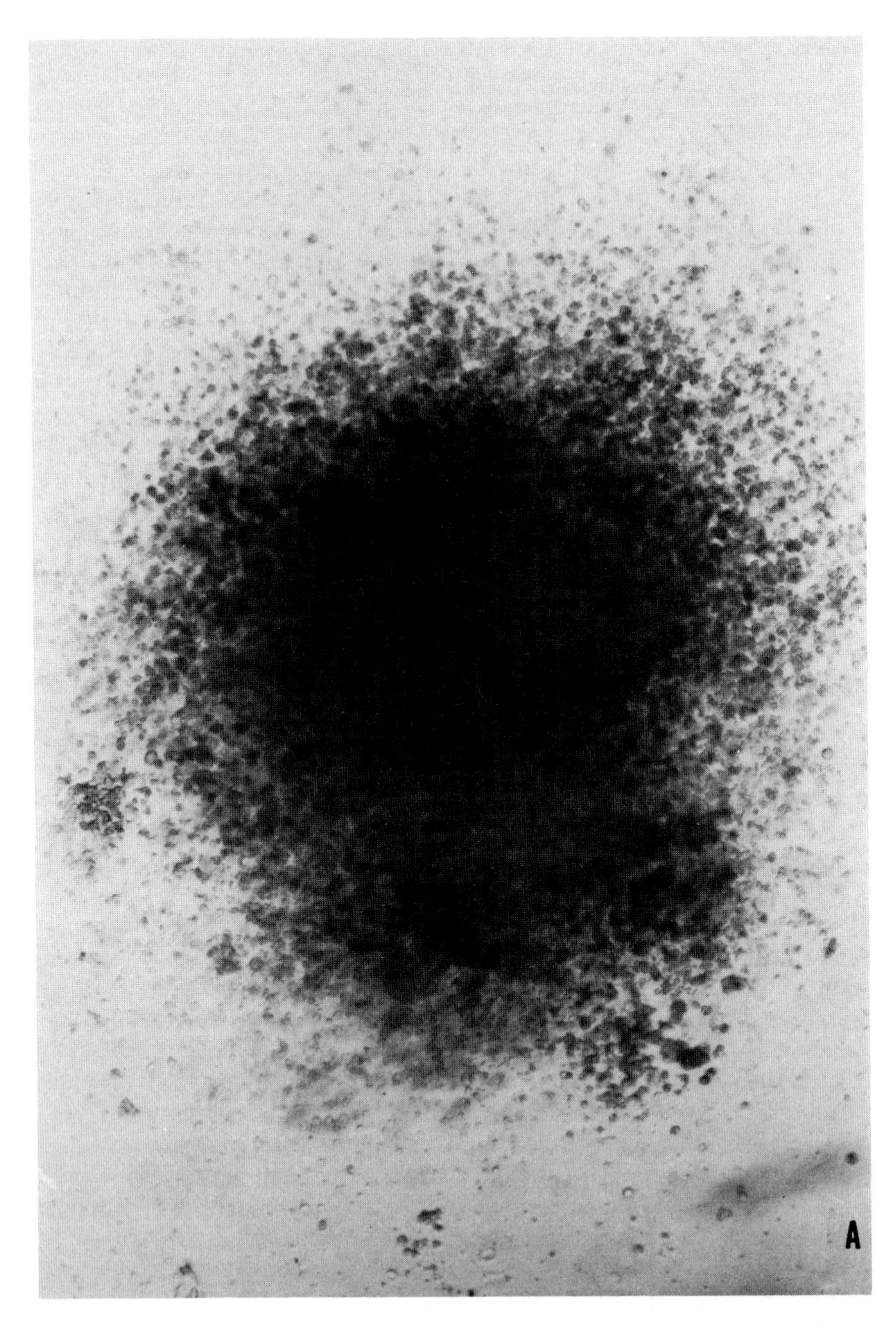

Fig. 2A. A large erythroid burst after 14 days of incubation. Low magnification view showing the entire colony.

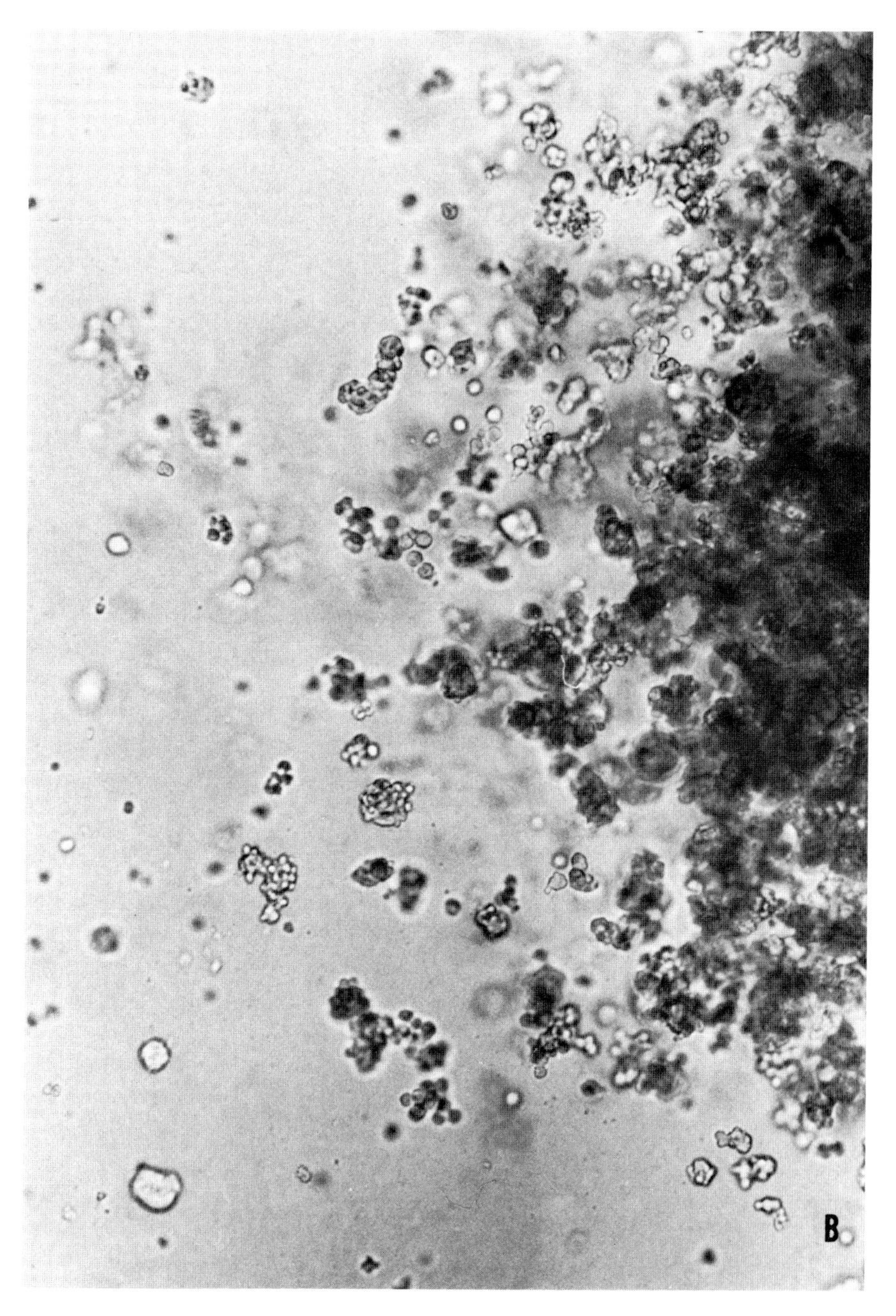

FIG 2B. A large erythroid burst after 14 days of incubation. A higher magnification view of the periphery of such a burst. The clusters of small, tightly agglutinated cells are characteristic and help to identify the colony as erythroid.

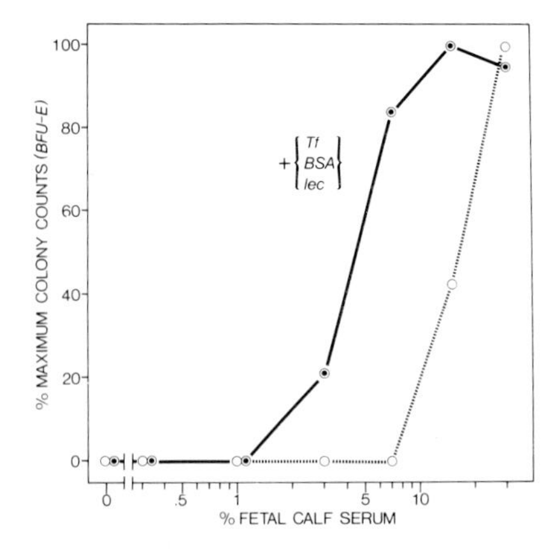

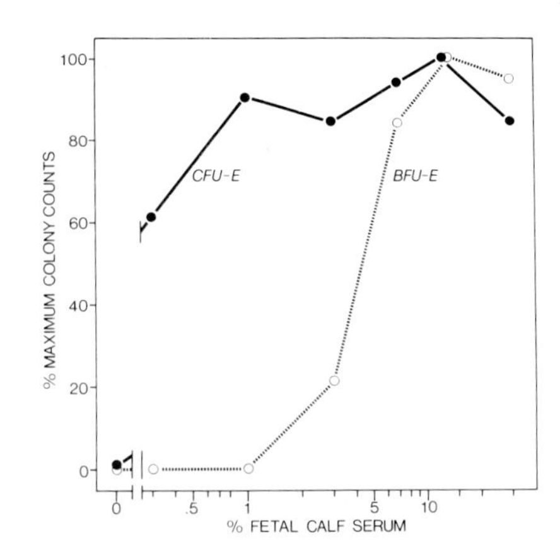

FIG. 3, left. Dependence of burst formation in culture on the concentration of fetal calf serum. The requirement for serum was more than threefold reduced in cultures containing added human transferrin (Tf, 3.4×10^{-6}M, 1/3 iron-saturated), bovine serum albumin (BSA, 1%), and soybean lecithin (lec, 50 μg/ml) (points joined by solid lines). Sodium selenite (10^{-7}M) was present in all cultures.

FIG. 4, right. Comparison of the serum dependence of CFU-E (points joined by solid lines) and BFU-E (10 day) for colony formation in culture. All cultures contained sodium selenite, transferrin, bovine serum albumin, and soybean lecithin. Concentrations were as in legend to Figure 3.

of CFU-E (Figure 4). Efforts are in progress to identify the additional requirements.

SIZE AND DENSITY OF CFU-E AND BFU-E

BFU-E can be almost completely separated from CFU-E by isopyknic centrifugation or velocity sedimentation. The measurements confirm that bursts do not originate from the cells that give rise to two-day colonies. As determined in isopyknic albumin gradients, BFU-E are relatively light, whereas CFU-E are somewhat denser (G. Wagemaker and N. Williams, work in progress). BFU-E sediment with a modal velocity of 3.8 mm/hour at unit gravity, CFU-E in a broader distribution between 5.0 and 7.5 mm/hour (Heath *et al.* 1976, N. Iscove, unpublished data). Sedimentation and density measurements combine (Miller 1973) to give size estimates of 7.8 μm in diameter for BFU-E and 8.2–10 μm in diameter for CFU-E.

The density and size of BFU-E are similar to those of the pluripotential stem cells (CFU-S) detected by spleen colony formation (Worton *et al.* 1969, Messner *et al.* 1972). CFU-E, on the other hand, occupy a size range similar to that of the recognizable proerythroblasts.

PHYSIOLOGICAL PROPERTIES OF CFU-E AND BFU-E

Erythropoietin Requirement in Culture

The generation of recognizable erythroid colonies by CFU-E and by BFU-E is normally strictly erythropoietin dependent. Only two exceptions are known: polycythemia vera in man (Prchal and Axelrad 1974) and mouse cells infected with Friend or related RNA tumor viruses (Liao and Axelrad 1975, Clarke *et al.* 1975). These instances will not be discussed here.

In methyl cellulose cultures containing 10^5 nucleated adult mouse marrow cells per milliliter, colonies from CFU-E first appear at .005 erythropoietin units/ml and reach maximum numbers at .5 units/ml. The former concentration would correspond to 2×10^{-12}M, assuming that erythropoietin preparations of 60,000 units/mg are pure. The plateau response extends to 10 units/ml or further, provided the erythropoietin preparations are relatively free of inhibitors. Under the same conditions of culture, recognizable colonies from BFU-E at 10 days of culture first appear at .3 units and reach maximum numbers at 5 units/ml (Iscove and Sieber 1975). Intermediate BFU-E, forming small bursts recognizable at three days of culture, respond to erythropoietin concentrations intermediate between the requirements of CFU-E and 10-day BFU-E (Gregory 1976).

It has been shown that erythropoietin activity does not decline in the cultures during the time required for BFU-E to generate colonies (Iscove and Sieber 1975). The differing erythropoietin requirement therefore appears to reflect a significant difference in sensitivity between these two populations.

Erythropoietin sensitivity can be influenced in vitro by a number of factors. Preliminary data suggest that sensitivity of CFU-E and BFU-E is affected by the number of marrow cells plated, the sensitivity being higher at higher seeding densities. Androgenic steroids or their derivatives can enhance sensitivity (Singer and Adamson 1976), as can dimethylsulfoxide and neuraminidase (Sieber 1976). It can therefore be expected that differing culture conditions, including variations in the serum constituent of medium as well as cell numbers and composition, will give rise to differing sensitivity estimates. While it may be safe to compare sensitivities of populations under rigidly equivalent conditions of culture, it would appear hazardous to make quantitative extrapolations to sensitivities in vivo or to draw comparisons between sensitivities of cells from mice in different physiological states in which cell composition could vary.

Restriction of CFU-E and BFU-E to Erythroid Differentiation

Pluripotential hemopoietic stem cells (CFU-S) of mice are capable of forming spleen colonies in vivo containing cells of the erythroid, granulocytic, monocytic-macrophage, and megakaryocyte-platelet lineages (Wu *et al.* 1967). In contrast, the clones derived from CFU-E in culture contain erythroid cells exclusively.

TABLE 1. *Independence of colony formation by BFU-E and CFU-C*

Stimulus	10-day bursts /10^5 cells	Granulocyte/ macrophage colonies/10^5 cells	Mixed colonies*
None	0	44	0
G-CSF	0	210	0
Ep†	13	34	0
G-CSF + Ep	11	220	.4‡

* Colonies containing hemoglobin-positive cells as well as granulocytes or macrophages.
† Ep: erythropoietin.
‡ Not significantly greater than the number expected to occur by random coincidence in the same region of the plate.
(Reproduced, with permission of Cell and Tissue Kinetics, from Iscove 1977a)

Clones derived from BFU-E also appear restricted in the sense that they have not been demonstrated to contain granulocytes or macrophages or their precursors. The experiment detailed in Table 1 illustrates this point. Adult mouse bone marrow cells were cultured with erythropoietin or with a source of granulocyte/macrophage colony-stimulating factor (G-CSF), or both. (G-CSF's are glycoproteins required for proliferation of granulocyte and macrophage precursors [CFU-C] in culture.) The populations recruited into specific colony formation by erythropoietin (BFU-E) and G-CSF (CFU-C) appear to be independent of one another. More important, when both factors are present, i.e., under conditions of culture permitting growth of granulocytes and macrophages as well as erythroid cells, virtually all the colonies remain either exclusively erythroid or exclusively granulocyte/macrophage.

Although it therefore appears that BFU-E cannot give rise to cells of granulocyte or macrophage lineage in culture, colonies derived from BFU-E can contain megakaryocytes in addition to erythoid cells (McLeod *et al.* 1976). The same observation has been made with a minority of granulocyte/macrophage colonies (Metcalf *et al.* 1975). It appears that the transition from pluripotential to unipotential cells may take place in a stepwise fashion, with potentialities being shed sequentially.

Temporal Proximity to CFU-S

Two observations already mentioned suggest that BFU-E are at a stage in erythroid maturation closer to CFU-S than to CFU-E. First, whereas CFU-E are limited to only two days of proliferation in culture before terminal maturation, the progeny of BFU-E can proliferate for as many as eight to ten days before synthesizing hemoglobin. Second, while BFU-E share identical density and size with CFU-S, CFU-E are denser and larger.

A third, independent experiment supports this view. Within individual spleen colonies, the number of CFU-S varies over a wide range (Siminovitch *et al.*

TABLE 2. *Correlation between number of CFU-S and later cells in individual spleen colonies**

Cell type	Correlation with CFU-S†
BFU-E (day 8‡)	.81
CFU-C	.76
BFU-E (day 3‡)	.57
CFU-E	.13§

* Adapted from Gregory and Henkelman 1977.
† Pearson product-moment correlation coefficients.
‡ Determined by scoring bursts at eight or three days of culture.
§ Not significantly different from 0.

1963). It is therefore possible to measure correlations between CFU-S and other precursor classes. At any given time in the development of a colony, the number of CFU-S will depend on the number of parent CFU-S that was present in the colony in the previous generation. If BFU-E and CFU-C also derive directly from CFU-S, then their numbers should correlate closely with the number of sister CFU-S in their own generation. Conversely, if many generations of maturing cells intercede between CFU-S and a particular cell class, no correlation would be expected between that class and CFU-S. Data from examination of individual spleen colonies have been obtained and analyzed by Gregory and Henkelman (1977) and are summarized in Table 2. High positive correlations were found between the granulocyte/macrophage precursors (CFU-C) and CFU-S, and between BFU-E and CFU-S, whereas CFU-E and CFU-S did not correlate. There is, therefore, persuasive evidence for considering BFU-E as early cells, closely related to their pluripotential precursors but restricted to erythroid and megakaryocytic development.

Proliferative Activity

The proliferative activity of colony-forming cells can be estimated by measuring their sensitivity to the lethal effect of tritiated thymidine (^{3}HTdR) at high specific activity and concentration. The percentage of colony-forming cells inactivated by a 20-minute pulse in vitro is taken as a measure of the proportion of the population engaged in DNA synthesis at the time of exposure. In marrow taken from normal mice, 30% of BFU-E and 75% of CFU-E are sensitive to killing by ^{3}HTdR (Iscove 1977a). Thus, virtually all CFU-E are normally in active cell cycle. The 25% that are resistant are likely to be in G_1 or G_2 at the time of exposure.

If CFU-E are actively cycling at the time they are plated, then their dependence on erythropoietin for colony formation must involve more than a simple "triggering" role of erythropoietin. Preliminary experiments have been done in which cultures were established without erythropoietin, and the hormone was added at variable intervals thereafter (Figure 5). CFU-E responded fully when erythro-

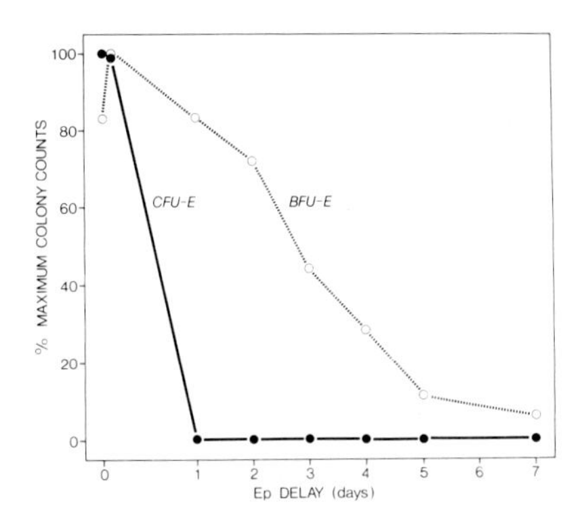

FIG. 5. Effect of delayed addition of erythropoietin (Ep) on colony formation by CFU-E (points joined by solid lines) and BFU-E in culture.

poietin was added after only two hours of culture, but no colony response was obtained when it was added after a delay of 24 or more hours. It therefore appears that if CFU-E are not exposed to erythropoietin during a critical time interval in culture, they become incapable of further cell division if they are later re-exposed to it. In contrast, delay of addition of erythropoietin by 48 hours had little effect on the number of bursts identifiable at ten days of culture. When the addition of erythropoietin was prolonged beyond the third day, the number of bursts declined, even in plates examined as late as 20 days after initiation of the cultures. The observations suggest that persistence of BFU-E, and perhaps even their proliferation, can occur independently of erythropoietin. However, as the progeny of BFU-E continue to mature they reach a stage at which they are dependent on erythropoietin for the survival of their capacity to divide further or mature.

REGULATION OF BFU-E AND CFU-E IN VIVO

Role of Erythropoietin

Circulating erythropoietin concentrations in mice can be raised by bleeding and suppressed by hypertransfusion with excess red cells. The resulting erythropoietin levels differ by at least 100-fold between bled and hypertransfused animals (for references, see Iscove 1977a). The responses of BFU-E and CFU-E are outlined in Table 3. During the first three days after bleeding, the number of marrow CFU-E increased almost threefold over normal, while after hypertransfusion they decreased about threefold below normal (confirming the observations of Gregory *et al.* 1973). BFU-E showed a contrasting pattern in only slowly declining during the three days after bleeding and in slightly increasing after hypertransfusion.

Sensitivity to ^{3}HTdR was also monitored in these experiments (Table 4). No change was noted in the sensitive proportion of BFU-E (30%) after bleeding or hypertransfusion, whereas the sensitive proportion of CFU-E was slightly decreased after hypertransfusion.

TABLE 3. *Effect of anemia and plethora on the numbers of BFU-E* and CFU-E in marrow*

Treatment	Days after treatment	Hematocrit†	Colony-forming cells/two femurs BFU-E	CFU-E
None	—	42	4,000	85,000
Bled	3	24	3,500	230,000
Hypertransfused	8–11	67	5,400	26,000

* Scored at 10 days of culture.
† A measure of the number of circulating erythrocytes.
(Reproduced, with permission of Cell and Tissue Kinetics, from Iscove 1977a)

These findings indicate that while CFU-E are likely to derive from precursors which are erythropoietin dependent, the precursors of BFU-E are clearly not responsive to physiological erythropoietin concentrations. Furthermore, the ^{3}HTdR data strongly indicate that even BFU-E are not targets for regulation by erythropoietin. It therefore appears that erythropoietin sensitivity is first acquired at a stage of maturation intermediate between BFU-E and CFU-E. Combined with kinetic data obtained in vivo (Lajtha *et al.* 1971) and the results of experiments involving delayed addition of erythropoietin to cultures, the findings suggest that the earliest phases of erythroid development involve proliferation which is not under the control of erythropoietin.

Proliferative Stimuli in Irradiated Mice

There is, however, evidence that regulation of proliferation of BFU-E does occur in vivo. When heavily irradiated mice are given isogeneic bone marrow cells, rapid growth of the grafted cells occurs until marrow populations are restored. When regenerating marrow was sampled and exposed to ^{3}HTdR, the kill of BFU-E was substantially higher than that observed with steady-state marrow (see Table 4), and this higher degree of sensitivity was unaffected by hypertransfusion of the regenerating mice. Similar increases in ^{3}HTdR sensitivity

TABLE 4. *Proliferative activity of BFU-E* and CFU-E in anemic, plethoric, and regenerating mice*

Treatment	Days after treatment	Hematocrit	Percent killed by ^{3}HTdR BFU-E	CFU-E
None	—	43	30	76
Bled	2–4	23	29	76
Hypertransfused	8–10	67	37	61†
Regenerating hypertransfused	5	63	63	80

* Scored at 10 days of culture.
† Significantly different from controls, $P < .01$.
(Reproduced, with permission of Cell and Tissue Kinetics, from Iscove 1977a)

occur among CFU-S (Becker *et al.* 1965), as well as CFU-C (Iscove *et al.* 1970). Thus, erythropoietin-independent stimuli exist, in the irradiated mouse, that elevate the proliferative activity of committed as well as uncommitted hemopoietic precursors. It is possible that the early committed progeny of the stem cells retain responsiveness to the same stimuli controlling proliferation of the stem cells themselves.

Effect of Mutant Genes at the *w* and *sl* Loci

Mice bearing mutant alleles at the *w* or *sl* loci are affected by an erythropoietin-resistant macrocytic anemia (Russell 1970). In W/W^v mice, the hemopoietic cells are defective, most strikingly apparent in the failure of W/W^v stem cells to generate macroscopic spleen colonies in normal irradiated hosts (McCulloch *et al.* 1964). In Sl/Sl^d mice, on the other hand, the hemopoietic cells are normal, but the hemopoietic environment is defective, so that normal +/+ cells fail to generate macroscopic spleen colonies in irradiated Sl/Sl^d hosts (McCulloch *et al.* 1965).

The results of measurements of CFU-C, BFU-E, and CFU-E in these mutants and their normal littermates are presented in Table 5. The bursts derived from BFU-E from both mutants in culture were normal in appearance and erythropoietin sensitivity. However, the number of 10-day BFU-E in both was low, compared to the number of granulocyte/macrophage progenitors (CFU-C). This observation could hint at a deficient rate of production of BFU-E from pluripotential precursors. However, the ratio of CFU-C to BFU-E is also subject to change in normal mice after bleeding (see Table 5). As already stressed elsewhere (Iscove 1977b, in press), conclusions on this point await studies comparing mutants to normal litermates in equivalent steady-states.

In Sl/Sl^d mice, the ratio of CFU-E to BFU-E was low, particularly when compared with anemic +/+ mice. The finding is strongly suggestive of a defect in proliferation or maturation between BFU-E and CFU-E and provides a cellular

TABLE 5. *Number of BFU-E and CFU-E in* W/W^v *and* Sl/Sl^d *mice*

Source	Hematocrit	Colony-forming cells/two femurs CFU-C	BFU-E*	CFU-E	CFU-C/BFU-E	CFU-E/BFU-E
+/+	43	64,000	4,000	97,000	16	24
Bled +/+						
4th day	22	58,000	1,400	275,000	42	196
7th day	37	50,000	3,300	122,000	15	37
W/W^v	40	45,000	1,300	63,000	35	49
Sl/Sl^d	31	25,000	890	12,000	28	13

* Scored at 10 days of culture.
(Reproduced, with permission of Jichi Medical School [Proceedings of the International Symposium on Aplastic Anemia], from Iscove 1977b, in press.)

basis for explaining the severe anemia and virtually complete erythropoietin resistance of these mice.

SUMMARY AND DISCUSSION

The observations with colony assays now make possible a tentative outline of erythroid differentiation in the mouse before the proerythroblast stage (Figure 6). The initial event is the transition from multipotential stem cells to cells that have lost the capacity for granulocyte/macrophage differentiation while retaining potential for erythroid and megakaryocytic differentiation. Cells at this early stage predominate among the population defined operationally as BFU-E, forming large colonies after 14 days of culture. Although these "determined" cells have presumably undergone changes in the organization of their chromatin, they remain similar to their parent pluripotential stem cells in size, density, low proliferative activity in the steady-state, and responsiveness to regulatory influences in vivo.

The next stage covers a spectrum characterized by the gradual onset of sensitivity to high concentrations of erythropoietin. Such cells would predominate among those BFU-E forming medium-sized colonies that cease growing before eight days of culture. With progression through this intermediate stage, sensitivity to erythropoietin increases, and the cells respond by increasing their proliferative activity. In addition, they become increasingly dependent on erythropoietin for maintenance of their further capacity to proliferate. Transition through this intermediate stage is influenced by an environmental cue related to the *sl* gene product.

CFU-E represent the final stage in this series. Their formation, beginning from the earliest BFU-E, can take as many as 8 to 12 days. Now larger and denser than the earliest cells, they are also highly erythropoietin sensitive and proliferate maximally. In the absence of erythropoietin, they lose their ability to respond to it when re-exposed. CFU-E precede the onset of hemoglobin synthesis by about 24 to 30 hours.

The key advances have yet to be made in understanding determination and proliferation control at the initiation of hemopoiesis. The humoral requirements for determination and early growth are unknown. Regulation at short range, although known to play a role, remains substantially unclarified. Defined markers, antigenic or biochemical, are not yet available for early cells. The identifica-

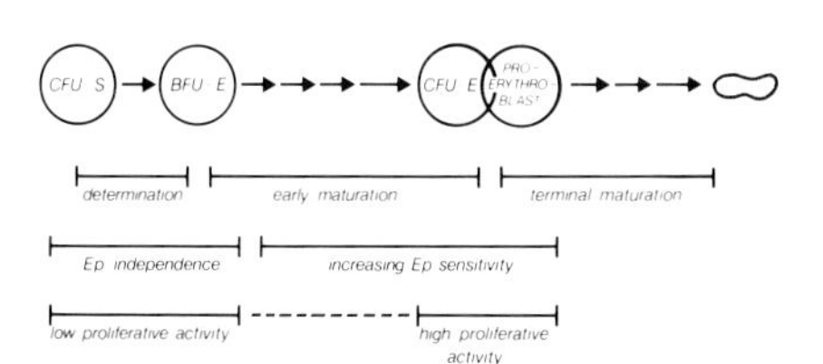

FIG. 6. Stages in erythroid differentiation. See text for details.

tion of such markers is an urgent task, because they should make possible direct study of precursor cells and provide a highly specific basis for their purification. Finally, no tool equivalent to the Friend virus-transformed lines is yet available as a model for events that precede the CFU-E stage. Because of the rarity of early cells in normal hemopoietic tissue, the development of such lines may offer the only realistic hope for understanding determination and early maturation at the molecular level.

In the development of the required tools, the approaches and the conceptual framework reviewed in this paper should provide a base upon which to build. The colony assays allow direct and selective detection of the relevant early precursors. Specific factors influencing later proliferation and maturation—G-CSF and erythropoietin—are at hand in largely defined form. The serum requirement for later stages of proliferation and maturation is now largely understood, permitting selective assay for factors important at early stages. Finally, a culture system has recently emerged (Dexter *et al.,* 1977), providing conditions for stem cell replication and determination in vitro. The stage, therefore, appears set for a rapid expansion of understanding of these critical early processes.

ACKNOWLEDGMENT

The author is grateful to Ms. M. Schweizer for excellent technical assistance.

REFERENCES

Axelrad, A. A., D. L. McLeod, M. M. Shreeve, and D. S. Heath. 1974. Properties of cells that produce erythrocytic colonies in vitro, *in* Hemopoiesis in Culture, W. A. Robinson, ed. U.S. Government Printing Office, Washington, D.C., pp. 226–234.

Becker, A. J., E. A. McCulloch, L. Siminovitch, and J. E. Till. 1965. The effect of differing demands for blood cell production on DNA synthesis by hemopoietic colony-forming cells of mice. Blood 26:296–308.

Broome, J. D., and M. W. Jeng. 1973. Promotion of replication in lymphoid cells by specific thiols and disulfides in vitro. J. Exp. Med. 138:574–592.

Clarke, B. J., A. A. Axelrad, M. M. Shreeve, and D. L. McLeod. 1975. Erythroid colony induction without erythropoietin by Friend leukemia virus in vitro. Proc. Natl. Acad. Sci. USA 72:3556–3560.

Click, R. E., L. Benck, and B. J. Alter. 1972. Immune responses in vitro. I. Culture conditions for antibody synthesis. Cell. Immunol. 3:264–276.

Dexter, T. M., T. D. Allen, and L. G. Lajtha. 1977. Conditions controlling the proliferation of haemopoietic stem cells in vitro. J. Cell. Physiol. 91:335–344.

Gregory, C. J. 1976. Erythropoietin sensitivity as a differentiation marker in the hemopoietic system: Studies of three erythropoietic colony responses in culture. J. Cell. Physiol. 89:289–302.

Gregory, C. J., and R. M. Henkelman. 1977. Relationships between early hemopoietic progenitor cells determined by correlation analysis of their numbers in individual spleen colonies, *in* Experimental Hematology Today, Springer-Verlag, N.Y.

Gregory, C. J., E. A. McCulloch, and J. E. Till. 1973. Erythropoietic progenitors capable of colony formation in culture: State of differentiation. J. Cell. Physiol. 81:411–420.

Guilbert, L. J., and N. N. Iscove. 1976. Partial replacement of serum by selenite, transferrin, albumin and lecithin in haemopoietic cell cultures. Nature 263:594–595.

Heath, D. S., A. A. Axelrad, D. L. McLeod, and M. M. Shreeve. 1976. Separation of the erythropoietin-responsive progenitors BFU-E and CFU-E in mouse bone marrow by unit gravity sedimentation. Blood 47:777–792.

Iscove, N. N. 1977a. The role of erythropoietin in regulation of population size and cell cycling of early and late erythroid precursors in mouse bone marrow. Cell Tissue Kinet. 10:323–334.

Iscove, N. N. 1977b. Committed erythroid precursor populations in genetically anemic W/W^v and Sl/Sl^d mice, *in* Proceedings of the International Symposium on Aplastic Anemia, Kyoto, 1976. (In press).

Iscove, N. N., and F. Sieber. 1975. Erythroid progenitors in mouse bone marrow detected by macroscopic colony formation in culture. Exp. Hematol. 3:32–43.

Iscove, N. N., J. E. Till, and E. A. McCulloch. 1970. The proliferative states of mouse granulopoietic progenitor cells. Proc. Soc. Exp. Biol. Med. 134:33–36.

Lajtha, L. J., C. W. Gilbert, and E. Guzman. 1971. Kinetics of haemopoietic colony growth. Br. J. Haematol. 20:343–354.

Liao, S.-K., and A. A. Axelrad. 1975. Erythropoietin-independent erythroid colony formation in vitro by hemopoietic cells of mice infected with Friend virus. Int. J. Cancer 15:467–482.

McCulloch, E. A., L. Siminovitch, and J. E. Till. 1964. Spleen-colony formation in anemic mice of genotype WW^v. Science 144:844–846.

McCulloch, E. A., L. Siminovitch, J. E. Till, E. S. Russell, and S. E. Bernstein. 1965. The cellular basis of the genetically determined hemopoietic defect in anemic mice of genotype Sl/Sl^d. Blood 26:399–410.

McLeod, D. L., M. M. Shreeve, and A. A. Axelrad. 1976. Induction of megakaryocyte colonies with platelet formation in vitro. Nature 261:492–494.

Messner, H., J. E. Till, and E. A. McCulloch. 1972. Density distributions of marrow cells from mouse and man. Ser. Haematol. 5:22–36.

Metcalf, D., H. R. MacDonald, N. Odartchenko, and B. Sordat. 1975. Growth of mouse megakaryocyte colonies in vitro. Proc. Natl. Acad. Sci. USA 72:1744–1748.

Miller, R. G. 1973. Separation of cells by velocity sedimentation, *in* New Techniques in Biophysics and Cell Biology, R. H. Pain and B. J. Smith, eds., Wiley, London, pp. 87–112.

Prchal, J. F., and A. A. Axelrad. 1974. Bone marrow responses in polycythemia vera. (Letter) N. Engl. J. Med. 290:1382.

Russell, E. S. 1970. Abnormalities of erythropoiesis associated with mutant genes in mice, *in* Regulation of Hematopoiesis, A. S. Gordon, ed., Vol. 1. Appleton-Century-Crofts, New York, pp. 649-675.

Sieber, F. 1976. Erythroid colony growth in culture: Effects of desialated erythropoietin, neuraminidase, dimethylsulfoxide and Amphotericin B, *in* Progress in Differentiation Research, N. Muller-Berat, ed., North Holland, Amsterdam, pp. 521–528.

Siminovitch, L., E. A. McCulloch, and J. E. Till. 1963. The distribution of colony-forming cells among spleen colonies. J. Cell. Comp. Physiol. 62:327–336.

Singer, J. W., and J. W. Adamson. 1976. Steroids and hematopoiesis. II. The effect of steroids on in vitro erythroid colony growth: Evidence for different target cells for different classes of steroids. J. Cell. Physiol. 88:135–144.

Stephenson, J. R., A. A. Axelrad, D. L. McLeod, and M. M. Shreeve. 1971. Induction of colonies of hemoglobin-synthesizing cells by erythropoietin in vitro. Proc. Natl. Acad. Sci. USA 68:1542–1546.

Worton, R. G., E. A. McCulloch, and J. E. Till. 1969. Physical separation of hemopoietic stem cells from cells forming colonies in culture. J. Cell. Physiol. 74:171–182.

Wu, A. M., J. E. Till, L. Siminovitch, and E. A. McCulloch. 1967. A cytological study of the capacity for differentiation of normal hemopoietic colony-forming cells. J. Cell. Physiol. 69:177–184.

Cell Differentiation and Neoplasia, edited by Grady F. Saunders. Raven Press, New York

Cellular Differentiation in the Myeloblastic Leukemias of Man

E. A. McCulloch, R. N. Buick, and J. E. Till

Institute of Medical Science, University of Toronto, and the Ontario Cancer Institute, Toronto, Ontario, Canada

Compelling genetic evidence exists that certain diseases of the hemopoietic system originate in pluripotent stem cells (clonal hemopathies). On the basis of studies of the isoenzymes of glucose-6-phosphate dehydrogenase, clonal cellular proliferation has been suggested as the cellular basis for polycythemia vera (Adamson *et al.* 1976), idiopathic myelofibrosis (Jacobson and Fialkow 1976), and chronic myelogenous leukemia (CML) (Fialkow *et al.* 1967). The identification of marker chromosomes in both myelopoietic and erythropoietic cells provides evidence for the origin of both acute (Blackstock and Garson 1974) and chronic (Whang *et al.* 1963) myeloblastic leukemia as transformations occurring in pluripotent stem cells. Continuing differentiation is obvious in the chronic disease; the finding of committed leukemic progenitors of granulopoiesis (Duttera *et al.* 1973, Moore *et al.* 1972, Aye 1974) and erythropoiesis (S. Lan, personal communication) in the marrow of patients with acute myeloblastic leukemia (AML) confirms the continuation of at least some myelopoietic differentiation. Much work has been reported on granulopoiesis in AML using culture methods for the study of committed progenitor cells (Pike and Robinson 1970, Moore *et al.* 1973, Brown and Carbone 1971, Greenberg *et al.* 1971, Cowan *et al.* 1972). Evidence has been presented for (Moore *et al.* 1974, Spitzer *et al.* 1976) and against (Curtis *et al.* 1975) the value of such culture methods as prognostic indicators. However, close parallels between granulopoiesis and leukemic proliferation usually have not been found (McCulloch and Till 1971), as might be expected if the disease originates in stem cells more undifferentiated than committed granulopoietic progenitors.

In parallel with studies of myelopoiesis in culture, we have developed methods for studying the major abnormal cell populations in AML leukemic blast cells (Aye *et al.* 1974a, Buick *et al.* 1977). The assays have provided evidence that this population is maintained by the proliferative activity of a minority subpopulation, regulated by cellular interactions based on mechanisms similar to those known to control granulopoiesis in culture. It is the purpose of this paper to review these studies of leukemic blast cell populations and to comment briefly on their possible significance for differentiation in AML.

LEUKEMIC BLAST CELLS IN SUSPENSION CULTURE

The peripheral blood of certain AML patients contains a large number of blast cells contaminated with relatively few other cell types; blasts from this source were considered to be appropriate for the development of new assays designed to measure leukemic cell proliferation and to determine the mechanism of its regulation. In our first procedure (Aye *et al.* 1974a), leukocytes from the peripheral blood of patients selected because of high peripheral blast cell concentrations were cultivated in liquid growth media consisting of fetal calf serum and a synthetic culture medium (alpha medium) to which putative stimulator cells or molecules could be added, as dictated by specific experimental regimens. After varying periods of incubation, the cultures were pulse labeled with tritiated thymidine (^{3}HTdR), and incorporation of radioisotope into acid-insoluble material was determined. In most studies, ^{3}HTdR incorporation increased with time in culture (Aye *et al.* 1975, McCulloch and Till 1977). This increase was abolished by 1,000 rads of ionizing radiation, and detailed radiation survival curves yielded parameters in the range of those associated with cell proliferation (McCulloch and Till 1977). These data were considered to provide adequate evidence for a proliferative basis of increases in ^{3}HTdR incorporation; the proliferation was considered to be leukemic since chromosome markers could be identified in cells recovered from cultures at the time of peak ^{3}HTdR incorporation (Aye *et al.* 1975). Detailed analysis of this cell population was possible because the cells retained their growth potential following storage at —70°C in dimethylsulfoxide (DMSO) (Aye *et al.* 1974a, 1975, McCulloch and Till 1977) and could be tested repeatedly with a variety of experimental designs.

As the system was used, marked patient-to-patient variation was encountered. The peak ^{3}HTdR incorporation varied from 2- to 50-fold and occurred from day 3 to day 9 of culture. The requirement for stimulators also varied. In a minority of specimens, proliferation was dependent upon the addition of leukocyte-conditioned media (LCM) to the cultures, whereas in most, proliferation occurred in the absence of stimulators but could be increased modestly by their addition. In some instances, no conditioned-media affect was noted. Where stimulation could be achieved with conditioned media, the mitogen phytohemagglutinin (PHA) also had an effect. Culturing cells over an extensive range of different cell concentrations proved useful in examining the basis for this variation. With this approach, at appropriately low cell numbers, dependence on stimulators could always be demonstrated. It became apparent that absent or modest stimulator dependence found in some kinetic experiments was a consequence of the presence in cultures at high cell densities (2×10^5 cells/ml) of adequate endogenous sources of stimulation (McCulloch and Till 1977).

By using low cell concentrations, it was feasible to examine some of the properties of the stimulator. Two of these are particularly relevant: first, the activity effective in stimulating ^{3}HTdR incorporation could be separated on hydroxylapatite from the activity essential for granulopoietic colony formation

(Aye 1974). Second, LCM active in stimulating leukemic blast cell proliferation were obtained regularly only if normal and/or leukemic leukocytes were incubated in the presence of PHA (Till *et al.* 1974). Thus, PHA-leukocyte conditioned media (PHA-LCM) were used routinely in experiments on leukemic blast cell proliferation.

By using the suspension culture method, a model was developed to account for heterogeneity in the blast cell population. On the basis of function in culture, three subpopulations could be defined: The first, a minority (1 in 10^2 to 1 in 10^4 cells) is capable of proliferation. The second acts to promote this proliferation. In culture, the interaction between these two populations requires PHA acting as an intermediary. Finally, the majority of morphologically recognized blast cells is functionally inert.

Much of the experimental basis for these conclusions can be derived from data obtained using a single experimental design. Figure 1 depicts the data of such an experiment; the underlying design was to culture varying numbers of cells for an appropriate time (9 days) and then to determine ^{3}HTdR incorporation. The figure is a log:log plot of these parameters. Three growth conditions were compared: In the first, cells were cultured in growth medium alone, with little evidence of proliferation. In the second, cells were cultured in the presence of PHA-LCM. For the same cell number, a 2- to 300-fold increase in ^{3}HTdR incorporation was observed over cells cultured in growth medium alone. In the third, an equivalent or greater stimulation was seen when cells were cultured with 10^5 irradiated cells in the presence of PHA. The controls were irradiated cells, which by themselves did not show any significant ^{3}HTdR incorporation and when added to cultures of cells in growth medium without PHA did not stimulate proliferation. Perhaps the most revealing control was the addition of PHA alone. At low cell numbers, for which either PHA-LCM or PHA plus irradiated cells had proved active stimulators, PHA alone was ineffective. With increasing cell numbers, PHA became increasingly effective. Finally, at cell numbers yielding maximum ^{3}HTdR incorporation, PHA by itself was as

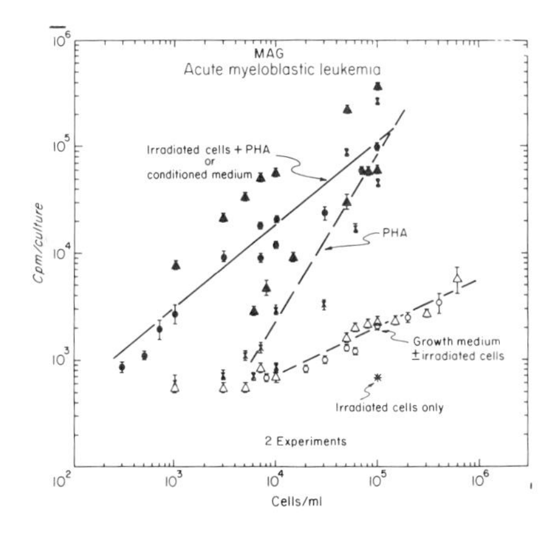

FIG. 1. The results of two experiments using cells from a patient, MAG, with acute myeloblastic leukemia. The data are depicted as a log : log plot of cell concentration vs. counts/minute/culture. Open circles, cultures in growth medium alone; open triangles, growth medium alone with 10^5 irradiated cells; closed circles, growth medium supplemented with leukocyte-conditioned medium; closed triangles, growth medium supplemented with 10^5 irradiated cells and 0.5% PHA; X, growth medium and 0.5% PHA. (Reprinted, with permission of Grune and Stratton, Inc., from *Blood,* McCulloch and Till 1977).

effective as irradiated cells in the presence of lectin. Under the conditions in which PHA alone appears to be effective as a stimulator, the total cell numbers were similar to those in cultures with irradiated cells, PHA, and a very small number of intact cells. These results are consistent with an indirect action of the lectin working to facilitate an interaction between a proliferative subpopulation and regulatory cells.

On the basis of the cell numbers required to give significant ^{3}HTdR incorporation over background, "proliferative units" are a minority population. Titration of irradiated "stimulator" cells against a fixed number of intact proliferative cells yielded data consistent with a small stimulator subpopulation (McCulloch and Till 1977). However, the liquid culture methodology provided only semi-quantitative descriptions of functional subpopulations of leukemic blasts. A colony assay was required not only to obtain better quantitation but also to permit characterization.

COLONY FORMATION BY A PROLIFERATIVE SUBPOPULATION OF LEUKEMIC BLAST CELLS

The development of a colony assay for proliferative blast cells was based on the role of PHA in the suspension culture system. The lectin not only facilitated interactions between proliferative and stimulatory subpopulations but also increased the stimulatory activity released by leukocytes into conditioned media. We found that bioactive species in such PHA-LCM preparations stimulated colony formation by proliferative blasts, even though conventionally prepared LCM was not effective (Buick *et al.* 1977) (Figure 2). Leukocytes were obtained from a patient with CML (patient COR). Fresh or cryopreserved aliquots were plated in methyl cellulose at 10^5 cells/ml. For the fresh specimens, colony formation was observed in the absence of exogenous stimulator, but it increased greatly with added PHA-LCM, reaching a peak at 5%. For the cryopreserved cells, colony formation was reduced (20% survival) and was observed only in the presence of PHA-LCM, again with a peak response at 5%. For the cryopreserved cells only, colony formation decreased with increasing concentration of PHA-LCM in excess of 5%.

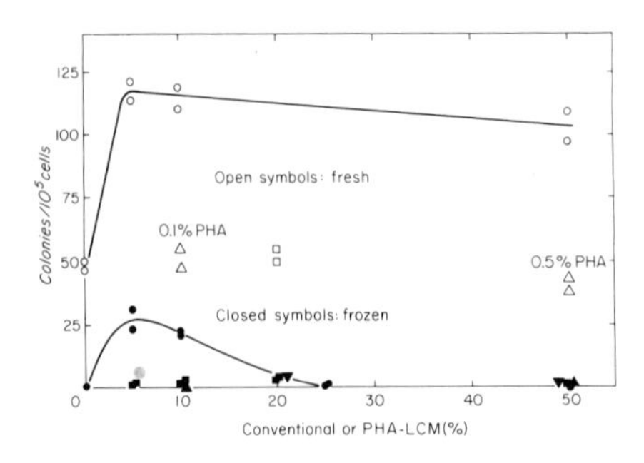

FIG. 2. Dose-response curve for the stimulation of colony formation. Open symbols are for cells freshly obtained. Closed symbols are for cryopreserved cells. Circles (○ ●) are experimental points with PHA-LCM. Controls consist of conventional LCM (squares □ ■). PHA at concentrations equivalent to the amount of lectin present in PHA-LCM, at concentrations indicated on the abscissa (triangles △ ▲), and conventional LCM mixed with equivalent amounts of PHA (inverted triangles ▼)

None of the controls (see Figure 2) stimulated colony formation. These included: (1) media conditioned in the absence of PHA by leukocytes immobilized in agar (this conventional leukocyte-conditioned medium was highly active in stimulating granulopoietic colony formation); (2) PHA at concentrations equivalent to those in PHA-LCM; and (3) PHA mixed with conventional LCM at the same concentrations. It is apparent that PHA-LCM has a proliferative target cell specificity different from the granulopoietic colony-stimulating activity in conventional LCM and from PHA itself.

Cells from three patients with AML and from one additional CML patient survived freezing and gave PHA-LCM dose response curves similar to that of the cryopreserved COR cells (Figure 2, lower curve). Of the fresh cell specimens from 13 patients, cells from only one additional AML patient were capable of colony formation without added stimulator. Four patients with AML and two with CML yielded PHA-LCM dose response curves similar to fresh COR cells (Figure 2, upper curve). Cells from two AML patients formed colonies only at low concentrations of PHA-LCM. In contrast, growth increased with increasing PHA-LCM for cells from four other AML patients. Leukocytes from only 1 of 13 AML patients tested at the time of initial diagnosis failed to form colonies.

Morphologically, the colonies were compact, contained 20 to 100 cells, and "broke up," becoming diffuse after long incubation periods. Both the morphology and the cellular composition of the colonies were different from granulopoietic colonies. In Giemsa-stained preparations, cells within colonies resembled blast cells. Other characteristics are summarized in Tables 1 and 2. The major finding is that the majority of such cells do not have characteristics permitting their confident assignment to any of the known cell lineages, although a few colonies from the cells of some patients contained one or two cells with peroxidase-positive granules localized in a single area, close to a nuclear cleft. It should be noted that colonies observed in cultures containing high concentrations of PHA-LCM (20%) were different from those seen at lower concentrations (5%). These colonies floated in the medium and contained a varying number of cells capable of E-rosette formation.

TABLE 1. *Tests for B- and T-cell surface markers on cells of PHA-LCM-stimulated colonies*

Tests	AML no.	CML no.	From single colonies no. pos/no. neg	From pooled colonies no. pos/no. neg	Normal leukocytes* no. pos/no. neg
E-rosettes†	5	3	0/176	8/1290	402/1017
Surface Ig‡	3	1	0/151	3/825	117/1300

* Total leukocytes or mononuclear fraction from Ficoll/Hypaque density separation.
† Modified from the method of Wybran *et al.* (1975).
‡ Determined using fluorescence-labeled goat anti-human Ig obtained from Meloy.

TABLE 2. *Peroxidase stain for cells in PHA-LCM-stimulated colonies**

Patient	Disease	No. of colonies examined	No. with peroxidase-positive cell or cells†
MAG	AML	16	10
R	AML	15	0
DY	AML	10	4
DI	AML	10	0
COR	CML	12	0
VAL	CML	26	12
DU	CML	20	0

* Graham-Knoll peroxidase reaction.
† Colonies containing 1 or 2 peroxidase-positive cells in colonies of 50 or more cells.

The relationship to leukemia of the cells of origin of the colonies appearing in cultures with 5% PHA-LCM was probed by chromosome analysis. Metaphase preparations were obtained from individual colonies from one patient with AML and two with CML (patients VAL and COR); these patients were chosen because chromosome markers were identified in fresh preparations and because cells survived freezing and were therefore available for study. The results are summarized in Table 3. It is evident that some colonies contained cells with markers characteristic of a leukemic clone. The unmarked cells might have derived from pre-existing normal clones or from leukemic clones that had not yet evolved to the stage of chromosomal abnormality.

An hypothesis that blast cell proliferation is not autonomous but rather is regulated by cell interactions was suggested by the earlier experiments on ^{3}HTdR incorporation. This hypothesis was tested using the colony assay method. A constant number (10^5) of intact peripheral leukocytes was mixed with increasing numbers of irradiated autologous cells and 0.5% PHA and then plated for colony formation. The results of a typical experiment are presented in Figure 3. A linear increase in colony formation was observed with increasing numbers of irradiated cells, provided PHA was present. Irradiated cells alone or PHA alone were ineffective. The data, like those based on ^{3}HTdR incorporation (McCulloch and Till 1977), are consistent with a stimulation of colony formation by a cell interaction requiring mediation by PHA, rather than a direct action of the lectin.

The proliferative populations detected by the two assays were compared by velocity sedimentation (Figure 4). In this figure, the upper panel is the nucleated cell profile and the lower 2 panels the profiles for ^{3}HTdR incorporation and colony formation, respectively. It is evident that the progenitors of colonies are more homogeneous populations in respect to sedimentation velocity than the proliferative population detected by ^{3}HTdR incorporation, although the colony-forming progenitors are included within the more heterogeneous population detected by liquid suspension assay. This experiment does not permit a firm conclusion about the relationship between the two populations. Nonetheless,

TABLE 3. *Karyotypic analysis of cells from individual colonies in culture*

Patient	No. of colonies examined	No. of metaphases obtained	No. of colonies with leukemia marker*	No. of colonies without recognizable abnormality
MAG	9	11	4	5
VAL	19	21	13	6
COR	2	2	2	0

* For MAG, abnormal karyotypes from freshly obtained marrow or individual colonies consisted of 1 additional acrocentric (G-group-like) and 1 additional metacentric chromosome (C-group-like). For VAL and COR, the marker was the Philadelphia chromosome.

it is reasonable to think that the colony assay might select for a population with high proliferative capacity, while cells capable of fewer divisions would also be detected by an assay less demanding of cell proliferation.

To test whether or not the colony assay using PHA-LCM yielded data that could be correlated with the major abnormal cell populations characteristic of acute leukemia, we compared the concentrations of morphologically identified blast cells with colony formation by leukocytes from 13 patients with AML taken at the time of diagnosis (Figure 5). A correlation significant at the 1% level was observed, and the Spearman rank correlation coefficient was 0.86.

COMMENTS

Three lines of evidence support the view that the assay for PHA-LCM-dependent colony formation detects cell classes closely related to leukemia: (1) Cells within colonies resemble leukemic blasts morphologically and markers of differentiation have not been found regularly in association with them. (2) At least some colonies contain chromosomal markers characteristic of leukemic clones. (3) A highly significant correlation exists between blast cell concentration and colony-forming capacity in leukemic peripheral blood.

The features of the PHA-LCM responsive colonies are not identical with any of the previously reported colonies. The colonies do not exhibit definite granulopoietic differentiation and their formation in culture is dependent upon the presence of stimulators other than those active on granulopoietic progenitors. Unlike erythropoietic colony formation in culture, the PHA-LCM colonies are erythropoietin independent and do not contain recognizable erythroblasts. Unlike colonies of either B or T cells, cells within the colonies (formed at low PHA-LCM concentrations) do not form E-rosettes and do not have surface immunoglobulins.

Dicke and co-workers (1976) have reported growth of colonies containing leukemic blasts in culture. However, both the culture procedures and the cellular composition of the colonies are different from those described here for PHA-LCM-dependent colony formation. In the system of Dicke *et al.,* leukemic mar-

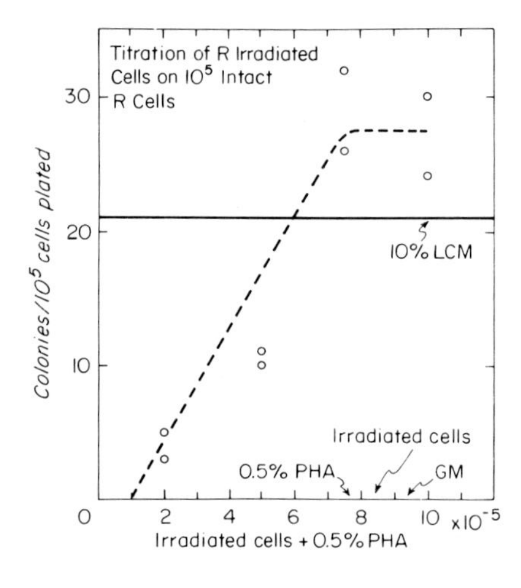

FIG. 3. Colony formation stimulated by increasing numbers of autologous irradiated cells in the presence of PHA. Controls consisting of growth medium, irradiated cells alone, or PHA alone, showed no colony formation. The dotted line marks the level of colony formation obtained with a potent PHA-LCM.

row but not peripheral blood yields colonies. Preincubation with PHA was required and LCM was not an effective stimulator. Although the cells of the colonies were blast-like in morphology, they were strongly peroxidase positive. These features of the system of Dicke *et al.* distinguished it from PHA-LCM-dependent colony formation, in which cells from the peripheral blood respond to conditioned media to yield predominantly peroxidase-negative cells.

The distinction between PHA-LCM colony formation and granulopoiesis in culture was first made on the basis of stimulator specificity (see Figure 2). As

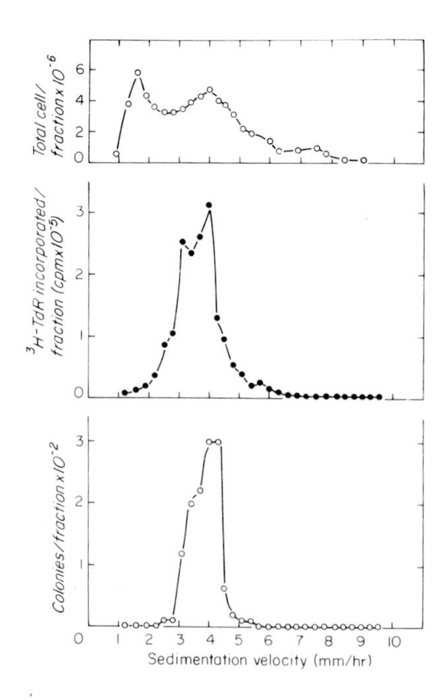

FIG. 4. Velocity sedimentation profiles from the cryopreserved peripheral leukocytes of a patient (R) with AML. Top panel, total cells; the very slowly sedimenting peak (2 mm/hour) is debris resulting in the main part from the freezing and thawing procedure. Middle panel, ^{3}HTdR incorporation per fraction. Bottom panel, PHA-LCM-dependent colony formation per fraction.

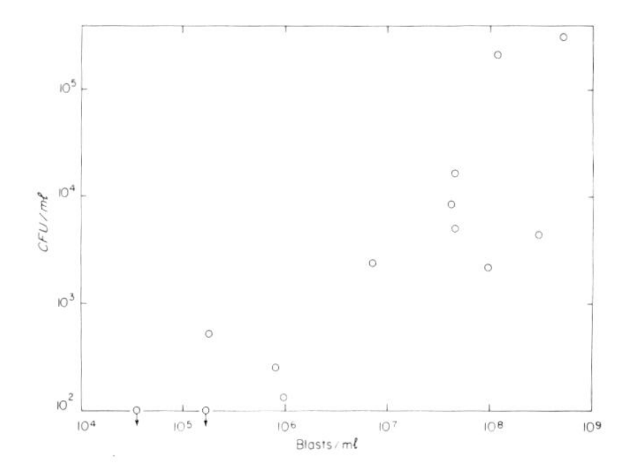

FIG 5. Correlation between the concentration of morphologically recognizable blast cells and colony formation in the peripheral blood of 13 patients with AML, studied at the time of diagnosis. Spearman's rank correlation = 0.86, and the correlation is significant ($p < 0.01$).

the bioactive molecules active in both assays are derived from peripheral leukocytes and obtained in conditioned media by similar methodologies, great specificity should not be expected of crude LCM preparations. The spectrum of responding cell classes might be even broader under conditions in which whole leukocytes rather than conditioned media were used as stimulator sources. Indeed, it is conceivable that interactions between homologous cells in the top and bottom layers used in the agar technique could cause release of bioactive proteins by mechanisms similar to those responsible for the PHA effect.

It is reasonable to consider that some of the differences observed between methods using methyl cellulose and stimulated with LCM, compared to methods using feeder cells immobilized in the bottom layer, might derive from just such variation in stimulator release. If such a broadened spectrum of stimulated proliferative classes expressing their growth potential in the double-layer method were to include cells analogous to the progenitors of PHA-LCM-dependent colonies, correlations between growth patterns and response to chemotherapy (Moore *et al.* 1974, Spitzer *et al.* 1976) might possibly relate to the detection of colonies from progenitors other than those of granulopoiesis.

We have suggested that the assay procedures applied to blast cells from the peripheral blood detect populations in differentiation pathways other than myelopoiesis and that these populations are directly related to leukemic proliferation. The question arises as to the nature of these nonmyelopoietic progenitors. In the absence of markers on the cells within colonies, even a tentative assignment is premature. Nonetheless, recent studies on the blast transformation of CML suggests that leukemic lymphopoiesis may be occurring, at least in some instances (Boggs 1974, Beard *et al.* 1976, Sarin and Gallo 1974, Schlossman *et al.* 1976). PHA-LCM-dependent colony formation is observed from peripheral blood cells of patients with CML before they reach blast crisis, and the colonies are not distinguishable from those obtained from the peripheral blood of patients with AML. Janossy *et al.* (1976) have recently advanced an hypothesis of the interrelationship between progenitors in the various forms of leukemia, based on heterogeneity within the stem cell population. It is attractive to consider that PHA-

LCM may contain bioactive molecules specific for a leukemic cell subpopulation common to the myeloblastic leukemias generally.

ACKNOWLEDGMENTS

The work reported in this paper is supported by the Ontario Cancer Treatment and Research Foundation, the Medical Research Council of Canada, and the National Cancer Institute of Canada. RNB is a special fellow of the Leukemia Society of America.

The authors are grateful to Mrs. Rita Chan for her excellent technical assistance.

REFERENCES

Adamson, J. W., P. J. Fialkow, S. Murphy, J. F. Prchal, and L. Steinmann. 1976. Polycythemia vera: Stem cell and probable clonal origin of the disease. N. Engl. J. Med. 295:913–916.

Aye, M. T. 1974. Studies of hemic cells from patients with leukemia. Ph.D. thesis, University of Toronto.

Aye, M. T., Y. Niho, J. E. Till, and E. A. McCulloch. 1974a. Studies of leukemic cell populations in culture. Blood 44:205–219.

Aye, M. T., J. E. Till, and E. A. McCulloch. 1974b. Cytological studies of colonies in culture derived from the peripheral blood cells of two patients with acute leukemia. Exp. Hematol. 2:362–371.

Aye, M. T., J. E. Till, and E. A. McCulloch. 1975. Interacting populations affecting proliferation of leukemic cells in culture. Blood 45:485–493.

Beard, M. E. J., J. Durrant, D. Catovsky, E. Wiltshaw, J. L. Amess, R. L. Bearly, B. Kirk, P. Wrigley, G. Janossy, M. F. Greaves and D. Galton.1976. Blast crisis of chronic myeloid leukaemia (CML). I. Presentation simulating acute lymphoid leukaemia (ALL). Br. J. Haematol. 34:167–178.

Blackstock, A. M., and O. M. Garson. 1974. Direct evidence for involvement of erythroid cells in acute myeloblastic leukemia. Lancet 2:1178–1179.

Boggs, D. R. 1974. Hematopoietic stem cell theory in relation to possible lymphoblastic conversion of chronic myeloid leukemia. Blood 44:449–457.

Brown, C. H., and P. P. Carbone. 1971. *In vitro* growth of normal and leukemic human bone marrow. J. Natl. Cancer Inst. 46:989–1000.

Buick, R. N., J. E. Till, and E. A. McCulloch. 1977. Colony assay for proliferative blast cells circulating in myeloblastic leukemia. Lancet 1: 862–863.

Cowan, D. H., A. Clarysse, H. Abu-Zahra, J. S. Senn, and E. A. McCulloch. 1972. The effect of remission induction in acute myeloblastic leukemia on colony formation in culture. Ser. Hematol. 5:179–188.

Curtis, J. E., D. H. Cowan, D. E. Bergsagel, R. Hasselback, and E. A. McCulloch. 1975. Acute leukemia in adults: Assessment of remission induction with combination chemotherapy by clinical and cell-culture criteria. Can. Med. Assoc. J. 113:289–294.

Dicke, K. A., G. Spitzer, and M. J. Ahearn. 1976. Colony formation in vitro by leukaemic cells in acute myelogenous leukaemia with phytohaemagglutinin as stimulating factor. Nature 259:129–130.

Duttera, J. J., J. M. Bull, J. Northup, E. S. Henderson, E. D. Stashick, and P. P. Carbone. 1973. Serial in vitro bone marrow culture in acute lymphocytic leukemia. Blood 42:687–700.

Fialkow, P. J., S. M. Gartler, and A. Yoshida. 1967. Clonal origin of chronic myelogenous leukemia in man. Proc. Natl. Acad. Sci. USA 58:1468–1473.

Greenberg, P. L., W. C. Nichols, and S. L. Schrier. 1971. Granulopoiesis in acute myeloid leukemia and preleukemia. N. Engl. J. Med. 284:1225–1232.

Jacobson, R. J., and P. J. Fialkow. 1976. Idiopathic myelofibrosis: Stem cell abnormality and probable neoplastic origin. (Abstract) Clin. Res. 24:439A.

Janossy, G., M. Roberts, and M. F. Greaves. 1976. Target cell in chronic myeloid leukaemia and its relationship to acute lymphoid leukaemia. Lancet 2:1058–1060.

McCulloch, E. A., and J. E. Till. 1971. Regulatory mechanisms acting on hemopoietic stem cells: Some clinical implications. Am. J. Pathol. 65:601–619.

McCulloch, E. A., and J. E. Till. 1977. Interacting cell populations in cultures of leukocytes from normal or leukemic peripheral blood. Blood 49:269–280.

Moore, M. A. S., N. Williams, and D. Metcalf. 1972. Characterization of in vitro colony forming cells in acute and chronic myeloid leukemia, *in* The Nature of Leukaemia, P. Vincent, ed. (Proceedings of the International Cancer Conference, Sydney, Australia), V. C. N. Blight, Sydney, pp. 135–249.

Moore, M. A. S., N. Williams, and D. Metcalf. 1973. In vitro colony formation by normal and leukemic human hematopoietic cells: Interaction between colony forming and colony stimulating cells. J. Natl. Cancer Inst. 50:591–602.

Moore, M. A. S., G. Spitzer, N. Williams, D. Metcalf, and J. Buckley. 1974. Agar culture studies in 127 cases of untreated acute leukemia: The prognostic value of reclassification of leukemia according to in vitro growth characteristics. Blood 44:1–18.

Pike, B. L., and W. A. Robinson. 1970. Human bone marrow colony growth in agar-gel. J. Cell. Physiol. 76:77–84.

Sarin, P. S., and R. C. Gallo. 1974. Terminal deoxynucleotidyltransferase in chronic myelogenous leukemia. J. Biol. Chem. 249:8051–8053.

Schlossman, S. F., L. Chess, R. E. Humphreys, and J. L. Strominger. 1976. Distribution of Ia-like molecules on the surface of normal and leukemic human cells. Proc. Natl. Acad. Sci. USA 73:1288–1292.

Spitzer, G., K. A. Dicke, E. A. Gehan, T. Smith, K. B. McCredie, B. Barlogie, and E. J Freireich. 1976. A simplified in vitro classification for prognosis in adult acute leukemia: The application of in vitro results in remission predictive models. Blood 48:795–807.

Till, J. E., T. W. Mak, G. B. Price, J. S. Senn, and E. A. McCulloch. 1974. Cellular and molecular approaches to the study of myeloproliferative disorders. Advances in the Biosciences 16:57–75.

Whang, J., E. Frei, J. H. Tjio, P. P. Carbone, and G. Brecher. 1963. The distribution of the Philadelphia chromosome in patients with chronic myelogenous leukemia. Blood 22:664–673.

Wybran, J., A. S. Levin, H. H. Fudenberg, and A. L. Goldstein. 1975. Thymosine: Effects on normal human blood T cells. Ann. NY Acad. Sci. 249:300–307.

Cell Differentiation and Neoplasia, edited by Grady F. Saunders. Raven Press, New York © 1978.

Control of Normal Cell Differentiation in Leukemic White Blood Cells

Leo Sachs

Department of Genetics, Weizmann Institute of Science, Rehovot, Israel

DIFFERENTIATION IN VITRO OF NORMAL MACROPHAGES, GRANULOCYTES, AND LYMPHOCYTES

We have studied the growth and differentiation of normal white blood cells and used these studies to elucidate the mechanism that controls cell differentiation in leukemia. In order to carry out such studies, we first developed experimental systems in which the growth and differentiation of normal mammalian white blood cells could be studied in culture. We have shown that differentiation of normal granulocytes, macrophages, mast cells, and lymphocytes can be induced in mass cultures in liquid medium (Sachs 1964, Ginsburg and Sachs 1963, 1965) and that normal macrophages, granulocytes (Figure 1) (Pluznik and Sachs 1965, 1966, Ichikawa *et al.* 1966), and lymphocytes (Figure 2) (Gerassi and Sachs 1976, Fibach *et al.* 1976) can be cloned in semisolid medium such as agar (Pluznik and Sachs 1965) or methylcellulose (Ichikawa *et al.* 1966). In contrast to the formation of lymphocyte colonies (Gerassi and Sachs 1976, Fibach *et al.* 1976, Kincade *et al.* 1976), the formation of granulocyte and macrophage colonies does not seem to require addition of a foreign antigen.

Normal, undifferentiated hematopoietic cells can be induced to form colonies with mature macrophages and granulocytes by a protein inducer (Pluznik and Sachs 1965, 1966, Ichikawa *et al.* 1966) that we now call MGI (macrophage and granulocyte inducer) (Landau and Sachs 1971, Sachs 1974a,b). It is secreted by various types of cells, including fibroblasts, and can be found in serum. This inducer, which has also been referred to as mashran gm (Ichikawa *et al.* 1967) CSF or CSA (Metcalf 1969, Austin *et al.* 1971), is specific for the induction of macrophage and granulocyte colonies. There are different requirements for macrophage and granulocyte colonies (Ichikawa *et al.* 1966, Sachs 1947b), and the type of colony induced may be due to different co-factors for MGI (Landau and Sachs 1971, Sachs 1974b). Another specific inducer, erythropoietin, induces the formation of erythroid colonies (Stephenson *et al.* 1971), and colonies of lymphocytes can be induced by using an appropriate lectin, such as phytohemagglutinin, pokeweed mitogen (Gerassi and Sachs 1976, Fibach *et al.* 1976), or concanavalin A. In all cases, the inducer has to be present until the formation of colonies with the differentiated cells is completed. Unfractionated mouse

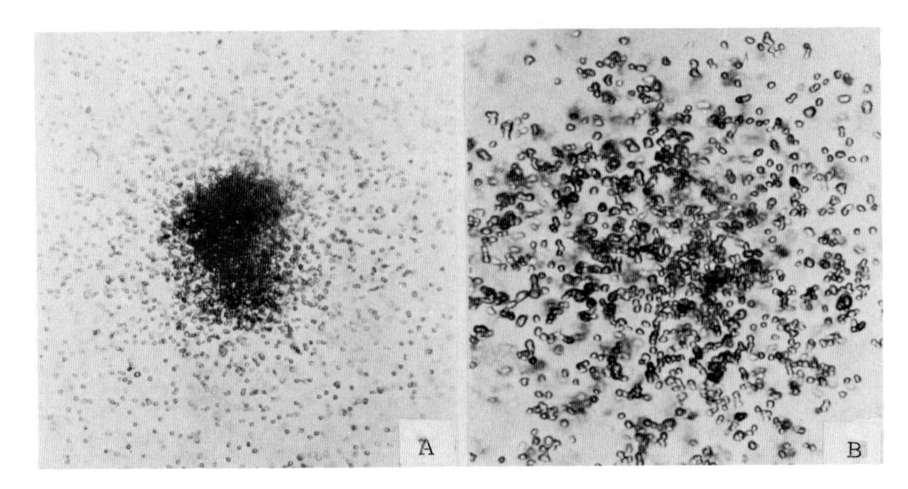

FIG. 1. A, Granulocyte colony; **B,** macrophage colony from normal mouse hematopoietic cells cloned in agar with MGI.

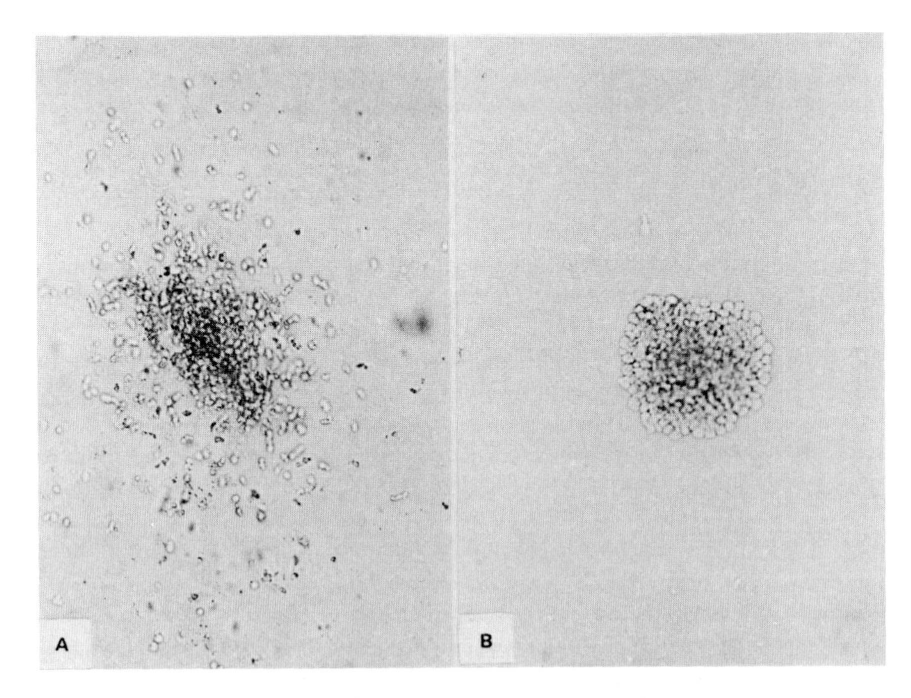

FIG. 2. Colonies of normal lymphocytes from human peripheral blood cells cloned in agar with the lectin pokeweed mitogen. **A,** colony inside the agar; **B,** colony on the agar.

lymphocytes can be induced to form colonies of B-lymphocytes in the presence of mercaptoethanol (Metcalf *et al.* 1975), and this induction also seems to require the addition of a mitogen (Kincade *et al.* 1976). With macrophages (Pluznik and Sachs 1966), granulocytes (Paran and Sachs 1969), and T-lymphocytes (Gerassi and Sachs 1976), we have shown that colonies can be derived from single cells and are, therefore, clones.

INDUCTION OF NORMAL CELL DIFFERENTIATION IN MYELOID LEUKEMIC CELLS

The finding of differentiation inducers raised the question of whether an inducer required for normal cell differentiation could induce the normal differentiation of leukemic cells. In order to test this possibility, we have studied myeloid leukemic cells from different sources (Paran *et al.* 1970, Fibach *et al.* 1972, 1973, Lotem and Sachs 1974, Krystosek and Sachs 1976). In contrast to the normal cells that require MGI for cell viability, growth, and differentiation, these myeloid leukemic cells do not require MGI for cell viability and growth (Fibach and Sachs 1976).

Our results have shown that there is one type of myeloid leukemic cell, which we will call $Fc^+C3^+D^+$, that can be induced by the protein inducer MGI to form surface membrane receptors for the Fc portion of immunoglobulin G and the C3 component of complement (Figure 3) (Lotem and Sachs 1974), to synthesize and secrete lysozyme (Figure 4) (Krystosek and Sachs 1976), and

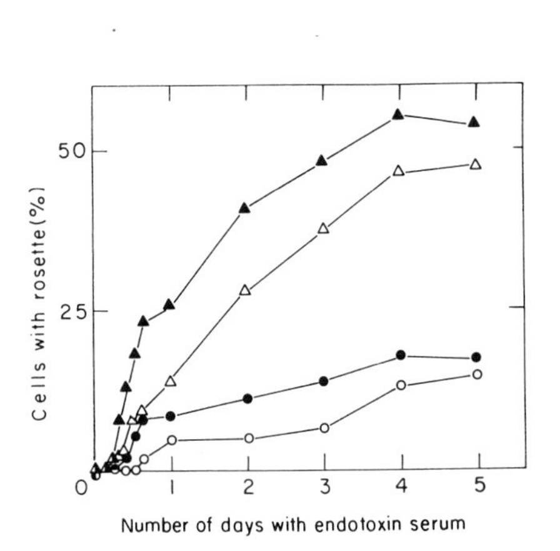

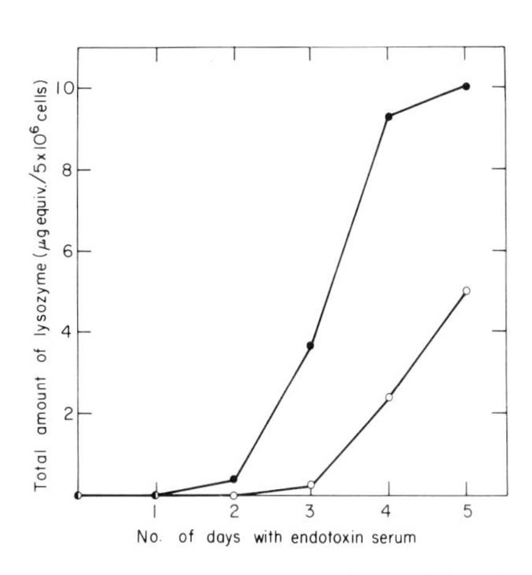

FIG. 3, left. Induction of Fc and C3 rosettes on $Fc^+C3^+D^+$ and $Fc^+C3^+D^-$ cells at different times after incubation with serum from mice injected with bacterial endotoxin (endotoxin serum). This serum contains MGI. Fc rosettes on $Fc^+C3^+D^+$ (●---●) and $Fc^+C3^+D^-$ (○---○) cells. C3 rosettes on $Fc^+C3^+D^+$ (▲---▲) and $Fc^+C3^+D^-$ (△---△) cells.

FIG. 4, right. Induction of lysozyme in $Fc^+C3^+D^+$ and $Fc^+C3^+D^-$ cells at different times after incubation with endotoxin serum that contains MGI. $Fc^+C3^+D^+$ (●---●), $Fc^+C3^+D^-$ (○---○).

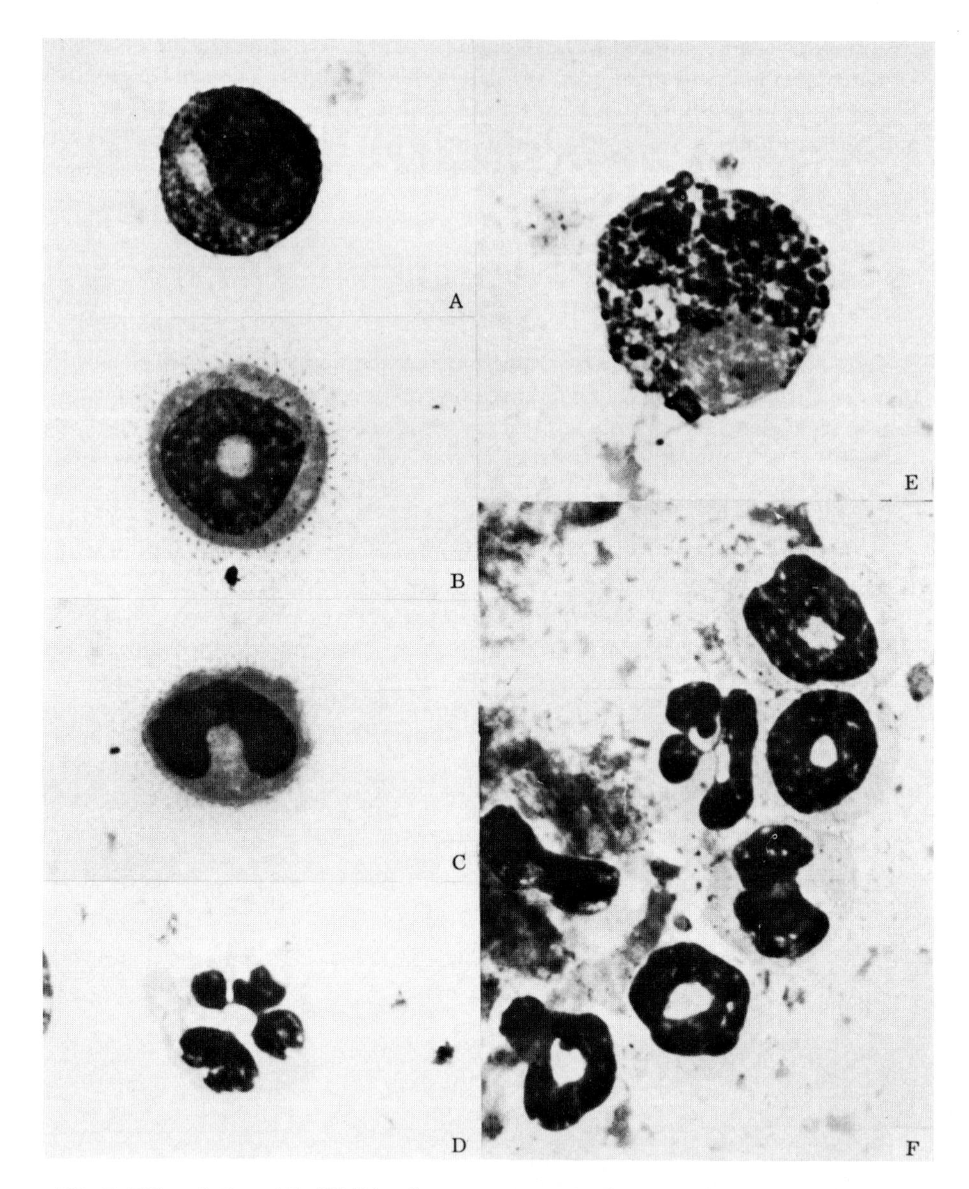

FIG. 5. Differentiation of $Fc^+C3^+D^+$ cells to mature macrophages and granulocytes by MGI. **A,** undifferentiated blast cell; **B-D,** stages in the differentiation to mature granulocytes; **E,** macrophage; **F,** group of granulocytes in different stages of differentiation.

to differentiate to mature granulocytes and macrophages (Figure 5) (Paran *et al.* 1970, Fibach *et al.* 1972, 1973). The surface receptors induced on these myeloid leukemic cells are also found on normal mature granulocytes and macrophages and can be detected by rosette formation with sheep erythrocytes coated

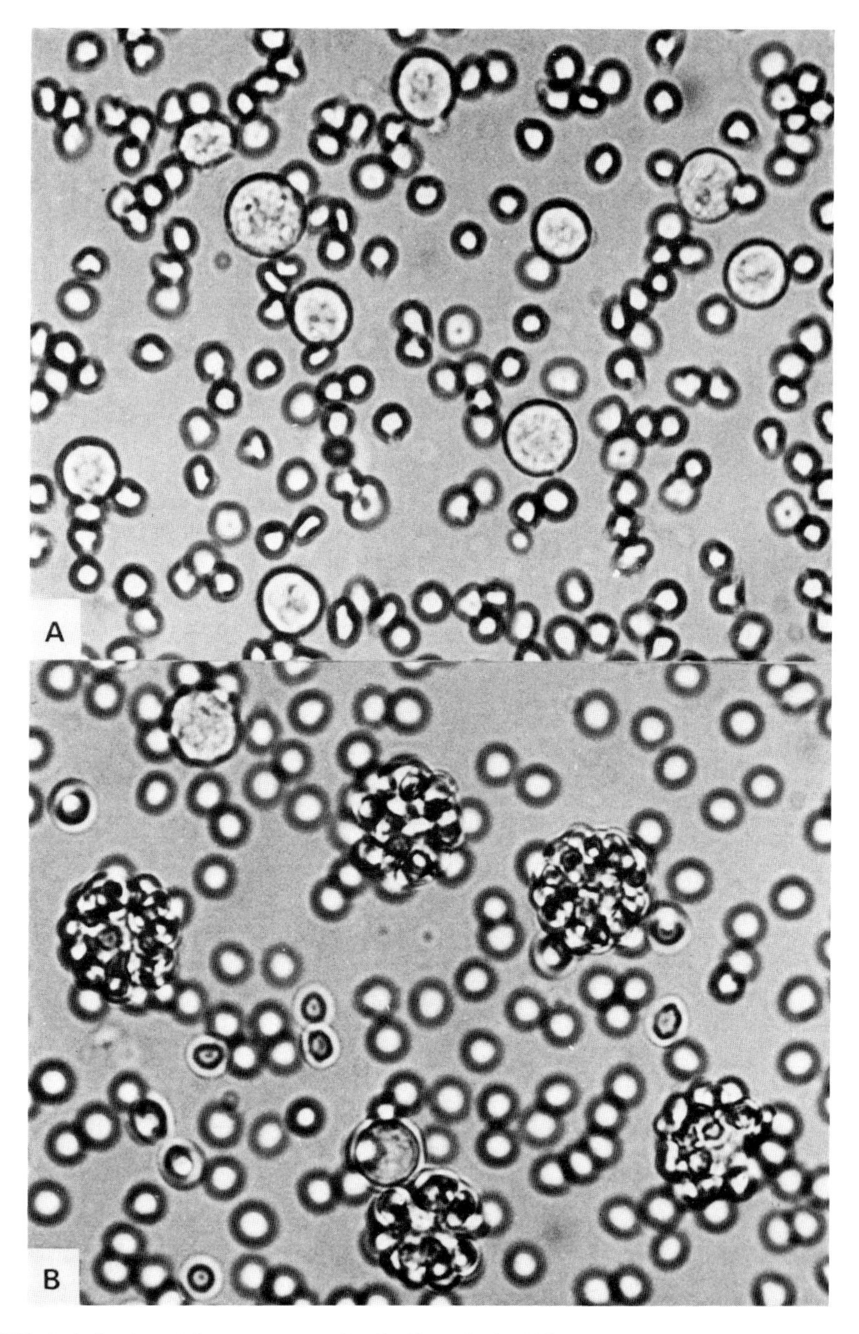

FIG. 6. Induction of C3 rosettes on $Fc^{+}C3^{+}D^{+}$ cells by MGI. **A,** untreated cells; **B,** rosettes induced by MGI.

with antibody or with antibody and complement (Figure 6). Induction of differentiation to mature granulocytes in these myeloid leukemic cells resulted in the loss of their leukemogenicity. The mature granulocytes induced by MGI from $Fc^+C3^+D^+$ leukemic cells behaved like normal granulocytes (Fibach and Sachs 1975, Vlodavsky *et al.* 1976). These myeloid leukemic cells differ from Friend erythroid leukemic cells in that they still respond to their normal differentiation-inducing protein, MGI, whereas the Friend cells no longer respond to the normal differentiation-inducing protein, i.e., erythropoietin (Kluge *et al.* 1974).

DIFFERENT BLOCKS IN THE DIFFERENTIATION OF MYELOID LEUKEMIC CELLS

In addition to $Fc^+C3^+D^+$ myeloid leukemic cells, there is another type of myeloid leukemic cell, $Fc^+C3^+D^-$, that can be induced to form Fc and C3 rosettes and lysozyme, with a lower inducibility than $Fc^+C3^+D^+$ cells but that could not be induced to differentiate to mature granulocytes or macrophages. A third type of cell, $Fc^-C3^-D^-$, could not be induced by MGI for rosettes, lysozyme, or mature cells (Figure 7). We have obtained tissue culture clones of these different types of myeloid leukemic cells (Fibach *et al.* 1973, Lotem and Sachs 1974, Krystosek and Sachs 1976). Blast cells from all three types induced myeloid leukemia in vivo and, as mentioned above, induction of granulocyte differentiation in $Fc^+C3^+D^+$ cells resulted in the loss of leukemogenicity. After the induction of differentiation, D^+ clones can also be distinguished from D^- clones by migration of the differentiated cells (Figure 8).

For comparison with the protein inducer MGI, we have studied the effects of various compounds, including steroid hormones (Lotem and Sachs 1975, 1976). These studies have shown that $Fc^+C3^+D^+$ cells were induced to form C3 but not Fc rosettes by dexamethasone, prednisolone, and estradiol (Figure 9). Induction required protein synthesis and was not inhibited by cordycepin or vinblastine; optimum induction required the continued presence of the hormone. These hormones also induced $Fc^+C3^+D^+$ cells to form macrophages but not granulocytes. In contrast to MGI, these hormones induced some but not all the changes associated with normal cell differentiation (Lotem and Sachs 1975). Incubation with dibutyryl-cyclic AMP or GMP, theophylline, aminophyl-

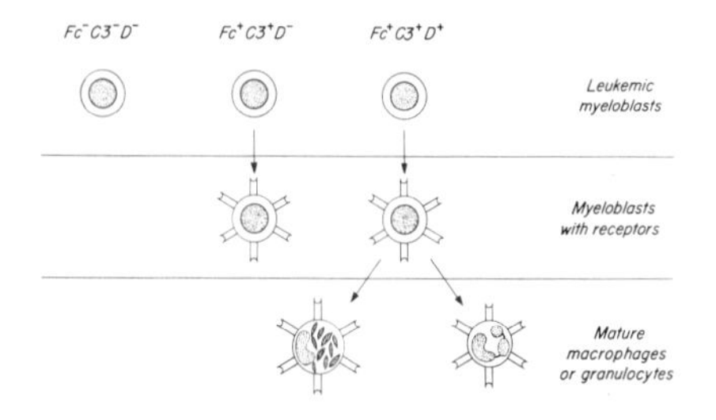

FIG. 7. Different blocks in the differentiation of myeloid leukemic cells. D, differentiation to mature macrophages and granulocytes; Receptors, receptors for Fc and C3 rosettes.

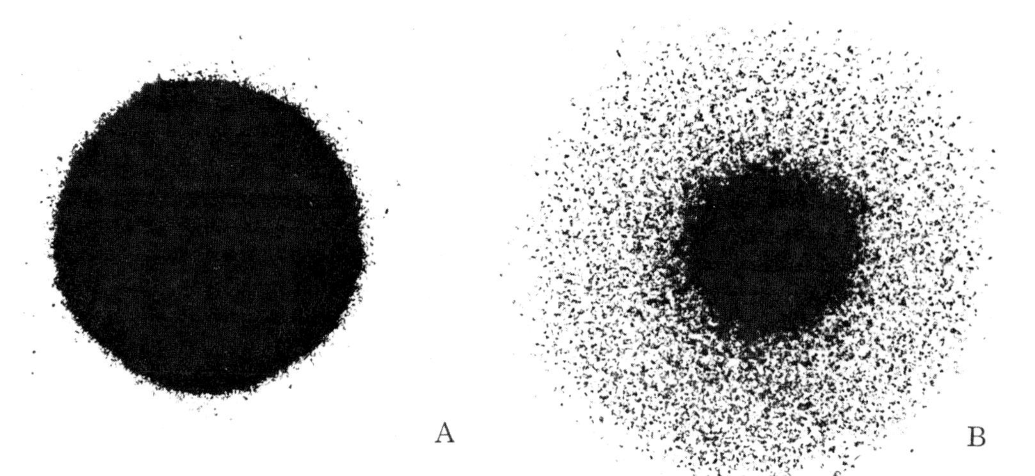

FIG. 8. A, $Fc^+C3^+D^-$ colony; **B,** $Fc^+C3^+D^+$ colony in agar. The cells were grown with MGI.

line, or prostaglandin E_1 or E_2 did not induce the formation of Fc or C3 rosettes, macrophages, or granulocytes (Lotem and Sachs 1976).

The results with steroid hormones have also shown differences in the response of the different types of myeloid leukemic cells. $Fc^+C3^+D^-$ cells showed a lower steroid inducibility for C3 rosettes than $Fc^+C3^+D^+$ cells, and no induction of macrophages. There was no induction of rosettes or macrophages with $Fc^-C3^-D^-$

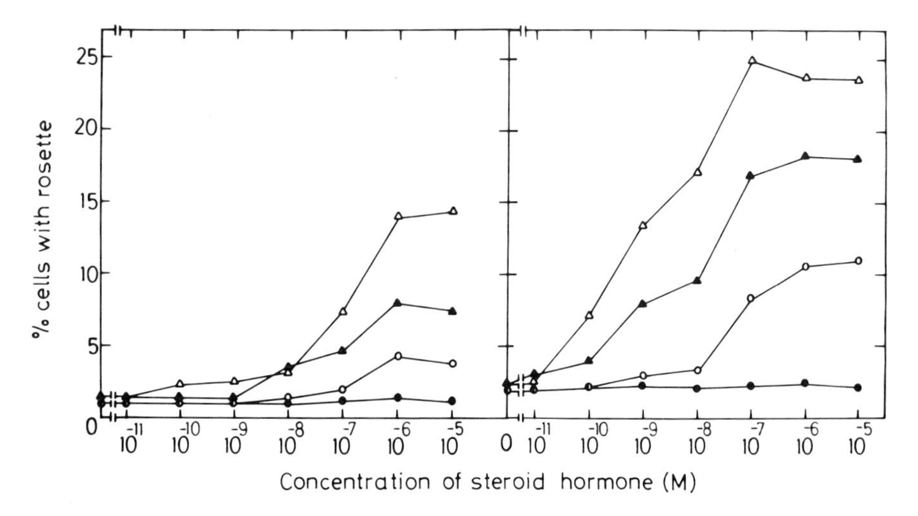

FIG. 9. Induction of C3 rosettes on $Fc^+C3^+D^+$ cells by the steroids prednisolone, dexamethasone, and estradiol at different times after incubation with 1 μM steroid. Untreated control (●---●), estradiol (○---○), prednisolone (▲---▲), dexamethasone (△---△). Progesterone, testosterone, and cortisone gave results similar to the untreated control.

cells. Progesterone, testosterone, or cortisone did not induce the formation of rosettes or macrophages and inhibited induction by the inducing hormones. Our results indicate that the steroid induction of surface changes in myeloid leukemia cells involves specific steroid hormone receptors (Lotem and Sachs 1975, 1976, Krystosek and Sachs, in press).

Other clones have been isolated that cannot be induced to form rosettes by MGI but can be induced to form Fc or Fc and C3 rosettes by dexamethasone. The results indicate that there are different cellular sites for MGI and the steroid hormones (Lotem and Sachs 1976). Dimethylsulfoxide, which can induce some of the stages of differentiation in erythroid leukemic cells (Friend *et al.* 1971, Reuben *et al.* 1976), induced C3 rosettes and lysozyme in only one $Fc^+C3^+D^+$ clone but in none of the $Fc^+C3^+D^-$ or $Fc^-C3^-D^-$ clones tested (Krystosek and Sachs 1976).

The inducing hormones prednisolone and dexamethasone can also induce the activity of alkaline phosphatase (Sela and Sachs 1974) and tyrosine aminotransferase (Rousseau *et al.* 1972) in some types of mammalian cells, enhance the induction of murine leukemia virus by 5-iododeoxyuridine, and induce the synthesis of B-type virus from mammary adenocarcinoma cells (Lowy *et al.* 1971, Paran *et al.* 1975). The enhancement of induction of murine leukemia virus by these steroid hormones seems to differ from the induction of changes in myeloid leukemic cells. Cordycepin, which inhibits the synthesis of poly A, inhibited virus induction but did not inhibit the induction of changes in the myeloid leukemic cells. Estradiol induced changes in the myeloid leukemic cells and did not enhance the induction of murine leukemia virus by 5-iododeoxyuridine.

Hormones like prednisolone are used in combination chemotherapy with compounds such as cytosine arabinoside for treatment of human myeloid leukemia. This compound, and other compounds such as actinomycin D, can induce Fc and C3 rosettes on $Fc^+C3^+D^+$ and $Fc^+C3^+D^-$, but not on $Fc^-C3^-D^-$ cells (Lotem and Sachs 1974). The therapeutic effects of prednisolone-like hormones and compounds such as cytosine arabinoside may, in part, be due to their induction of changes associated with differentiation. It will, therefore, be of interest to study the inducibility of Fc and C3 rosettes and other changes associated with differentiation in myeloid leukemic cells from different patients (Paran *et al.* 1970) in relation to their response to chemotherapy.

SIMILAR SEQUENCE OF DIFFERENTIATION IN NORMAL AND MYELOID LEUKEMIC CELLS

In order to compare the sequence of differentiation in normal and myeloid leukemic cells, we first developed a technique for enrichment of the myeloid precursor cells from normal bone marrow. An enriched population consisting of early myeloid cells (myeloblasts and promyelocytes) was obtained from normal bone marrow by injecting mice with sodium caseinate and removing cells with C3 rosettes using Ficoll-Hypaque density centrifugation. The enriched popula-

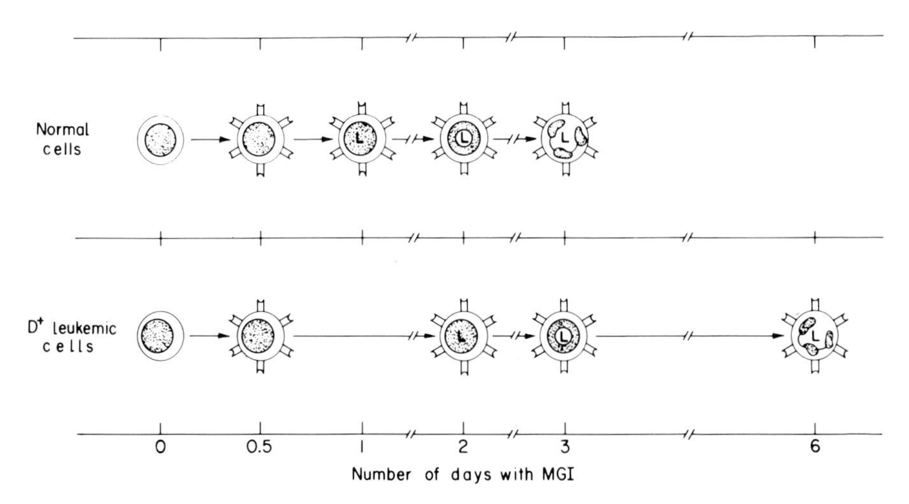

FIG. 10. The sequence and timing of differentiation in normal and $Fc^+C3^+D^+$ leukemic cells after incubation with MGI. Fc and C3 rosettes (⊓), lysozyme (L), intermediate stage of granulocyte differentiation (◉), mature granulocyte ( ).

tion had no C3 or Fc rosettes and contained about 90% early myeloid cells, compared to about 15% in the unfractionated bone marrow. Nearly all these early myeloid cells were stained for myeloperoxidase. After seeding with MGI, the enriched population showed a cloning efficiency of about 15%, compared to 0.3% in the unfractionated bone marrow, and both the enriched and the unfractionated cells gave rise to macrophage and granulocyte colonies. The increased cloning efficiency of the enriched population did not seem to be due to the removal of inhibitory cells in the bone marrow (Lotem and Sachs 1977).

The normal early myeloid cells were induced to differentiate by MGI to mature granulocytes and macrophages. The sequence of granulocyte differentiation was the formation of Fc and C3 rosettes, followed by the synthesis and secretion of lysozyme, and then morphological differentiation to mature cells. The $Fc^+C3^+D^+$ myeloid leukemic cells with no Fc or C3 rosettes before induction had a similar morphology to normal early myeloid cells and showed the same sequence of differentiation. The induction of Fc and C3 rosettes occurred at the same time in both the normal and $Fc^+C3^+D^+$ leukemic cells, but lysozyme synthesis and the formation of mature granulocytes was induced later in the leukemic than in the normal cells (Figure 10) (Lotem and Sachs 1977). The results indicate that normal and $Fc^+C3^+D^+$ myeloid leukemic cells have a similar sequence of differentiation but that the normal cells had a greater sensitivity for the formation of mature cells by MGI.

SURFACE MEMBRANE-CYTOSKELETON INTERACTIONS AND THE CONTROL OF DIFFERENTIATION

In order to study the association between surface membrane–cytoskeleton interactions and the inducibility of differentiation of myeloid leukemic cells by

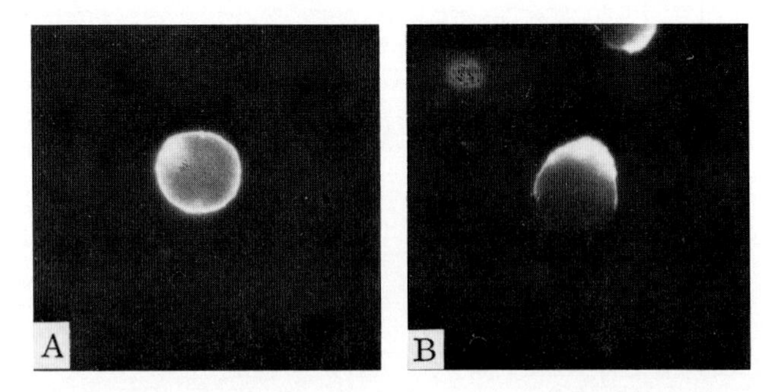

Fig. 11. A, $Fc^+C3^+D^+$ cells without a cap; and **B,** with a cap, induced by fluorescent Con A.

MGI, we have used, as a probe, the formation of caps by the lectin concanavalin A (Con A) (Sachs 1974a) and the effect of vinblastine on cap formation (Edelman 1976).

$Fc^+C3^+D^+$, $Fc^+C3^+D^-$, and $Fc^-C3^-D^-$ cells have a similar number of Con A receptors; before induction, they showed 50%, 5%, and 0% cells, respectively, with a Con A–induced cap (Figure 11) (Sachs 1974a, Lotem *et al.* 1976). Treatment with vinblastine or colchicine, but not with lumicolchicine, increased the frequency of cap formation from 50 to 100% in $Fc^+C3^+D^+$ cells, from 5 to 95% in $Fc^+C3^+D^-$, and from 0 to 50% in $Fc^-C3^-D^-$ cells (Figure 12). The increased ability to form a cap produced by vinblastine did not change the inducibility of cells for rosettes. Our results indicate that although free surface receptors for Con A and receptors anchored to tubulin can form a cap on myeloid leukemic cells, there are also receptors that may be anchored to struc-

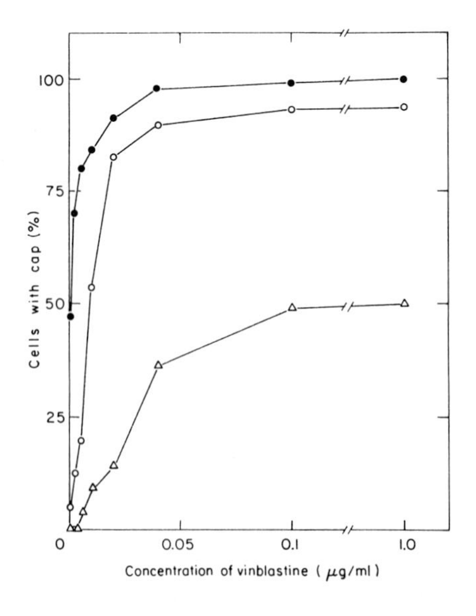

FIG. 12. Frequency of Con A–induced caps on mouse myeloid leukemic cells in the presence and absence of vinblastine. $Fc^+C3^+D^+$ (●---●), $Fc^+C3^+D^-$ (○---○), $Fc^-C3^-D^-$ (△---△).

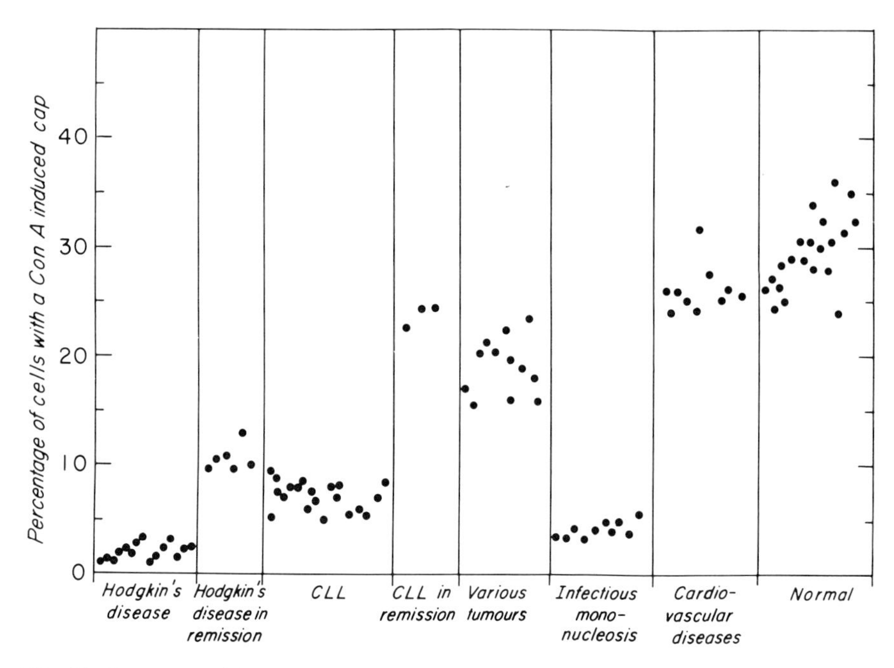

FIG. 13. Frequency of Con A–induced caps on human peripheral blood lymphocytes from normal persons, patients with different diseases, and patients in remission. (CLL, chronic lymphocytic leukemia.)

tures other than tubulin and that do not form a cap. The data suggest that the ability of myeloid leukemic cells to be induced by MGI to differentiate is associated with the frequency of Con A surface receptors that are free or have specific types of anchorage (Lotem *et al.* 1976). This association between cap formation and the ability to respond to a differentiation inducer may be useful as an aid in clinical diagnosis in various diseases (Figure 13) (Mintz and Sachs 1975, 1977).

CHROMOSOME MAPPING OF THE GENES THAT CONTROL DIFFERENTIATION AND MALIGNANCY IN MYELOID LEUKEMIC CELLS

The clonal origin and hereditability of the differences in inducibility by MGI suggested that there may be genetic differences between the different cell types and that it may be possible to map the chromosome location of the genes involved. The mouse myeloid leukemic cell clones used, even those that had been in culture for several years, still had a diploid or near diploid modal chromosome number.

An analysis of the chromosome banding pattern has shown that none of the cells had a completely normal diploid banding pattern. The $Fc^+C3^+D^+$ cells

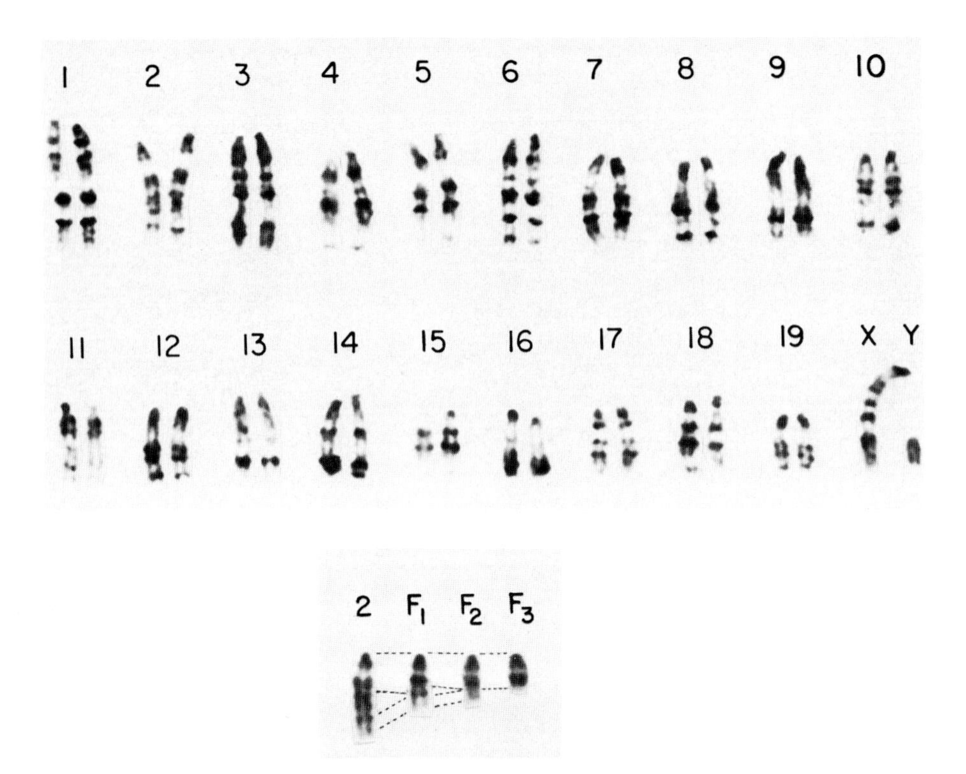

FIG. 14. Chromosome banding pattern of the normal male mouse and the loss of a piece in one chromosome number 2 found in the $Fc^-C3^-D^-$ clones (F_1, F_2, F_3).

can be induced to undergo normal cell differentiation even though the cells do not have a completely normal diploid chromosome banding pattern (Hayashi *et al.* 1974, Azumi and Sachs 1977). This shows that these malignant cells can be induced by MGI to have a normal differentiation phenotype without a completely normal genotype. There were specific chromosome differences between $Fc^+C3^+D^+$ and $Fc^+C3^+D^-$ cells (Hayashi *et al.* 1974). In addition, six clones of $Fc^-C3^-D^-$ cells derived from six independently produced myeloid leukemias showed a loss of a piece of one chromosome 2 (Figure 14). This loss was not found in $Fc^+C3^+D^+$, $Fc^+C3^+D^-$, or lymphoid leukemias. Five $Fc^+C3^+D^+$ mutants derived from an $Fc^-C3^-D^-$ clone with a loss of a piece of one chromosome 2, one normal chromosome 12, and two translocated chromosomes 12 maintained the abnormal chromosome 2 but had lost either the one normal or one translocated chromosome 12. These results indicate that chromosomes 2 and 12 carry genes that control the induction of differentiation of myeloid leukemic cells by MGI. The data suggest that there is a suppressor gene(s) on chromosome 12 that can suppress the inducing gene(s) on chromosome

2. The results also suggest that chromosomes 2 and 12 carry genes that control the malignancy of these myeloid leukemic cells (Azumi and Sachs 1977). The data support the suggestion that there is more than one gene that controls malignancy and differentiation of a specific cell type and that phenotypic expression depends on the balance between these different genes (Sachs 1974a, Azumi and Sachs 1977, Hitotsumachi *et al.* 1971, Yamamoto *et al.* 1973, Bloch-Shtacher and Sachs 1976, and in press).

CONCLUSIONS

The development of experimental systems for the culture and cloning of cells has made it possible to study the controls that regulate hematopoietic cell differentiation and the blocks that can occur in leukemia. All the main types of hematopoietic cells can be cloned in culture. It has been shown that the specific macrophage- and granulocyte-inducer MGI can induce normal cell differentiation in some types of myeloid leukemic cells ($Fc^+C3^+D^+$) and that in other types of myeloid leukemic cells there are different blocks in the process of normal cell differentiation. Unlike erythroleukemic cells that no longer respond to the normal regulator erythropoietin, the $Fc^+C3^+D^+$ myeloid leukemic cells can still respond to the normal regulator MGI.

The ability of the different types of myeloid leukemic cells to be induced to differentiate by MGI was associated with differences in the membrane-cytoskeleton interactions of specific surface membrane receptors. The sequence of differentiation in the $Fc^+C3^+D^+$ leukemic cells is the same as that in normal myeloid cells, and the leukemic cells studied appear to be less sensitive than normal cells to the induction of mature granulocytes by MGI. The normal cells require MGI for cell viability, growth, and differentiation, whereas the $Fc^+C3^+D^+$ leukemic cells are viable and can multiply in the absence of MGI. The change in the cells resulting in a decrease or lack of requirement of this protein for viability and growth may be one of the causes of myeloid leukemia.

Some of the stages of differentiation in myeloid leukemic cells can be induced by steroid hormones such as dexamethasone and other compounds such as cytosine arabinoside and actinomycin D. The therapeutic effects of these compounds in human leukemia, therefore, may in part be due to their induction of some stages of differentiation. There are different cellular sites for MGI and the steroids, and induction does not seem to be mediated by cyclic AMP.

Chromosome banding studies have shown that $Fc^+C3^+D^+$ cells can be induced to undergo normal cell differentiation by MGI even though the cells do not have a completely normal genotype. The results indicate that differentiation and malignancy of myeloid leukemic cells is controlled by the balance between different genes located on specific chromosomes which have been identified. The experimental system used in these studies appears to be particularly favorable for further elucidation of the molecular controls that regulate normal cell differentiation and the reversion of a malignant to a nonmalignant phenotype.

REFERENCES

Austin, P. E., E. A. McCulloch, and J. E. Till. 1971. Characterization of the factor in L-cell conditioned medium capable of stimulating colony formation by mouse marrow cells in culture. J. Cell. Physiol. 77:121–134.

Azumi, J., and L. Sachs. 1977. Chromosome mapping of the genes that control differentiation and malignancy in myeloid leukemic cells. Proc. Natl. Acad. Sci. USA 74:253–257.

Bloch-Shtacher, N., and L. Sachs. 1976. Chromosome balance and the control of malignancy. J. Cell. Physiol. 87:69–100.

Bloch-Shtacher, N., and L. Sachs. 1977. Identification of a chromosome that controls malignancy in Chinese hamster cells. J. Cell. Physiol. (In press).

Edelman, G. M. 1976. Surface modulation in cell recognition and cell growth. Science 192:218–226.

Fibach, E., E. Gerassi, and L. Sachs. 1976. Induction of colony formation in vitro by human lymphocytes. Nature 259:127–129.

Fibach, E., M. Hayashi, and L. Sachs. 1973. Control of normal differentiation of myeloid leukemic cells to macrophages and granulocytes. Proc. Natl. Acad. Sci. USA 70:343–346.

Fibach, E., T. Landau, and L. Sachs. 1972. Normal differentiation of myeloid leukemic cells induced by a differentiation-inducing protein. Nature New Biol. 237:276–278.

Fibach, E., and L. Sachs. 1975. Control of normal differentiation of myeloid leukemic cells. VIII. Induction of differentiation to mature granulocytes in mass culture. J. Cell. Physiol. 86:221–230.

Fibach, E., and L. Sachs. 1976. Control of normal differentiation of myeloid leukemic cells. XI. Induction of a specific requirement for cell viability and growth during the differentiation of myeloid leukemic cells. J. Cell. Physiol. 89:259–266.

Friend, C., W. Scher, J. G. Holland, and T. Sato. 1971. Hemoglobin synthesis in murine virus–infected leukemic cells in vitro: Stimulation of erythroid differentiation by dimethylsulfoxide. Proc. Natl. Acad. Sci. USA 68:378–382.

Gerassi, E., and L. Sachs. 1976. Regulation of the induction of colonies in vitro by normal human lymphocytes. Proc. Natl. Acad. Sci. USA 73:4546–4550.

Ginsburg, H., and L. Sachs. 1963. Formation of pure suspensions of mast cells in tissue culture by differentiation of lymphoid cells from the mouse thymus. J. Natl. Cancer Inst. 31:1–40.

Ginsburg, H., and L. Sachs. 1965. Destruction of mouse and rat embryo cells in tissue culture by lymph node cells from unsensitized rats. J. Cell. Comp. Physiol. 66:199–220.

Hayashi, M., E. Fibach, and L. Sachs. 1974. Control of normal differentiation of myeloid leukemic cells. V. Normal differentiation in aneuploid leukemic cells and the chromosome banding pattern of D^+ and D^- clones. Int. J. Cancer. 14:40–48.

Hitotsumachi, S., Z. Rabinowitz, and L. Sachs. 1971. Chromosomal control of reversion in transformed cells. Nature 231:511–514.

Ichikawa, Y., D. H. Pluznik, and L. Sachs. 1966. In vitro control of the development of macrophage and granulocyte colonies. Proc. Natl. Acad. Sci. USA 56:488–495.

Ichikawa, Y., D. H. Pluznik, and L. Sachs. 1967. Feedback inhibition of the development of macrophage and granulocyte colonies. I. Inhibition by macrophages. Proc. Natl. Acad. Sci. USA 58:1480–1486.

Kincade, P. W., P. Ralph, and M. A. S. Moore. 1976. Growth of B-lymphocyte clones in semisolid culture is mitogen dependent. J. Exp. Med. 143:1265–1270.

Kluge, N., G. Gaedicke, G. Steinheider, S. Dube, and W. Ostertag. 1974. Globin synthesis in Friend-erythroleukemia mouse cells in protein and lipid-free medium. Exp. Cell Res. 88:257–262.

Krystosek, A., and L. Sachs. 1976. Control of lysozyme induction in the differentiation of myeloid leukemic cells. Cell 9:675–684.

Krystosek, A., and L. Sachs. 1977. Steroid hormone receptors and the differentiation of myeloid leukemic cells. J. Cell. Physiol. (In press).

Landau, T., and L. Sachs. 1971. Characterization of the inducer required for the development of macrophage and granulocyte colonies. Proc. Natl. Acad. Sci. USA 68:2540–2544.

Lotem, J., and L. Sachs. 1974. Different blocks in the differentiation of myeloid leukemic cells. Proc. Natl. Acad. Sci. USA 71:3507–3511.

Lotem, J., and L. Sachs. 1975. Induction of specific changes in the surface membrane of myeloid leukemic cells by steroid hormones. Int. J. Cancer. 15:731–740.
Lotem, J., and L. Sachs. 1976. Control of Fc and C3 receptors on myeloid leukemic cells. J. Immunol. 117:580–586.
Lotem, J., and L. Sachs. 1977. Control of normal differentiation of myeloid leukemic cells. XII. Isolation of normal myeloid colony-forming cells from bone marrow and the sequence of differentiation to mature granulocytes in normal and D^+ myeloid leukemic cells. J. Cell. Physiol. 92:97–108.
Lotem, J., I. Vlodavsky, and L. Sachs. 1976. Regulation of cap formation by concanavalin A and the differentiation of myeloid leukemic cells: Relationship to free and anchored surface receptors. Exp. Cell Res. 101:323–330.
Lowy, D. R., W. P. Rowe, N. Teich, and J. W. Hartley. 1971. Murine leukemia virus: High frequency activation in vitro by 5-iododeoxyuridine and 5-bromodeoxyuridine. Science 174:155–156.
Metcalf, D. 1969. Studies on colony formation in vitro by mouse bone marrow cells. I. Continuous cluster formation and relation of clusters to colonies. J. Cell. Physiol. 74:323–332.
Metcalf, D., G. J. V. Nossal, N. L. Warner, J. F. A. P. Miller, T. E. Mandel, J. E. Layton, and G. A. Gutman. 1975. Growth of B lymphocyte colonies in vitro. J. Exp. Med. 142:1534–1549.
Mintz, U., and L. Sachs. 1975. Membrane differences in peripheral blood lymphocytes from patients with chronic lymphocytic leukemia and Hodgkin's disease. Proc. Natl. Acad. Sci. USA 72:2428–2432.
Mintz, U., and L. Sachs. 1977. Surface membrane changes in lymphocytes from patients with infectious mononucleosis. Int. J. Cancer 19:345–350.
Paran, M., R. C. Gallo, L. S. Richardson, and A. M. Wu. 1975. Adrenal corticosteroids enhance production of type C virus induced by 5-iodo-2-deoxyuridine from cultured mouse fibroblasts. Proc. Natl. Acad. Sci. USA 70:2391–2395.
Paran, M., and L. Sachs. 1969. The single cell origin of normal granulocyte colonies in vitro. J. Cell. Physiol. 73:91–92.
Paran, M., L. Sachs, Y. Barak, and P. Resnitzky. 1970. In vitro induction of granulocyte differentiation in hematopoietic cells from leukemic and nonleukemic patients. Proc. Natl. Acad. Sci. USA 67:1542–1549.
Pluznik, D. H., and L. Sachs. 1965. The cloning of normal "mast" cells in tissue culture. J. Cell. Comp. Physiol. 66:319–324.
Pluznik, D. H., and L. Sachs. 1966. The induction of clones of normal "mast" cells by a substance from conditioned medium. Exp. Cell Res. 43:553–563.
Reuben, R. C., R. L. Wife, R. Breslow, R. Rifkind, and P. A. Marks. 1976. A new group of potent inducers of differentiation in murine erythroleukemia cells. Proc. Natl. Acad. Sci. USA 73:862–866.
Rousseau, G. G., J. D. Baxter, and G. M. Tomkins. 1972. Glucocorticoid receptors: Relations between steroid binding and biological effects. J. Mol. Biol. 67:99–115.
Sachs, L. 1964. The analysis of regulatory mechanisms in cell differentiation, *in* New Perspectives in Biology. Elsevier, Amsterdam, pp. 246–260.
Sachs, L. 1974a. Regulation of membrane changes, differentiation and malignancy in carcinogenesis. Harvey Lect. 68:1–35.
Sachs, L. 1974b. Control of growth and differentiation in normal hematopoietic and leukemic cells, *in* Control of Proliferation in Animal Cells, R. Baserga and B. Clarkson, eds., Cold Spring Harbor Laboratory, Cold Spring Harbor, New York, pp. 915–925.
Sela, B., and L. Sachs. 1974. Alkaline phosphatase activity and the regulation of growth in transformed mammalian cells. J. Cell. Physiol. 83:27–34.
Stephenson, J. R., A. A. Axelrad, D. L. McLeod, and M. M. Shreeve. 1971. Induction of colonies of hemoglobin-synthesizing cells by erythropoietin in vitro. Proc. Natl. Acad. Sci. USA 68:1542–1546.
Vlodavsky, I., E. Fibach, and L. Sachs. 1976. Control of normal differentiation of myeloid leukemic cells. X. Glucose utilization, cellular ATP and associated membrane changes in D^+ and D^- cells. J. Cell Physiol. 87:167–177.
Yamamoto, T., Z. Rabinowitz, and L. Sachs. 1973. Identification of the chromosomes that control malignancy. Nature New Biol. 243:247–250.

Cell Differentiation and Neoplasia, edited by Grady F. Saunders. Raven Press, New York © 1978.

In Vitro Growth of Normal and Malignant Hemopoietic Cells in Man

K. A. Dicke, G. Spitzer, D. Verma, and K. B. McCredie

Department of Developmental Therapeutics, The University of Texas System Cancer Center M. D. Anderson Hospital and Tumor Institute, Houston, Texas 77030

Evaluation of the hemopoietic function of the bone marrow on the basis of morphology is severely limited, especially in malignant hematological disorders. These types of studies are not of value for predicting recurrence of disease or for detecting low numbers of residual tumor cells. Any tumor population lower than 1 to 5% of the total cell number is difficult to detect morphologically.

It has only been for the past seven years that methods other than morphology have been available to evaluate the hemopoietic function of human bone marrow. Robinson in 1970 described an in vitro technique with which he could clone human bone marrow cells in much the same way that Bradley *et al.* (1969) cloned mouse bone marrow. Robinson and Pike (1970) used a stimulus for colony formation of human bone marrow cells, i.e., leukocytes from peripheral blood as a feeder layer. It became evident that the cell population in the bone marrow responsible for colony formation in vitro belonged to the myeloid compartment (Moore and Williams 1972), although it can not be excluded that a fraction of the colonies are derived from pluripotent hemopoietic stem cells (Dicke *et al.* 1972). Using this system, normal human bone marrow forms colonies (40 cells or more) of granulocytes and macrophages after approximately seven days of incubation and clusters (3 to 40 cells) with the same morphology in a frequency of 2 to 17 times greater. This is depicted schematically in the upper part of Figure 1.

Robinson and many other investigators, such as Moore (1975) and his coworkers (1973,1974), Bull *et al.* (1973), Chervenick and Boggs (1971), Greenberg *et al.* (1971), and Spitzer *et al.* (1976a,b), used this technique for cloning leukemic cells and marrow cells from other hematological disorders, and it appeared that this system, referred to as the Robinson system, not only supports colony formation of normal myeloid precursors but also malignant cells. In other words, this assay is not specific for growth of hemopoietic cells. Such a lack of specificity has its advantages, but this also can be a disadvantage, as can be seen from the following analysis of the Robinson system. Because of the lack of specificity of the Robinson system, early detection of leukemic cell proliferation is not possible, due to the overwhelming growth of normal hemopoietic cells. Since

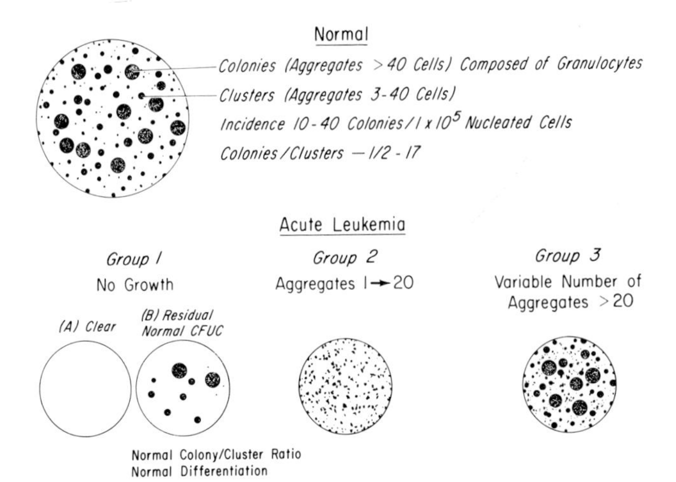

FIG. 1. Schematic representation of the pattern of normal bone marrow and the three subdivisions of acute leukemic growth. Reprinted, with permission of Blood, from Spitzer *et al.* 1976b.

the presence of residual acute leukemic proliferative cells in remission of acute leukemia or in the benign phase of CML (chronic myelogenous leukemia) may determine the clinical outcome of the disease, it is essential to investigate other methods of detection that are more specific. In our laboratory, we have recently developed an in vitro system, called the PHA assay, in which colony growth by leukemic cells is supported frequently without the necessity of leukocyte feeder layers. This is substantially different from growth of leukemic cells in the Robinson assay, since in the Robinson system leukemic cells predominantly grow as clusters containing 5 to 20 cells. Colonies in the PHA assay contain 50 to 500 cells. It is important to prove both the specificity of this PHA assay for leukemic cell growth and the sensitivity of the assay for the detection of leukemia. Another fundamental question concerns the nature of the subpopula-

The Robinson Assay System

Values:

1. Prognosis of response to chemotherapy in acute leukemia
2. Early detection of the outcome in AML by sequential cultures
3. Diagnosis of leukemic nature of refractory anemia
4. Quantification of viability of bone marrow after storage

Limitations:

1. Lack of early detection of proliferative leukemic cells in benign CML
2. Lack of early detection of proliferative disease in smoldering leukemia
3. Lack of detection of proliferative leukemic cells in remission of acute leukemia
4. Uncertainty of whether the leukemia cells populated that proliferate in this system are representative of the leukemic cell population that causes proliferation of the disease in vivo

tion that forms colonies in the PHA system, i.e., is this a true reflection of proliferation in vivo? In this paper we will demonstrate the results of our studies and the impact of these two in vitro assays on several clinical situations in hematological malignant disorders.

THE ROBINSON SYSTEM

Culture Procedure

The culture procedure used was similar to that previously described (Spitzer *et al.* 1976a,b). Briefly, bone marrow aspiration was performed from the posterior iliac spine, and the aspirate was placed into tissue culture tubes (Falcon 3033, Oxnard, Calif.) with 200 units of preservative-free heparin. The buffy coat was aspirated after centrifugation at 1200 X g for 10 minutes and then washed twice in phosphate-buffered saline. After the final wash, the cells were suspended in an approximate volume of 2 ml, and the total nucleated cell count was determined by counting the Turk's solution in a hemocytometer. The cell suspension was mixed with 0.3% agar-alpha medium and then 1 ml aliquots layered on top of previously prepared leukocyte underlayers (1×10^6 leukocytes/ml/dish in 0.5% agar). The preparation of the leukocyte feeder layer obtained from healthy individuals has been described extensively elsewhere (Moore *et al.* 1974). After gelling, the cultures were placed for seven days in a Forma CO_2 incubator with 7.5% CO_2 and air. All samples were cultured at two cell concentrations: 1×10^4 and 1×10^5 nucleated cells/ml. Each specimen at both cell concentrations was plated on two different leukocyte underlayers in replicates.

Culture Scoring

Cultures were evaluated visually at seven days using an Olympus dissecting microscope at magnification, 25–40X. They were analyzed for the following: (1) total number of aggregates of three cells or greater, (2) largest aggregate size, (3) presence or absence of single cells, and (4) the total number of colonies per plate. The total incidence of aggregates was the mean of four counts in dishes with 1×10^5 nucleated cells/ml, or, if the plating efficiency was high, the estimate was obtained from counts of dishes with 1×10^4 nucleated cells/ml. The other parameters were assessed from all eight plates. To assure constant assessment of adequate culture conditions, marrow from patients with solid tumors was simultaneously cultured and the number of colonies scored. The most important criterion we have used for assessment of adequate culture conditions was the percentage of colonies equal to or greater than 100 cells.

THE IN VITRO PHA ASSAY

Basically, the technique consists of two phases: an initial liquid phase of 15 hours at 37°C and a semi-solid phase of seven days incubation at 37°C (Dicke *et al.* 1976a,b). In the liquid phase, 2 X 10^6 cells/ml of medium (Dulbecco's MEM + 20% serum) were cultured in Pyrex glass tubes to which 0.05 ml phytohemagglutinin (PHA) (Difco, PHA-M)/ml medium was added. After 15 hours of incubation, the cells were washed twice using Hank's balanced salt solution (HBSS) (305 mOsm), and then 1 X 10^5 cells/ml were cultured in a system identical to that described above. Simultaneously, the cells were plated in petri dishes containing agar underlayers without leukocytes. After seven days of incubation in a 7.5% CO_2 gas-controlled humidified incubator at 37°C, colonies were visible microscopically. These colonies were counted using an inverted light microscope. Aggegrates containing 50 cells or more were considered colonies. Aggregates containing 50 cells or less were considered to be clusters. The number of colonies and clusters presented in this paper is the mean value of triplicate petri dish cultures.

LIGHT MICROSCOPE PROCEDURE

Bone marrow cells drawn by aspiration from the iliac crest were stained using Wright-Giemsa to establish the diagnosis of AML (acute myelogenous leukemia). A differential count of at least 500 cells was undertaken. Cells obtained from in vitro colonies using a fine Pasteur pipette were either stained individually with acetoorecin or pooled and stained with Giemsa, as previously described (Spitzer *et al.* 1976b).

ELECTRON MICROSCOPE PROCEDURE

Soft agar colonies for morphological observation were fixed in their petri dishes for 12 hours at 37°C with 2.5% Sorensen's buffered glutaraldehyde, pH 7.2. The petri dishes were rinsed in three changes of phosphate buffer, pH 7.2, for 30 minutes each and then incubated in the dark with 3–3′ diaminobenzidine tetrahydrochloride reagent for two hours at room temperature. The staining solution of endogenous peroxidase was removed with three additional rinses of Sorensen's phosphate buffer and postfixed in 1% tetroxide, pH 7.2, for one hour at 4°C. Following three rinses with distilled H_2O, the agar dishes were removed to flat covered glass dishes for the acetone dehydration and Epon polymerization at 80°C; the aluminum foil was removed from the specimen, and the colonies of interest marked under a dissecting microscope. Those colonies selected for observation were cut from the specimen disk and mounted on plastic rods for ultramicrotomy. Alternate thick and thin serial sections were cut for light and electron microscopic study of the entire colony. Epon sections for light microscopy were stained with Paragon's stain for frozen sections. For ultrastructural observation, thin sections were stained with 0.5% uranyl acetate and Reynold's Lead Citrate prior to their examination with a Siemen's Elmiskop IA at 80 kv.

CYTOGENETIC PROCEDURE

Chromosome studies were performed on cells from both the liquid phase and from the colonies that had formed after seven days in culture. The cells obtained from the liquid phase, following the 15-hour incubation with PHA, were placed in 10 ml Ham's F-10 tissue culture medium supplemented with 20% fetal calf serum and incubated overnight at 37°C. The following day the cells were arrested in metaphase using 0.01 mg/ml colchicine, submitted to hypotonic treatment with 0.075 M KCl, and fixed in methanol:acetic acid (3:1) mixture. Air-dry slide preparations were made, stained with Giemsa, and scanned for well-spread metaphases using a Zeiss microscope.

To obtain dividing cells from the colonies that had formed after seven days in culture, 0.1 g Colcemid was added to the petri dishes, and the dishes were incubated for an additional three hours. The colonies were collected with a fine Pasteur pipette and pooled in 0.2 ml HBSS. The harvesting and slide preparations were completed as above and the slides scanned for analyzable metaphases.

STUDIES WITH THE ROBINSON SYSTEM

As has been documented in the foregoing outline of the Robinson assay, the Robinson system is used by our group in four areas of clinical practice:

1) Prognosis of response to chemotherapy in acute leukemia. We have analyzed the growth patterns and their prognostic value in untreated adult leukemia. In the lower part of Figure 1, the three in vitro growth patterns in acute leukemia are presented: Group 1 reflects the group of patients in which the marrow cultured in vitro does not show leukemic proliferation in vitro, with its two subdivisions: (a) no growth, i.e., an empty plate and (b) a low incidence of normal colonies. Groups 2 and 3 are patients whose marrow demonstrates leukemic cell growth in vitro. In Group 2, the aggregates in the culture are small, up to 20 cells, and in Group 3, the aggregates contain 20 cells or more. Leukemic origin of these aggregates or clusters in Groups 2 and 3 is suggested because of the primitive morphology of the cells, the cytogenetic markers, and the abnormal buoyant density of the cluster-forming cells. In Table 1 it can be seen that patients whose cells form large leukemic

TABLE 1. *Response rate by in vitro subdivisions*

		Leukemic Growth		
	Group 1 No leukemic growth	Group 2 Aggregates 1–20	Group 3 Aggregates >20	Total
Complete remission	16(76%)	27(75%)	4(21%)	47(62%)
Not complete remission	5(24%)	9(25%)	15(79%)	29(38%)
Total	21	36	19	76

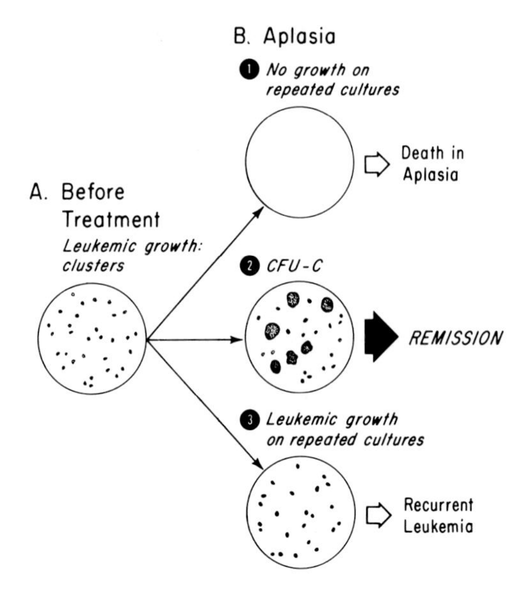

FIG. 2. The in vitro changes with remission induction therapy in AML.

aggregates in vitro (Group 3) have a poor response to chemotherapy. Patients in Group 2, in which the marrow proliferates in vitro but only as small aggregates, do respond to chemotherapy and, therefore, have a good prognosis. The same holds true for Group 1 patients. It appears, by utilizing this system, that extensive in vitro growth of acute leukemia may be associated with a poor prognosis (Spitzer *et al.* 1976b).

2) Early detection of clinical outcome of leukemia after chemotherapy. In Figure 2, three characteristic patterns are depicted that were recognized during treatment of AML (Spitzer *et al.* 1977). On the left side of the figure, the classical pattern of AML in vitro (small clusters with varying frequency) has been depicted. On the right side, the following three patterns (after chemotherapy treatment) have been listed: Pattern 1, a persistent absence of colony-forming units (CFU-C); Pattern 2, presence of colonies in the culture; and Pattern 3, persistence of cluster formation. Table 2 shows that those patients who show an early return of colony formation go into remission in the majority of the cases. However, patients showing persistent absence of colony formation on subsequent cultures one week apart died with irreversible aplasia, and patients showing persistent leukemia growth patterns in vitro were still leukemic after chemotherapy. Patients destined to achieve complete remission already showed return of colony formation in the bone marrow at the time that morphological evaluation is difficult because of morphologically unidentifiable blast cells.

3) Detection of leukemia in pre-leukemic disorders. In Figure 3, the three growth patterns that we have recognized in pre-leukemia and oligoblastic leukemia

TABLE 2. *Relationship of clinical outcome and the bone marrow in vitro culture change in AML*

In vitro culture pattern	Clinical outcome	Number of patients	Remission rate
Persistent absence of leukemic or CFU-C* growth	Persistent aplasia and death	5	0/5
Return of CFU-C	Remission—same course of chemotherapy	22	
	Remission—next course of chemotherapy	5	
	Death—day 8—first course	1	27/28
Persistent leukemia in vitro growth	Leukemic regrowth and death	3	0/3
TOTAL		36	27/36

* CFU-C: colony-forming unit culture

(smoldering leukemia) have been schematically presented. In Category I, no leukemic growth is present, just residual colony formation by normal hemopoiesis. Category II and Category III demonstrate leukemic growth in vitro. Category III has the classical leukemic pattern in vitro. Category II, besides having clusters of leukemic origin, also shows colonies of varying morphology.

Marrow from 10 patients with refractory anemia demonstrated leukemic growth in vitro (Categories II and III). These patients progressed to leukemia within one year after the initial culture. However, none of the six patients without leukemia growth (Category I) have yet progressed to acute leukemia.

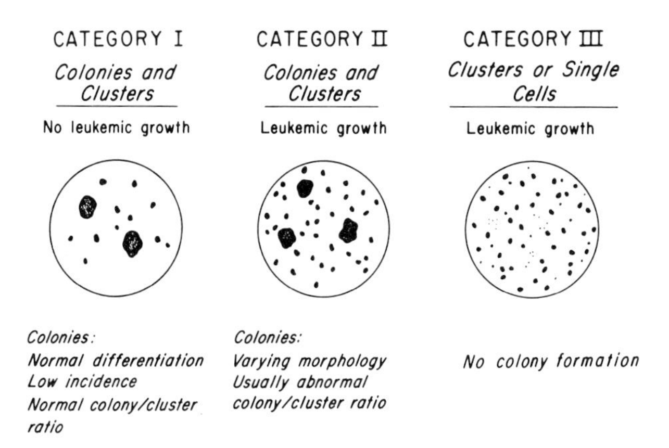

FIG. 3. The in vitro culture patterns in pre-leukemic and oligoblastic leukemia.

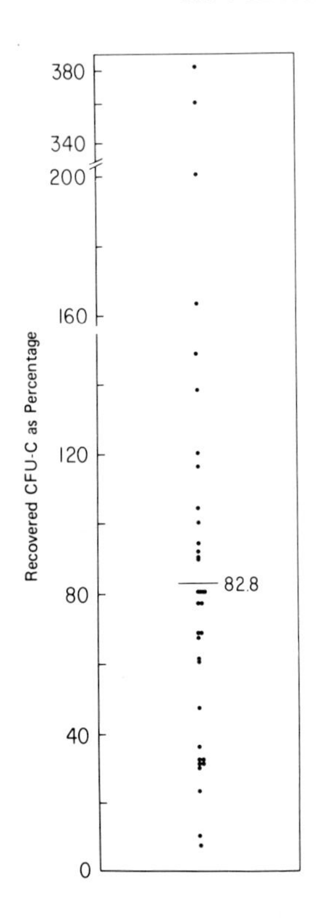

FIG. 4. CFU-C variability/10^5 plated cells after freezing expressed as percentage of the number of CFU-C of fresh marrow samples.

Similarly, in 37 patients with oligoblastic leukemia, we have noticed that some patients with the Category I pattern have their original diagnosis altered (e.g., leukemia or lymphoma or have not shown a progressive course). In patients with leukemia, growth sequential cultures were performed to study the significance of an increase in clusters. In 8 of 16 patients, a marked increase in cluster incidence occurred in advance of morphological evidence of progression.

4) Quantification of viability of bone marrow after storage. In previous studies, a correlation between the number of CFU-C in monkey bone marrow suspensions and engraftment in lethally irradiated recipients has been demonstrated (Dicke *et al.* 1971); therefore, this assay has been utilized to determine the optimal storage procedure of human bone marrow. We used the freezing method of Schaefer and Dicke (1971) for storage of bone marrow cells collected from acute leukemia patients in remission and from CML patients in the benign phase of the disease. In Figure 4, the CFU-C recovery after freezing bone marrow samples from 33 patients have been documented. The lowest CFU-C recovery is from the CML patients. There is an enormous variation in CFU-C recovery. From our preliminary experience of autologous bone marrow transplantation in relapsed acute leukemia, we have noticed that the degree and rapidity of restoration of hemopoiesis after transplantation

correlates with the concentration of CFU-C in the marrow after thawing.

Also we have been able to predict early engraftment after transplantation (day 7 after transplantation) by culturing mononuclear marrow cells after concentration by Ficoll-Hypaque gradients.

STUDIES WITH THE PHA ASSAY

Before application of the PHA assay to those clinical areas in which the Robinson system is of limited value, it has to be demonstrated that the PHA assay is specific for growth of leukemic cells. We have tested marrows from at least 60 untreated acute leukemia patients on colony formation in the PHA system, of which results on 43 AML patients are depicted in Figure 5. It can be noted in this figure that colonies can be grown without the presence of leukocytes in the underlayer. There is an identical response when peripheral blood cells of untreated AML patients containing over 90% blast cells are cultured (Figure 6). To prove the leukocyte independence of the colonies, a linear regression analysis has been done between numbers of colonies obtained from marrow cells from untreated AML with and without the presence of leukocyte underlayers, and it appears that there exists a strict correlation between those two parameters ($r = 0.979$), as can be noted in Figure 7. In Table 3, the results of colony formation of bone marrow cells from hematologically normal patients and from patients in remission have been documented. It can be noted

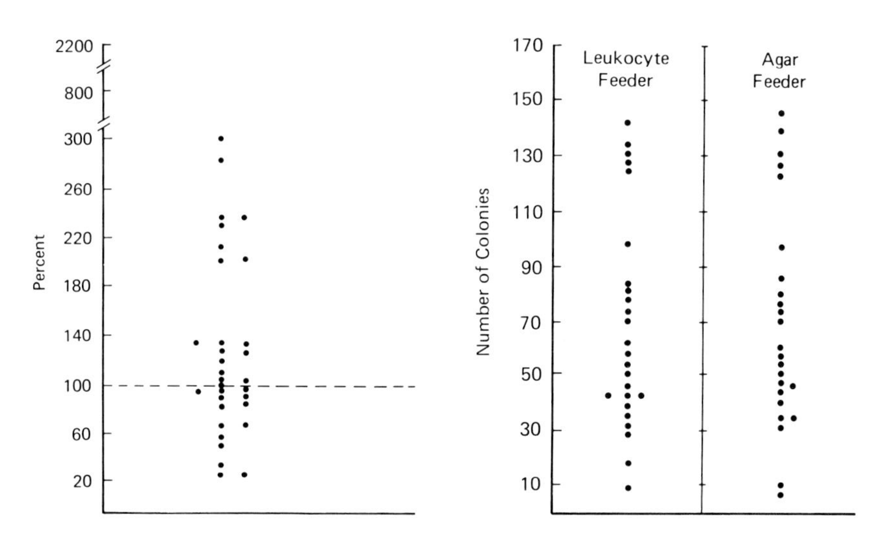

FIG. 5, left. The incidence of colony formation from untreated AML bone marrow in the PHA assay with and without leukocyte stimulus. The colonies on the leukocyte feeder layer are expressed as a percentage of those in cultures without leukocytes.

FIG. 6, right. The incidence of PHA-induced colonies from the peripheral blood of untreated AML patients.

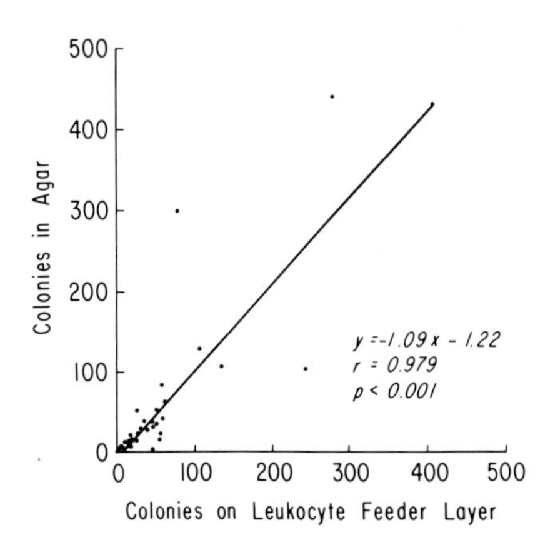

FIG. 7. Relationship between number of AML colonies with and without leukocyte feeder layer in the PHA assay.

TABLE 3. *Comparison between number of CFU-C in the Robinson assay and the PHA assay of marrow from hematologically normal patients and from AML patients in remission*

	Patient	Robinson Assay Leukocyte feeder	Robinson Assay Agar	PHA Assay Leukocyte feeder	PHA Assay Agar
AML	KNT	1	n.d.*	75	n.d.
	SAN	0	n.d.	68	n.d.
	SAL	0	n.d.	290	n.d.
	SE	0	0	280	440
	HE	0	0	29	30
	KA	0	0	59	43
	NE	0	0	50	54
	PA	0	0	14	22
	PO	0	0	22	16
NORMAL	1	30	n.d.	25	n.d.
	2	25	n.d.	30	n.d.
	3	40	1	35	0
	4	44	2	34	0
	5	32	1	22	0

*n.d. = not done

TABLE 4. *Number of CFU-C in the Robinson assay and the PHA assay of marrow from AML relapse*

	Patient	Robinson assay Leukocyte feeder	Robinson assay Agar*	PHA assay Leukocyte feeder	PHA assay Agar*
	Wo	2†	n.d.‡	52	0
AML	Be	0	0	150	130
RELAPSE	El	0	0	140	130
	Fe	0	0	160	103

* Agar underlayer without leukocytes
† Figure represents number of colonies per 10^5 cells plated
‡ n.d. = not done

that colony formation is leukocyte dependent and PHA independent, which is in contrast with colony formation of marrows from untreated leukemia patients. Marrow from relapsed leukemia patients behaved as untreated AML; colonies can be grown in the PHA system that are dependent on the presence of PHA and independent of leukocyte underlayers (Table 4).

The nature of the colonies obtained from marrow of untreated AML in the PHA system was analyzed by means of electron microscopy and cytogenetics. Individual colonies were removed from 7-day cultures, fixed and stained for electron microscope analysis. Individual cells were analyzed and were found to be characteristic of myeloblasts. Thick cross-sections of the colonies were stained with peroxidase, and it was demonstrated by this method that the cells in the colonies from untreated AML bone marrow were peroxidase positive, confirming the single-cell analysis by electron microscopy.

MECHANISM OF COLONY FORMATION IN THE PHA ASSAY

The mechanism of induction of colony formation by PHA is not yet known. It could either be a direct effect of PHA on the leukemic cell or an indirect effect via cells producing factors that trigger leukemic cells to form colonies. A linear relationship found in the solid phase between the number of colonies observed and the number of cells plated may indicate a direct action of PHA on the clonogenic leukemic cell (Dicke *et al.* 1976b). The number of colonies/10^5 plated cells did not vary by varying the number of cells/ml liquid culture from 0.5×10^6 to 2×10^6. However, using cell concentrations lower than 0.2×10^6/ml in the liquid phase, the colony formation was reduced by up to 20%.

We have attempted to produce factors that may trigger leukemic cells to form colonies, by incubation of either leukemic cells or normal peripheral blood cells in the presence of PHA for up to seven days. Media prepared in this way has been shown to stimulate thymidine incorporation in leukemic cells

TABLE 5. *Comparison of effect of PHA and conditioned medium added to the liquid culture phase prior to the semi-solid phase on leukemic colony formation*

	Robinson assay		Preincubation							
			PHA* (0.05)		NCM+†		LCM+†		PHA‡ (0.005)	
Patient	agar	leukocyte	agar	leukocyte	agar	leukocyte	agar	leukocyte	agar	leukocyte
DO	0	0	8	27	0	0	0	0	0	0
CA	0	0	40	34	8	14	0	0	8	13
DO	0	0	16	27	0	0	0	0	0	0
STE	0	0	0	16	0	0	0	0	0	0
NE	0	0	54	52	0	0	0	0	0	0
TU	0	0	144	166	—	—	0	0	0	0
FRE	0	0	140	300	0	0	0	0	0	0
AN	0	0	51	46	0	0	0	0	0	0

* PHA 0.05: PHA, 0.05 ml/ml of culture. This concentrate of PHA is normally used in the PHA assay (see Materials and Methods).

† NCM+: PHA-conditioned medium from peripheral blood cells of normal individuals (4 batches tested).

‡ PHA 0.005 ml (Difco-M)/ml of medium. This concentration has to be used as a control for the conditioned medium, since this amount of PHA is also present in the NCM- and LCM-supplemented cultures.

(Aye *et al.* 1974). Until now, our attempts to stimulate colony formation by conditioned medium have been negative, as can be seen in Table 5.

We have also attempted to grow marrow cells from nonmyelogenous leukemia in the PHA assay. Thus far, our results have been negative. This implies that the PHA phenomenon is specific for AML cells.

STUDIES CORRELATING THE PHA ASSAY WITH CLINICAL PROLIFERATION OF LEUKEMIA

In pre-clinical settings, we have attempted to detect early proliferative disease using the PHA assay in: (a) chronic myelogenous leukemia (CML), benign phase (Spitzer *et al.* 1976c), (b) smoldering leukemia (Spitzer *et al.* 1976c), and (c) remission of acute leukemia.

In Table 6 the results of the PHA assay of seven patients in CML blast crisis have been documented, which indicate that in the acute phase of CML, colonies are being found in the PHA system. In 12 of 22 CML patients in the benign phase of the disease, colonies in the PHA system were found. Five of these 12 patients developed blastic transformation within four months after assay, whereas the clinical status of the other seven patients has remained unchanged. Ten of 22 CML patients in the benign phase were negative in the PHA system and are still clinically unchanged nine months after culture (Table 7).

Five patients with smoldering leukemia, positive in the PHA assay, developed rapidly progressive leukemia two months after assay. The other patients, whose marrow did not form colonies in the PHA assay, are still clinically stable up to six months after assay.

We have tested marrows from four patients in remission using a slightly modified PHA assay. This modified assay system is depicted in Figure 8. Basically, the cells are incubated with PHA for 24 to 48 hours, after which they are examined electron microscopically and by cytogenetic methods.

The leukemia cell population of the four patients under study demonstrated chromosomal abnormalities detected at the time before chemotherapy remission

TABLE 6. *Colonies in CML blast crisis*

Patient	Robinson assay Leukocyte feeder	Agar	PHA assay Leukocyte feeder	Agar
1	0	0	10	0
2	0	0	33	8
3	0	0	22	20
4	0	0	0	0
5	0	0	20	38
6	1	0	205	117
7	3	0	100	85

TABLE 7. *CML—Benign phase summary of PHA response data on agar underlayers*

	Group 1	Group 2	Group 3
Number of patients	10	7	5
PHA response	—ve*	+ve†	+ve
Clinical behavior	All still benign	Benign	Progression to blast crisis

* —ve: negative.
† +ve: positive.

induction treatment (Table 8). In the four patients under study, the same chromosome abnormalities could be found as those observed before treatment. It is remarkable that in the cells of three patients cultured without the presence of PHA, no cell with the leukemic chromosome patterns were found; whereas after PHA stimulation, abnormal chromosome markers were observed. The electron microscopic analysis confirmed the cytogenetic results—patients with high numbers of leukemic cells after PHA stimulation had a tendency to early relapse. Four patients died 3 weeks after assay from cardiotoxicity due to chemotherapy, so that a correlation with the clinical outcome was impossible.

In conclusion, the PHA assay forms a valuable tool, especially for those areas in which the Robinson system can not be used because of its aspecificity. However, the technical problem of agglutination of cells in the PHA assay still exists. This prohibits exact quantitation of the colonies in 25% of the cases. Experiments are underway to eliminate agglutination. Nevertheless, this system has enormous potential in the clinic, as has been demonstrated in this paper. In addition, sophisticated separation methodology based on the differential affin-

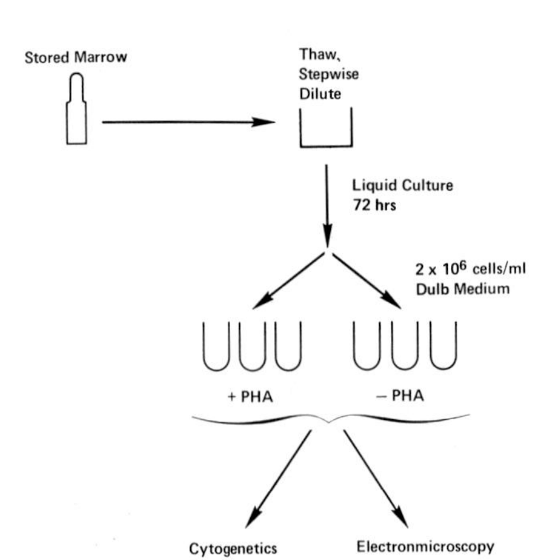

FIG. 8. Detection of PHA-sensitive leukemic cells in the remission phase of acute leukemia.

TABLE 8. *Cytogenetic and electron microscope analysis of marrow cells from AML remission patients before and after PHA stimulation. Relationship between CFU-C formation and relapse*

Patient	Cytogenetic analysis		Electron microscope study		PHA assay* CFU-C formation	Relapse	
	−PHA	+PHA	−PHA	+PHA		Time interval	Progression
Ha	0/50	6/50	1/1000	1/1000	0	—	—
We	2/200	2/49	0/1000	7/1000	5	2m	Fast
Hu	0/50	1/50	3/1000	0/1000	5	2m	Fast
Br	0/34	1/50	0/1000	1/1000	0	2m	Slow

* CFU-C formation in agar without leukocyte feeder after incubation with PHA for 15 hours in liquid suspension.

ity of leukemic cells and normal hemopoietic stem cells to PHA could be developed, which would be valuable for the improvement of autologous bone marrow transplantation in acute leukemia.

ACKNOWLEDGMENTS

This research was supported in part by Grants CA-11520 and CA-14528 from the National Institutes of Health.

REFERENCES

Ahearn, M. J., J. M. Trujillo, A. Cork, A. Fowler, and J. S. Hart. 1974. The association of nuclear blebs with aneuploidy in human acute leukemia. Cancer Res. 34:2887–2896.

Aye, M. T., Y. Niho, J. E. Till, and E. A. McCulloch. 1974. Studies of leukemic cell populations in culture. Blood 4:205–219.

Bradley, T. R., D. Metcalf, M. A. Sumner, and E. R. Stanley. 1969. Characteristics of in vitro colony formation by cells from haemopoietic tissues, *in* Hemic Cells In Vitro, P. Farnes, ed., Vol. 4. Williams and Wilkins Co., Baltimore, p. 22.

Bull, J. M., J. J. Cuttera, J. D. Northrup, E. S. Henderson, E. Statshick, and P. P. Carbone. 1973. Serial in vitro marrow culture in acute myelocytic leukemia. Blood 42:679–686.

Chervenick, P. A., and D. R. Boggs. 1971. In vitro growth of granulocytic and mononuclear cell colonies from blood of normal individuals. Blood 37:131.

Dicke, K. A., M. G. C. Platenburg, and D. W. van Bekkum. 1971. Colony formation in agar: *In vitro* assay for haemopoietic stem cells. Cell Tissue Kinet. 4:463–477.

Dicke, K. A., G. Spitzer, and M. J. Ahearn. 1976a. In vitro colony formation by leukemic cells in acute myelogenous leukemia using PHA as stimulating factor. Nature 259:129–130.

Dicke, K. A., G. Spitzer, A. Cork, and M. J. Ahearn. 1976b. In vitro colony growth of acute myelogenous leukemia. Blood Cells 2:125–137.

Dicke, K. A., M. J. van Noord, B. Maat, U. W. Schaefer, and D. W. van Bekkum. 1972. Attempts at morphological identification of the haemopoietic stem cells in primates and rodents, *in* CIBA Foundation Symposium on Haemopoietic Stem Cells, G. E. W. Wolstenholme, ed., Associated Scientific Publishers, Amsterdam, pp. 47–69.

Greenberg, P. L., W. C. Nichols, and S. L. Schrier. 1971. Granulopoiesis in acute myeloid leukemia and pre-leukemia. N. Engl. J. Med. 284:1225.

Moore, M. A. S. 1975. In vitro studies in the myeloid leukemias, *in* Advances in Acute Leukemia, F. J. Cleton, D. Crowther, and J. S. Malpas, eds., North-Holland Publishing Co., Amsterdam, pp. 161–227.

Moore, M. A. S., G. Spitzer, N. Williams, D. Metcalf, and J. Buckley. 1974. Agar culture studies in 127 cases of untreated acute leukemia: The prognostic value of reclassification of leukemia according to in vitro growth characteristics. Blood 44:1–18.

Moore, M. A. S., and N. Williams. 1972. Physical separation of colony-stimulating cells from in vitro colony forming cells in monkey hemopoietic tissue. J. Cell Comp. Physiol. 80:195.

Moore, M. A. S., N. Williams, and D. Metcalf. 1973. In vitro colony formation by normal and leukemic human hematopoietic cells. J. Natl. Cancer Inst. 50:603–723.

Robinson, W. A., and B. L. Pike. 1970. Colony growth of human bone marrow cells in vitro, *in* Hemopoietic Cellular Proliferation, F. Stohlman, Jr., ed., Grune and Stratton, New York, p. 249.

Schaefer, U. W., and K. A. Dicke. 1973. Preservation of hematopoietic stem cells. Transplantation potential and CFU-C activity of frozen marrow tested in mice, monkeys and man, *in* The Cryopreservation of Normal and Neoplastic Cells (Proceedings of the International Conference held at l'Institut de Cancerologie et d'Immunogenetique, Villejuif, France), R. S. Weiner, R. K. Oldham, and L. Schwarzenberg, eds., Paris, Inserm, pp. 63–69.

Spitzer, G., K. A. Dicke, E. A. Gehan, T. Smith, and K. B. McCredie. 1976a. The use of the Robinson in vitro agar culture assay in adult acute leukemia. Blood Cells 2:139–148.

Spitzer, G., K. A. Dicke, E. A. Gehan, T. Smith, K. B. McCredie, B. Barlogie, and E. J Freireich. 1976b. A simplified in vitro classification for prognosis in adult acute leukemia: The application of in vitro results in remission-predictive models. Blood 48:795–806.

Spitzer, G., K. A. Dicke, K. B. McCredie, and B. Barlogie. 1977. The early detection of remission in acute myelogenous leukaemia by in vitro cultures. Br. J. Haematol. 35:411–418.

Spitzer, G., M. A. Schwarz, K. A. Dicke, J. M. Trujillo, and K. B. McCredie. 1976c. Significance of PHA induced clonogenic cells in chronic myeloid leukemia and early acute myeloid leukemia. Blood Cells 2(1–2):149–156.

Control of Differentiation and Neoplasia—Molecular Mechanisms of Gene Expression

Cell Differentiation and Neoplasia, edited by Grady F. Saunders. Raven Press, New York

Experimental Study of Alpha-Fetoprotein Re-Expression in Liver Regeneration and Hepatocellular Carcinomas

G. I. Abelev

Laboratory of Tumor Immunochemistry and Diagnosis, N. F. Gamaleya Institute of Epidemiology and Microbiology, USSR Academy of Medical Sciences, Moscow, USSR

Alpha-fetoprotein (AFP) is the most extensively studied embryonic antigen, synthesis of which is resumed in certain malignancies, e.g., in germ cell teratocarcinomas and hepatocellular carcinomas (Abelev 1971, Hirai and Miyaji 1973, Masseyeff 1974, Hirai and Alpert 1975, Fishman and Sell 1976). It is an embryo-specific serum protein that comprises one polypeptide chain with a molecular weight of about 70,000 and that contains a small carbohydrate moiety (Ruoslahti *et al.* 1974). AFP is characterized in the primary structure by well-pronounced homology with serum albumin (Ruoslahti and Terry 1976), by shared antigenic determinants revealed after denaturation (Ruoslahti and Engvall 1976), and by similar physicochemical properties (Abelev 1971).

During ontogenesis, AFP first becomes synthesized in the endoderm of the yolk sac and then in the hepatocytes of the embryonic liver. The site and time of AFP synthesis are in strong correlation with embryonic hemopoiesis. In adults, AFP is synthesized only in trace amounts; its level rises to 10^{-5} to 10^{-6}, as compared with the level in the embryonic serum (Ruoslahti *et al.* 1974).

AFP synthesis is temporarily resumed during regeneration of the liver in mice and rats, induced by partial hepatectomy or hepatotoxic poisoning (Abelev 1971, 1974, Ruoslahti *et al.* 1974, Watanabe *et al.* 1976b). More intense synthesis of AFP is observed during the acute phase of chemical hepatocarcinogenesis, still *preceding* the appearance of tumor cells (Watabe 1971, Kroes *et al.* 1972, Becker *et al.* 1975). Stable AFP production takes place in germ cell teratocarcinomas and hepatocellular cancer; moreover, this protein is synthesized by the tumor cells (Abelev 1971).

The mechanisms that may be responsible for AFP expression in ontogenesis, termination of its synthesis in postnatal life and its re-expression in tumors, are, as yet, under only preliminary investigation. From the available information, elucidation of factors that give rise to AFP synthesis by germ cell teratocarcinomas seems to be the most promising.

Germ cell teratocarcinomas have been shown to synthesize AFP with the elements that are analogs of the embryonic yolk sac endoderm responsible for AFP synthesis in normal ontogenesis (Engelhardt *et al.* 1973, Norgaard-Pedersen 1976, Teilum 1976). Polypotent teratocarcinomas are regularly able to differentiate into various structures, including the structures corresponding to the yolk sac of the embryo. Some of their forms ("yolk-sac tumors") undergo exclusively these differentiations (Teilum 1976).

The mechanisms responsible for AFP regulation in the liver and re-expression of this protein in hepatomas appear to be much more complicated and obscure. The present paper is intended to consider this problem and to analyze those approaches that are under way in our laboratory.

AFP IN NORMAL ONTOGENESIS OF THE LIVER

Immunofluorescence permits detection of AFP in the embryonic liver of rats from the beginning of its formation in the hepatic bud; in this case, the adjoining epithelial areas of the foregut do not contain AFP (Shipova and Goussev 1976). At this point, embryonic hepatocytes are responsible for AFP production. Human hepatocytes, studied by immunofluorescence and local hemolysis-in-gel method (see below) in an early stage of liver ontogenesis (6- to 12-week-old development), have shown that all or nearly all hepatocytes are active AFP and also serum albumin producers (Eraizer *et al.* 1977, Abelev 1976). The same picture is observed in the embryonic liver of mice and rats.

The picture of the early postnatal liver in mice and rats, the time when fading and termination of AFP synthesis takes place, seems to be of special interest. Fading of AFP synthesis is of uneven character; it is most pronounced at the time when liver lobules make their appearance (Stratil *et al.* 1976). The AFP-containing cells remain around the central venae to form "cylinders" surrounding a vessel. The volume of the "cylinders" and fluorescence intensity of cells therein show rapid decrease (Shipova *et al.* 1974, Shipova and Goussev 1976, Stratil *et al.* 1976) (Figure 1). One may suggest that differences in the peripheral and central areas of the lobule constitute the key factor that is responsible for transition of hepatocytes into the "adult" type of synthesis.

It is very interesting that differences in central and peripheral areas of the lobule are retained in the first weeks after termination of AFP synthesis and disappearance of "AFP+" cells. According to the findings of V. S. Poltoranina in our laboratory, during regeneration of the liver (induced by partial hepatectomy in 6-week-old mice*), AFP synthesis is mainly resumed in cells adjoining the central venae (Figure 2). One may suggest that the decisive condition responsible for transition of cells from embryonal into "adult-type" synthesis is the establishment of specific intercellular contacts in the definitive hepatic cord that is characteristic of the liver of the adult animals. In any case, one of the

* "AFP+" cells are no longer detectable by the fourth week of postnatal development.

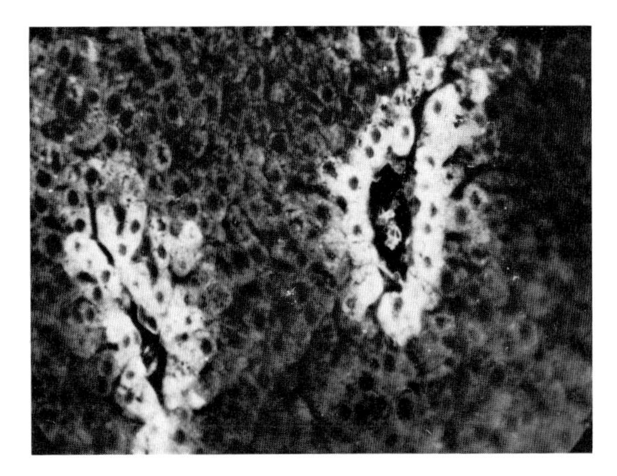

FIG. 1. Alpha-fetoprotein in the area of the central venae during the early postnatal development in the liver of two-week-old-mice. (Immunofluorescence technique of L. Ya. Shipova.) (obj. X 20.)

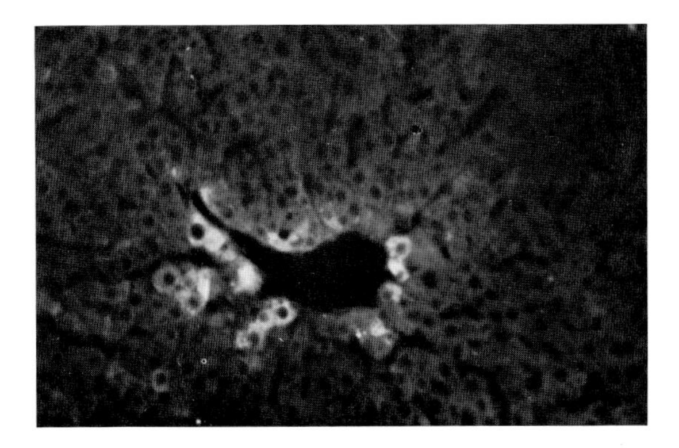

FIG. 2. Alpha-fetoprotein in the area of the central venae in the regenerating liver of six-week-old mice. (72 hours after partial hepatectomy.) (Immunofluorescence technique of V. S. Poltoranina.) (obj. X 20.)

possibilities for establishing factors responsible for regulation of AFP production in normal ontogenesis lies, apparently, in comparative study on the peculiarities of the cells bordering the central venae and the portal tract.

AFP SYNTHESIS DURING REGENERATION OF THE LIVER AND HEPATOTOXIN ACTION

AFP synthesis is resumed in the liver of adult animals poisoned with CCl_4 and other hepatotoxins or after partial hepatectomy (Abelev 1971, Watanabe *et al.* 1976b). This phenomenon is especially pronounced in mice.

A most important and interesting feature of AFP synthesis lies in the fact

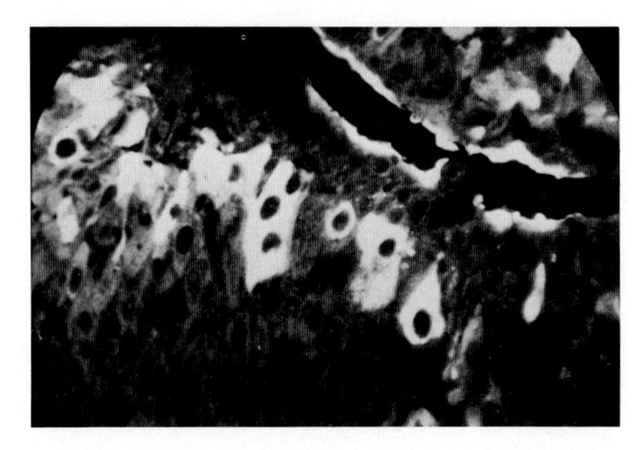

FIG. 3. Alpha-fetoprotein containing hepatocytes, bordering on the centrolobular necrosis, in the liver of mice poisoned with CCl_4. (72 hours after treatment.) (Immunofluorescence technique of N. V. Engelhardt and M. N. Lazareva.) (obj. X 40.)

that AFP production is resumed only in *single* cells, the number of which does not exceed 3 to 5% of the viable cells of hepatic parenchyma (Engelhardt *et al.* 1976c). "AFP^{+}" cells stand sharply apart from the surrounding cells that do not contain AFP (Figure 3). Their number and fluorescence intensity rise in parallel with the AFP level in the blood of the animal. These cells make their appearance 24 to 30 hours after CCl_4 action, reach their peak by the third day, and then a few days later are no longer detectable in the liver.

This discrete involvement of single hepatocytes into AFP synthesis takes place in mice after partial hepatectomy or poisoning with CCl_4, amile alcohol, and paracetamol (acetaminophen), each of which were investigated by the immunofluorescence technique (Lazareva 1977).

What, then, makes the difference between cells responsible for resumption of AFP synthesis and the nonsynthesizing hepatocytes? It has been suggested that aside from the main process of rehabilitation, due to proliferation of mature hepatocytes, regeneration of the liver also gives rise to the poorly expressed process of new hepatocyte formation from the precursor cells, that is, the process which resembles hepatocyte "ontogenesis," and, hence, involves synthesis of AFP. This process is quite likely to occur in the acute stage of hepatocarcinogenesis (Onoe *et al.* 1973, 1975), and one might assume that a reduced form of this process would take place during regeneration of the liver (Abelev 1974).

It is also possible that the earliest hepatocytes, those that have not yet progressed through the main stages of differentiation, are responsible for induction of AFP synthesis. These immature hepatocytes might be expected to belong to a low ploidy cell class, because during the postnatal growth of the liver the population of hepatocytes becomes almost exclusively polyploid (Carriere 1969).

However, all these suggestions have not yet been confirmed. "AFP^{+}" hepatocytes did not differ from their neighboring "AFP⁻" cells in size, ploidy, or shape (Engelhardt *et al.* 1976b,c). Moreover, repeated regeneration cycles, induced by several consecutive injections of animals with CCl_4 (from four to nine times) and accompanied by parallel rise in hepatocyte ploidy class (including even the formation of "giant" forms), have shown AFP also to be expressed in such cells (Lazareva 1977) (Figure 4). The question arose as to whether *pre-existing* hepatocytes were responsible for AFP induction or whether the cell would need to begin preliminary progress through the cell cycle, or, at least, DNA synthesis, in order to begin AFP synthesis. In order to elucidate an answer to this question, the following study was undertaken. Mice were poisoned with CCl_4, and in either three- or six-hour intervals were repeatedly injected with ^3H-thymidine (S-phase duration under these conditions is about seven hours). The animals were sacrificed either before or at the beginning of the first wave of DNA synthesis, i.e., 42 or 48 hours after CCl_4 injection. "AFP^+" cells on liver sections were identified (and photographed), and this was followed by radioautographic study of the same sections. From these analyses we discovered that from 18% to 94% of "AFP^+" cells proved to have not yet even entered the S phase (Engelhardt *et al.* 1976a). Hence, it appears clear that AFP induction in the regenerating murine liver takes place in pre-existing, mature, differentiated hepatocytes before they enter the S phase. These cells will certainly enter the cell cycle, since the percentage of labeled hepatocytes under these conditions is approximately 100, and the presence of "AFP^+" cells hardly makes a difference in this respect from the main hepatocyte population. On the other hand, it follows from the same data that the mere fact that mature hepatocytes enter the cell cycle is not sufficient to induce AFP synthesis therein. The hepatocyte that enters the cell cycle during regeneration appears to require

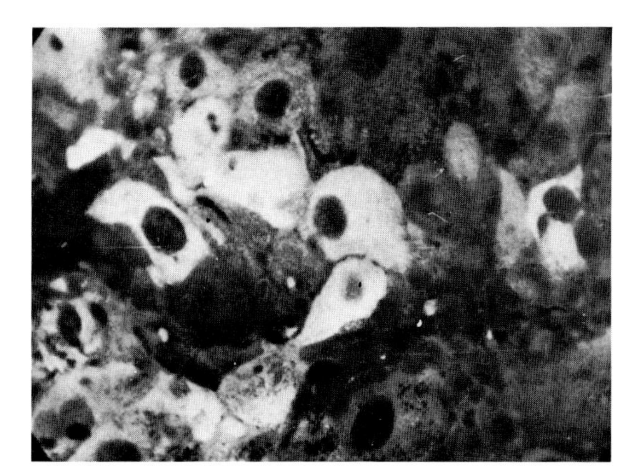

FIG. 4. Alpha-fetoprotein in large hepatocytes of a mouse repeatedly poisoned with carbon tetrachloride. (48 hours after the seventh poisoning.) (Immunofluorescence technique M. N. Lazareva.) (obj. × 40.)

some additional factors for induction of AFP synthesis. In what way is it possible to define these additional critical factors?

The only difference between the AFP-containing hepatocytes and the cells in the remainder population, according to our data, lies in their characteristic *topographical distribution* with regard to the necrosis zone. In the study of CCl_4 poisoning in mice, the AFP-containing hepatocytes are the first layer of cells, in close proximity to the zone of centrolobular necrosis (see Figure 3); in necrosis zones of other localization, as in the studies involving paracetamol and amile alcohol, the AFP-containing cells are localized in the same "peri-necrotic" layer (Figure 5). Peri-necrotic localization of "AFP^{+}" cells is of strictly regular character; it shows stability, irrespective of size of necrosis (CCl_4 dose), age of animals, ploidy of hepatocytes, or nature of hepatotoxin (Lazareva 1977). This characteristic arrangement of "inducible" cells suggests their sublethal damage (necrosis border) as the factor required for AFP induction. However, this suggestion does not appear to be probable. Endogeneous serum immunoglobulin can not penetrate "AFP^{+}" cells. They normally enter the cycle, scarcely different in this respect from the remainder population, and do not exhibit morphological features of degeneration.

In our opinion, the data obtained on the regenerating liver may well provide us with a plausible explanation, proceeding from the hypothesis formulated in the previous section. The hepatocyte in all the stages of its development has the capacity to synthesize AFP. During the formation of the definitive hepatic cord, it establishes specific and strictly defined contacts with neighboring hepatocytes and Kupffer's cells. Establishment of these contacts is a specific signal for *reversible* "repression" of AFP synthesis, which actually takes place

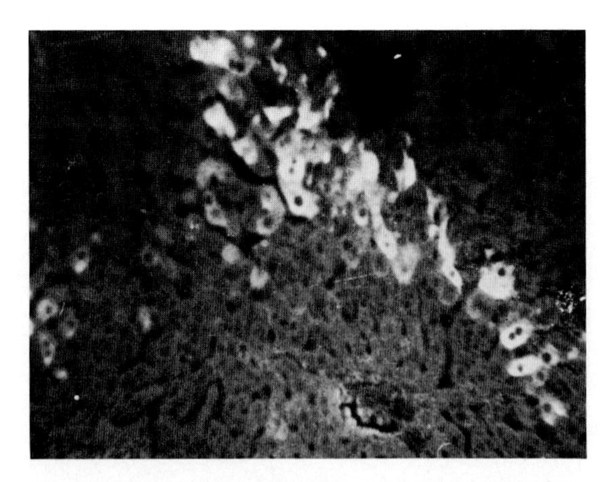

FIG. 5. Alpha-fetoprotein-containing hepatocytes in the perinecrotic area in the liver of mice poisoned with paracetamol (acetaminophen). (48 hours after the poisoning.) (Immunofluorescence technique of M. N. Lazareva.) (obj. × 20.)

during the postnatal development. Partial hepatectomy in very young animals that, at the time, leads to regeneration "loosens" the liver trabeculae in its most vulnerable spot, i.e., in the area adjoining the central venae. Upon formation of necrosis, there are exposed ends of the liver cords, and bordering edge cells lose the contacts with the neighboring cells of the necrotic area. As a result, these cells become responsible for AFP synthesis, which takes place until which time the cell is again included in the composition of the normal cord. This hypothesis provides a rather plausible explanation based on the available data and permits experimental testing (which is of primary importance). To study intercellular contacts, membrane hepatocyte antigens that are characteristically distributed on the cell surface, for instance, glycoprotein antigen of the bile capillaries (Khramkova and Beloshapkina 1974), lipid hapten of the hepatic sinuses (Yakimenko *et al.* 1977), and protein antigen of the hepatocyte membrane, evenly localized throughout the cell membrane can be used. In any case it is also necessary to find out whether induction of AFP synthesis is characterized by strictly specific activation, or if the same cells also give rise to synthesis of other serum proteins, such as albumin and transferrin. Experiments to determine this are presently under way in our laboratory.

Of critical importance to these studies is the immunofluorescent examination of hepatotoxin action, which is not accompanied by necrosis, as is the case with ethyonine poisoning in rats (Watanabe *et al.* 1976a). Thus, AFP studies in mouse liver regeneration have distinctly shown AFP synthesis to be induced in mature, differentiated hepatocytes, i.e. their temporary *dedifferentiation.*

Is this phenomenon in any way related to carcinogenesis? Is it possible for "AFP^{+}" cells to become tumor cell precursors? This problem is recently under study in our laboratory, using the model of carcinogenesis induced by $CC1_4$ poisoning, since this system permits us to follow all the events, proceeding from the acute toxic stage of this carcinogenic action and ending with the establishment of the tumor nodules.

However, it follows from analysis of histogenesis of hepatomas and AFP synthesis during chemical hepatocarcinogenesis in rats that mature differentiated hepatocytes hardly seem to be responsible for the formation of tumors. Hepatomas are most likely to arise in hyperplastic nodules that represent the clonal loci of liver regeneration, built up by newly formed hepatocytes (Farber 1973, 1976). The pre-existing hepatocytes are replaced by the newly formed ones and degenerate.

According to the findings of a number of authors (Kitagawa *et al.* 1972, Onoe *et al.* 1973, 1975, Bannikov *et al.* 1977), intense AFP synthesis in the liver of rats during the acute stage of chemical carcinogenesis is due to the process of new hepatocyte formation and is performed by the so-called "transitional" cells. These small basophilic hepatocyte-like cells, which commonly develop into ductal structures, are regarded as transitional forms from the oval cells to the typical hepatocytes (Inaoka 1967). It is a far better substantiated suggestion that such cells may be the precursors of hepatomas or are, at least, closely related to the tumor precursor cells. However, this type of hepatic cell has not actually been studied in detail.

AFP PRODUCTION BY HEPATOMAS

The main difficulty in interpretation of AFP production by hepatomas lies in the fact that the serum AFP levels vary in individual cases from the background (10^{-5} mg/ml) up to the level exceeding the embryonal—over 10 mg/ml, without clear-cut correlation either with morphological structure or biological properties of the tumor. Hepatoblastoma (Ishak and Glunz 1967)—the discrete type of hepatocellular carcinoma—composed of the cells that are analogs to the embryonic or fetal hepatocytes is the most "productive." But even this group is characterized by a wide variety of AFP levels (Elgort *et al.* 1976). Study on correlations in a number of transplantable hepatomas in mice and rats has allowed us to establish only very general and not strictly pronounced dependencies; on the rate of growth and the degree of differentiation (Abelev 1971, Sell and Morris 1974), low-differentiated strains were far stronger AFP-producers than high-differentiated strains and were aneuploid rather than diploid. However, tumors of the same type, according to this feature, also show appreciable variance; it is difficult to establish any strict regularities in human tumors of the same histogenesis (Masseyeff 1972, Purves *et al.* 1970).

Hence, regular reappearance of AFP in hepatomas may clearly suggest specific relationship of this phenomenon with the very process of hepatocyte malignant transformation, but quantitative expression of this character does not in any way permit us to "bind" it to any characteristic feature of the malignant tissue. It seems to us that *heterogeneity* of neoplastic cells in their ability to produce AFP may constitute the key point in the solution of this problem. This heterogeneity appears to be generally expressed in AFP-producing hepatomas. According to immunofluorescence assays, AFP is revealed only in a part of tumor cells; moreover, the number of "AFP^{+}" cells is in correspondence with serum AFP levels (Figure 6) (Engelhardt *et al.* 1974, Goussev and Shipova 1975). Hepatoblastomas, in this respect, appear to stand clearly apart from the tumors of the adult type and are more homogeneous in AFP contents (Norgaard-Pedersen *et al.* 1974). It stands to reason that if one can elucidate the cause of the observed heterogeneity and the specific peculiarities of "AFP^{+}" cells, as compared with "AFP^{-}" cells, we might have a better understanding of AFP re-expression in the neoplastic cell.

Several possibilities are under investigation to explain this heterogeneity:

1) Clonal heterogeneity of hepatomas, which has received quite conclusive demonstration for rat ascitic hepatomas (Isaka *et al.* 1976). Individual clones, isolated from the same ascitic tumor and maintained in vitro, varied over 1,000-fold with respect to AFP production. However, this approach makes it difficult to evaluate both the actual contribution of different clones in

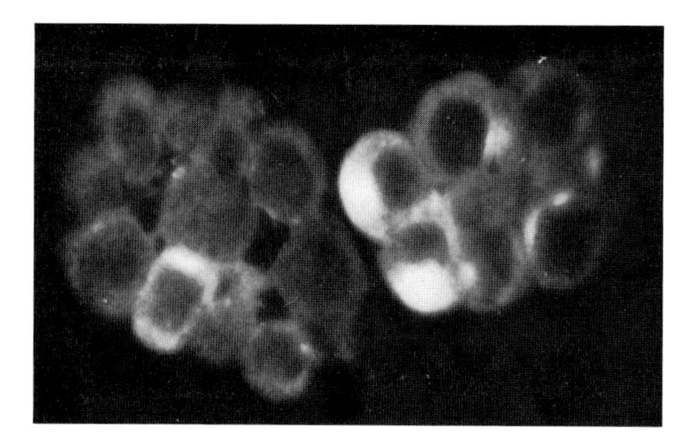

FIG. 6. Alpha-fetoprotein in the Zajdela rat ascitic hepatoma. (Immunofluorescence technique of A. I. Goussev and L. Ya. Shipova.)

the original tumor and secondary alterations due to cell adaptation to in vitro growth, which is certain to take place when the clones are established. Also, there are no data as to whether a cell population inside the clone is homogeneous in AFP production.

2) Dependency of AFP production on cell cycle has been clearly demonstrated on synchronized rat hepatoma cultures. AFP was secreted into the medium in the late G_1 and S phases (Tsukada and Hirai 1975). However, there are no indications as to whether all hepatoma cells in G_1 and S phases are AFP-producers and to what extent the observed heterogeneity of cells in the tumor is related to "cycle-dependency."
3) Intratumoral and intraclonal heterogeneity of hepatoma cells is due to the degree of their differentiation (Abelev 1971, 1974). This approach has, as yet, no direct experimental confirmation. It proceeds from the hypothesis that the tumor stem line is represented by a predetermined, although nondifferentiated, "precursor" cell line that has the ability to differentiate. "AFP^+" cells belong only to this "side" branch of cells, diverging from the stem line. The main problem here lies in finding experimental approaches to characterize the ability of the stem cells to produce AFP and other serum proteins.

Hence, an attempt to analyze the basis of intratumoral heterogeneity in AFP production by hepatomas involves an obviously multifactor system, and the necessity for consideration of each factor; however, we lack an adequate experimental approach for solution on this point. In this respect, in vitro analysis of individual cells in their population seems to be especially promising. Based on our view and experience, the method of local hemolysis-in-gel, suggested by Molinaro *et al.* (1975) for detecting the antigens secreted by single cells in vitro (for example, serum albumin by hepatocytes), appears to be especially attractive in this respect. In our laboratory this method was applied, with slight modifications, to the primary culture of human embryonal hepatocytes. The method of local hemolysis-in-gel, employed in this system, has permitted us to analyze quantitatively the population of embryonal hepatocytes with regard to their production of AFP and serum albumin (Eraizer *et al.* 1977).

The method offers unique possibilities. In our modification, study is made on liver or hepatoma cells explanted in vitro to achieve their adherence to the bottom of the plate. The adherent cells are then coated with a thin layer of agarose, wherein are suspended sheep erythrocytes, chemically conjugated with the antibodies against AFP or serum albumin. After four-hour incubation, the antiserum to the antigen under study is added for one hour into the system, and then the complement in succession. A zone of hemolysis—plaque—is formed around antigen-secreting cells or a colony of cells, almost in the same way as in the Jerne technique (Figure 7).

In studies with Drs. Elgort, Perova, and Eraizer, we have obtained direct indications that this method is perfectly applicable for analyzing Zajdela rat ascitic hepatoma, both for analysis of the adherent cells or for the cells in suspension (Figure 8). The method clearly reveals pronounced heterogeneity of hepatoma cell population and permits the study of its nature. The method allows one to evaluate the contribution of clonal heterogeneity, since an ascitic hepatoma forms with high frequency most likely in the islands of clonal origin. The method permits large-scale study of microcolonies, immediately taken from the ascitic fluid. Analysis of clones, isolated from gel, after their minimal in vitro growth is expected to characterize intraclonal heterogeneity of hepatoma cells. Since the test can be carried out on the adherent cells, it is quite possible to subsequently treat the preparations with ordinary staining and radioautographic techniques. This will permit one to decide to what extent the cells that are in the same stage of cycle are homogeneous in their AFP production.

The unique characteristic of the local hemolysis-in-gel technique is that it allows us to work with the living cells, which appear to resist the conditions of the assay. Hence, it appears possible to investigate the clonogeneic potency

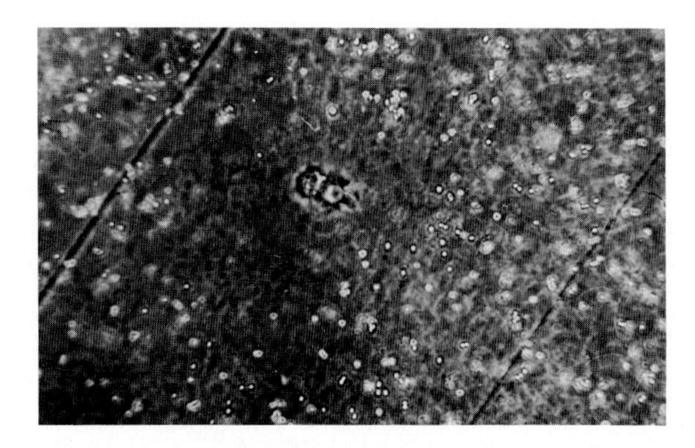

FIG. 7. Hemolysis plaque around the hepatocytes from human embryonic liver.

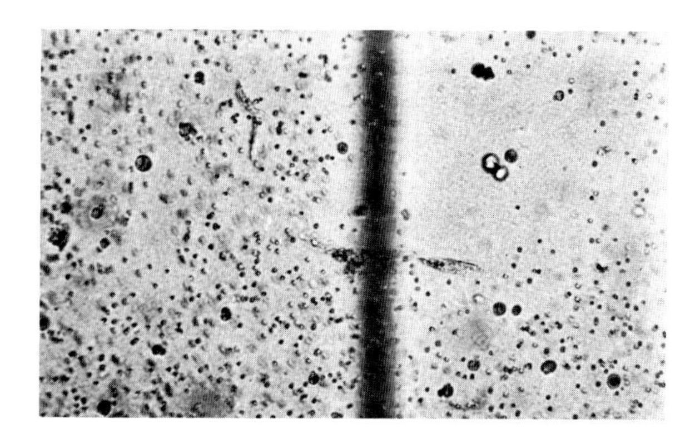

FIG. 8. Hemolysis plaque around the cells of the Zajdela hepatoma.

of AFP-producing and nonproducing cells and, thereby, to elucidate whether AFP production is characteristic of the stem cells, or only of a part of its progeny that is on the way to differentiation. Studies on this problem are in progress.

In our opinion, the analysis of cell heterogeneity in hepatomas, as well as heterogeneity of hepatocytes in the postnatal and regenerating liver, may constitute a key point for gaining further insight into the mechanism responsible for regulation of alpha-fetoprotein during ontogenesis and the re-expression of this protein in tumors.

ACKNOWLEDGMENTS

I would like to express my sincere gratitude to Ms. M. Berdichevskaya for her help in the English translation of this manuscript and to Dr. N. Engelhardt, Ms. M. Lazareva, and Mrs. L. Shipova for providing the illustrations.

The experimental studies presented in this paper were partially supported by the World Health Organization Immunology Unit.

REFERENCES

Abelev, G. I. 1971. Alpha-fetoprotein in ontogenesis and its association with malignant tumors. Adv. Cancer Res. 14:295–358.

Abelev, G. I. 1974. α-fetoprotein as a marker of embryospecific differentiations in normal and tumor tissues. Transpl. Rev. 20:3–37.

Abelev, G. I. 1976. Cellular aspects of alpha-fetoprotein synthesis, *in* Onco-developmental Gene Expression, W. Fishman and S. Sell, eds., Academic Press, New York, pp. 191–202.

Bannikov, G. A., V. I. Gelstein, and T. A. Chipisheva. 1977. Study of evolution of liver cell population during chemical carcinogenesis with the use of antigen differentiation markers. Voprosi Onkologii 23(4):39–44.

Becker, F. F., A. A. Horland, A. Shurgin, and S. Sell. 1975. A study of α-fetoprotein levels during exposure to 3′methyl-4-dimethylaminoazobenzene and its analogs. Cancer Res. 35:1510–1513.

Carriere, R. 1969. The growth of liver parenchemal nuclei. Int. Rev. Cytol. 25:201–277.

Elgort, D. A., G. I. Abelev, L. A. Durnov, Yu. V. Pashkov, D. M. Levina, A. V. Laskina,

M. A. Izrailskaya, I.O. Harrit, E. N. Ter-Grigorova, and A. P. Reizis. 1976. Alpha-feroprotein in diagnosis of tumors in children. Vestn. Acad. Med. Nauk. USSR (2):31–44.

Engelhardt, N. V., M. N. Lazareva, G. I. Abelev, I. V. Uryvaeva, V. M. Factor, and V. Ya. Brodsky. 1976a. Detection of α-foetoprotein in mouse liver differentiated hepatocytes before their progression through S phase. Nature 263:146–148.

Engelhardt, N. V., M. N. Lazareva, I. V. Uryvaeva, V. M. Factor, V. S. Poltoranina, A. S. Gleiberman, V. Ya. Brodsky, and G. I. Abelev. 1976b. Alpha-fetoprotein in adult differentiated hepatocytes of the regenerating liver, *in* Onco-developmental Gene Expression, W. Fishman and S. Sell, eds., Academic Press, New York, pp. 533–540.

Engelhardt, N. V., V. S. Poltoranina, M. N. Lazareva, S. D. Perova, and A. S. Gleiberman. 1976c. Synthesis and localization of α-fetoprotein in the regenerating liver of mice. Bull. Exp. Biol. Med. 82:1251–1254.

Engelhardt, N. V., V. S. Poltoranina, L. Y. Shipova, A. I. Goussev, and A. K. Yazova. 1974. Cellular distribution of AFP in mice during normal ontogenesis and in transplantable teratocarcinoma and hepatomas, *in* Alpha-Feto-Protein, R. Masseyeff, ed., Inserm, Paris, pp. 217–229.

Engelhardt, N. V., V. S. Poltoranina, and A. K. Yazova. 1973. Localization of alpha-fetoprotein in transplantable murine teratocarcinomas. Int. J. Cancer 11:448–459.

Eraizer, T. L., D. A. Elgort, and G. I. Abelev. 1977. Detection of the number of alpha-fetoprotein and albumin-producing cells in the embryonic liver with the use of local hemolysis-in-gel method. Bull. Exp. Biol. Med. 83(6):711–713.

Farber, E. 1973. Carcinogenesis-cellular evolution as a unifying thread. Cancer Res. 33:2537–2550.

Farber, E. 1976. Hyperplastic areas, hyperplastic nodules, and hyperbasofilic areas as putative precursor lesions. Cancer Res. 36:2532–2533.

Fishman, W., and S. Sell, eds. 1976. Onco-developmental Gene Expression. Academic Press, New York.

Goussev, A. I., and L. Ya. Shipova. 1975. Distribution of α-fetoprotein and albumin on paraffine sections of the cells of Zajdela's ascitic hepatoma. Bull. Exp. Biol. Med. 81:80–82.

Hirai, H., and E. Alpert, eds. 1975. Carcinofetal Proteins: Biology and Chemistry. Annals of the New York Academy of Science, New York.

Hirai, H., and T. Miyaji, eds. 1973. Alpha-fetoprotein and Hepatoma (Gann Monograph on Cancer Research #14), University Park Press, Baltimore.

Inaoka, Y. 1967. Significance of the so-called oval cell proliferation during azo-dye hepatocarcinogenesis. Gann 58:355–366.

Isaka, H., S. Umehara, H. Yoshii, Y. Tsukada, and H. Hirai. 1976. α-fetoprotein and albumin produced by subclonal cell population of the ascites hepatoma AH-66 in vitro. Gann 67:131–135.

Ishak, K., and P. Glunz. 1967. Hepatoblastoma and hepatocarcinoma in infancy and childhood. Cancer 20:396.

Khramkova, N. I., and T. D. Beloshapkina. 1974. Antigen of mouse bile capillaries and cuticule of intestinal mucosa. Nature 251:627–628.

Kitagawa, T., T. Yokochi, and H. Sugano. 1972. α-fetoprotein and hepatocarcinogenesis in rats fed 3′-methyl-4-(dimethylamino)-azobenzene and N-2-fluorenylacetamide. Int. J. Cancer 10:368–381.

Kroes, R., G. M. Williams, and J. H. Weisburger. 1972. Early appearance of serum α-fetoprotein during hepatocarcinogenesis as a function of age of rats, and extent of treatment with 3′-methyl-4-dimethylamino-azobenzene. Cancer Res. 32:1526–1532.

Lazareva, M. N. 1977. Characterization of the alpha-fetoprotein-containing hepatocytes in the liver of mice, poisoned with $CC1_4$ and other hepatotoxins. Bull. Exp. Biol. Med. 83(7):97–101.

Masseyeff, R. 1972. Human alpha-feto-protein (Review). Pathol. Biol. 20:703–727.

Masseyeff, R., ed. 1974. Alpha-Feto-Protein. Inserm, Paris.

Molinaro, G. A., E. Maron, W. C. Eby, and S. Sray. 1975. A general method for enumerating single cells secreting antigen: Albumin-secreting hepatocytes detected as plaque-forming cells. Eur. J. Immunol. 5:771–774.

Norgaard-Pedersen, B. 1976. Human alpha-fetoprotein. Scand. J. Immunol. (Suppl. N4):1–45.

Norgaard-Pedersen, B., E. Dabelsteen, and C. J. Edeling. 1974. Localization of human α-fetoprotein synthesis in hepatoblastoma cells by immunofluorescence and immunoperoxidase methods. Acta Pathol. Microbiol. Scand. 82:169–174.

Onoe, T., K. Dempo, A. Kaneko, and H. Watabe. 1973. Significance of α-fetoprotein appearance in the early stage of azo-dye carcinogenesis, *in* Alpha-Fetoprotein and Hepatoma, H. Hirai and T. Miyaji, eds. (Gann Monograph on Cancer Research #14), University Park Press, Baltimore, pp. 233–247.

Onoe, T., A. Kaneko, K. Dempo, K. Ogawa, and M. Takashi. 1975. α-fetoprotein and early histological changes of hepatic tissue in DAB-hepatocarcinogenesis. Ann. NY Acad. Sci. 259:168–180.

Purves, L. R., I. Bersohn, and E. W. Geddes. 1970. Serum alpha-feto-protein in primary cancer of the liver in man. Cancer 25:1261–1270.

Ruoslahti, E., and E. Engvall. 1976. Immunological cross-reaction between alpha-fetoprotein and albumin. Proc. Natl. Acad. Sci. USA 73:4641–4644.

Ruoslahti, E., H. Pihko, and M. Seppala. 1974. Alpha-fetoprotein: Immunochemical purification and chemical properties. Expression in normal state and in malignant and non-malignant liver diseases. Transplant. Rev. 20:38.

Ruoslahti, E., and W. D. Terry. 1976. α-foetoprotein and serum albumin show sequence homology. Nature 260:804.

Sell, S., and P. Morris. 1974. Relationship of rat α_1-fetoprotein to growth rate and chromosome composition of Morris hepatomas. Cancer Res. 34:1413–1417.

Shipova, L. Ya., and A. I. Goussev. 1976. Alpha-fetoprotein in the liver of embryonic and newborn rats. Ontogenez 7:392–395.

Shipova, L. Ya., A. I. Goussev, and N. V. Engelhardt. 1974. Immunohistochemical study of α-fetoprotein and serum albumin in the early postnatal period in mice. Ontogenez 5:53–60.

Stratil, P., V. Dolezalova, J. Feit, and A. Kocent. 1976. Localization of α-fetoprotein in liver tissue of rats during postnatal development: Comparison of the immunofluorescent and autoradiographic methods. Neoplasma 23:1–10.

Teilum, G. 1976. Special Tumors of Ovary and Testis and Related Extragonadal Lesions: Comparative Pathology and Histological Identification. 2nd ed., Munksgaard, Copenhagen.

Tsukada, Y., and H. Hirai. 1975. α-fetoprotein and albumin synthesis during the cell cycle. Ann. NY Acad. Sci. 259:37–44.

Watabe, H. 1971. Early appearance of embryonic α-globulin in rat serum during carcinogenesis with 4-dimethylaminoazobenzene. Cancer Res. 31:1192–1194.

Watanabe, A., M. Miyazaki, and K. Taketa. 1976a. Prompt elevation of serum α-fetoprotein by acute liver injury following a single injection of ethyonine to rat. Int. J. Cancer 17:518–524.

Watanabe, A., K. Taketa, K. Kosaka, and M. Miyazaki. 1976b. Mechanisms of increased alpha-fetoprotein production by hepatic injury and its pathophysiological significance, *in* Onco-developmental Gene Expression, W. Fishman and S. Sell, eds., Academic Press., New York, pp. 209–217.

Yakimenko, E. F., N. Y. Khramkova, and T. D. Rudinskaya. 1977. Antigenic structure of plasmatic membrane of mouse liver cells. III. Lipid haptens. Cytologia 19:545–551.

Cell Differentiation and Neoplasia, edited by Grady F. Saunders. Raven Press, New York

Biosynthesis of Rauscher Leukemia Virus Reverse Transcriptase and Structural Proteins: Evidence for Translational Control

Ralph B. Arlinghaus, Ghazi A. Jamjoom, John Kopchick, and Robert B. Naso

Department of Biology, The University of Texas System Cancer Center M. D. Anderson Hospital and Tumor Institute, Houston, Texas 77030

The genetic complexity of the genome of avian and murine RNA tumor viruses has been estimated to be about 3.0×10^6 daltons (Beemon *et al.* 1976, Billetter *et al.* 1974), which would have the capacity to code for proteins with total molecular weight of about 300,000 daltons. This limited size increases the hope for characterization of all the translational products of the viral genome. Genetic and immunological studies (Baltimore 1974) have indicated that the viral genome codes for the viral core proteins, the envelope proteins, the RNA-dependent DNA polymerase, and the proteins that are thought to be involved in the expression of the oncogenic properties of these viruses.

Studies of virus-specific polypeptides in cells infected with RNA tumor viruses have indicated that mature viral proteins* are formed by the cleavage of high molecular weight precursors (Vogt and Eisenman 1973, Naso *et al.* 1975a). In Rauscher leukemia virus (RLV), several virus-specific precursors of the core, and envelope proteins have been characterized (Arcement *et al.* 1976, Naso *et al.* 1976, Van Zaane *et al.* 1975). However, it remained to study the synthesis of the viral polymerase and to detail the steps in the synthesis pathway of the mature viral proteins. In addition, more studies were needed to illustrate the overall scheme of translation of the viral genome and the relation between the translational products of the main viral genes.

In the present study, we have attempted to shed more light on these aspects of translation of the viral genome, by searching in virus-infected cell extracts for unstable precursor polypeptides that can be recognized by monospecific antisera made against the viral structural components or the viral polymerase. In addition, we have tried to interfere with the proteolytic cleavage of viral

* We have used the nomenclature of mature virion proteins of oncogenic RNA viruses as proposed by August *et al.* 1974. For virus-specific intracellular precursor polypeptides, however, we are temporarily using our own nomenclature system, because the estimated molecular weights of these polypeptides are only approximate. Once these molecular weights are determined with more accuracy, a system similar to the one agreed upon by the above-mentioned authors can be easily adopted (e.g., Pr4 may be called Pr 70).

precursor polypeptides by adding inhibitors of serine proteases and amino acid analogues. Both of these agents have been extensively used in other viral systems (Jacobson *et al.* 1970, Summers *et al.* 1972, Korant 1975). Our results suggest the existence of translational control in the expression of the viral genome and point to specific models of control, which are amenable to future evaluation and tests.

MATERIALS AND METHODS

Cells and Virus

Rauscher murine leukemia virus-infected N.I.H. Swiss mouse embryo cells (JLS-V16) were used in these studies (Naso *et al.* 1975a). The culture medium contained a modified Eagle's amino acid formula and 10% fetal calf serum, as described previously (Syrewicz *et al.* 1972). Cells were grown in 2 oz. prescription glass bottles or quart glass bottles. Cells were subcultured 1 to 3 days before use and were used when subconfluent.

Labeling of Cells and Virus

Cells were rinsed in warm Hank's solution and pulse labeled for 15 minutes in Hank's solution containing ^{35}S-methionine, 25 μCi/ml (288 Ci/mM, Amersham Searle). For chase incubations, the radioactive medium was removed, and the cells were rinsed with Hank's solution and incubated in complete growth medium. Labeling of virus (6 to 8 hours) with ^{35}S-methionine was carried out in growth medium containing 5% dialyzed calf serum, 1/10th Eagle's concentration of unlabeled methionine, and no tryptose phosphate.

Immune Precipitation and Gel Electrophoresis

Cytoplasmic extracts and anti-RLV serum were prepared as previously described (Naso *et al.* 1975a). Monospecific goat antisera prepared against RLV p30, p15, p12, p10, and gp69/71 and anti-reverse transcriptase (RT) sera were obtained through the Office of Program Resources and Logistics, Viral Oncology, National Institutes of Health. The anti-RT serum inhibits the in vitro RT reaction. It was prepared with a polypeptide with RT activity, which was purified by affinity chromatography. The antisera to the group antigens of the virus *(gag)* were prepared from proteins purified by guanidine-HCl agarose chromatography and gel filtration (p12 and p30). The anti-gp69/71 serum was prepared with gp69/71 purified by ion exchange chromatography and gel filtration. Additional anti-RT sera were also kindly provided by Dr. George Vande Woude (prepared by Dr. Strickland) and Dr. Takis Papas of N.I.H. Anti-RT sera were in some cases preabsorbed with 100 μg of virus proteins. Marker RLV-RT ($\approx$ 72,000 daltons) was also generously supplied by Dr. Papas. For direct immune precipitation with anti-RLV serum, cytoplasmic extracts of cells from a 2 oz. bottle ($\approx 2 \times 10^7$ cells) were mixed with 0.3 ml of anti-RLV serum. For indirect immune precipitation, 12 μl of monospecific antisera were mixed with cytoplasm extracts, as above, and incubated for 15 minutes at room temperature and overnight at 4°C. For the second immune reaction, 0.25 ml of rabbit anti-goat IgG serum was then added and further incubated for 15 minutes at room temperature and 4 hours at 4°C. In all experiments, the antisera (including the rabbit anti-goat IgG) were preabsorbed with excess uninfected JLS-V16 cytoplasmic extracts (Jamjoom *et al.* 1975) by incubating the antisera with two or three times its volume of a solution of uninfected cell cytoplasmic extract containing 10 to 30 mg/ml of cell protein for 15 minutes at room temperature and overnight at 4°C. The antisera were then clarified at 10,000 × g for 10 minutes, and used as above, after accounting for its dilution. The immune pellets of indirect precipitates were collected by centrifugation at 10,000 × g for 10 minutes in a Sorvall HB-4 rotor over a 2 ml cushion of 1 M sucrose in immune buffer (0.02 M Tris, 0.05 M NaCl, 0.5% sodium deoxycholate [DOC], 0.5% NP-40) containing 1% Triton X-100.

Sodium dodecyl sulfate–polyacrylamide gel electrophoresis (SDS-PAGE) of the immune pellets was performed using the buffer system described by Laemmli (1970). The gels were processed for fluorography, as described by Bonner and Laskey (1974). To obtain a linear response to radioactivity, the X-ray films were preflashed (Laskey and Mills 1975). To quantitate the relative amounts of radioactivity in different bands of a certain slot, the film was scanned at 590 nm in a Gilford spectrophotometer, and the relative areas of different peaks in the resulting curve were measured in a DuPont 310 curve resolver. Similar amounts of radioactivity in the samples were applied to each gel unless otherwise stated.

RESULTS

Effect of Protease Inhibitors on Pulse Labeling of Viral Precursor Polypeptides

Mature RLV particles are composed of several proteins, the major ones being p30, p15, p12, and p10, and the glycoproteins gp69/71 and gp45 (August *et al.* 1974). The p12 class consists of two different proteins (Naso *et al.* 1976, Karshin *et al.* 1977), one of which is related to an envelope protein termed p15E (Ikeda *et al.* 1975). In addition, viral particles also contain variable amounts of an uncleaved p30 precursor, termed Pr4, of molecular weight of 67,000 to 70,000 daltons. In virus produced by some cell lines, it is one of the major proteins (Jamjoom *et al.* 1975). Virus-specific polypeptides in infected cells have been studied by immune precipitation of cytoplasmic extracts with antisera made against detergent-disrupted virions or purified viral proteins (Naso *et al.* 1975a, Arcement *et al.* 1976, Arlinghaus *et al.* 1976). Such studies have indicated that viral proteins are made by cleavage of rapidly synthesized, high molecular weight precursors that can be detected by short pulse labeling with radioactive amino acids. Figure 1 shows the pattern of labeling of virus-specific polypeptides that were immune precipitated from infected cells by anti-RLV serum and analyzed by SDS-PAGE. After a 15-minute pulse labeling (Figure 1, slot B), virus-specific polypeptides include the major species, designated as follows: Pr1*a*+*b* ($\simeq$ 200,000 daltons), Pr2*a*+*b* ($\simeq$ 90,000 daltons), Pr3 ($\simeq$ 80,000 daltons), and Pr4 ($\simeq$ 70,000 daltons). None of the mature viral polypeptides, however, are labeled in such a short pulse. Pr1*a*+*b*, Pr3, and Pr4 are immune precipitable with antisera made against purified p30 (Naso *et al.* 1976) and have been shown by tryptic fingerprinting to contain p30 peptide sequences (Arcement *et al.* 1976). Polypeptides Pr2*a*+*b*, which are glycosylated, are immune precipitable with anti-gp69/71 serum and share tryptic peptides with this glycoprotein (Naso *et al.* 1976).

When a 60-minute chase is allowed after a short pulse, mature viral proteins can be detected in the infected cell cytoplasm (Figure 1, slot A). Some of the viral proteins (p15, p10) are not seen after labeling with ^{35}S-methionine, which was used here, because they are deficient in this amino acid.

It is apparent from such studies that viral precursor polypeptides are built up and processed at different rates. Thus, the buildup of Pr2*a*+*b* is slower than that of Pr3 and Pr4. In addition, the cleavage of Pr1*a* and Pr3 occurs much faster than the cleavage of Pr1*b*, Pr2*a*+*b*, and Pr4.

Figure 1, slot C shows the effect of 0.1 mM tolylsulfonyl-phenylalanyl chloromethyl ketone (TPCK) on the pulse labeling of virus-specific polypeptides. It is evident that TPCK altered the pattern of labeling of these polypeptides. Thus, Pr1*a*+*b* were increased (Pr1*a* more so than Pr1*b*) in comparison to Pr3 and Pr4. The ratio of Pr1*a*+*b* to Pr3 plus Pr4 was estimated from densitometer tracings of SDS gel patterns such as the ones shown in Figure 1. This ratio was found to increase on treatment with TPCK from about 0.3 to approximately 0.54 to 0.94. In addition, the ratio of Pr3 to Pr4 was reversed, and the amount of Pr2*a*+*b* was reduced. Figure 1, slot D is a control for nonspecific immune precipitation, which shows that if the antiserum were absorbed with mature virus proteins, none of the viral precursor polypeptides were immune precipitated.

Figure 1, slots E and F show the pattern of virus-specific polypeptides after 5 minutes of pretreatment and 15 minutes of pulse labeling in the presence of leupeptin (150 μg/ml), and antipain (150 μg/ml). Leupeptin inhibits plasmin, trypsin, papain, and cathepsin B, while antipain inhibits trypsin, papain, and cathepsin A (Aoyagi and Umezawa 1975). Under the conditions that we used in this experiment, these inhibitors did not have any noticeable effect on the pattern of labeling of viral polypeptides.

Since TPCK is a site-specific inhibitor of α-chymotrypsin (Shaw 1967), the results mentioned above suggest that a chymotrypsin-like enzyme is involved in the initial cleavage steps of the viral precursor polypeptides. However, it is

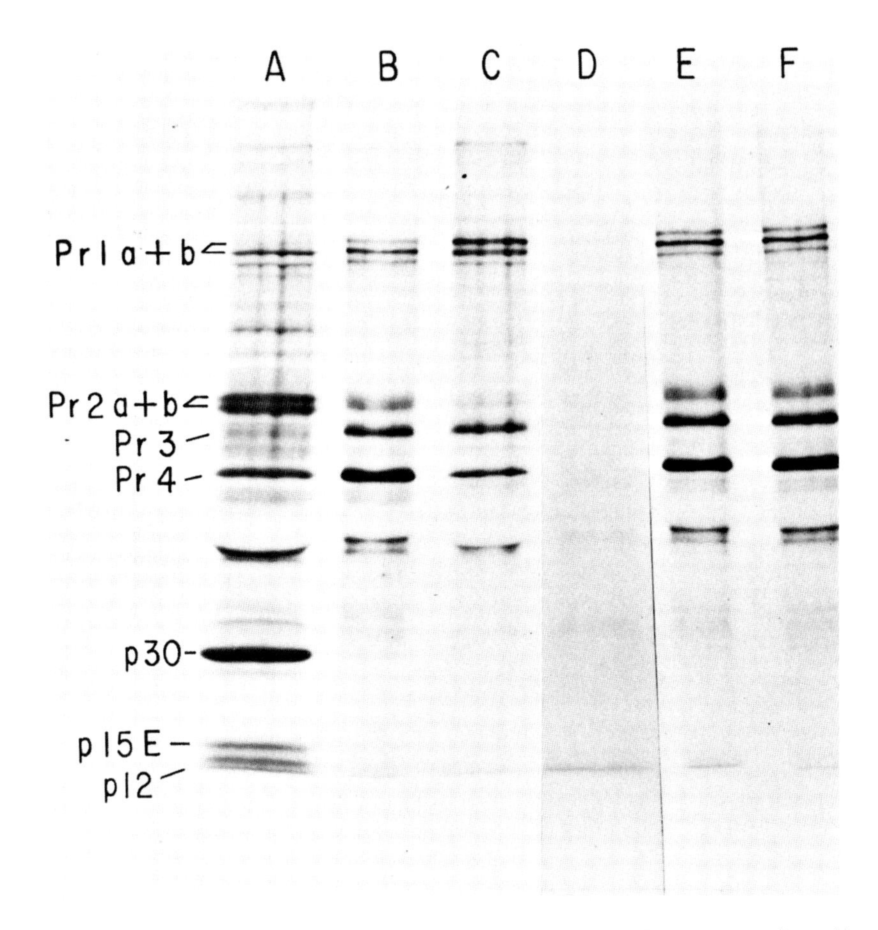

FIG. 1. Effect of TPCK on the pulse labeling of intracellular virus-specific precursor polypeptides. Monolayers of virus-infected cells were rinsed and then pulsed for 15 minutes in warm Hank's solution containing ^{35}S-methionine. Cells in slot A were then chased in complete growth medium for 60 minutes, while those in slots B, C, D, E, and F were lysed directly after the pulse. During the pulse, cells were treated as follows: slots A and B, DMSO; slots C and D, 0.1 mM TPCK (dissolved in DMSO); slot E, 150 μg/ml leupeptin; slot F, 150 μg/ml antipain. Treatment was initiated 5 minutes before addition of radioactive label. The final concentration of DMSO in slots A, B, C, and D was 0.06%.

After the above treatments, cells were rinsed, homogenized in a buffer (30 mM Tris, pH 7.5, 3.6 mM $CaCl_2$, 6 mM $MgCl_2$, 125 mM KCl, 0.5 mM EDTA, and 3 mM 2-mercaptoethanol) containing 0.5% NP-40 and 0.5% DOC. Cytoplasmic extracts were prepared and immune precipitated with anti-RLV serum. The anti-RLV serum used in slot D was preabsorbed with excess purified RLV and excess uninfected cell proteins. The immune pellet was washed with immune buffer and analyzed by SDS polyacrylamide gel electrophoresis on a 6% to 12% linear gradient gel. Similar amounts of radioactivity were applied for all samples except D, for which only about 1/6th of the counts were precipitated and applied. All samples were exposed for one day, except D, which was exposed for approximately four days.

possible that the effect of TPCK is nonspecific, since the compound is a strong alkylating agent. To judge TPCK's specificity, we tested the effect of other chloromethyl ketones on the pulse labeling of virus-specific precursor polypeptides. The other compounds tested were tolylsulfonyl-lysyl chloromethyl ketone (TLCK), an inhibitor of trypsin (Shaw 1967), and carbobenzyloxyl-phenylalanyl chloromethyl ketone (L-ZPCK), another inhibitor of chymotrypsin, and D-ZPCK, the isomer of L-ZPCK, which is inactive against chymotrypsin (Shaw and Ruscia 1971).

TLCK had no effect on the pattern of labeling; TPCK and D- and L-ZPCK caused a relative enrichment of Prl*a*+*b* and a reversal in the Pr3 to Pr4 ratio. That D-ZPCK, which is inactive against chymotrypsin, had an effect identical to TPCK and L-ZPCK indicates that the effect is not specific for chymotrypsin inhibitors but is more likely the result of other properties of these compounds.

Small and variable amounts of high molecular weight polypeptides, larger than Pr1*a*+*b*, were also observed after treatment with TPCK or ZPCK, among the proteins that are precipitable with anti-RLV serum. The identity and significance of these polypeptides are not known.

Taken together, our results indicated that the chloromethyl ketone derivatives TPCK and D- and L-ZPCK cause a relative enrichment of Pr1*a*+*b* and a reversal of the ratio of Pr3 to Pr4. These results differ from those reported by Vogt *et al.* (1975) and Van Zaane *et al.* (1975), who obtained no effects with these compounds. The reversal of the ratio of Pr3 to Pr4 was consistent with the idea that Pr3 is cleaved to Pr4. This idea was initially suggested by pulse-chase studies, which indicated that Pr3 has a shorter half-life than Pr4, and by precipitation with anti-p30 serum and tryptic fingerprinting, which indicated that both Pr3 and Pr4 contain p30 sequences (Arcement *et al.* 1976).

Since Pr1*a*+*b* are also precipitated with anti-p30 serum and contain p30 tryptic peptides (Arcement *et al.* 1976), the relative enrichment of Pr1*a*+*b* after treatment with TPCK and ZPCK is consistent with the idea that Pr1*a*+*b* are early precursors of p30. If this is the case, the intermediate p30 precursors Pr3 and Pr4 would be formed by cleavage of Pr1*a*+*b*. However, relatively large amounts of Pr3 are synthesized in the presence of inhibitors. Moreover, the amount of Pr1*a*+*b* accumulated in the presence of these inhibitors is not as large as would be expected if these polypeptides were the major precursors in the synthesis of p30. This, together with the effect of protease inhibitors on protein synthesis, which results in more inhibition of the smaller rather than larger proteins (as discussed below), makes it difficult to suggest, on the basis of the inhibitors effect, whether p30 precursors such as Pr3 are formed by the cleavage of Pr1*a*+*b*.

The effect of TPCK and ZPCK on the glycoprotein precursor Pr2*a*+*b* can not yet be assessed. The effect could be due to the inhibition of cleavage of a larger precursor polypeptide or to a nonspecific effect on the synthesis or modification of Pr2*a*+*b*. Our studies indicate that Pr1*a*+*b*, unlike Pr2*a*+*b*, are not precipitated with anti-gp69/71 serum (Naso *et al.* 1976).

Since Pr1*a*+*b* are the largest major precursor polypeptides that accumulate in the presence of TPCK and ZPCK, it is possible either that polypeptides Pr2*a*+*b* are made by way of Pr1*a*+*b* but that processing and glycosylation are essential in determining the gp69/71 cross reactivity or that Pr2*a*+*b* sequences are not present in Pr1*a*+*b*. The latter explanation appears to be more likely, since a major methionine-containing tryptic peptide of glycopolypeptides Pr2*a*+*b* and its nonglycosylated cleavage product p15E is not found in Pr1*a*+*b* (Arcement *et al.* 1976). Thus, the effect of TPCK on Pr2*a*+*b* is probably due to a nonspecific effect on the synthesis or modification (e.g., glycosylation) of these glycoproteins.

One problem with the use of such protease inhibitors is that they inhibit protein synthesis. In our studies, TPCK at 0.1 mM inhibited total protein synthesis by 80% to 90% and the synthesis of virus-specific polypeptides by 50% to 80% in a 15-minute pulse labeling after 5 minutes of pretreatment with the drug. TLCK (at 0.3 to 1.0 mM) caused a 10% to 25% inhibition of total protein synthesis. Pong *et al.* (1975) have indicated that one effect of TPCK probably occurs in the initiation of translation. In this case, the labeling of large polypeptides would be less sensitive to inhibition by the compound than the labeling of small polypeptides. This might result in an artifactual enrichment of large polypeptides. It would thus be even more difficult to determine whether the enrichment of Pr1*a*+*b* relative to Pr3 and Pr4 by treatment with TPCK and ZPCK is due to an inhibition of a cleavage step from Pr1*a*+*b* to Pr3 and Pr4. However, it is clear from our studies that while its specificity is unclear, the effect of the chloromethyl ketone derivatives used can not be totally attributed to their inhibition of protein synthesis. This was demonstrated by pulse labeling cells in the absence of protease inhibitors and then adding the inhibitors during a chase period (Jamjoom *et al.* 1977).

TPCK, D-ZPCK, and L-ZPCK at 0.1 mM and TLCK at 0.3 to 1.0 mM inhibited the cleavage of Pr3 and Pr4 and the formation of p30. On the other hand, cycloheximide at 200 μg/ml, which inhibits protein synthesis maximally, did not inhibit the formation of p30 initially, although it reduced the formation of p30 as the chase period was continued (Jamjoom *et al.* 1976). Similarly, pactamycin and puromycin did not inhibit the formation of p30. Finally, our results indicated that TPCK (0.1 mM), but not pactamycin (5×10^{-7} M), inhibits the processing of Pr1*a*+*b* to two intermediate precursors of the reverse transcriptase, termed Pr RT2 and Pr RT3.

These results indicate that the effects of the protease inhibitors are, at least in part, more likely to be a result of their effect on the cleavage of precursor polypeptides rather than inhibition of protein synthesis. Additional control experiments have been published that distinguish the effects of TPCK from those of pactamycin, an inhibitor of the initiation of translation (Jamjoom *et al.* 1977).

Pr1*a*+*b* as Early Precursors Containing Determinants of the Reverse Transcriptase

Figure 2, slots B, C, and D show the pattern of polypeptides that are immune precipitable with three different preparations of anti-RT sera, after pulse labeling cells for 15 minutes in the presence of TPCK to enrich for the high molecular weight polypeptides Pr1*a*+*b*. It is evident that Pr1*a*+*b* constitute the major polypeptides that are precipitable with the anti-RT sera. Anti-p30 serum also precipitated Pr1*a*+*b* (Figures 2, slot E), but in addition, it precipitated Pr3 and Pr4, which are the major components in this case. The glycoprotein precursors Pr2*a*+*b* were precipitated with antisera made against whole virions (Figure 2, slot A) but not with anti-RT or anti-p30 sera. Small amounts of Pr3 and Pr4 were brought down with two preparations of anti-RT sera (Figure 2, slot C,D). Results will be presented below which indicate that this is due to contamination of anti-RT sera with antibodies directed against the viral group-specific

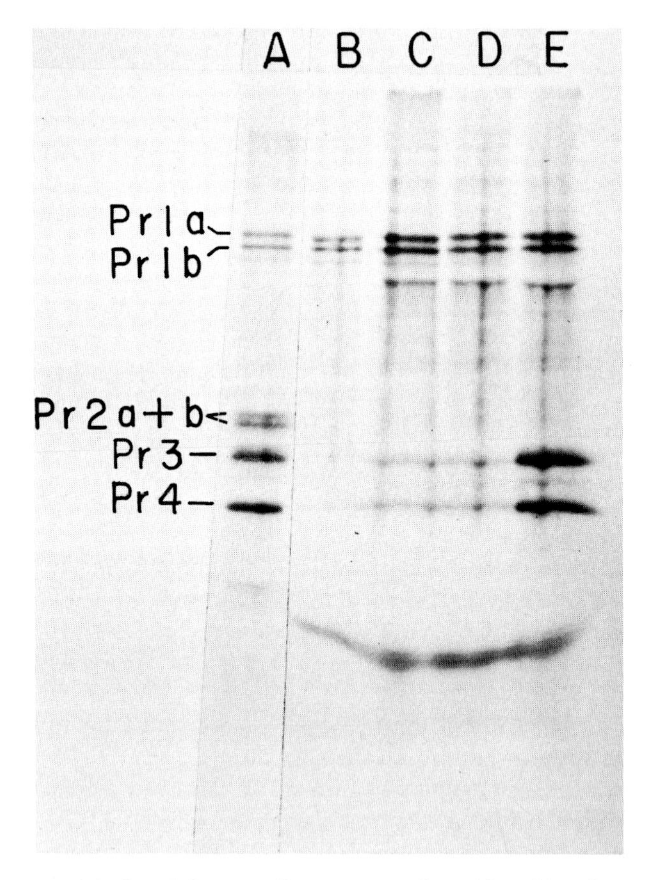

FIG. 2. Immune precipitation of virus-specific precursor polypeptides with anti-reverse transcriptase sera after pulse labeling in the presence of TPCK. Virus-infected cells were pulse labeled for 15 minutes with ^{35}S-methionine in Hank's solution containing 0.1 mM TPCK, after 5 minutes of pretreatment with this inhibitor. Precipitation was with the following sera: slot A, anti-RLV, slots B, C, and D, anti-reverse transcriptase (three independent antisera preparations); slot E, anti-p30. Precipitation in slot A was direct and in slots B-E, indirect.

antigens. It should be noted that while the use of inhibitors such as TPCK has facilitated the characterization of Pr1*a*+*b* as early precursors containing both p30 and RT determinants, the precipitation of Pr1*a*+*b* with anti-RT or anti-p30 sera can be clearly demonstrated without the use of these inhibitors (see below). Thus, any objection to the use of these inhibitors does not affect our results: that Pr1*a*+*b* contain RT and p30 determinants.

Pr1*a*+*b* Contain Determinants of the *gag* Gene Products

To determine whether the high molecular weight polypeptides Pr1*a*+*b* contain all of the viral group-specific antigens p30, p15, p12, and p10, we precipitated cytoplasmic extracts of cells pulsed labeled for 45 minutes with antisera made against each of these proteins. It is evident that each of these antisera can recognize and precipitate Pr1*a*+*b* (Figure 3, slots C-F). This indicates that Pr1*a*+*b* share antigenic determinants with all the viral group-specific antigens. The amount of Pr1*a* is reduced in the pattern of immune precipitate obtained with anti-p15. The reason for this abnormality is not understood. Antisera made against the viral reverse transcriptase precipitated the two polypeptides termed Pr RT1 and Pr RT2, as well as Pr1*a*+*b*. As in the previous experiments, the anti-RT preparation used here precipitated a small amount of Pr3 and Pr4 (Figure 3, slot B), most likely because the anti-RT serum was contaminated with antibodies directed against the viral group-specific antigens. These contaminating antibodies can be selectively absorbed by adding unlabeled viral proteins to the anti-RT serum before immune precipitation (Figure 3, slot A, and see below).

Characterization of Intermediate Precursors of the Reverse Transcriptase

Since it was clear from the studies described above that Pr1*a*+*b* contain determinants of the reverse transcriptase, it was of interest to follow the processing of these precursors into the mature viral enzyme. To do this, we pulse labeled cells for 15 minutes or pulse labeled them and then added unlabeled medium for a chase period of 2.5 hours. Figure 4 shows the pattern of polypeptides that are immune precipitable with anti-RT serum after a pulse (slot B) or a pulse chase (slot E). It is evident that Pr1*a*+*b* constitute the major polypeptides that are labeled in a pulse, while Pr RT1 and Pr RT2 are labeled to a smaller extent. After the chase, however, Pr1*a*+*b* disappeared, Pr RT1 ($\approx$ 145,000 daltons) was reduced, and the majority of label was shifted to Pr RT2 ($\approx$ 135,000 daltons) and to a newly formed band, named Pr RT3, of molecular weight of about 80,000 to 85,000 daltons. Since most preparations of marker RT isolated from RLV have a molecular weight of about 72,000 daltons, the relationship of the presumptive RT precursors Pr RT2 and Pr RT3 to the mature enzyme remains unclear.

The possibility exists that Pr RT3 may represent the mature form of the

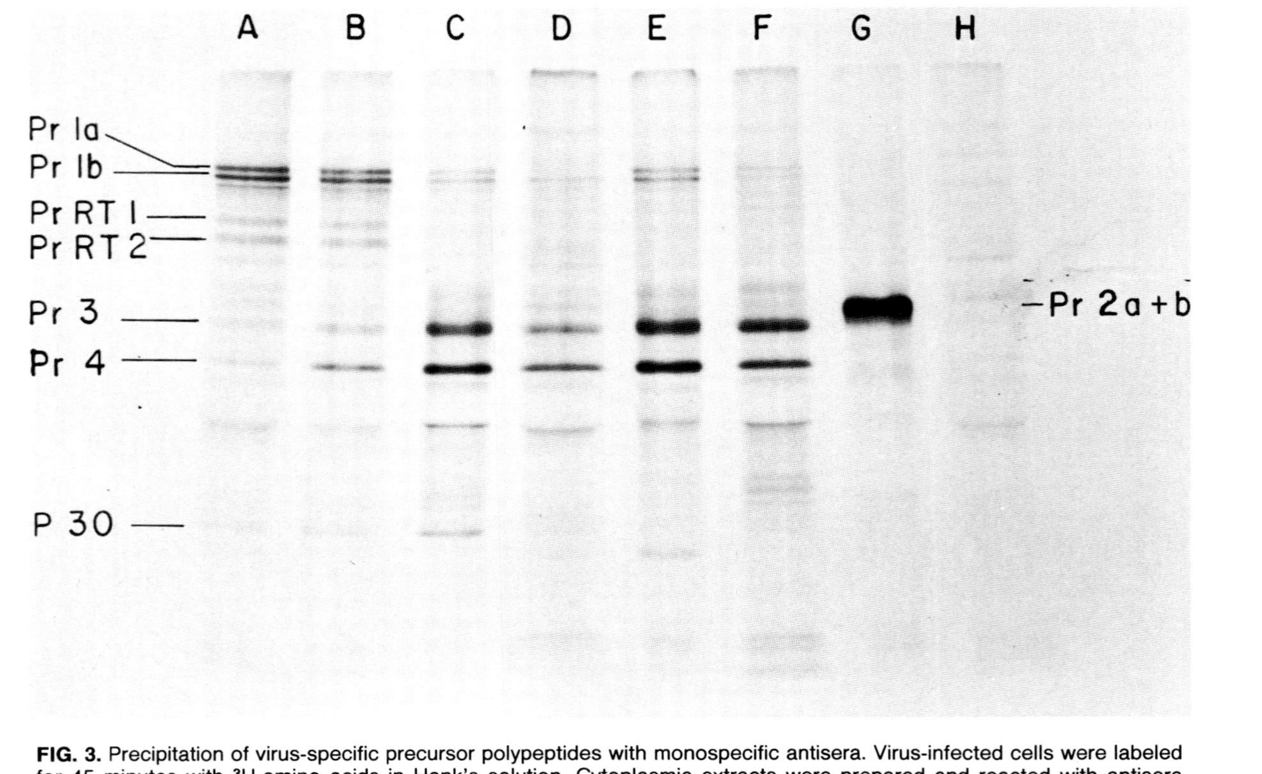

FIG. 3. Precipitation of virus-specific precursor polypeptides with monospecific antisera. Virus-infected cells were labeled for 45 minutes with ^{3}H-amino acids in Hank's solution. Cytoplasmic extracts were prepared and reacted with antisera monospecific for the following viral proteins: In slots A and B, reverse transcriptase; C, p30; D, p15; E, p12; F, p10; G, gp69/71; and H, rabbit anti-goat globulin only. Anti-RT serum in slot A was preabsorbed with 100 μg of RLV suspension in .01 M Tris. The immune precipitate was analyzed as in Figure 1.

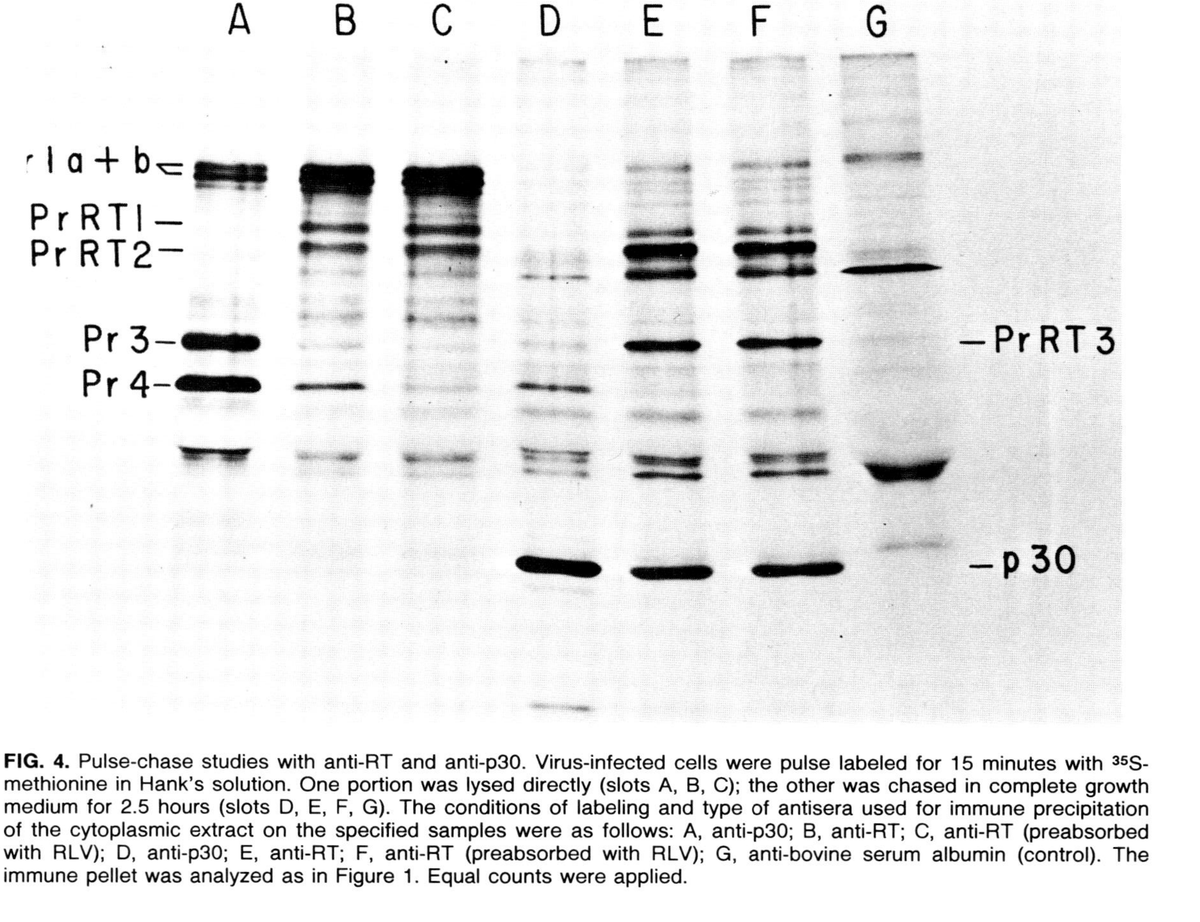

FIG. 4. Pulse-chase studies with anti-RT and anti-p30. Virus-infected cells were pulse labeled for 15 minutes with ^{35}S-methionine in Hank's solution. One portion was lysed directly (slots A, B, C); the other was chased in complete growth medium for 2.5 hours (slots D, E, F, G). The conditions of labeling and type of antisera used for immune precipitation of the cytoplasmic extract on the specified samples were as follows: A, anti-p30; B, anti-RT; C, anti-RT (preabsorbed with RLV); D, anti-p30; E, anti-RT; F, anti-RT (preabsorbed with RLV); G, anti-bovine serum albumin (control). The immune pellet was analyzed as in Figure 1. Equal counts were applied.

virion enzyme. If this is the case, Pr RT3 would be similar to the enzymes of Friend and Moloney leukemia viruses, which have a molecular weight of 80,000 to 84,000 daltons (Moelling 1976, Verma 1975). Purified RLV enzyme preparations of lower molecular weight could possibly then arise by a slower process of cleavage, which may occur in virus particles or by degradation during the purification procedure. Moelling (1976) indicated the occurrence of such degradation during routine enzyme purification. Studies are in progress to characterize the early form of the virion enzyme and to test, by fingerprinting, the relationship of Pr RT1, Pr RT2, and Pr RT3 to the isolated viral enzyme, either on the basis of its function or by precipitation with antisera.

Figure 4 also shows the polypeptides obtained after immune precipitation of the same cell extract preparation with anti-p30 after pulse (slot A) and chase (slot D). Pr1*a*+*b*, Pr3, and Pr4 are precipitated with anti-p30 serum in the pulse, while mainly p30 and a small amount of Pr4 are present after the chase. Pr RT1, Pr RT2, and Pr RT3 are not precipitated with anti-p30, which indicates that they lack determinants of this protein and strengthens the specificity of the precipitation of these polypeptides with anti-RT sera. Note the similar mobility of Pr3, the precursor to *gag* genes, and the anti-RT specific polypeptide Pr RT3. We emphasize that Pr RT3 is a stable polypeptide formed during chase incubations and is precipitated by anti-RT serum but not by anti-p30 serum. In contrast, Pr3 is a rapidly labeled, unstable polypeptide, which quickly disappears in chase incubations (Arcement *et al.* 1976) and is precipitated by anti-*gag* sera but not by anti-RT sera preabsorbed with viral proteins (see below).

Apparent Association between RT Precursors and p30

One of the problems encountered in our studies on the characterization of the precursors of the reverse transcriptase is the contamination of most of the available preparations of anti-RT with antibodies that precipitate *gag* gene polypeptides. Thus anti-RT sera precipitated small amounts of Pr3 and Pr4 in pulse-labeled extracts and quite significant amounts of p30 in chase extracts. Because we are heavily dependent on the specificity of the antisera to characterize viral precursor polypeptides, particularly Pr1*a*+*b*, which contain both *gag* and *pol* (the RNA-directed DNA polymerase) gene products, it was essential to determine which polypeptides are precipitated by RT-specific antibodies and which by contaminating antibodies to *gag* gene products. Fortunately, this was not difficult in the case of Pr1*a*+*b*, since these polypeptides are the major ones precipitated by anti-RT sera and since they were precipitated by relatively clean anti-RT preparations that did not bring down detectable amounts of *gag* gene products (see Figure 2, slot B).

However, we have found a simple method of selectively absorbing out the contaminating anti-*gag* antibodies in anti-RT sera by adding unlabeled virus to sera or to the cell extract prior to immune precipitation. While the addition of virus scarcely affects the precipitation of Pr1*a*+*b* or Pr RT1, 2, and 3 with anti-RT serum (see below), it effectively prevents the precipitation of Pr1*a*+*b* and Pr3 and Pr4 with anti-RLV (see Figure 1, slot D) or anti-p30 (not shown). This selective absorption results from the fact that the amount of RT present in viral preparations is much less than the amount of the structural *gag* gene products. This method of selective absorption is effective in removing the contaminating Pr3 and Pr4 present in precipitates obtained with anti-RT serum after pulse labeling (compare Figure 3, slots A and B, Figure 4, slots B and C). Surprisingly, however, this method, although effective in absorbing anti-p30 reactivity (Figure 5, slots A, B), was ineffective in reducing the amount of

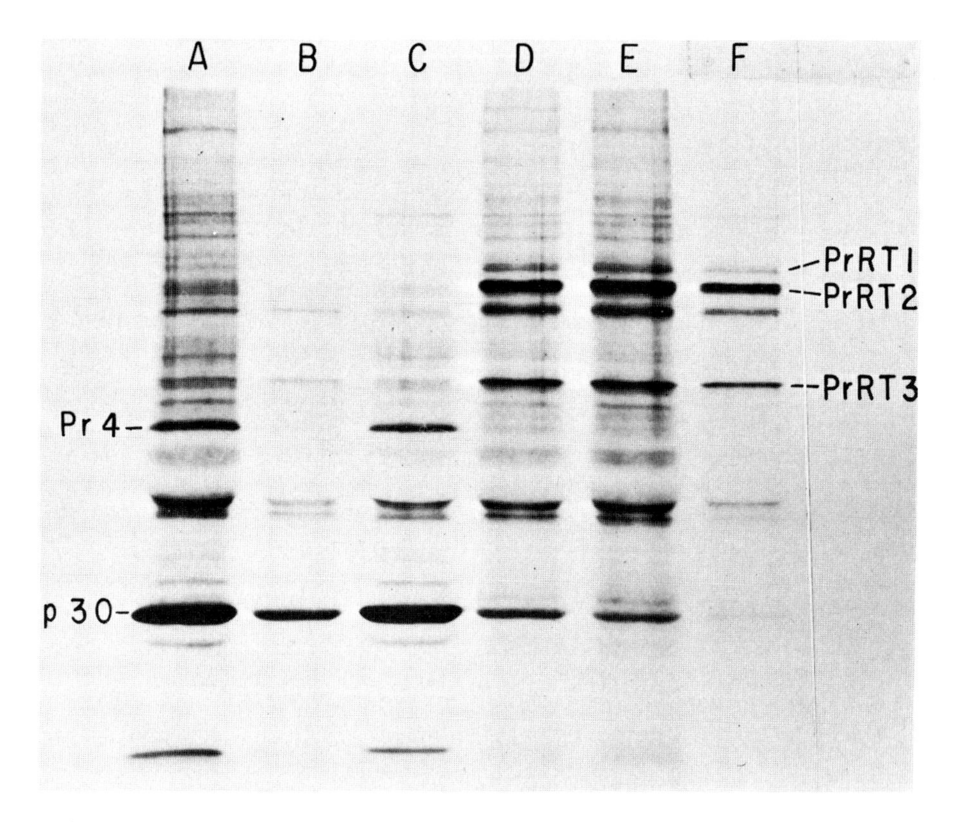

FIG. 5. Effect of competing RLV or sodium dodecyl sulfate on chase polypeptides precipitable with anti-p30 and anti-RT. Virus-infected cells were pulsed for 15 minutes with ^{35}S-methionine in Hank's solution and chased for 2.5 hours in complete growth medium. Cytoplasmic extracts were prepared and divided into equal portions. Samples A and D were treated directly with antisera; 100 μg of RLV were added to B and D, while C and F were made 0.3% in SDS, prior to addition of antisera. The antisera used were as follows: slot A, no treatment, anti-p30; slot B, RLV added, anti-p30; slot C, SDS added, anti-p30; slot D, no treatment, anti-RT; slot E, RLV added, anti-RT; slot F, SDS added, anti-RT. The anti-RT serum was previously absorbed with RLV. The total amount of the immune precipitate was applied in each case.

TABLE 1. *Effect of SDS on the amount of p30 associated with RT precursors*

Sample	Treatment with SDS	PrRT1*	PrRT2	PrRT3	p30
(D)	−	1	2.4	1.2	1.1
(E)	−	1	2.2	1.3	1.0
(F)	+	1	3.0	1.4	0.4

* The numbers represent areas of the indicated bands relative to PrRT1. The filmstrips containing Figure 5, slots D, E, F were scanned, and the areas measured as described under Methods.

p30 that is brought down with anti-RT serum in pulse-chase experiments (see Figure 4, slots E, F; Figure 5, slots D, E). This suggests that the precipitation of p30 by anti-RT antisera is not due to contaminating p30 antibodies in the sera. The main alternative possibility that would explain this phenomenon is the existence of an association between p30 and RT precursors. This would mean that p30 molecules are brought down by genuine anti-RT antibodies, by virtue of their association with RT precursors. One approach that we took to test this possibility was to try to break up this possible complex by adding SDS at a final concentration of 0.3% to the cytoplasmic extract before immune precipitation. Figure 5, slot C, shows that addition of this detergent does not affect the pattern of peptides that are precipitated with anti-p30. Moreover, it does not affect precipitation of Pr RT1, Pr RT2, or Pr RT3 with anti-RT serum; rather, it selectively reduces the amount of associated p30 by approximately 2.5-fold (Figure 5, slot F, Table 1). This result is in accordance with the presence of an association between RT precursors and p30 molecules.

However, the objection may be raised that if p30 is associated with RT precursors, then precipitation with anti-p30 should also bring down these RT precursors, which is not the case. This dilemma, in our opinion, can be explained by the finding that far fewer RT precursor molecules are present in the cell than p30 molecules (Panet *et al.* 1975). Thus, most of the RT precursor molecules may be associated with p30, but only a minor proportion of p30 molecules are associated with RT precursors. In immune precipitates with anti-p30, therefore, the small amount of associated RT molecules would be difficult to detect.

Finally, we have observed that the addition of SDS to cell extract before immune precipitation, while decreasing nonspecific precipitation, does not affect the precipitation of Pr1*a*+*b* with anti-RT or anti-p30. This indicates that the precipitation is probably not the result of association of Pr1*a*+*b* with other molecules that are recognized by these antisera.

Comparison of Peptide Maps of Pr1*a*+*b* and the RT-Specific Precursors

In order to determine the relationship of the intermediate RT precursors (Pr RT1, 2, and 3) to the Pr1*a*+*b*, tryptic digests of these proteins were compared

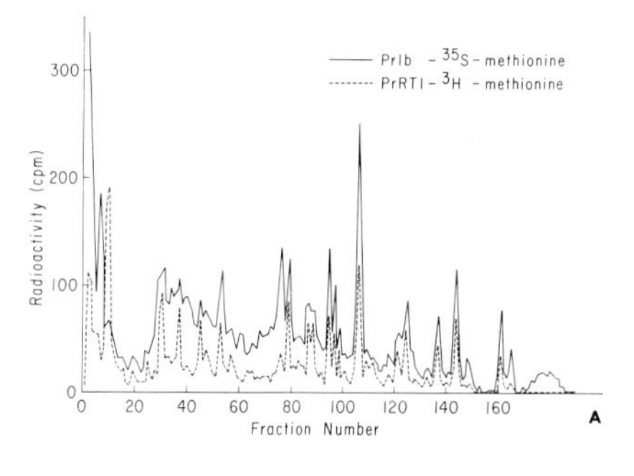

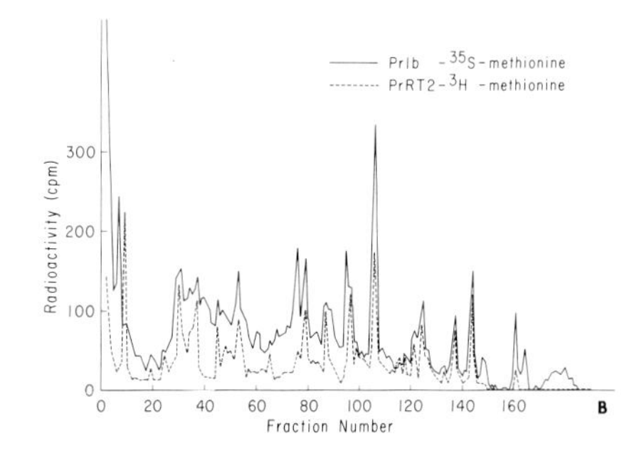

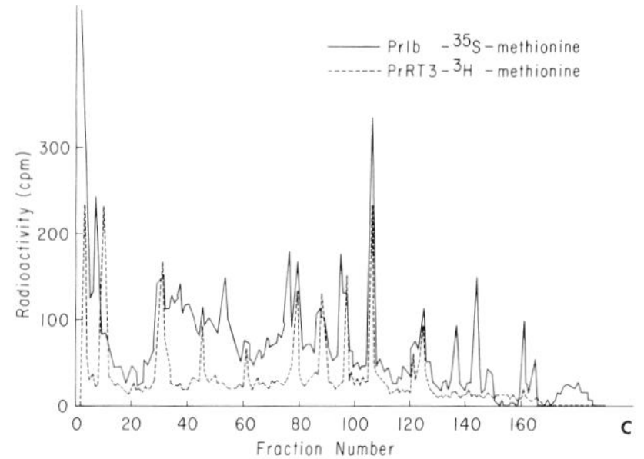

FIG. 6. Tryptic digest patterns of Pr1*b* and *pol-* specific precursors. Tryptic digests of the purified proteins were fractionated on a cation exchange column as previously described (Arcement *et al.* 1976). All profiles represent dual label experiments using ^{3}H-methionine and ^{35}S-methionine-labeled proteins.

(Figure 6). In these experiments, the proteins were labeled with either ^{35}S-methionine or ^{3}H-methionine. The labeled proteins were purified by SDS gel electrophoresis, and the tryptic digests were fractionated on ion exchange columns. The results showed that Pr1*b* ($\simeq$ 200,000 daltons) shares methionine-containing tryptic peptide fractions with Pr RT1 ($\simeq$ 145,000 daltons), Pr RT2 ($\simeq$ 135,000 daltons), and Pr RT3 ($\simeq$ 85,000 daltons). The complexity of the tryptic digest chromatograms clearly decreases as the proteins decrease in molecular weight, from Pr1*b* to Pr RT3. Thus, Pr RT1 and Pr RT2 are not dimeric forms of Pr RT3.

The major methionine-containing tryptic peptide fraction in fractions 105 to 107 from Pr1*b* contains a p30–specific tryptic peptide that comigrates with an RT-specific peptide. Consistent with this is the observation that the relative molar amounts of this fraction in Pr1*b* are twice that found in Pr RT1, 2, and 3.

Effect of the Arginine Analogue Canavanine on the Formation of *pol*-Specific Precursor Polypeptides

Figure 7 indicates that in the presence of the arginine analogue, only Pr1*a*, among the anti-RT precipitable polypeptides, is detected. The absence of formation of Pr RT1 and Pr RT2 in the presence of canavanine is a strong indication that these polypeptides are formed by cleavage of Pr1*a*, as similarly suggested by the effect of TPCK (see Figures 2 and 8A).

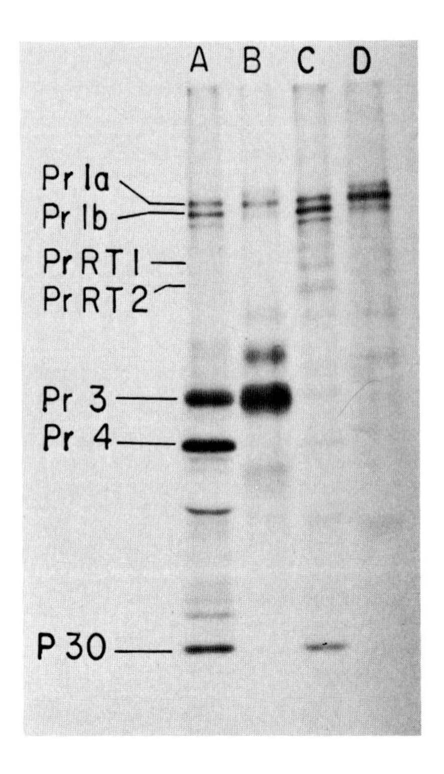

FIG. 7. Anti-RT precipitable polypeptides labeled in the presence of canavanine. Virus-infected cells were labeled for 45 minutes with ^{3}H-amino acids in Hank's solution in the absence (slots A, C) or presence (slots B, D) of 4 mM canavanine. Canavanine was added 10 minutes prior to labeling. Slots A, B, anti-p30; slots C, D, anti-RT. Before anti-RT was added, 300 μg of unlabeled RLV were added to each sample of cytoplasmic extract.

Relationship of Pr1*a*+*b* to Precursors of the Group-Specific Antigens and the Reverse Transcriptase

The results presented above indicate that Pr1*a*+*b* contain determinants of both *gag* and *pol* gene products. The obvious consideration, then, is that in the expression of the *gag-pol* gene segment, Pr1 is translated as the primary gene product, which is then cleaved into separate *gag* and *pol* intermediate precursors. Such a mode of translation would predict the formation of equimolar amounts of *gag* and *pol* gene products. To test this possibility, we have compared the amount of RT-specific precursors to the amount of p30 precursors after pulse labeling of infected cells (Table 2). Pr1*a*+*b*, which is precipitated with both anti-RT and anti-p30, serves as a standard to make this comparison possible. These measurements indicated that for the equivalent of one mole of Pr1*a*+*b*, about 12 moles of Pr3 and Pr4 exist but only 0.8 moles of Pr RT1 and Pr RT2. If we make the assumption that no selective degradation of either RT or p30 sequences takes place during the time of the pulse (see below), our results would be a measure of the relative rate of synthesis of these polypeptide sequences. Thus, it can be estimated that approximately 1/15th as many RT sequences are made in the cell as p30 sequences. Measurements by other investigators have previously indicated that, in the steady state in infected cells, far fewer RT-specific peptide sequences are present than are sequences of viral structural proteins (Panet *et al.* 1975, Chen and Hanafusa 1974). Thus, both the rate of formation and the steady-state level of RT sequences are considerably less than those of p30 sequences. This makes it unlikely that the sole pathway of formation of both *gag* and *pol* gene products is by cleavage of the common precursors Pr1*a+b*.

It should be noted that in the quantitative measurements of the type presented here, the assumption is made that the antisera precipitate precursor polypeptides in the same ratio as they exist in the cell. This was verified by experiments in which extracts from pulsed cells were mixed with extracts from pulse-chased cells in different ratios before immune precipitation (not shown). The ratio of pulse polypeptides, such as Pr1*a+b*, were compared to chase polypeptides, such as p30. The experimental results indicated that as the proportion of pulse polypeptides increased, their relative amount in the immune precipitate proportionately increased in relation to chase polypeptides.

Another approach that can be taken to determine the role of Pr1*a+b* in the formation of RT and p30 intermediate precursors is to compare the kinetics of formation of these precursors and the effect of different agents on their formation. Figure 8 shows the profile of anti-RT and anti-p30 precipitable polypeptides after pulse (C, D), pulse in the presence of TPCK (A, B), and chase (E, F). Comparison of RT-specific polypeptides in a pulse (C) and a chase (E) shows that there is a simple conversion from Pr1*a+b* to Pr RT1, 2, and 3. Thus, Pr1*a+b* are the major bands present in a pulse. Pr RT1 and Pr RT2 are present in smaller amounts. After a 2.5-hour chase, Pr1*a+b* have completely

TABLE 2. *Estimate of the relative amounts of RT and p30 precursor polypeptides*

Labeling conditions	Pr1a + b (moles)	PrRT1 + PrRT2 (moles)	Pr3 + Pr4 (moles)
15-minute pulse, ^{35}S-methionine	1	0.73	7.8
	1	0.87	—
	1	—	6.9
	1	—	7.8
45-minute pulse, ^{3}H amino acids	1	0.50	12
	1	0.71	14
	1	0.67	19
		0.48	—
		0.53	—
45-minute pulse, ^{35}S-methionine	1	—	15

Patterns of viral specific polypeptides immune precipitated with anti-p30 or anti-RLV sera were used to estimate the ratio of Pr3 + Pr4 to Pr1*a* + *b*. Immune precipitates with anti-RT were used to estimate the ratio of PrRT1 plus PrRT2 to Pr1*a* + *b*. All patterns were obtained from SDS-polyacrylamide gel electrophoresis, similar to and including some of those in Figures 1 through 8. Films were scanned and the peak areas were quantitated as described under Methods. For calculations of molar amounts, the relative area of a particular peak was divided by the estimated molecular weight of the polypeptide. The following average molecular weights were used: Pr1*a* + *b*, 380; PrRT1 + PrRT2, 280; Pr3 + Pr4, 155. Only relative areas from the same scan are compared.

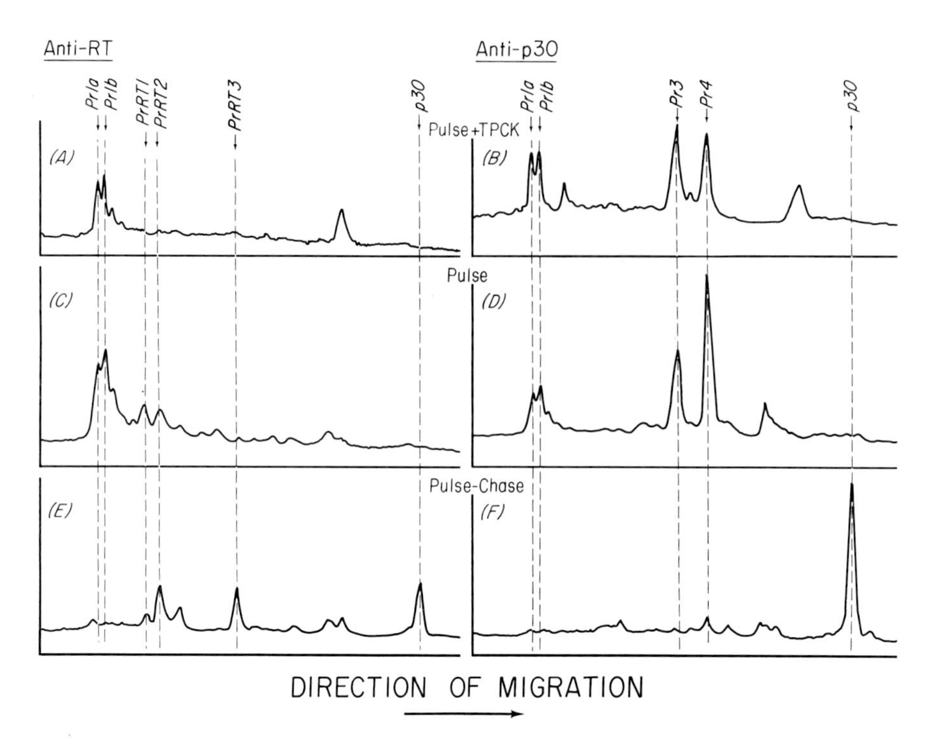

FIG. 8. Densitometer tracings of anti-RT and anti-p30 immune precipitates obtained from fluorographs of SDS gels. RLV-infected JLS-V16 cells were incubated with ^{35}S-methionine as described below. Cytoplasmic extracts were immune precipitated with either anti-RT or anti-p30 sera. Anti-RT profiles in panels A, C, and E were taken from the following figures: Panel A from Figure 2B, 15-minute pulse in the presence of 0.1 mM TPCK; panel C from Figure 4C, 15-minute pulse; panel E from Figure 4F, 15-minute pulse, 2.5 hour chase. Anti-p30 profiles in panels B, D and F were taken from the following figures: panel B from Figure 2E, 15-minute pulse in the presence of 0.1 mM TPCK; panel D from Figure 4A, 15-minute pulse; Panel F from Figure 4D, 15-minute pulse, 2.5 hour chase.

disappeared, while Pr RT2 increases and a new band, Pr RT3, appears. The efficiency of processing RT precursors and p30 precursors can be compared from these profiles because equal amounts of radioactivity were applied to the gels and because the ratio of radioactivity that is immune precipitable with anti-RT to those precipitable with anti-p30 remains relatively unchanged in a pulse or a pulse-chase (the ratio of anti-RT to anti-p30 counts in immune precipitates was decreased in the chase incubation by 20% to 30% in this experiment). The ratio of p30 in the chase (F) to Pr3 plus Pr4 (D) in the pulse is about 0.48. The ratio of Pr RT1, Pr RT2, and Pr RT3 in the chase (E) to Pr1*a*+*b* plus Pr RT1 and Pr RT2 in the pulse (C) is 0.59, or 0.80 if the amount of p30 in the anti-RT precipitate (E) is corrected for. Such results indicate that the processing of RT-specific polypeptides is probably as efficient as the processing of p30-specific polypeptides. This rules out the possibility, mentioned above, of selective degradation in the cell of RT-specific peptide sequences, at least during a 2.5-hour chase incubation.

The cleavage of Pr1*a+b* to RT-specific intermediate precursors is inhibited by TPCK (see Figure 8A) and also by canavanine (see Figure 7, slot D). This result, together with the results mentioned above on the amounts of Pr RT1, Pr RT2, and Pr RT3, indicates that the formation of these precursors occurs by cleavage of preformed Pr1*a+b*, rather than by cleavage of unfinished (nascent) Pr1*a+b*.

The formation of p30 precursors follows different kinetics. Thus, after a 15-minute pulse (Figure 8D) most of the radioactivity is present in Pr3 and Pr4, while Pr1*a+b* are present as minor polypeptides. A similar profile is obtained in pulses as short as 1 to 2 minutes. This is difficult to reconcile with a precursor-product relationship between Pr1*a+b* and the majority of Pr3 without hypothesizing an extremely fast cleavage or, if Pr3 is near the N-terminus, a nascent chain cleavage, before completion of Pr1*a+b*. Both of these hypotheses contradict the kinetics of formation of RT-specific polypeptides. In addition, neither TPCK (see Figure 8B) nor canavanine (see Figure 7, slot B) inhibit the formation of Pr3, although both of these compounds exert a noticeable effect on the processing of the Pr1*a+b* precursors.

DISCUSSION

The data presented in this study provide evidence of the existence in RLV-infected cells of high molecular weight polypeptides that contain determinants of both the viral group antigens and the reverse transcriptase. This indicates that in RLV the *gag* and *pol* genes are adjacent and can be translated as one polypeptide. Moreover, it is consistent with the large sizes (35S and 21S) of the intracellular viral mRNA species thought to function as messenger in avian and murine tumor virus-infected cells (Fan and Baltimore 1973, Schincariol and Joklik 1973, Shanmugam *et al.* 1974).

In the map of avian sarcoma virus genome, which has been constructed by oligonucleotide fingerprints (Wang *et al.* 1976, Coffin and Billetter 1976, Duesberg *et al.* 1976, Joho *et al.* 1975, 1976), the *gag* and *pol* genes have been tentatively assigned neighboring locations, in the order 5′..*gag*..*pol*..*env*..*src*..3′. Thus, the murine RNA tumor viruses seem to be similar to the avian sarcoma virus at least in the adjacent location of the *gag* and *pol* genes. A few observations suggest that the *gag* gene is located near the N-terminal end of the viral mRNA. Thus, Pawson *et al.* (1976), using viral RNA in cell-free translation, have demonstrated the incorporation of N-formyl methionine in the major translational product, which is precipitable with anti-p30, indicating that it is a *gag* gene precursor. We have also noted that after synchronization of translation by treatment of cells with high salt (Saborio *et al.* 1974), the group antigen protein precursors Pr3 and Pr4 are quickly detected, which is consistent with a location near the N-terminus (Naso, unpublished observations).

The detection of both *pol*-specific and *gag*-specific antigenic determinants in the high molecular weight polypeptides Pr1*a*+*b* raised the following issue concerning the translation of the viral genome. If the synthesis of the *gag* and *pol* gene products proceeds by synthesis and cleavage of this common precursor only, then *gag* and *pol* gene products must be synthesized in equimolar amounts. Virus particles contain much less polymerase than structural proteins (Panet *et al.* 1975, Stromberg *et al.* 1974). Moreover, measurement of the steady-state amount of *pol*-specific sequences in virus-infected cells has indicated that these sequences are present in much smaller amounts than the viral structural proteins (Panet *et al.* 1975, Chen and Hanafusa 1974). This indicates that if *pol*-specific polypeptide sequences are indeed made in equimolar amounts, compared to structural proteins, they must be quickly degraded. In the present study, we were able to shed more light on this point by comparing the amount of *pol*-precursors to the amount of p30 precursors after pulse labeling of the cells. These conditions should more closely approximate the rate of synthesis of these polypeptides. Pr1*a*+*b* was used as the basis of comparison since it is common to both types of precursors. Our measurements revealed that roughly 1/10 to 1/25 as many *pol*-specific precursors as p30 precursors are present in the cell after pulse labeling. Moreover, intermediate *pol*-precursors appear to be formed by the cleavage of preformed Pr1*a*+*b*, whereas this mechanism of formation is possible for p30 precursors only if very fast or nascent cleavage of Pr1*a*+*b* takes place, which is not supported by our results. Thus, it is unlikely that the pathways of synthesis of *gag* and *pol* gene products involve only the cleavage of a common precursor.

Two possibilities are theoretically compatible with these findings: The mRNA coding for Pr1*a*+*b* constitutes a minor population of the messenger involved in the synthesis of *gag* gene products or only one species of mRNA codes for both *gag* and *pol*, but a translational control mechanism allows the synthesis on the messenger of more *gag* than *pol* gene products. Consideration of other available information favors the second possibility. Thus, the major species of viral mRNA in murine and avian RNA tumor virus-infected cells has a size of 35S, but 21S and 16S species have also been detected (Fan and Baltimore 1973, Gielkens *et al.* 1976, Schincariol and Joklik 1973, Shanmugam *et al.* 1974). Moreover, recent findings indicate that the 35S RNA is the main species present in polysomes precipitable with anti-p30 monospecific serum (Mueller-Lantzsch *et al.* 1976). These observations make it unlikely that there is a small mRNA on which the majority of the *gag* gene products are synthesized and disfavors the possibility that large species of viral mRNA, such as the one able to code for Pr1*a*+*b*, constitute a minor population.

The possibility of a translational control that regulates the amount of *gag* and *pol* gene products made on a large messenger constituting a major fraction

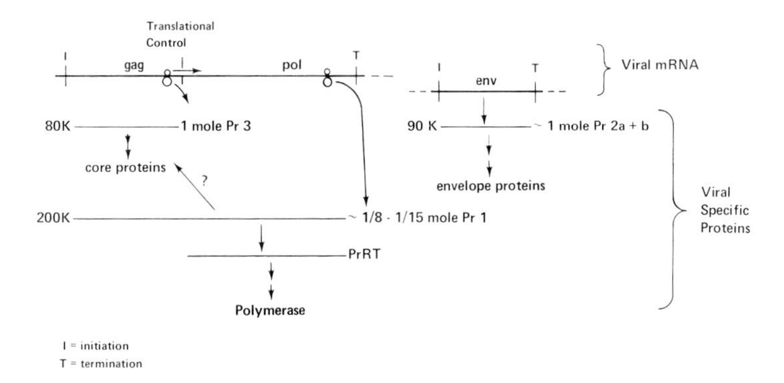

FIG. 9. A model for the initial stages of synthesis of Rauscher leukemia virus proteins.

of the viral mRNA is depicted in the model presented in Figure 9. In this model, translation is initiated at the beginning of the *gag* gene and proceeds until the end of this gene. At that point on the gene, most of the ribosomes are released. However, few of the ribosomes manage to bypass this release point and continue on to produce a complete *gag pol* common product (Pr1*a*).

Preliminary work from cell-free translation of viral genomic RNA has also suggested the possible existence of control in the translation of RNA tumor virus genomes. Thus, while the products of such translation include high molecular weight ($\simeq$ 180,000 daltons) polypeptides (Naso *et al.* 1975b), which have been shown to contain sequences of some of the major viral group-specific antigens (Kerr *et al.* 1976), most have a size of 60,000 to 80,000 daltons (Naso *et al.* 1975a, Von Der Helm and Duesberg 1975, Salden *et al.* 1976a, Kerr *et al.* 1976). The latter polypeptides have also been found to represent precursors of *gag* gene products (Pawson *et al.* 1976, Salden *et al.* 1976b), which are probably equivalent to Pr3 and Pr4. If the high molecular weight polypeptides turn out to contain RT or *pol* determinants, cell-free synthesis would be similar to synthesis within the cell. The fact that translation of genomic 35S results in the synthesis of the *gag* gene products is consistent with the finding that viral 35S mRNA is the major species coding for these proteins in the cell. Moreover, it suggests the presence of control in the translation of the viral genome (Salden *et al.* 1976b). However, further characterization of high molecular weight products of cell-free translation are required to support such control on the basis of the cell-free work.

One possible variation on the model in Figure 9 is that a factor may exist which blocks the translation of most viral mRNA molecules at the end of the *gag* gene. Only mRNA molecules that have escaped interaction with this factor can translate Pr1*a*+*b*. However, it is more difficult to envisage this possibility because the amount of the factor must be carefully regulated so as not to shut off the synthesis of the polymerase completely. If future studies on cell-free translation of intact 35S viral RNA indicate that *pol* sequences are indeed translated in vitro, and in lower molar amounts than *gag* gene products, then the possible translational control between *gag* and *pol* genes may be dependent on the structure of the mRNA rather than on an associated protein factor. The factor would presumably be lost during purification of the viral RNA.

Recent studies by Davidson and his colleagues (Bender and Davidson 1976) have shown that the 60S to 70S RNA of type C viruses contains novel structural features. One such feature is a loop, which in RD-114 virus is approximately 2.4 kilo-bases from the 5′ end of the 35S subunit RNA. The loop has a coding capacity expected for protein of about 120,000 daltons, whereas the 2.4 kilobase stretch at the 5′ end of the RNA could conceivably code for a precursor the size of Pr3 ($\approx$ 80,000 daltons). This type of loop structure could serve as a structural block, which would frequently aid in ribosomal termination but occasionally permit a read through to the end of the *pol* gene.

The idea of translational control in animal viruses, while still novel, has been previously suggested in other systems. Lucas-Lenard (1974) and Paucha *et al.* (1974) have proposed similar models of control based on specific termination to account for the molar excess in the production of capsid proteins in mengovirus (Lucas-Lenard 1974, Paucha *et al.* 1974) and polio virus (Paucha and Colter 1975). However, McLean *et al.* (1976) attributed a similar excess of rhinovirus 1A capsid proteins to selective degradation and ambiguous processing of the noncapsid proteins. Clarification of these differences in picornaviruses awaits further studies.

So far, we have not detected any polypeptide that carries both *gag* and *env* determinants in RLV-infected cells. This suggests that these two polypeptides are translated independently. However, it does not rule out the possibility that they are translated from a single initiation site but undergo nascent cleavage. If the *env* gene in RLV is located at the 3′ end of the *pol* gene, as is the case with Rous sarcoma virus, then our results would favor an independent initiation site for the translation of the *env* gene to allow for the synthesis of the envelope proteins at a higher molar yield than the polymerase precursors, so as to represent the situation in the cell. Earlier results on the differential expression of the group antigens and envelope proteins in high and low leukemia mouse strains (Strand *et al.* 1974) have also favored the independent translation of the envelope and group antigens. If translation of the envelope proteins does proceed from a separate initiation site than the one used for *gag* and *pol* gene products, then it remains to find out whether that initiation site is functional on the 35S RNA or is only activated when present on a smaller species of subset viral mRNA. Since no internal initiation sites have yet been found to function in animal viruses with polycistronic mRNAs, the second possibility seems more likely.

ACKNOWLEDGMENTS

This work was supported in part by Public Health Service contract CP-61017 from the National Cancer Institute, and a grant from The Robert A. Welch

Foundation (G-429). G. A. J. is supported by the Riyadh University, Saudi Arabia. We thank Drs. T. Papas and G. VandeWoude for generous gifts of anti-RT sera and Dr. E. Shaw for his gift of D-ZPCK. We also thank Drs. L. J. Arcement and E. Murphy, Jr., for reviewing the manuscript and Mr. J. Syrewicz and Mrs. M. E. Leroux for technical assistance.

REFERENCES

Aoyagi, T., and H. Umezawa. 1975. Structures and activities of protease inhibitors of microbial origin, *in* Protease and Biological Control, E. Reich, D. B. Rifkin, and E. Shaw, eds., Vol. 2. Cold Spring Harbor Laboratory, Cold Spring Harbor, New York, pp. 429–454.

Arcement, L. J., W. L. Karshin, R. B. Naso, G. A. Jamjoom, and R. B. Arlinghaus. 1976. Biosynthesis of Rauscher leukemia viral proteins: Presence of p30 and envelope p15 sequences in precursor polyproteins. Virology 60:763–774.

Arlinghaus, R. B., R. B. Naso, G. A. Jamjoom, L. J. Arcement, and W. L. Karshin. 1976. Biosynthesis and processing of Rauscher leukemia viral precursor polyproteins, *in* Animal Virology: ICN-UCLA Symposia on Molecular and Cellular Biology, D. Baltimore, A. S. Huang, and C. F. Fox, eds., Vol. 4. Academic Press, New York, pp. 689–716.

August, J. T., D. P. Bolognesi, E. Fleissner, R. V. Gilden, and R. C. Nowinski. 1974. A proposed nomenclature for the virion proteins of oncogenic RNA viruses. Virology 60:595–601.

Baltimore, D. 1974. Tumor viruses: 1974. Cold Spring Harbor Symp. Quant. Biol. 39:1187–1200.

Beemon, K. L., A. J. Faras, A. T. Haase, P. H. Duesberg, and J. E. Maisel. 1976. Genomic complexities of murine leukemia and sarcoma, reticuloendotheliosis, and visna viruses. J. Virol. 17:525–537.

Bender, W., and N. Davidson. 1976. Mapping of poly (A) sequences in the electron microscope reveals unusual structures of type C oncornavirus RNA molecules. Cell 7:595–607.

Billeter, M. A., J. T. Parsons, and J. M. Coffin. 1974. The nucleotide sequence complexity of avian tumor virus RNA. Proc. Natl. Acad. Sci. USA 71:3560–3564.

Bonner, W. M., and R. A. Laskey. 1974. A film detection method for tritium-labelled proteins and nucleic acids in polyacrylamide gels. Eur. J. Biochem. 46:83–88.

Chen, J. H., and H. Hanafusa. 1974. Detection of a protein of avian leukoviruses in uninfected chicken cells by radioimmunoassay. J. Virol. 13:340–346.

Coffin, J. M., and M. A. Billeter. 1976. A physical map of the Rous sarcoma virus genome. J. Mol. Biol. 100:293–318.

Duesberg, P. H., L. H. Wang, K. Beemon, P. Mellon, P. K. Vogt, S. Kawai, and H. Hanafusa. 1976. Towards a complete genetic map of Rous sarcoma virus, *in* Animal Virology: ICN-UCLA Symposia on Molecular and Cellular Biology, D. Baltimore, A. S. Huang, and C. F. Fox, eds., Vol. 4. Academic Press, New York, pp. 107–125.

Fan, H., and D. Baltimore, 1973. RNA metabolism of murine leukemia virus: Detection of virus-specific RNA sequences in infected and uninfected cells and identification of virus-specific messenger RNA. J. Mol. Biol. 80:93–117.

Gielkens, A. L. J., D. Van Zaane, H. P. J. Bloemers, and H. Bloemendal. 1976. Synthesis of Rauscher murine leukemia virus-specific polypeptides in vitro. Proc. Natl. Acad. Sci. USA 73:356–360.

Ikeda, H., S. Hardy, Jr., E. Tress, and E. Fleissner. 1975. Chromatographic separation and antigenic analysis of proteins of the oncornaviruses. V. Identification of a new murine viral protein, p 15(E). J. Virol. 16:53–61.

Jacobson, M. F., J. Asso, and D. Baltimore. 1970. Further evidence on the formation of poliovirus proteins. J. Mol. Biol. 49:657–669.

Jamjoom, G., W. L. Karshin, R. B. Naso, L. J. Arcement, and R. B. Arlinghaus. 1975. Proteins of Rauscher murine leukemia virus: Resolutions of a 70,000-dalton, non-glycosylated polypeptide containing p30 peptide sequences. Virology 68:135–145.

Jamjoom, G. A., R. B. Naso, and R. B. Arlinghaus. 1976. A selective decrease in the rate of cleavage of an intracellular precursor to Rauscher leukemia virus p30 by treatment of infected cells with actinomycin D. J. Virol. 19:1054–1072.

Jamjoom, G. A., R. B. Naso, and R. B. Arlinghaus. 1977. Further characterization of intracellular precursor polyproteins of Rauscher leukemia virus. Virology 78:11–34.
Joho, R. H., M. A. Billeter, and D. Wessmann. 1975. Mapping of biological functions of RNA of avian tumor viruses: Location of regions required for transformation and determination of host range. Proc. Natl. Acad. Sci. USA 72:4772–4776.
Joho, R. H., E. Stoll, R. R. Friis, M. A. Billeter, and C. Weissmann. 1976. A partial genetic map of Rous sarcoma virus RNA: Location of polymerase, envelope and transformation markers, *in* Animal Virology: ICN-UCLA Symposia on Molecular and Cellular Biology, D. Baltimore, A. S. Huang, and C. F. Fox, eds., Vol. 4. Academic Press, New York, pp. 127–245.
Karshin, W. L., L. J. Arcement, R. B. Naso, and R. B. Arlinghaus. 1977. Common precursor for Rauscher leukemia virus gp 69/71, p15(E), and p12(E). J. Virol. 23:787–798.
Kerr, I. M., U. Olsherski, H. F. Lodish, and D. Baltimore. 1976. Translation of murine leukemia virus RNA in cell-free systems from animal cells. J. Virol. 18:627–635.
Korant, B. D. 1975. Regulation of animal virus replication by protein cleavage, *in* Cold Spring Harbor Conferences on Cell Proliferation, Vol. 2. Proteases and Biology Control, E. Reich, D. B. Rifkin, and E. Shaw, eds., Cold Spring Harbor Laboratory, Cold Spring Harbor, New York, pp. 621–644.
Laemmli, U. K. 1970. Cleavage of structural proteins during the assembly of the head of bacteriophage T4. Nature 227:680–685.
Laskey, R. A., and A. D. Mills. 1975. Quantitative film detection of ^{3}H and ^{14}C in polyacrylamide gels by fluorography. Eur. J. Biochem. 56:335, 341.
Lucas-Lenard, J. 1974. Cleavage of mengovirus polyproteins in vivo. J. Virol. 14:261–269.
Mc Lean, C., T. J. Matthews, and R. R. Rueckert. 1976. Evidence of ambiguous processing and selective degradation in the noncapsid proteins of rhinovirus 1A. J. Virol. 19: 903–914.
Mueller-Lantzsch, N., L. Hatlen, and F. Hung. 1976. Immunoprecipitation of murine leukemia virus-specific polyribsomes: Identification of virus-specific messenger RNA, *in* Animal Virology: ICN-UCLA Symposia on Molecular and Cellular Biology, Vol. 4, D. Baltimore, A. S. Huang, and C. F. Fox, eds., Academic Press, New York, pp. 37–53.
Moelling, K. 1976. Further characterization of the Friend murine leukemia virus reverse transcriptase-RNase H complex. J. Virol. 18:418–425.
Naso, R. B., L. J. Arcement, and R. B. Arlinghaus. 1975a. Biosynthesis of Rauscher leukemia viral proteins. Cell 4:31–36.
Naso, R. B., L. J. Arcement, T. G. Wood, T. E. Saunders, and R. B. Arlinghaus. 1975b. The cell-free translation of Rauscher leukemia virus RNA into high molecular weight polypeptides. Biochim. Biophys. Acta 383:195–206.
Naso, R. B., L. J. Arcement, W. L. Karshin, G. A. Jamjoom, and R. B. Arlinghaus. 1976. A fucose-deficient glycoprotein precursor to Rauscher leukemia virus gp69/71. Proc. Natl. Acad. Sci. USA 73:2325–2330.
Panet, A., D. Baltimore, and T. Hanafusa. 1975. Quantitation of avian RNA tumor virus reverse transcriptase by radioimmunoassay. J. Virol. 16:146–152.
Paucha, E., and J. S. Colter. 1975. Evidence for control of translation of the viral genome during replication of mengovirus and poliovirus. Virology 67:300–305.
Paucha, E., J. Seehaefer, and J. S. Colter. 1974. Synthesis of viral-specific polypeptides in mengovirus-infected L cells: Evidence for asymmetric translation of the viral genome. Virology 61:315–326.
Pawson, T., G. S. Martin, and A. E. Smith. 1976. Cell-free translation of virion RNA from nondefective and transformation-defective Rous sarcoma viruses. J. Virol. 19:950–967.
Pong, S-S., D. L. Nuss, and G. Koch. 1975. Inhibition of initiation of protein synthesis in mammalian tissue culture cells by L-1-Tosylamino-2-phenylethyl chloromethyl ketone. J. Biol. Chem. 250:240–245.
Saborio, J. L., S. Pong, and G. Koch. 1974. Selective and reversible inhibition of initiation of protein synthesis in mammalian cells. J. Mol. Biol. 85:195–211.
Salden, M., F. Asselbergs, and H. Bloemendal. 1976a. Translation of oncogenic virus RNA in *Xenopus laevis* oocytes. Nature 259:696–699.
Salden, M., H. L. Selten-Versteegen, and H. Bloemendal. 1976b. Translation of Raucher murine leukemia viral RNA. A model for the function of virus-specific messenger. Biochem. Biophys. Res. Commun. 72:610–618.
Schincariol, A., and W. Joklik. 1973. Early synthesis of virus-specific RNA and DNA in cells rapidly transformed with Rous sarcoma virus. Virology 56:532–548.

Shanmugam, G., S. Bhaduri, and M. Green. 1974. The virus-specific RNA species in free and membrane bound polyribosomes of transformed cells replicating murine sarcoma-leukemia virus. Biochem. Biophys. Res. Commun. 56:697–702.

Shaw, E. 1976. Site-specific reagents for chymotrypsin and trypsin, *in* Methods in Enzymology, S. P. Colowick and N. O. Kaplan, eds., Vol. 11. Academic Press, New York, pp. 677–686.

Shaw, E., and J. Ruscica. 1971. The reactivity of His-57 in chymotrypsin to alkylation. Arch. Biochem. Biophys. 145:485–489.

Strand, M., F. L. Lilly, and J. T. August. 1974. Host control of endogenous murine leukemia virus gene expression: Concentrations of viral proteins in high and low leukemia virus strains. Proc. Natl. Acad. Sci. USA 71:3682–3686.

Stromberg, K., N. E. Hurley, N. L. Davis, R. R. Rueckert, and E. Fleissner. 1974. Structural studies of avian myeloblastosis virus: Comparison of polypeptides in virion and core component by dodecyl sulfate-polyacrylamide gel electrophoresis. J. Virol. 13:513–528.

Summers, D. F., E. N. Shaw, M. L. Steward, and J. V. Maizel, Jr. 1972. Inhibition of cleavage of large polio-virus-specific precursor proteins in infected HeLa cells by inhibitors of proteolytic enzymes. J. Virol. 10:880–884.

Syrewicz, J. J., R. B. Naso, C. S. Wang, and R. B. Arlinghaus. 1972. Purification of large amounts of murine ribonucleic acid tumor viruses in roller bottle cultures. Appl. Microbiol. 24:488–498.

Van Zaane, D., A. L. J. Gielkens, M. J. A. Dekker-Michielsen, and H. P. J. Bloemers. 1975. Virus-specific precursor polypeptides in cells infected with Rauscher leukemia virus. Virology 67:544–552.

Verma, J. 1975. Studies on reverse transcriptase of RNA tumor viruses: III. Properties of purified Moloney murine leukemia virus DNA polymerase and associated RNase H. J. Virol. 15:843–854.

Vogt, V. M., and R. Eisenman. 1973. Identification of a large polypeptide precursor of avian oncornavirus proteins. Proc. Natl. Acad. Sci. USA 70:1934–1938.

Vogt, V. M., R. Eisenman, and H. Diggelmann. 1975. Generation of avian myeloblastosis virus structural proteins by proteolytic cleavage of a precursor polypeptide. J. Mol. Biol. 96:471–493.

Von der Helm, L., and P. H. Duesberg. 1975. Translation of Rous sarcoma virus RNA in cell-free systems from ascites Krebs-2 cells. Proc. Natl. Acad. Sci. USA 72:614–618.

Wang, L. H., P. H. Duesberg, S. Kawai, and H. Hanafusa. 1976. Location of envelope-specific and sarcoma-specific oligonucleotides on RNA of Schmidt-Ruppin Rous sarcoma virus. Proc. Natl. Acad. Sci. USA 73:447–451.

Cell Differentiation and Neoplasia, edited by Grady F. Saunders. Raven Press, New York

The Dual 5S RNA Gene System in *Xenopus*

Donald D. Brown and Nina V. Fedoroff

Department of Embryology, Carnegie Institution of Washington, Baltimore, Maryland 21210

Those of us who purify genes of known function believe that we can use them to understand gene control without relying on traditional genetics. The idea is to have available large amounts of two genes of known and related function but under different controls. Can we reconstruct in vitro their faithful transcription and their biological control? The dual 5S RNA gene system in *Xenopus* has advantages for such an analysis.

BIOLOGY OF THE DUAL 5S RNA GENES

The low molecular weight RNA of ribosomes termed 5S RNA is coded for by 5S DNA. In their genomes amphibians maintain two kinds of 5S RNA genes that are controlled separately (Brown and Sugimoto 1973). Both kinds of genes are present in multiple copies—a feature characteristic of genes that code for abundant stable RNA species such as ribosomal RNAs and tRNAs. The reason that there are two kinds of 5S RNA genes is related to the process of gene amplification in amphibian oocytes. These oocytes synthesize as many ribosomes in a month as a somatic cell could in many years. This capability is supported by massive amplifications of the genes (rDNA) for 18S and 28S rRNA in the oocyte nucleus (germinal vesicle) (Brown and Dawid 1968, Gall 1968). In most eukaryotes, including *Xenopus,* the 5S RNA genes are not physically linked with the 18S and 28S rRNA genes. Rather than amplify the 5S RNA genes in oocytes to keep up with the massive synthesis of rRNA, evolution has chosen to introduce a second set of 5S RNA genes into the *Xenopus* genome. Somatic cells express one kind of 5S RNA gene (somatic-type 5S RNA) (Ford and Brown 1976). Most of the 5S RNA synthesized by oocytes differs by six bases from the somatic 5S RNA in *Xenopus laevis.* We refer to this RNA as oocyte-type 5S RNA, but it is not the only type synthesized by oocytes. Somatic-type 5S RNA is also synthesized by oocytes. The control mechanism seems to be that all kinds of 5S RNA genes are functioning in oocytes, but only the somatic-type 5S RNA genes function in somatic cells. If we are to construct faithful transcription of 5S DNA in vitro we will have to explain how the oocyte-type 5S RNA genes are kept silent in somatic cells.

ADVANTAGES OF THE DUAL 5S RNA GENE SYSTEM

The "transcription" unit of 5S DNA is simple, and it is known. Evidence from several sources suggest that 5S RNA is itself the true transcription unit of 5S DNA, i.e., there is no obligatory precursor form of 5S RNA in *Xenopus* (Brown and Sugimoto 1973). This means the 5′ nucleotide of mature 5S RNA is the first one inserted by the polymerase, and the 3′ nucleotide is the last one added before termination of transcription. Proper reconstruction means simply the synthesis of 5S RNA. The oocyte-type and somatic-type 5S RNA isolated from two species of *Xenopus, X. laevis* and *X. borealis** have been sequenced (Ford and Brown 1976). Thus, the transcription units for the two kinds of 5S DNA in *Xenopus* are known exactly.

RNA polymerase form III transcribes 5S DNA. Roeder and his colleagues have shown that RNA polymerase form III transcribes 5S and 4[illegible]NA in living eukaryotic cells (Roeder *et al.* 1978, see pages 305 to 32[illegible] me). Recently, Parker and Roeder (1977) caused chromatin from [illegible] cytes to synthesize 5S RNA by the addition of purified enzyme III. O[illegible] karyotic RNA polymerases and *Escherichia coli* polymerase were unfaithful in their transcription. We know which polymerase form to use for reconstruction of faithful, in vitro transcription.

The 5S DNA has been purified from the *Xenopus* genome. We have isolated two kinds of 5S DNA from the genomic DNA of *X. laevis* (Brown *et al.* 1971) and *X. borealis* (Brown and Sugimoto 1973). The major component in each species is the oocyte-type 5S DNA (Brownlee *et al.* 1974, R. D. Brown and D. D. Brown, unpublished results). The minor component in *X. borealis* DNA contains somatic-type 5S RNA genes (J. Doering and D. D. Brown, unpublished results). The minor *X. laevis* 5S DNA is neither of the two; it may code for a trace 5S DNA synthesized by oocytes (Brown *et al.* 1975). In this paper, we will summarize what we know about the structure of these 5S DNAs, with emphasis on *X. laevis* oocyte-type 5S DNA. Finally, we will summarize our strategy for reassembling their control in vitro.

THE STRUCTURE OF *XENOPUS* 5S DNAs

The two kinds of 5S DNA in the genome of these two species behave as satellites in actinomycin D-CsCl gradients. By a series of density gradients, each of the four kinds of 5S DNA has been purified, starting with DNA from nucleated erythrocytes of *Xenopus.* Table 1 summarizes what we know about their structure. Each type of 5S DNA is present in multiple copies, the principal oocyte-types being the most abundant. The two oocyte-type 5S DNAs are located at the ends of most of the chromosomes (Pardue *et al.* 1973). This means that they are in clusters of several hundred to several thousand repeating units at

* *Xenopus borealis* is the correct name for the species referred to in our previous publications as *X. mulleri.*

TABLE 1. *Characteristics of four kinds of* Xenopus *5S DNA*

	Oocyte-type		Minor 5S DNA	
	X. laevis	*X. borealis*	*X. laevis**	*X. borealis*†
Density (g/cm³)	1.692	1.695	1.701	1.712
Average no. of repeats/1C DNA	24,000	9,000	2,000	700
Repeat length (base pairs)	700	1,400	350	900
Length heterogeneity	Yes	Great	No	No
Strand separation in alkaline CsCl	Yes	Yes	No	Yes
5S RNA coding strand	Light	Light	—	Light

* Contains neither the principal oocyte-type nor the somatic-type 5S RNA genes.
† Identified as the somatic-type 5S DNA (J. Doering and D. D. Brown, unpublished results).

these locations. Each kind of 5S DNA comprises a gene family in which the gene alternates with a spacer region. The 5S RNA genes (the transcription units) are highly conserved in evolution, but the spacers evolve rapidly and are species and family specific. This fact has centered attention on the problem of "tandem gene" evolution or "horizontal" evolution (Brown and Sugimoto 1973).

The Structure of *X. laevis* Oocyte-Type 5S DNA

Xenopus laevis oocyte-type 5S DNA was the first one isolated from *Xenopus* (Brown *et al.* 1971), and we know the most about its structure. Figure 1 is a circular diagram of an average repeat of *X. laevis* oocyte-type 5S DNA (Xlo 5S DNA). Although the DNA does not exist in circles, tandem repeats are conveniently represented this way. Some of the structure of the spacer and the location of the transcription unit (the gene) and the pseudogene are shown. The repeat has been sequenced recently using the HaeIII restriction enzyme fragments (Fedoroff and Brown, in press). This has been accomplished using

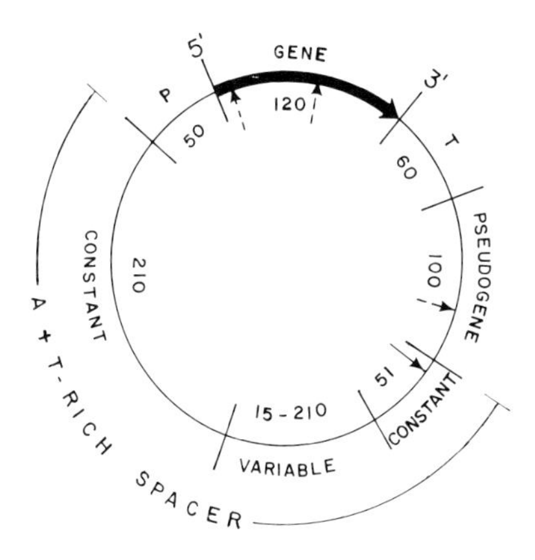

FIG. 1. A diagram of one repeating unit of *X. laevis* oocyte-type 5S DNA. The DNA is not circular but is most conveniently represented as such. The various regions are discussed in the text.

5S DNA isolated from the frog by the traditional CsCl methods and by sequencing cloned fragments of 5S DNA (Sanger and Coulson 1975, Maxam and Gilbert 1977).

Sequencing shows a great deal about the structure of the spacer. It consists of several discrete regions, which are diagrammed in the figure. The sequence adjacent to the 3′ end of the gene (called T) is rich in adenine-thymine (A + T) residues. At seventy-three nucleotides from the end of the transcription unit begins a region that resembles the gene. It was found by George Brownlee and his associates, who named it the "pseudogene." It differs from the gene in several nucleotides throughout the first 101 residues. The principal part of the A + T-rich spacer begins at this point in the repeat and proceeds up to 49 nucleotides from the 5′ end of the transcription unit. The A + T-rich spacer is the longest part of the repeat, ranging from 300 to 500 nucleotides in length. It endows 5S DNA with those physical characteristics that enabled us to purify the 5S DNA from the total genomic DNA.

The 24,000 repeats of Xlo 5S DNA in a frog differ slightly from each other in length, and this variation lies mainly in the A + T-spacer region. Earlier work showed that this region is not a unique sequence but is composed of simple sequences in more than one copy per repeat (Brownlee *et al.* 1974). In particular, a 15-nucleotide repeat was sequenced. Furthermore, restriction enzyme analysis has demonstrated that individual repeats can vary in length by units of 15 nucleotides (Carroll and Brown 1976). Sequencing studies of individual, cloned monomers of 5S DNA have localized this variable region within the A + T-rich spacer. The entire A + T-rich spacer consists of related, but not identical, simple sequences, which undoubtedly were derived from some ancestral duplication event. Yet, the A + T-rich spacer has an important structural arrangement. It has one variable region surrounded by two regions that are mainly constant. The variable region consists of adjacent polymers of 15 nucleotides which are the same or contain one or more base substitutions. The A + T-rich regions on either side of the variable region are usually constant in length, with some sequence heterogeneity. Some repeats vary in length in the A + T-rich region between the variable (V) region and the transcription unit, but many repeats have the same length. The variable and constant (C) regions clearly represent two different parts of the spacer. The fact that repeat lengths can differ by 15 nucleotide amounts means that the V region has something to do with the repeat length heterogeneity. If unequal crossing over causes tandem evolution, then the V region either enhances this process or is generated by it.

The A + T-rich region changes abruptly 50 nucleotides from the 5′ end of the transcription unit. This part of the spacer, which we optimistically have termed "P" for promotor region, is rich in guanine-cytosine (G + C). Is control region? We must devise an experimental design to determine this.

We know a great deal about the other three kinds of 5S DNA. Some of this information is summarized in Table 1. In *X. borealis* oocyte-type 5S DNA

(Xbo 5S DNA), there is clearly a simple sequence in the spacer detected by the restriction enzyme "Hha" (J. Doering and D. D. Brown, unpublished). It is 25 nucleotides long and occurs in clusters. We do not know its sequence. Preliminary information suggests that the two smaller 5S DNA components have simple sequence structures in their spacers also. Sequencing of these three DNAs is underway.

HOW CAN WE TEST FOR GENE FUNCTION?

If we cannot use traditional genetics, we will need a new means of determining gene function. Numerous attempts to transcribe *Xenopus* 5S DNA in vitro with bacterial or eukaryotic polymerases have been unsuccessful. Recently, a promising new method has given us new optimism. When 5S DNA is injected into *Xenopus* oocytes, it is transcribed faithfully (Brown and Gurdon 1977). Earlier experiments using unfertilized eggs showed low-level transcription of injected DNA (Gurdon and Brown 1977). The oocyte is a better candidate for RNA synthesis. Its enormous nucleus, the germinal vesicle (GV), actively synthesizes RNA, whereas unfertilized eggs normally have low, perhaps absent, RNA-synthesizing activity. Furthermore, we know that the GV contains large amounts of stored polymerases and histones much in excess of what is required for the functions of the genomic DNA in the GV. These substances are stored for future use in embryogenesis.

When purified *Xenopus* 5S DNA is injected into the GV, along with a radioactive precursor, the response is dramatic (Brown and Gurdon 1977). Normally, 5S RNA is about 0.3% of endogeneous RNA synthesis. Depending on the kind of 5S DNA injected, the amount of 5S RNA synthesized ranges from 10 to 60% of the total RNA made. The newly made 5S RNA can be shown to be a product of the injected DNA, as determined by fingerprint analysis of the 5S RNA. All four kinds of purified 5S DNA support 5S RNA synthesis in the oocyte, and each gives the expected fingerprint.

GENETICS WITH "HOMEMADE" MUTANTS

The oocyte injection system provides a means of testing a simple gene, such as 5S DNA, for its ability to function in vitro. A scheme for producing and testing the effect of base changes in a presumptive control region of 5S DNA is as follows: A single repeat of 5S DNA is cloned and shown to make 5S RNA in the oocyte system. Base changes are introduced into the repeat, their location determined by sequencing, and their effect on faithful transcription assayed by the oocyte system. By this means we can determine which nucleotides in the "promotor" region are required for proper transcription. We also hope to introduce 5S DNA into the GV and then reisolate it with the molecules involved in this control still attached to it.

Even with these new tools, we recognize the complexities of prokaryotic gene regulation and the essential role that mutants played in research to elucidate these controls. The question must be asked, is it possible to reconstruct a system of at least comparable complexity in the absence of genetic mutants? We have the following substitutes for functional mutations:

1) Genetics by evolution. We make the assumption that functional nucleotide sequences evolve more slowly than nonfunctional ones. If there is a region adjacent to the gene that has some control function we might expect it to be the same or at least similar in the two related *Xenopus* species. We have already compared the sequences adjacent to the 3′ end of the principal oocyte-type genes of *X. laevis* and *X. borealis* (Brown and Brown 1975). In this study, it was found that the first two nucleotides in the spacer were T residues in both kinds of 5S DNA, and then the spacers of the two kinds of 5S DNA diverge. We concluded, tentatively, that it is the 3′ end of the gene itself, plus the two additional T residues which comprise the termination signal.

2) Genetics by differentiation. In order for us to truly reconstruct control we must understand why the oocyte-type 5S genes are "turned off" in somatic cells. This cell specificity is equivalent to having available a process like enzyme induction or enzyme repression in prokaryotes. Because we know the biological activity of these two kinds of 5S genes, it helps us to ask the right questions about the control of 5S RNA genes.

These are some of the tools of genetics by gene isolation and some of the proposed experiments that we hope can be used in the absence of traditional genetics.

ACKNOWLEDGMENTS

This research has benefitted from the expert assistance provided by Mrs. E. Jordan. We are supported in part by Grant No. 1 R01 GM22395-02 from The National Institutes of Health.

REFERENCES

Brown, D. D., and I. B. Dawid. 1968. Specific gene amplification in oocytes. Science 160:272–280.

Brown, D. D., and J. B. Gurdon. 1977. High fidelity transcription of 5S DNA injected into Xenopus oocytes. Proc. Natl. Acad. Sci. USA 74:2064–2068.

Brown, D. D., E. Jordan, R. D. Brown, and D. Carroll. 1975. Purification and characterization of new 5S DNA from *Xenopus laevis* and *Xenopus mulleri.* Carnegie Institution of Washington Yearbook 74:13.

Brown, D. D., and K. Sugimoto. 1973. 5S DNAs of *Xenopus laevis* and *Xenopus mulleri:* Evolution of a gene family. J. Mol. Biol. 78:397–415.

Brown, D. D., and K. Sugimoto. 1974. The structure and evolution of ribosomal and 5S DNAs in *Xenopus laevis* and *Xenopus mulleri.* Cold Spring Harbor Symp. Quant. Biol. 38:501–505.

Brown, D. D., P. C. Wensink, and E. Jordan. 1971. Purification and characteristics of 5S DNA from *Xenopus laevis.* Proc. Natl. Acad. Sci. USA 68:3175–3179.

Brown, R. D., and D. D. Brown. 1976. The nucleotide sequence adjoining the 3′ end of the genes coding for oocyte-type 5S RNA in *Xenopus.* J. Mol. Biol. 102:1–14.

Brownlee, G. G., E. M. Cartwright, and D. D. Brown. 1974. Sequence studies of the 5S DNA of *Xenopus laevis*. J. Mol. Biol. 89:703–718.

Carroll, D., and D. D. Brown. 1976. Repeating units of *Xenopus laevis* oocyte-type 5S DNA are heterogeneous in length. Cell 7:467–475.

Federoff, N. V., and D. D. Brown. 1977. The nucleotide sequence of the repeating unit in the oocyte 5S ribosomal DNA of *Xenopus laevis*. Cold Spring Harbor Symp. Quant. Biol. (In press).

Ford, P. J., and R. D. Brown. 1976. Sequences of 5S ribosomal RNA from *Xenopus mulleri* and the evolution of 5S gene-coding sequences. Cell 8:485–493.

Gall, J. G. 1968. Differential synthesis of the genes for ribosomal RNA during amphibian oogenesis. Proc. Natl. Acad. Sci. USA 60:553–560.

Gurdon, J. G., and D. D. Brown. 1977. Toward an in vivo assay for the analysis of gene-control and function, *in* Molecular Biology of the Genetic Apparatus, P. T'so, ed., North-Holland Publishing Co., Amsterdam. (In press).

Maxam, A. M., and W. Gilbert. 1977. A new method for sequencing DNA. Proc. Natl. Acad. Sci. USA 74:560–564.

Parker, C. S., and R. G. Roeder. 1977. Selective and accurate transcription of *Xenopus laevis* 5S RNA genes in isolated chromatin by purified RNA polymerase III. Proc. Natl. Acad. Sci. USA 74:44–48.

Pardue, M. L., D. D. Brown, and M. L. Birnstiel. 1973. Location of the genes for 5S ribosomal RNA in *Xenopus laevis*. Chromosoma 42:191–203.

Roeder, R. G., C. S. Parker, J. A. Jaehning, S-Y. Ng, and V. E. F. Sklar. 1978. Structure and function of animal RNA polymerases and selective gene transcription in reconstructed systems, *in* Cell Differentiation and Neoplasia (The University of Texas System Cancer Center 30th Annual Symposium on Fundamental Cancer Research, 1977), G. F. Saunders, ed., Raven Press, New York, pp. 305–324.

Sanger, F., and A. R. Coulson. 1975. A rapid method for determining sequences in DNA by primed synthesis with DNA polymerase. J. Mol. Biol. 94:441–448.

Cell Differentiation and Neoplasia, edited by Grady F. Saunders. Raven Press, New York

Structure and Function of Animal RNA Polymerases and Selective Gene Transcription in Reconstructed Systems

Robert G. Roeder, Carl S. Parker, Judith A. Jaehning, Sun-Yu Ng, and Virgil E. F. Sklar

Department of Biological Chemistry, Division of Biology and Biomedical Sciences, Washington University, St. Louis, Missouri 63110

The precise nature and mechanism of action of the components that regulate transcription of the eukaryotic genome are largely unknown, although transcription appears to be regulated at several levels. Considerable evidence suggests that structural modifications of the chromosomal matrix (chromatin) govern the accessibility of specific genes and, hence, their ability to be transcribed by an RNA polymerase. In addition, the presence of multiple molecular species of RNA polymerase in eukaryotic cell nuclei suggests that the cell may in part regulate the transcription of specific genes or classes of genes through distinct RNA polymerases, possibly through alterations in the intracellular activities or in the specificities of individual RNA polymerases. Studies of the structure, function, and regulation of these enzymes can be expected to enhance our understanding of transcriptional control. Moreover, a detailed understanding of various transcriptional controls, whatever their nature, will require the reconstruction of these controls in cell-free systems from various components, and specific eukaryotic enzymes may be necessary for these controls to be manifested in vitro as well as in vivo.

In this report we will briefly review the diversity, structure, and general functions of eukaryotic RNA polymerases and observations relevant to the regulation of these enzymes in the cell. Subsequently, we will present more recent observations from this laboratory on the transcription of specific genes in reconstructed systems. These later studies suggest that both chromatin-associated components and specific RNA polymerases are required for the faithful transcription of specific genes, and they describe systems suitable for more detailed studies of transcriptional regulatory mechanisms.

DIVERSITY AND STRUCTURE OF EUKARYOTIC RNA POLYMERASES

Eukaryotic cells contain multiple forms of nuclear RNA polymerases, which can be separated by chromatographic and electrophoretic procedures. These enzymes fall into three distinct classes (designated I, II, and III, or A, B, and C), distinguished on the basis of their enzymatic properties and subunit structures (reviewed in Roeder 1976a, Chambon 1975). Two major class I enzymes (I_A and I_B), three major class II enzymes (II_O, II_A, and II_B), and two major class III enzymes (III_A and III_B) have been identified in mammalian cell types. The diversity of enzymes detected within each enzyme class in lower eukaryotes is less.

The most comprehensive structural studies of the class I, II, and III RNA polymerases from animal cells are those of the mouse plasmacytoma enzymes (Schwartz and Roeder 1974, 1975, Sklar *et al.* 1975, Sklar and Roeder 1976, Roeder *et al.* 1976). The molecular structures of these enzymes are summarized in Table 1. An analysis of these data reveals several important features. Each of the nuclear RNA polymerases contains two polypeptides with molecular weights greater than 100,000 and several smaller polypeptides. The total number of distinct polypeptides in each enzyme form is five to six for class I enzymes ($M_r \simeq 400{,}000$), nine for class II enzymes ($M_r \simeq 500{,}000$), and ten for class III enzymes ($M_r \simeq 650{,}000$). Similar complexities have been reported for the enzymes from several other eukaryotes (Chambon 1975, Roeder 1976a), including lower eukaryotes such as yeast (Valenzuela *et al.* 1976, Sentenac *et al.* 1976).

The various polypeptides present in the highly purified enzymes appear to be tightly associated, as evidenced by their co-purification through a variety of procedures, including ion-exchange chromatography, gel electrophoresis, and sucrose gradient sedimentation at high ionic strength. They are also present in roughly equimolar proportions and have their counterparts in the cognate enzymes from different organisms (Sklar *et al.* 1976, Roeder 1976a, Chambon 1975). Hence, they are tentatively regarded as enzyme subunits or transcription components, although this must ultimately be confirmed by more rigorous experimentation involving enzyme dissociation and reconstitution. Most of the polypeptides present in each enzyme class (including the two largest subunits) differ in size from those present in the other enzyme classes. These observations suggest that the various classes of enzymes are assembled primarily from distinct gene products, have distinct functions, and are subject to independent regulation. Two or three smaller subunits appear to be common to the three classes of enzymes, suggesting that these enzymes utilize some similar catalytic mechanisms or that they recognize some common transcription elements.

The enzymes within each class of mammalian enzymes show only minor structural differences. Thus, the class I enzymes appear to differ by the presence or absence of a specific polypeptide (Matsui *et al.* 1976, Gissinger and Chambon 1975). The class II enzymes each have a unique high molecular weight polypeptide, which in each case is the largest subunit (Kedinger *et al.* 1974, Krebs and Chambon 1976). The class III enzymes each have a unique low molecular weight polypeptide. These observations raise the possibility of distinct, but related, functions for intraclass enzymes. However, it should be emphasized

TABLE 1. *Subunits of mouse plasmacytoma Class I, II and III RNA polymerases*

Subunit	I_A MW	I_B MW	Subunit	II_O MW	II_A MW	II_B MW	Subunit	III_A MW	III_B MW
			IIo	240					
			IIa		205				
Ia	195	195							
			IIb			170			
							IIIa	155	155
			IIc	140	140	140			
							IIIb	138	138
Ib	117	117							
							IIIc	89	89
							IIId	70	70
Ic	61								
							IIIe1	53	53
Id	49	49					IIIe2	49	49
			IId	41	41	41	IIIf	41	41
							$IIIg_B$		33
							$IIIg_A$	32	
Ie	29	29	IIe	29	29	29	IIIh	29	29
			IIf	27	27	27			
			IIg	22	22	22			
			IIh1	19.5	19.5	19.5			
If	19	19	IIh2	19	19	19	IIIi	19	19
			IIi	16.5	16.5	16.5			

Numbers indicate approximate molecular weights in daltons $\times 10^{-3}$. In some cases, these values differ slightly from previously reported values. Molar ratios are approximately unity for all subunits except the heterogeneous subunit IIIi (ratio $\simeq 3$). Subunit IIIf also contains at least two subcomponents. Only polypeptides that remain associated with the enzymes during electrophoresis under nondenaturing conditions are indicated. Data from Schwartz and Roeder 1974, 1975, Sklar *et al.* 1975, Sklar and Roeder 1976, Roeder *et al.* 1976. For further details and for similarities to other eukaryotic RNA polymerases, see text and Roeder (1976a).

that the biological significance of the multiple intraclass mammalian enzymes has not been established. It is possible that the multiple enzyme forms within a class may be generated by the loss or modification (e.g., proteolytic cleavage) of specific subunits during enzyme isolation, as has been reported for the enzymes from lower eukaryotes (Dezelee *et al.* 1976, Greenleaf *et al.* 1976, Huet *et al.* 1975, Hager *et al.* 1976). Other RNA polymerase modifications are discussed below. The possible functional significance of these modifications and the various intraclass forms must be assessed by more direct means (see below).

GENERAL AND SPECIFIC FUNCTIONS OF RNA POLYMERASES

The class I, II, and III enzymes from animal cells are readily distinguished by their differential sensitivities to α-amanitin (Schwartz *et al.* 1974, reviewed in Roeder 1976a). This property has been used in investigations of the functions of the various RNA polymerases, which have employed isolated nuclei that are active in the synthesis of defined classes of RNA and that are freely permeable to the toxin. Table 2 summarizes the results of these studies (reviewed in Roeder 1976a) and also indicates the subcellular localizations of the RNA polymerases. In animal cells, the class I enzymes transcribe the genes that encode the 18S and 28S rRNAs, the class III enzymes transcribe the genes that encode the tRNA and 5S rRNA genes, and the class II enzymes transcribe those DNA sequences that encode the heterogeneous-nuclear RNAs (and presumably, therefore, most mRNAs). Host RNA polymerases (or possibly modified forms of them) have also been shown to be involved in the transcription of viral genes. In cells infected with DNA tumor viruses, a class II enzyme(s) transcribes viral genes that encode nuclear precursors to viral mRNA (reviewed in Roeder 1976a). In human cells infected with adenovirus 2, a class III enzyme(s) transcribes viral genes that encode a group of low molecular weight viral RNAs 140–200 nucleotides in length (Weinmann *et al.* 1974, 1976a). The studies of viral gene transcription provide convincing support for the hypothesis that class

TABLE 2. *Localization and general functions of animal cell RNA polymerases*

Enzyme class	Subnuclear localization	Cellular gene transcripts	Viral gene transcripts
I	nucleolus	18S, 28S rRNAs	none identified
II	nucleoplasm	HnRNAs	mRNA precursors
III	nucleoplasm	tRNAs, 5S rRNA	low molecular weight RNAs (V_{200}, V_{156}, V_{140})

The functions of the RNA polymerases in transcribing cellular genes were analyzed in mouse plasmacytoma cells, in amphibian cells, and in HeLa cells. The functions of the RNA polymerases in transcribing viral genes were analyzed in adenovirus 2–infected KB cells. The viral V_{200}, V_{156}, and V_{140} species are approximately 200, 156, and 140 nucleotides in length, respectively. For references and discussion, see text and Roeder (1976a).

II and III enzymes have general functions in the synthesis of mRNA or mRNA precursors and low molecular weight RNAs, respectively. The studies of low molecular weight RNA synthesis in isolated nuclei are particularly significant, because they indicate that at least some (i.e., class III) eukaryotic RNA polymerases can initiate and terminate synthesis efficiently at the correct gene sites in cell-free systems (Marzluff *et al.* 1974, Price and Penman 1972, Weinmann *et al.* 1974). Hence, one can be optimistic about the possibility that these reactions can be duplicated in systems reconstructed from more purified components.

Since the individual RNA polymerases within a given class of mammalian enzymes have identical α-amanitin sensitivities, it has not yet been possible to ascertain from these types of studies whether the intraclass enzymes have distinct functions (e.g., III_A vs. III_B in the synthesis of tRNA vs. 5S RNA). It may be possible to answer this question through the use of antibodies to the intraclass enzyme-specific subunits, if they prove to be antigenically distinct, or the use of transcription systems that respond effectively to specific exogenous RNA polymerases (see below). It is apparent, however, that at least some class I, II, and III enzymes must be present in most all cell types. In fact, no consistent tissue specificities have yet been convincingly demonstrated for individual forms of class I, II, or III RNA polymerases (Krebs and Chambon 1976, Roeder 1976b).

REGULATION OF TRANSCRIPTION IN EUKARYOTIC CELLS

The existence of structurally and functionally distinct RNA polymerases suggests that synthesis of the major classes of RNA could be regulated in part through the direct regulation of individual enzymes. It is presumed that each enzyme recognizes specific structural features of the template common to the various genes which that enzyme transcribes. The intracellular activities of specific enzymes could then be regulated independently and in a direct fashion through alterations in enzyme concentrations or regulatory modifications of the enzymes. Regarding the first possibility, altered RNA polymerase concentrations have been reported in a number of cell types in response to certain physiological stimuli (Roeder 1976a) and may be responsible for effecting gross changes in the rates of synthesis of the major classes of RNA. However, such changes are not characteristic of all physiological transitions that result in altered rates of RNA synthesis. In these situations (e.g., early embryonic development, lytic virus infection) other factors must clearly modulate the intracellular activity and selectivity of the RNA polymerases. With respect to RNA polymerase modifications, some possible functional modifications have already been mentioned. In addition, other laboratories have reported the phosphorylation of specific RNA polymerase subunits both in vivo and in vitro (Hirsch and Martelo 1976, Bell *et al.* 1976) and the isolation of proteins that stimulate the transcription of various DNA templates by specific RNA polymerases (reviewed in Chambon 1975, Roeder 1976a). It is not clear whether the phosphorylation events or the various stimulatory proteins modulate enzyme activity or selectivity within the cell. However, these observations suggest additional mechanisms by which the intracellular enzyme activities and the overall rates of transcription of the major classes of genes could be regulated. Obviously, chromatin structural alterations could also effect overall rates of transcription.

Because of the limited number of structurally distinct RNA polymerases, it is apparent that additional determinants are required to effect the differential transcription of genes transcribed by a common enzyme. Although the various components responsible for the differential transcription within a specific class of genes have yet to be elucidated, considerable evidence suggests that chromatin structural changes are involved, at least in the regulation of some genes. Thus, the globin and ovalbumin genes (or more precisely, regions of them) can be transcribed by an exogenous bacterial RNA polymerase in chromatin from cells actively expressing the respective genes but not in chromatin from most other cell types (Axel *et al.* 1973, Gilmour and Paul 1973, Tsai *et al.* 1976); and the components determining whether or not the gene can be transcribed apparently reside in the nonhistone protein fraction (O'Malley *et al.* 1978, see pages 473 to 485, this volume). Studies of the sensitivity of specific genes in chromatin to nuclease digestion also indicate that chromatin structural modification accompanies functional changes in gene activity (Weintraub and Groudine 1976, Garel and Axel 1976). These results suggest an indirect role for the RNA polymerase in the selective activation of these genes. However, it has not yet been shown that the chromatin transcripts of these genes accurately reflect the natural primary gene transcripts. Furthermore, some genes that do not appear to be transcribed in vivo are transcribed in isolated chromatin by exogenous bacterial RNA polymerases (Reeder 1973). Thus, other components besides those present in isolated chromatin may be necessary to effect the same transcription patterns observed in vivo.

SELECTIVE GENE TRANSCRIPTION IN RECONSTRUCTED SYSTEMS

Although the readily available bacterial RNA polymerases have proved to be useful probes for tissue-specific, functional alterations within isolated chromatin, it seems probable that homologous eukaryotic enzymes will be necessary in any reconstituted system to reproduce precisely the transcriptional controls operating in vivo. Several studies of the transcription of specific genes in isolated chromatin by animal RNA polymerases have been reported previously (reviewed in Roeder 1976a). In no case were the class I and II RNA polymerases found to transcribe specific genes any more accurately or efficiently than prokaryotic RNA polymerases, possibly reflecting the use of impure or damaged RNA polymerases or templates or the absence of other necessary transcription components. Similarly, no animal RNA polymerase has yet been shown to faithfully transcribe specific genes in purified DNA templates, including simple viral genomes (reviewed in Roeder 1976a).

In the studies presented below, we have undertaken a similar approach but have focused on those genes transcribed by the class III RNA polymerases. These genetic systems are advantageous for such studies for the following reasons: (1) the genes are present in many copies per cell; (2) the gene products are simple and readily characterized; (3) the gene products are not extensively processed; (4) the genes are in some cases available in purified form and serve both as defined templates and as hybridization probes; (5) well-characterized class III enzymes are available in purified form; and (6) unlike the class I and II enzymes, the class III enzymes in isolated nuclei readily initiate synthesis.

Transcription of the 5S RNA Genes Expressed in *Xenopus laevis* Oocytes

The DNAs that encode the 5S RNAs have been isolated in pure form from *X. laevis* and from *X. borealis* (incorrectly designated *X. mulleri* in several previous publications) and have been shown to consist of tandem assays of genes interspersed with nontranscribed spacer regions (Brown and Sugimoto 1973). The major class of *X. laevis* 5S RNA genes numbers about 25,000 per haploid amount of DNA. These genes are present in all cell types but are expressed only in oocytes (Ford and Southern 1973, Brownlee *et al.* 1974). In immature previtellogenic oocytes, the transcription of these genes and of the tRNA genes accounts for the vast majority of newly synthesized RNA (Denis 1974). Although *X. laevis* oocytes contain all three classes of RNA polymerases, RNA polymerase III constitutes an unusually high fraction (about 30%) of the total oocyte RNA polymerase activity (Roeder 1974). This enzyme has a subunit structure (10 distinct polypeptides) remarkably similar to that of mammalian RNA polymerases (Sklar *et al.* 1975). In the following studies, both homologous class III RNA polymerases and heterologous RNA polymerases have been employed to study the transcription of 5S RNA genes in oocyte chromatin and in purified DNA templates. Transcripts of the 5S gene region have been assayed by hybridization to *X. borealis* 5S DNA, since only the gene regions of *X. laevis* and *X. borealis* 5S DNAs are homologous. This permits an assessment of the absolute amounts of 5S RNA synthesis, as well as the detection of both sense and antisense strand transcripts of the gene region (see Table 3).

An analysis of the transcription of oocyte chromatin by homologous RNA polymerase III is shown in Table 3. This chromatin exhibits a low level of 5S RNA synthesis by endogenous RNA polymerase III. This RNA accounts for about 30% of the total RNA synthesized. However, the absolute rate of 5S RNA synthesis is stimulated about 40-fold by exogenous RNA polymerase III. Under these conditions, 5S RNA synthesis accounts for about half of the total RNA synthesized, and the gene region is transcribed in a highly asymmetric fashion (sense:antisense ratio of 44 in the experiment shown). A further analysis by polyacrylamide gel electrophoresis of the RNA products synthesized under these conditions is shown in Figure 1. In the presence of RNA polymerase III, a discrete RNA product, indistinguishable from natural 5S RNA, is synthe-

TABLE 3. *Transcription by homologous and heterologous RNA polymerases of the 5S RNA genes in* X. laevis *chromatin*

Added RNA polymerase	Input $[^3H]$(CPM)	$[^3H]$ Hybridized: 5S gene (sense strand)	$[^3H]$ Hybridized: 5S gene (antisense strand)	$[^{32}P]$5S RNA hybridization efficiency	$[^3H]$(CPM) in 5S RNA
None (endogenous)	2,800	103	0	14.4%	722
X. laevis III (265 units)	65,140	3,982	91	13.9%	27,993
MOPC II (130 units)	6,440	99	7	13.0%	710
E. coli holoenzyme (210 units)	56,010	138	42	14.4%	660
E. coli holoenzyme (730 units)	157,000	457	185	15.4%	1,760

Chromatin was isolated from purified immature oocytes and transcribed with the indicated units (based on activity with native DNA) of various RNA polymerases, including RNA polymerase III from *X. laevis* oocytes. The input $[^3H]$(CPM) represents the total amount of RNA synthesized in each case. The purified RNAs were hybridized to *X. borealis* 5S DNA alone and to *X. borealis* 5S DNA presaturated with 5S RNA. Only the 5S gene regions of the *X. laevis* and *X. borealis* 5S DNA are homologous (Brown and Sugimoto 1973). The radioactive RNA that hybridized to *X. borealis* 5S DNA presaturated with 5S RNA was scored as 5S gene antisense strand hybridization. The difference between the amount of radioactivity that hybridized to *X. borealis* 5S DNA alone and that which hybridized to the 5S DNA presaturated with 5S RNA was scored as gene sense strand hybridization. The total amount of $[^3H]$ radioactivity in 5S RNA was calculated from the amount of $[^3H]$(CPM) hybridized to the 5S gene sense strand and from the hybridization efficiency of 5S RNA. The latter was determined by inclusion of a bonafide $[^{32}P]$5S RNA internal standard in each hybridization reaction. Data from Parker and Roeder (1977).

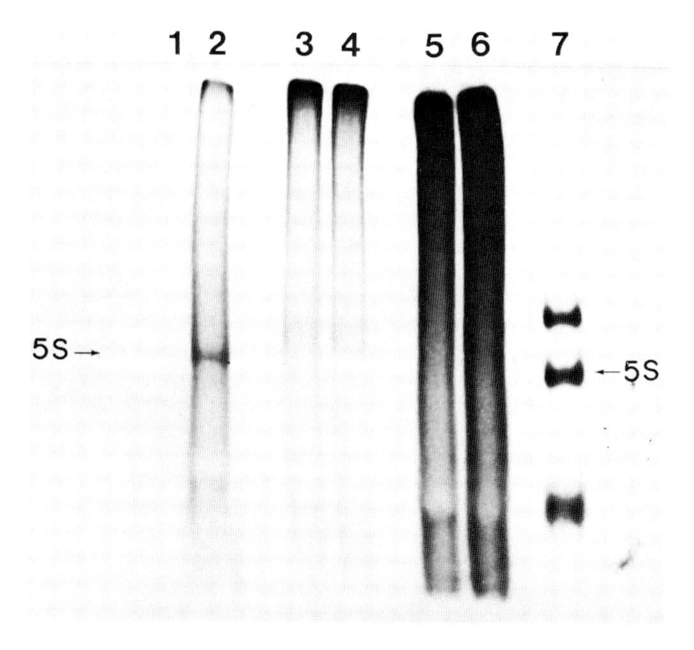

FIG. 1. Electrophoretic analysis of *X. laevis* oocyte chromatin transcripts. Chromatin was isolated and transcribed with purified RNA polymerases as described in Table 3. The ^{3}H-labeled transcripts purified from reactions containing equivalent amounts of chromatin were subjected to electrophoresis on a 12% polyacrylamide slab gel. Radioactive RNAs were localized by fluorography. Equivalent amounts of chromatin were transcribed under the following conditions: Slot 1, no exogenous RNA polymerase; slot 2, 250 units of *X. laevis* RNA polymerase III; slots 3 and 4, 700 units of *E. coli* core RNA polymerase; slots 5 and 6, 375 units of *E. coli* RNA polymerase (holoenzymes). The reactions analyzed in slots 4 and 6 contained sufficient α-amanitin (200 μg/ml) to completely inhibit endogenous RNA polymerase III activity. Marker RNAs (5.5S, 5S, and 4S) were run in slot 7. For details, see Parker and Roeder (1977). (From C. Parker, unpublished data.)

sized (slot 2). The identity of the in vitro 5S RNA transcript with natural 5S RNA has also been confirmed by fingerprint analysis (C. S. Parker and R. G. Roeder, manuscript in preparation). These results demonstrate a highly selective transcription (3,000-fold above random) of the 5S RNA genes by the RNA polymerase III (Parker and Roeder 1977). They further suggest that the RNA polymerase recognizes natural initiation and termination sites within the template.

The transcription of 5S genes in oocyte chromatin by heterologous RNA polymerases has also been examined. Saturating amounts of a murine (MOPC) RNA polymerase II (see Table 3) and a homologous oocyte RNA polymerase I (Parker and Roeder 1977) were found to stimulate total RNA synthesis severalfold (above the endogenous level), but in neither case was the endogenous level of 5S RNA synthesis increased. *Escherichia coli* RNA polymerase (holoenzyme) was found to markedly stimulate total RNA synthesis (see Table 3). The lower level of enzyme used in the experiments shown stimulated total RNA synthesis about 20-fold (the same as RNA polymerase III) with no apparent stimulation of 5S RNA synthesis. A higher level of *E. coli* holoenzyme, which stimulated total RNA synthesis 56-fold, did increase the level of 5S RNA synthesis about two to threefold. However, under these conditions both strands of the 5S RNA gene were transcribed, and the fractional level of 5S RNA synthesis (1%) approximated that observed when native DNA templates were transcribed by the *E. coli* RNA polymerase (Parker and Roeder 1977). Moreover, no discrete 5S RNA products were observed when the in vitro *E. coli* RNA polymerase transcripts were analyzed by polyacrylamide gel electrophoresis (see Figure 1, slots 5 and 6). In other experiments (C. S. Parker and R. G. Roeder, manuscript in preparation), the *E. coli* core RNA polymerase has also been found to stimulate by severalfold 5S RNA synthesis in chromatin, although discrete 5S RNA transcripts are not synthesized in response to the exogenous core enzyme (see Figure 1, slots 3 and 4). From these studies it appears that an RNA polymerase III is necessary for effecting a preferential and accurate transcription of the 5S RNA genes in isolated chromatin.

The ability of RNA polymerase III to selectively transcribe the 5S RNA genes in purified DNA templates has also been examined. These studies have employed intact recombinant DNAs that contain bacterial DNA sequences and *X. laevis* 5S DNA sequences (Carroll and Brown 1976). The recombinant DNAs used for these experiments contained five repeat units (genes plus spacers) of *X. laevis* 5S DNA, which comprise about one third of the total plasmid DNA. The use of these intact templates avoids complications due to artificial initiation sites (e.g., nicks), which are present in most high molecular weight cellular DNA templates (Roeder 1976a). As shown in Table 4, both the relaxed and supercoiled forms of the recombinant DNA plasmids (Parker *et al.* 1976) are active templates for *X. laevis* RNA polymerase III. However, two observations strongly suggest that the recombinant DNA templates are randomly transcribed. First, both the sense and antisense strands of the 5S RNA genes are transcribed with equal frequency. Second, the apparent ratios of 5S gene transcripts to bacterial DNA (pSC134) transcripts (0.12–0.19) are only a fewfold higher than the ratio expected if the DNA were transcribed randomly (0.08). In related studies, the 5S RNA genes in total cellular DNA were shown to be aberrantly transcribed by RNA polymerase III (Parker and Roeder 1977). These and the preceding studies suggest, therefore, that both chromatin-associated components and an RNA polymerase III are required for the selective and asymmetric transcription of the 5S RNA genes in amphibian oocytes.

Transcription of Mouse Plasmacytoma tRNA and 5S RNA Genes

Related studies have examined the transcription of those genes that encode the 5S RNA and the 4.5S RNAs (tRNA precursors) in mouse plasmacytoma cells. In these studies, purified RNA polymerases have been employed to tran-

TABLE 4. *Transcription by RNA polymerase III of a recombinant DNA containing* Xenopus laevis *5S DNA*

DNA form	KCl (M)	Input (CPM)	CPM Hybridized			Hybridization ratio (5S gene DNA:Plasmid DNA)
			Plasmid DNA (pSC134)	5S gene (sense strand)	5S gene (antisense strand)	
Supercoil	0.018	533,000	87,047	8,129	8,300	0.19
	0.098	295,000	48,176	3,914	4,876	0.18
	0.218	124,000	10,414	1,168	801	0.19
Relaxed	0.018	167,000	18,683	2,037	1,130	0.17
	0.098	71,000	7,800	506	945	0.19
	0.218	63,000	6,106	145	612	0.12

Both supercoiled and relaxed forms of a recombinant DNA containing *X. laevis* 5S DNA (5 repeat units) and plasmid pSC134 DNA (Parker *et al.* 1976) were transcribed at the indicated salt concentrations by an RNA polymerase III purified from *X. laevis* oocytes. The input radioactivity in the third column represents the amount of RNA synthesized in each case. The newly synthesized transcripts were hybridized to pSC134 DNA and to *X. borealis* 5S DNA alone or presaturated with 5S RNA. The amounts of radioactivity hybridized to the sense and antisense strands of the 5S gene were determined as described in Table 3. The mass ratio of 5S gene DNA to pSC134 DNA in the template is 0.043. For further details see text and Parker *et al.* (1976).

scribe the genes in isolated nuclei rather than in chromatin. The rationale for this approach is that nuclei contain nucleoprotein templates in a minimally damaged form and that specific transcription events in response to purified RNA polymerases may be more readily detected. Isolated mouse plasmacytoma nuclei readily synthesize discrete 5S RNA and 4.5S RNAs via an endogenous RNA polymerase III (Weinmann and Roeder 1974, Sklar and Roeder 1977). However, both RNA polymerase III_A and III_B from MOPC cells stimulate significantly the synthesis of 5S RNA and 4.5S RNA (Figure 2), as well as other discrete low molecular weight RNAs somewhat larger than 5S RNA (Sklar and Roeder 1977). Hybridization studies have shown that the 5S RNA genes are transcribed asymmetrically under these conditions (Sklar and Roeder

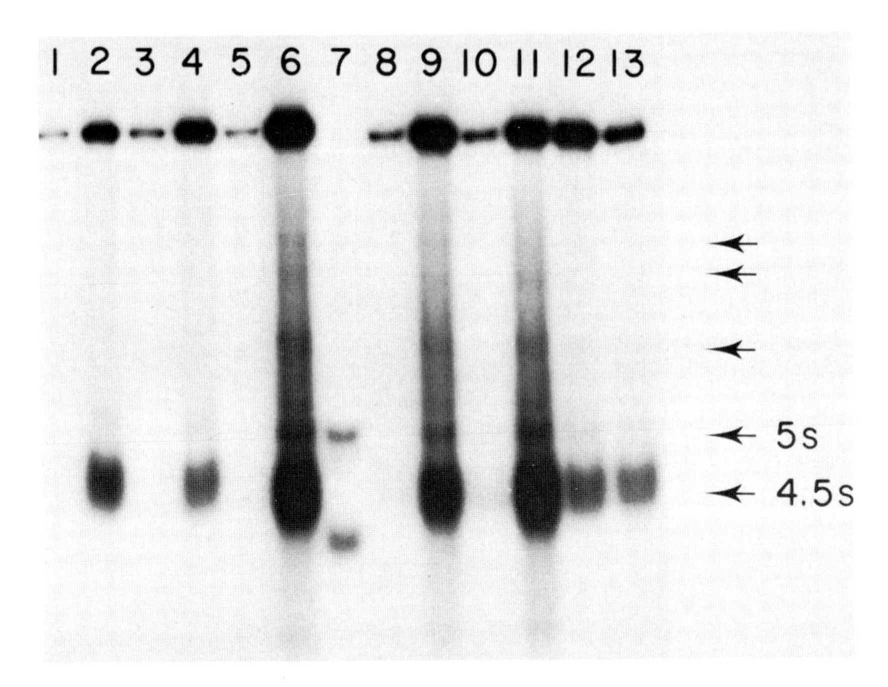

FIG. 2. Electrophoretic analysis of mouse plasmacytoma nuclear transcripts synthesized by exogenous RNA polymerases. Nuclei isolated from cultured MOPC 460 cells were incubated under RNA synthesis conditions in the presence of different purified RNA polymerases. The purified ^{32}P-labeled transcripts were analyzed by autoradiography following electrophoresis in 12% polyacrylamide gels. Nuclei were transcribed under the following conditions: Slots 1 and 2, 250 units of MOPC RNA polymerase I; slots 3 and 4, endogenous RNA polymerase alone; slots 5 and 6, 500 units of MOPC RNA polymerase III_A; slots 8 and 9, 500 units of MOPC RNA polymerase III_B; slots 10 and 11, 450 units of *X. laevis* RNA polymerase III; slots 12 and 13, 400 units of MOPC RNA polymerase II. The reactions analyzed in slots 2, 4, 6, 9, 11, and 13 contained sufficient α-amanitin (0.5 μg/ml) to specifically inhibit RNA polymerase II. The reactions analyzed in slots 1, 3, 5, 8, and 10 contained sufficient α-amanitin (200 μg/ml) to completely inhibit RNA polymerase III. ^{32}P-labeled marker RNAs (5S and 4.5S) were run in lane 7. The upper, unlabeled arrows denote the migration positions of minor 6.6S, 6.3S, and 5.8S RNAs synthesized by RNA polymerase III. From Sklar and Roeder (1977).

1977). These studies suggest that the RNA polymerases III_A and III_B are not functionally distinct, although possible interconversions of these enzymes in the reconstructed systems have not been ruled out.

The transcription of nuclear templates by heterologous RNA polymerases is also shown in Figure 2. RNA polymerases I and II from MOPC cells do not significantly stimulate the synthesis of 4.5S RNA and 5S RNA. In contrast, the heterologous RNA polymerase III from *X. laevis* oocytes stimulates 5S and 4.5S RNA synthesis as effectively as do the homologous class III enzymes. This result is somewhat surprising since these enzymes are isolated from very different cell types and from different organisms. However, the MOPC and *X. laevis* class III enzymes are structurally very similar, as mentioned above (Sklar *et al.* 1975) and apparently are capable of recognizing heterologous transcription components (signals).

Transcription of Specific Adenovirus Genes

During the infection of human KB cells by adenovirus 2, the viral genome is transcribed by both class II and class III RNA polymerases, as discussed above. Those viral genes transcribed by RNA polymerase III are clustered near map position 0.3 of the viral genome (Weinmann *et al.* 1976a, Soderlund *et al.* 1976) and encode several low molecular weight RNAs, the most prominent of which is the V_{156} or 5.5S RNA. While the functions of these viral gene products are unknown, this genetic system offers advantages for an analysis of transcriptional controls. The present studies have focused on transcription of the viral 5.5S gene in purified DNA templates and in nuclear templates.

As reported previously (Weinmann *et al.* 1974, 1976a), the viral 5.5S RNA is actively synthesized in nuclei from adenovirus-infected cells via an endogenous RNA polymerase III (Figure 3, slot 1). The addition of exogenous RNA polymerase III isolated from infected cells significantly stimulates (up to fivefold) the synthesis of the 5.5S RNA as well as the V_{200} RNA species (Figure 3, slot 2). To assess more directly the effects of exogenous RNA polymerases on viral gene transcription, attempts were made to irreversibly inactivate the endogenous RNA polymerase III. This was effectively accomplished by preincubation of the isolated nuclei with low concentrations of N-ethylmaleimide (Figure 3, slot 3). The addition of exogenous RNA polymerase III from human KB cells to these preincubated nuclei has been found to stimulate 5.5S RNA synthesis by as much as 80- to 100-fold (Figure 3, slot 4) (J. A. Jaehning and R. G. Roeder, manuscript in preparation).

A further analysis of this system is shown in Figure 4. RNA polymerase III_B from uninfected KB cells (slot 3) and RNA polymerase III_B from adenovirus-infected KB cells (slot 4) are equally effective in transcribing the viral 5.5S gene. Exactly the same results have been obtained with RNA polymerase III_A from uninfected and infected cells (see Figure 3). Moreover, as shown in Figure 4, the class III RNA polymerases from mouse plasmacytoma cells (slot 5) and

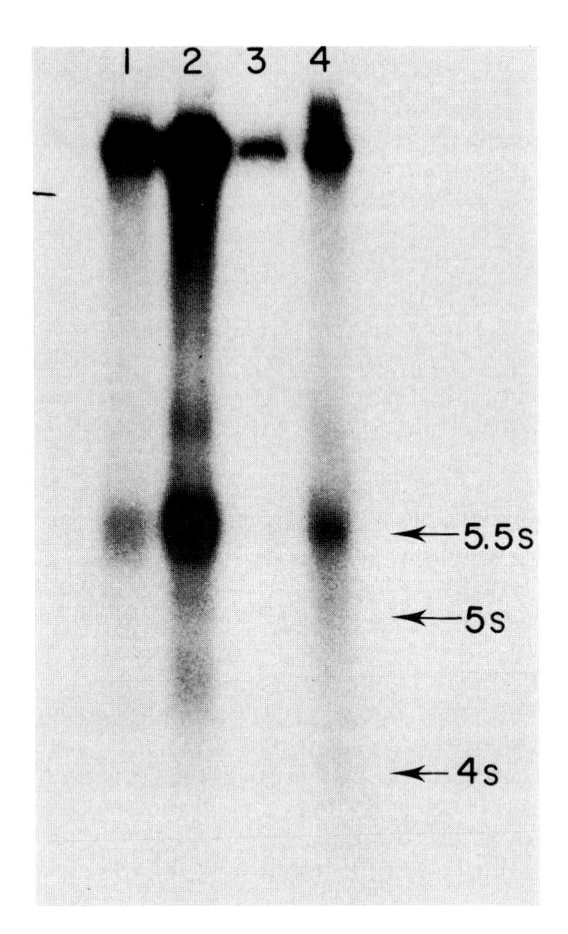

FIG. 3. Electrophoretic analysis of low molecular weight adenovirus RNAs synthesized in isolated nuclei. Nuclei were isolated from human KB cells 14 hours after infection with adenovirus-2 and incubated under RNA synthesis conditions as described previously (Weinmann *et al.* 1976a). ^{32}P-labeled transcripts were analyzed by autoradiography following electrophoresis in 12% polyacrylamide gels. Nuclei were incubated in the absence (slots 1 and 3) or in the presence (slots 2 and 4) of 130 units of RNA polymerase III_A from KB cells. The nuclei used for the analyses in slots 2 and 4 were preincubated with N-ethylmaleimide just prior to incubation under RNA synthesis conditions. All reactions contained 0.5 μg α-amanitin per milliliter. Positions of ^{32}P-labeled marker RNAs are indicated with arrows. The band just above the 5.5S RNA in slot 2 is the V_{200} viral RNA species (see footnote to Table 2). (From J. Jaehning, unpublished data.)

X. laevis oocytes (slot 6) also transcribe the 5.5S gene very effectively. In contrast, neither RNA polymerase II from adenovirus 2–infected human KB cells (slot 8) nor *E. coli* RNA polymerase (slot 7) appears to transcribe the viral 5.5S RNA gene, even though total RNA synthesis is significantly stimulated in each case (particularly in the latter). These conclusions have been confirmed by hybrid-

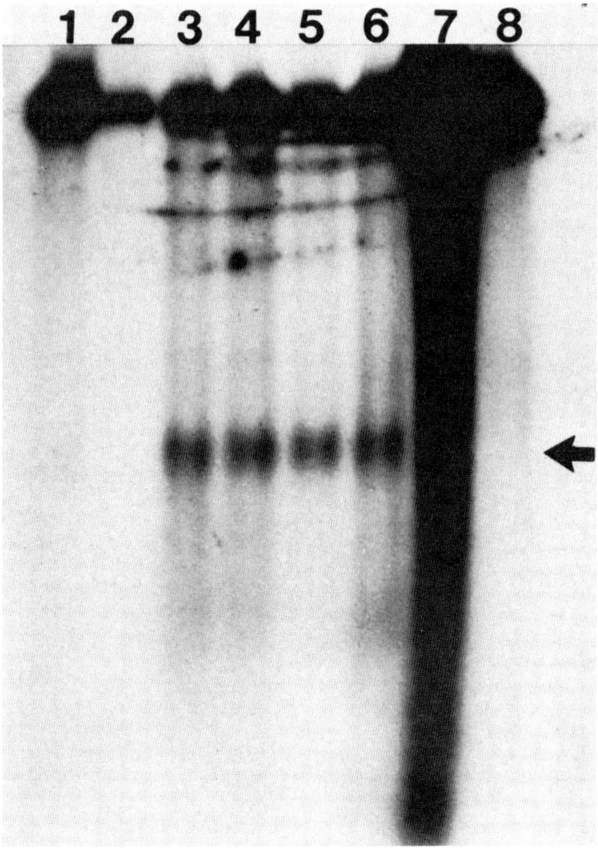

FIG. 4. Electrophoretic analysis of low molecular weight adenovirus RNAs synthesized in isolated nuclei. Nuclei were isolated from adenovirus-infected KB cells, preincubated with N-ethylmaleimide, and incubated as indicated in Figure 3. Nuclei were incubated under the following conditions: Slots 1 and 2, no exogenous RNA polymerase; slot 3, 160 units uninfected KB cell RNA polymerase III_B; slot 4, 130 units RNA polymerase III_B from adenovirus-infected KB cells; slot 5, 200 units MOPC RNA polymerase III_A; slot 6, 280 units *X. laevis* oocyte RNA polymerase III; slot 7, 640 units *E. coli* RNA polymerase (holoenzyme); slot 8, 200 units RNA polymerase II from adenovirus-infected KB cells. The reactions analyzed in slots 1 and 8 contained no α-amanitin, while those analyzed in slots 2–7 contained 0.5 μg α-amanitin per milliliter. The arrow indicates the position of the 5.5S RNA marker. No discrete 5.5S RNA band was evident in the *E. coli* RNA polymerase reaction (slot 7) following shorter autoradiographic exposure times. (From J. Jaehning, unpublished data.)

ization-competition studies (J. A. Jaehning and R. G. Roeder, manuscript in preparation).

These studies demonstrate the preferential and accurate transcription of specific viral genes by purified class III RNA polymerases. It appears from these studies that modifications of host RNA polymerases are not a prerequisite for viral gene transcription, although they might still be involved in modulating the intracellular RNA polymerase III activity, which has been shown to increase dramatically during virus infection (Weinmann *et al.* 1976b). Moreover, possible modifications of the uninfected KB cell or the heterologous class III enzymes in the crude reconstructed systems have not yet been ruled out. It is presumed that the human class III enzymes are structurally similar to the murine and amphibian class III enzymes, since all these enzymes appear capable of recognizing common transcription components (signals).

Other studies have examined the transcription of purified viral DNA templates by RNA polymerase III from infected cells. As shown by the hybridization analyses in Table 5, the purified RNA polymerase III appears to transcribe the viral genome randomly, as indicated by the following: the in vitro transcripts hybridize equally well to the *r* and *l* strands of the viral genome and to three specific DNA fragments produced by digestion of the viral genome with the restriction endonuclease *Sal*I. In contrast, the natural RNA polymerase III transcripts are apparently all derived from the r strand of the *Sal*I C fragment (Weinmann *et al.* 1976a, B. Harris and R. G. Roeder, unpublished observations). Thus, it appears that other components present in nuclei from infected cells are also required for the selective transcription of the viral genome by RNA polymerase III. These components could be directly associated with the viral DNA, although the nature of the viral template is not known.

TABLE 5. *Transcription of adenovirus-2 DNA by purified RNA polymerase III*

Input (CPM)	Adenovirus DNA hybridization probe (μg)	CPM hybridized	CPM hybridized fraction of genome/fragment
72,000	0.50 (Total)	14,072	—
	0.25 (*l* strand)	6,831	—
	0.25 (*r* strand)	8,702	—
14,044	0.35 (Total)	2,261	—
	0.70 (e.g., *Sal*I A)	1,789	3,312
	0.70 (e.g., *Sal*I B)	1,564	6,256
	0.70 (e.g., *Sal*I C)	951	4,655

Intact adenovirus-2 DNA was transcribed by RNA polymerase III from infected KB cells harvested 14 hours postinfection. The nuclease-free RNA polymerase (a mixture of III_A and III_B) was purified as previously described (Sklar and Roeder 1976). Incubation conditions were as described previously (Weinmann *et al.* 1976a), except for the presence of 5 mM Mg^{++} (in place of Mn^{++}), purified adenovirus DNA (70 μg/ml), and RNA polymerase III (3,000 units/ml). The input (CPM) represents the total amount of RNA synthesized in each experiment. Purified transcription products were hybridized with adenovirus-2 DNA, or with *l* or *r* strands of adenovirus-2 DNA, or with *Sal*I restriction endonuclease fragments. As discussed in the text, restriction endonuclease fragment C contains the DNA sequences that are transcribed by RNA polymerase III in intact cells. In the last column, the radioactivity hybridized is normalized to the fraction of total adenovirus DNA present in each restriction endonuclease fragment. Results similar to those reported here were obtained with Mn^{++} as the divalent metal ion. (From J. Jaehning and R. Roeder, unpublished data.)

Implications

These studies have demonstrated the selective and accurate transcription of specific genes (those encoding 5S RNA, pre-tRNA, and viral 5.5S RNA) in response to purified RNA polymerases. Only those RNA polymerases (class III enzymes) known to transcribe these genes in vivo selectively transcribe the same genes in vitro. Moreover, this selective and accurate transcription is seen only when chromatin or nuclear templates are employed and not when intact, defined DNA templates are used. These studies suggest, therefore, that nuclear or chromatin-associated proteins are required for the selective and accurate transcription of these genes in vivo. Thus, these genes may be regulated, at least in part, by template restriction and activation mechanisms analogous to those postulated for other genes (e.g., those encoding globin and ovalbumin). Our studies also emphasize the importance of the homologous RNA polymerases in effecting the selective transcription of these genes in vitro as well as in vivo. The studies further suggest that a purified RNA polymerase and a chromatin template are sufficient for selective transcription of specific genes in vitro. They do not rule out the possibility that these genes are also subject to regulation by other components not active in these fractions. However, the systems described here seem appropriate for investigation of this question and for analysis of the transcriptive functions of specific components and structural modifications within the complex chromatin (see Weintraub and Groudine 1976) and RNA polymerase macromolecular assemblies. It is also expected that similar approaches (the reconstruction of transcription systems from completely homologous components) will be useful for the analysis of more complex gene systems, such as those which produce mRNAs.

CONCLUSIONS

Distinct classes of genes (those encoding the major classes of RNA) are transcribed by three structurally distinct RNA polymerases and may be regulated, in part, through the direct regulation of these enzymes. Altered enzyme concentrations, enzyme structural modifications, and enzyme-specific stimulatory proteins have all been documented and may be directly responsible for physiological modulations of gene activity. The only available evidence suggests that the differential transcription of genes transcribed by a common enzyme is mediated via alterations of chromatin structure, although the involvement of other components has not been excluded.

In considering the possible mechanisms involved in transcriptional control, it is appropriate to consider the extreme structural complexities of the eukaryotic RNA polymerases, which appear considerably more complex than necessary solely for the recognition of specific initiation and termination sites on the DNA. One possible explanation for the enzyme complexity is that the constituent polypeptides are necessary not only for base sequence recognition and nucleotide polymerization but also for a variety of other catalytic functions associated with the transcription of complex nucleoprotein templates. An alternative explanation is that the enzymes play more than a passive role in the regulation of transcription and that many of the diverse polypeptides are involved in regulatory interactions with other cellular components, effecting modes of control in addition to those imposed by chromatin structure alone.

Presently, the diversity of transcriptional regulatory mechanisms in eukaryotes is a matter of conjecture. However, it seems probable that these will be elucidated primarily through an analysis of specific genes in reconstructed transcription systems and that the homologous enzymes will be necessary to achieve a detailed understanding of these controls. The apparent ability of some highly purified class III RNA polymerases to effect a highly selective and accurate transcription of specific genes from nucleoprotein templates is encouraging in this regard. Given these systems and the development of similar systems for other genes, the appropriate dissociation and reconstitution experiments should facilitate meaningful studies of the functions of individual RNA polymerase and chromatin components (and modified forms of them), as well as other potential transcription components.

ACKNOWLEDGMENTS

This work was supported by research grants from the National Science Foundation (BMS 74-24657), the National Institutes of Health (CA-16640 and CA-16217), and the American Cancer Society, Inc. (VC-159). R.G.R. is a Research Career Development Awardee (GM 70661); C.S.P. is a Predoctoral Trainee (GM 01311) of the National Institutes of Health, and J.A.J. is supported by a predoctoral fellowship from Sigma Chemical Company.

REFERENCES

Axel, R., H. Cedar, and G. Felsenfeld. 1973. Synthesis of globin ribonucleic acid from duck-reticulocyte chromatin in vitro. Proc. Natl. Acad. Sci. USA 70:2029–2032.

Bell, G. I., P. Valenzuela, and W. J. Rutter. 1976. Phosphorylation of yeast RNA polymerases. Nature 261:429–431.

Brown, D. D., and K. Sugimoto. 1973. 5S DNA's of *Xenopus laevis* and *Xenopus mulleri:* Evolution of a gene family. J. Mol. Biol. 78:397–415.

Brownlee, G. G., E. M. Cartwright, and D. D. Brown. 1974. Sequence studies of the 5S DNA of *Xenopus laevis.* J. Mol. Biol. 89:703–718.

Carroll, D., and D. D. Brown. 1976. Adjacent repeating units of *Xenopus laevis* 5S DNA can be heterogeneous in length. Cell 7:477–486.

Chambon, P. 1975. Eukaryotic nuclear RNA polymerases. Ann. Rev. Biochem. 43:613–638.

Denis, H. 1974. Nucleic acid synthesis during oogenesis and early embryonic development of the amphibians, *in* Biochemistry of Cell Differentiation, J. Paul, ed., University Park Press, Baltimore, pp. 95–125.

Dezelee, S., F. Wyers, A. Sentenac, and P. Fromageot. 1976. Two forms of RNA polymerase B in yeast. Proteolytic conversion in vitro of enzyme BI into BII. Eur. J. Biochem. 65:543–552.

Ford, D. J., and E. M. Southern. 1973. Different species for 5S RNA in kidney cells and ovaries of *Xenopus laevis.* Nature New Biol. 241:7–12.

Garel, A., and R. Axel. 1976. Selective digestion of transcriptionally active ovalbumin genes from oviduct nuclei. Proc. Natl. Acad. Sci. USA 73:3966–3970.

Gilmour, R. S., and J. Paul. 1973. Tissue specific transcription of the globin gene in isolated chromatin. Proc. Natl. Acad. Sci. USA 70:3440–3442.

Gissinger, F., and F. Chambon. 1975. Subunit SA3 is not required for the activity of calf thymus DNA-dependent RNA polymerase AI. FEBS Lett. 58:53–56.

Greenleaf, A. L., R. Haars, and E. K. F. Bautz. 1976. In vitro proteolysis of a large subunit of *Drosophila melanogaster* RNA polymerase B. FEBS Lett. 71:205–208.

Hager, G., M. Holland, P. Valenzuela, and W. J. Rutter. 1976. RNA polymerases and transcriptive specificity in *Saccharomyces cerevisiae, in* RNA Polymerases, R. Losick and M. Chamberlin, eds. Cold Spring Harbor Laboratory, Cold Spring Harbor, New York, pp. 745–762.

Hirsch, J., and O. J. Martelo. 1976. Phosphorylation of rat liver RNA polymerase I by nuclear protein kinases. J. Biol. Chem. 251:5408–5413.

Huet, J., J. Buhler, A. Sentenac, and P. Fromageot. 1975. Dissociation of two polypeptide chains from yeast RNA polymerase A. Proc. Natl. Acad. Sci. USA 72:3034–3038.

Kedinger, C., F. Gissinger, and P. Chambon. 1974. Animal DNA-dependent RNA polymerases. Molecular structures and immunological properties of calf thymus enzyme A1 and calf thymus and rat liver enzymes B. Eur. J. Biochem. 44:421–436.

Krebs, G., and P. Chambon. 1976. Animal DNA-dependent RNA polymerases. Purification and molecular structure of hen-oviduct and liver class B RNA polymerases. Eur. J. Biochem. 61:15–25.

Marzluff, W. F., Jr., E. C. Murphy, Jr., and R. C. C. Huang. 1974. Transcription of the genes for 5S ribosomal RNA and transfer RNA in isolated mouse myeloma cell nuclei. Biochemistry 13:3689–3696.

Matsui, T., T. Onishi, and M. Muramatsu. 1976. Nucleolar DNA-dependent RNA polymerases from rat liver. 1. Purification and subunit structure. Eur. J. Biochem. 71:351–360.

O'Malley, B. W., S. Y. Tsai, M-J. Tsai, and H. Towle. 1978. Regulation of gene expression in eukaryotes, *in* Cell Differentiation and Neoplasia (The University of Texas System Cancer Center 30th Annual Symposium on Fundamental Cancer Research, 1977), G. F. Saunders, ed., Raven Press, New York, pp. 473–485.

Parker, C. S., and R. G. Roeder. 1977. Selective and accurate transcription of the *Xenopus laevis* 5S RNA genes in isolated chromatin by purified RNA polymerase III. Proc. Natl. Acad. Sci. USA 74:44–48.

Parker, C. S., S. Ng, and R. G. Roeder. 1976. Selective transcription of the 5S RNA genes in isolated chromatin by RNA polymerase III, *in* Molecular Mechanisms in the Control of Gene Expression, D. P. Nierlich, W. J. Rutter, and C. F. Fox, eds., Academic Press, New York, pp. 223–242.

Price, R., and S. Penman. 1972. A distinct RNA polymerase activity, synthesizing 5.5S, 5S and 4S RNA in nuclei from adenovirus 2–infected HeLa cells. J. Mol. Biol. 70:435–470.

Reeder, R. H. 1973. Transcription of chromatin by bacterial RNA polymerase. J. Mol. Biol. 80:229–241.

Roeder, R. G. 1974. Multiple forms of deoxyribonucleic acid-dependent ribonucleic acid polymerase in *Xenopus laevis:* Levels of activity during oocyte and embryonic development. J. Biol. Chem. 248:249–256.

Roeder, R. G. 1976a. Eukaryotic nuclear RNA polymerases, *in* RNA Polymerases, R. Losick and M. Chamberlin, eds., Cold Spring Harbor Laboratory, Cold Spring Harbor, New York, pp. 285–329.

Roeder, R. G. 1976b. Selective transcription of genes by eukaryotic RNA polymerase, *in* Organization and Expression of Chromosomes, V. G. Allfrey, E. K. F. Bautz, B. J. McCarthy, R. T. Schimke, and A. Tissieres, eds., Abakon Verlagsgesellschaft, Berlin, pp. 285–300.

Roeder, R. G., L. B. Schwartz, and V. E. F. Sklar. 1976. Structure, function, and regulation of eukaryotic RNA polymerases, *in* Hormones and Molecular Biology (The 34th Symposium of the Society for Developmental Biology, 1975), J. Papaconstantinou, ed., Academic Press, New York. pp. 29–52.

Schwartz, L. B., and R. G. Roeder. 1974. Purification and subunit structure of deoxyribonucleic acid-dependent ribonucleic acid polymerase I from the mouse myeloma, MOPC 315. J. Biol. Chem. 249:5898–5906.

Schwartz, L. B., and R. G. Roeder. 1975. Purification and subunit structure of deoxyribonucleic acid-dependent ribonucleic acid polymerase II from the mouse plasmacytoma, MOPC 315. J. Biol. Chem. 250:3221–3228.

Schwartz, L. B., V. E. F. Sklar, J. Jaehning, R. Weinmann, and R. G. Roeder. 1974. Isolation and partial characterization of the multiple forms of deoxyribonucleic acid-dependent ribonucleic acid polymerase in mouse myeloma, MOPC 315. J. Biol. Chem. 249:5889–5897.

Sentenac, A., S. Dezelee, F. Iborrea, J. M. Buhler, J. Huet, F. Wyers, A. Ruet, and P. Fromageot. 1976. Yeast RNA polymerases, *in* RNA Polymerases, R. Losick and M. Chamberlin, eds., Cold Spring Harbor Laboratory, Cold Spring Harbor, New York, pp. 763–778.

Sklar, V. E. F., and R. G. Roeder. 1976. Purification and subunit structure of deoxyribonucleic acid-dependent ribonucleic acid polymerase III from the mouse plasmacytoma, MOPC 314. J. Biol. Chem. 251:1064–1073.

Sklar, V. E. F., and R. G. Roeder. 1977. Transcription of specific genes in isolated nuclei by exogenous RNA polymerases. Cell 10:405–414.

Sklar, V. E. F., J. A. Jaehning, L. P. Gage, and R. G. Roeder. 1976. Purification and subunit structure of deoxyribonucleic acid-dependent ribonucleic acid III from the posterior silk gland of *Bombyx mori.* J. Biol. Chem. 251:3794–3800.

Sklar, V. E. F., L. B. Schwartz, and R. G. Roeder. 1975. Distinct molecular structures of nuclear class I, II, and III DNA-dependent RNA polymerases. Proc. Natl. Acad. Sci. USA 72:348–352.

Soderlund, H., U. Pettersson, B. Vennstrom, L. Phillipson, and M. B. Matthews. 1976. A new species of virus-coded low molecular weight RNA from cells infected with adenovirus type 2. Cell 7:585–593.

Tsai, M., H. Towle, S. Harris, and B. O'Malley. 1976. Effect of estrogen on gene expression in the chick oviduct. Comparative aspects of RNA chain initiation in chromatin using homologous versus *Escherichia coli* RNA polymerase. J. Biol. Chem. 251:1960–1968.

Valenzuela, P., G. L. Hager, F. Weinberg, and W. J. Rutter. 1976. Molecular structure of yeast RNA polymerase III. Demonstration of the tripartite transcriptive system in lower eukaryotes. Proc. Natl. Acad. Sci. USA 73:1024–1028.

Weinmann, R., and R. G. Roeder. 1974. Role of DNA-dependent RNA polymerase III in transcription of the tRNA and 5S RNA genes. Proc. Natl. Acad. Sci. USA 71:1790–1794.

Weinmann, R., H. Raskas, and R. G. Roeder. 1974. Role of DNA-dependent RNA polymerases II and III in the transcription of the adenovirus genome in KB cells. Proc. Natl. Acad. Sci. USA 71:3426–3430.

Weinmann, R., J. A. Jaehning, H. J. Raskas, and R. G. Roeder. 1976a. Viral RNA synthesis and levels of DNA-dependent RNA polymerases during replication of adenovirus 2. J. Virol. 17:114–126.

Weinmann, R., T. Brendler, H. J. Raskas, and R. G. Roeder. 1976b. Identification of small viral-coded RNAs transcribed by RNA polymerase III in adenovirus 2 infected cells. Cell 7:557–566.

Weintraub, H., and M. Groudine. 1976. Chromosomal subunits in active genes have an altered conformation. Science 193:848–856.

Cell Differentiation and Neoplasia, edited by Grady F. Saunders. Raven Press, New York

Chromosomal Proteins in Differentiation

Lubomir S. Hnilica, Jen-Fu Chiu, Kenneth Hardy, and Hideo Fujitani

Department of Biochemistry, Vanderbilt University School of Medicine, Nashville, Tennessee 37232

Although all somatic cells in a given animal or plant species contain the same genetic information in the form of DNA, individual cells within the same organism may differ in the expression of their genome. During cellular differentiation, large segments of DNA are transcriptionally inactivated, and only genetically active parts of the differentiated genome can be expressed. The process of differentiation is assumed to proceed by selective transcriptional inactivation of DNA in differentiating cells. The exact mechanisms of this gene selection are not known. It is believed that certain macromolecular components of chromatin and chromosomes (especially the proteins) play an important role in the biochemistry of differentiation, organogenesis and genetic regulation.

There are two main categories of chromosomal proteins. The structurally simple and relatively small histones are basic proteins found essentially in the chromatin of all eukaryotes. Most higher organisms contain only five principal kinds of histones, which exhibit limited tissue and species specificity (Hnilica 1972, Elgin and Weintraub 1975). Although the histones appear to play important roles in the structure and function of chromatin, they do not possess sufficient heterogeneity and specificity, which would be expected from gene regulatory macromolecules.

The other principle group of chromosomal proteins is categorized only under "umbrella" names such as chromosomal nonhistone proteins, acidic nuclear proteins, etc. Because of their excessive heterogeneity, poor solubility, and the lack of suitable specific assay procedures, the chromosomal nonhistone proteins are much less biochemically characterized than histones. Indeed, only polyacrylamide gel electrophoresis in the presence of sodium dodecyl sulfate permits the realization of the extreme heterogeneity of this protein group.

Since it is reasonable to expect that the macromolecules that operate and maintain the differentiated state of cells will reflect the specificity of this phenomenon, the past several years witnessed an intensive effort to correlate changes in chromosomal nonhistone proteins with transcriptional activities of selected cellular systems. In search for their biological significance, the tissue specificity and quantitative or qualitative effects on gene transcription of chromosomal nonhistone proteins were studied by many investigators. Presently, enzymatic phosphorylation of chromosomal nonhistone proteins, their effects on the transcription of specific genes, DNA-binding properties, and immunological tissue specificity are the principal "handles" lending themselves to biochemical and biological investigations on this interesting group of proteins. We will discuss here a select group of chromosomal nonhistone proteins that bind to homologous DNA and exhibit exceptional tissue specificity, detectable immunologically.

NUCLEAR PROTEINS AS ANTIGENS

Immunological techniques are perhaps the most powerful tools available for probing the heterogeneity, diversity, and specificity of macromolecular components present in differentiated cells. Histones of higher animals are relatively poor antigens, and immunological studies show that they are not tissue or species specific (Sandberg *et al.* 1967, Stollar and Ward 1970). These findings are in a very good agreement with comparative studies on amino acid sequences of histones isolated from various sources. The evolutionary conservation of the amino acid sequences of histones, and especially of the arginine-rich fractions, is truly amazing (Hnilica 1972, Elgin and Weintraub 1975). The very lysine-rich histones, H_1, present a notable exception. They are heterogeneous, and the H_1 histone subfractions were shown to be tissue and species specific in respect to their amino acid sequences and immunological properties (Bustin and Cole 1968, Bustin and Stollar 1972, 1973).

The poor solubility and extensive heterogeneity of chromosomal nonhistone proteins apparently discouraged serious studies on their antigenic specificity. Early reports on immunological tissue specificity of nuclear proteins and nucleohistones (Henning 1962, Messineo 1961) did not stimulate much interest, and only recently are nuclear antigens receiving due attention.

Chytil and Spelsberg (1971) showed that dehistonized chromatin can elicit the formation of tissue-specific antibodies when injected into rabbits. Their results were quickly confirmed and extended (Wakabayashi and Hnilica 1972, 1973, Zardi *et al.* 1973). It was shown that the specific antisera react well with dehistonized or native chromatin and that the exceptional tissue specificity of the nuclear protein antigens depends on their association with homologous DNA. The antigenicity of chromosomal nonhistone proteins in *Drosophila melanogaster* was demonstrated by Silver and Elgin (1976) who studied the in situ distribution of nuclear antigens on polytene chromosomes.

Other nuclear antigens were reported to result from oncogenic virus-induced transformation such as the SV40 T-antigen (Todaro *et al.* 1965), the Epstein-Barr virus-associated antigens (Suzuki and Hinuma 1974), or appear in tumors of nonvirus etiology (Messineo 1961, Morris *et al.* 1975, Perez-Cuadrado *et al.* 1965, Gaffar *et al.* 1967, Carlo *et al.* 1970). Some of these antigens may be related to growth and proliferation (Chytil *et al.* 1974, Chiu *et al.* 1974, 1976a); others have been found in embryonic tissues as well as tumors (Chiu *et al.* 1975, 1976a, Yeoman *et al.* 1976). The relationships between all these antigens are not known at present. It can be speculated that many of these antigens represent products of genes activated by the process of carcinogenesis.

In our experiments, chromatins from rat liver, calf thymus, and Novikoff hepatoma were isolated, dehistonized, and used as antigens to produce antibodies in New Zealand rabbits (Wakabayashi *et al.* 1974, Chytil and Spelsberg 1971). The antigenicity of the individual preparations and isolated chromatins was

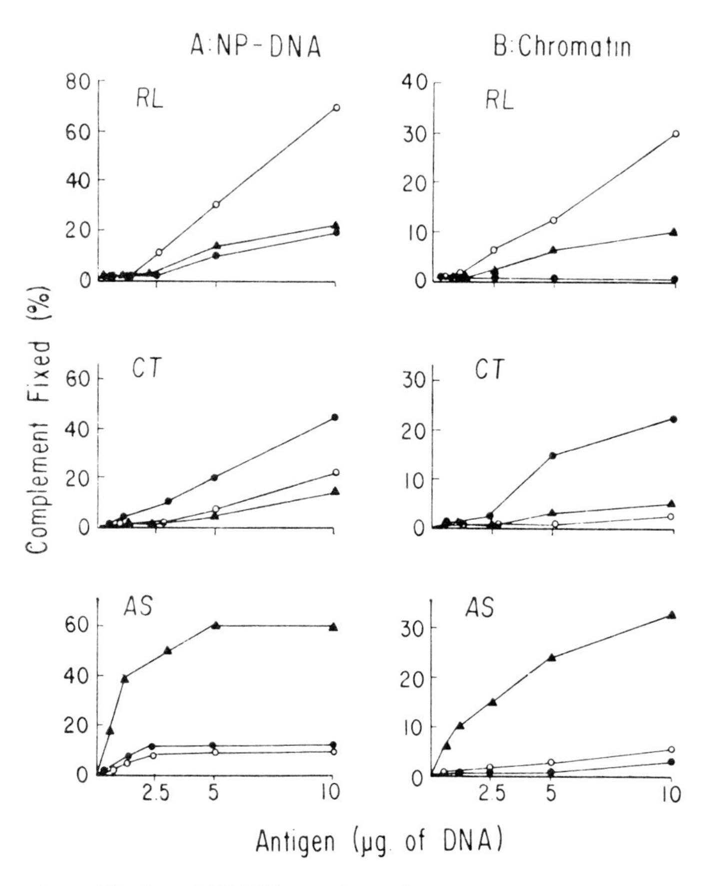

FIG. 1. Complement fixation of NP-DNA complexes (A) and chromatin (B) from rat liver, calf thymus, and Novikoff hepatoma in the presence of antiserum against the individual NP-DNA complexes. The reaction mixtures, each containing, in a total volume of 0.8 ml, various amounts of NP-DNA complexes or chromatin, antiserum (0.1 ml of 200 X diluted rabbit antiserum), and complement (0.2 ml of 50 X diluted guinea pig serum), were incubated at 4°C overnight. Then 0.2 ml of activated sheep erythrocytes was added, and the mixture was incubated at 37°C for 20 minutes. The extent of hemolysis was determined by measuring the absorbancy at 413 nm in a spectrophotometer. All experimental points were corrected for anticomplementarity: RL, CT, and AS, antisera against NP-DNA pellets from rat liver, calf thymus, and Novikoff hepatoma, respectively. The NP-DNA complexes (A) or chromatin (B) were prepared from rat liver (○), calf thymus (●), and Novikoff hepatoma (▲). The NP-DNA pellets were prepared according to the schedule in Figure 14. (Reprinted, with permission of Biochemistry, from Wakabayashi *et al.* 1974.)

compared by the microcomplement fixation method of Wasserman and Levine (1961). The individual antisera were purified by chromatography on DEAE cellulose (Rapp 1964). In agreement with Chytil and Spelsberg (1971), each antiserum fixed the complement in the presence of its corresponding antigen (chromatin) but not in the presence of chromatin from another tissue (Figure 1). The differences between the individual tissues were reproducible in other species, indicating that certain chromosomal nonhistone proteins may correlate directly to the process of differentiation and organogenesis. Perhaps most interesting was the observation that the immunological specificity of chromosomal nonhistone proteins of Novikoff hepatoma differed significantly from that of normal rat liver.

NUCLEAR ANTIGENS IN DIFFERENTIATION AND EXPERIMENTAL NEOPLASIA

The programmed selective expression of individual genes during cellular differentiation is perhaps the most fascinating aspect of higher organisms. Mechanisms similar to those responsible for cell differentiation may also operate the restriction and activation of genes in response to dietetic, hormonal, environmental, and perhaps carcinogenic stimuli. Because of the mounting experimental evidence that chromosomal nonhistone proteins play an important role in genetic regulations, their specificity in differentiating tissues was investigated. Using antisera

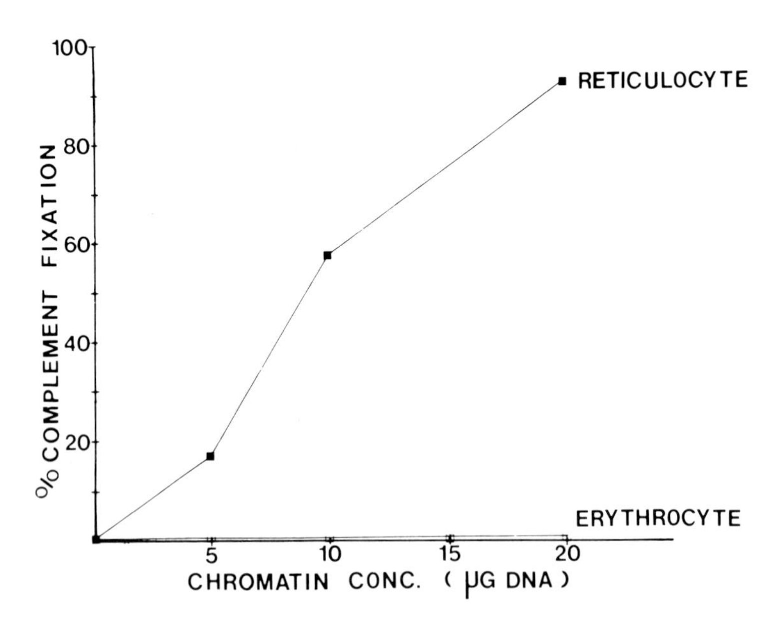

FIG. 2. Complement fixation of chicken reticulocyte (■——■) and erythrocyte (□——□) chromatin in the presence of antiserum against dehistonized reticulocyte chromatin. All experimental values were corrected for anticomplementarity.

obtained against dehistonized chromatin from adult rat liver, Chytil *et al.* (1974) showed gradual changes in the antigenicity of chromosomal nonhistone proteins in developing embryonic and postpartum livers of rats. Similar changes were seen in the immunoreactivity of oviduct chromatins of chicks stimulated with diethylstilbestrol, as compared with the unstimulated, seven-day-old controls (Spelsberg *et al.* 1973). In both instances, the immunological specificity of the dehistonized chromatin changed with differentiation. This indicates either qualitative or structural changes in chromatin of the differentiating tissues.

To investigate the contribution of chromatin structure to its antigenicity, we have initiated experiments with maturing reticulocytes. Adult male Leghorn chickens (DeKalb strain) were made anemic by daily injections of 1% neutralized phenyl hydrazine (10 mg/kg). The polychromatic primitive erythrocyte (reticulocyte) count was determined daily, and after reaching 60 to 70%, daily bleeding (12 ml/day) was administered to each animal until the reticulocytes represented 95 to 97% of the total red cell population. Mature erythrocytes were obtained from untreated animals. Chromatin was isolated, dehistonized, and used in studies on immunological specificity. As is shown in Figure 2, antibodies against dehistonized reticulocyte chromatin reacted with reticulocyte but not erythrocyte chromatins.

Because of the progressive nuclear condensation (heterochromatization) in maturing erythrocytes, both reticulocyte and erythrocyte chromatin was fractionated by limited shearing, centrifugation, and precipitation with a mixture of divalent cations (Hardy *et al.*, manuscript in preparation). This procedure, outlined in Figure 3, yields three principal fractions of chromatin which differ in their capacity to transcribe globin messenger-like RNA (Hardy *et al.*, in preparation). A small, transcriptionally active fraction CAS, which represented less than 1% of the total nuclear DNA, contained essentially all the reticulocyte-specific antigens (Figure 4). A similar fraction isolated by shearing erythrocyte

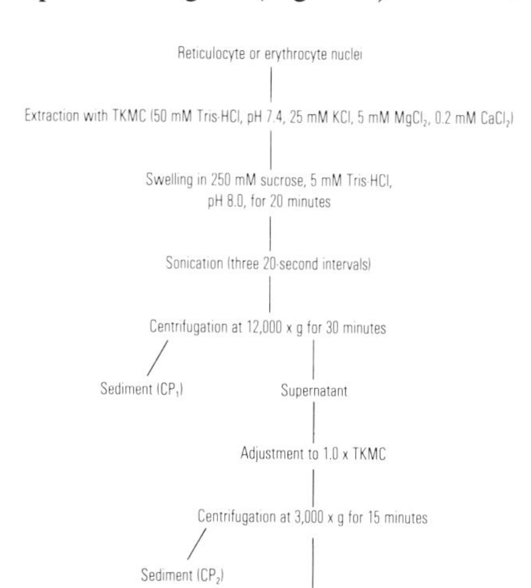

FIG. 3. Flow diagram of procedure employed to fractionate reticulocyte and erythrocyte nuclei.

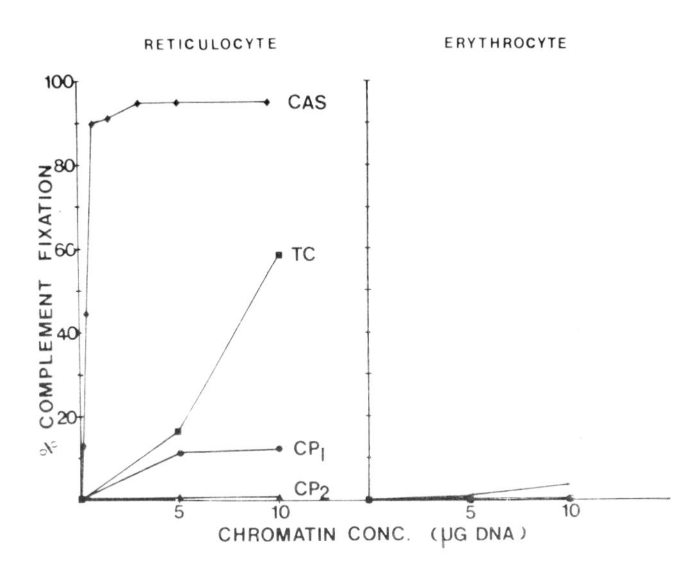

FIG. 4. Complement fixation of chicken reticulocyte and erythrocyte chromatin fractions isolated according to the schedule in Figure 3. The assays were performed in the presence of antiserum against dehistonized chicken reticulocyte chromatin. The individual fractions are identified by their symbols. TC = unfractionated (total) chromatin. CP_1, CP_2, and CAS are fractions identified in Figure 3. All experimental points were corrected for anticomplementarity.

nuclei did not exhibit any affinity for reticulocyte-specific antiserum. This indicates that the immunological tissue specificity of the nuclear antigens probably reflects qualitative changes in chromosomal nonhistone proteins rather than altered conformation of chromatin. It is of considerable interest that the small fraction of reticulocyte chromatin (CAS) which contains almost all the reticulocyte-specific antigens also transcribed the globin gene much more efficiently than unfractionated chromatin or the fractions CP_1 and CP_2 (Hardy *et al.*, in preparation). Digestion with various enzymes showed that the immunological specificity of reticulocyte chromatin depends on the presence of both proteins and DNA (Figure 5).

Neoplastic growth frequently reverts some of the phenotypical characteristics of the affected cells to those similar in embryonic and differentiating tissues (Criss 1971, Knox 1967, Pitot 1966, Walker and Potter 1972, Weber and Lea 1966, Weinhouse *et al.* 1972). This suggests the presence of abnormal gene expression, i.e., activation of some embryonic genes by the process of carcinogenesis. It can be anticipated that such transcriptional modifications are accompanied by parallel changes in macromolecules that comprise and regulate the genetic apparatus. Indeed, specific nuclear antigens were recently reported to be present in both embryonic and malignant tissues (Chiu *et al.* 1975, 1976a, Yeoman *et al.* 1976).

Assuming that recognition and characterization of tissue-specific antigens that may be characteristic for the process of malignant transformation could contrib-

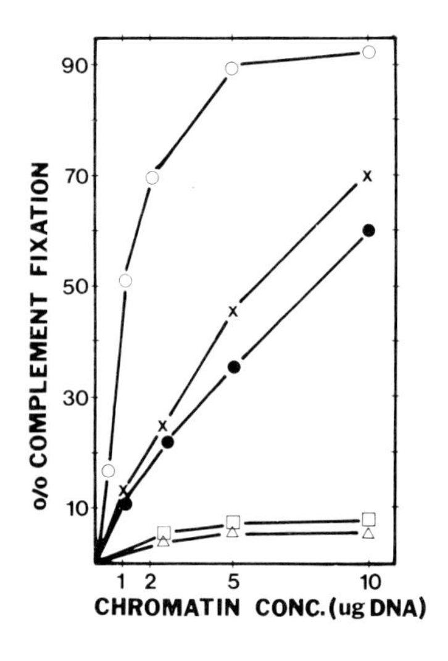

FIG. 5. Effects of enzymatic digestions on the complement fixation of reticulocyte chromatin active fraction CAS in the presence of antiserum against dehistonized chicken reticulocyte chromatins. Control (undigested); ○——○, RNase digested; ●——●, DNase I digested; X——X, DNase II digested; □——□, and pronase digested; △——△ reticulocyte CAS fraction. Individual CAS samples were incubated with 50 μg/ml of RNase (Worthington), 50 μg/ml DNase (RNase-free, Worthington), or 100 μg/ml Pronase (Calbiochem) in appropriate buffers for 60 minutes at 37°C.

ute to the diagnosis, treatment, and perhaps prevention of cancer, we have investigated the immunogenic properties of nuclear nonhistone proteins and chromatin in cancer cells. Following the procedure of Chytil and Spelsberg (1971), we have produced rabbit antisera against normal rat liver and Novikoff hepatoma chromatins. As can be seen in Figure 1, each antiserum reacted with its own antigen, while the reactivity of Novikoff hepatoma chromatin with rat liver antiserum and vice versa was only marginal.

Encouraged by these results, we wanted to determine possible differences between individual transplantable tumors. Chromatins of several transplantable rat tumors were assayed with the antiserum against Novikoff hepatoma dehistonized chromatin. The results are presented in Figure 6. There was essentially no tissue specificity between the Novikoff hepatoma, 30 D ascites hepatoma, or Walker carcinosarcoma. On the other hand, these three tumor chromatins exhibited only marginal reactivity with the antiserum against rat liver dehistonized chromatin. It appears as if the neoplastic process changed the specificity of the original tissue into an immunochemically new type, common to at least all the three tumors compared in Figure 6.

Novikoff hepatoma is a very fast-growing ascites tumor. Although slower by comparison, the growth rates of 30 D ascites hepatomas and Walker tumor are also fast. Consequently, the antigenic similarity of these tumors and their summary difference from a slow-growing, well-differentiated rat liver may be associated with cellular proliferation rather than with the cancerous phenotype. To test this possibility, we compared the immunoreactivity of chromatins isolated from various Morris hepatomas (Figure 7). The ability of these chromatins to fix complement in the presence of Novikoff hepatoma antiserum increased with the growth rates of individual tumors. The poorly differentiated, fast-growing 7777 and 3924 A hepatomas were more immunoreactive than the better differentiated and slow-growing 7800 and 7787 tumors. However, the differences in

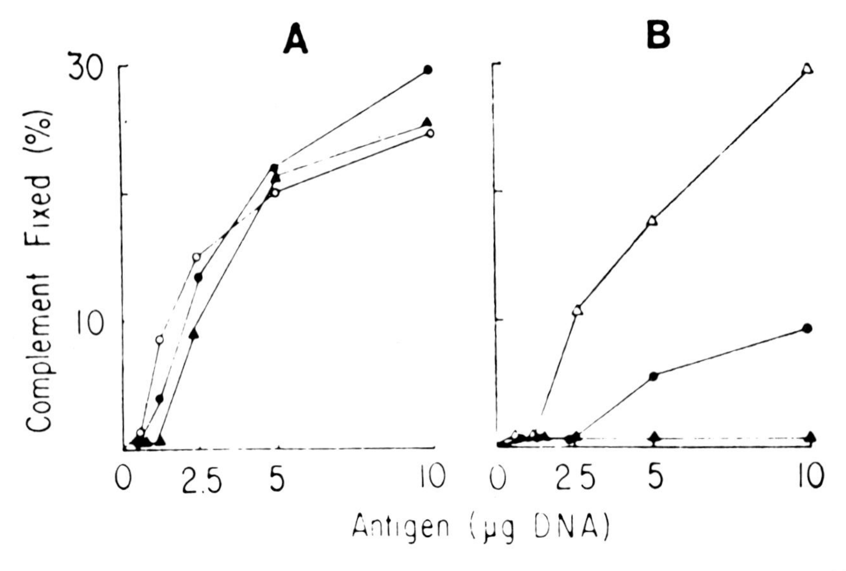

FIG. 6. Complement fixation of chromatin from various tumors. The assay was performed in the presence of antisera against (A) Novikoff hepatoma and (B) rat-liver NP-DNA complexes:rat liver chromatin (△); 30D ascites hepatoma chromatin (●); Walker carcinosarcoma chromatin (▲); Novikoff hepatoma chromatin (○). (Reprinted, with permission of Biochemistry, from Wakabayashi *et al.* 1974.)

complement fixation were not sufficient to accommodate the idea that the immunological tissue specificity of the tumors is caused by a growth-associated antigen.

This conclusion is further supported by our experiments with Fisher rats maintained on a diet containing a hepatocarcinogen, N, N-dimethyl-p-(m-tolylazo) aniline. As soon as 15 days after initiation of this diet, the immunological

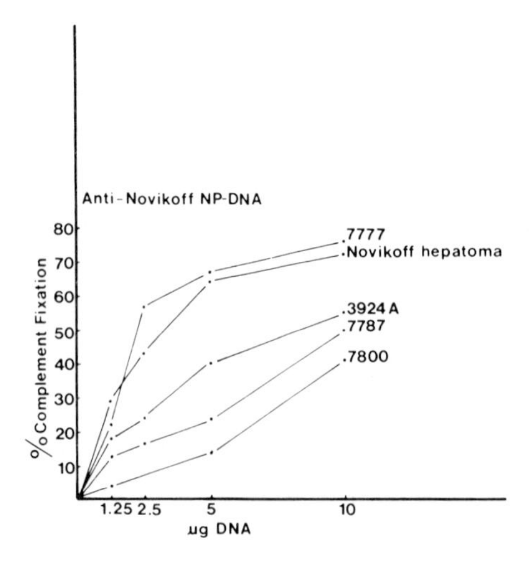

FIG. 7. Complement fixation of chromatin nonhistone protein-DNA complexes from Novikoff hepatoma and Morris hepatomas 7777, 7787, 7800, and 3924A in the presence of antiserum against Novikoff hepatoma NP-DNA. All experimental points were corrected for anticomplementarity. (Reprinted with permission of FEBS Lett., from Chiu *et al.* 1974.)

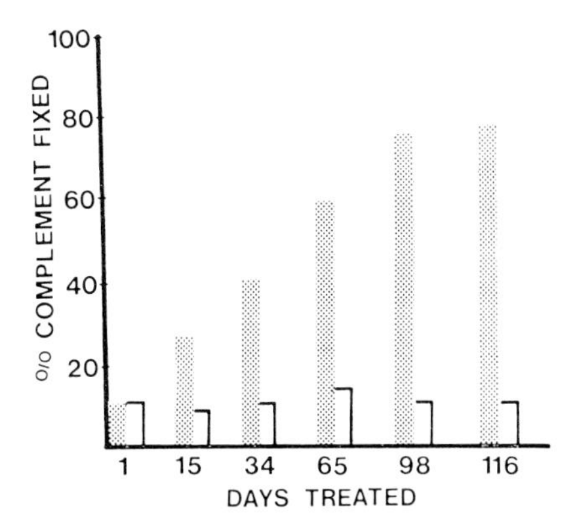

FIG. 8. Time course of increasing complement fixation of chromatins isolated from liver of rats maintained on 3′-MDAB-containing diet. The complement fixation assay was performed in the presence of antisera against Novikoff hepatoma dehistonized chromatin. ▭, rats maintained on 3′MDAB; ▭, rats maintained on α-NIT. (Reprinted, with permission of Academic Press, from Chiu *et al.* 1976a.)

specificity of the chromosomal nonhistone proteins in chromatin began to change from the type characteristic of normal tissue (liver) to a new type (Figure 8), common to the three experimental tumors shown in Figure 6. This change in immunological tissue specificity paralleled, to some extent, the increase of the α_1-fetoprotein in sera of the experimental animals. Liver chromatins of control animals or rats fed a diet containing α-naphtylisothiocyanate were also compared. This compound produces bile duct cell proliferation similar to that caused by the carcinogen, but it is not carcinogenic. As can be seen in Figure 8, the proliferation of bile duct cells did not result in any significant immunological changes of the isolated chromatins.

The animal tumors assayed in Figures 1, 6, 7, and 8 were produced by the administration of carcinogens and maintained through numerous transfers in normal animals. It was therefore desirable to compare the immunogenicity of spontaneous neoplasias, preferably of human origin. We have produced antibodies in rabbits against dehistonized chromatin preparations from either human lung or breast carcinomas. As is illustrated in Figures 9 and 10, these antisera are highly specific, in that they fix the complement only in the presence of chromatin isolated from the homologous tissue, i.e., antiserum against human breast carcinoma fixed the complement extensively in the presence of breast carcinoma chromatin, while the reaction was negative or marginal in the presence of chromatins from breast benign tumor, normal human lung, breast, or placenta (Figure 9). This exceptionally selective tissue recognition was also seen where antiserum against human lung carcinoma, dehistonized chromatin was assayed in the presence of lung carcinoma chromatin (extensive complement fixation) or other human tissues, including normal lung, breast, and human placenta (all negative). Complement fixation of chromatins from various tumors of two different animal species (human breast and lung carcinomas, HeLa cells, and rat Novikoff hepatoma) in the presence of antiserum against human lung carcinoma is shown in Figure 10. Again, strict tissue specificity is observed. This observation is somewhat surprising because our experiments with transplantable animal tumors could not demonstrate any significant differences in their reactivity

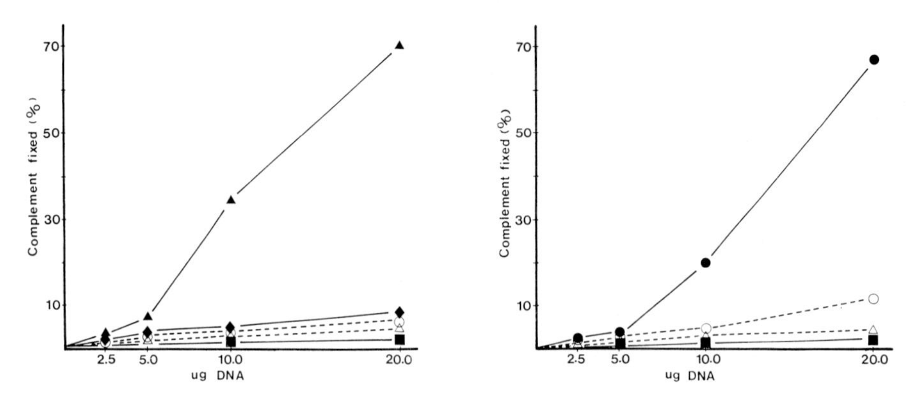

FIG. 9, left. Complement fixation of chromatins in the presence of antiserum against human breast carcinoma dehistonized chromatin. Human breast carcinoma, ▲——▲; human normal breast, △——△; benign tumor of human breast, ◆——◆; normal human lung, ○——○; and human placenta, ■——■.

FIG. 10, right. Complement fixation of chromatins in the presence of antiserum against human lung carcinoma dehistonized chromatin. Human lung carcinoma, ●——●; normal human lung, ○——○; normal human breast, breast carcinoma, placenta, and HeLa cells, △——△; rat Novikoff hepatoma, ■——■.

with serum against dehistonized Novikoff hepatoma chromatin (see Figures 6 and 7).

In an attempt to explain this difference, chromatins isolated from human lung and breast carcinomas were assayed in the presence of antiserum against Novikoff hepatoma dehistonized chromatin. Both human tumors fixed complement extensively (Figure 11), despite the species heterology. However, substantial tissue specificity was retained in this system since Novikoff hepatoma antiserum did not react with chromatins isolated from normal lung and breast tissues (Figure 11). The antiserum against Novikoff hepatoma, dehistonized chromatin also reacted with fetal rat liver chromatin (Figure 12). The reactivity rapidly diminished after birth, and the liver chromatin of three-week-old rats was essentially nonreactive (Figure 12). These experiments may indicate the presence of several different antigens present in chromatin of higher animals, some associated with the process of differentiation and others with growth or malignancy.

To ascertain that the tissue-specific antibodies are directed against the nuclear material and not against some cytoplasmic components which could have become absorbed to chromatin during its isolation, we have attempted intracellular localization of these antigens. Horseradish peroxidase-labeled antibody method (Chytil 1975, Nakane 1970) was used to localize the antigens of normal rat liver. As shown in Figure 13, the antibodies localized selectively in the nuclei and not in the cytoplasm.

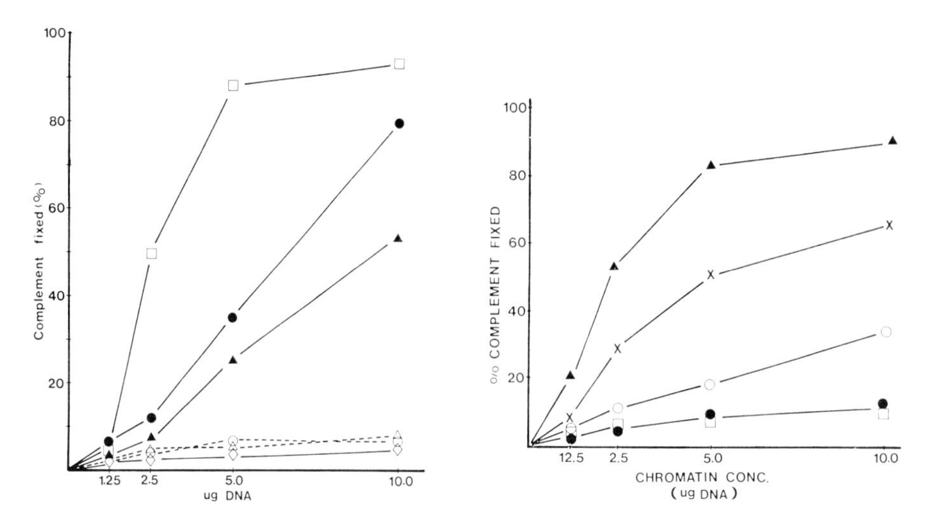

FIG. 11, left. Complement fixation of chromatins in the presence of antiserum against rat Novikoff hepatoma dehistonized chromatin. Novikoff hepatoma, □——□; human lung carcinoma, ●——●; human breast carcinoma, ▲——▲; normal human breast, △——△; normal human lung, ○——○; and normal rat liver, ◇——◇.

FIG. 12, right. Complement fixation of chromatins isolated from developing rat liver in the presence of antisera against Novikoff hepatoma dehistonized chromatin. Fetal rat liver, X——X; one-week-old rat liver, ○——○; three-week-old rat liver, □——□; normal adult rat liver, ●——●; and Novikoff hepatoma, ▲——▲. (Reprinted, with permission of Academic Press, from Chiu *et al.* 1976a.)

CHARACTERIZATION OF NUCLEAR ANTIGENS

The exceptional tissue specificity of the nuclear antigens points to their involvement in cellular differentiation. It is, therefore, very important to characterize these antigens and establish a biochemical base for studies on their biological significance. As is illustrated in Figure 5, the antigenicity of chicken reticulocyte chromatin was completely abolished by digestion with pronase or DNase. This indicates the involvement of specific interactions between DNA and nuclear proteins. Similar tissue-specific immunological inactivation of chromatin by DNase was also observed by Wakabayashi and Hnilica (1973) and Wakabayashi *et al.* (1974) who report studies in which dehistonized or native Novikoff or rat liver chromatin were digested with DNase I. In all instances, the digestion resulted in a complete loss of immunological tissue specificity. However, the nontissue-specific antigenicity of chromatin was still retained after this treatment.

Isolated chromatin can be dissociated into its components in solutions containing high concentrations of salt and urea. In 5.0 M urea and 2.5 M NaCl at pH 8, the dissociation is almost complete, and chromosomal DNA can be sepa-

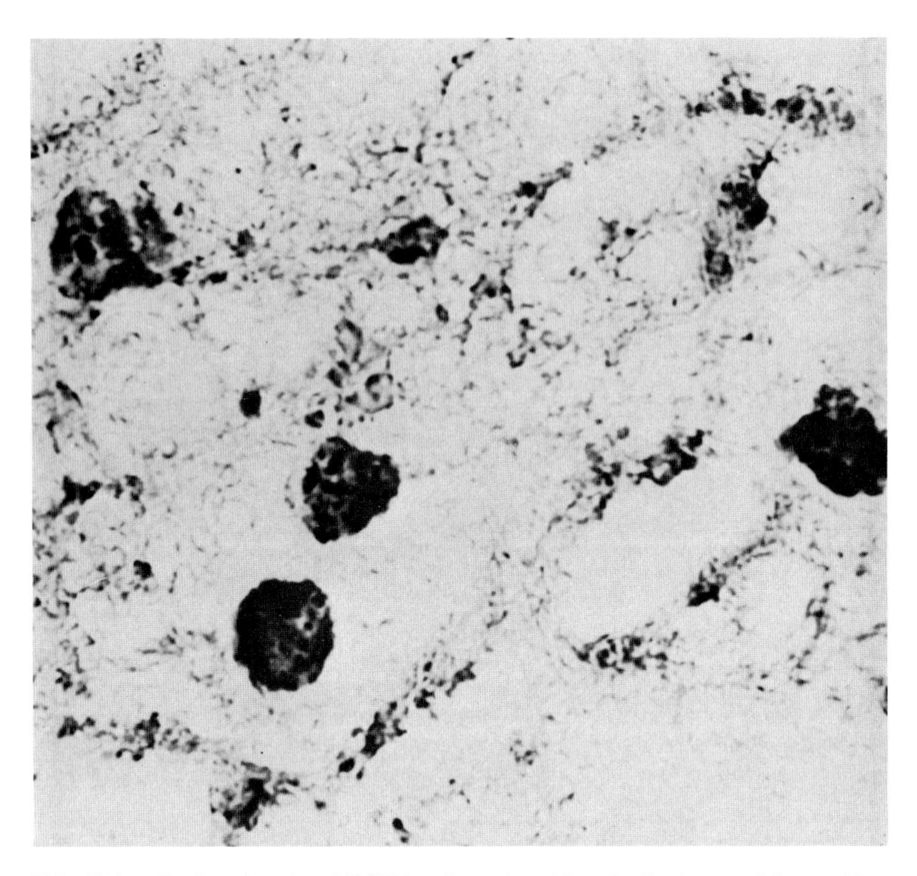

FIG. 13. Localization of nuclear NP-DNA antigens in rat liver by the horseradish peroxidase bridge technique in the presence of rabbit serum against dehistonized rat liver chromatin. (Reprinted, with permission of Academic Press, from *Chiu et al.* 1976a.)

rated from associated proteins by prolonged ultracentrifugation. Individual components of chromatins from various tissues can be brought together and reconstituted by a slow removal of first NaCl and then urea (Paul and Gilmour 1968, Bekhor *et al.* 1969, Spelsberg *et al.* 1971). Dissociation of isolated chromatin in the described manner and its reconstitution did not significantly affect its immunospecificity (Chiu *et al.* 1975). Removal of the histones by dissociation and centrifugation at pH 6.0 increased the specific immunoreactivity of chromatin (Chytil and Spelsberg 1971). When the dehistonized chromatin was dissociated and centrifuged at pH 8.0, the resulting nonhistone proteins (supernatant) and DNA (pellet) were not immunospecific by complement fixation. However, reconstitution of the DNA (pellet) with nonhistone proteins (supernatant) resulted in a full immunological activity and specificity of the complexes (Wakabayashi and Hnilica 1973). These experiments supplement and confirm the results

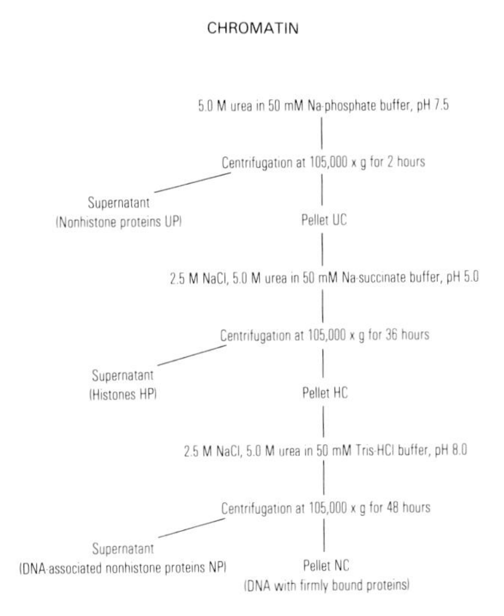

FIG. 14. Scheme for fractionation of chromosomal proteins.

obtained by protease and DNase digestion (see Figure 5). Obviously, a specific interaction between the DNA and some nonhistone protein(s) is necessary for the observed immunological tissue specificity of chromatin.

Based on the observation that the immunological tissue specificity of chromatin depends on selective associations between DNA and chromosomal nonhistone proteins, the fractionation schedule illustrated in Figure 14 was developed to seek a fraction enriched in this antigenic material. As can be seen in Figure 15, essentially all the immunological activity is associated with the last protein fraction, NP. Indeed, when this fraction from Novikoff hepatoma was reconstituted with DNA purified from liver chromatin or vice versa, the immunological tissue specificity followed the donor tissue of the proteins NP (Figure 16).

Protein and DNA recoveries from several fractionation experiments of chromosomal proteins are presented in Table 1. The immunologically tissue-specific proteins NP represent less than 5% of the total chromatin protein content in most tissues. The NP proteins are slightly acidic (Table 2), and polyacrylamide gel electrophoresis in the presence of sodium dodecyl sulfate revealed their heterogeneity. As can be seen in Figure 17, in Novikoff hepatoma NP there are three principal low molecular weight polypeptides and several relatively minor high molecular weight proteins.

To identify the tumor-specific protein fraction NP with the previously reported immunogenic DNA-binding proteins of chromatin (Wakabayashi *et al.* 1974), reconstituted mixtures of DNA and NP were sedimented through sucrose density gradients. Radioactive in vitro labeling of the NP proteins with ^{125}I facilitated the detection and quantitative evaluation of the interaction experiments. The distribution of DNA on the gradient was determined by its absorbance at 260 nm. Under these experimental conditions, free protein remained on top of the gradient, followed by the protein-DNA complexes. As can be seen in Figure

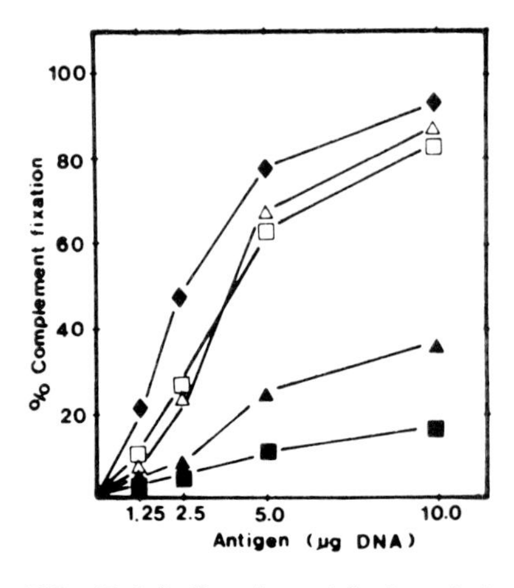

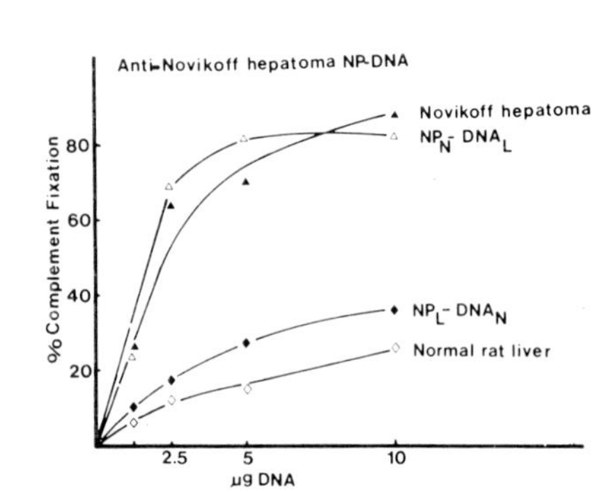

FIG. 15, left. Complement fixation of chromatin preparations performed in the presence of antiserum against dehistonized Novikoff hepatoma chromatin. All data were corrected for anticomplementarity. ♦——♦, Novikoff hepatoma native chromatin; △——△, Novikoff hepatoma UC pellet (i.e., chromatin devoid of UP proteins); □——□, Novikoff hepatoma HC pellet (i.e., chromatin devoid of UP and HP proteins); ▲——▲, Novikoff hepatoma NC pellet (i.e., chromatin devoid of UP, HP, and NP proteins); ■——■, normal rat liver chromatin. (Reprinted, with permission of J. Biol. Chem., from Wang *et al.* 1976.)

FIG. 16, right. Complement fixation of normal and reconstituted NP-DNA complexes from rat liver and Novikoff hepatoma in the presence of antiserum against Novikoff hepatoma NP-DNA. All experimental points were corrected for anticomplementarity. ▲——▲, Novikoff hepatoma chromatin (native); △——△, reconstituted complex of Novikoff hepatoma NP and normal rat liver DNA (NP_N–DNA_L); ◇——◇, normal rat liver chromatin (native); ♦——♦, reconstituted complex of rat liver NP and Novikoff hepatoma DNA (NP_L–DNA_N). (Reprinted, with permission of FEBS Lett., from Chiu *et al.* 1974.)

18A, approximately 1.5% (w/w) of the NP fraction from azo dye-produced hepatoma associated with purified rat spleen DNA under saturating conditions. Essentially identical ratios were observed for rat liver or Novikoff hepatoma DNA. Conversely, as shown in Figure 18B, there was only minimal interaction between the NP proteins from the azo dye-produced hepatoma and calf thymus

TABLE 1. *Distribution of DNA and protein in fractions resulting from the scheme shown in Figure 14*

Fraction	% of total DNA	% of total protein
Chromatin	100	100
UP	2–3	44–47
HP	1–2	50–52
NP	1–2	2–3
NC	93–96	3–5

The values are averages of several preparations of rat liver chromatin with DNA/protein ratios 1.0 : 1.6–1.9. (Data are from Chiu *et al.* 1975.)

TABLE 2. *Amino acid composition of the protein fractions obtained according to the schedule shown in Figure 14*

	Fraction		
Amino acid	UP	HP	NP
Lysine	5.5	14.1	7.8
Histidine	2.4	1.4	1.9
Arginine	5.8	7.4	8.0
Aspartic acid	9.9	4.9	7.5
Threonine	3.8	5.0	4.7
Serine	8.5	6.0	6.8
Glutamic acid	14.2	9.2	12.3
Proline	6.8	5.6	5.0
Glycine	12.5	9.5	10.7
Alanine	7.4	15.3	8.8
Valine	4.8	5.0	5.6
Methionine	1.8	0.6	1.4
Isoleucine	4.0	4.2	4.2
Leucine	7.6	8.0	8.8
Tyrosine	2.6	2.1	3.7
Phenylalanine	2.3	1.7	2.8
Acidic/basic	1.7	0.6	1.1

The amino acid concentration is expressed as mole percent of all amino acids recovered. Tryptophan was not determined. All serine values are corrected (10%) for hydrolytic losses. The figures are averages of three to four determinations. The amino acid analyses were kindly performed by Dr. D. N. Ward, Department of Biochemistry, The University of Texas System Cancer Center M. D. Anderson Hospital and Tumor Institute. (Data are from Chiu *et al.* 1975.)

DNA. Numerical evaluations of similar experiments with chicken erythrocyte and *E. coli* DNA in addition to the rat or calf DNA are presented in Table 3. The interactions are clearly species specific.

Because of the electrophoretic heterogeneity of NP proteins, we have attempted their further fractionation. Gel filtration on Sephadex G-100 separated the NP proteins into two peaks (Figure 19). Electrophoretically, the three major, low molecular weight polypeptides were contained in the second peak while the first peak consisted of the high molecular weight material (Figure 20). Essentially all the immunological activity was associated with the high molecular weight material (Figure 21).

The high molecular weight nature of the specific antigens was confirmed by preparative polyacrylamide gel electrophoresis. The NP proteins were separated on gel slabs, and prominent groups of bands were cut according to segments stained with Coomassie Brilliant Blue. The protein bands were eluted, their associated sodium dodecyl sulfate removed, and the recovered proteins were assayed after reconstitution with homologous DNA. Results of these experiments, shown in Figure 22, identify the antigenic material with a gel segment containing three protein bands of molecular weight between 50–60,000.

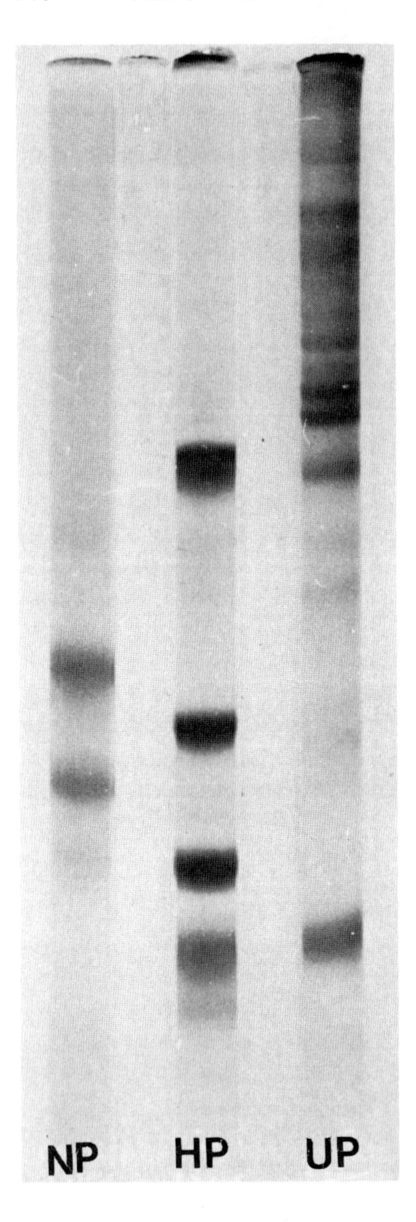

FIG. 17. Polyacrylamide gel electrophoresis of the UP, HP, and NP fractions from rat liver chromatin. The electrophoresis was formed in the presence of sodium dodecyl sulfate. The origin of migration is at the top of the gels.

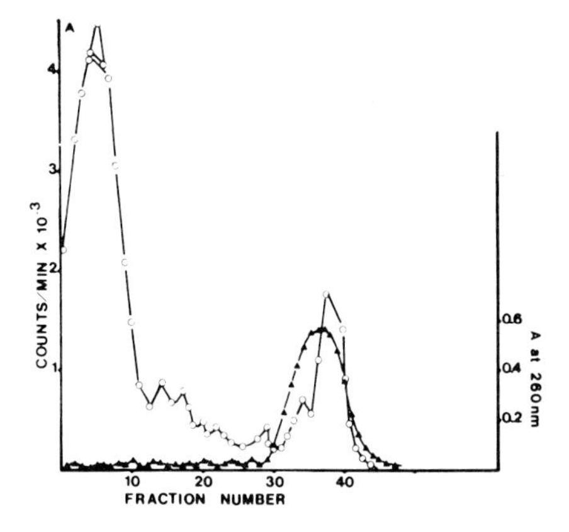

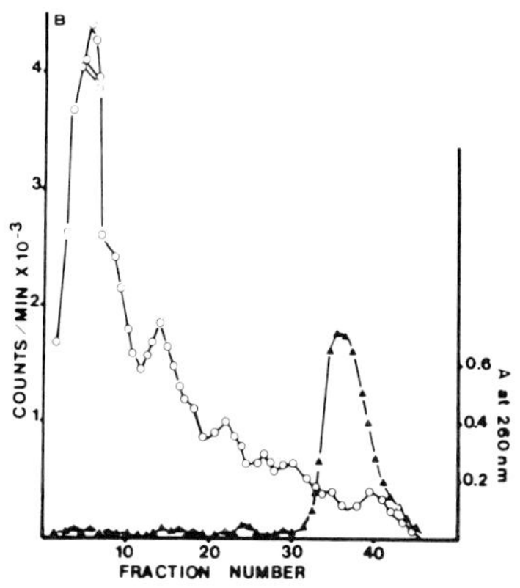

FIG. 18. Sucrose density gradient centrifugation of (A) rat spleen DNA and (B) calf thymus DNA reconstituted with NP proteins isolated from azo dye-produced hepatoma. Radioactivity of ^{125}I-labeled NP protein, ○——○; absorbance at 260 nm (DNA), ▲——▲. (Reprinted, with permission of Cancer Res., from Chiu *et al.* 1975.)

TABLE 3. *Interactions of rat liver NP fraction proteins with homologous native or fractionated and with heterologous DNA*

Source of DNA	DNA	Protein applied (μg)	Protein bound	Protein/DNA binding (ratio)
Rat spleen	400	40	5.8	0.0145
Rat liver	400	40	5.6	0.0141
Calf thymus	400	40	1.0	0.0025
Chicken erythrocyte	400	40	0.3	0.0008
Escherichia coli	400	40	0.1	0.0003
Rat spleen				
Single-stranded unique sequences	200	20	1.8	0.0089
Double-stranded unique sequences	200	20	3.7	0.0185

The formation of DNA-protein complexes was assayed by sucrose density gradient centrifugation or by nitrocellulose filter assay. Labeled (^{125}I) NP protein was used to determine the binding ratios. The binding ratios represent weight percentages of protein retained by the DNA. Rat spleen DNA was fractionated after shearing by hydroxylapatite chromatography. (Data are from Wang *et al.* 1976.)

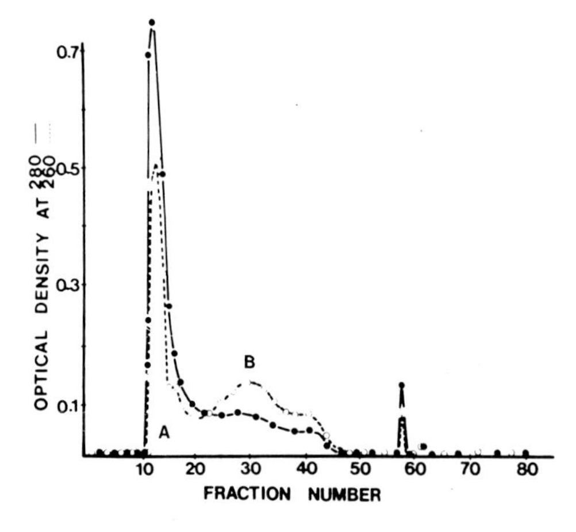

FIG. 19. Fractionation of Novikoff hepatoma NP proteins on Sephadex G-100. The solvent was 5 M urea in 50 mM Tris-HCl buffer, pH 8.0.

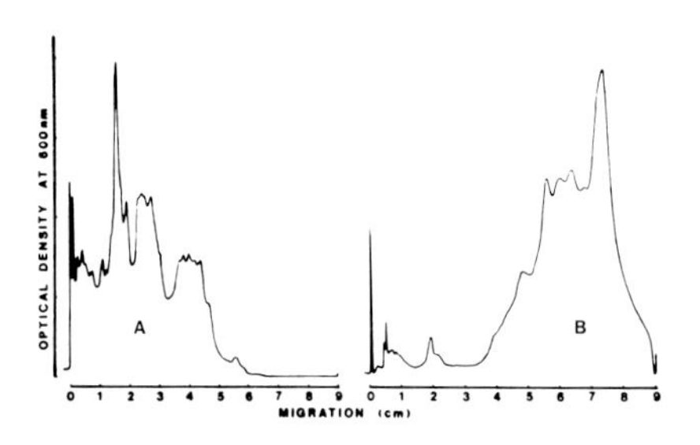

FIG. 20. Polyacrylamide gel electrophoresis in the presence of sodium dodecyl sulfate of the low and high molecular weight portions of the elution pattern shown in Figure 19.

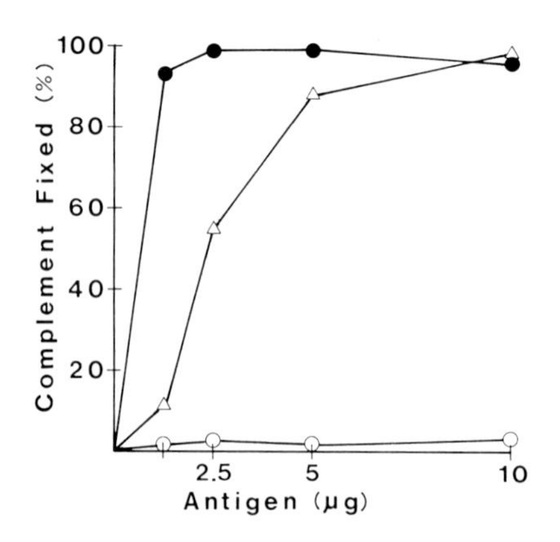

FIG 21. Complement fixation of the low molecular weight, ○——○, and high molecular weight, ●——●, proteins recovered from gel filtration on Sephadex G-100 (see Figure 19). For immunoassays, the proteins were reconstituted with purified rat spleen DNA. The control was unfractionated, NP protein fraction (△——△).

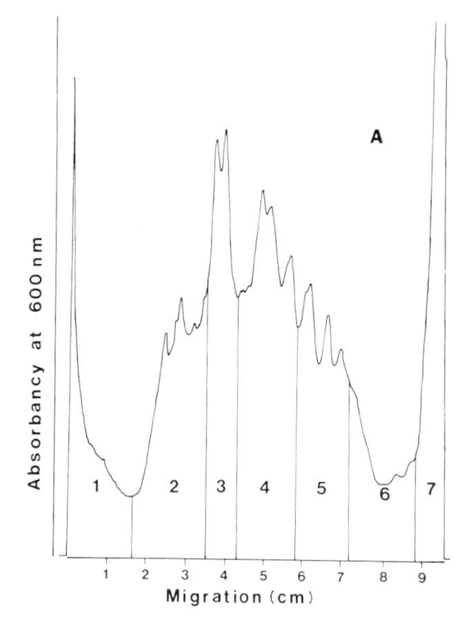

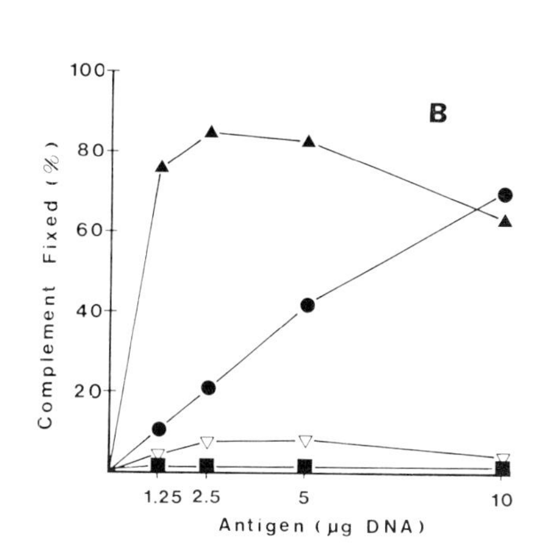

FIG. 22. Complement fixation of fractions obtained by preparative gel electrophoresis of Novikoff hepatoma NP protein fraction. The electrophoresis was performed in the presence of sodium dodecyl sulfate (A). The slabs were cut according to the insert in this figure, and the proteins were eluted by diffusion from broken-up gels. After the removal of sodium dodecyl sulfate, the proteins were reconstituted with purified rat spleen DNA. Unfractionated NP proteins, ●——●; gel fraction 4, ▲——▲; gel fraction 7, ▽——▽; all other gel fractions, ■——■. The complement fixation was performed in the presence of antiserum against Novikoff hepatoma dehistonized chromatin (B).

DISCUSSION

Experiments presented here point to the presence in chromatin of protein species which interact with homologous DNA in a highly specific manner. This interaction specificity changes with cellular differentiation (Chytil *et al.* 1974, Spelsberg *et al.* 1973, Chiu *et al.* 1976a), chemical (Chiu *et al.* 1975), or viral (Zardi *et al.* 1973) carcinogenesis and phases of the cell cycle (Chiu *et al.*, manuscript in preparation). It can be suggested that alterations in the composition, and perhaps the structure, of nuclear nonhistone proteins are closely associated with the process of cytodifferentiation. Chytil and Spelsberg (1971) as well as Chiu *et al.* (1976a) have shown that tissue-specific antisera will inhibit the in vitro transcription of chromatin. This phenomenon is tissue specific in that individual chromatins will be inhibited only by their homologous antisera. This points to the possible presence of gene regulatory proteins in the NP fraction. Recently, we have employed the chicken reticulocyte chromatin system to analyze the effects of NP proteins on in vitro transcription of globin genes. The NP fraction was found to be essential for the in vitro transcription of globin mRNA by chromatins reconstituted from DNA and isolated chromosomal protein components (Chiu *et al.* 1976b).

It can be speculated that if the immunologically specific proteins NP represent a part of the mechanism by which cells express and maintain their phenotype, our experiments may open the way for detailed biochemical and immunological studies of cancer.

If specific localization techniques can be developed to detect the cancer-specific antigens, this may provide a very sensitive procedure for differentiation of normal and cancerous cells in tissue biopsies. Preliminary experiments in our laboratory indicate that it may be possible to distinguish immunochemically the cells that became transformed to malignancy but whose morphological phenotype appears still normal. We are confident that studies on chromatin-associated specific antigens may lead to a better understanding of the mechanisms of cell differentiation, growth, and carcinogenesis.

ACKNOWLEDGMENTS

Original research reported in this communication was supported by National Cancer Institute Contracts N01-CP-65730, N01-CB-53896, and U. S. Public Health Service Grant CA-18389.

REFERENCES

Bekhor, I., G. M. Kung, and J. Bonner. 1969. Sequence-specific interaction of DNA and chromosomal protein. J. Mol. Biol. 39:351–364.

Bustin, M., and R. D. Cole. 1968. Species and organ specificity in very lysine-rich histones. J. Biol. Chem. 243:4500–4505.

Bustin, M., and B. D. Stollar. 1972. Immunochemical specificity in lysine-rich histone subfractions. J. Biol. Chem. 247:5716–5721.

Bustin, M., and B. D. Stollar. 1973. Immunological relatedness of thymus and liver F1 histone subfractions. J. Biol. Chem. 248:3506–3510.

Carlo, D. J., N. J. Bigley, and Q. Van Winkle. 1970. Immunochemical differences in deoxyribonucleoproteins from normal and malignant canine tissues. Immunology 19:879–889.

Chiu, J. F., C. Craddock, H. P. Morris, and L. S. Hnilica. 1974. Immunospecificity of chromatin nonhistone protein-DNA complexes in normal and neoplastic growth. FEBS Lett. 42:94–97.

Chiu, J. F., M. Hunt, and L. S. Hnilica. 1975. Tissue-specific DNA-protein complexes during azo dye heptocarcinogenesis. Cancer Res. 35:913–919.

Chiu, J. F., F. Chytil, and L. S. Hnilica. 1976a. Onco-fetal antigens in chromatin of malignant cells, *in* Onco-Developmental Gene Expression, W. H. Fishman and S. Sell, eds., Academic Press, New York, pp. 271–280.

Chiu, J. F., Y. H. Tsai, K. Sakuma, and L. S. Hnilica. 1976b. Regulation of *in vitro* mRNA transcription by a fraction of chromosomal proteins. J. Biol. Chem. 250:9431–9433.

Chytil, F., and T. C. Spelsberg. 1971. Tissue differences in antigenic properties of non-histone protein-DNA complexes. Nature New Biol. 233:215–218.

Chytil, F., S. R. Glasser, and T. C. Spelsberg. 1974. Alterations in liver chromatin during perinatal development of the rat. Dev. Biol. 37:295–305.

Chytil, F. 1975. Immunochemical characteristics of chromosomal proteins, *in* Methods in Enzymology, B. W. O'Malley and J. G. Hardman, eds., Vol. 40. Academic Press, New York, pp. 191–198.

Criss, W. E. 1971. A review of isozymes in cancer. Cancer Res. 31:1523–1542.

Elgin, S. C. R., and H. Weintraub. 1975. Chromosomal proteins and chromatin structure. Ann. Rev. Biochem. 44:725–774.

Gaffer, A., N. J. Bigley, J. P. Minton, and M. C. Dodd. 1967. Serological analysis of nuclear antigens from normal and malignant human tissues. Fed. Proc. 26:753.

Henning, N., W. Frenger, F. Scheiffarth, and A. Assaf. 1962. Immunologische studien über die Antigenität von Zellkernsubstanzen und über Zellkern-Antikörper. Zeitschrift fur Rheumaforschung 21:13–20.

Hnilica, L. S. 1972. The Structure and Biological Functions of Histones. CRC Press, Cleveland, Ohio.

Knox, W. E. 1967. The enzymic pattern of neoplastic tissue. Adv. Cancer Res. 10:117–161.

Messineo, L. 1961. Immunological differences of deoxyribonucleoproteins from white blood cells of normal and leukemic human beings. Nature 190:1122–1123.

Morris, A. D., C. Littleton, L. C. Corman, J. Esterly, and G. C. Sharp. 1975. Extractable nuclear antigen effect on the DNA-anti-DNA reaction and NZB/NZW mouse nephritis. J. Clin. Invest. 55:903–907.

Nakane, P. K. 1970. Classification of anterior pituitary cell types with immunoenzyme histochemistry. J. Histochem. Cytochem. 18:9–20.

Paul, J., and R. S. Gilmour. 1968. Organ-specific restriction of transcription in mammalian chromatin. J. Mol. Biol. 34:305–316.

Perez-Cuadrado, S., S. Haberman, and G. T. Race. 1965. Fluorescent antibodies of human cancer-specific DNA and nuclear proteins. Cancer 18:193–200.

Pitot, H. C. 1966. Some biochemical aspects of malignancy. Ann. Rev. Biochem. 35:335–368.

Rapp, H. J. 1964. The nature of complement and the design of a complement fixation test, *in* Immunochemical Methods, J. F. Ackboyd, ed., A. Davis Co., Philadelphia, pp. 1–23.

Sandberg, A. L., M. Liss, and B. D. Stollar. 1967. Rabbit antibodies induced by calf thymus histone-serum albumin complexes. J. Immunol. 98:1182–1189.

Silver, L. M., and S. C. R. Elgin. 1976. A method for determination of the *in situ* distribution of chromosomal proteins. Proc. Natl. Acad. Sci. USA 73:423–427.

Spelsberg, T. C., L. S. Hnilica, and A. T. Ansevin. 1971. Proteins of chromatin in template restriction. III. The macromolecules in specific restriction of the chromatin DNA. Biochim. Biophys. Acta 228:550–562.

Spelsberg, T. C., W. M. Mitchell, F. Chytil, E. M. Wilson, and B. W. O'Malley. 1973. Chromatin of the developing chick oviduct: Changes in the acidic proteins. Biochim. Biophys. Acta 312:765–778.

Stollar, D., and M. Ward. 1970. Rabbit antibodies to histone fractions as specific reagents for preparative and comparative studies. J. Biol. Chem. 245:1261–1266.

Suzuki, M., and Y. Hinuma. 1974. Evaluation of Epstein-Barr virus-associated nuclear antigen with various human cell lines. Int. J. Cancer 14:753–761.

Todaro, G. J., K. Habel, and H. Green. 1965. Antigenic and cultured properties of cells doubly transformed by polyoma virus and SV40. Virology 27:179–185.

Wakabayashi, K., and L. S. Hnilica. 1972. Immunochemical and transcriptional specificity of chromatin. J. Cell Biol. 55:271a.

Wakabayashi, K., and L. S. Hnilica. 1973. The immunospecificity of nonhistone protein complexes with DNA. Nature New Biol. 242:153–155.

Wakabayashi, K., S. Wang, and L. S. Hnilica. 1974. Immunospecificity of nonhistone proteins in chromatin. Biochemistry 13:1027–1032.

Walker, P. R., and V. R. Potter. 1972. Isoenzyme studies on adult, regenerating, precancerous and developing liver in relation to findings in hepatomas. Adv. Enzyme Regul. 10:339–364.

Wang, S., J. F. Chiu, L. Klyszejko-Stefanowicz, H. Fujitani, and L. S. Hnilica. 1976. Tissue-specific chromosomal nonhistone protein interactions with DNA. J. Biol. Chem. 251:1471–1475.

Wasserman, E., and L. Levine. 1961. Quantitative microcomplement fixation and its use in the study of antigenic structure by specific antigen-antibody inhibition. J. Immunol. 87:290–295.

Weber, G., and M. A. Lea. 1966. The molecular correlation concept of neoplasia. Adv. Enzyme Regul. 4:115–145.

Weinhouse, S., J. B. Shalton, W. E. Criss, and H. P. Morris. 1972. Molecular forms of enzymes in cancer. Biochimie 54:685–693.

Yeoman, L. C., J. J. Jordan, R. K. Busch, C. W. Taylor, H. E. Savage, and H. Busch. 1976. A

fetal protein in chromatin of Novikoff hepatoma and Walker 256 carcinosarcoma tumors that is absent from normal and regenerating rat liver. Proc. Natl. Acad. Sci. USA 73:3258–3262.
Zardi, L. 1975. Chicken antichromatin antibodies: Specificity to different chromatin fractions. Eur. J. Biochem. 55:231–238.
Zardi, L., J. C. Lin, and R. Baserga. 1973. Immunospecificity to non-histone chromosomal proteins of antichromatin antibodies. Nature New Biol. 245:211–213.

Cell Differentiation and Neoplasia, edited by Grady F. Saunders. Raven Press, New York © 1978.

Gene Regulation in Mammalian Cells: Some Problems and the Prospects for Their Solution

James E. Darnell, Jr.

The Rockefeller University, New York, New York 10021

The question underlying most of the sessions at this Symposium is this: How do eukaryotic cells (actually mammalian and, even more particularly, human cells) govern the production of specific proteins? This question is obviously basic to the understanding of determination and differentiation of specific cells and of those proteins (or enzymatic products) in all cells whose concentrations are responsible for growth control.

Cell biologists studying animal cells have had their pattern of thinking about the question of protein-synthesis regulation (or gene expression, as it is commonly called) shaped mainly by the discoveries in bacterial genetics and physiology, which quite precisely define the "rules" used by bacteria to regulate specific protein synthesis (for example, see Watson 1976). The primary assumption descended from bacterial physiology is that the concentration of specific mRNA is directly related to the rate of production, and probably the cellular concentration, of the homologous protein. Another frequently made assumption is that the regulation of transcription (generally used to mean control of initiation of transcription) is solely responsible for controlling the level of mRNA.

To begin a consideration of eukaryotic gene regulation with these assumptions is quite understandable, because they have been thoroughly proved for bacteria. In the future, when we know how gene regulation is accomplished in human cells, it may prove true that the only models we ever needed were the bacterial models. In the meantime, however, in case more complicated cells have more complicated mechanisms of regulation, perhaps there is some profit in trying to describe more precisely how regulation of gene expression might differ between prokaryotic and eukaryotic cells.

Table 1 gives a list of four possible levels of gene regulation: transcription, posttranscription (nuclear), posttranscription (cytoplasmic, pretranslational), and translation. Experimental results from a variety of sources suggest the possibility that each of these levels may be used at some time. This paper will briefly review some of these cases and describe the type of measurements that can now be made which should allow a clear demonstration of the true level of regulation in eukaryotic cells.

TABLE 1. *Four possible levels of gene regulation*

Possible levels of regulation	Step involved	Exact locus	
Transcription	DNA → hnRNA or mRNA	Initiation:	Prevent, start, stimulate
		Termination:	Attentuate, antiterminate
Posttranscription (nuclear)	hnRNA → mRNA +cap +m6A +poly A	Cleavage and modifications	
	mRNA (modified) → Cyto	Transport	
Posttranscription (cyto, pretranslation)	mRNA (modified) → Polysomes	Stabilization	
Translation	Polysomal mRNA → Protein	Initiations:	Few or many
		Half-life:	Long or short

TRANSCRIPTIONAL REGULATION

Transcriptional regulation in bacteria should be understood to be a much broader and more diverse means of gene regulation than the original model described by Jacob and Monod (1961). Certainly, the most critical decision in the regulation of any gene is whether or not to allow RNA polymerase access to the promoter site to begin to transcribe a given region. However, at this critical step the decision can either be to prevent or to stimulate the entry of RNA polymerase. Thus, the classic repressor model that accurately describes "negative" transcriptional control for β-galactosidase and lambda bacteriophage has had to be expanded because of the discovery of positive transcriptional elements that operate by increasing the rate of RNA polymerase initiation at particular genes (Englesberg *et al.* 1969). These positive regulatory proteins can affect transcription as dramatically as negative regulatory proteins (i.e., many 100-fold for the ara-C gene, Englesberg *et al.* 1969, Greenblatt and Schlief 1971) or can simply stimulate more efficient transcription (about 10-fold changes in enzyme levels, e.g., cyclic-AMP-binding protein, Zubay *et al.* 1970).

Another significant division in the mechanisms of transcriptional control has come to light recently from the studies of RNA chain termination (Roberts 1975) in the transcription of lambda bacteriophage and tryptophan genes, for example. The N gene of lambda phage is responsible for "antitermination," which allows RNA polymerase to transcribe beyond the region at which it normally terminates (Lozeron *et al.* 1976). Likewise, in severe tryptophan starvation, *Escherichia coli* RNA polymerase transcribes past a site 166 nucleotides from the promoter of the tryptophan operon, whereas normally, chain termination occurs at this point (Bertrand *et al.* 1976). Even in the presence of tryptophan, at least a part of this "leader" sequence is transcribed, so that control of this gene clearly resides at the level of allowing this "attenuating" site to be "read through" (Kano *et al.* 1976).

Thus, even if *all* gene regulation in eukaryotes finally proves to take place at the level of transcription, it is not obvious *a priori* which bacterial transcriptional regulatory mechanism—positive or negative, initiation or termination—most likely applies in the case of any eukaryotic gene.

Table 1 lists several levels of regulation in addition to transcription. No candidates (or very few) for such regulatory sites have been identified in bacteria, and essentially no cases of eukaryotic regulation have been completely settled. Therefore, naturally enough, discussions of models for gene regulation, other than transcription, in eukaryotes have been diffuse, speculative, and, while occasionally provocative, largely unconvincing. Enough has been learned recently about RNA biosynthesis and mRNA function in eukaryotic cells to at least make discussion about levels of regulation other than transcription quite explicit prior to decisive experimental evidence.

Before describing possible points at which gene regulation might operate, a clear definition of regulation must be established. An individual step in the biosynthesis of functional mRNA that occurs at less than 100% efficiency but that always occurs at the same relative efficiency is *not* a regulatory step. Only a step in the formation of mRNA that changes relative efficiency in accord with physiological change should be recognized as a regulatory step.

REGULATION ASSOCIATED WITH POSTTRANSCRIPTIONAL MODIFICATIONS: NUCLEAR

RNA Chain Cleavage

Much of the evidence that justifies the consideration of possible differences in gene expression in eukaryotic cells arises from studies on nuclear RNA synthesis and analysis of structures found in eukaryotic mRNA. Nuclear RNA synthesis in mammalian cells (including human cells) vastly exceeds the amount of RNA that ever appears in the cytoplasm (Soeiro *et al.* 1968). This situation does not exist in bacteria, in which it is believed that all (or at least most) transcription results in products that are used. A great deal of this rapidly "turned over" nuclear RNA belongs to the class termed hnRNA; these molecules range in length from about 1,000 nucleotides to perhaps 30,000 or so nucleotides. Recent analyses using new approaches to the transcription unit size of the hnRNA confirms that more than 90% are longer than mRNA molecules, more than 50% are longer than 5,000 nucleotides, and 20% are longer than 20,000 nucleotides (Derman *et al.* 1976, Giorno and Sauerbier 1976). Some of the sequences in these long molecules are found as mRNA in the cell cytoplasm as detected by molecular hybridization (Herman *et al.* 1976). Recently, hemoglobin-specific sequences also have been found in nuclear molecules about 2,000 to 3,000 nucleotides long, whereas the functional hemoglobin mRNA is less than 700 nucleotides long (Curtiss and Weissman 1976, Ross 1976, Lingrel *et al.* 1978, see pages 361 to 368, this volume).

Evidence from our laboratory (Bachenheimer and Darnell 1975, Weber *et al.* 1977, Goldberg *et al.* 1977) demonstrates that late in adenovirus-2 (Ad-2) infection of HeLa cells, a single, large RNA molecule arises as the predominant RNA product from a region that encodes at least six smaller mRNA molecules (Sharp *et al.* 1974, Philipson *et al.* 1974). Thus, it appears highly likely that specific enzyme cleavage occurs in making mRNA in eukaryotic cells. Such posttranscriptional RNA cleavage is well established in the production of animal cell (Darnell 1975) and bacterial tRNA and rRNA and in bacteriophage T7 mRNA production (Smith 1975, Dunn and Studier 1973). In the face of a large amount of nuclear RNA turnover, it seems quite possible that only portions of such large transcripts survive as mRNA, but it remains a total mystery whether such cleavages could be a *regulatory* point. *Does the cell ever decide how much of a given mRNA to ship to the cytoplasm by regulating the cleavage of a potential mRNA precursor?*

Terminal Additions to mRNA and Transport of mRNA

In the past six years, it has become clear that the greatest biochemical distinction between eukaryotic and prokaryotic mRNA lies at the ends of the mRNA (Shatkin 1976, Darnell 1975). In bacteria, after transcription has progressed a few hundred nucleotides or less, ribosomes begin translating even before mRNA chain completion. Moreover, the 5′ terminus is an unmodified pppApX--- or pppGpX---. In eukaryotic cells, both ends of the mRNA are modified in the cell nucleus before the mRNA enters the cytoplasm; the 5′ terminus on most, if not all, cellular mRNA molecules is m^7GpppXmp--- or m^7GpppXmpXmp---, a blocked (5′-5′ linked), methylated structure (Shatkin 1976). The 3′ terminus in most mRNAs is a segment of polyadenylic acid originally over 200 nucleotides long (Darnell 1975, Brawerman 1976). The presumed function of the cap is to promote the proper initiation of protein synthesizing complexes. No cytoplasmic function has yet been identified for poly(A). It is clear, however, that without the addition of the poly(A) segment of about 235 nucleotides, the mRNA does not appear from the nucleus as polyadenylated cytoplasmic RNA, as it otherwise would (Darnell 1975, Sawicki *et al.* 1977).

Furthermore, proteins that bind to poly(A) have been identified in both the nucleus and the cytoplasm, suggesting a possible mechanism for participation of poly(A) in the transport of mRNA from the nucleus to the cytoplasm. This, of course, does not imply that all mRNA requires poly(A) for transport. In fact, mRNA lacking poly(A), notably histone mRNA, is known to exist (Adesnik and Darnell 1972).

Does the cell use "cap" or poly(A) addition as decision points in making mRNA? Without a change in transcription rates, can the cell decide to send more or less of a particular mRNA through the cap or poly(A) pathway and thus change the rate of appearance of competent mRNA? Such a nuclear posttranscription level of control, at first glance, seems very different and more complex than in bacteria. Formally, however, no more complicated recognition system is needed for posttranscriptional control than for transcriptional control. In the latter, a protein recognizes a sequence in DNA; in the former, recognition would simply be transferred to RNA. A system based only on posttranscriptional control would, of course, require constant transcription of the gene in question and, thus, seems highly unlikely.

POSTTRANSCRIPTIONAL REGULATION: STABILIZATION OF CYTOPLASMIC mRNA BEFORE TRANSLATION

Myoblast cultures that have not yet begun to differentiate, as evidenced by myosin production, nevertheless synthesize a 26S poly(A)-terminated, presumptive myosin mRNA (Buckingham *et al.* 1976, Whalen *et al.* 1976). This is detected as a cytoplasmic, nonpolysome-associated, potential mRNA species. Later in the course of development, the same cells accumulate label in 26S RNA at a somewhat similar rate, but most of the 26S RNA enters the polyribosomes.

May cells make regulatory decisions about gene expression at the cytoplasmic level by selective and differential initiation of protein synthesis with consequent stabilization of mRNA? Whether the terminal differentiation of myoblasts is based solely on this apparent myosin mRNA stabilization or not, such a general mechanism for gene expression must be borne in mind when considering eukaryotic cell regulation.

TRANSLATION AS A LEVEL OF REGULATION

Stopping Ongoing Protein Synthesis

During vaccinia virus infection, a program of synthesis of virus-specific enzymes occurs. Thymidine kinase increases for several hours and then abruptly ceases to be made. Over 10 years ago, McAuslan (1963) showed that simply stopping RNA synthesis did not stop thymidine kinase formation. Paradoxically, it made it continue. This apparent requirement for continuing RNA synthesis and perhaps for protein synthesis in order to stop the ongoing synthesis of an induced protein has been found repeatedly. Interferon (Tann *et al.* 1970) and tyrosine amino transferase (Tomkins 1974) are examples of cell proteins whose "shut-off" apparently requires ongoing RNA synthesis. This subject has been much discussed largely because of details in interpretation about rates of protein synthesis during and after drug treatments, but all such experiments leave no doubt that under conditions where normally there is a decrease in the synthesis of a specific protein, such a decrease does not occur as rapidly or completely when RNA synthesis is inhibited (Lodish 1978, see pages 369 to 388, this volume). *Does the eukaryotic cell have a mechanism (proteins? RNA?) in the cell cytoplasm to regulate ongoing mRNA utilization?*

Changing mRNA Half-Life

Finally, the role of mRNA half-life, independent of any hypothetical, direct cytoplasmic regulatory mechanisms, must be considered. Mammalian cell mRNA exists with half-lives of an hour or so to many hours (Puckett *et al.* 1975), perhaps as long as the generation time of the cell (Singer and Penman 1973). *Does the cell have the capacity to change the half-life of a given mRNA?* It has recently been argued, in contrast to earlier conclusions (Tomkins 1974), that the mRNA of certain hormone-dependent proteins has the same half-life whether in the presence or the absence of inducing hormone (Stiles *et al.* 1976). Moreover, from these studies it appeared that the synthesis of two enzymes, tyrosine amino transferase (TAT) and alanine amino transferase (AAT), had fixed half-lives; but TAT synthesis decayed with about a 2- to 3-hour half-life, while AAT synthesis had a much longer half-life of about 10 to 12 hours. This point is so important that as new methods of direct measurement of mRNA become available, the half-life determinations should be repeated.

Aviv and colleagues (Bastos *et al.* 1977) have been studying the relationship of hemoglobin mRNA stability to the mechanism of differentiation that finally results in a reticulocyte in which hemoglobin constitutes 95% of the new protein synthesized. They found that hemoglobin mRNA apparently has a fixed half-life of between 15 and 20 hours in *all* cells tested. They also found that induced Friend erythroleukemia cells (inducible by dimethyl sulfoxide [DMSO] to form hemoglobin) accumulate hemoglobin mRNA to a level of about 10 to 15% of the total mRNA, but an mRNA class with a half-life longer than hemoglobin's precludes further accumulation and thus terminal differentiation. Red blood cell precursors from the spleens of anemic mice exhibit the same phenomenon. Aviv postulates that final differentiation requires a destabilization of the long-lived mRNA, which in turn allows hemoglobin mRNA, with its relatively long half-life, to become the major mRNA during terminal differentiation.

Whether cells can change mRNA half-lives of specific mRNAs or classes of mRNAs is an extremely important point to know in considering levels of mammalian gene regulation. Even if each mRNA normally has a fixed half-life, removal of mRNAs *en bloc* (e.g., *all* long-lived mRNA, as Bastos *et al.* [1977] suggest) could be an example of dramatic fluctuations in mRNA lifetime.

SUMMARY OF POSSIBLE LEVELS OF REGULATION

It seems profitable to speculate about which possible mode(s) of gene regulation likely holds for eukaryotic cells only if an incisive, previously unrecognized pathway of exploration can be suggested that will yield the truth. It has been clear all along to many cell biologists that no satisfactory picture of eukaryotic gene regulation would develop unless a representative set of genes were available to measure the flow of specific RNA from template through to cytoplasmic function. The recent revolutionary advances in cloning genes within *E. coli* promises to provide such a battery of pure genes (see, for example, Wensink *et al.* 1974). Using the purified DNA to hybridize labeled cellular RNA, it should be possible within a few years to describe, under a variety of physiological conditions, transcription units containing a sufficiently broad variety of genes, the rate of transcription, the extent of mRNA conservation from primary transcripts, and the rate of mRNA turnover from these transcription units. Thus, by "brute force" biochemistry (i.e., experiments not based on classical genetics), the problem of levels of gene expression may be solved.

Speculation about the probable outcome of these now possible experiments is perhaps both idle and useless, but the temptation offered by a symposium paper (which, in fact, is a required exercise) nevertheless calls forth a few musings. Is the basically flexible, gene regulation system(s) at the transcriptional level that evolved to solve all types of problems for prokaryotes likely to have been discarded in the evolution of eukaryotes? It seems unlikely. Transcriptional regulation in its many ramifications is certainly theoretically sufficient for all gene regulation. A potentially vexing problem in attempting to transfer the logic of bacterial physiology to higher cells is the lack of clustered related genes (operons) in higher cells. If each scattered gene within the huge genomes of higher eukaryotic cells must be controlled by a regulatory gene product, a problem described early by Jacob and Monod (1961) arises: What will control the regulatory genes? The discovery of autogenous regulation of transcription of the mRNA for regulatory proteins in lambda bacteriophage (Dottin *et al.* 1975) essentially removes the logical restraint on how to control the controllers: They control themselves. In addition, the discovery of positively acting, transcriptional regulatory proteins (Englesberg *et al.* 1969, Greenblatt and Schlief 1971) greatly broadens the mechanisms. In particular, positive regulatory elements (like the ara-C gene) have great appeal as a mechanism in cell differentiation. It is widely believed that a cell which is committed to a pathway usually stays on this pathway through many generations until the trigger for differentiation is supplied. Positive regulatory elements that are themselves autogenously regulated would suffice to endow committed cells with their chief characteristics: Once a cell started to express a gene under the influence of such a positive regulatory system, it would always continue making that product. The view that a variety of transcriptional controls remain the still unproved "best bet" to explain eukaryotic gene regulation would therefore seem easy to accept.

If transcriptional regulation rules all in the kingdom of gene expression, then all the intricate steps of mRNA formation—cleavage of precursors, capping, poly(A) addition, m^6A methylation, transport to the cytoplasm—must all be regarded as necessary and important mechanical events in mRNA production but *not* as steps used for regulation. This would remain the correct view even if these posttranscriptional steps occurred to only a fraction of the potential mRNA molecules synthesized, so long as that fraction were unchanging for a given type of transcript. This may, in fact, be the case, and all the "extras" that eukaryotic mRNAs feature and prokaryotic mRNAs lack may serve functions other than gene regulation, e.g., protection against translation of accidentally introduced but improperly modified mRNA. Or, finally, it may be that at least some part of the control of gene expression in mammalian cells is posttranscriptional, perhaps as a "fine-tuning" or ultimate "quantity-control." The problems are much more clearly defined now than 10 years ago and means are at hand for answering the questions. Further guesswork seems out of place at this point.

THE MEASUREMENTS NECESSARY TO EXPLORE LEVELS OF GENE CONTROL

If it is granted that the new recombinant DNA technology will allow decisive experiments on how gene regulation is accomplished in eukaryotes, we should be certain that the experimental questions are properly phrased to make best use of the new techniques.

Definition of Transcription Units

The recent work in our laboratory has used adenovirus-infected cells and adenovirus-transformed cells in experiments that will be prototypes for the study of any cellular genes whose RNA products we can measure. The first issue to settle when studying regulation of RNA production from any gene or group of contiguous genes is the nature of the primary transcript from that region. We have argued that since RNA processing may be rapid, experiments which study RNA at its earliest stage, namely nascent RNA, are the best experiments to define transcription units (Bachenheimer and Darnell 1975). We have further suggested that perhaps the best way to study nascent RNA is to allow isolated nuclei to elongate RNA chains already begun in the cell (Weber *et al.* 1977). To illustrate these points, we have examined RNA synthesis late in Ad-2 infection, in which 98% of RNA synthesis occurs by transcription of one strand. Labeled, nascent, viral RNA molecules at this time of infection form a series of increasing size, the shortest being those nearest the promoter, the longest being most distal to the promoters. The order of such increasingly long molecules follows the order from 0.1 to 1.0 on the physical map of Ad-2 virus. Thus, the transcription unit begins around 0.15 and proceeds to 1.0.

An independent means of mapping the origin of RNA synthesis and the length of a transcription unit was provided by the development of ultraviolet (UV) transcription mapping (Giorno and Sauerbier 1976). For any transcription unit, the most UV-sensitive region is the most distal from the promoter, and the most resistant is the region proximal to the promoter. With an available restriction enzyme map of any putative transcription unit, hybridization of RNA labeled after UV irradiation to various restriction fragments will reveal how transcription normally occurs in that region. Even if RNA processing occurred simultaneously with transcription so that labeled, nascent RNA representing the entire transcription unit could not be observed, hybridization of labeled RNA produced after UV irradiation would show both the direction of transcription and boundaries of a single transcription unit (Goldberg *et al.* 1977). With this approach it was shown that the most UV-sensitive region was from 0.9 to 1.0 on the map, and the sensitivity decreased exponentially back to approximately the region of 0.2. From the UV data it could be concluded that over 95% of the RNA transcribed from the rightmost region was initiated 20,000 to 25,000 nucleotides to the left.

A recent approach to locating more precisely the origin of transcription is an extension of experiments involving isolated nuclei (Evans *et al.*, unpublished data). If promotion of RNA chains is from a specified site and processing does not immediately cleave growing chains from all regions of a transcriptional unit to small RNA chains (about 500 nucleotides or less), then the only RNA labeled during incubation of isolated nuclei that is shorter than, say, 500 nucleotides should be the proximal promoter RNA segment. We have found that late in Ad-2 infection, almost all short, labeled, nascent RNA is complementary to a DNA fragment that lies between 11.3 and 18.1 on the physical map. This observation is in agreement with the earlier projections of promoter locations based on analyzing the increasing sizes of the large, ordered, nascent RNA molecules and with the relative UV sensitivity of RNA synthesis from various regions of the Ad-2 genome (Evans *et al.* 1977).

The definition of the late Ad-2 transcription unit will be completely rigorous if it can be shown that the presently suspected origin is the only region to bind large RNA molecules containing pppGp--- or pppAp---.

The definition of this large, late Ad-2 transcript, which appears to be the only (>95%) mode of transcription for 80 to 85% of the rightward reading DNA strand, appears to be strong evidence for the capacity of cells to cleave primary RNA products in the production of mRNA. At least six mRNAs of smaller size are contained within this large transcript (Sharp *et al.* 1974, Philipson *et al.* 1974).

Given a means of defining transcriptional units, what must be questioned in any future definition of cell transcription units is whether a given gene is always transcribed in the same fashion, i.e., that changes in gene expression are not accompanied by shifting frames of transcription.

Instantaneous Rates of Transcription

Assuming the definition of the transcription unit, the next problem in eventually understanding regulation is to measure the differential rate of primary transcript formation from that unit compared to other units when physiological changes occur. This is an almost impossible task in whole cells, in which the complications of differential pool equilibrations and potentially differential changes in the turnover of RNA from any transcription unit make measurements of total accumulation of radioactivity in different RNAs suspect as a means of measuring differential rates of synthesis. If isolated nuclei do not initiate chains, but only add a few nucleotides onto *all* already started chains, then labeled RNA from isolated nuclei afford an accurate means of measuring instantaneous rates of synthesis from any available pure DNA from which transcription is occurring.

We have approached this problem by measuring the fraction of labeled RNA that is Ad-2 specific in both isolated nuclei as well as in very brief pulse-labeled cells under three conditions—early and late in the lytic cycle and in transformed cells. The fraction of virus-specific RNA labeled in the cells and in the nuclei was about the same in each case: 0.5% early in infection, 10 to 20% late in infection, and 0.01% in the transformed cells (Wilson *et al.*, unpublished data). Thus, not only does the isolated nucleus add label to virus-specific molecules approximately the same size as in the cell, but the isolated nucleus appears to label these molecules at the same rate as the whole cell. Instantaneous rates of RNA synthesis will be very difficult to measure for cellular genes by any technique other than labeling isolated nuclei because sufficient labeled RNA to measure single RNA species may only be produced in this manner.

If the instantaneous rates of nuclear transcription of well-described transcription units could be measured under various conditions for a variety of purified genes and compared to the levels of mRNA existing in all these conditions, a giant step toward locating the locus of regulation obviously would have been made.

mRNA Transport, Stabilization, and Turnover

The area that will need more research effort if it is found that mRNA levels do not vary in proportion to transcription rates is mRNA transport, stabilization, and turnover. Such studies on specific mRNA molecules from cellular genes again will come with the availability of cloned DNA. Prior to such studies, we have recently endeavored to study two specific mRNAs that are produced from integrated Ad-2 DNA in transformed cells. We had earlier noted that the labeling of total Ad-2-specific RNA fluctuated about threefold around the cell cycle (Hoffman and Darnell 1975) and when it was demonstrated that the two adenovirus mRNAs were not overlapping, i.e., represented two independent polynucleotides, we realized that we should be able to observe differences in mRNA metabolism, if they existed, between two mRNAs arising from the same type of integrated DNA. In fact, evidence for differential behavior of the two mRNAs was easy to obtain. Simply determining the time of arrival of the two, 14S and 19S, in the cell cytoplasm revealed that the 14S mRNA exited within a few minutes of the onset of labeling, while only after 30 to 60 minutes was the larger mRNA observed (Wilson *et al.,* in press). The smaller mRNA remained dominant for the first 100 minutes, but after long label times the longer mRNA was the dominant molecule by an amount that appeared greater than the molecular weight difference of the two. Thus, the shorter mRNA may also have a shorter half-life. Experiments of this type (and further analysis of this phenomenon) should provide valuable models for the later study of the cytoplasmic metabolism of specific cell mRNAs.

CONCLUSION

The conceptual framework and adequate technology have developed in the last few years to learn how gene expression is regulated in eukaryotes. As studies are performed in the next 10 years or so, a deeper understanding of the results, as well as facility in their communication, will be aided by a comparison with already understood models. The case is made that transcriptional regulation in bacteria and phages, especially in its recently broadened perspective, might conceivably be sufficient to explain all gene regulation in eukaryotes. However, decision points for gene regulation might also exist in eukaryotes at those points where a biochemical departure in mRNA metabolism already has been detected between eukaryotes and prokaryotes.

In constructing experiments to determine at which level gene regulation truly functions, emphasis should be given to careful description of transcriptional units and the constancy of their boundaries and instantaneous rates of transcription, as well as improvements in our knowledge of how best to measure the metabolic fate of mRNA.

REFERENCES

Adesnik, M., and J. E. Darnell. 1972. Biogenesis and characterization of histone mRNA in HeLa cells. J. Mol. Biol. 67:397–406.

Bachenheimer, S., and J. E. Darnell. 1975. Adenovirus-2 mRNA is transcribed as part of a high molecular-weight precursor RNA. Proc. Natl. Acad. Sci. USA 72:4445–4449.

Bastos, R. N., Z. Volloch, and H. Aviv. 1977. Messenger RNA population analysis during erythroid differentiation: A kinetical approach. J. Mol. Biol. 110:191–203.

Bertrand, K., C. Squires, and C. Yanofsky. 1976. Transcription termination in vivo in the leader region of the tryptophan operon of *Escherichia coli.* J. Mol. Biol. 103:319–337.

Brawerman, G. 1976. Characteristics and significance of the polyadenylate sequence in mammalian RNA. Prog. Nucleic Acid Res. Mol. Biol. 17:118–148.

Buckingham, M., A. Cohen, and F. Gros. 1976. Cytoplasmic distribution of pulse-labeled poly(A)-containing RNA, particularly 26S RNA, during myoblast growth and differentiation. J. Mol. Biol. 103:611–626.

Curtiss, P. J., and C. Weissman. 1976. Purification of globin mRNA from DMSO-induced Friend cells and detection of a putative globin mRNA precursor. J. Mol. Biol. 106:1061–1075.

Darnell, J. E. 1975. The origin of mRNA and the structure of the mammalian chromosome. Harvey Lect. 69:1–47.

Derman, E., S. Goldberg, and J. E. Darnell. 1976. hnRNA in HeLa cells: Distribution of transcript sizes estimated from nascent chain profile. Cell 9:465–475.

Dottin, R. P., L. S. Cutler, and M. L. Pearson. 1975. Repression and autogenous stimulation *in vitro* by bacteriophage lambda repressor. Proc. Natl. Acad. Sci. USA. 72:804–808.

Dunn, J. J., and W. Studier. 1973. T7 early RNAs and *E. coli* ribosomal RNAs are cut from large precursor RNAs in vivo by ribonuclease III. Proc. Natl. Acad. Sci. USA 70:3296–3300.

Englesberg, E., D. Sheppard, C. Squires, and F. Meronik, Jr. 1969. An analysis of "revertants" of a deletion mutant in the C gene of the L-arabinose gene complex in *E. coli* B/r: Isolation of initiation constitutive mutants (I^c). J. Mol. Biol. 43:281–293.

Evans, R., N. Fraser, E. Zigg, J. Weber, M. Wilson, and J. E. Darnell. 1977. The initiation sites for RNA transcription in AD-2 DNA. Cell (In press).

Giorno, R., and W. Sauerbier. 1976. A radiological analysis of the transcription unit for heterogeneous nuclear RNA in cultured murine cells. Cell 9:775–784.

Goldberg, S., J. Weber, and J. E. Darnell. 1977. The definition of a large viral transcription unit late in AD-2 infection of HeLa cells: Mapping by effects of UV irradiation. Cell 10:617–621.

Greenblatt, J., and R. Schlief. 1971. Arabinose C protein: Regulation of the arabinose operon in vitro. Nature New Biol. 273:166–169.

Herman, R. H., J. G. Williams, and S. Penman. 1976. Message and non-message sequences adjacent to poly(A) in steady state heterogeneous nuclear RNA of Hela cells. Cell 7:429–438.

Hoffman, P. R., and J. E. Darnell. 1975. Differential accumulation of virus-specific RNA during the cell cycle of adenovirus-transformed rat embryo cells. J. Virol. 15:806–811.

Jacob, F., and F. Monod. 1961. Genetic regulatory mechanisms in the synthesis of proteins. J. Mol. Biol. 3:318–356.

Kano, Y., M. Kuwano, and F. Imamoto. 1976. Initial *trp* operon sequence in *E. coli* is transcribed without coupling to translation. Mol. Gen. Genet. 146:179–188.

Lingrel, J. B., T. G. Wood, S-P. Kwan, P. Rosteck, Jr., and K. Smith. 1978. Isolation of a precursor to globin messenger RNA, *in* Cell Differentiation and Neoplasia (The University of Texas System Cancer Center 30th Annual Symposium on Fundamental Cancer Research, 1977), G. F. Saunders, ed., Raven Press, New York, pp. 361–368.

Lodish, H. F., J. E. Bergmann, and T. H. Alton. 1978. Regulation of messenger RNA translation, *in* Cell Differentiation and Neoplasia (The University of Texas System Cancer Center 30th Annual Symposium on Fundamental Cancer Research, 1977), G. F. Saunders, ed., Raven Press, New York, pp. 369–388.

Lozeron, H. A., J. E. Dahlberg, and W. Szybalski. 1976. Processing of the major leftward mRNA of coliphage lambda. Virology 71:262–277.

McAuslan, B. R. 1963. The induction and repression of thymidine kinase in the poxvirus infected HeLa cell. Virology 21:383–389.

Philipson, L., U. Pettersson, U. Lindberg, C. Tibbetts, B. Vennstrom, and T. Persson. 1974. RNA synthesis and processing in adenovirus-infected cells. Cold Spring Harbor Symp. Quant. Biol. 34:447–456.

Puckett, L., S. Chambers, and J. E. Darnell. 1975. Short-lived mRNA in HeLa cells and its impact on the kinetics of accumulation of cytoplasmic polyadenylate. Proc. Natl. Acad. Sci. USA 72:389–393.

Roberts, J. 1975. Transcription termination and late control in phage lambda. Proc. Natl. Acad. Sci. USA 72:3300–3304.

Ross, J. 1976. A precursor of globin mRNA. J. Mol. Biol. 106:403–415.

Sawicki, S., W. Jelinek, and J. E. Darnell. 1977. 3′ terminal addition to HeLa cell nuclear and cytoplasmic poly(A). J. Mol. Biol. 113:219–235.

Sharp, P. A., P. H. Gallimore, and S. J. Flint. 1974. Mapping of adenovirus-2 RNA sequences in lytically infected cells and transformed cell lines. Cold Spring Harbor Symp. Quant. Biol. 39:457–474.

Shatkin, A. 1976. Capping of eukaryotic mRNA's. Cell 9:645–654.

Singer, R. H., and S. Penman. 1973. Messenger RNA in HeLa cells: Kinetics of formation and decay. J. Mol. Biol. 78:321–331.

Smith, J. 1975. Transcription and processing of transfer RNA precursors. Prog. Nucleic Acid Res. Mol. Biol. 16:25–73.

Soeiro, R., M. H. Vaughan, J. R. Warner, and J. E. Darnell. 1968. The turnover of nuclear DNA-like RNA in HeLa cells. J. Cell Biol. 39:112–118.

Stiles, C. D., K-L. Lee, and F. T. Kenney. 1976. Differential degradation of mRNAs in mammalian cells. Proc. Natl. Acad. Sci. USA 73:2634–2638.

Tann, Y. H., J. Armstrong, J. H. Ke, and M. Ho. 1970. Regulation of cellular interferon production: Enhancement by antimetabolites. Proc. Natl. Acad. Sci. USA 67:464–471.

Tomkins, G. M. 1974. Regulation of gene expression in mammalian cells. Harvey Lect. 68:37–67.
Watson, J. D. 1976. Molecular Biology of the Gene. 3rd ed. Benjamin, New York.
Weber, J., W. Jelinek, and J. E. Darnell. 1977. The definition of a large viral transcription unit late in AD-2 infection of HeLa cells: Mapping of nascent RNA molecules labeled in isolated nuclei. Cell 10:611–616.
Wensink, P. C., D. J. Finnegan, J. E. Donelson, and D. S. Hogness. 1974. A system for mapping DNA sequences in the chromosomes of *D. melanogaster.* Cell 3:315–325.
Whalen, R. G., M. E. Buckingham, and F. Gros. 1976. Protein and mRNA synthesis in cultured muscle cells. Prog. Nucleic Acid Res. Mol. Biol. 19:485–489.
Wilson, M. C., S. Sawicki, M. Salditt-Georgieff, and J. E. Darnell. 1977. AD-2 in RNA in transformed cells: Map positions and difference in transport time. J. Virol. (In press).
Zubay, G., D. Schwartz, and J. Beckwith. 1970. Mechanism of activation of catabolite-sensitive genes: A positive control system. Proc. Natl. Acad. Sci. USA 66:104–110.

Cell Differentiation and Neoplasia, edited by Grady F. Saunders. Raven Press, New York © 1978.

Isolation of a Precursor to Globin Messenger RNA

Jerry B. Lingrel, T. Gordon Wood, Sau-Ping Kwan, Paul Rosteck, Jr., and Kate Smith

Department of Biological Chemistry, University of Cincinnati Medical Center, Cincinnati, Ohio 45267

Regardless of the etiology of cancer, it is clear that the genetic information of the cell is involved in the neoplastic state. This supposition is based on the fact that cancerous cells confer the neoplastic state on their progeny. It is likely that in cancerous cells the program for the expression of genetic information is altered in an irreversible or nearly irreversible way. This could be the result of a somatic mutation, a viral product that alters gene expression, or reprogramming of the genetic information (as is thought to occur in differentiation). These considerations lead us to the conclusion that a detailed understanding of gene expression in higher organisms is important to an understanding of cancer.

If the regulation of gene expression is to be studied, it will be necessary to identify the initial transcription product of specific genes. Once this is accomplished, factors that affect the expression of these genes can be isolated and studied.

Much of the work to date in this area has centered around the study of heterogeneous nuclear RNA (HnRNA) (see review by Darnell 1974, Lewin 1975a,b, and Darnell, pages 347 to 359, this volume). It has been proposed that this RNA is a precursor to cytoplasmic RNA, but the evidence has been indirect. Unequivocal proof that a precursor to mRNA exists will require that an RNA larger than the mRNA in question be identified and that it be kinetically related to the mature form. Both cell-free translational assays and complementary DNA (cDNA) probes have been used in an attempt to identify precursors to one mRNA, i.e., the one for globin (Imaizumi *et al.* 1973, Williamson *et al.* 1973, Ostertag *et al.* 1973, Ruiz-Carrillo *et al.* 1973, Macnaughton *et al.* 1974, Knöchel and Tiedemann 1975, Melli and Pemberton 1972). Several problems exist with these studies, including the sensitivity of the assay and the possible presence of aggregates between mature globin mRNA and itself or other RNAs. Furthermore, because unlabeled RNA is used, these studies neither lend themselves to precursor-product analysis nor do they provide a method for isolating the precursor species.

Present Addresses: T. G. Wood, The University of Texas System Cancer Center M. D. Anderson Hospital and Tumor Institute, Houston, Texas 77030; S-P. Kwan, Albert Einstein College of Medicine, Yeshiva University, Bronx, New York 10461.

Our approach to this problem has been to prepare an affinity column that specifically binds globin mRNA and to use this material to isolate globin mRNA containing sequences from pulse-labeled cells (Kwan *et al.* 1977). The affinity matrix has been prepared by copying globin mRNA with reverse transcriptase while it is attached to oligo(dT)-cellulose (Wood and Lingrel 1977). The cDNA cellulose selectively retains globin mRNA. In this paper we describe the isolation of a 16S RNA using the globin cDNA column and prove by hybridization kinetic analysis that it contains globin mRNA sequences. We also show that the 16S RNA is kinetically related to mature globin mRNA.

EXPERIMENTAL PROCEDURES

Nucleated erythroid cells were obtained from the spleens of anemic mice and then incubated in the presence of [5-^{3}H] uridine, as previously described (Merkel *et al.* 1976, Kwan *et al.* 1977). Following incubation, the cells were washed with isotonic saline, and the total nucleic acid isolated by the proteinase K procedure (Kwan *et al.* 1977). The DNA was removed by DNase, as previously described (Kwan *et al.* 1977), except that the DNase was not treated with iodoacetate. RNA containing globin mRNA sequences was isolated from the total RNA by globin cDNA cellulose chromatography, as previously described (Wood and Lingrel 1977, Kwan *et al.* 1977).

Polacrylamide gel electrophoresis in formamide was carried out according to the procedure of Orkin *et al.* (1975), with the following modifications. Four percent polyacrylamide gels containing 3.64% acrylamide and 0.36% N, N-methylenebisacrylamide were polymerized in formamide containing 20 mM phosphate and 0.1 M acetic acid. The electrode buffer was 20 mM phosphate (pH 6.5) and was recirculated during electrophoresis. Samples were applied to the gels in buffered formamide containing 20% sucrose and 0.05% bromophenol blue. Electrophoresis was for 30 minutes at 1 mA/gel, then for three hours at 2.5 mA/gel at 20 to 25°C.

Hybridizations were carried out in 0.12 M phosphate, 1 mM EDTA, with 100 μg/ml of tRNA as carrier. Following denaturation at 95°C for 5 minutes, an excess of cDNA was annealed to 6,000 dpm of globin mRNA and 18,000 dpm of 16S RNA for 15 minutes or for 10 hours at 65°C. Volumes ranged from 24 ml to 0.05 ml. The extent of duplex formation was determined by digestion with RNase A (20 μg/ml) and RNase T_1 (10 units/ml) for one hour at 25°C, and the acid precipitable material was collected and counted.

The globin mRNA used in these studies was labeled to high specific radioactivity ($>$100,000 cpm/μg) with ^{3}H-uridine (Merkel *et al.* 1976). Complementary DNA to reticulocyte 9S mRNA was prepared essentially according to Wood and Lingrel (1977), except that the reaction mixture contained 20 μg/ml of reverse transcriptase, 100 μg/ml of mRNA, 20 μg/ml of oligo $(dT)_{12\text{-}18}$, and 0.5 mM nucleotide triphosphates.

TABLE 1. *Percent total—10-minute labeled RNA bound to globin cDNA cellulose*

Experiment	% Bound
1	0.50
2	0.66
3	0.68
4	0.49
5	0.79
L-Cell RNA	<0.10

Nucleated erythroid cells (4 ml of packed cells in 60 ml of tissue culture media) were incubated for 10 minutes in the presence of 500 μCi of [5-^{3}H]uridine/ml, and the RNA isolated as described under Experimental Procedures. The percent bound by the globin cDNA column was calculated from the total number of counts originally applied to the column and the amount retained after the second application.

RESULTS

We have shown previously that our globin cDNA cellulose column specifically retains globin mRNA sequences (Wood and Lingrel 1977). Neither rRNA nor the poly(A)-containing RNA of L-cells is retained. Table 1 shows that when the RNA isolated from nucleated erythroid cells labeled for 10 minutes in the presence of ^{3}H-uridine is applied to the column, approximately 0.5% is retained following the second application. The second application to the column is necessary to remove the large amount of nonglobin mRNA that is present in these cells. Less than 0.1% of L-cell RNA is retained under identical conditions. When the bound RNA is analyzed using polyacrylamide gels run in the presence of formamide, two species of RNA are observed (Figure 1). The faster moving component comigrates with α- and β-globin mRNAs. The slower moving component migrates at approximately 16S, which is equivalent to a molecular weight of approximately 600,000. The 16S RNA can also be resolved from globin

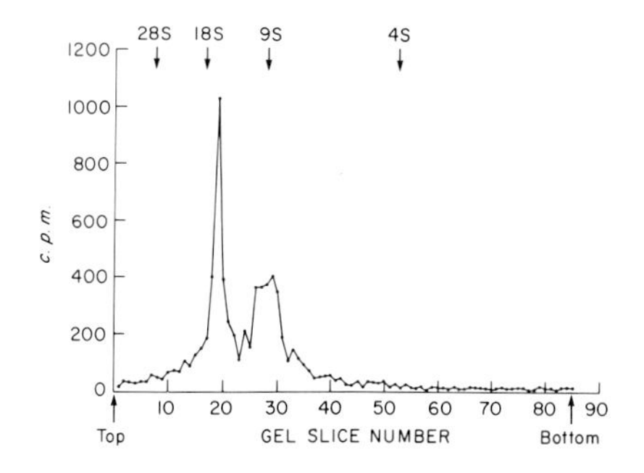

FIG. 1. Formamide-polyacrylamide gel electrophoresis of the RNA bound to the globin cDNA column. Two gels were run, one containing 18,000 dpm of globin cDNA bound RNA and the other containing 4S, 9S, 18S, and 28S reticulocyte marker RNAs. Marker RNA bands were visualized by staining and radioactivity determined by slicing the gels and counting the fractions. The results of both gels are superimposed.

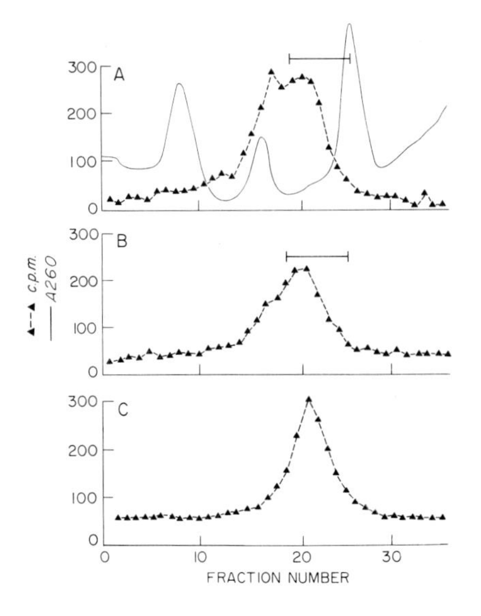

FIG. 2. Sucrose density gradient analysis of RNA bound to the globin cDNA column. The RNA was heated to 80°C for five minutes prior to layering on 10 to 30% linear sucrose gradients. The centrifugation was for 17 hours in an SW-40 rotor run at 17°C and 40,000 rpm. A, Total globin cDNA bound RNA; B, Recentrifugation of the material under the bar of (A); C, Recentrifugation of the material under the bar of (B). (———) A_{260}; (▲——▲) cpm.

mRNA on sucrose density gradients (Figure 2a). When material migrating in the region of 16S (designated by the bars) is recovered and recentrifuged, the pattern shown in Figure 2b is obtained. Recentrifugation of the 16S RNA from this gradient gives a single component (Figure 2c).

While the binding of the 16S RNA to the globin cDNA cellulose suggests that it contains globin mRNA sequences, this might not be the case. For example, it is possible that the globin mRNA used to prepare the cDNA cellulose contained a minor impurity that was copied by the reverse transcriptase. Because the cDNA cellulose column is run under conditions in which the cDNA is in excess and thus, driving the hybridization reaction, annealing between nonglobin cDNA and its complementary RNA could occur. It should be pointed out that whenever cDNA is used at high Dot's (Dot is defined as the concentration of cDNA in moles of nucleotide per liter times time in seconds), to assure complete hybridization, this possibility exists. In order to determine if the 16S RNA actually contains globin mRNA sequences, 16S RNA was purified by sucrose density gradient centrifugation (Figure 2c) and hybridized to nonradioactive cDNA at various Dot values (Figure 3). The conversion of radioactive RNA to RNase-resistant hybrids by cDNA was followed. If the RNA were complementary to a minor species present in the cDNA, it would hybridize at a much higher Dot½ than the globin cDNA. The Dot½ of the 16S RNA is 1.0×10^{-3}, which compares favorably with 0.75×10^{-4} for highly purified globin mRNA. The plateau value for the 16S RNA is approximately one-third that of globin mRNA, which would be expected if the 16S RNA contains only one copy of the globin mRNA sequence and is approximately three times its size. As little additional RNA hybridizes at the high Dot values, sequences

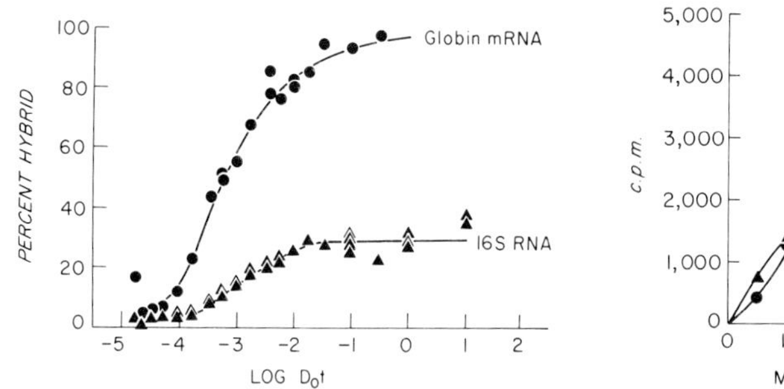

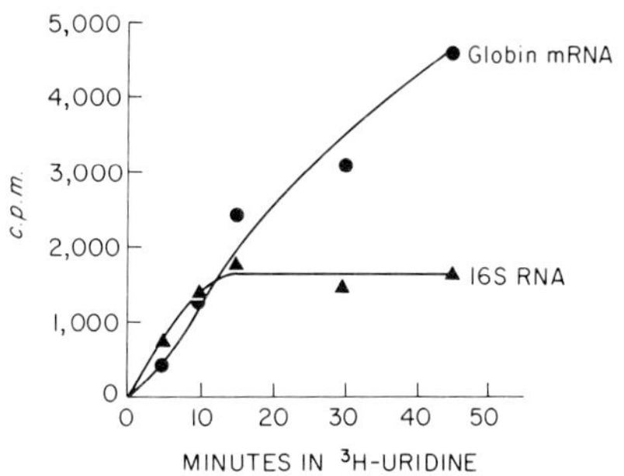

FIG. 3, left. A comparison of the kinetics of hybridization of globin mRNA and 16S RNA to excess nonlabeled globin cDNA. Hybridizations were carried out as described under Experimental Procedures. (●——●) globin mRNA; (▲——▲) 16S RNA.

FIG. 4, right. The accumulation of radioactivity in 16S RNA and globin mRNA during continuous labeling. Aliquots of cells were removed at various times after the introduction of label, the RNA isolated, and the amount of radioactivity in the 16S and mature globin mRNA determined. (●——●) globin mRNA; (▲——▲) 16S RNA.

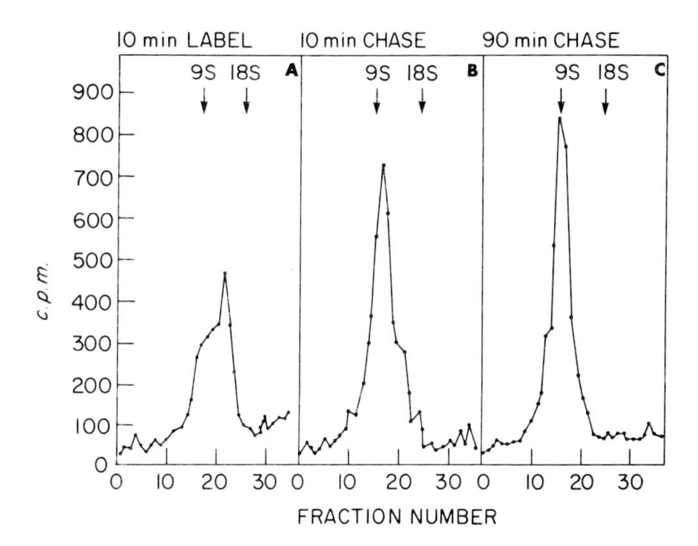

FIG. 5. Pulse-chase analysis of 16S and globin mRNA. Cells were incubated for 10 minutes in the presence of ^{3}H-uridine; actinomycin D and daunomycin were added to a final concentration of 40 μg/ml and 56 μg/ml, respectively, and the incubation continued. Cells were removed, the RNA isolated and applied to the globin cDNA cellulose column. The RNA retained by the cDNA cellulose was analyzed by sucrose density gradient centrifugation as described in Figure 2. A, RNA from cells labeled for 10 minutes in the presence of ^{3}H-uridine; B, RNA from cells labeled for 10 minutes in the presence of ^{3}H-uridine, followed by 10 minutes in the presence of actinomycin D and daunomycin; C, RNA from cells labeled for 10 minutes in the presence of ^{3}H-uridine, followed by 90 minutes in the presence of actinomycin D and daunomycin.

complementary to minor species of cDNA must be virtually absent in the 16S RNA preparation. It is therefore concluded that the 16S RNA does indeed contain globin mRNA sequences.

In order for the 16S RNA to be a precursor to globin mRNA its synthesis must be kinetically related to cytoplasmic globin mRNA. Both a time course of synthesis and a pulse-chase experiment were performed. The data of Figure 4 show that the 16S RNA is rapidly labeled and plateaus after approximately 15 minutes, while globin mRNA is initially labeled more slowly and continues to be labeled throughout the time course. This is as expected if the 16S RNA were a precursor to globin mRNA.

A pulse-chase study was also carried out. Cells were prelabeled for 10 minutes with ^{3}H-uridine at which time new RNA synthesis was inhibited by the addition of actinomycin D and daunomycin. The incubation was continued, and at 10 and 90 minutes after the addition of the drugs, globin mRNA-containing sequences were isolated and analyzed by sucrose density gradient centrifugation. At 10 minutes there is more radioactivity in the 16S RNA than in mature globin mRNA (Figure 5a). During the chase (Figure 5b and c), the radioactivity in the 16S RNA disappears, with a concomitant increase in radioactivity in mature globin mRNA. Again, these results are compatible with a precursor-product relationship between the 16S RNA and globin mRNA.

DISCUSSION

It has been proposed that mRNA in eukaryotic cells is synthesized as a larger precursor molecule (see reviews by Darnell 1974, Lewin 1975a,b); however, the identification and isolation of precursors to specific mRNAs has been a difficult task. In the study reported here we have shown that a highly specific globin mRNA affinity column can be used to isolate RNAs that contain globin mRNA sequences. Radioactive RNA from nucleated erythroid cells exposed to ^{3}H-uridine for 10 minutes was applied to the column, and the bound RNA was analyzed by denaturing polyacrylamide gels. Both mature globin mRNA and a larger globin mRNA-containing molecule were observed. However, because the isolation involved globin cDNA excess it was possible that the 16S RNA was complementary to a small contaminant in the cDNA. Kinetics of hybridization of the 16S RNA with nonlabeled excess cDNA revealed a Dot½ similar to that for authentic globin mRNA, thus proving that the 16S RNA contains globin mRNA sequences. The kinetics of labeling of the 16S RNA and globin mRNA both during continuous label and in pulse-chase studies are compatible with a precursor-product relationship between these two RNAs.

A similar RNA has been detected in mouse fetal liver cells by liquid cDNA hybridization (Ross 1976). This RNA is kinetically related to cytoplasmic globin mRNA. Erythroleukemia cells induced to synthesize hemoglobin also produce a larger RNA which contains globin mRNA sequences (Curtis and Weissmann 1976). These workers hybridized globin cDNA containing a poly(C) tail to total RNA in liquid and selected for the hybrid by applying the material to a poly(I) column. The finding of this RNA in such diverse erythroid cells as those from fetal mouse liver, anemic mouse spleens, and induced erythroleukemia cells lends support to the physiological significance of this RNA.

The finding of a precursor that is approximately three times the size of mature globin mRNA requires that the size of the globin gene be reconsidered. Rather than only being large enough to code for the approximately 660 transcribed nucleotides in globin mRNA, the globin gene must contain at least 2,000 nucleotides. Even this is a minimum estimate as there is no compelling evidence at the present time to suggest that the 16S RNA is the primary transcript of the globin genes. Shorter labeling times may reveal larger precursors. Proof that an RNA is a primary transcript requires that it contain a triphosphate at the 5′ terminus. Such an analysis has not yet been performed on the 16S globin mRNA precursor.

ACKNOWLEDGMENTS

This work was supported by Grant GM-10999 from the National Institutes of Health, Grant BMS 7401783 AO1 from the National Science Foundation, and Grant NP 59H from the American Cancer Society. T. G. W. was a postdoctoral fellow of the National Cancer Institute, and K. S. is a postdoctoral fellow of the Damon Runyon-Walter Winchell Cancer Fund. The reverse transcriptase used in these studies was obtained from the Office of Program Resources and Logistics, Viral Cancer Program, Viral Oncology, National Cancer Institute. The authors wish to thank Drs. Armentrout and Brown for their many helpful suggestions.

REFERENCES

Curtis, P. J., and C. Weissmann. 1976. Purification of globin messenger RNA from dimethylsulfoxide-induced Friend cells and detection of a putative globin messenger RNA precursor. J. Mol. Biol. 106:1061–1075.

Darnell, J. E., Jr. 1974. The origin of mRNA and the structure of the mammalian chromosome, *in* The Harvey Lecture Series #60. Academic Press, New York, pp. 1–47.

Darnell, J. E., Jr. 1978. Gene regulation in mammalian cells: Some problems and the prospects for their solution, *in* Cell Differentiation and Neoplasia (The University of Texas System Cancer Center 30th Annual Symposium on Fundamental Cancer Research, 1977), G. F. Saunders, ed., Raven Press, New York, pp. 347–359.

Imaizumi, T., H. Diggelmann, and K. Scherrer. 1973. Demonstration of globin messenger sequences in giant nuclear precursors of messenger RNA of avian erythroblasts. Proc. Natl. Acad. Sci. USA 70:1122–1126.

Knöchel, W., and H. Tiedemann. 1975. Size distribution and cell-free translation of globin-coding HnRNA from avian erythroblasts. Biochim. Biophys. Acta 378:383–393.

Kwan, S.-P., T. G. Wood, and J. B. Lingrel. 1977. Purification of a putative precursor of globin messenger RNA from mouse nucleated erythroid cells. Proc. Natl. Acad. Sci. USA 74:178–182.

Lewin, B. 1975a. Units of transcription and translation: The relationship between heterogeneous nuclear RNA and messenger RNA. Cell 4:11–20.

Lewin, B. 1975b. Units of transcription and translation: Sequence components of heterogeneous nuclear RNA and messenger RNA. Cell 4:77–93.

Macnaughton, M., K. B. Freeman, and J. O. Bishop. 1974. A precursor to hemoglobin mRNA in nuclei of immature duck red blood cells. Cell 1:117–125.

Melli, M., and R. E. Pemberton. 1972. A new method of studying the precursor-product relationship between high molecular weight RNA and messenger RNA. Nature New Biol. 236:172–173.

Merkel, C. G., T. G. Wood, and J. B. Lingrel. 1976. Shortening of the poly(A) region of mouse globin messenger RNA. J. Biol. Chem. 251:5512–5515.

Orkin, S. H., D. Swan, and P. Leder. 1975. Differential expression of α- and β-globin genes during differentiation of cultured erythroleukemia cells. J. Biol. Chem. 250:8753–8760.

Ostertag, W., G. Gaedicke, N. Kluge, H. Melderis, B. Weimann, and S. K. Duke. 1973. Globin messenger in mouse leukemic cells: Activity associated with RNA species in the region of 8 to 16S. FEBS Lett. 32:218–222.

Ross, J. 1976. A precursor of globin messenger RNA. J. Mol. Biol. 106:403–420.

Ruiz-Carrillo, A., M. Beato, G. Schutz, P. Feigelson, and V. G. Allfrey. 1973. Cell-free translation of the globin message within polydisperse high-molecular-weight ribonucleic acid of avian erythrocytes. Proc. Natl. Acad. Sci. USA 70:3641–3645.

Williamson, R., C. E. Drewienkiewicz, and J. Paul. 1973. Globin messenger sequences in high molecular weight RNA from embryonic mouse liver. Nature New Biol. 241:66–68.

Wood, T. G., and J. B. Lingrel. 1977. Purification of biologically active globin mRNA using cDNA-cellulose affinity chromatography. J. Biol. Chem. 252:457–463.

Cell Differentiation and Neoplasia, edited by Grady F. Saunders. Raven Press, New York

Regulation of Messenger RNA Translation

Harvey F. Lodish, John E. Bergmann, and Thomas H. Alton

Department of Biology, Massachusetts Institute of Technology, Cambridge, Massachusetts 02139

Regulation at the translational level—the interaction of ribosomes and messenger RNA—is an important means by which cells and viruses modulate synthesis of specific proteins. In the case of certain RNA viruses, such as the RNA phages, this is the only level at which control of protein synthesis is exerted (Zinder 1975). Prokaryotic and eukaryotic cells, and also DNA viruses, regulate gene expression primarily at the levels of synthesis, processing, or degradation of mRNA; translational controls have been shown to be important, however, in several systems.

The literature of the past fifteen years is replete with claims of translational control of protein synthesis in a large number of systems. Many of these involve studies in which cellular RNA synthesis is inhibited by actinomycin D or some other antibiotic, and synthesis of proteins is followed. These studies are, for the most part, impossible to evaluate because one is never certain if the drug inhibits synthesis of the mRNA in question or if it inhibits protein synthesis by means other than by blocking mRNA biogenesis. Many such side effects of actinomycin D have been documented (c.f. Lodish 1976). Other types of experiments on translational control involve the effects of various compounds and reaction conditions on the translation of different messenger RNAs in various cell-free systems. The problem with many such studies is that most of the cell-free systems in current use, such as crude extracts from cultured mammalian cells, are very inefficient; typically they synthesize less than one molecule of protein per molecule of added mRNA. The factors that limit the rate of protein synthesis in these extracts are not necessarily those which are limiting in the intact cell. All of the experiments indicating the existence of protein factors or RNA molecules required for the translation of only a specific class of mRNAs have utilized these types of extracts, and the existence of mRNA-specific translational elements remains unproved (Lodish 1974, 1976). In this paper we will review current concepts and recent experiments on translational control of protein synthesis. Our recent work on differentiation of the slime mold *Dictyostelium discoideum* has elucidated a particularly clear example of control at this level, and we will discuss it in some detail.

Present Address: T. H. Alton, Department of Biochemistry, Stanford University School of Medicine, Stanford, California 94305.

At which of the many steps of polypeptide biosynthesis are controls operative? How do they work? At the outset, five summary points can be made; the evidence for these generalities is found in a recent, rather extensive review (Lodish 1976) and in the discussion of more recent results that follows.

1) Control is exerted primarily at the level of polypeptide chain initiation, not elongation or termination.

2) Cells regulate the overall rate of polypeptide chain initiation affecting the translation of all mRNAs.

3) The same ribosomes, initiation factors, and elongation and termination factors are apparently required and are utilized for translation of all mRNAs in a cell. There is no compelling evidence that these ribosomes or factors change in a significant way during development of an organism. Nor is there strong evidence for the existence of components with absolute mRNA specificity that might result in preferential or exclusive translation of a particular class of mRNAs by a particular class of cells (Lodish 1976).

4) Different mRNAs have different affinities for ribosome subunits or other factors involved in the initiation process; thus, over a period of time, certain mRNAs (those with higher rate constants for chain initiation) will direct the synthesis of more molecules of protein than will those with lower initiation constants.

5) Changes in a cell of the concentration or activity of a component required for polypeptide chain initiation or elongation will not necessarily affect translation of all mRNAs to the same extent. As is demonstrated by the kinetic model of protein synthesis discussed below, any reduction in the concentration or activity of a component of the initiation process which acts at steps at or before binding of mRNA will result in preferential inhibition of translation of mRNAs with lower rate constants for polypeptide chain initiation (the poorer mRNAs). Hence, in cases where the overall rate of initiation of cellular protein synthesis is either increasing or decreasing with time, one would expect that translation of different mRNAs would be affected to different extents. Such changes would not require any alterations in the specificity of the protein synthetic apparatus.

KINETIC MODEL FOR PROTEIN SYNTHESIS

Before presenting the results of our recent computer modeling, it is perhaps appropriate to ask why such a model is useful or necessary. The complexity inherent in the multistep reactions of protein synthesis, together with the steric hinderance known to exist between translating ribosomes (MacDonald *et al.* 1968), has made it difficult to assess quantitatively the role any particular parameter might play in the overall rate of translation. In theory, each individual step in protein synthesis can be assigned an intrinsic rate constant (which, of course, may vary in different cells or physiological states). The rate of addition, say, of a leucine residue to a growing polypeptide chain in response to a particular codon should have a defined, determinable value. Likewise, the rate constant for binding of the initiation site on an mRNA to an initiation factor or to a met-tRNA$_f$-40S ribosome complex is also definable. The problem is that all of these rate constants and rates interact in a complex manner. Two examples will suffice to illustrate this. First, if the density of ribosomes on an mRNA is high enough (as it is, for instance, in normal reticulocytes), "collisions" become appreciable; a ribosome may have reached a particular codon on the mRNA but be physically blocked from further elongation until its downstream neighbor ribosome has moved along (Lodish 1974). Second, one may consider the case in which a ribosome had just initiated protein synthesis on an mRNA. Ribosomes protect a region of mRNA much larger than the codons being read. Another ribosome cannot initiate protein synthesis at this site until the first has moved along the mRNA a sufficient number of nucleotides to vacate and expose the initiation site. The number of ribosomes that can initiate protein synthesis on a given mRNA per unit of time (the flux) is governed not only by the intrinsic ability of the mRNA to bind to ribosomes but also, in a complex way, by the rate of addition of each of the internal amino acids (the elongation rate constants). Prediction of the effects on protein synthesis caused by alterations in the rates of individual initiation or elongation stages is not a simple matter; complete solution of the complex rate equations requires the use of a digital computer.

Polypeptide synthesis on ribosomes can be described most simply by the following scheme:

$$m + R^* \xrightarrow{K_1} P_1 \xrightarrow{K_2} P_2 \xrightarrow{K_3} P_3 \xrightarrow{K_4} \cdots \xrightarrow{K_S} P_S \xrightarrow{K_R} P_R \qquad (1)$$

The binding of met-tRNA$_f$ to the 40S subunit is the first step in protein synthesis in eukaryotic cells (Darnbrough *et al.* 1973, Schrier and Staehelin 1973); in this case, R* is the concentration of the met-tRNA$_f$-40S ribosome complex able to bind an mRNA. The concentration of this component, R*, is assumed to be rate limiting for chain initiation. The concentration of the mRNA is m, and K_1 is the overall rate constant for the binding of the 40S and 60S ribosomal subunit to the initiation region of an mRNA when it is unblocked by a ribosome. Clearly, there are several factors and steps involved in the initiation process (Weissbach and Ochoa 1976). The use of an overall rate constant K_1, possibly different for different mRNAs, is sufficient to establish the main conclusions of the model. Later we will consider the case in which K_1 is itself a complex function of the concentrations of various factors. The polypeptide contains S amino acids and the message S codons; P_n (n = 1, 2, 3 . . . S) represents a nascent chain of n amino acids. P_R is the released polypeptide chain. K_n (n = 2, 3 . . . S) is the rate constant for addition of the nth amino acid; K_R is the rate constant for release of the completed polypeptide chain. Ribosomes are known to cover many more codons on a message than those being decoded at any instant; the value L is the number of codons covered by one ribosome.

Solution of these equations (Bergmann and Lodish, in press) is done with a

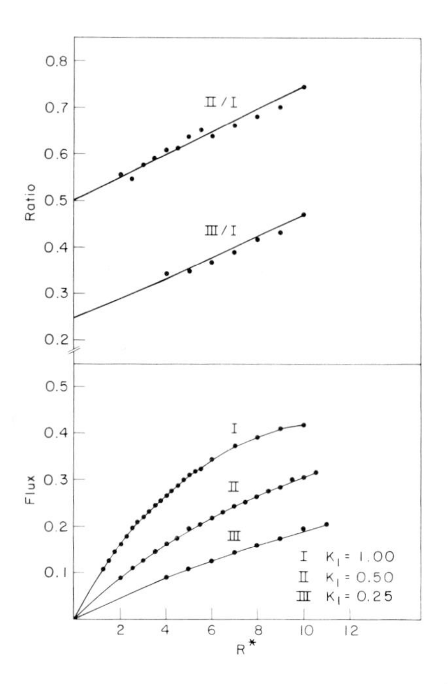

FIG. 1. Theoretical calculation of parameters of protein synthesis as a function of R*, the concentration of met-tRNA$_f$-40S ribosome subunits capable of initiating protein synthesis. A value of 100 is assumed for K_e (rate constant for elongation); 10 for K_R (rate constant for termination); and 12 is taken for L, the number of codons covered by one ribosome. Rate constants for chain initiation, K_1, are taken as 1.00 (curve I), 0.50 (curve II) and 0.25 (curve III). The bottom panel shows the amount of protein synthesized per mRNA per unit time (the flux) plotted as a function of R*, and the top panel shows the ratio of the amount of protein synthesized by the different mRNAs, plotted as a function of R*. Details of the calculation are given in Bergmann and Lodish (in press). Since a stochastic method is employed (see text), the data points will not necessarily fall on a smooth curve.

Monte Carlo model approach. Each of the rate constants is treated as a probability. The computer starts with 100 naked mRNAs and determines stochastically, during each interval of time, whether a ribosome initiates synthesis on any given mRNA or whether a nascent peptide is elongated by one amino acid during that interval.

It is easiest to illustrate our findings by modeling a greatly simplified translation process. Later we will introduce some of the complexities. We begin by assuming that the rate constant for addition of any but the first amino acid is the same, that is, K_n ($n = 2, 3 \ldots S$) $= K_e$, the single elongation rate constant. Also, $K_R = 0.1\ K_e$, that is, the time for release of a completed chain is that required for addition of 10 amino acids. This is the situation which obtains in reticulocytes in which the termination rate has been measured directly (Lodish and Jacobsen 1972). Figure 1B shows a theoretical plot as a function of R* of the synthesis of three proteins, the mRNAs for which differ in K_1 but have the same value of K_e. The upper panel makes the important point that changes in R* affect differently the translation of these different mRNAs. Reduction in the value of R*—*nonspecific* reduction in the overall rate of chain initiation—results in preferential inhibition of translation of the mRNAs with lower values of K_1. The implications of this conclusion (c.f., Lodish 1974) are discussed in a later section of this review.

This model can be extended in several productive directions. It can easily encompass the case in which binding of an initiation factor (possibly IF-3, see below) to mRNA is a prerequisite for attachment to ribosomes. If the concentration of the free factor is f, and the equilibrium constant for this binding is K_f, then we have

$$m + F \rightleftarrows (mf) \quad \text{message + factor} \overset{K_f}{\rightleftarrows} \text{message-factor complex} \qquad (2)$$

$$(mf) + R^* \xrightarrow{K_a} P_1 \quad \text{message-factor complex} + R^* \xrightarrow{K_a} P_1 \qquad (3)$$

Suppose two mRNAs differ only in K_f—the affinity for an essential rate-limiting initiation factor—but otherwise have identical values for the other rate constants (K_a, K_e, etc.). The fraction of mRNA bound to factor, as a function of the concentration of free factor, F, is given in Figure 2. (Very similar results are obtained if K_f is considered as a forward rate constant, when reaction [3] follows so closely that [2] is not in equilibrium.) The major conclusion from this graph is that changing the concentration of F affects differently the amount of protein made by the two mRNAs; reduction in F inhibits preferentially the translation of the mRNA with the lower constant for binding to F. Note that this conclusion is very similar to those seen in Figure 1 and is in fact expected; comparing equation (1) with (2) and (3), it is seen that K_1 and R^* in equation (1) are quite equivalent to K_f and F in equations (2) and (3). Reducing the overall level of chain initiation (R^*) in Figure 1 is equivalent in effect to lowering the concentration of a rate-limiting initiation factor (Figure 2).

This kinetic model can also be used to study the effects of limitations in specific tRNAs or termination factors on the overall rate of protein synthesis. As will be seen, the effects are very slight. First, Figure 3 shows a calculated distribution of ribosomes along an mRNA the size of globin. As expected from experimental results on globin (Hunt *et al.* 1968, Luppis *et al.* 1970), the distribution is quite uniform. However, there is a preponderance of ribosomes with a completed but unreleased protein chain; this is a direct consequence of setting

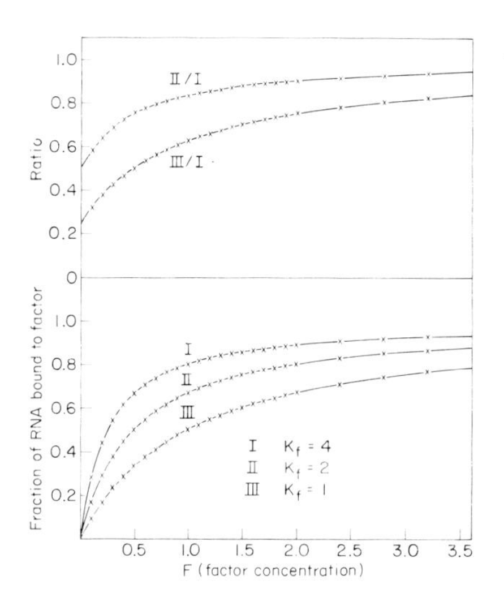

FIG. 2. Theoretical calculation of the effect of a rate-limiting initiation factor, F. It is assumed that a complex of mRNA and factor is a prerequisite stage in initiation of protein synthesis and that different mRNAs have different equilibrium rate constants, K_f, for binding to this factor. Values of K_f are taken as 4.0 (mRNA I), 2.0 (II), and 1.0 (III). The bottom panel shows the fraction of mRNA bound to factor as a function of F, and the top panel shows the ratio of the amount of two mRNAs bound to factor. The latter value will be proportional to the relative rate of polypeptide chain initiation.

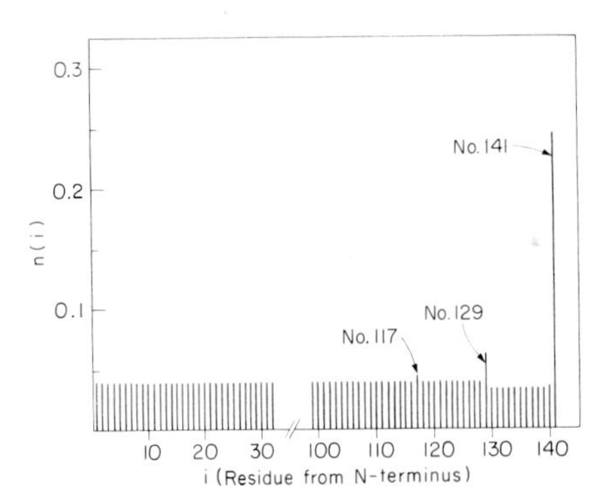

FIG. 3. Theoretical plot of the distribution of ribosomes along a globin mRNA. Plotted is the probability, n(i), that a given mRNA has on it a ribosome at the ith codon or, equivalently, that it is being translated by a ribosome with a nascent chain of i amino acids in length. The completed chain is assumed to be 141 amino acids long, and any ribosome is assumed to cover 12 codons. It is also assumed that $K_1 = 7.0$; $K_e = 100$; and $K_R = 15$.

$K_R = 0.15\ K_e$. Note also an excess number of ribosomes 12 codons upstream from the COOH-terminal codon. This arises because, with a certain probability, a ribosome at that codon cannot advance because a downstream ribosome with an unreleased completed chain is protecting 12 codons and thus is blocking the path of the upstream ribosome.

Table 1 shows that reducing the rate of chain termination even below what is here assumed has little effect on the overall rate of protein synthesis. As long as the rate constant for termination is greater than that for chain initiation, the queue of ribosomes does not extend back to the initiation site. The overall rate of chain initiation remains unchanged, and the flux of ribosomes moving

TABLE 1. *Calculated effect of reducing the rate of polypeptide chain termination on the overall rate of protein synthesis and on the average size of polyribosomes*

K_R Rate constant for chain termination	Flux	Polysome size (ribosomes per mRNA)
1	0.80	11.4
2	1.60	9.3
3	2.10	6.1
4	2.14	4.9
5	2.16	4.3
6	2.16	3.7
10	2.16	3.4

In this calculation, the size of the protein is assumed to be 141 amino acids. The elongation rate constant, K_e, is set equal to 100 for all residues: $K_1 = 1$; $R^* = 3$ (i.e., the total rate constant for initiation is $1 \times 3 = 3$). Shown is the effect of varying K_R, the rate constant for chain termination, on the overall rate of protein synthesis per mRNA (the flux) and on the average number of ribosomes bound to each mRNA in the steady state. Note that there is a reduction in the flux only when K_R is less than the overall rate constant for chain initiation (here equal to 3.0). However, as the value of K_R is increased above 3.0, there is a marked decrease in the number of ribosomes bound to each mRNA (last column) due to the reduction in queuing of ribosomes behind a ribosome bearing an unreleased, completed nascent chain. This reduction in ribosome density is not accompanied by a change in the overall flux of ribosomes along the mRNA.

TABLE 2. *Calculated effect of a single rate-limiting codon on overall rate of globin synthesis*

Position of codon	Control flux (%)
2	0.68
13	0.78
25	0.85
37	0.90
49	0.91
61	0.91
73	0.92

In this calculation it is assumed that the mRNA encodes a protein of 141 amino acids and that a ribosome covers 12 codons. It is assumed that $K_1 = 1$; $R^* = 5$ (i.e., the overall rate constant for initiation is $1 \times 5 = 5$); and $K_R = 10$. Each elongation rate constant except one is set at 100. Shown is the effect on the overall flux, or rate of protein synthesis per mRNA, by assuming a value of $K_e = 6$ for a single amino acid codon. The data in column 2 are normalized to the control case where all values of K_e are set to 100.

along the mRNA will be unchanged. Thus, the termination step is not likely to be a point of regulation of protein synthesis.

As might be expected from the above result, specific internal codons are also unlikely to be possible points of regulation of protein synthesis. As an illustration, Table 2 shows the case in which all of the elongation constants are set equal to K_e except one, which is assigned the value $0.06K_e$. If this slow or rate-limiting codon is more than 12 codons from the initiation site, negligible reduction in the rate of protein production is observed. Even if this codon is near the beginning of the mRNA, the effect is slight; this low effect is simply due to blocking of the initiation site by a ribosome "stuck" a fraction of the time at this codon. Although not shown here, the cumulative effect of several such rate-limiting codons is additive, but it is still unlikely that differential control at the level of chain elongation could be effective.

REGULATION OF mRNA TRANSLATION OCCURS AT THE INITIATION STEP

It is generally assumed that elongation and termination of all nascent polypeptides in a cell occur at the same rate. This is the case for the α- and β-globins and for other polypeptide chains in reticulocytes and for certain proteins in the hen oviduct, the only instances in which this has been measured directly (Lodish and Jacobsen 1972, Hunt 1974, Palmiter 1972). The rate of elongation of the same mRNA could differ in different cells. Different cell types often contain different amounts of iso-accepting tRNA species, which could be reflected in changes in the rates of reading of certain mRNA codons (e.g., see Smith

1975). There is no evidence as to whether these differences between cells affect the *overall rate* of translation of different mRNAs. As mentioned above, our kinetic modeling studies suggest that controls at the termination stage cannot be an efficient regulatory mechanism and that modulation at the level of a single internal tRNA can be important only if the codon is located very near the amino terminus of the protein. Even if there are specific points at which elongation of globin chains is reduced (Protzel and Morris 1974), our calculations indicate that these would not serve to affect the *overall* rate of α or β chain synthesis.

Rabbit reticulocytes do contain a spectrum of tRNAs that appear specialized for the synthesis of hemoglobin. Compared to a liver cell, they contain a relatively large amount of $tRNA_{his}$ and a relatively small amount of $tRNA_{ileu}$; hemoglobin is particularly rich in histidine residues and poor in isoleucine (Smith and McNamara 1971). In such a case, where reading of many codons in an mRNA is affected, it is clear that a slight increase in the rate of production of hemoglobin molecules can result. Changes in tRNA composition to match utilization can also be ascribed to overall cellular optimization of translational efficiency. A ribosome that is "struck" at some codon not only slows elongation on that message but also reduces the number of ribosomes that are available to translate any other message in the cell.

REGULATION OF THE OVERALL LEVEL OF POLYPEPTIDE CHAIN INITIATION

A particularly clear example of this type of control is the hemin requirement for globin synthesis in reticulocytes (reviewed in Hunt 1976, London *et al.* 1976). When deprived of iron (a component of hemin), intact reticulocytes stop synthesizing globin. Likewise, when a reticulocyte lysate is incubated in the absence of hemin, synthesis of globin is normal for about 10 minutes, but then ceases. This inhibition is manifest only at the level of initiation of protein synthesis. There is build-up of an inhibitor that blocks the attachment of met-$tRNA_f$ to the 40S subunit; apparently, this is mediated by the phosphorylation of one of the subunits of IF-MP, the initiation factor that binds met-$tRNA_f$ to the small ribosome subunit and thus inactivating it (Hunt 1976, Levin *et al.* 1976, Traugh *et al.* 1976, Ranu and London 1976). This inhibitor can be overcome by addition of excess factor IF-MP. At high levels, this inhibitor blocks initiation of synthesis of α- and β-globin and of all other reticulocyte proteins. As another example, during mitosis in cultured mammalian cells, the overall rate of polypeptide chain initiation decreases threefold (Fan and Penman 1970). In this case, however, the nature of the factor(s) affected is not clear.

If rabbit reticulocytes are deprived of a specific amino acid, such as isoleucine, protein synthesis is inhibited. As might be expected, the average size of β-globin polyribosomes increases, as ribosomes are queued behind the now rate-limiting ileu codons (Kazazian and Freedman 1968). By contrast, when cultured mammalian cells are deprived of an essential amino acid or when synthesis of any aa-tRNA is inhibited, the polyribosomes are found to become much smaller. This is indicative of a block in polypeptide chain initiation much greater than that at the elongation step (Vaughan and Hansen 1973). How deprivation of an amino acid results in specific block at the chain initiation step is unclear, but it is possible that inactivation of a specific initiation factor is involved.

A similar phenomenon occurs in the first few minutes of differentiation in the cellular slime mold *Dictyostelium discoideum* (Alton and Lodish, in press). *Dictyostelium* grow as single-celled amoebae. Development is initiated by removal of exogenous nutrients; recent work has shown that removal of any of several amino acids from the growth medium suffices to initiate the developmental cycle (Marin 1976). Beginning at six to eight hours after starvation, the cells aggregate, forming a multicellular organism containing about 10^5 cells, which ultimately differentiate into spore cells and stalk cells (reviewed in Jacobson and Lodish 1975, Loomis 1975). Gradually, during the first six to eight hours, a large number of new enzymes appear in the cells, along with several new

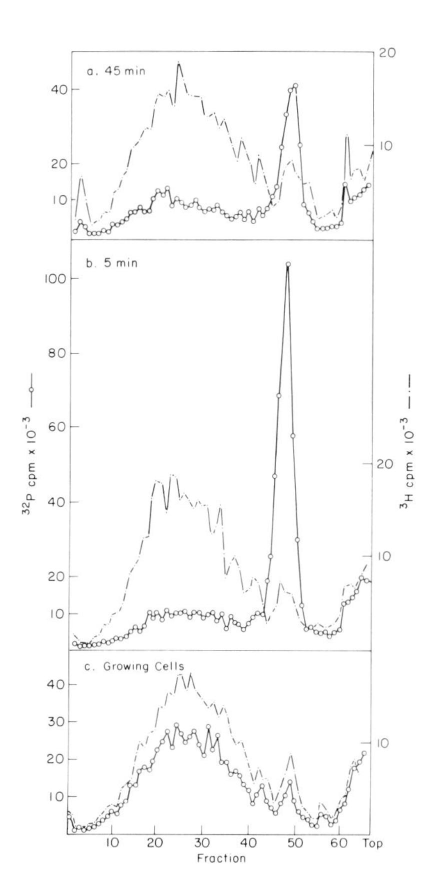

FIG. 4. Polysomes in differentiating *Dictyostelium* cells. Vegetative cells were labeled with either ^{32}P (40 μCi/ml) or ^{3}H-uracil (70 μCi/ml) for seven generations. Cells labeled with ^{32}P were plated for differentiation; at the time indicated in panels a and b, they were mixed with ^{3}H-labeled cells in ratio of 2:3. Polysomes were prepared as described by Alton and Lodish (in press). Cells were lysed by rapid agitation at 0° C for one minute in a buffer containing 50 mM Hepes, pH 7.5, 40 mM $MgAc_2$, 20 mM KCl, 1 mg/ml heparin, 5% sucrose, and 4% Cemulsol NP 12. Cellular debris was removed by rapid high-speed centrifugation, and the resulting cytoplasmic extract layered onto linear gradients of sucrose from 7% to 50% (w/v) made up in the above buffer, but without the Cemulsol and heparin. The gradients were centrifuged at 26,500 rpm at 1° C for three hours in a Beckman SW27 rotor. Gradients were collected and fractionated from the bottom; aliquots were precipitated in 6% ice cold TCA, filtered, dried and counted. a, ^{32}P vegetative cells mixed with ^{3}H vegetative cells; b, ^{32}P cells after five minutes of differentiation mixed with ^{3}H vegetative cells; c, ^{32}P cells plated 45 minutes mixed with ^{3}H vegetative cells. Sedimentation is from right to left; the peak of 80S monoribosomes is at fraction 48.

surface molecules concerned with aggregation and chemotaxis. But the very earliest biochemical change noted is an immediate threefold reduction in the rate of polypeptide chain initiation; this is reflected in a threefold drop in the average size of polyribosomes within five minutes after initiation of differentiation (Figures 4, 5). This effect is reversible; polysomes re-form quickly if the cells are returned to growth medium (Figure 5F), suggesting that there is no irreversible inactivation of ribosomes in these cells. Starvation is the key factor in inducing polysome runoff; manipulations associated with plating the cells are not important (c.f. Figure 5D). A variety of hybridization and cell-free translation studies have established that the amount and nature of the mRNA species in these early differentiating cells is identical to that of growing cells (Alton and Lodish, in press). As is later discussed in more detail, this reduction in the overall rate of chain initiation is accompanied by a selective inhibition of translation of certain of these mRNA species.

A related question concerns the existence and function of inactive or masked mRNA molecules in cells. Perhaps the best example of this is in sea urchin eggs, in which following fertilization or artificial activation, there is a fivefold

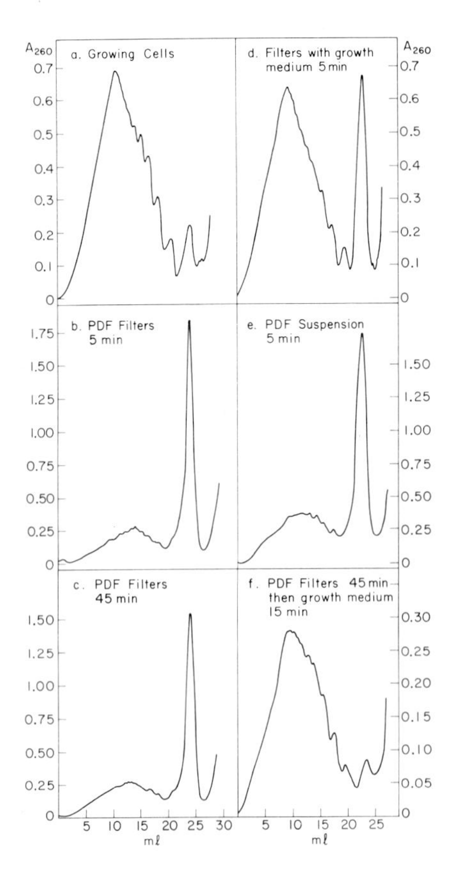

FIG. 5. Profiles of polysomes in *Dictyostelium* cells. Polysomes were prepared and analyzed as described in Figure 4; the gradients were pumped through a 5 mm flow cell and the absorbance at 260 nm monitored continuously. Sedimentation is from right to left. a, growing axenic cells; b, cells plated for 5 minutes on Millipore filters in PDF (phosphate buffered saline); c, cells plated for development for 45 minutes as in b; d, cells plated on Millipore filters containing growth medium; e, cells shaken in PDF suspension for 5 minutes; f, cells plated for development 45 minutes (as in c) then returned to growth medium for 15 minutes.

or greater rise in protein synthesis and a concomitant rise in the proportion of ribosomes engaged in protein synthesis. This increase does not depend on synthesis of new mRNAs—it occurs in activated enucleated eggs or in embryos treated with amounts of actinomycin sufficient to block new RNA synthesis (Humphreys 1971, Gurdon 1974). The majority of protein synthesis during the early cleavage stages is supported by mRNA transcribed during oogenesis and stored in the cytoplasm in inactive ribonucleoprotein particles. Of considerable importance is the finding that, as analyzed by two-dimensional gel electrophoresis, the pattern of proteins made at all of the early stages of egg differentiation is the same (Brandhorst 1976). Thus, there does not appear to be a sequential activation of different classes of mRNAs after fertilization; rather, all classes of mRNAs seem to be activated or released from RNPs (ribonucleoproteins) at about the same rate.

There is some evidence that mRNAs for actin and other muscle proteins are stored in inactive RNPs prior to fusion of myoblasts (Buckingham *et al.* 1974, Bag and Sarkar 1975); these mRNPs are believed to be activated following the cell fusion, which generates a muscle-like myotube. What is not clear is if all classes of cellular mRNAs are sequestered in these RNPs and if different RNPs are activated at different times after fusion. This point is crucial with respect to the possibility of message-specific proteins (or RNAs) that hold only certain mRNA species in inactive RNPs.

RELATIVE RATES OF INITIATION OF TRANSLATION OF DIFFERENT EUKARYOTIC mRNAs

It appears that different eukaryotic mRNAs may differ in the rate of attachment to ribosomes or in other stages of initiation of protein synthesis, but the number of systems studied in this regard is small. Direct comparison of the rates of protein synthesis with the amounts of mRNA have shown that the ten monogenic mRNAs of reovirus in infected cells differ at least tenfold in the rate of chain initiation (Joklik 1973).

In rabbit reticulocytes each molecule of α-globin mRNA initiates protein synthesis only 60% as frequently as does each β-globin mRNA (Lodish and Jacobsen 1972, Hunt 1974). This conclusion derives primarily from the fact that each β-globin mRNA contains about one and a half times as many ribosomes as does each α-globin mRNA and that the rate of polypeptide chain elongation and release are the same for the two mRNAs. Since reticulocytes make equal amounts of α- and β-globin, there must be one and a half times as much α-globin as β-globin mRNA (reviewed in Lodish 1976). It is possible to estimate the relative amounts of different mRNAs by measuring the relative amounts of protein made under conditions in which chain elongation is slowed by antibiotics (e.g., cycloheximide or sparsomycin). Under such conditions, the initiation step is not the limiting one, and the amount of protein made is proportional only to the amount of mRNA, not to its intrinsic affinity for ribosomes. Such experiments on reticulocytes show that the ratio of α-globin mRNA to β-globin mRNA is about 1.4:1 (Lodish 1971).

Possibly, these differences can be ascribed to intrinsic differences in the RNA structure or sequences, which perhaps involve RNA secondary structure, as is the case in the RNA phages. An alternative, but not mutually exclusive, possibility is that these variations result from differences in the mode of interaction with an initiation factor or the 40S ribosome subunit. A recent study suggests that translation of β-mRNA has a lower requirement for initiation factor IF-M3 or M4 than does α-mRNA; addition of IF-M3 or M4 to cell-free systems will preferentially enhance translation of α-mRNA (Kabat and Chappell, in press).

Many other eukaryotic mRNA have been claimed to differ in their ability to initiate protein synthesis, but in most cases, the experiments are less complete. One potentially useful system was devised by Koch and his associates. They showed that treatment of mammalian cells with hypertonic medium results in an inhibition of initiation of protein synthesis (Saborio *et al.* 1974). Whereas the mechanism of this shut-off is unknown, it is of interest that the hypertonic treatment affects differently the translation of different mRNAs. Those viral and cellular mRNAs whose translation is relatively resistant to the high salt treatment are believed to have the higher rate constants for chain initiation, an assumption that is consistent with the results of the kinetic model discussed earlier. In at least one case, these results could be checked by an independent means. Translation of light chain immunoglobin mRNA is more resistant to hypertonic salt treatment than that of average myeloma cell mRNA, suggesting that it has a higher rate constant for chain initiation than does the average mRNA in myeloma cells (Nuss and Koch 1976, Sonenshein and Brawerman 1976a). A number of studies both on cell-free translation of myeloma mRNA and on the effects of inhibitors of polypeptide chain elongation on synthesis of immunoglobin proteins by intact myeloma cells are in agreement with this conclusion (Nuss and Koch 1976, Sonenshein and Brawerman 1976b). Whether hypertonic salt treatment is a valid method for ascertaining initiation rate constants of all mRNAs is not clear; it is possible that the results relate only to the relative translatability of mRNAs in solutions of high ionic strength and that the correlations with immunoglobin mRNAs is fortuitous.

DIFFERENTIAL TRANSLATION OF mRNAs

There are a number of rather well-studied cases in which developmental changes or physiological treatments affect the relative translation of different mRNAs in a cell. For a number of these, it has been proposed that changes in the amount of message-specific translation elements is involved, but as we have argued previously, the evidence for the existence of such mRNA-specific factors is not compelling (Lodish 1974, 1976). In several cases, differential translation of mRNAs is consistent with, and explained by, results of our kinetic model of protein synthesis. Although in some cases the biochemical evidence supporting this interpretation is incomplete, it does seem worthwhile discussing these examples and their interpretation.

Perhaps the best example is the inhibition of globin synthesis in intact reticulocytes or in cell-free extracts by the deprivation of hemin. Although translation of both α- and β-globin mRNA is inhibited, at limiting concentrations of hemin, translation of α-mRNA is inhibited preferentially (Lodish 1974, Beuzard and London 1974). Given the fact that α-mRNA has a lower efficiency of chain initiation relative to β-mRNA, this preferential inhibition of α synthesis is an explicit prediction of our kinetic model.

A similar mechanism has been proposed to explain the mechanism by which infection of animal cells by the picornaviruses, poliovirus, or EMC virus results in a rapid drop in the rate of host cell-protein synthesis. Cellular mRNA remains intact following infection and can be translated in cell-free systems as efficiently as can RNA from infected cells (Leibowitz and Penman 1971, Lawrence and Thach 1974, Colby *et al.* 1974, Fernandez-Munoz and Darnell 1976). Poliovirus inhibition of HeLa cell protein synthesis does not require replication of poliovirus RNA because the inhibition still occurs in the presence of inhibitors of RNA replication such as guanidine. Inhibition of host synthesis is followed by a period of poliovirus or EMC protein synthesis; this is dependent on synthesis of polio RNA (Leibowitz and Penman 1971). One hypothesis is that poliovirus infection results in a change in the specificity of the protein synthesis apparatus for the 5′ end of mRNA such that polio mRNA, which contains the 5′ sequence pUp . . . can be translated in preference to cellular mRNA with the usual 5′ sequence $m^7G^{5'}ppp^{5'}NmpNp$ . . . (Fernandez-Munoz and Darnell 1976). Alternatively, it is possible that there is a nonspecific reduction in the rate of polypeptide chain initiation resulting in a reduced rate of initiation of translation on both cellular and poliovirus mRNA. Poliovirus RNA would be translated preferentially if it had a lower requirement for (higher affinity for) whatever initiation component is rate limiting (Lawrence and Thach 1974, Nuss *et al.* 1975). Supporting this interpretation are cell-free translation studies which show that translation of EMC RNA requires much less initiation factor IF-M3 than does the average cellular mRNA. When IF-M3 is limiting, EMC is translated in preference to cellular mRNA, and addition of IF-M3 stimulates preferentially translation of cellular mRNA (Golini *et al.* 1976). It is postulated that infection of cells by EMC or poliovirus results in inactivation of initiation factor IF-M3, but this has not been shown directly.

A possibly similar example concerns the changing pattern of protein synthesis that occurs during the first half hour of differentiation in the slime mold *Dictyostelium discoideum* (Alton and Lodish, in press). As analyzed by two-dimensional gel electrophoresis, growing cells synthesize about 100 definable proteins in amounts greater than 0.07% of the cells' total protein synthesis (Figure 6). During the first 15 to 20 minutes of differentiation, at a time when the overall rate of chain initiation is reduced threefold (see above), synthesis

FIG. 6. Two-dimensional gel of proteins synthesized by vegetative cells. Cells growing at 2×10^6/ml were labeled with 300 μCi/ml ^{35}S-methionine for 20 minutes. Total cellular protein was dissolved by boiling in a buffer containing 80 mM Tris, pH 6.8, 2% sodium dodecyl sulfate, and 0.1 M dithioerythrotol. The procedure for two-dimensional gel analysis was essentially that of O'Farrell (1975). The first dimension (left to right) is isoelectric focusing between pH 2 (left) and 10 (right). The second dimension (top to bottom) is electrophoresis in a gel containing sodium dodecyl sulfate and an exponential 9–20% gradient of acrylamide. Shown is an autoradiogram of the dried gel. The largest proteins migrate near the top of the gel; the molecular weight of actin is 43,000. Details are given in Alton and Lodish (in press).

of no new proteins can be detected on two-dimensional gels. Strikingly, at least six proteins made by growing cells are not synthesized by cells at 15, 30, or 60 minutes of differentiation. These are labeled spots 1, 6, 7, 8, 9, and 10 in Figures 6 and 7. These polypeptides, made by growing cells, are not turned over preferentially during differentiation.

Two explanations of this result are possible: The mRNAs for proteins 1, 6, 7, 8, 9, and 10 are unstable in early developing cells, or the mRNAs are present and intact but are not being translated. Our recent experiments suggest that the latter explanation is correct.

The total amount of translatable mRNA that can be extracted from growing cells or cells at 15 or 60 minutes of differentiation is the same. Fractionation by two-dimensional gel electrophoresis of the translation products of cytoplasmic mRNA isolated from growing cells in a wheat germ, cell-free extract shows the synthesis of a large number of authentic proteins, including actin and proteins 1, 6, 7, 8, 9, and 10 (Figure 8, c.f. Figure 6). Since the amounts of these proteins made are proportional to the amount of added *Dictyostelium* mRNA, this system provides an assay of the amount of translatable mRNA for these proteins. Figure 9 shows that the amount of translatable mRNA for polypeptides 1, 6, 7, 8, 9, and 10 present in cells at 15 minutes of differentiation is unchanged from that in growing cells, in spite of the fact that these proteins are not made in detectable amounts by cells after 15 minutes of development. Even when mRNA is prepared from cells four hours after initiation of the differentiation cycle, the amount of translatable mRNAs for species 1, 6, 7, 8, 9, and 10 is at least 50% that in mRNA preparations from growing cells.

It is concluded that early in differentiation of *Dictyostelium,* cells contain several species of mRNA which are specifically not being translated. As noted previously, the overall rate of protein synthesis drops precipitously early in differentiation; translation of all proteins is reduced (about threefold) while that of species such as 1, 6, 7, 8, 9, and 10 is reduced even more. The following explanation is speculative, but testable; it does not invoke mRNA-specific translation elements. It is proposed that the activity of one initiation factor, say IF-M3, is reduced, possibly as a direct consequence of the starvation of amino acids, and that translation of mRNAs for polypeptides 1, 6, 7, 8, 9, and 10 requires a higher level of this factor than does that of other *Dictyostelium* mRNAs. This would result in preferential reduction of synthesis of proteins 1, 6, 7, 8, 9, and 10. Or the concentration of met-$tRNA_f$-40S complexes could be reduced, possibly due to inactivation of initiation factor IF-MP. This would result in preferential inhibition of synthesis of proteins 1, 6, 7, 8, 9, and 10, if the mRNAs encoding these proteins have a lower than average rate constant for binding to met-$tRNA_f$-40S ribosome subunit complexes. Obviously, proof of this explanation will require additional experimentation, similar to the approaches used for the study of α- and β-globin mRNA.

Early differentiation in *Dictyostelium* exhibits a very clear example of differentiation translation of mRNAs. The value of the above explanations, based on

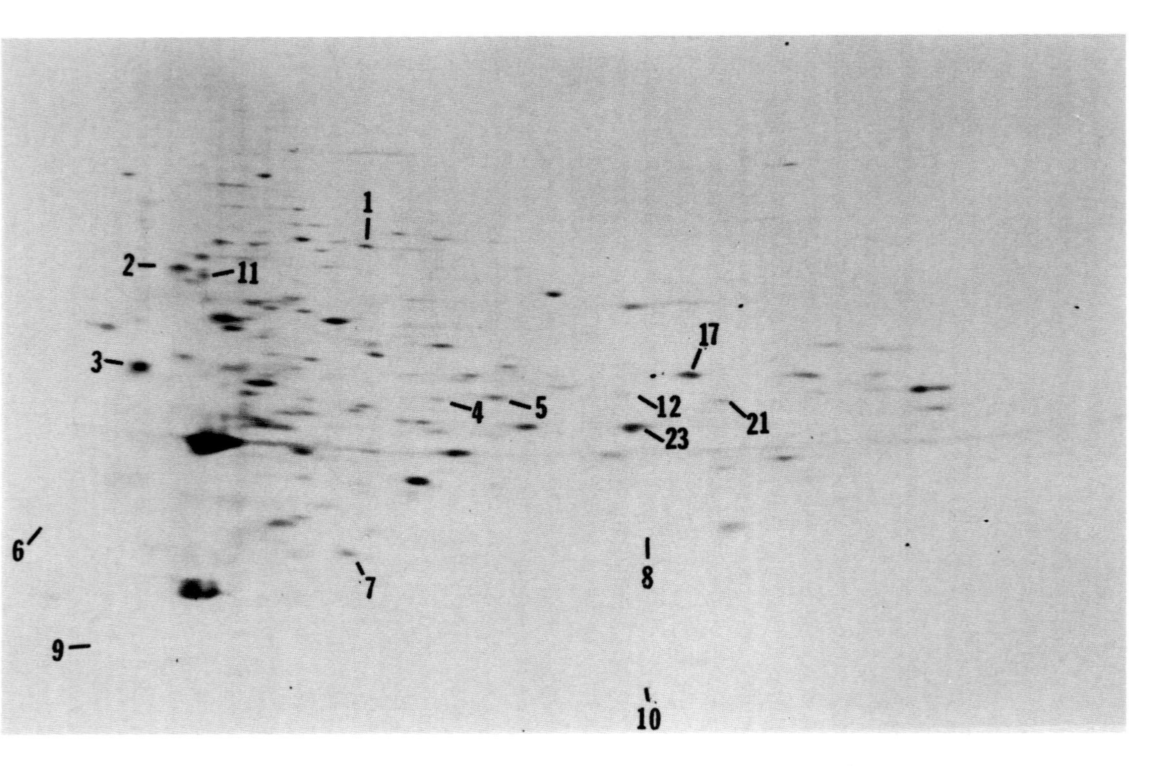

FIG. 7. Two-dimensional gel analysis of proteins synthesized by cells immediately after plating. 6×10^6 cells were labeled for 15 minutes with 50 μCi of ^{35}S-methionine immediately after plating for differentiation. Analysis was as in Figure 6.

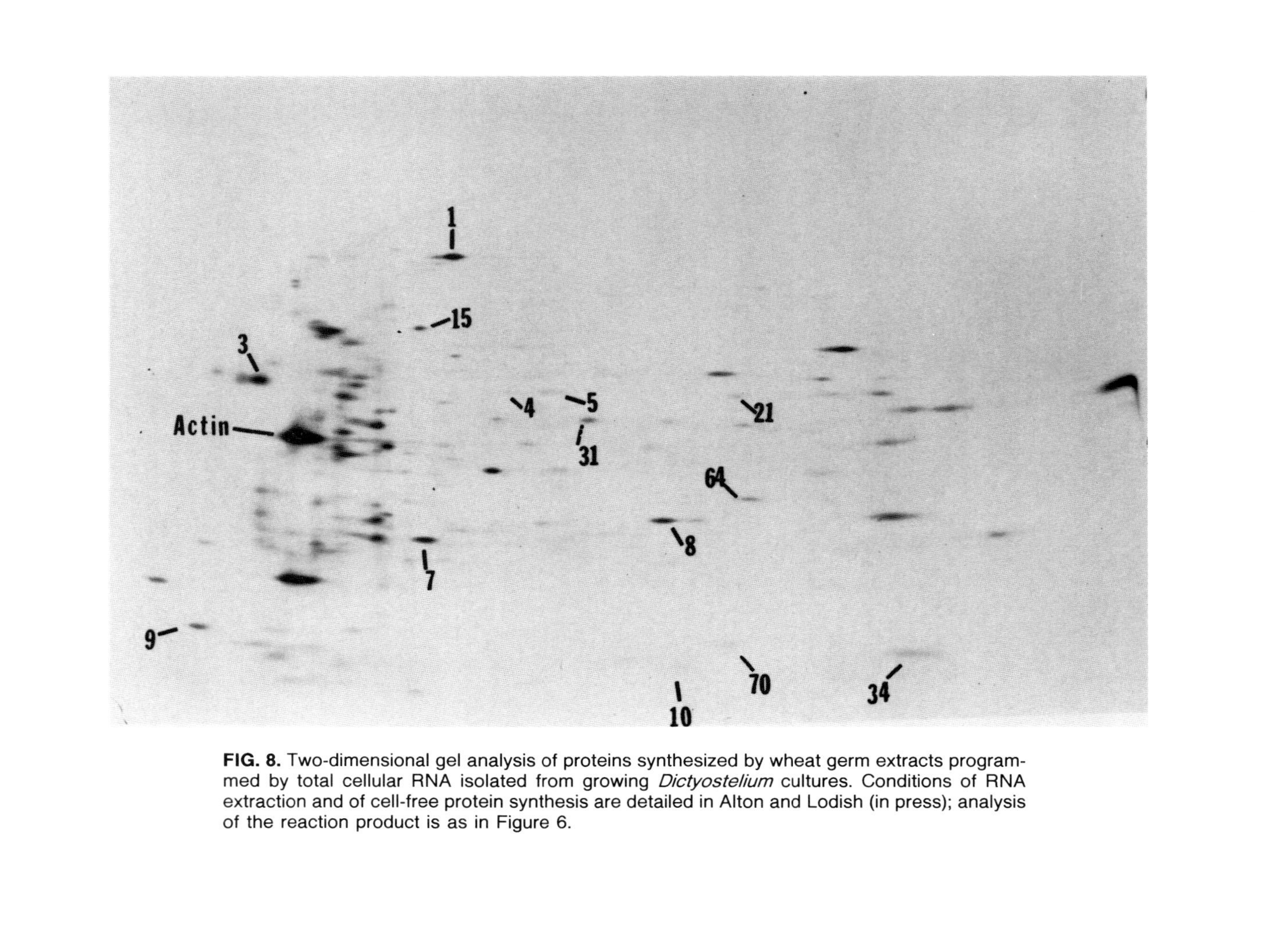

FIG. 8. Two-dimensional gel analysis of proteins synthesized by wheat germ extracts programmed by total cellular RNA isolated from growing *Dictyostelium* cultures. Conditions of RNA extraction and of cell-free protein synthesis are detailed in Alton and Lodish (in press); analysis of the reaction product is as in Figure 6.

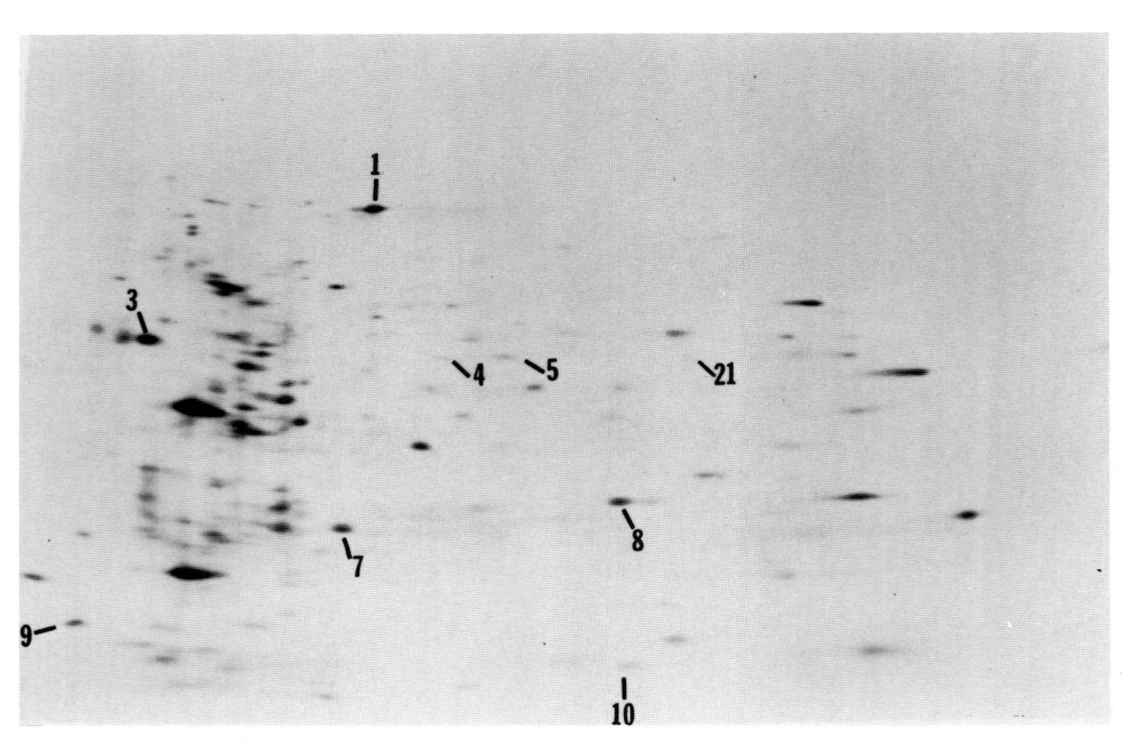

FIG. 9. Two-dimensional gel analysis of proteins synthesized by wheat germ extracts programmed by total cellular RNA isolated from cultures at 15 minutes of differentiation. Analysis is as in Figure 6.

results of our kinetic modeling of protein synthesis, is that it shows how such changes can be brought about without invoking changes in the specificity of the protein synthetic apparatus.

ACKNOWLEDGMENTS

This work was supported by Grants AI-08814 and AM 15929 from the U.S. National Institutes of Health, PCM 74–04869 A02 from the National Science Foundation, and NP 180 from the American Cancer Society.

REFERENCES

Alton, T., and H. F. Lodish. 1977. Translational control of protein synthesis during the initial stages of differentiation in the slime mold *Dictyostelium discoideum.* (In press).

Bag, J., and S. Sarkar. 1975. Cytoplasmic nonpolysomal messenger ribonucleoprotein containing actin messenger RNA in chicken embryonic muscles. Biochemistry 14:3800–3807.

Bergmann, J., and H. F. Lodish. 1977. Kinetic modeling of initiation, elongation, and termination of protein synthesis. (In press).

Beuzard, Y., and I. M. London. 1974. The effects of hemin and double-stranded RNA on α and β globin synthesis in reticulocyte and Krebs II ascites cell-free systems. Proc. Natl. Acad. Sci. USA 71:2863–2866.

Brandhorst, B. P. 1976. Two-dimensional gel patterns of protein synthesis before and after fertilization of sea urchin eggs. Dev. Biol. 52:310–317.

Buckingham, M., D. Caput, A. Cohen, R. G. Whalen, and F. Gros. 1974. The synthesis and stability of cytoplasmic messenger RNA during myoblast differentiation in culture. Proc. Natl. Acad. Sci. USA 71:1466–1470.

Colby, D., V. Finnerty, and J. Lucas-Lennard. 1974. Fate of mRNA in L-cells infected with mengovirus. J. Virol. 13:858–869.

Darnbrough, C., S. Legon, T. Hunt, and R. Jackson. 1973. Initiation of protein synthesis: Evidence for mRNA-independent binding of met-tRNA to the 40S ribosomal subunit. J. Mol. Biol. 76:379–403.

Fan, H., and S. Penman. 1970. Regulation of protein synthesis in mammalian cells. II. Inhibition of protein synthesis at the level of initiation during mitosis. J. Mol. Biol. 50:655–679.

Fernandez-Munoz, R., and J. E. Darnell. 1976. Structural difference between the 5′ termini of viral and cellular mRNA in poliovirus-infected cells. J. Virol. 18:719–726.

Golini, F., S. S. Thach, C. H. Birge, B. Safer, W. C. Merrick, and R. E. Thach. 1976. Competition between cellular and viral mRNAs *in vitro* is regulated by a messenger discriminatory factor. Proc. Natl. Acad. Sci. USA 73:3040–3044.

Gurdon, J. B. 1974. The Control of Gene Expression in Animal Development. Harvard University Press, Cambridge, pp. 37–74.

Humphreys, T. 1971. Measurement of messenger RNA entering polysomes upon fertilization of sea urchin eggs. Dev. Biol. 26:201–208.

Hunt, T. 1974. The control of globin synthesis in rabbit reticulocytes. Ann. NY Acad. Sci. 241:223–231.

Hunt, T. 1976. Control of globin synthesis. Br. Med. Bull. 32:257–261.

Hunt, T., A. R. Hunter, and A. J. Munro. 1968. Control of hemoglobin synthesis: Distribution of ribosomes on the messenger RNA for α and β chains. J. Mol. Biol. 36:31–45.

Jacobson, A., and H. F. Lodish. 1975. Genetic control of development of the cellular slime mold *Dictyostelium discoideum.* Ann. Rev. Genet. 9:145–186.

Joklik, W. 1973. The transcription and translation of reovirus RNA, *in* Virus Research, C. F. Fox and W. S. Robinson, eds., Academic Press, New York and London, pp. 105–126.

Kabat, D., and M. R. Chappell. 1977. Competition between globin messenger ribonucleic acids for a discriminating initiation factor. J. Biol. Chem. (In press).

Kazazian, H., and M. Freedman. 1968. The characterization of separated α and β polyribosomes in rabbit reticulocytes. J. Biol. Chem. 253:6446–6450.

Lawrence, C., and R. E. Thach. 1974. Encephalomyocarditis virus infection of mouse plasmacytoma cells. I. Inhibition of cellular protein synthesis. J. Virol. 14:598–610.
Leibowitz, R., and S. Penman. 1971. Regulation of protein synthesis in HeLa cells. III. Inhibition during poliovirus infection. J. Virol. 8:661–668.
Levin, D. H., R. J. Ranu, V. Ernst, M. A. Fiter, and I. M. London. 1976. Regulation of protein synthesis in rabbit reticulocytes: Phosphorylation of methionyl-$tRNA_f$ binding factor by protein kinase activity of translational inhibitor isolated from heme-deficient lysates. Proc. Natl. Acad. Sci. USA 73:3112–3116.
Lodish, H. F. 1971. Alpha and beta globin messenger ribonucleic acid: Different amounts and rates of initiation of translation. J. Biol. Chem. 246:7131–7138.
Lodish, H. F. 1974. Model for the regulation of mRNA translation applied to hemoglobin biosynthesis. Nature 251:385–388.
Lodish, H. F. 1976. Translational control of protein synthesis. Ann. Rev. Biochem. 45:39–72.
Lodish, H. F., and M. Jacobsen. 1972. Regulation of hemoglobin synthesis: Equal rates of translation and termination of α and β globin chains. J. Biol. Chem. 247:3622–3629.
London, I. M., M. J. Clemens, R. S. Ranu, D. Levin, L. Cherbas, and V. Ernst. 1976. The role of hemin in the regulation of protein synthesis in erythroid cells. Fed. Proc. 35:2218–2222.
Loomis, W. F., Jr. 1975. *Dictyostelium discoideum,* A Developmental System. Academic Press, New York.
Luppis, B., A. Bargellesi, and F. Conconi. 1970. Control of hemoglobin synthesis at the translational level. Biochemistry 9:4175–4179.
MacDonald, C., J. Gibbs, and A. Pipkin. 1968. Kinetics of biopolymerization on nucleic acid templates. Biopolymers 6:1–25.
Marin, F. T. 1976. Regulation of development in *Dicytostelium discoideum.* I. Initiation of the growth to development transition by amino acid starvation. Dev. Biol. 48:110–117.
Nuss, D. L., and G. Koch. 1976. Variation in the relative synthesis of immunoglobin G and non-immunoglobin G proteins in cultured MPC-11 cells. J. Mol. Biol. 102:601–612.
Nuss, D. L., H. Opperman, and G. Koch. 1975. Selective blockage of initiation of host protein synthesis in RNA virus infected cells. Proc. Natl. Acad. Sci. USA 72:1258–1262.
O'Farrell, P. H. 1975. High resolution two-dimensional electrophoresis of proteins. J. Biol. Chem. 250:4007–4021.
Palmiter, R. D. 1972. Regulation of protein synthesis in the chick oviduct. II. Modulation of polypeptide elongation and initiation rates by estrogen and progesterone. J. Biol. Chem. 247:6770–6780.
Protzel, A., and A. J. Morris. 1974. Gel chromatographic analysis of nascent globin chains. Evidence for non-uniform size distribution. J. Biol. Chem. 249:4594–4600.
Ranu, R., and I. M. London. 1976. Regulation of protein synthesis in rabbit reticulocyte lysates: Purification and initial characterization of the cyclic 3′5′-AMP independent protein kinase of the heme-regulated translational inhibitor. Proc. Natl. Acad. Sci. USA 73:4349–4353.
Saborio, J. L., S-S. Pong, and G. Koch. 1974. Selective and reversible inhibition of initiation of protein synthesis in mammalian cells. J. Mol. Biol. 85:195–211.
Schrier, M., and T. Staehelin. 1973. Initiation of eucaryotic protein synthesis: [Met $tRNA_f \cdot$ 40S ribosome] Initiation complex catalyzed by purified initiation factors in the absence of mRNA. Nature New Biol. 242:35–38.
Smith, D. W. E. 1975. Reticulocyte transfer RNA and hemoglobin synthesis. Science 190:529–535.
Smith, D. W. E., and A. L. McNamara. 1971. Specialization of rabbit reticulocyte transfer RNA content for hemoglobin synthesis. Science 171:577–581.
Sonenshein, G., and G. Brawerman. 1976a. Regulation of immunoglobin synthesis in mouse myeloma cells. Biochemistry 15:5497–5501.
Sonenshein, G., and G. Brawerman. 1976b. Differential translation of mouse myeloma messenger RNAs in wheat germ cell-free system. Biochemistry 15:5501–5506.
Traugh, J., S. M. Tahara, S. B. Sharp, B. Safer, and W. Merrick. 1976. Factors involved in initiation of hemoglobin synthesis can be phosphorylated *in vitro.* Nature 263:163–165.
Vaughan, M. H., and B. S. Hansen. 1973. Control of protein synthesis in human cells. Evidence for a role of uncharged transfer RNA. J. Biol. Chem. 248:7087–7096.
Weissbach, H., and S. Ochoa. 1976. Soluble factors required for eucaryotic protein synthesis. Ann. Rev. Biochem. 45:191–216.
Zinder, N. D., ed. 1975. RNA Phages. Cold Spring Harbor Laboratory, Cold Spring Harbor, New York.

Control of Differentiation and Neoplasia—Gene Expression in Early Development and Cancer

Cell Differentiation and Neoplasia, edited by Grady F. Saunders. Raven Press, New York

Synthesis of Stage-Specific Proteins in Early Embryogenesis

David A. Wright

Department of Biology, The University of Texas System Cancer Center M. D. Anderson Hospital and Tumor Institute, Houston, Texas 77030

Evidence from a variety of sources, including the work of several participants in this Symposium, indicates that in amphibians, the time just before gastrulation is a time of activation of embryonic genes directing the synthesis of RNA (Brown 1964, Bachvarova and Davidson 1966, Gurdon 1974). Prior to that time, metabolic activity is dependent on gene products made during oogenesis. Studies of hybrid embryos also indicate the gastrula stage as a time of embryonic gene expression. Many species hybrids arrest at the gastrula stage due to a nuclear-cytoplasmic incompatability (Brachet 1974, Subtelny 1974). Viable hybrids between those with different developmental rates show a maternal rate prior to gastrulation but an intermediate rate thereafter (Moore 1941).

It is the purpose of this paper to examine the proteins made in frog embryos by using electrophoretic variants of proteins to determine if that protein is made under the direction of templates present in the maternal cytoplasm or produced by the expression of nuclear genes.

What types of proteins should be examined? In this paper I will consider two classes of proteins: (1) enzymes of known function, and (2) proteins synthesized during brief periods in early development. First, I shall describe how we have examined the expression of electrophoretic phenotypes in frog embryos heterozygous for genes coding for enzymes of known function. I do this partly for its intrinsic interest and partly because it demonstrates rather simply the rationale for using genetic variants of proteins in the analysis of gene expression in embryonic development.

EXPRESSION OF GENES FOR ENZYMES IN HYBRID EMBRYOS

Figure 1 illustrates four different kinds of frog embryos that result when experimental crosses are made using artificial means to induce ovulation and fertilize eggs (Rugh 1934). These embryos are all the same age (tail-bud stage) and are the result of combinations of gametes from four animals. The parents are from two different populations of the *Rana pipiens* complex—*Rana pipiens* from Vermont (V) and *Rana berlandieri* (T) from Texas, a frog that is now

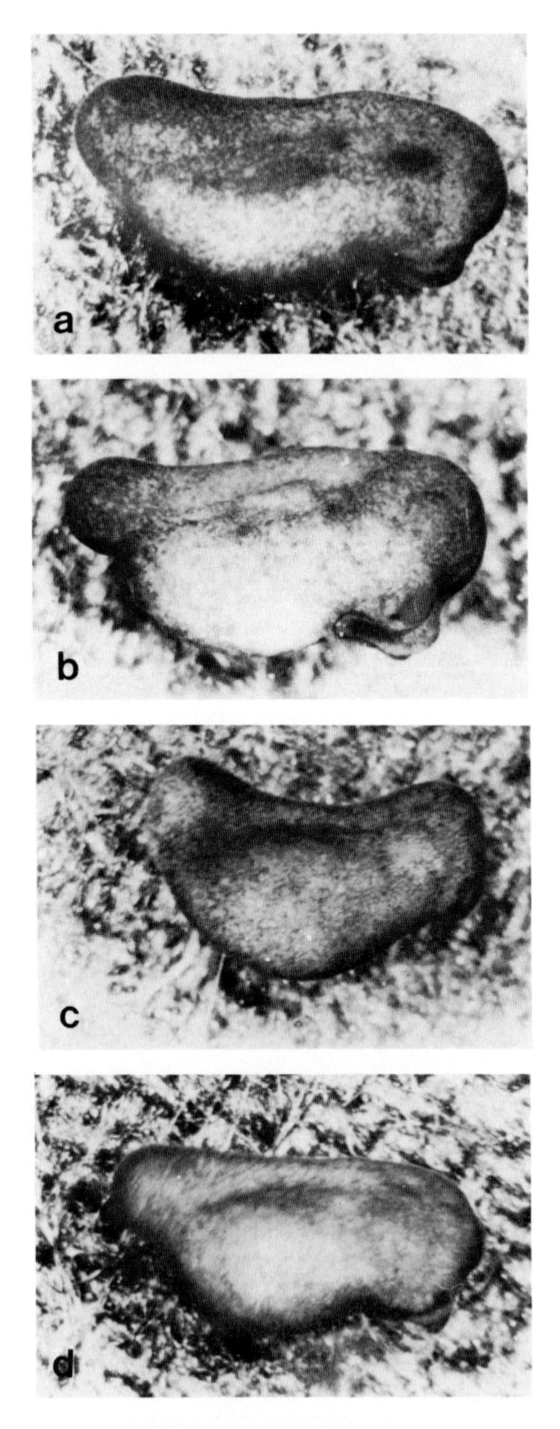

FIG. 1. Photographs of tail-bud frog embryos resulting from an experimental cross. a, Vermont female × Vermont male (VV); b, Vermont female × Texas male (VT); c, Texas female × Vermont male (TV); d, Texas female × Texas male (TT).

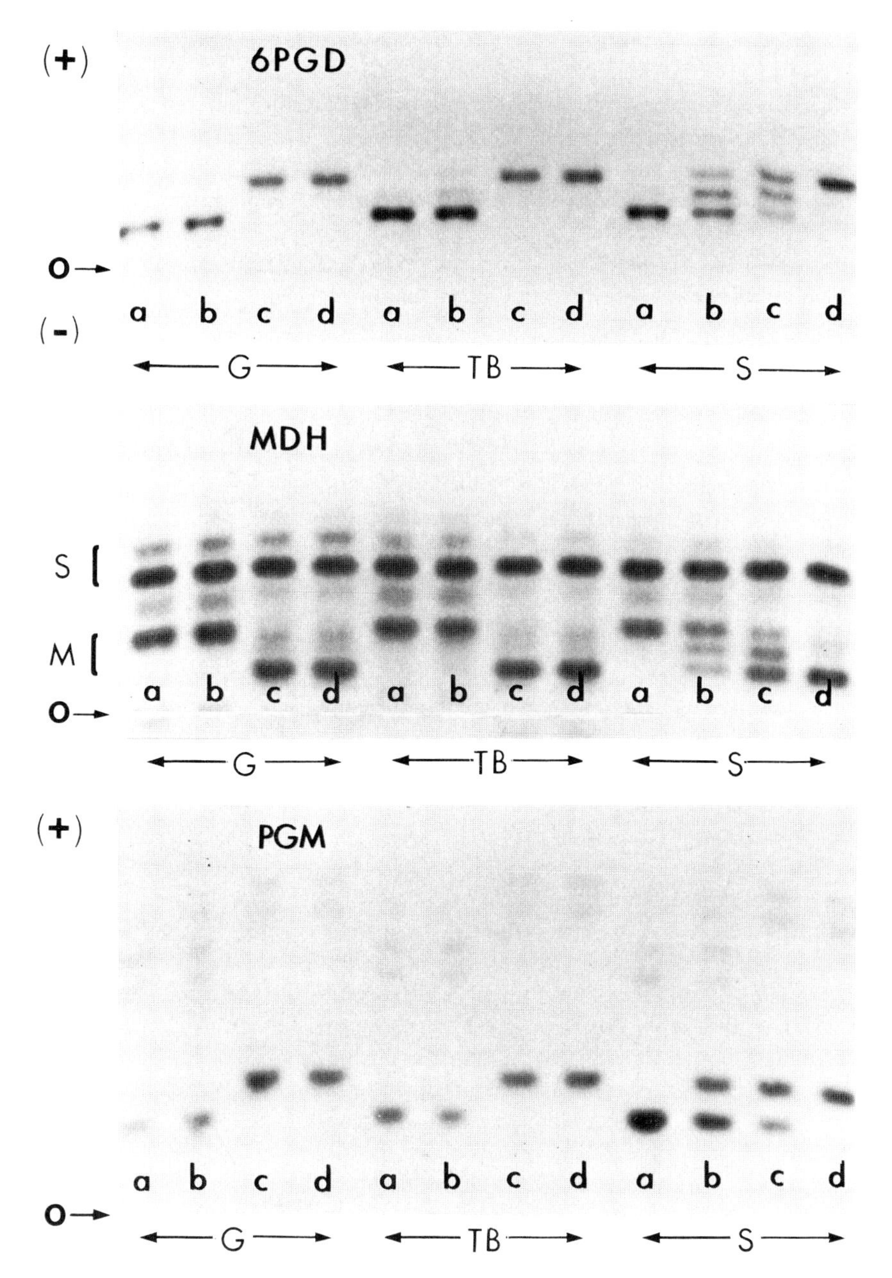

FIG. 2, top. Photograph of a starch gel slice stained for 6-phosphogluconate dehydrogenase (6PGD). Electrophoretic patterns of extract of embryos at the gastrula stage, "G," tail-bud, "TB," and swimming, "S," of the four different cross-types of Vermont (V) and Texas (T) frogs. a, VV; b, VT; d, TV; d, TT are shown. Note hybrid enzyme pattern in tail-bud VT hybrids (TB-b) and in both reciprocal hybrids at swimming stage (S-b, c).

FIG. 3, middle. Photograph of a different slice of the same starch gel as illustrated in Figure 2 but stained for malate dehydrogenase (MDH). The supernatant MDH (S) is the same in the Vermont and Texas frogs, while the mitochondrial MDH (M) is different. Note hybrid pattern for this enzyme in the reciprocal hybrids at swimming stages (S-b, c).

FIG. 4, bottom. Photograph of a slice of the same starch gel shown in Figures 2 and 3, but stained for phosphoglucomutase (PGM). Both maternal and paternal alleles for this enzyme are expressed in hybrids by the swimming stage (S-b,c).

regarded as a separate species. There are abnormalities in the reciprocal hybrids. In the V × T hybrids, the embryos are macrocephalic, while the T × V embryos are microcephalic. These kinds of hybrids have been described in detail by others (Moore 1969, Fowler 1961). For examples of expression of embryonic genes and maternal effects on enzyme phenotype we have considered three enzymes. Methods for starch gel electrophoresis and enzyme staining are from Siciliano and Shaw (1976).

Figure 2 illustrates a slice of a starch gel stained specifically for 6-phosphogluconate dehydrogenase (6PGD). The two parental types differ in the electrophoretic mobility of this enzyme. The first four samples are extracts of gastrula embryos. Patterns of a and d are the pure V and T types; b and c are the reciprocal hybrids. The maternal-type pattern of the hybrids is obvious here. The next four samples repeat the order for tail-bud stages (the stage illustrated in Figure 1). A hybrid enzyme pattern consisting of both parental species and a band of intermediate electrophoretic mobility in one of the hybrid combinations is present. Hybrid patterns again are evident in both reciprocal hybrids at the swimming stages. The three-band pattern is typical for a dimer molecule. The hybrid molecule, consisting of a subunit from each parental type, is an indication that both alleles are being expressed.

The same type pattern is shown for malate dehydrogenase (MDH) in which the supernatant form is invariant, but a difference is seen in the mitochondrial enzyme (Figure 3). Again, the phenotype is maternal in the reciprocal hybrids until about the hatching stage, at which time the nuclear gene for this enzyme is expressed. A similar time of expression of nuclear phenotype is seen in Figure 4 for the enzyme phosphoglucomutase (PGM). In the case of phosphoglucomutase, the enzyme is a monomer. The hybrids have the parental types with no intermediate hybrid form.

It seems clear from these studies that the genes for a variety of what might be considered "housekeeping proteins" are not among the genes activated at the gastrula stage. On the contrary, the embryo has a store of these kinds of proteins in the maternal egg cytoplasm. The evidence from nuclear cytoplasmic hybrids suggests that there are no maternal cytoplasmic templates for the synthesis of these enzymes functioning in early stages. The genome of the embryo does not direct synthesis of these proteins until relatively late in embryonic development (Wright and Moyer 1966, 1968, Wright and Subtelny 1971, Wright 1975). It was the use of genetic variants that made possible this analysis of nuclear vs. cytoplasmic effects on molecular phenotypes.

EXPRESSION OF GENES FOR PROTEINS SYNTHESIZED IN EARLY EMBRYOS

What proteins are synthesized in the early embryo? Are nuclear genes directing the synthesis of those proteins? A combination of techniques is now available to study proteins synthesized in embryos, helping us to answer these questions.

Microinjection of radioactive precursors allows us to label macromolecules synthesized by intact amphibian embryos, which are rather impermeable to such precursors in solution. High resolution polyacrylamide gel electrophoresis combined with autoradiography of the gel (Laemmli 1970) allows us to identify changing patterns in protein synthesis. When these procedures are carried out without detergents such as sodium dodecyl sulfate, charge differences between allelic variants of a protein are also detectable. The details of procedures used in these studies are published elsewhere (Wright, in press).

By using these techniques, I have attempted to demonstrate the following: (1) that there are differences in the proteins synthesized in different embryonic stages; (2) that genetic variants can be found in these stage-specific proteins; and (3) that the genetic differences can be used in hybrid embryos as an indication of the expression of nuclear genes either of the embryo or of the maternal genome active during oogenesis.

Figure 5 shows a picture of an autoradiogram of one of our earliest gels. It

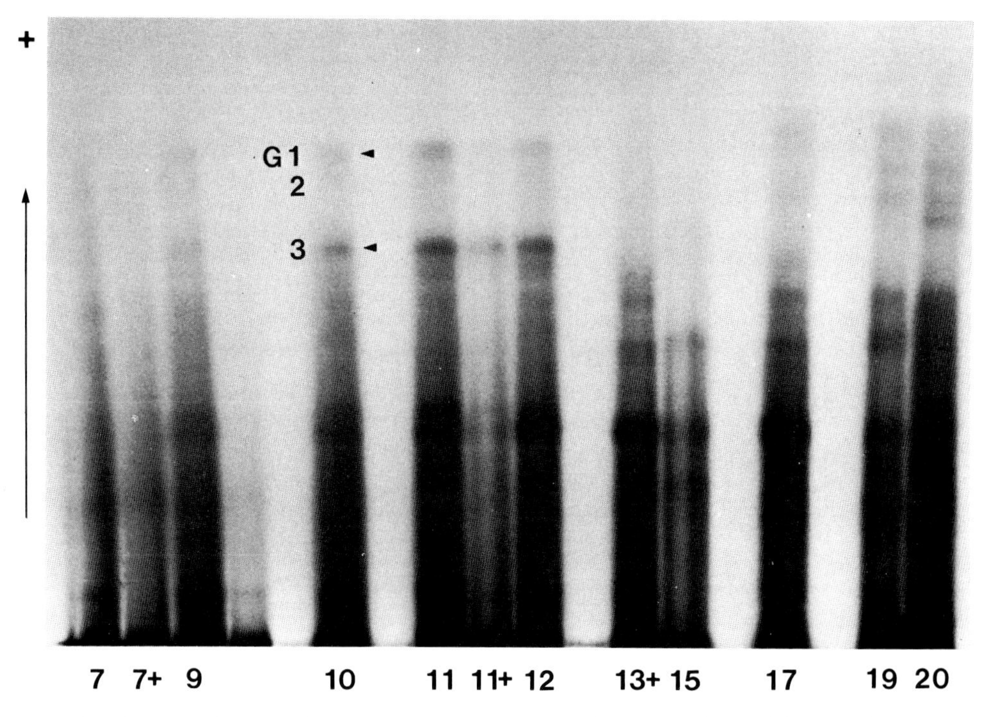

FIG. 5. Autoradiogram of polyacrylamide gel showing changing pattern of incorporation of ^{14}C amino acids into proteins during embryonic development. The labeled amino acid mixture was injected into embryos of different stages and incubated for three hours. Soluble proteins were separated on a polyacrylamide gel slab. Proteins synthesized were detected by placing a sheet of X-ray film over the dried gel. Note the synthesis of certain proteins in the late blastula stage and gastrula stages (9–12) not synthesized earlier and whose synthesis has stopped by the neurula, tail-bud, or hatching stages (13–20).

illustrates some of the problems in resolution and inadequate incorporation of label in early developmental stages, especially in cleavage and early blastula stages. It also illustrates that late blastula and gastrula stages synthesize certain proteins not seen in cleavage. Some of these gastrula proteins are not made by neurula or older embryos. One protein in particular that is not made in cleavage, synthesized in late blastula and gastrula, and absent in neurula and later stages is G3.

After obtaining this initial information it seemed reasonable to look for differences between different populations of frogs and their hybrids. Figure 6 illustrates

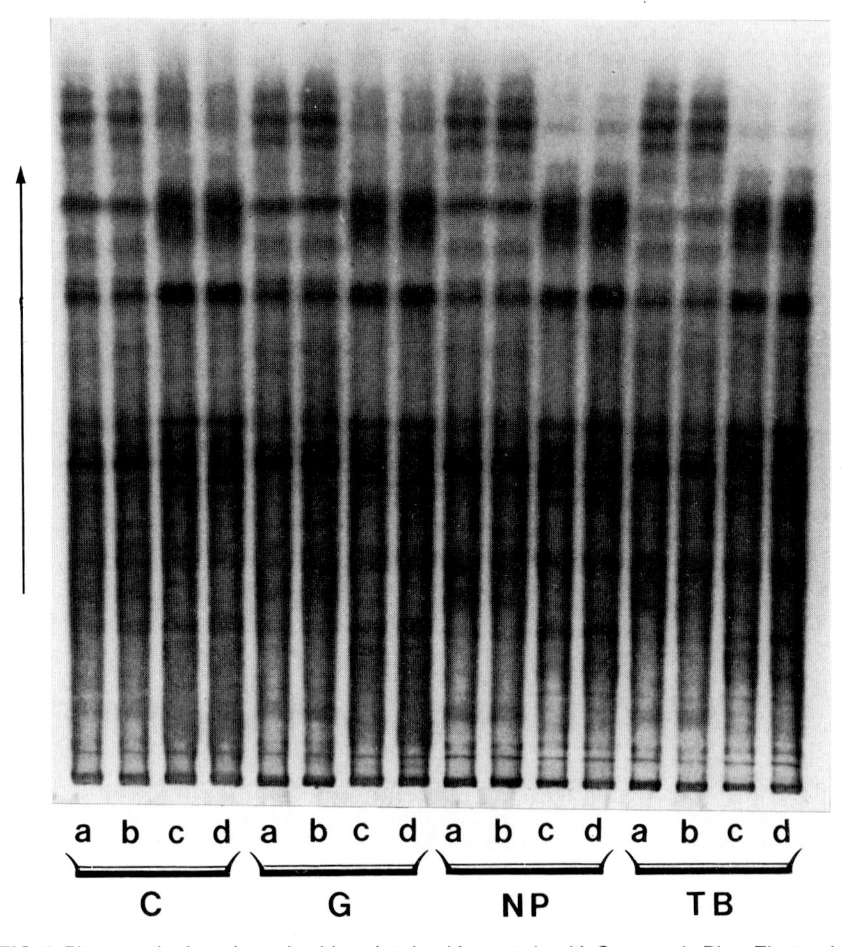

FIG. 6. Photograph of a polyacrylamide gel stained for protein with Coomassie Blue. Electrophoretic patterns of the bulk proteins are shown for frog embryos derived from crosses. a, Texas female X Texas male (TT); b, Texas female X Vermont male (TV); c, Vermont female X Texas male (VT); d, Vermont female X Vermont male (VV); at C, cleavage; G. gastrula; NP, neural plate and TB, tail-bud stages. Note the consistent pattern of bulk proteins characteristic of the source of female gamete in the cross, with very little change in pattern with development.

bulk protein patterns of extracts of embryos from Texas frogs (TT), Vermont frogs (VV), and their reciprocal hybrids (TV, VT) from cleavage, gastrula, neurula, and tail-bud stages. Differences in the proteins from the different egg source, maternal pattern of F_1 hybrids, and the relative unchanged pattern throughout early development can be seen. In the autoradiogram of this gel (Figure 7) the cleavage and gastrula proteins are especially important. There are obvious maternal effects on proteins synthesized by cleavage stages. These maternal

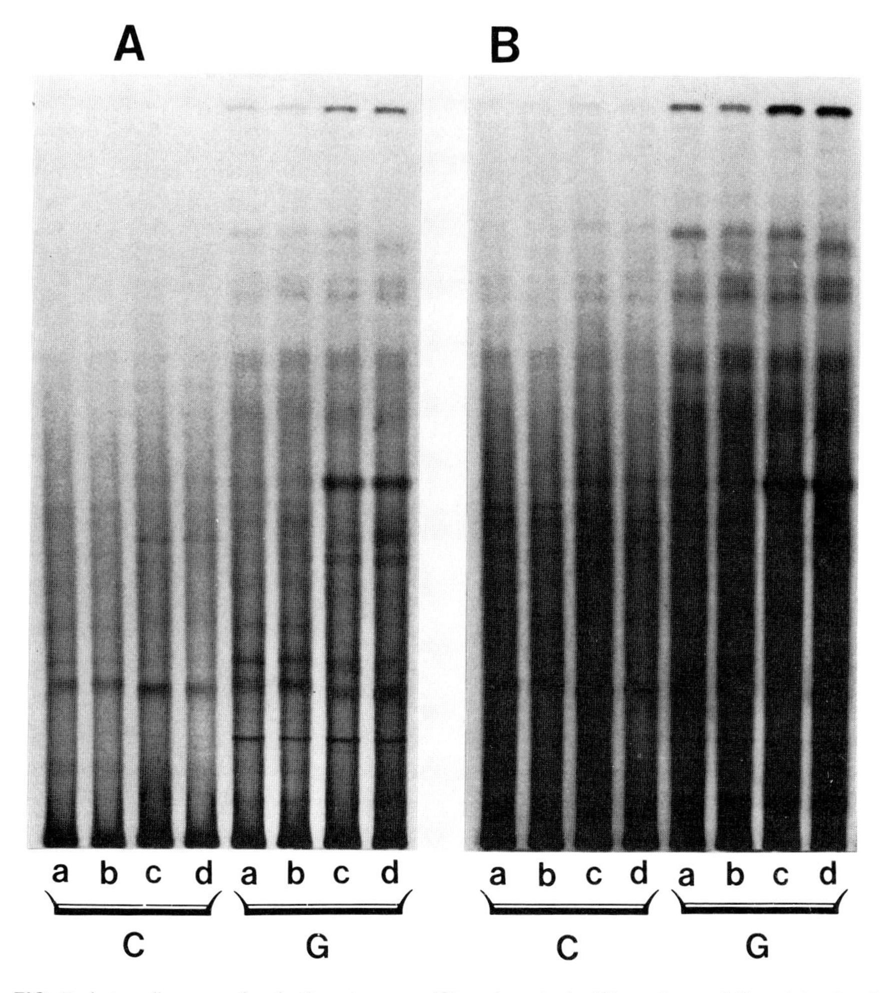

FIG. 7. Autoradiogram of only the cleavage (C) and gastrula (G) portions of the stained gel shown in Figure 6. a, TT; b, TV; c, VT; d, VV. A is a two-week exposure, and B is a six-week exposure. A maternal effect, best seen in Section A, is demonstrated for several proteins synthesized during cleavage and gastrula stages, i.e., the hybrids have the same pattern of protein synthesis as the maternal species. Expression of the nuclear gene for the G3 protein is evident in section B in the gastrula stages in which the Vermont and Texas frogs differ in the electrophoretic mobility of this protein, and the hybrids have two bands.

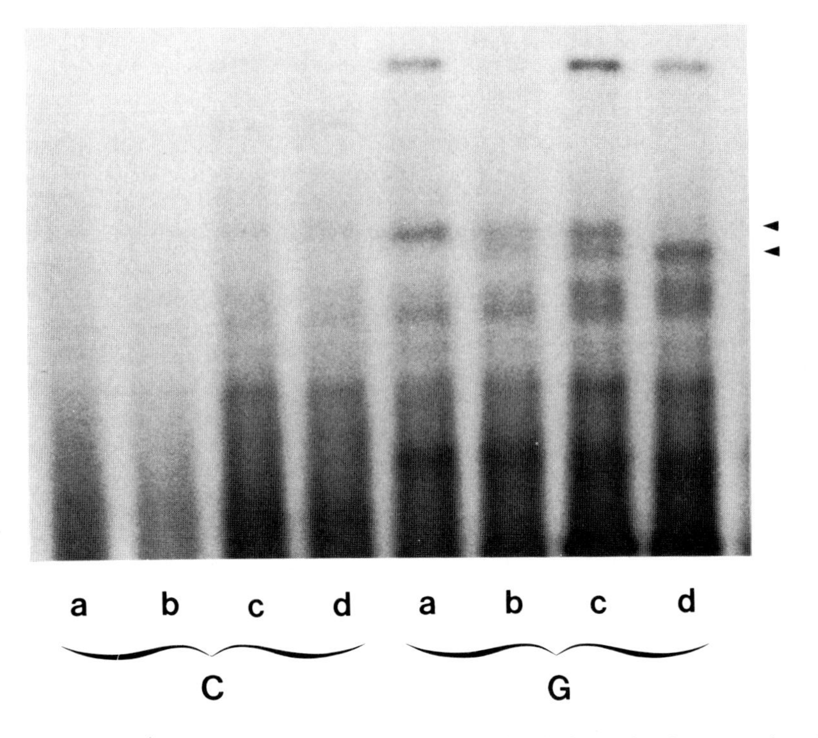

FIG. 8. Portion of an autoradiogram of a gel containing labeled proteins from a series of crosses of Texas and Wisconsin frogs. C, cleavage stages; G, gastrula stages; a, Texas female × Texas male (TT); b, Texas female × Wisconsin male (TW); c, Wisconsin female × Texas male (WT); d, Wisconsin female × Wisconsin male (WW). Note differences in the G3 protein of the Texas and Wisconsin gastrulae and the presence of both parental types in the hybrids.

effects on protein synthesis are detected only because of electrophoretic differences between proteins of the parental species. While some similarities exist, frogs from the different localities have characteristic patterns of protein synthesis in the cleavage stages. The hybrid embryos have patterns identical to those of the pure maternal species (Wright 1976, and in press). This is as expected if there is no mRNA synthesis in cleavage nuclei and if protein synthesis occurs on maternal cytoplasmic templates.

Examination of autoradiograms for labeled gastrula stages reveals that there are many more kinds of proteins synthesized. The overall quantity of protein synthesis is also increased. Among the proteins synthesized by the gastrula are most of the proteins made in cleavage stages. These proteins show the same kind of maternal effects in the reciprocal hybrids as they did in cleavage stages. This suggests that long-lived cytoplasmic mRNA is responsible for continued synthesis of these proteins.

As seen in Figures 5 and 7, several proteins are made by late blastula and gastrula stages not made in cleavage stages or in later neurula stages. Are these

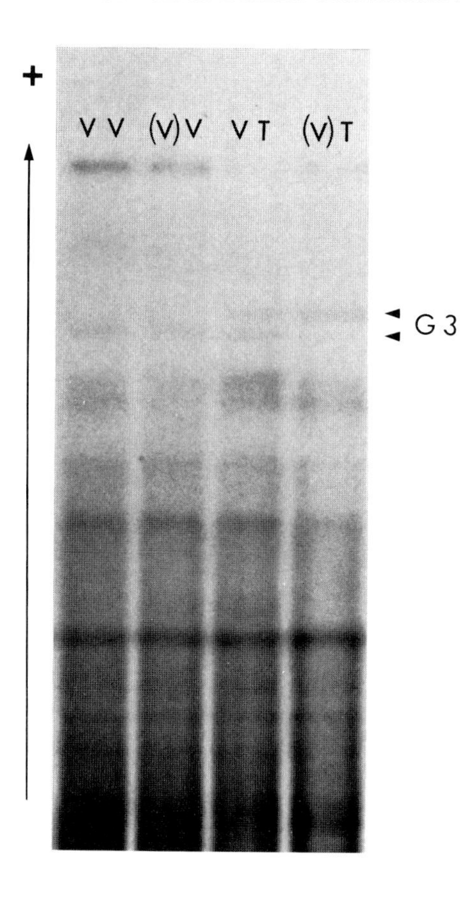

FIG. 9. Patterns of gastrula proteins synthesized in androgenetic haploid hybrids. VV, diploid gastrulae resulting from a cross of two Vermont frogs; (V)V, homologous haploids in which the egg nucleus was removed in a cross of two Vermont frogs; VT, diploid hybrids between a Vermont female and Texas male (note double band for the G3 protein); (V)T, haploid hybrid in which the Vermont egg nucleus was removed (note the production of only the single G3 band, characteristic of the nuclear type).

new proteins made by mRNAs produced by activation of nuclear genes? It is possible that some maternal cytoplasmic mRNAs are stored in an inactive form and are not used for protein synthesis until the embryo prepares for gastrulation.

Because there are electrophoretic differences between the parental types for one of these proteins, we can distinguish between the two possibilities—nuclear gene activation or activation of stored cytoplasmic mRNA unused until the gastrula stage. In hybrids, both forms of the G3 protein are found, indicating nuclear gene activity (Wright 1976, and in press). This is most clearly seen in Figure 8 in which gastrulae resulting from a series of crosses involving Texas and Wisconsin frogs are analyzed. The hybrids seem to have equal amounts of both parental forms in these crosses of Texas with Wisconsin frogs.

In frogs, androgenetic haploid hybrids having only the paternal genome can be produced by removing the egg nucleus shortly after fertilization, when it becomes visible on the egg surface in preparation for the second meiotic division (Porter 1939). A verification of the expression of a nuclear gene for this particular protein is that in androgenetic haploid gastrulae only the paternal form is seen (Figure 9.).

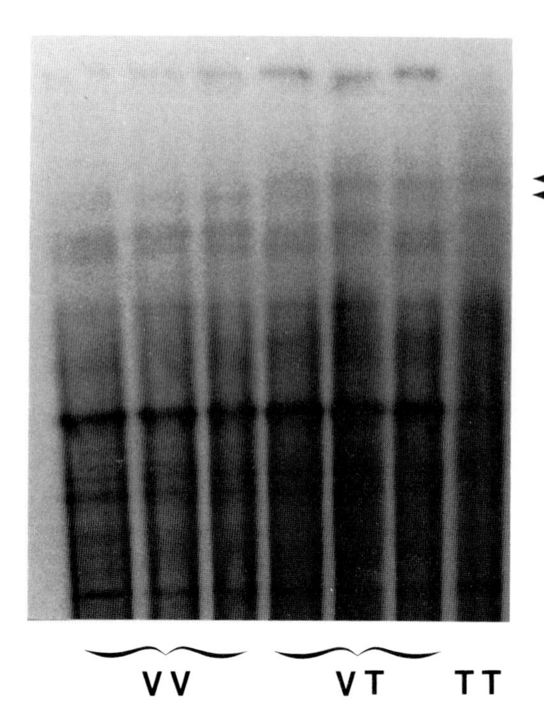

FIG. 10. Autoradiograms showing variation in patterns of the G3 protein in individual embryos (see text).

GENETIC BASIS OF A GASTRULA PROTEIN VARIATION

In one series of crosses involving Vermont frogs, some peculiarities were found in the position and quantities of bands in the G3 region of the autoradiogram. Further analysis of this phenomenon provides evidence that we are dealing with a genetic variation and that variation can occur not only between widely separated populations but also within a population. In Figure 7 we can see that in comparing the V, VT, TV, and TT gastrulae, the bands do not quite match up in position or expected relative intensity. But these samples were prepared by pooling four different individual embryos. In Figure 10 are shown

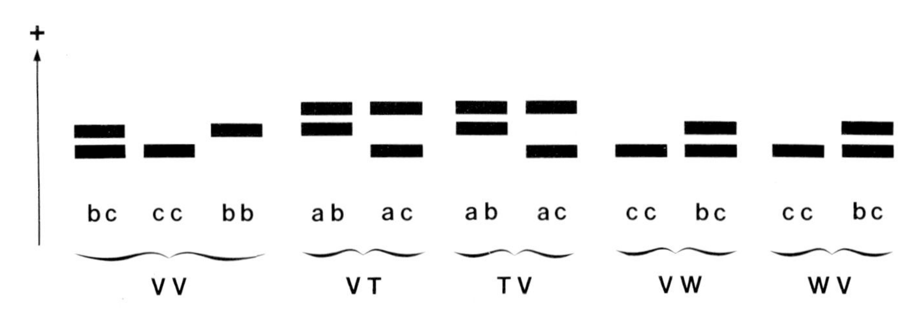

FIG. 11. Diagram of G3 protein variation seen in crosses involving two Vermont frogs.

autoradiograms from gels in which extracts of single embryos were run. The first three are from the V × V cross; the next three are from the V × T cross. In both cases, two different patterns are seen for the G3 protein.

A summary of analyses of individual embryos from different crosses made with the two Vermont frogs is seen in Figure 11. While very limited numbers were analyzed (only five in each cross) this indicates that both of the Vermont frogs used in these crosses were heterozygous.

DISCUSSION

Whether other proteins in the gastrula pattern are also synthesized under the direction of nuclear genes awaits further study and may depend on finding electrophoretic variants of these proteins. Some experiments have been initiated using actinomycin D, an inhibitor of RNA synthesis, in concentrations that block gastrulation. When treatment starts in mid-blastula, we have found that synthesis of gastrula-specific proteins is prevented, while the blocked embryos continue to synthesize cleavage-type proteins (Kubo and Wright 1977).

In considering the value of these studies in the context of a symposium on cell differentiation and neoplasia, it seems a long distance from hybrid frog embryos to human suffering in cancer. But as can be seen from the papers included in this Symposium, the mechanisms of cellular differentiation through control of gene expression in normal and neoplastic cells is a single problem having many branches. In that regard, cellular differentiation itself is a branching process involving a series of steps leading to each particular cell type. I believe that the description of a gastrula-specific protein controlled by nuclear genes may be an important step in understanding how certain gene products may be expressed for only short periods in the life cycle of an organism but may provide the basis for further differentiation of progeny cells. A cellular system in which all genes are "off," then certain genes peculiar to that development stage become expressed for a short period of time and then are no longer expressed may provide the needed material for understanding the mechanisms that control cellular differentiation.

ACKNOWLEDGMENTS

This work was supported by Institutional Research Grant RR5511 and by Grant HD07021 from the National Institutes of Health.

REFERENCES

Bachvarova, R., and E. H. Davidson. 1966. Nuclear activation at the onset of amphibian gastrulation. J. Exp. Zool. 163:285–295.

Brachet, J. 1974. Introduction to Molecular Embryology. Springer-Verlag, New York.

Brown, D. D. 1964. RNA synthesis during amphibian development. J. Exp. Zool. 157:101–113.

Fowler, J. A. 1961. Anatomy and development of racial hybrids of *Rana pipiens.* J. Morphol. 109:251–268.
Gurdon, J. B. 1974. The Control of Gene Expression in Animal Development. Harvard University Press, Cambridge.
Kubo, K., and D. A. Wright. 1977. Actinomycin D treated blastula fail to synthesize gastrula specific proteins. Am. Zool. 17 (Abstract).
Laemmli, U. K. 1970. Cleavage of structural proteins during the assembly of the head of bacteriophage T4. Nature 227:680–685.
Moore, J. A. 1941. Developmental rate of hybrid frogs. J. Exp. Zool. 86:405–422.
Moore, J. A. 1969. Interrelations of the populations of the *Rana pipiens* complex, *in* Biology of Amphibian Tumors, M. Mizell, ed., Springer-Verlag, New York, pp. 26–33.
Porter, K. R. 1939. Androgenetic development of the egg of *Rana pipiens.* Biol. Bull. 77:233–257.
Rugh, R. 1934. Induced ovulation and artificial fertilization in the frog. Biol. Bull. 66:22–29.
Siciliano, M. J., and C. R. Shaw. 1976. Separation and localization of enzymes on gels, *in* Chromatographic and Electrophoretic Techniques, I. Smith, ed., Vol. 2. 4th ed., Interscience Publ., New York, pp. 184–209.
Subtelny, S. 1974. Nucleocytoplasmic interactions in development of amphibian hybrids. Int. Rev. Cytol. 39:35–88.
Wright, D. A. 1975. Expression of enzyme phenotypes in hybrid embryos, *in* Isozymes, IV. Genetics and Evolution, C. Markert, ed., Academic Press, New York, pp. 649–664.
Wright, D. A. 1976. Genetic control of stage specific proteins synthesized in early frog embryos. Am. Zool. 16:231.
Wright, D. A. 1978. Genetic control of protein synthesis in early frog embryos. J. Exp. Zool. (In press).
Wright, D. A., and F. H. Moyer. 1966. Parental influences on lactate dehydrogenase in the early development of hybrid frogs in the genus *Rana.* J. Exp. Zool. 163:215–230.
Wright, D. A., and F. H. Moyer. 1968. Inheritance of frog lactate dehydrogenase patterns and the persistance of maternal isozymes during development. J. Exp. Zool. 167:197–206.
Wright, D. A., and S. Subtelny. 1971. Nuclear and cytoplasmic contributions to dehydrogenase phenotype in hybrid frog embryos. Dev. Biol. 24:119–140.

Cell Differentiation and Neoplasia, edited by Grady F. Saunders. Raven Press, New York

Control of Synthesis of Specific Gene Products during Spermatogenesis

Marvin L. Meistrich, William A. Brock, Sidney R. Grimes, Jr.,* Robert D. Platz, and Lubomir S. Hnilica*

*Section of Experimental Radiotherapy, and *Department of Biochemistry, The University of Texas System Cancer Center M.D. Anderson Hospital and Tumor Institute, Houston, Texas 77030*

At the biochemical level, normal cellular differentiation can be viewed as the regulated synthesis of products required for specialized cell function, cell division, and cell migration. Cancer cells, on the other hand, often differentiate and synthesize some of the same specialized products as normal cells, but the regulation of this synthesis is abnormal. In order to understand this aspect of cancer, we must elucidate the control mechanisms of biochemical differentiation in normal cells, i.e., the control of appearance of the proteins of differentiation. The appearance of these proteins may be regulated at both the transcriptional and posttranscriptional levels. In this paper we shall summarize the results of studies of specific protein synthesis during spermatogenesis in the rat and mouse. Several modes for control of gene expression are demonstrated, and methods for studying the level of control are suggested.

Spermatogenesis is a particularly favorable system for studying the biochemical events of cellular differentiation; the process yields only one cell type, the spermatozoon, is precisely regulated, is localized in a specific organ, and yields sufficient quantities of cells for biochemical analysis. Furthermore, a large number of tissue-specific genes are active only during parts of this process. All of these events can be correlated to the morphological changes that occur in spermatogenesis and have been beautifully described (Leblond and Clermont 1952, Oakberg 1956a). These changes are shown in Figure 1. Briefly, the type A spermatogonial stem cells pass through a series of mitotic divisions and differentiations to produce spermatocytes. During the long meiotic prophase, the synaptonemal complex forms to hold the homologous chromosomes together in pairs so that genetic recombination can occur (Moses 1968). At the same time, the mitochondria begin to show changes in both their morphology and their complement of enzymes (De Domenech *et al.* 1972). The cells then pass through the two meiotic

Present addresses: S. R. Grimes, Veterans Administration Hospital, Shreveport, Louisiana 71130; R. D. Platz, Frederick Cancer Research Center, Frederick, Maryland 21701; L. S. Hnilica, Department of Biochemistry, Vanderbilt University School of Medicine, Nashville, Tennessee 37232.

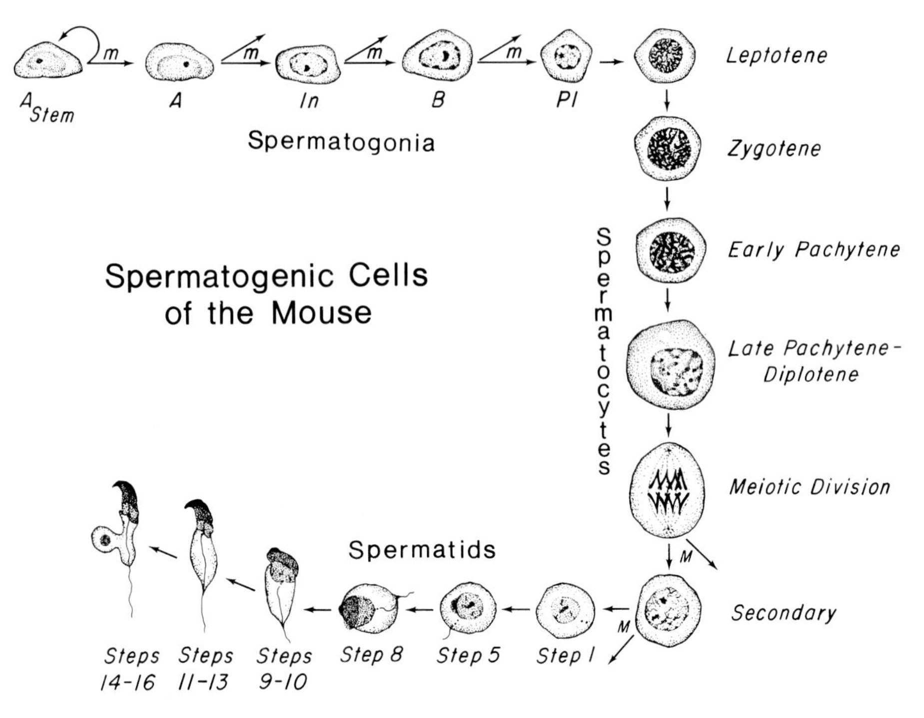

FIG. 1. Sequence of spermatogenic cell differentiation in the mouse.

divisions forming haploid spermatids that initially possess round nuclei. Beginning at step 9 of spermiogenesis, the nuclei of these cells condense and assume a precisely defined shape. As differentiation continues, the acrosome and tail are formed; the mitochondria become aligned along the midpiece, and, finally, most of the residual cytoplasm is lost.

The macromolecular events associated with the morphological changes of spermatogenesis have been most thoroughly characterized by the use of autoradiography. The results demonstrate that protein synthesis occurs throughout spermatogenesis, although at certain stages the synthesis of arginine- and lysine-rich nuclear proteins predominates (Monesi 1965). DNA synthesis is essentially limited to the spermatogonia and preleptotene spermatocytes (Monesi 1962), although a small amount of DNA synthesis does occur during the zygotene and pachytene spermatocyte stages (Meistrich *et al.* 1975). RNA synthesis increases dramatically during the spermatocyte stage, reaching a maximum at mid-pachytene (Figure 2). Although RNA synthesis declines after mid-pachytene, appreciable amounts of RNA are synthesized in the round spermatids. All detectable RNA synthesis ceases in the spermatid when nuclear elongation commences at step 8 in the rat or step 10 in the mouse.

With one exception, very little is known about the synthesis of specific messenger RNA during spermatogenesis. Dixon and co-workers have isolated the

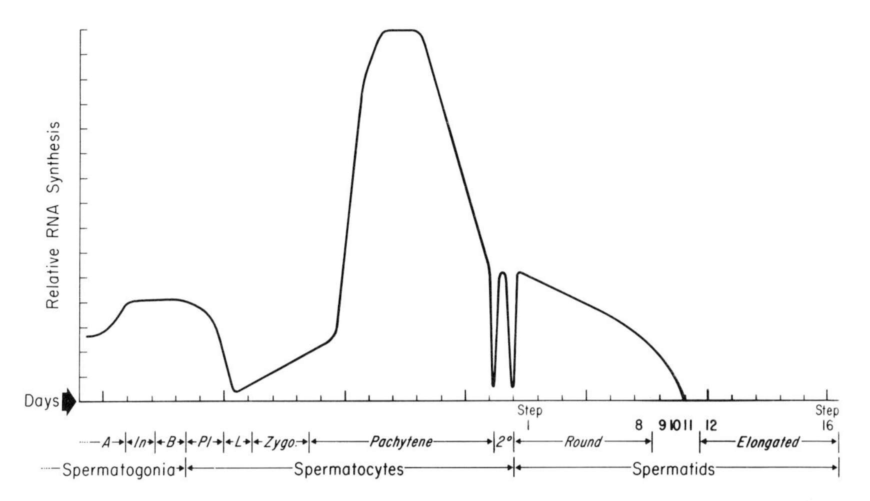

FIG. 2. Relative levels of RNA synthesis during mouse spermatogenesis as measured by ^{3}H-uridine incorporation and autoradiography. (Data taken from Monesi 1971, Moore 1971, Kierszenbaum and Tres 1975).

mRNA for protamine, the basic protein of trout sperm, which is synthesized in late spermatids (Gedamu and Dixon 1976a, b). Their results indicate that this mRNA is stored in RNP particles prior to its translation. This result suggests that the final expression of the protamine genes is controlled by a posttranscriptional event. In rodents, the peak of RNA synthesis in the late pachytene spermatocyte corresponds to heterogeneous nuclear RNA (HnRNA) (Utakoji 1970, Murumatsu *et al.* 1968, Soderstrom and Parvinen 1976a), and the turnover of this RNA is quite slow. No biochemical analysis of the RNA synthesized in spermatids has been performed, but electron microscopy of spread preparations indicates that the RNA may be HnRNA. However, there is only one example of biochemical expression of a gene in haploid cells (Yanagisawa *et al.* 1974).

In order to further resolve the macromolecular events occurring at specific stages of spermatogenesis, it was necessary to develop methods for bulk separation of testicular cells and nuclei. These methods, which are summarized elsewhere (Meistrich 1977), include separation of cells by velocity sedimentation in an elutriator rotor, separation of nuclei by velocity sedimentation in a Staput apparatus, and preparation of elongated spermatid nuclei that are resistant to sonication. These methods provide enriched populations of cells and nuclei at specific stages of spermatogenesis in quantities sufficient for biochemical analysis.

Regulation of Histone Synthesis

The control of histone synthesis in somatic mammalian cells is unique. With the possible exception of histone H1 (Gurley and Hardin 1970), histone synthesis occurs only during the S phase (Spalding *et al.* 1966) and is tightly coupled

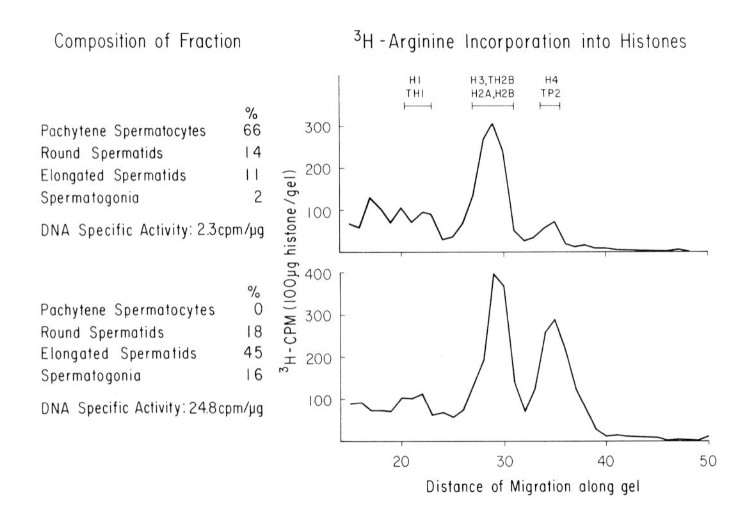

FIG. 3. Histone and DNA synthesis in different fractions of testis cells. Rats were injected with either ^{3}H-arginine or ^{3}H-thymidine and sacrificed 24 hours later. Cells were separated at unit gravity by the Staput method of velocity sedimentation. Acid-soluble nuclear proteins were extracted and then separated by polyacrylamide gel electrophoresis. Methods are given in more detail by Platz *et al.* 1975.

to DNA replication (Borun *et al.* 1967). Regulation of histone synthesis is primarily at the level of mRNA synthesis, although some translational modulation may exist (Gallwitz and Mueller 1969). As an exception, histone mRNA in sea urchin eggs is stored in ribonucleoprotein particles and is subsequently translated during early embryogenesis (Skoultchi and Gross 1973).

In the male germ line, histone synthesis occurs in the preleptotene primary spermatocyte concurrent with the last S phase DNA replication (Liapunova and Babadjanian 1973). There is only a small amount of DNA synthesis (roughly 0.1% of the genome) occurring during the zygotene and pachytene spermatocyte stages (Hotta *et al.* 1966, Meistrich *et al.* 1975, Soderstrom and Parvinen 1976b), but large amounts of histone synthesis do occur during these stages (Bogdonov *et al.* 1973). We have analyzed histone synthesis in purified populations of testis cells. Spermatocytes can be separated from other rat testes cells by velocity sedimentation with minimal contamination by S phase cells (Platz *et al.* 1975, Meistrich 1977). The cells were labeled by an in vivo injection of ^{3}H-amino acids; the nuclear proteins were then separated by electrophoresis, and the following patterns of radioactivity were observed. Round spermatids showed essentially no histone synthesis (Grimes *et al.* 1975a); a fraction containing elongated spermatids, round spermatids, and spermatogonia showed synthesis of all histones, and a fraction enriched in pachytene spermatocytes showed synthesis of a histone migrating in the region of H3-H2B-H2A (Figure 3). Contamination of this fraction by S phase cells, as determined by measurement of ^{3}H-thymidine incorporation, could not account for this high level of histone synthesis (Platz *et al.* 1975). Contamination by spermatids would only serve to lower the specific activity of the histones. Similarly, by using sodium dodecyl sulfate (SDS) slab gel electrophoresis and fluorography, it was determined that the maximum incorporation of ^{3}H-amino acids occurred in the region of histone H2B in fractions containing late pachytene spermatocytes.

Although we have not yet identified this histone, it is quite likely that it represents the testis-specific form of histone H2B, which has been designated X3 (Branson *et al.* 1975) or TH2B (Shires *et al.* 1976). This protein is similar in amino acid composition to the somatic H2B, but it has several differences in amino acid composition. The appearance of TH2B during development of the rat testis is associated with the first appearance of pachytene spermatocytes (Grimes *et al.* 1975a). TH2B is also clearly present in purified late pachytene spermatocytes and round spermatids. From these data we conclude that TH2B synthesis is turned on during the primary spermatocyte stage and turned off prior to the second meiotic division. TH2B is present in greater amounts than H2B (Grimes *et al.* 1975), and in spermatocytes and spermatids, TH2B accounts for more than 13% of the total histone. In contrast, meiotic DNA synthesis is less than 0.1% of the total genome. Thus, we conclude that, unlike somatic histone synthesis, meiotic histone synthesis is not coupled to DNA synthesis. Therefore, the mode of regulation of meiotic histone must be different from that of somatic histone synthesis. We speculate that synthesis of the mRNA for TH2B occurs during the burst of RNA synthesis at mid-pachytene and that this message is immediately translated and then degraded after a predetermined length of time. This would place the turning on of TH2B synthesis under transcriptional control.

Synthesis of LDH-X

Spermatogenic cells and spermatozoa differ from somatic cells in that they contain a unique isozymic form of lactate dehydrogenase, LDH-X. During spermatogenesis, this isozyme, which is the product of the LDH-c gene, replaces the somatic isozymes. LDH-X becomes associated with the mitochondria and is believed to be important in sperm metabolism (Wheat and Goldberg 1975). LDH-X is not present in cells prior to the mid-pachytene stage of spermatogenesis. Its synthesis begins at mid-pachytene, reaches a maximum at late pachytene, and continues throughout spermiogenesis, although at a reduced rate (Table 1) (Meistrich *et al.*, in press).

The turning on of LDH-X synthesis may be correlated with the RNA synthesis at mid-pachytene and may be under transcriptional control. It is not known whether additional LDH-X mRNA is synthesized after meiosis, but genetic data suggest that this is unlikely (Zinkham *et al.* 1964). Since LDH-X synthesis continues after the cessation of all RNA synthesis, elongated spermatid LDH-X synthesis must be directed by stable mRNA. Because of the gradually declining levels of LDH-X synthesis after pachytene, we speculate that there is no posttran-

TABLE 1. *Incorporation of ^{3}H-valine into LDH-X in different testicular cell types*

Cell type	Cellular composition of fraction	^{3}H-LDH-X (dpm/mg protein)
Pachytene spermatocytes	80%	14.2 ± 1.3
Round spermatids	60%	
Elongated spermatids	20%	6.3 ± 2.9
Spermatid cytoplasmic fragments	10%	
Elongated spermatids	75%	
Spermatid cytoplasmic fragments	25%	2.3 ± 0.2

Testis cell suspensions were prepared from mice two hours after injection of ^{3}H-valine and were then separated by centrifugal elutriation. LDH-X was extracted from these cells, purified by ammonium sulfate precipitation and immunoprecipitation, and then counted for radioactivity (Meistrich *et al.,* in press).

scriptional regulation of LDH-X synthesis. If the rate of LDH-X synthesis is proportional to the concentration of LDH-X mRNA, then our data suggest that LDH-X mRNA has a half-life of 4.3 days.

Synthesis of Basic Nuclear Proteins in Elongated Spermatids

Coincident with nuclear condensation and elongation, dramatic changes occur in the protein composition of the rat spermatid nucleus. Round spermatids in steps 1–8 of development contain both somatic and testis-specific histones, as well as nonhistone proteins, but they contain none of the spermatidal basic proteins (Platz *et al.* 1975, Grimes *et al.* 1975b). Spermatids in steps 13–15 contain no somatic histones, have reduced amounts of nonhistones, but now contain two testis-specific proteins that have been designated TP and TP2 (Table 2) (Grimes *et al.,* in press). The details of this transition are not known, but the histones are completely removed during steps 9–12 and then are replaced by these two basic proteins. Synthesis of TP and TP2 does not occur in the round spermatids or after step 15. The bulk of TP and TP2 synthesis occurs during steps 13–15. At step 16, synthesis of S1, the basic protein of spermatozoan nuclei, commences. This protein completely replaces TP and TP2, and it is the major basic protein of step 16–19 spermatids.

Since RNA synthesis drops below detectable levels at about step 9 of spermatid development, the synthesis of TP, TP2, and S1 begins after cessation of RNA synthesis. Because the time interval between the cessation of RNA synthesis and the beginning of S1 synthesis is five days, the mRNA must be stabilized in some form prior to its translation. The observation that protamine mRNA in trout is associated with ribonucleoprotein particles suggests a possible mechanism of RNA storage in the rat testis. If the mRNA for these proteins is

TABLE 2. *Summary of properties of basic proteins from elongated spermatids*

	TP	TP2	S1
Molar composition of amino acids (%)			
Arginine	20	14	65
Lysine	20	10	3
Cysteine	0	4	9
Molecular weight	6200	11,000	6400
Cellular localization of protein	step 13–15 spermatids	step 13–15 spermatids	step 16–19 spermatids spermatozoa
Cellular localization of synthesis	step 13–15 spermatids	step 13–15 spermatids	step 16–18 spermatids

Data compiled from Kistler *et al.* 1975, Marushige and Marushige 1975, Grimes *et al.* 1975b, and Grimes *et al.*, in press.

synthesized in the spermatocyte, then there would be an interval of at least 10 days between transcription and translation of these mRNA species. There must be at least two specific control mechanisms: one for activating the translation of the mRNAs for TP and TP2 at about step 12 and another for activating the translation of S1 at step 16.

The turning off of the synthesis of these proteins is also highly stage specific. There is no synthesis of TP or TP2 after step 15 and no synthesis of S1 after step 18. It is not known what mechanism is responsible for the turning off of the synthesis of specific proteins at specific stages. Possible mechanisms include: (1) rapid exponential degradation of the mRNA, (2) specific degradation of mRNA after translation a certain number of times, or (3) appearance or loss of a "factor" responsible for respectively blocking or promoting the further translation of a given class of mRNA at a particular stage.

DISCUSSION AND CONCLUSIONS

The data presented here suggest that the synthesis of meiotic histone, LDH-X, and basic nuclear proteins of elongated spermatids is regulated in quite different ways. Possible modes of regulation are outlined in Table 3. Since we have only studied protein synthesis, these models are speculative, but experiments to test them are suggested.

One way of testing these models is by the use of inhibitors of nucleic acid synthesis. Seminiferous tubules or isolated testicular cells can be incubated in vitro with radioactive precursors for up to one day, without affecting cell viability (Soderstrom and Parvinen 1976a). DNA synthesis may be inhibited in these experiments (Hotta and Stern 1971) to determine whether meiotic histone synthe-

TABLE 3. *Comparison of postulated modes of control of protein synthesis during spermatogenesis*

	Turning on	Turning off
Testis histone TH2B	Transcriptional	Abrupt (Rapid turnover of mRNA after translation)
LDH-X	Transcriptional	Gradual (Gradual exponential decay of mRNA)
Spermatidal basic proteins TP, TP2, S1	Posttranscriptional	Abrupt (Rapid turnover of mRNA after translation)

sis is directly coupled to DNA synthesis. RNA synthesis may be inhibited to determine whether these proteins are synthesized from newly transcribed RNAs.

A more direct test would require assays of the messenger RNA species for some testis-specific proteins. This may be done by transcribing mRNA from spermatogenic cells in a cell-free system or by hybridization with specific cDNA probes. The latter would, of course, first require the purification of the mRNA species. It should be feasible to purify some of the testis messengers because of the following considerations: (1) LDH-X synthesis represents 3% of the total testicular protein synthesis, (2) the cells that are most active in synthesizing TH2B, TP, TP2, and S1 can be highly enriched by cell separation methods, (3) the basic protein mRNA should be short molecules found on small polysomes, probably disomes in the case of TP and S1 (Ling and Dixon 1970), (4) the basic protein mRNAs may not be polyadenylated, (5) the S1 DNA and mRNA should be G-C rich because of the high arginine content of S1, and (6) all of these proteins have been purified, and, in the case of LDH-X, pure antibody has been made, permitting the purification of LDH-X mRNA by immunoprecipitation of polyribosomes containing nascent polypeptides.

Once we are able to assay the mRNA molecules, it would be possible to determine in which cell types these molecules are present. Using this technique, along with cell separation, we could determine if the mRNA for a particular protein is present in cell stages prior to the turning on of its synthesis. If this is indeed found to be true, it would be a direct example of posttranscriptional control of protein synthesis. In addition, the intracellular localization of the stored mRNA in separated cell types could be studied. We could ask questions such as whether mRNA is stored in the nucleus or cytoplasm and what is the nature of the RNA particles. With an assay for specific mRNAs we could also examine the mechanisms for turning off the synthesis of specific proteins and how this relates to the message half-life.

In conclusion, the study of macromolecular synthesis during rodent spermatogenesis using methods for cell separation and biochemical analysis provides a valuable system for investigating the mechanisms involved in control of gene expression during cellular differentiation.

ACKNOWLEDGMENTS

We thank Patricia Trostle, Gwen Hord, and Betty Reid for their excellent technical assistance. We also thank Dr. Robert P. Erickson for his valuable collaboration on the LDH-X studies and Dr. Gordon H. Dixon for his results on the synthesis of protamine message. This study was supported in part by the Department of Health, Education, and Welfare, National Institutes of Health, National Cancer Institute Grants HD-05803, CA-17364, and CA-06294, and National Science Foundation Grant PCM 76–08836.

REFERENCES

Branson, R. E., S. R. Grimes, G. Yonuschot, and J. L. Irvin. 1975. The histones of rat testis. Arch. Biochem. Biophys. 168:403–412.

Bogdanov, Yu. F., A. A. Strokov, and S. A. Reznickova. 1973. Histone synthesis during meiotic prophase in *Lilium*. Chromosoma 43:237–245.

Borun, T. W., M. D. Scharff, and E. Robbins. 1967. Rapidly labelled, polyribosome associated RNA having the properties of histone messenger. Proc. Natl. Acad. Sci. USA 58:1977–1983.

De Domenech, E. M., C. E. Domenech, A. Aoki and A. Blanco. 1972. Association of the testicular lactate dehydrogenase isozyme with a special type of mitochondria. Biol. Reprod. 6:136–147.

Gallwitz, D., and G. C. Mueller. 1969. Histone synthesis *in vitro* on HeLa cell microsomes. J. Biol. Chem. 244:5947–5952.

Gedamu, L. and G. H. Dixon. 1976a. Assay of protamine messenger RNA from rainbow trout testis. J. Biol. Chem. 251:1446–1454.

Gedamu, L., and G. H. Dixon. 1976b. Purification and properties of biologically active rainbow trout testis protamine mRNA. J. Biol. Chem. 251:1455–1463.

Grabske, R. J., S. Lake, B. L. Gledhill, and M. L. Meistrich. 1975. Centrifugal elutriation: Separation of spermatogenic cells on the basis of sedimentation velocity. J. Cell. Physiol. 86:177–190.

Grimes, S. R., C. B. Chae, and J. L. Irvin. 1975a. Effects of age and hypophysectomy upon relative proportions of various histones in rat testis. Biochem. Biophys. Res. Commun. 64:911–917.

Grimes, S. R., R. D. Platz, M. L. Meistrich, and L. S. Hnilica. 1975b. Partial characterization of a new basic nuclear protein from rat testis elongated spermatids. Biochem. Biophys. Res. Commun. 67:182–189.

Grimes, S. R., M. L. Meistrich, R. D. Platz, and L. S. Hnilica. 1977. Nucleoprotein transitions in rat testis spermatids. Exp. Cell Res. (In press).

Gurley, L. R., and J. M. Hardin. 1970. The metabolism of histone fractions. III. Synthesis and turnover of histone fl. Arch. Biochem. Biophys. 136:392–401.

Hotta, Y., M. Ito, and H. Stern. 1966. Synthesis of DNA during meiosis. Proc. Natl. Acad. Sci. USA 56:1184–1191.

Hotta, Y., and H. Stern. 1971. Analysis of DNA synthesis during meiotic prophase in *Lilium*. J. Mol. Biol. 55:337–355.

Kierszenbaum, A. L., and L. L. Tres. 1975. Structural and transcriptional features of the mouse spermatid genome. J. Cell. Biol. 65:258–270.

Kistler, W. S., C. Noyes, R. Shu, and R. L. Heinrikson. 1975. The amino acid sequence of a testis specific basic protein that is associated with spermatogenesis. J. Biol. Chem. 250:1847–1853.

LeBlond, C. P., and Y. Clermont. 1952. Definition of the stages of the cycle of the seminiferous epithelium in the rat. Ann. NY Acad. Sci. 55:548–573.

Liapunova, N. A., and D. P. Badadjanian. 1973. A quantitative study of histones in meiocytes. Chromosoma 40:387–399.

Ling, V., and G. H. Dixon. 1970. The biosynthesis of protamine in trout testis. II. Polysome patterns and protein synthetic activities during testis maturation. J. Biol. Chem. 245:3035–3042.

Marushige, Y., and K. Marushige. 1975. Transformation of sperm histone during formation and maturation of rat spermatozoa. J. Biol. Chem. 250:39–45.

Meistrich, M. L. 1977. Separation of spermatogenic cells and nuclei from rodent testes, *in* Methods in Cell Biology, D. M. Prescott, ed., Vol. 15. Academic Press Inc., New York, pp. 15–54.

Meistrich, M. L., B. O. Reid, and W. J. Barcellona. 1975. Meiotic DNA synthesis during mouse spermatogenesis. J. Cell. Biol. 64:211–222.

Meistrich, M. L., P. K. Trostle, M. L. Frapart, and R. P. Erickson. 1977. Localization of LDH-X biosynthesis to spermatocytes and spermatids in mouse testes. Dev. Biol. (In press).

Monesi, V. 1962. Autoradiographic study of DNA synthesis and the cell cycle in spermatogonia and spermatocytes of the mouse testis, using tritiated thymidine. J. Cell. Biol. 14:1–18.

Monesi, V. 1965. Synthetic activities during spermatogenesis in the mouse—RNA and protein. Exp. Cell Res. 39:197–224.

Monesi, V. 1971. Chromosome activities during meiosis and spermiogenesis. J. Reprod. Fertil. (Suppl.) 13:1–14.

Moore, G. P. M. 1971. DNA-dependent RNA synthesis in fixed cells during spermatogenesis in mouse. Exp. Cell Res. 68:462–465.

Moses, M. J. 1968. Synaptonemal complex. Ann. Rev. Genet. 2:363–412.

Oakberg, E. F. 1956a. A description of spermiogenesis in the mouse and its use in analysis of the cycles of the seminiferous epithelium and germ cell renewal. Am. J. Anat. 99:391–413.

Oakberg, E. F. 1956b. Duration of spermatogenesis in the mouse and timing of the stages of the cycle of the seminiferous epithelium. Am. J. Anat. 99:507–516.

Platz, R. D., S. R. Grimes, M. L. Meistrich, and L. S. Hnilica. 1975. Changes in nuclear proteins of rat testis cells separated by velocity sedimentation. J. Biol. Chem. 250:5791–5800.

Shires, A., M. P. Carpenter, and R. Chalkley. 1976. A cysteine-containing H2B-like histone found in mature mammalian testes. J. Biol. Chem. 251:4155–4158.

Skoultchi, A., and P. R. Gross. 1973. Maternal histone messenger RNA: Detection by molecular hybridization. Proc. Natl. Acad. Sci. USA 70:2840–2844.

Soderstrom, K. O., and M. Parvinen. 1976a. RNA synthesis in different stages of rat seminiferous epithelial cycle. Mol. Cell Endocrinol. 5:181–200.

Soderstrom, K. O., and M. Parvinen. 1976b. DNA synthesis during male meiotic prophase in the rat. Hereditas 82:25–28.

Spalding, J., K. Kajiwara, and G. C. Mueller. 1966. The metabolism of basic proteins in HeLa cell nuclei. Proc. Natl. Acad. Sci. USA 56:1535–1542.

Utakoji, T. 1970. Isolation of the pachytene stage nuclei from the Syrian hamster testis, *in* Methods in Cell Physiology, D. M. Prescott, ed., Vol. 4. Academic Press, Inc., New York, pp. 1–17.

Utakoji, T., M. Murumatsu, and H. Sugano. 1968. Isolation of pachytene nuclei from the Syrian hamster testis. Exp. Cell Res. 53:447–458.

Wheat, T. E., and E. Goldberg. 1975. LDH-X: The sperm-specific C_4 isozyme of lactate dehydrogenase, *in* Isozymes, C. L. Markert, ed., Vol. 3. Academic Press, Inc., New York, pp. 325–345.

Yanagisawa, K., D. R. Pollard, D. Bennett, L. C. Dunn, and E. A. Boyse. 1974. Transmission ratio distortion at the T-locus: Serological identification of two sperm populations in t-heterozygotes. Immunogenetics 1:91–96.

Zinkham, W. H., A. Blanco, and L. Kupchyk. 1964. Lactate dehydrogenase in pigeon testes: Genetic control by three loci. Science 144:1353–1354.

Cell Differentiation and Neoplasia, edited by Grady F. Saunders. Raven Press, New York

Cytoplasmic Control of Gene Expression in Oogenesis

J. B. Gurdon, E. M. De Robertis, R. A. Laskey, J. E. Mertz, G. A. Partington, and A. D. Wyllie

MRC Laboratory of Molecular Biology, Cambridge CB2 2QH, England

It is commonly supposed that the cytoplasm of animal eggs contains molecules which determine the first steps in embryonic development (Davidson 1968, Gurdon 1977). This effect could be understood if such molecules were to act as controllers of gene activity and if they were asymmetrically arranged in fertilized eggs. Nuclei in different regions of an early embryo would accumulate different gene-control molecules, and in this way regional differences in gene activity would arise.

The gene-controlling molecules presumed, by this hypothesis, to be present in eggs would almost certainly be synthesized during oogenesis. In amphibia, oogenesis lasts for several weeks or months; in contrast, oocyte maturation into an egg and fertilization take only a few hours. The experiments described here have been designed to reveal the existence of, and to lead eventually to the isolation of, gene-controlling molecules in oocytes of the South African frog, *Xenopus laevis.*

NUCLEI INJECTED INTO OOCYTES

Mammalian Nuclei

Our original method for introducing somatic nuclei into oocytes (Gurdon 1968) involved the disruption of whole cells by sucking them into a micropipette. However, this procedure is cumbersome, and we have tried to find a method of preparing a nuclear suspension that is suitable for injection. Cell nuclei are commonly isolated from cells by procedures involving the use of such agents as Triton or Nonidet P40. Nuclei prepared with these agents are synthetically active in vitro for a few hours but are unsatisfactory for injection into living oocytes. After investigating various procedures, we devised a method that involves the controlled lysis of cells by lysolecithin (Gurdon 1976). A suspension of about 200 nuclei prepared by this means can be conveniently injected into

---

Present address: J. E. Mertz, McArdle Laboratory for Cancer Research, University of Wisconsin, Madison, Wisconsin 53706.

each oocyte. The injected nuclei survive for up to a month in the injected oocytes and during this period remain continuously active in RNA synthesis (Gurdon *et al.* 1976a). Most injected nuclei enlarge very substantially in volume, sometimes by over 100 times. The amount and rate of their enlargement is increased if they are injected into the oocyte nucleus (germinal vesicle) or into the region of the dispersed contents of a germinal vesicle (Gurdon 1976). The RNA synthesizing activity of the injected nuclei and their accumulation of cytoplasmic proteins increases nearly in proportion to their increase in volume (Gurdon *et al.* 1976b).

To find out whether the RNA synthesized by mammalian nuclei in oocytes includes translatable messenger RNA, we have examined labeled proteins using two-dimensional electrophoresis. We have found that oocytes injected three or more days before with HeLa nuclei synthesize some proteins which are not synthesized by uninjected oocytes (Gurdon *et al.* 1976a, De Robertis *et al.* 1977). There are several reasons for believing that the synthesis of these new proteins depends on transcription of messenger RNA by nuclei after their injection into oocytes and that it is not attributable to "carried-over" messenger RNA which might contaminate the nuclear suspension: (1) The new proteins are not seen immediately after injection, as would be expected if carried-over messenger RNA were responsible; (2) the synthesis of the new proteins is suppressed by α-amanitin; (3) the injection of a tenfold excess of RNA isolated from our HeLa cell nuclear suspension does not lead to the synthesis of the new proteins. At least one of the new proteins has been proved by peptide analysis to be HeLa coded (De Robertis *et al.* 1977), and it is synthesized after the injection of HeLa nuclei into enucleated oocytes (De Robertis *et al.* 1977). These experiments show that somatic nuclei may be transplanted to living oocytes where they are transcriptionally active for many days. We are now in a position to use this experimental system to ascertain if oocytes contain conditions or components capable of controlling the activity of genes.

Amphibian Nuclei

We have transplanted nuclei from cultured *Xenopus* somatic cells (a line derived from adult kidney) into oocytes of a Urodele amphibian, *Pleurodeles waltlii.* We first established that differences in gene activity between *Xenopus* cultured cells, *Xenopus* oocytes, and *Pleurodeles* oocytes could be detected by two-dimensional electrophoresis of proteins. The proteins seen fall into the following two categories: (1) Cultured cell-specific proteins, synthesized by *Xenopus* cultured cells but not by *Xenopus* oocytes; (2) oocyte-specific proteins, synthesized by *Xenopus* oocytes but not by *Xenopus* cultured cells. In addition, many proteins were observed that are synthesized both in cultured cells and in oocytes of *Xenopus.* A few readily detectable proteins that belong to categories (1) and (2), and which were not represented by proteins of similar size and charge in *Pleurodeles* oocytes, were selected for examination in the following experiments, which are described in detail by De Robertis and Gurdon (1977).

Xenopus-cultured cell nuclei, prepared by the lysolecithin method, were injected into *Pleurodeles* oocytes. The oocytes were labeled with ^{14}C-amino acids for six hours immediately after injection, or several days later. The proteins were analyzed on fluorographed gels, and the following conclusions were reached. First, and most important, some of the *Xenopus* oocyte-specific proteins were synthesized in the *Pleurodeles* oocytes that had been injected three or more days before with *Xenopus*-cultured cell nuclei. Second, the cultured cell specific proteins were not synthesized at a detectable rate. Appropriate controls showed that these effects were dependent on transcription in the injected oocytes. The first result shows that oocytes possess components that can reactivate genes which had evidently become inactive in the course of normal development. The second result indicates that this "switching-on" effect is selective, since not all genes in the injected nuclei are expressed. We cannot exclude the possibility that the nonexpression of the cultured cell specific genes is at the translational level, all genes being transcribed. It is clear, however, that if we consider gene expression at the level of protein synthesis, then oocytes contain components which have selective effects and in this sense, at least, have gene-controlling molecules of the kind referred to in the introduction.

DNA INJECTED INTO OOCYTES

Heterologous DNAs

If gene-controlling substances exist in oocyte cytoplasm, we must consider how to identify them and how to determine the molecular basis of their activity. The idea that we have been pursuing in recent years is to inject purified genes of a known kind, as purified DNA molecules, in such a way that they function normally in oocytes and then to reisolate these pieces of DNA with the gene-controlling molecules of oocytes attached to them. Before summarizing the progress that we have made toward this eventual aim, it will be helpful to outline the extent to which other purified nucleic acids can be introduced into eggs or oocytes in such a way that they function normally.

Our first experiments of this general type have involved the injection of purified DNA into eggs and oocytes to test the proposition that eggs contain, in their cytoplasm, all molecules required for the proper replication of DNA (Gurdon *et al.* 1969). The results of these initial experiments were much extended by subsequent work (Laskey and Gurdon 1973); together they argue strongly that semi-conservative replication of purified polyoma DNA takes place in eggs but not in oocytes. Injected DNA, which probably never exists uncomplexed with protein in living cells, was evidently associated with cytoplasmic molecules in such a way as to adopt the normal replicative activity of DNA in eggs.

The next important step in the microinjection of purified nucleic acids into cells involved the use of messenger RNA. This is another type of macromolecule that is believed never to exist in cells uncomplexed with proteins and that is, furthermore, exceptionally sensitive to RNase activity. Oocytes and eggs of *Xenopus* contain a very substantial activity of RNase when homogenized. It was therefore surprising to discover that globin mRNA was very efficiently translated into the proteins for which it codes, when injected into the cytoplasm of eggs and oocytes (Gurdon *et al.* 1971, Lane *et al.* 1971). It has now been established from the work of this and many other laboratories that nearly all vertebrate messenger RNAs are very efficiently translated when injected into eggs or oocytes. Again we may conclude that purified macromolecules may be introduced into living cells in such a way that they become associated with normal cell components, and then function efficiently.

With this background, it seemed worth investigating the possibility that injected genes might be transcribed in eggs and oocytes. The first series of experiments (Gurdon and Brown 1977) involved the injection of purified *Xenopus* ribosomal DNA (DNA coding for 18S and 28S RNA) and 5S DNA (that which codes for 5S RNA), as well as mouse satellite DNA, into fertilized eggs of *Xenopus.* The very substantial endogenous synthesis of 18S, 28S, and 5S RNA at all post-blastula stages made it difficult to recognize transcripts from the injected DNA at advanced embryonic stages. During early cleavage, however, it was clear that transcripts from injected 18S, 28S, and 5S DNA were synthesized. Of special interest was the finding that the mouse satellite DNA, which is not transcribed in mouse cells, is also not transcribed in frog eggs, even though it is remarkably stable.

In the nuclear injection experiments outlined above, we had discovered a way of injecting DNA into the nucleus of oocytes. The procedure is described by Gurdon (1976). We soon found that purified SV40 DNA is transcribed to a substantial extent if it is injected into the nucleus of oocytes, but not if injected into oocyte cytoplasm (Mertz and Gurdon 1977). The injected molecules can code for up to 50% of all transcripts synthesized by the oocytes. The injected DNA appears stable, since transcription from it continues at about the same rate for at least a week. The results of further experiments with SV40 DNA injected into oocytes are as follows (J. Mertz and J. Gurdon, unpublished data). The stable transcripts of SV40, which are synthesized in oocytes, are similar in size, relative abundance, and map location to those found in lytically infected monkey cells. The transcription of SV40 DNA in oocytes is very sensitive to low concentrations of α-amanitin, a result which indicates that the process depends on the activity of RNA polymerase II. Some of the SV40 transcripts formed in oocytes are translated into correct viral proteins (De Robertis and Mertz, in press), a further indication that they include correct transcripts. A further conclusion from this early stage of the work was that most kinds of DNA injected are transcribed; in the case of double-stranded DNA from the replicative form of phage ϕX174, 95% of transcripts are from the minus strand, the strand that is transcribed in vivo (Mertz and Gurdon 1977).

Homologous DNAs

It was appropriate, with this background, to investigate the use of *Xenopus* oocytes for transcribing *Xenopus* genes. It was of particular interest to know whether the fidelity of transcription of 5S DNA would be better in injected living oocytes than it is when 5S DNA is transcribed in vitro with added RNA polymerase. The results (Brown and Gurdon 1977) show that this is indeed the case. Thus, transcripts from 5S DNA injected into oocytes are nearly all from the naturally transcribed strand, whereas in vitro transcripts are mostly from the wrong strand. Most of the transcripts made in injected oocytes are from the gene, not spacer, region of 5S DNA. Most important, the majority of transcripts are of the correct size (a sharp 5S RNA band on an electrophoresis gel), and these molecules are of the correct oligonucleotide composition. This implies that the injected 5S DNA is transcribed starting and ending at the right positions or processed so as to give transcripts of this composition. We may conclude that purified genes can be introduced into living cells in such a way that they function nearly as well as in the cells from which they are purified, and much better than in in vitro cell-free transcription systems.

Our current work involving the injection of purified DNA molecules into oocyte nuclei is directed toward the recovery of the injected DNA, so that it is still complexed with the oocyte components with which it becomes associated in the living cell.

CONCLUSIONS

We have seen that nuclear injection into oocytes indicates the existence of gene-controlling molecules. These certainly include those that activate previously unexpressed genes and probably include molecules that inhibit the activity of other, previously expressed, genes.

We have also seen that the injection of purified DNA molecules into the nucleus of living oocytes provides an experimental system with high transcriptional fidelity, of a type not currently obtainable with cell-free in vitro systems. The prospect, therefore, seems favorable for eventually using this experimental system to "fish out" gene-controlling molecules from living cells. A more detailed discussion of this experimental approach is presented elsewhere (Gurdon 1977). If this experimental procedure for extracting gene-controlling molecules by injecting genes proves successful, the results seem likely to be of general relevance to the mechanism of gene control in normal as well as abnormal cells.

ADDENDUM

We have limited this account to work from this laboratory. Appropriate reference to the work of other laboratories is made in the original papers cited below. However, it may be helpful to add that we are aware of only one other paper concerning gene expression by nuclei injected into oocytes (Etkin 1976). At the time of writing, no other work has been published on the transcription of purified DNA molecules injected into oocyte nuclei.

REFERENCES

Brown, D. D., and J. B. Gurdon, 1977. High fidelity transcription of 5S DNA injected into *Xenopus* oocytes. Proc. Natl. Acad. Sci. USA 74:2064–2068.

Davidson, E. H. 1968. Gene Activity in Early Development. Academic Press, New York.

De Robertis, E. M., and J. B. Gurdon. 1977. Gene activation in somatic nuclei after injection in amphibian oocytes. Proc. Natl. Acad. Sci. USA 74:2470–2474.

De Robertis, E. M., and J. E. Mertz. 1977. Cell (In press).

De Robertis, E. M., G. A. Partington, R. F. Longthorne, and J. B. Gurdon. 1977. Somatic nuclei in amphibian oocytes: Evidence for selective gene expression. J. Embryol. Exp. Morphol. 40:199–214.

Etkin, L. 1976. Regulation of lactate dehydrogenase (LDH) and alcohol dehydrogenase (ADH) synthesis in liver nuclei, following their transfer into oocytes. Dev. Biol. 52:201–209.

Gurdon, J. B. 1968. Changes in somatic cell nuclei inserted into growing and maturing amphibian oocytes. J. Embryol. Exp. Morphol. 20:401–414.

Gurdon, J. B. 1976. Injected nuclei in frog oocytes: Fate, enlargement and chromatin dispersal. J. Embryol. Exp. Morphol. 36:523–540.

Gurdon, J. B. 1977. Egg cytoplasm and gene control in development. Proc. Roy. Soc. (B) 198:211–247.

Gurdon, J. B., M. L. Birnstiel, and V. A. Speight. 1969. The replication of purified DNA introduced into living egg cytoplasm. Biochim. Biophys. Acta 174:614–628.

Gurdon, J. B., and D. D. Brown. 1977. Towards an *in vivo* assay for the analysis of gene control and function, *in* Symposium on Molecular Biology of the Genetic Apparatus. P. T'so, ed., North-Holland Publishing Co., Amsterdam 2:111–123.

Gurdon, J. B., E. M. De Robertis, and G. A. Partington. 1976a. Injected nuclei in frog oocytes: A living cell system for the study of transcriptional control. Nature 260:116–120.

Gurdon, J. B., C. D. Lane, H. R. Woodland, and G. Marbaix. 1971. The use of frog eggs and oocytes for the study of messenger RNA and its translation in living cells. Nature 233:177–182.

Gurdon, J. B., G. A. Partington, and E. M. De Robertis. 1976b. Injected nuclei in frog oocytes: RNA synthesis and protein exchange. J. Embryol. Exp. Morphol. 36:541–553.

Lane, C. D., G. Marbaix, and J. B. Gurdon. 1971. Rabbit haemoglobin synthesis in frog cells: The translation of reticulocyte 9S RNA in frog oocytes. J. Mol. Biol. 61:73–91.

Laskey, R. A., and J. B. Gurdon. 1973. Induction of polyoma DNA synthesis by injection into frog-egg cytoplasm. Eur. J. Biochem. 37:467–471.

Mertz, J. E., and J. B. Gurdon. 1977. Purified DNAs are transcribed after microinjection into *Xenopus* oocytes. Proc. Natl. Acad. Sci. USA 74:1502–1506.

Cell Differentiation and Neoplasia, edited by Grady F. Saunders. Raven Press, New York

Cytoskeletal Changes in Cell Transformation to Malignancy

B. R. Brinkley, C. L. Miller, J. W. Fuseler, D. A. Pepper, and L. J. Wible

Department of Human Biological Chemistry and Genetics, Division of Cell Biology, The University of Texas Medical Branch, Galveston, Texas 77550

Transformation of cells in vitro by viral or chemical agents is accompanied by a variety of physiological, morphological, and growth-related changes, including altered lectin-binding characteristics and changes in surface proteins (Burger 1973, Pollack and Burger 1969). In many forms of neoplasia, striking changes occur in cell morphology, including the transition from flattened anisotropic forms that grow as monolayers to rounded pleomorphic forms that pile up as multilayered foci in culture (Temin and Rubin 1958, Stoker and Abel 1962). Changes in growth properties include loss of density-dependent control of growth (Todaro *et al.* 1964), loss of contact-inhibited mobility (Gail and Boone 1971), and acquisition of anchorage-independent growth (MacPherson and Montagnier 1964, Stoker and MacPherson 1961, Freedman and Shin 1974, Benedict *et al.* 1975, Evans and DiPaolo 1975, Risser and Pollack 1974). Transformed cells grow well in media containing reduced serum content and are often characterized by reduced cyclic AMP levels (Smith *et al.* 1971, Holly and Kiernan 1968, Anderson *et al.* 1973, Sheppard 1972, Monahan *et al.* 1973) and altered calcium requirements (Balk *et al.* 1973). Although it is known that various properties in this list are dissociable and are not linked as a group to malignant transformation, taken collectively these findings attest to the molecular complexity of transformation. Moreover, it can be concluded that many properties of transformed cells are attributable to discrete alterations at the cell surface. In this regard, one of the most important concepts to emerge in the field of membrane biology in recent years is that proteins of the cell surface are dynamically continuous with and, to some extent, regulated by an underlying assemblage of cytoplasmic filaments and tubules composing the cytoskeleton. Thus, the mobility of the specific lectins and immunoglobulin receptors can be altered by drugs such as colchicine, vinca alkaloid, and cytochalasin B, which selectively dissociate cy-

Present Addresses: B. R. Brinkley, D. A. Pepper, and L. J. Wible, Department of Cell Biology, Baylor College of Medicine, Houston, Texas 77030; C. L. Miller, Department of Biology, University of Pennsylvania, Philadelphia, Pennsylvania 19104; J. W. Fuseler, Department of Cell Biology, The University of Texas Health Science Center, Dallas, Texas 75235.

toskeletal components (DePetris 1974, 1975, Edelman *et al.* 1973, Yahara and Edelman 1973, 1975, Yin *et al.* 1972, Berlin *et al.* 1974, Poste *et al.* 1975b, Wessels *et al.* 1971, Taylor *et al.* 1971). Through the use of electron microscopy and immunofluorescent probes, we have begun to define more clearly the structure and organization of cytoskeletal elements in normal and transformed cells in vitro. This paper will review the structure, assembly, and distribution of two major cytoskeletal components—microtubules and microfilaments—and describe specific changes that occur in these structures in transformed cells. Evidence will be presented that correlates alterations in the cytoskeleton with the loss of anchorage-dependent regulation of growth and related properties of transformed cells.

MATERIALS AND METHODS

Cell Culture

Nontransformed cell lines used in this study were PA-2 (a generous gift to Dr. Brinkley from Dr. Uta Franke), HSF-CF (both lines of human skin fibroblasts), 3T3 (murine fibroblasts), and PtK_1 (rat kangaroo kidney). Transformed lines observed included CHO (Chinese hamster ovary), SV3T3 (SV40-transformed 3T3), and RAG (murine renal adenocarcinoma). Somatic cell hybrids (CRAG clones) are the progeny of a Sendai virus-induced fusion between HSF-CF and RAG cells. Both the CRAG clones and the RAG cells were given to Dr. Brinkley by Dr. R. J. Klebe.

Cells are maintained as monolayer cultures. Every two to three days, the stock cultures are subcultivated by removing the cells from the substrate with 0.1% or 0.025% trypsin (Gibco) and replating an appropriate number of single cells as the new culture. Growth media used are Dulbecco's Modified Eagle's (DMEM) (Gibco), Hsu's modified McCoy's (Gibco), or Ham's F-10 (Schwarz-Mann), supplemented with 10% fetal calf serum (nonheat inactivated, Gibco). Hybrids are maintained in DMEM supplemented with 10% fetal calf serum, 14 mg/1 hypoxanthine, 0.2 mg/1 aminopterin, and 4 mg/1 thymidine (HAT medium). Incubation is at 37°C in a humidified atmosphere of 5% or 10% CO_2 in air.

Cells used in immunofluorescent observations were plated onto sterile 11 × 22 mm coverslips at least 24 hours prior to experimentation to insure exponential growth of the cultures.

Determination of In Vitro Growth Potential

Plating efficiency on plastic is determined by plating a predetermined number of cells into a petri dish containing growth medium and allowing the dishes to incubate undisturbed for one week. The colonies are then fixed with methanol, stained with crystal violet, and the number of colonies consisting of 50 cells or more on each dish are counted. Plating efficiency is determined by expressing the number of colonies per dish as a percentage of the number of cells originally plated.

Efficiency of plating in agar is determined through the use of a modification of the procedure of MacPherson and Montagnier (1964). Base layers of 0.6% Noble agar in growth medium are first poured into petri dishes and allowed to solidify. A predetermined number of cells are then suspended in 0.3% agar in growth medium and established in the dishes as a growth layer. The dishes are then incubated and allowed to grow undisturbed for 14 days. At the end of this time, the number of colonies in each dish are counted. Plating efficiency is calculated by expressing the number of cells per dish that grew into colonies as a percentage of the number of cells originally plated.

Saturation density of a cell line is determined by plating 250,000 cells into each of 10 petri dishes and incubating them at 37°C. At measured intervals thereafter, one dish is removed, and the total number of cells in the dish determined. Medium is replaced with fresh, prewarmed medium every third day. Saturation density is considered to have been reached when the number of cells in each dish no longer increases.

Drug and Enzyme Treatments

Chemicals used in this study were dibucaine hydrochloride (Chemicals Procurement Labs), 0.2 mM in DMEM; ionophore A23187 (Lilly), 5 μg/ml in DMEM supplemented with calcium to yield a final concentration of 4.7 mM; trypsin (Gibco), 60 μg/ml in Puck's saline; neuraminidase (Worthington), 20 μg/ml in DMEM; 0.3 mM N^6, 0^2-dibutyryl adenosine 3′:5′ cyclic monophosphoric acid monosodium salt (Bt_2cAMP) (Sigma) and 1 mM 1,3-dimethylxanthine (theophylline) (Sigma) in DMEM; and cytochalasin B (Aldrich), 10 μg/ml in DMEM. Dibucaine, ionophore, and cytochalasin B were stored as concentrated stock solutions in dimethylsulfoxide (DMSO). Appropriate aliquots of stock solutions were added to fresh media to give the final working concentrations listed above. To insure that none of the cellular effects observed were due to the presence of solvent, control medium containing DMSO was also used.

Two coverslip cultures were incubated in the appropriate chemical-containing medium at 37°C for the length of time indicated as follows:

Colcemid—60 minutes
Colcemid reversal—10, 20, 30, and 60 minutes
dibucaine—15 minutes
ionophore + Ca^{++}—60 minutes
trypsin—6 minutes
neuraminidase—30 minutes
cytochalasin B—15 minutes
DMSO control—60 minutes

Coverslip cultures of SV3T3 cells were incubated in the Bt_2cAMP + theophylline medium at 37°C for eight hours. A second culture of SV3T3 cells was incubated in normal medium alone under the same conditions. Following treatment, all coverslips were fixed with formaldehyde and processed for tubulin and actin immunofluorescence as described below.

Indirect Immunofluorescence

The cytoplasmic microtubules and microfilaments are visualized by indirect immunofluorescence using either rabbit antibovine tubulin antibody made against 6S tubulin monomers (Fuller *et al.* 1975a) or rabbit antiactin antibody, respectively. Coverslips with the attached monolayer of cells are rinsed briefly in phosphate-buffered saline (PBS), fixed in 3% Ultrapure formaldehyde (Tousimis) in PBS for 20 minutes, then dehydrated in acetone at −10°C for seven minutes. Following a second PBS wash, the fixed cells are incubated with the antibody (0.1 mg/ml in PBS) for 45 minutes at 37°C. After three gentle 10-minute washes in PBS, the cells are incubated in fluorescein-conjugated goat anti-rabbit immunoglobulin G (Meloy Laboratories) diluted 1:6 in PBS for 45 minutes at 37°C. The cells are finally rinsed three times in PBS for 30 minutes and mounted on glass slides in a drop of glycerol:PBS (9:1 by volume at pH 9.0–10.5).

Antiactin was prepared according to the procedure of Lazarides and Weber (1974) and was a generous gift from Dr. E. Lazarides to Dr. Brinkley.

Control slides consisted of cells prepared as above with rabbit albumin antiserum substituted for the primary antibody.

Cold Treatment

The conditions for cold treatment of cultured cells was precisely the same as those described by Brinkley and Cartwright (1975), except that cells were held in the cold for two hours.

Light Microscopy

Observations were made with a Leitz Orthoplan microscope adapted with a Leitz darkfield condenser for fluorescence microscopy. A Leitz 54X Fl Oel (N.A. 0.95) objective was used exclusively for viewing the cells. The output of an HBO 200W high-pressure mercury arc lamp used as an illumination source was filtered with a Leitz KG1 heat-absorbing filter, and a BG 38 red suppression filter. The excitation filter used was a Leitz KP 490, and the barrier filter was a Leitz K530. Photographs were made using the Leitz Orthomat camera with Kodak Tri-X Pan film. Magnification was calibrated by photographing a standard-stage micrometer through the optical system.

Electron Microscopy

Cells to be examined by electron microscopy were prepared according to the procedures described by Brinkley and Cartwright (1975).

RESULTS AND DISCUSSION

Nature of the Cytoskeleton in Nontransformed Cells

Examination of ultrathin sections of appropriately fixed cells in vitro by transmission electron microscopy reveals a network of 240Å microtubules and bundles of 60Å microfilaments, which course their way through the cytoplasm and frequently come in association with the plasma membrane (Figure 1). A number of investigators have described microtubules and microfilaments in the cytoplasm and consider these elements to be the major components of the cytoskeleton (Goldman *et al.* 1976). Unless whole cells are examined by high voltage electron microscopy (Porter 1976, Buckley and Porter 1975), however, it is difficult to visualize the extensive distribution of these structures by conventional thin-section electron microscopy. Recent preparation of antibodies directed against microtubule protein (tubulin) and proteins of the 60Å microfilaments (actin) has made it possible to visualize the cytoskeleton in large populations of cells by immunofluorescence procedures (Lazarides and Weber 1974, Fuller *et al.* 1975b, Brinkley *et al.* 1975, Weber *et al.* 1975).

The Cytoplasmic Microtubule Complex

Nontransformed cells (Figures 2–4) stained with antitubulin antibody display an elaborate complex of fine fluorescent filaments throughout the cytoplasm, when observed by indirect immunofluorescence. In most instances, these filaments associate with one or two specific foci in the cytoplasm near the nucleus and radiate out toward the cell periphery, where they either terminate at the plasma membrane or bend and extend along the cell surface (Figures 8 and 9). We have termed this array of fluorescent filaments the cytoplasmic microtubule complex (CMTC) and have identified it in a wide variety of nontransformed cells in vitro (Brinkley *et al.* 1975, Fuller and Brinkley 1976). In independent studies, Weber and co-workers (1975) have described essentially the same complex using antitubulin immunofluorescence on 3T3 cells in vitro.

The CMTC is present throughout interphase but is disassembled as the cells enter mitosis. As shown in Figure 11, the CMTC begins to disassemble in late G_2 or early prophase, and the fluorescent filaments disappear except for those that are associated with the two poles of the spindle. As the cells progress into prometaphase, they become more rounded in appearance, and fluorescence becomes entirely localized within the forming mitotic spindle and the astral

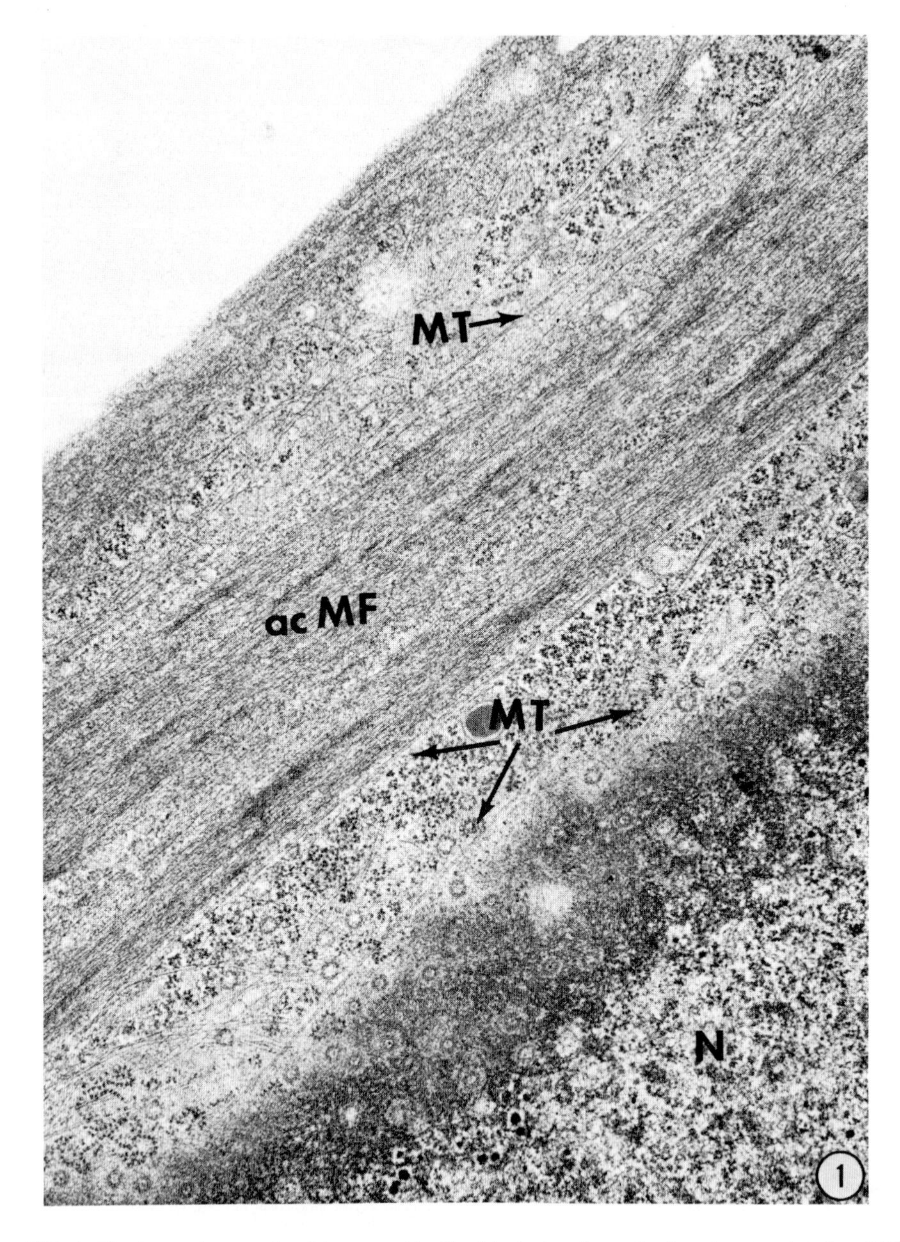

FIG. 1. Electron micrograph of human skin fibroblast showing cytoplasmic microtubules (MT) and bundles of actin filaments (acMF) forming actin stress fibers (X16,000)

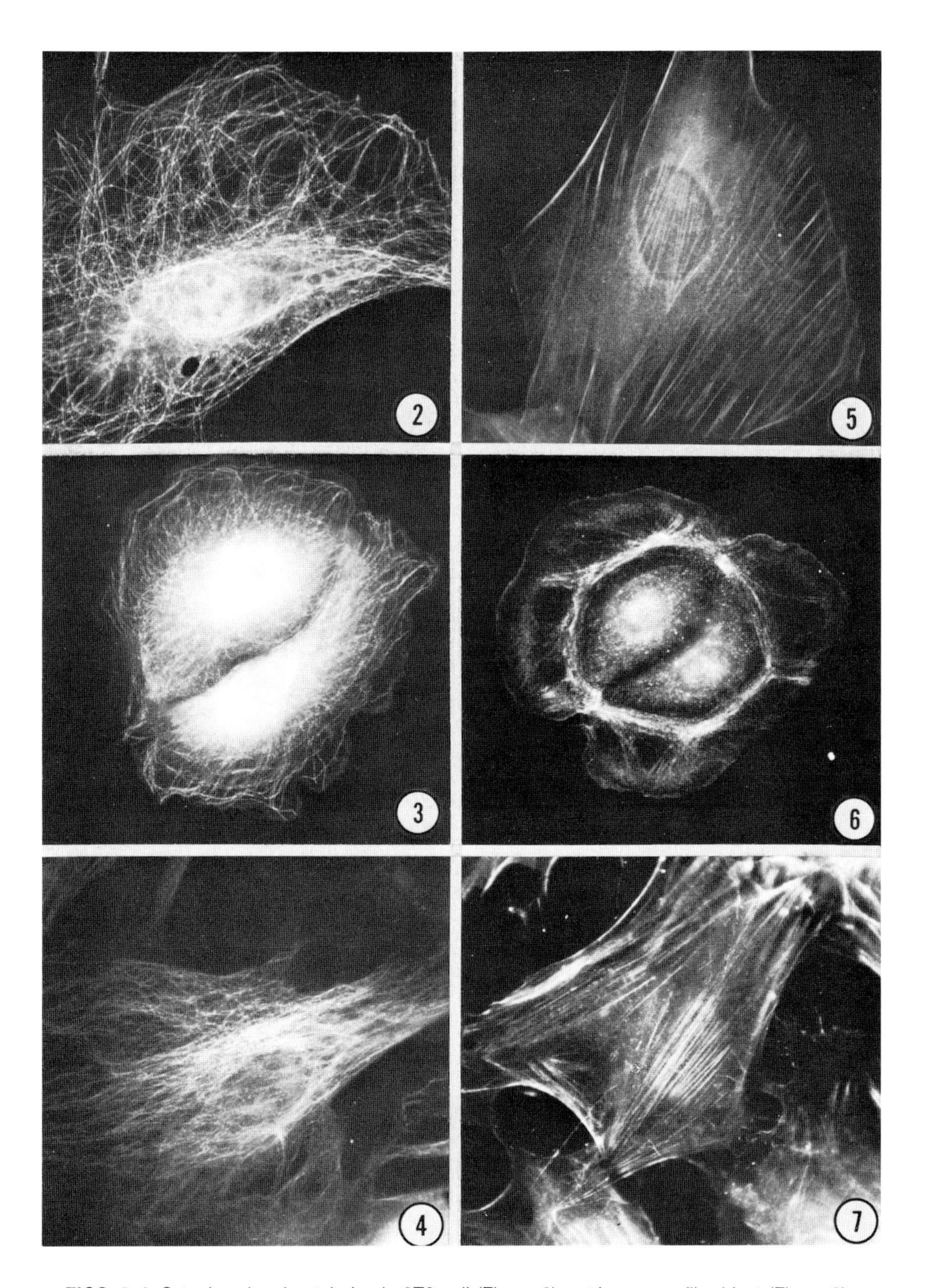

FIGS. 2–4. Cytoplasmic microtubules in 3T3 cell (Figure 2), rat kangaroo fibroblast (Figure 3), and human skin fibroblast (Figure 4).

FIGS. 5–7. Actin stress fiber patterns shown by antiactin immunofluorescence; 3T3, rat kangaroo, and human skin fibroblast, respectively.

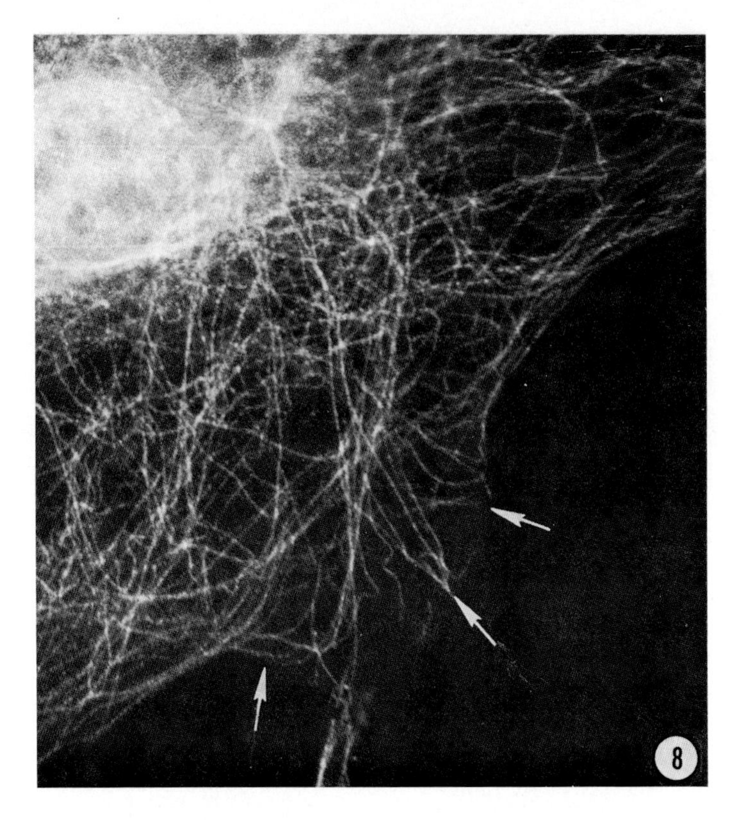

FIG. 8. Higher magnification of CMTC in 3T3 cells. Arrows point to position of the plasma membrane. Note some microtubules appear to terminate near the plasma membrane, while others bend and extend along the membrane.

fibers (Figure 12). At metaphase, the chromosomes become aligned on the equatorial plate, and a clearly defined spindle is present near the center of the cell (Figure 13). The cytoplasm surrounding the spindle is totally free of any filamentous components. As the cell progresses through anaphase and into telophase, the chromosomes move to the poles, and cytokinesis occurs (Figures 14 to 17). During cytokinesis, the fluorescence becomes concentrated within the midbody, which bridges the two daughter cells (Figure 18). This structure is known to contain numerous microtubules and probably represents the rudiments of the interpolar spindle fibers. During late telophase or early G_1, a large aster forms in association with the centriole region of each daughter cell (Figures 17 and 18). Subsequently, the complete CMTC is generated from each of these regions as the cells become flattened and assume a more fibroblastic appearance.

Thus, in all normal cells examined, the extensive cytoplasmic microtubule complex is disassembled at the beginning of mitosis and is replaced by a highly fluorescent mitotic apparatus that functions to segregate the chromosomes during

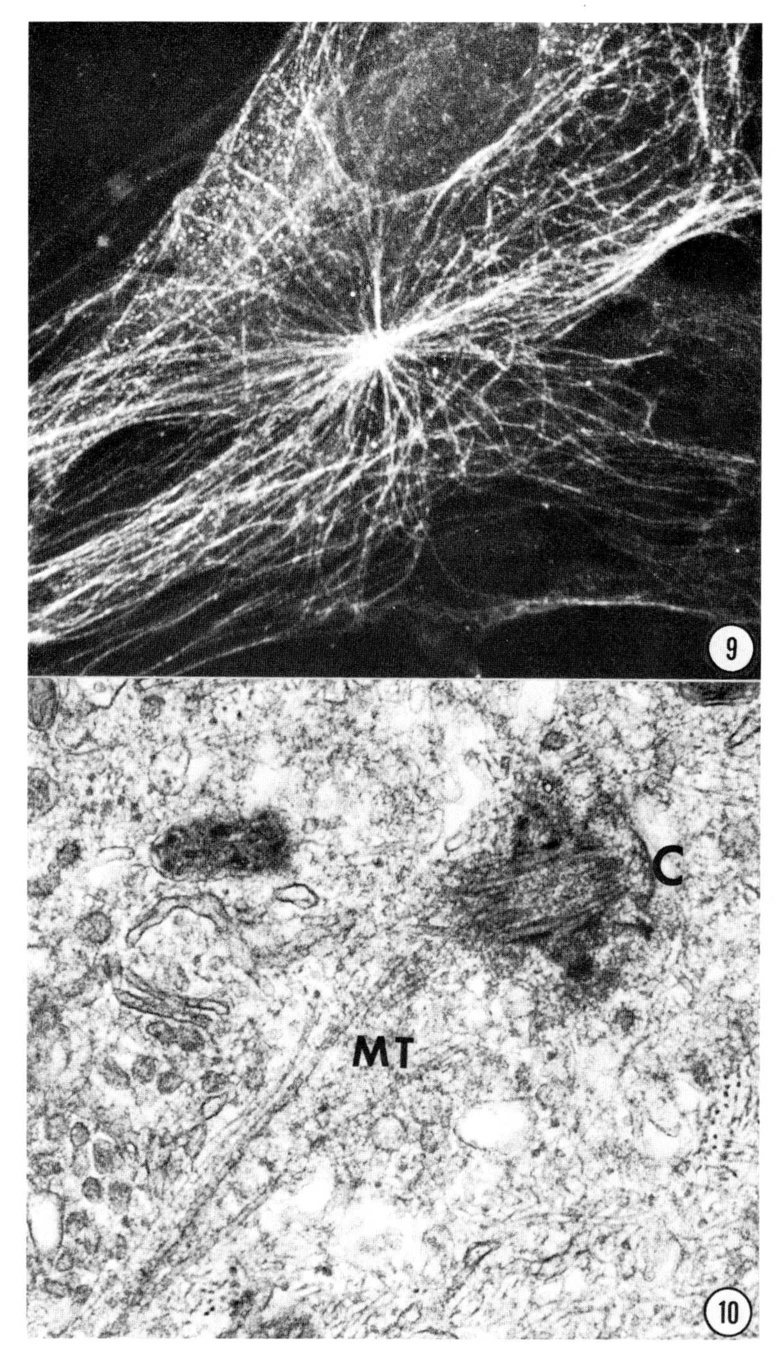

FIG. 9. CMTC of 3T3 cells showing central, bright fluorescent spot from which microtubules are extended radially.

FIG. 10. Electron micrograph of a cell through a region corresponding to the bright fluorescent spot shown in Figure 9. Mt, microtubules; C, centriole.

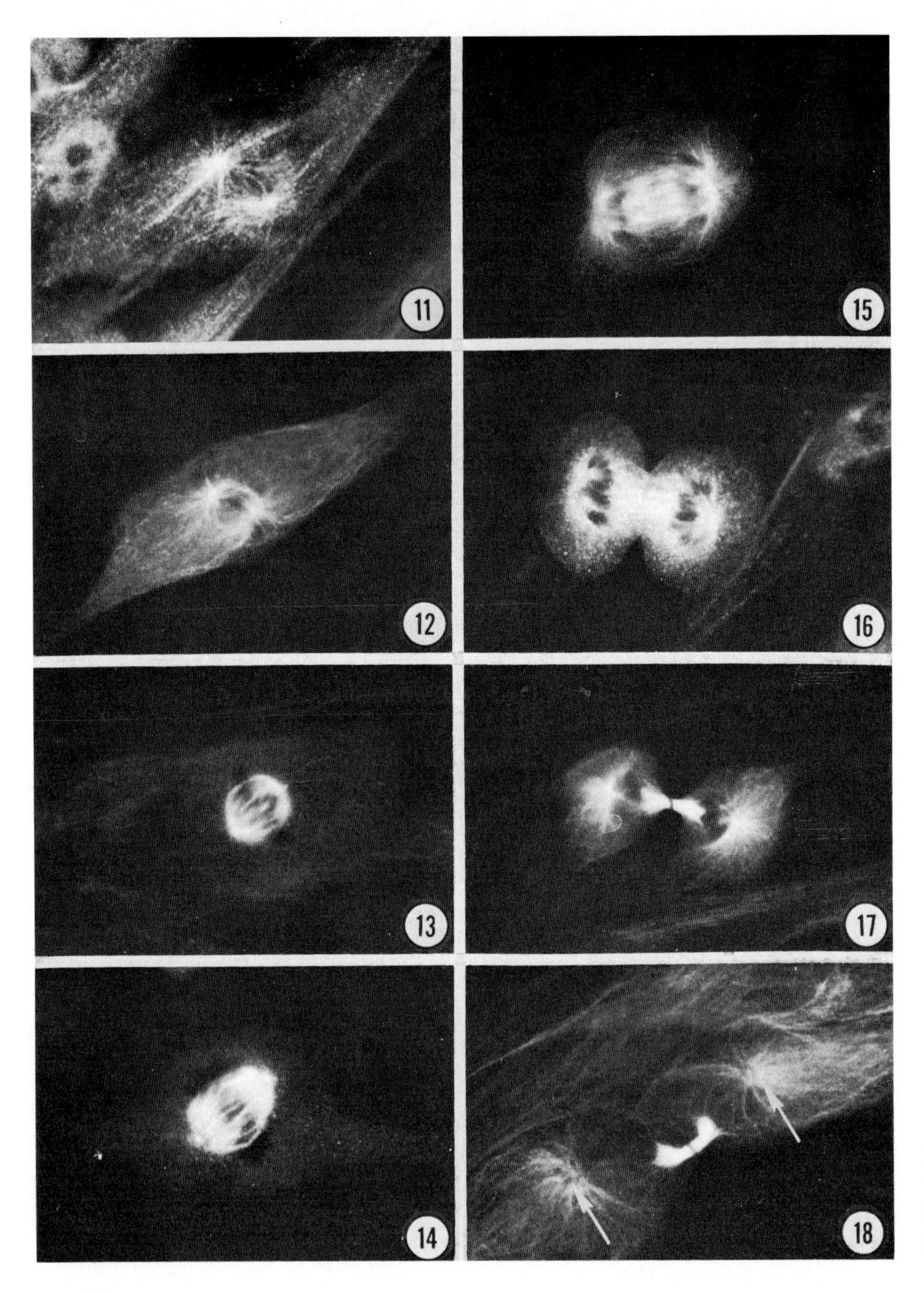

FIGS. 11–18. Stages of mitosis in human skin fibroblast (strain PA-2). Prophase (Figure 11), prometaphase (Figure 12), metaphase (Figure 13), early anaphase (Figure 14), late anaphase (Figure 15), telophase with cytokinesis (Figure 16), telophase with bright fluorescent midbody (Figure 17), and late telophase-early G_1 phase showing the return of the CMTC at each centrosome (arrows, Figure 18).

mitosis. The entrance of cells into mitosis and the disassembly of the cytoplasmic microtubule apparatus is accompanied by the major alterations in cell shapes from a flattened state to a more rounded appearance. When mitosis is completed, the mitotic apparatus disappears and is replaced by cytoplasmic microtubules that originate at the asters in each daughter cell. The formation of the CMTC in G_1 phase of the cell cycle is accompanied by the return of the cell to a flattened, more fibroblastic appearance. Since both systems are composed of microtubules of similar structure and properties, it appears likely that tubulin is recycled from one apparatus to the other during the mitotic cycle (Brinkley *et al.* 1975).

Actin Stress Fibers

Examination of many types of nonmuscle cells by phase contrast, Nomarski or polarizing optics reveals an array of "stress fibers" that exist as bundles or meshworks in the cytoplasm (Goldman *et al.* 1976). When glutaraldehyde-osmium fixed cells are examined by transmission electron microscopy, the stress fibers are seen to be composed of bundles of individual 60Å microfilaments (see Figure 1). Although they are more numerous near the region of cell-substrate contact, bundles of actin filaments are also found throughout the cytoplasm.

As originally shown by Lazarides and Weber (1974), the intracellular distribution of actin fibers is best demonstrated by immunofluorescence using antiactin antibodies. Figures 5 to 7 show actin fibers in flattened cells from three different cell lines, mouse 3T3 (Figure 5), rat kangaroo PtK_1 (Figure 6), and human PA-2 (Figure 7). The fluorescent fibers extend over the entire cell and are often arranged in parallel array. The cell is usually characterized by bright fluorescence near the plasma membrane. As described by Lazarides and Weber (1974), the pattern of actin fibers may vary with the cell and the culture. We find this to be true but have also noted that pairs of daughter cells in G_1 phase often display actin fibers arranged in mirror images (see Figure 6). Such observations suggest that discrete microfilament organizing centers may be transmitted to opposite poles of the spindle during mitosis and serve to regulate the geometry of actin fiber assembly in daughter cells. To some extent, the pattern of cytoplasmic microtubules in daughter cells also seems to exist in mirror images in flattened cells (see Figure 3), but such patterns are much more difficult to visualize due to the complexity of the CMTC.

During mitosis the actin stress fibers disappear in prophase and are absent in the cytoplasm throughout mitosis. The presence of actin filaments in the mitotic spindle have been described in some cell types (Forer 1976, Sanger and Sanger 1976), but the functional status of actin in the mitotic apparatus is still controversial. When mitosis is completed, the actin stress fibers reappear in the daughter cells and become numerous as the cell assumes a more flattened and anisotropic appearance.

In many respects, the CMTC and actin stress fibers exhibit similar behavior throughout the cell cycle of nontransformed cells. Both are present in the cytoplasm of flattened cells during interphase and appear to be involved in the maintenance of cell shape.

Regulation of Assembly of the Cytoskeleton

As indicated in the previous section of this report, the cytoskeleton is a labile structure that undergoes rapid assembly and disassembly at precise times in the cell cycle. It is likely that the intracellular motility as well as the motility of cell surface components require regulated assembly and disassembly of microtubules (Mt) (and perhaps microfilaments). Most of our knowledge concerning the kinetics of Mt assembly and disassembly has come from in vitro reassembly studies of tubulin derived from neurons (Weisenberg 1972, Borisy *et al.* 1975, Murphy and Borisy 1975, Shelanski *et al.* 1973). Relatively little is known of the factors that regulate microtubules and microfilaments in intact cells. Studies using immunofluorescence as molecular probes for localizing actin and tubulin are beginning to provide useful information on the mechanism of cytoskeletal regulation in vivo.

Effects of Colcemid

When nontransformed cells are incubated in medium containing Colcemid (0.06 μg/ml) for two hours, the CMTC is completely disassembled, as shown in Figure 26. The cells usually undergo major changes in morphology, becoming more rounded and pleomorphic in appearance. The cytoplasm is free of fluorescent filaments and staining is diffuse. One or two bright fluorescent spots may be seen near the nucleus (Figure 26), which represent the centriole region (see Figures 9 and 10). When the cells are washed free of Colcemid, reassembly of cytoplasmic microtubules proceeds from these regions (Figures 27 and 28). Within 30 minutes after reversal, an extensive CMTC is formed around the region, and long microtubules extend to the cell periphery (Figure 29). After 60 minutes, the CMTC is completely reassembled in the cytoplasm, and the cells reassume their flattened state (Figure 30).

Cold Temperatures

The cytoplasmic microtubules are also cold labile. Cells that have been chilled to 0–4°C for two hours are devoid of a CMTC. Warming the cells to 37°C initiates reassembly of the CMTC (Figures 19–24) in a manner similar to that observed for Colcemid reversal. Neither Colcemid nor cold treatment has any appreciable effect on actin stress fibers (Weber *et al.* 1975).

Role of Centrioles in CMTC Assembly

On the basis of the above evidence, we have concluded that most of the cytoplasmic microtubules are assembled at one or two specific foci within the

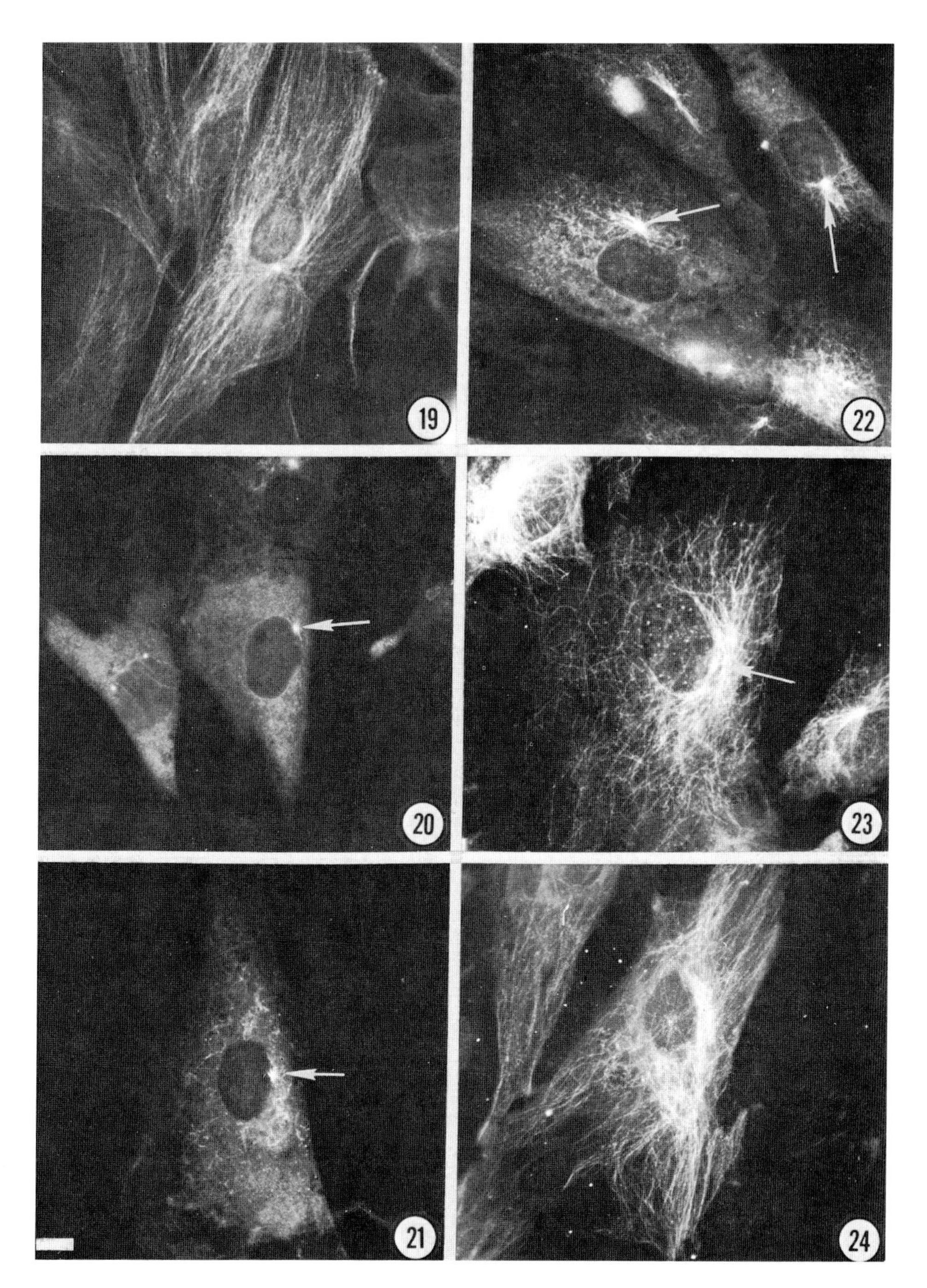

FIGS. 19–24. *Recovery of the CMTC from cold treatments in PA-2 fibroblasts.* Typical CMTC of interphase cell before cold treatment (Figure 19). Interphase cell after two hours at 0°C; no intact microtubules are observed (Figure 20). Five-minute temperature reversal at 33.8°C; CMTC begins re-forming from centrosphere-organizing center. A few fluorescent fibers are seen re-forming, but not associated with, the organizing centers (Figure 21). Ten-minute temperature reversal at 37°C; the CMTC is about 70% re-formed with much activity at the organizing center (Figure 22). Twenty-minute temperature reversal at 37°C (Figure 23). Thirty-minute temperature reversal at 37°C (Figure 24).

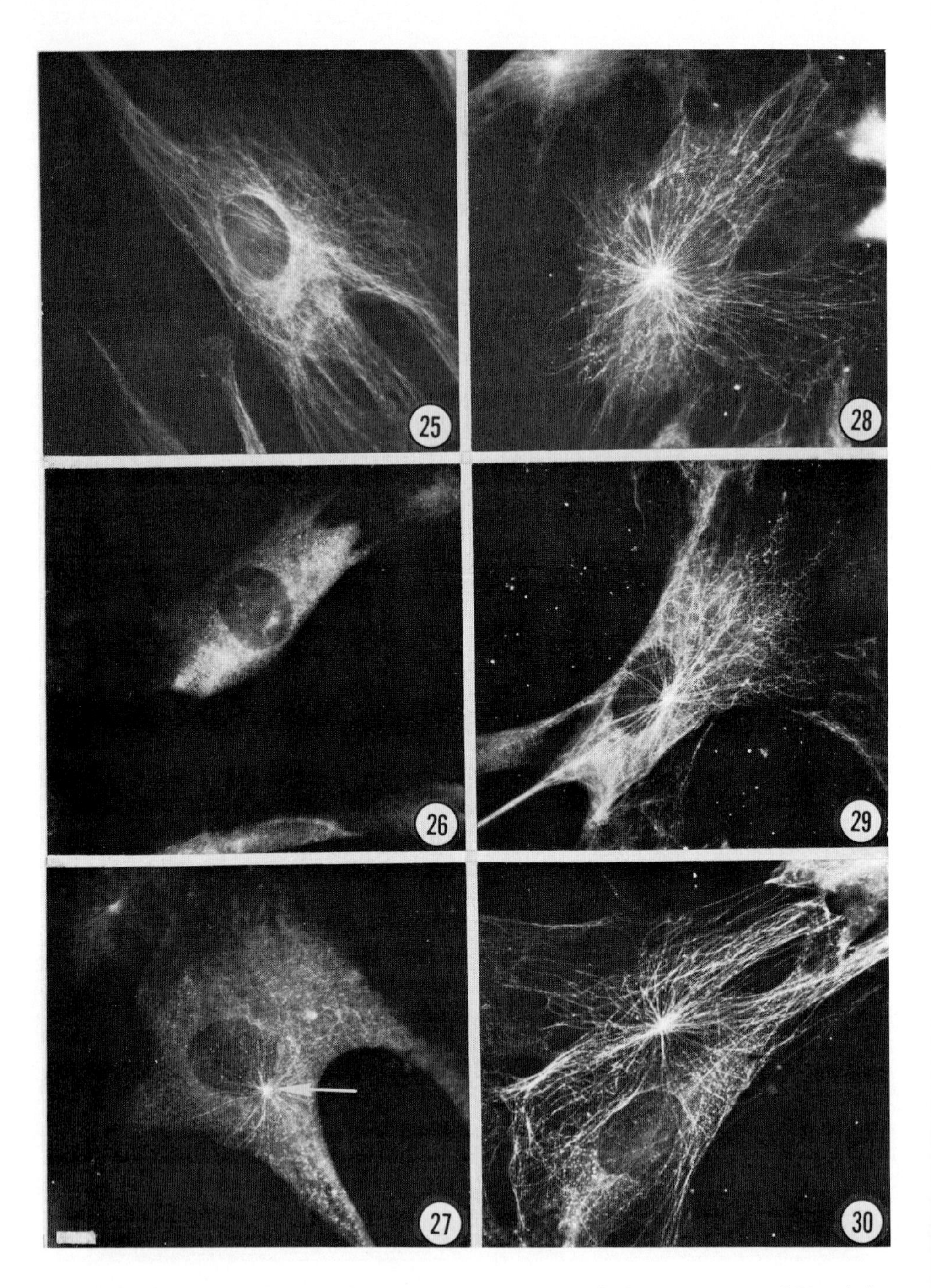

FIGS. 25–30. *Reversal of the CMTC of PA-2 interphase cells from exposure to 0.06* μg/ml *Colcemid.* Interphase PA-2 cell, no drug treatment (Figure 25). After incubation in Colcemid for 2 hours at 37°C (Figure 26). Ten-minute reversal from drug treatment; microtubules begin re-forming at bright fluorescent region (arrows, Figure 27). Twenty-minute reversal; activity continues, and microtubules radiate out into the cytoplasm (Figure 28). Thirty-minute reversal; the CMTC is about 70% re-formed (Figure 29). Sixty-minute reversal; the morphology of the CMTC and cell shape is returned to normal (Figure 30).

cytoplasm, which can be shown by electron microscopy to contain centrioles. Centrioles are also thought to be involved in regulating the assembly of the mitotic spindle (Brinkley and Stubblefield 1970). Therefore, it is not surprising that they are also found in the region where the CMTC is organized. It can be concluded that the major microtubule organizing center of mammalian cells is the centrosphere region of the cell containing the centrioles and associated organelles.

Ionophore A23187

It is now well known that the in vitro assembly of microtubules can be inhibited when excess calcium is present in the reassembly buffer, and several investigators have proposed that calcium ion flux may regulate the assembly of microtubules in intact cells (Fuller *et al.* 1976, Kirschner *et al.* 1974, Solomon 1976). Recently Fuller and co-workers (1975b) demonstrated that inhibiting levels of Ca^{++} could be overcome when isolated mitochondria were added to the in vitro microtubule reassembly system. These same investigators (Fuller and Brinkley 1976) demonstrated that the CMTC could be disassembled in cultured 3T3 cells when ionophore A23187 was present with elevated calcium in the culture medium. We have carried out similar studies and observed the effects of elevated calcium on both the CMTC as well as actin cables. As shown in Figure 33, 3T3 cells grown for one hour in media containing ionophore A23187 plus 4.7 mM calcium show a greatly diminished CMTC. In fact, most of the cells round up and float off the substrate after such treatment. A somewhat less dramatic effect was observed in the actin cables (Figure 35). Even in the more rounded cells, a few actin cables could be observed.

Local Anesthetics

Poste and co-workers (1975a) found that treatment of mouse and hamster cells in vitro with the local anesthetics dibucaine (a tertiary amine) and procaine led to enhanced cell agglutination by plant lectin. In a subsequent study (1975b), they found that such treatment also resulted in the disassembly of both microtubules and microfilaments in the cytoplasm. They concluded that cytoskeletal disassembly may have resulted from release of sequestered calcium caused by the anesthetic. The accumulation of intracellular calcium may be antagonistic to microtubule and microfilament organization.

Fuller (manuscript in preparation) found that dibucaine, procaine, and tetracaine treatment of mouse 3T3 cells led to rapid reversible disassembly of the CMTC. We further utilized these drugs to determine their effects on actin fibers, concurrently monitoring the CMTC. Incubation of 3T3 cells in medium containing 0.2 mM dibucaine for 15 minutes leads to a dramatic reduction of both actin fibers and the CMTC (see Figures 32 and 35). Control cells incubated in medium containing only DMSO showed no changes in the cytoskeleton after

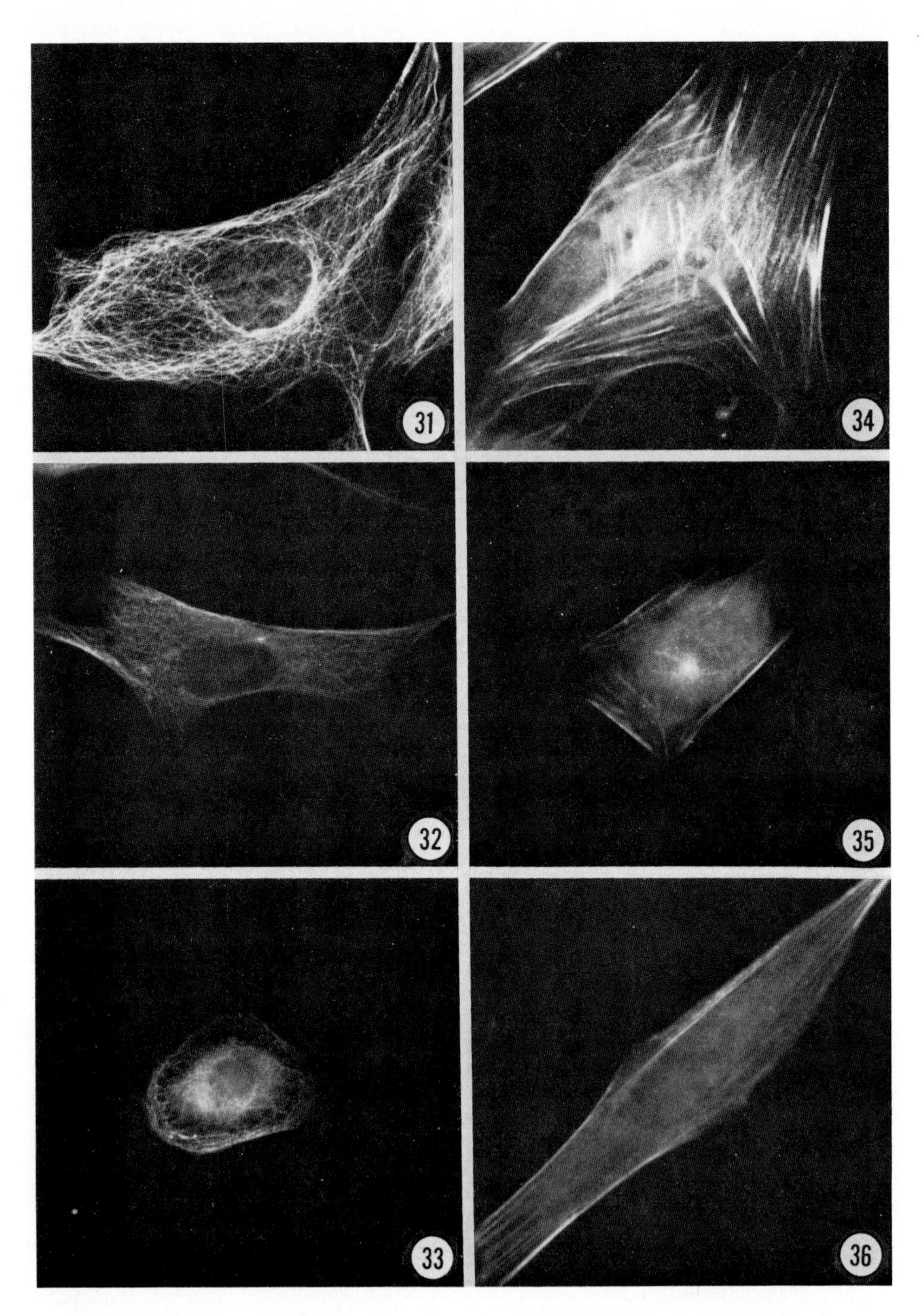

FIGS. 31–36. Effect of dibucaine and ionophore A23187 on CMTC and actin stress fibers. A control 3T3 shows normal CMTC patterns (Figure 31) and actin fibers (Figure 34). Dibucaine treatments result in striking diminution of CMTC (Figure 32) and actin cables (Figure 35). Treatment of 3T3 cells with ionophore A23187 produces diminution of CMTC (Figure 33) and decrease in the number of actin cables (Figure 36).

60 minutes of exposure (see Figures 31 and 34). The rapid diminution of microtubules and the subsequent alteration in cell shape are essentially the same as that observed in cells exposed to ionophore A23187 (Figures 33 and 36) and would lend support to the contention that the local anesthetics affect cytoskeletal organization through the release of bound calcium.

Trypsin and Neuraminidase

It has been proposed that mild exposure of the cell surface to proteolytic enzymes can lead to altered cell shape changes (Revel *et al.* 1974) and disassembly of cytoskeletal structure (Fuller and Brinkley 1976). As shown in Figures 37, 38, 40, and 41 both trypsin and neuraminidase treatment of 3T3 cells result in rapid disassembly of actin filaments and microtubules. The observation is consistent with the findings of Pollack and Rifkin (1975) who reported the dissociation of actin cables in rat embryo cells by treatment with plasmin and trypsin, and of Fuller and Brinkley (1976) who found that trypsin disrupts the CMTC.

Cyclic Nucleotides

It is well known that cyclic AMP and its derivative, 3′:5′-dibutyryl cyclic AMP can stimulate cells to undergo microtubule-related morphological changes (Hsie and Puck 1971). Moreover, cyclic GMP also stimulates microtubule assembly in some cells (Oliver *et al.* 1975). Untreated, transformed SV3T3 cells (Figure 39) display only diffuse staining with tubulin immunofluorescence. When they are treated with dibutyryl cyclic AMP, the cells become more flattened and fibroblastic in appearance. Moreover, they exhibit a more organized CMTC (Figure 42), suggesting that the cyclic nucleotide stimulates microtubule assembly. This observation is consistent with the finding that morphological changes induced by cAMP are inhibited by colchicine (Hsie and Puck 1971).

Although the mechanism of cyclic nucleotide stimulation of microtubule assembly by cAMP is unknown, it is conceivable that such treatments could affect movement of calcium ions in the cytoplasm (Borle 1974), producing conditions favoring microtubule assembly.

Cytochalasin B

As reported in an earlier section, treatment of cells with Colcemid produces a rapid, reversible disassembly of the cytoplasmic microtubules but has little effect on the actin stress fibers. Treatment of cells with cytochalasin B abolishes actin fibers, as Colcemid disrupts microtubules. As shown in Figure 43, the CMTC is still intact although somewhat disarrayed in a 3T3 cell treated with cytochalasin B (10 μg/ml) for 15 minutes. The actin filaments, however, are largely dissolved (Figure 44). This observation is not unexpected, due to the

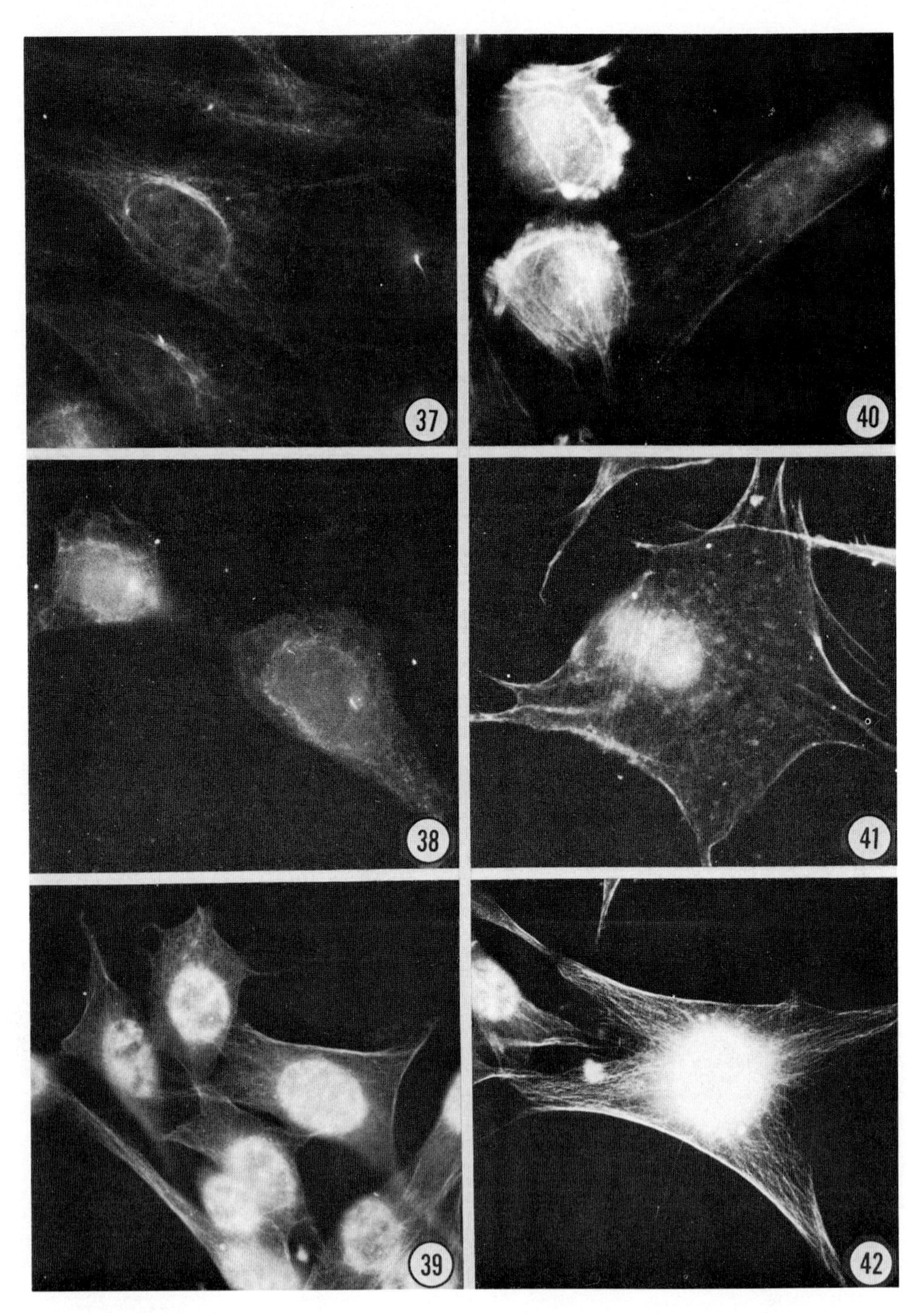

FIGS. 37–42. Effects of trypsin, neuraminidase, and dibutyryl cyclic AMP on cytoskeleton. After brief trypsin treatment, both CMTC (Figure 37) and actin fibers (Figure 40) are diminished. Neuraminidase digestion of the cell surface also disassembles CMTC (Figure 38) and actin (Figure 41). SV3T3 cells show only diffuse staining after antitubulin immunofluorescence (Figure 39). After treatment of the cells with dibutyryl cAMP, the cell becomes more fibroblastic, and an extensive CMTC appears (Figure 42).

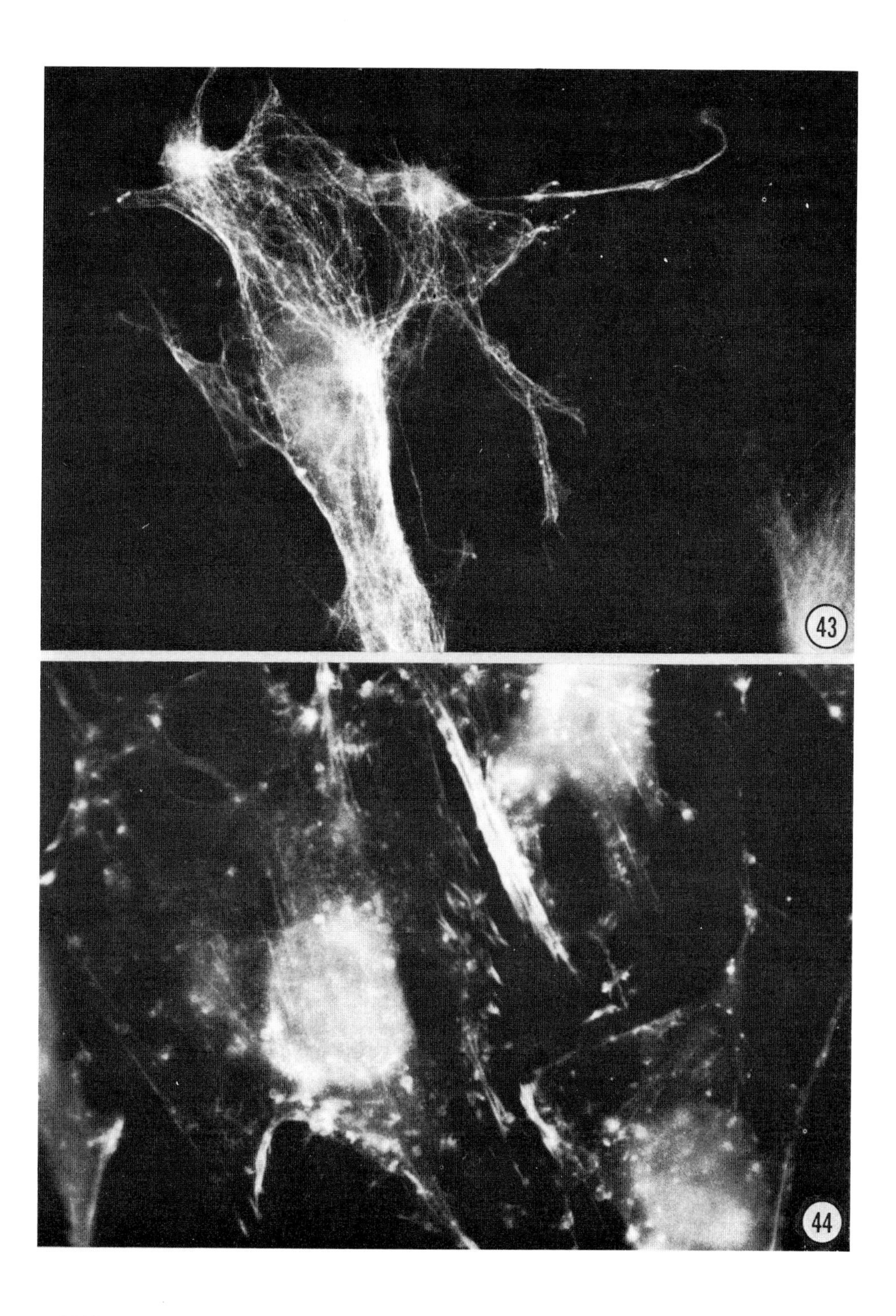

FIGS. 43–44. 3T3 cells after treatment with cytochalasin B. CMTC is still present, although somewhat disorganized after treatment (Figure 43). The actin stress fibers are diminished after cytochalasin B treatment (Figure 44).

well-known action of cytochalasin B on actin microfilaments. It does illustrate, however, that the CMTC can remain essentially intact in the absence of actin cables.

Generalizations Concerning the Assembly and Disassembly of the Cytoskeleton

Although any conclusions concerning the control of assembly of cytoskeletal elements would be premature, some generalizations can be drawn from the experiments described above. Many activities of the cell, including morphogenesis, cell motility, surface receptor mobility, cell-cell and cell-substrate interactions, and intracytoplasmic motility seem to require controlled assembly and disassembly of the cytoskeleton. We have observed that the primary site for the initiation of assembly of cytoplasmic microtubules is a centralized region of the cytoplasm containing centrioles and associated structures. We know far less about the factors that regulate the assembly of microtubules and microfilaments. Excellent progress is being made in the analysis of microtubule assembly using in vitro model systems with purified exogenous tubulin. However, our knowledge of factors that regulate cytoskeletal assembly and disassembly in intact cells is still relatively primitive. Nevertheless, experiments described in the preceding section strongly suggest that intracellular calcium flux may mediate both microtubule and microfilament assembly and disassembly. Other ions such as magnesium are probably involved (Olmsted 1976), and various molecular entities such as calcium-binding protein, microtubule-associated proteins (Borisy *et al.* 1975) or tau factors (Penningroth *et al.* 1976) may also influence assembly. As discussed in the remaining sections, knowledge of the regulation of cytoskeletal assembly could be of considerable importance in understanding the transformation of cells to malignancy.

Cytoskeletal Alterations in Transformed Cells

It has recently been reported by several investigators that virally or chemically induced transformation of cells in vitro is accompanied by conspicuous alteration of the cytoskeleton (Fonte and Porter 1974, Brinkley *et al.* 1975, Edelman and Yahara 1976). When we first examined transformed cells with antitubulin-immunofluorescent staining (Brinkley *et al.* 1975) images such as those shown in Figures 45, 46, and 47 were obtained. When compared with nontransformed cells (see Figures 2, 3, and 4), it was apparent that the cytoplasm of transformed cells contained fewer and more randomly oriented cytoplasmic microtubules. Even prior to the use of immunofluorescence procedures, Fonte and Porter (1974) reported that rat kidney cells transformed with Kirstan sarcoma virus showed fewer cytoplasmic microtubules than did their normal counterparts, when examined by conventional transmission electron microscopy. Edelman and Yahara (1976), using immunofluorescence, have recently reported similar differ-

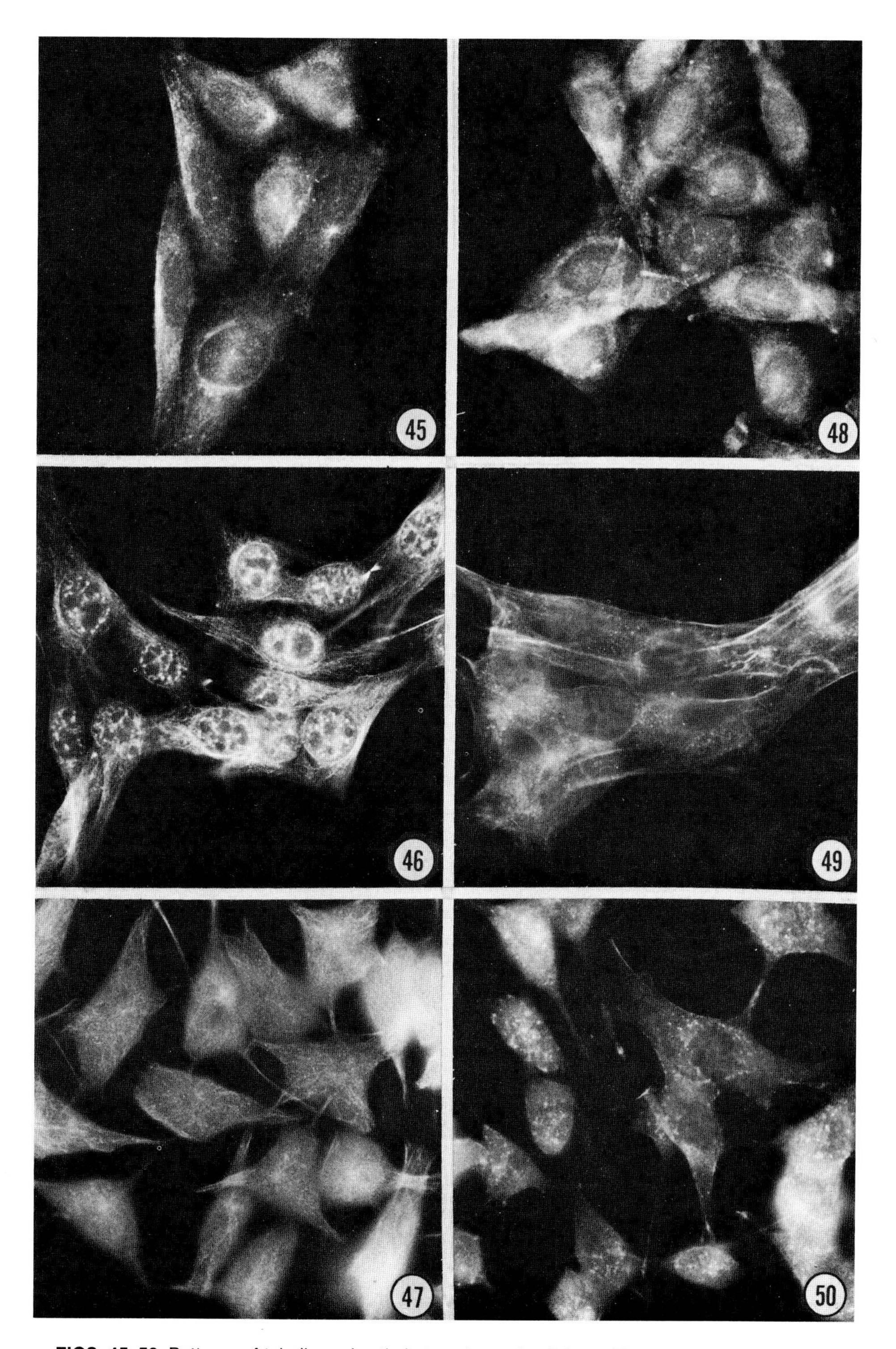

FIGS. 45–50. Patterns of tubulin and actin in transformed cell lines. Figures 45–47 show CHO, RAG, and SV3T3 cells, respectively, after staining with antitubulin. Note the diffuse staining and paucity of CMTC. Figures 48–50 show the same cell lines after staining for actin. Note the absence of actin stress fibers in these cells.

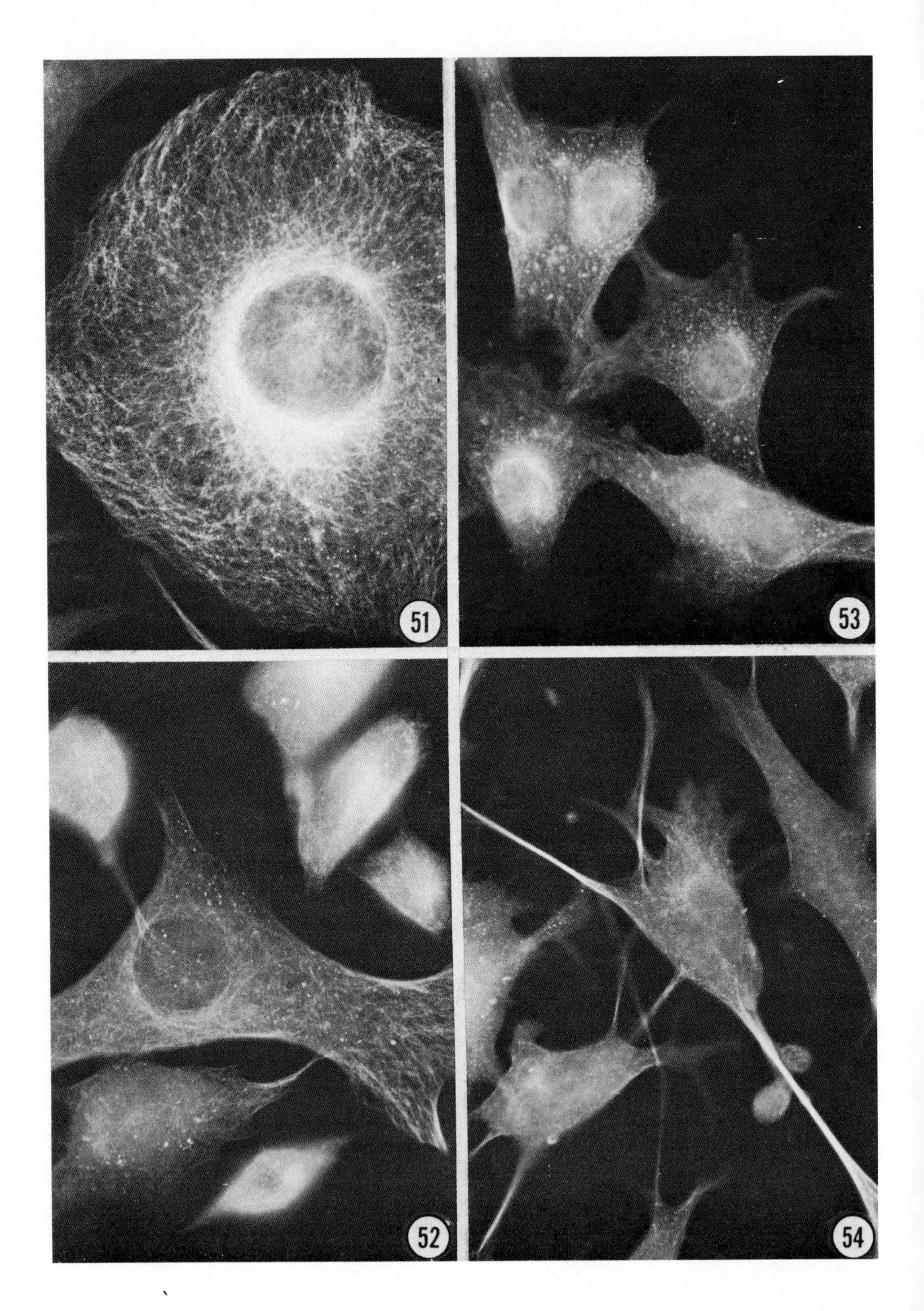

FIGS. 51–54. Cytoplasmic microtubule patterns in cultured cells. A full intricate pattern (FC-type) is characteristic of nontransformed cell populations (Figure 51). An extremely diminished complex (DC-type) is shown in Figure 53. Intermediate between these types are the sparse, random patterns or SR-type (Figure 52) and the random-oriented (SO-type) pattern (Figure 54).

ences in the cytoskeleton of SV3T3 and Rous sarcoma virus-transformed chick cells. The actin stress fiber pattern is also affected, and many transformed cells are essentially void of immunofluorescent fibers after incubation with antiactin antibody (Figures 48–50). Pollack and Rifkin (1975) were able to correlate the loss of actin-containing cables in virally transformed cultured cells with their acquisition of anchorage-independent growth.

CMTC Phenotypes

In more recent studies, in our laboratory, of transformed and nontransformed cells, we have identified four stable phenotypes with regard to the cytoplasmic microtubule complex. These are illustrated in Figures 51 to 54. The phenotype, which is predominantly found in nontransformed populations is termed the Full Microtubule Complex (FC-type). These cells have extensive cytoplasmic microtubules that radiate out from one or two bright foci in the cytoplasm (Figure 51). Three other phenotypes have also been described. One is the Diminished Microtubule Complex (DC-type), in which cells display mostly diffuse cytoplasmic staining with antitubulin. Occasionally a few short filaments are seen scattered throughout the cytoplasm (Figure 53). The Sparse Random Microtubule Complex (SR-type) is characterized by short, randomly oriented microtubules, which are clustered near the cell periphery, or a few lone microtubules extending the length of the cell (Figure 52). The Sparse Oriented Complex (SO-type) contains microtubules arrayed in loose bundles in the extended processes of the cell (Figure 54).

The distribution of CMTC phenotypes in populations of nontransformed and transformed cells is shown in Table 1. Although we have observed variation in the actin stress fiber patterns, we have not quantitated stable actin phenotypes.

Microtubules, Microfilaments, and Growth Properties of Cells

If the diminution of actin stress fibers and cytoplasmic microtubules observed in most transformed cell populations is a reliable indicator for transformation to neoplasia, a direct correlation should exist between the alteration of cytoskeletal structures and other properties of transformed cells. Pollack and Rifkin

TABLE 1. *CMTC phenotypes in nontransformed (—) and transformed (+) cell cultures*

Cell line	Growth	% FC	% SR	% SO	% DC	No. counted
PA-2	(—)	98.0	1.0	1.0	0.0	200
BALBc/3T3	(—)	86.7	12.0	2.0	0.0	500
RAG	(+)	0.0	1.0	4.0	95.0	200
LMTK$^-$	(+)	0.0	0.0	5.5	94.5	200

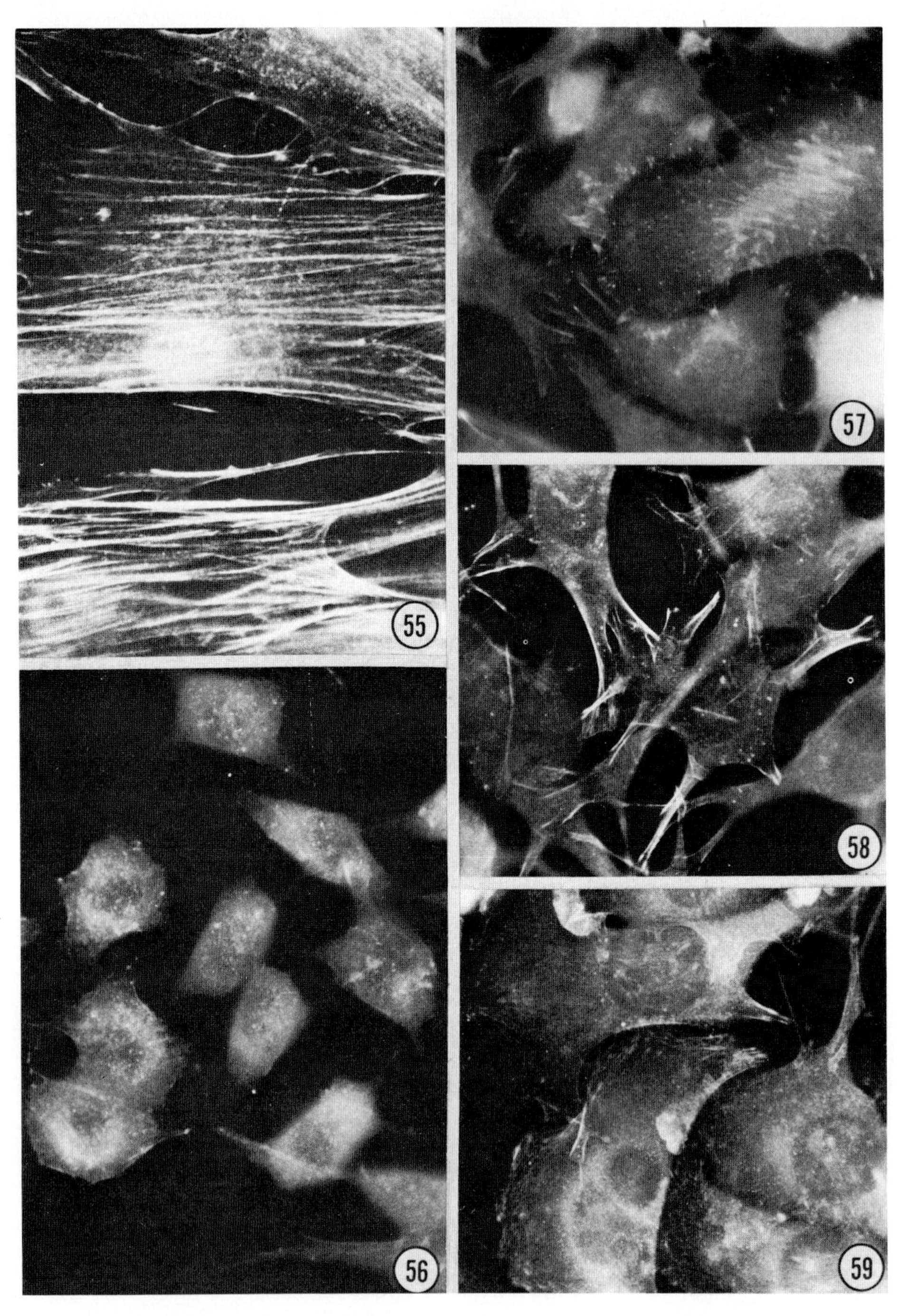

FIGS. 55–59. Actin patterns in somatic cell hybrids and parental cell lines. The nontransformed HSF-CF parent (Figure 55) has numerous parallel bundles of actin fibers, while the transformed RAG cell (Figure 56) shows only diffuse staining. All three hybrid clones CRAG-2 (Figure 57), CRAG-3, (Figure 58), and CRAG-9 (Figure 59) show a diffuse staining pattern very much like the RAG parent.

(1975) have observed a direct correlation between the loss of cytoplasmic actin cables from cells and their acquisition of anchorage-independent growth, as measured by colony formation in agar. Moreover, Freedman and Shin (1974) and Shin *et al.* (1975) have provided evidence that the capacity of cells to form colonies in agar is one of the most reliable indicators for transformation to malignancy.

Miller *et al.* (1977) have observed a similar correlation between cytoplasmic microtubule complex morphology and growth characteristics using somatic cell hybrids derived from the fusion of nontransformed human × transformed mouse cells. Human fibroblasts were fused with mouse renal adenocarcinoma cells (RAG), and hybrid clones were isolated in HAT selective medium. The actin and CMTC phenotypes of the parental cells are shown, respectively, in Figures 55, 56, 60, and 61. The normal parent cell population displayed 97% FC phenotypes. Similarly, a large majority of the nontransformed population contained extensive actin cables. The RAG parentals were the opposite, with 95% of the cells displaying a greatly diminished CMTC (DC-type) and a similarly high proportion without actin stress fibers (Figures 56 and 61).

The phenotypes of three representative clones are shown in Figures 57 to 59 for actin fibers and Figures 62 to 64 for CMTC patterns. Both CRAG-2 (Figure 62) and CRAG-3 (Figure 63) displayed high percentages of the intermediary CMTC patterns, the SR- and SO-types, while CRAG-9 (Figure 64) resembles the nontransformed parental. The actin fiber patterns in the three subclones (Figures 57 to 59) are greatly diminished, with only a few stress fibers visible over the nuclei or in the cell processes.

When we analyzed the capacity of the parentals and hybrids to grow in agar, a direct correlation was observed between the frequency of cells in the population with the DC-type CMTC pattern and the percentage of cells capable of colony formation in agar (Figure 65). When each of the hybrid clones was subcloned by selection from agar, we observed a large increase in the frequency of the DC phenotype in the surviving clones (Figure 66). These data suggest that there are both anchorage-dependent and anchorage-independent cells in the original hybrid clones but that only those cells which have lost their cytoplasmic microtubule organization, and thus anchorage-dependent growth control, form colonies in the agar. In future experiments, survival in agar and CMTC phenotype will be correlated with tumorigenicity in immunosuppressed animal hosts. The present data, however, are sufficient to permit a correlation to be made between the acquisition of anchorage-independent growth in vitro and the diminution of cytoskeletal elements.

Extent and Significance of Cytoskeletal Changes in Neoplastic Transformation

From the data presented here, and from numerous other studies, we can conclude that the cytoskeleton plays a major role in normal cell growth and

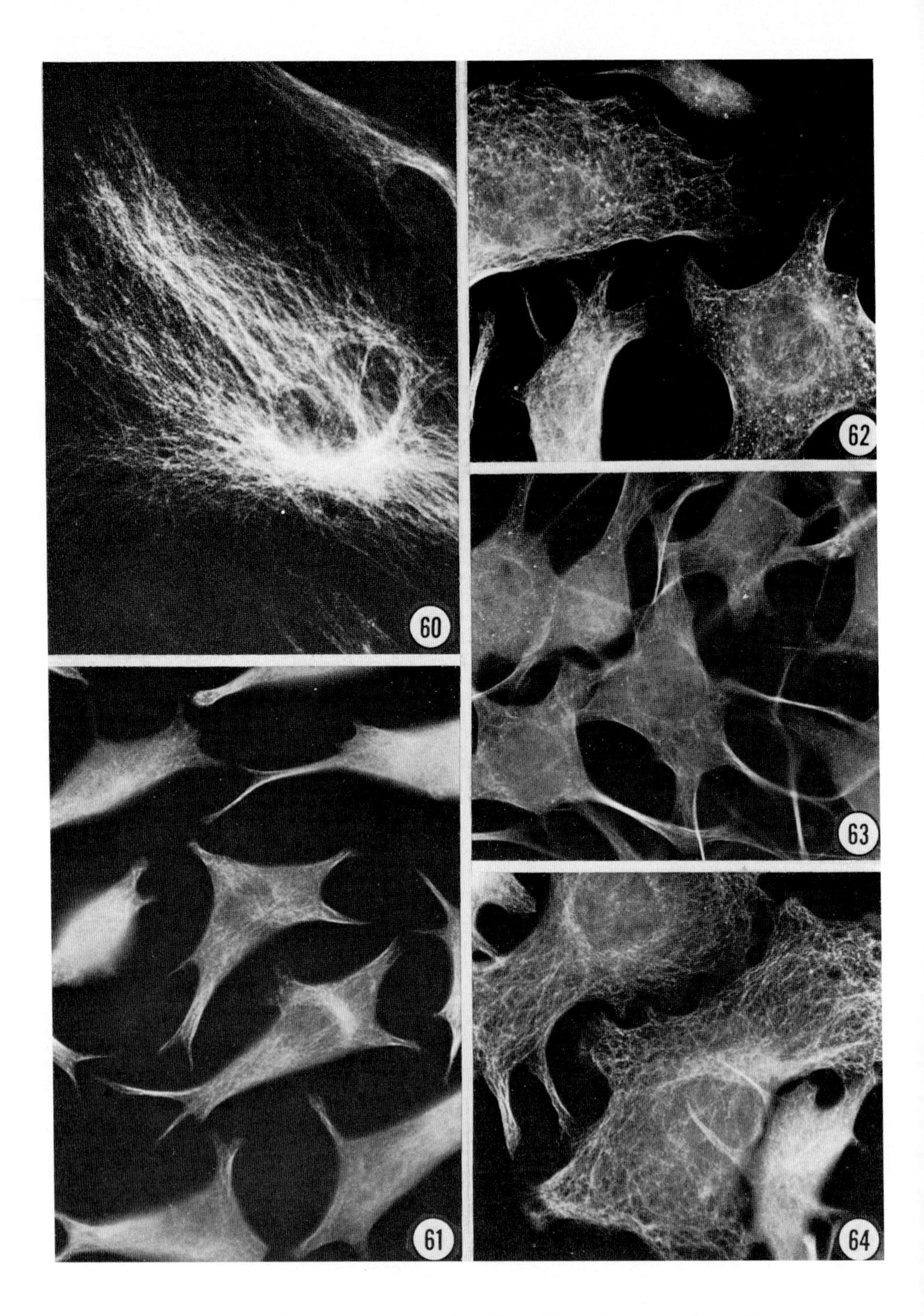

FIGS. 60–64. Microtubule patterns in somatic cell hybrids and parental lines. The nontransformed HSF-CF displays FC-type cytoplasmic phenotype (Figure 60), while the RAG parent shows diffuse staining, characteristic of DC phenotype (Figure 61). The clone CRAG-2 (Figure 62) displays an SR-type pattern; CRAG-3 is predominantly SO-type (Figure 63), and CRAG-9 has a high proportion of FC phenotypes, like the nontransformed parental line (Figure 64).

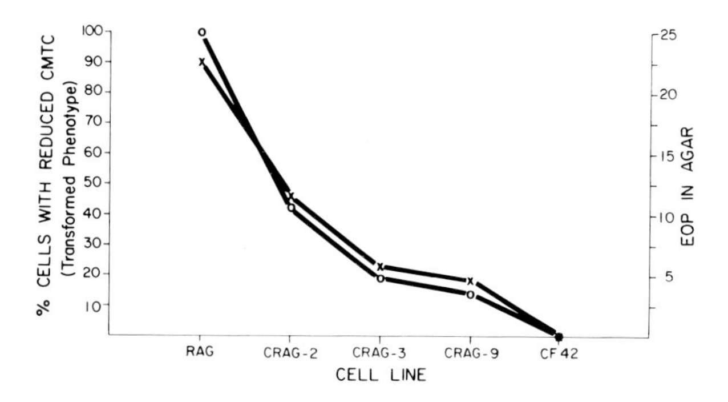

FIG. 65. This figure shows the percentage of cells in the parental and three somatic cell hybrids that display the transformed phenotype (X) and the efficiency of plating (EOP) in agar (0) of these same cell populations.

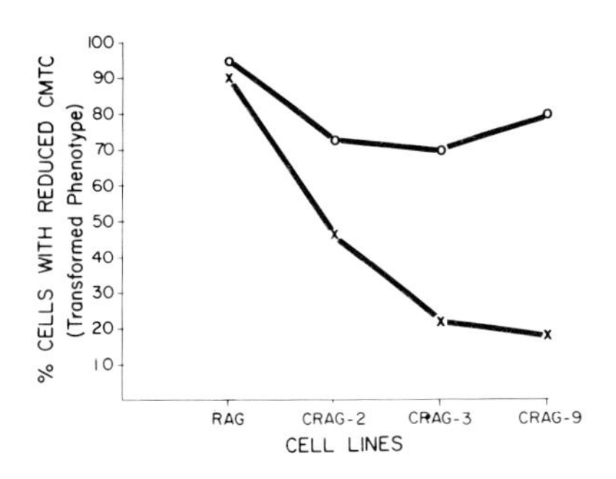

FIG. 66. After cloning the three somatic cell hybrids in agar, the percentage of cells with transformed phenotypes is significantly increased. (X) Phenotypes before cloning, (0) phenotypes after cloning.

differentiation. Microtubules and microfilaments are required for a variety of cell functions, including the maintenance of shape and form (Porter 1966), chromosome movement in mitosis and meiosis (Nicklas 1971), cell motility (Allison 1973), secretion, including the transport and release of hormones (Wolff and Bhattacharayya 1975) and plasma proteins (Redman *et al.* 1975), mobilization and release of lysosomes (Malawista 1975), and axonal transport in neurons (Smith *et al.* 1975). The recent evidence linking microtubules and microfilaments to cell surface receptors and the modulation of surface-receptor mobility and activity (Edelman 1976) adds an important new dimension to cytoskeletal involvement in cell regulation. The immunofluorescent analyses support the notion that cytoskeletal alterations are directly involved in the transformation of normal cells to malignancy. Obviously, many important points remain to be clarified. For example, a major criticism of studies on cytoskeletal alteration in transformed cells is that the studies are based on established cell lines of mesenchymal origin, most of which were derived from embryonic or neonatal tissues. To what extent do alterations seen in established cell lines relate to actual tumor cells of major importance to humans? A preliminary answer to this question has been given in the recent study of Asch *et al.* (submitted for publication) who observed significant alterations in cytoplasmic actin stress fibers and microtubules in primary cell cultures derived from murine mammary carcinoma and hyperplastic lesions of the mammary glands. Their study is significant in that it represents the first report of cytoskeletal changes in neoplastic and preneoplastic lesions of epithelial origin.

Another controversial point concerns the molecular mechanisms associated with cytoskeletal alterations in transformed cells. Do the alterations observed in microtubules and actin stress fibers relate to defects in actin and tubulin molecules or their capacity for assembly? It is unlikely that tubulin itself is greatly altered, since Fuller *et al.* (1975b), Edelman and Yahara (1976), and, more recently, Wiche *et al.* (1977) have all found that tubulin isolated from transformed cells is competent to polymerize in vitro. The size of the tubulin pool, however, may be significantly smaller in transformed cells. Edelman and Yahara reported approximately equal amounts of colchicine-binding activity and vinblastine-precipitated tubulin from homogenates of paired normal and virally transformed cells, and a similar finding was reported by Wiche *et al.* (1977). Two other laboratories, however, reported reduced colchicine-binding activity in the homogenates of transformed cells (Ostlund and Pastan 1975, Fine and Taylor 1976), and the question of pool size remains unresolved.

A suppression of tubulin assembly could be manifested in a defect in any one of several factors that control microtubule assembly, as discussed in an earlier section of this report. Much less is known about actin assembly in cells, but it is conceivable that conditions such as elevated calcium levels, which inhibit tubulin assembly, could also inhibit the assembly and function of actin-containing filaments.

Perhaps the most intriguing and challenging problem posed by these findings is whether the cytoskeletal alterations are a direct manifestation of a mutagenic or carcinogenic event or just one of many ancillary events in the pathway to neoplasia. From the findings presented in the present report and elsewhere, it is tempting to speculate that alteration in cytoskeletal structures represents early manifestations in the transformation of cells to malignancy. Edelman and Yahara (1976) have presented several models that could link the acquisition of altered growth control, morphological changes, and altered cytoskeleton with the direct action of the *src* gene. It is also possible that carcinogenic agents could interact directly with tubulin or actin or factors that regulate their assembly and disassembly and thereby lead to transformation without having initial interaction with the genome. Such effects might explain the action of those carcinogens that fail to elicit a mutagenic response.

Finally, it should be pointed out that we have only considered two cytoskeletal components in the present report. Although microtubules and actin filaments are major components of the cytoskeleton, other important fibrous elements are ubiquitous in the cytoplasm, including myosin, tropomyosin, troponin, and α-actinin. A recent diagram (Figure 67) published by Loor (1976) depicts possible relationships of several cytoskeletal elements with the plasma membrane proteins.

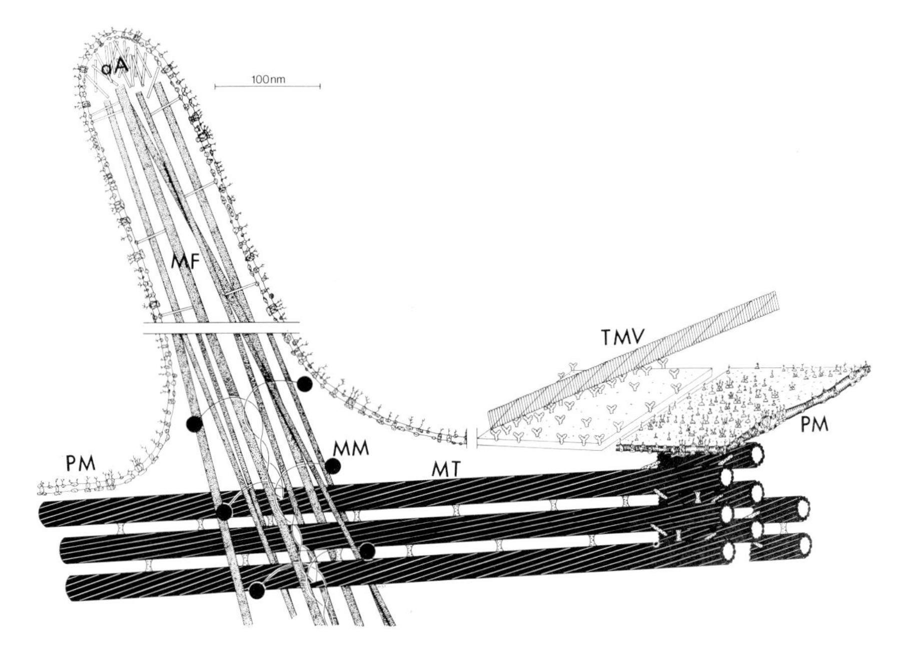

FIG. 67. Diagram illustrating the possible interaction of cytoskeletal components with the cell surface and receptor immunoglobulins. PM, plasma membrane; MT, microtubules; MF, actin microfilaments; MM, myosin filaments; aA, α-actinin. A tobacco mosaic virus (TMV) is shown as a marker for cell size. For more specific details, see Loor (1976). (Reproduced, with permission of Nature, from Loor 1976.)

The true structural relationship, however, is probably even more complex, since recent high-voltage electron microscope studies indicate that the cytoplasmic gel or ground substance in which the filaments and tubules are embedded is itself composed of a finer structured meshwork of microtrabeculae (Porter 1976).

Clearly, the cytoskeleton is complex, and its limits are difficult to define. Much is yet to be learned of its structure and function, but it would not be too surprising to find that it is intimately involved in the expression of malignancy in cells.

ACKNOWLEDGMENTS

The authors are grateful for the advice and collaboration of Dr. G. M. Fuller, and to Dr. E. Lazarides for the generous gift of antiactin antibody. Appreciation is extended to Dr. R. Klebe and Ms. Brenda Mayo for allowing us to use their CRAG hybrids. We thank Dr. Uta Franke for the generous gift of human PA-2 cells. Special appreciation is extended to Glenda Garza and Susan Cox for assistance with the manuscript.

This investigation was supported by a grant from the National Cancer Institute (CA 14675) and a grant from Dow Chemical Company to Dr. B. R. Brinkley.

REFERENCES

Allison, A. C. 1973. The role of microfilaments and microtubules in cell movement, endocytosis, and exocytosis, *in* Locomotion of Tissue Cells (CIBA Foundation Symposium), M. Abercrombie, ed., Vol. 14. Associated Scientific Publishers, New York, pp. 109–129.

Anderson, W. G., G. S. Johnson, and I. Pastan. 1973. Transformation of chick embryo fibroblasts by wild-type and temperature-sensitive Rous sarcoma virus alters adenylate cyclase activity. Proc. Natl. Acad. Sci. USA 70:1055–1059.

Balk, S. D., J. F. Whitfield, T. Youdale, and A. C. Braun. 1973. Roles of calcium, serum, plasma, and folic acid in the control of proliferation of normal and Rous sarcoma virus-infected chicken fibroblasts. Proc. Natl. Acad. Sci. USA 70:675–679.

Benedict, W., P. Jones, and W. Long. 1975. Characterization of human cells transformed *in vitro* by urethane. Nature 256:322–324.

Berlin, R. D., J. M. Oliver, T. E. Ukena, and H. H. Yin. 1974. Control of cell surface topography. Nature 247:45–46.

Borisy, G. C., J. M. Marcum, J. P. Olmsted, D. B. Murphy, and K. A. Johnson. 1975. Purification of tubulin and associated high molecular weight proteins from porcine brain and characterization of microtubule assembly *in vitro.* Ann. NY Acad. Sci. 253:107–132.

Borle, A. B. 1974. Cyclic AMP stimulation of calcium efflux from kidney, liver and heart mitochondria. J. Membr. Biol. 16:221–236.

Brinkley, B. R., and J. Cartwright. 1975. Cold-labile and cold-stable microtubules in the mitotic spindle of mammalian cells. Ann. NY Acad. Sci. 253:428–439.

Brinkley, B. R., G. M. Fuller, and D. P. Highfield. 1975. Cytoplasmic microtubules in normal and transformed cells in culture: Analysis by tubulin antibody immunofluorescence. Proc. Natl. Acad. Sci. USA 72:4981–4985.

Brinkley, B. R., and E. T. Stubblefield. 1970. Ultrastructure and interaction of the kinetochore and centriole in mitosis and meiosis, *in* Advances in Cell Biology, D. M. Prescott, L. Goldstein, and E. McConkey, eds., Appleton-Century-Crofts, New York, pp. 119–185.

Buckley, I. K., and K. R. Porter. 1975. Electron microscopy of critical point dried whole cultured cells. J. Microsc. 104:107.

Burger, M. M. 1973. Surface changes in transformed cells detected by lectins. Fed. Proc. 32:91–101.

De Petris, S. 1974. Inhibition and reversal of capping by cytochalasin B, vinblastine and colchicine. Nature 250:54–56.

De Petris, S. 1975. Concanavalin A receptors, immunoglobulins, and θ antigen of the lymphocyte surface. J. Cell Biol. 65:123–146.

Edelman, G. M., and I. Yahara. 1976. Temperature-sensitive changes in surface modulating assemblies of fibroblasts transformed by mutants of Rous sarcoma virus. Proc. Natl. Acad. Sci. USA, 73:2047–2051.

Edelman, G. M. 1976. Surface modulation in cell recognition and cell growth. Science 192:218–226.

Edelman, G. M., I. Yahara, and J. L. Wang. 1973. Receptor mobility and receptor-cytoplasmic interactions in lymphocytes. Proc. Natl. Acad. Sci. USA 70:1442–1446.

Evans, C. H., and J. A. DiPaolo. 1975. Neoplastic transformation of guinea pig fetal cells in culture induced by chemical carcinogens. Cancer Res. 35:1035–1044.

Fine, R. E., and L. Taylor. 1976. Decreased actin and tubulin synthesis in 3T3 cells after transformation by SV40 virus. Exp. Cell Res. 102:162–168.

Fonte, V., and K. R. Porter. 1974. Topographic changes associated with viral transformation of normal cells to tumorigenicity, *in* 8th International Congress on Electron Microscopy, Australian Academy of Science, Canberra, p. 334.

Forer, A. 1976. Actin filaments and brefringent spindle fibers during chromosome movements, *in* Cell Motility, R. Goldman, T. Pollard, and J. Rosenbaum, eds., Cold Spring Harbor Laboratory, Cold Spring Harbor, New York, pp. 1273–1293.

Freedman, V. H., and S. Shin. 1974. Cellular tumorigenicity in nude mice: Correlation with cell growth in semi-solid medium. Cell 3:355–359.

Fuller, G. M., C. S. Artus, and J. J. Ellison. 1976. Calcium as a regulator of cytoplasmic microtubule assembly and disassembly. J. Cell Biol. 70:68a.

Fuller, G. M., and B. R. Brinkley. 1976. Structure and control of assembly of cytoplasmic microtubules in normal and transformed cells. J. Supramol. Struct. 5:497–514.
Fuller, G. M., B. R. Brinkley, and J. M. Boughter. 1975a. Immunofluorescence of mitotic spindles by using monospecific antibody against bovine brain tubulin. Science 187:948–950.
Fuller, G. M., J. Ellison, M. McGill, L. Sordahl, and B. R. Brinkley. 1975b. Studies on the inhibitory role of calcium in the regulation of microtubule assembly *in vitro* and *in vivo*, *in* International Symposium on Microtubules and Microtubular Inhibitors, Beerse, Belgium, M. Borgers and M. de Brabander, eds. American Elsevier, New York, pp. 379–390.
Gail, M. H., and C. W. Boone. 1971. Cytochalasin effects on BALB/3T3 fibroblasts: Dose dependent, reversible alteration of motility and cytoplasmic cleavage. Exp. Cell Res. 68:226–228.
Goldman, R. D., J. A. Schloss, and J. M. Starger. 1976. Organizational changes of actin like microfilaments during animal cell movement, *in* Cell Motility, R. Goldman, T. Pollard, and J. Rosenbaum, eds., Cold Spring Harbor Laboratory, Cold Spring Harbor, New York, pp. 217–245.
Holley, R. W., and J. A. Kiernan. 1968. Contact inhibition of cell division in 3T3 cells. Proc. Natl. Acad. Sci. USA 60:300–304.
Hsie, A. W., and T. T. Puck. 1971. Morphological transformation of chinese hamster cells by dibutyryl adenosine cyclic 3′:5′-monophosphate and testosterone. Proc. Natl. Acad. Sci. USA 68:358–361.
Kirschner, M. W., R. C. Williams, M. Weingarten, and J. C. Gerhart. 1974. Microtubules from mammalian brain: Some properties of their depolymerization products and a proposed mechanism of assembly and disassembly. Proc. Natl. Acad. Sci. USA 71:1159–1163.
Lazarides, E., and K. Weber. 1974. Actin antibody: The specific visualization of actin filaments in non-muscle cells. Proc. Natl. Acad. Sci. USA 71:2268–2272.
Loor, F. 1976. Cell surface design. Nature 264:272–273.
MacPherson, I., and L. Montagnier. 1964. Agar suspension culture for the selective assay of cells transformed by polyoma virus. Virology 23:291–294.
Miller, C. L., J. W. Fuseler, and B. R. Brinkley. 1977. Cytoplasmic microtubules in transformed mouse × nontransformed human cell hybrids: Correlation with in vitro growth. Cell 12:319–331.
Monahan, T. M., R. R. Fritz, and C. W. Abell. 1973. Levels of cyclic AMP in murine L5178Y lymphoblasts grown in different concentrations of serum. Biochem. Biophys. Res. Commun. 55:642–646.
Murphy, D. M., and G. C. Borisy. 1975. Association of high-molecular-weight proteins with microtubules and their role in microtubule assembly *in vitro*. Proc. Natl. Acad. Sci. USA 72:2696–2700.
Nicklas, R. B. 1971. Mitosis, *in* Advances in Cell Biology, D. M. Prescott, L. Goldstein, and E. N. McConkey, eds., Vol. 2. Appleton-Century-Crofts, New York, pp. 225–298.
Oliver, J. M., R. B. Zurier, and R. D. Berlin. 1975. Concanavalin A cap formation on polymorphonuclear leukocytes of normal and beige (Chediak-Higashi) mice. Nature 253:471–473.
Olmsted, J. B. 1976. The role of divalent cations and nucleotides in microtubule assembly *in vitro*, *in* Cell Motility, R. Goldman, T. Pollard, and J. Rosenbaum, eds., Cold Spring Harbor Laboratory, Cold Spring Harbor, New York, pp. 1081–1092.
Osborn, M., and K. Weber. 1976. Cytoplasmic microtubules in tissue culture cells appear to grow from an organizing structure towards the plasma membrane. Proc. Natl. Acad. Sci. USA 73:867–871.
Ostlund, R., and I. Pastan. 1975. Fibroblast tubulin. Biochemistry 14:1064–1066.
Penningroth, S. M., D. W. Cleveland, and M. W. Kirschner. 1976. *In vitro* studies of the regulation of microtubule assembly, *in* Cell Motility, R. Goldman, T. Pollard, and J. Rosenbaum, eds., Cold Spring Harbor Laboratory, Cold Spring Harbor, New York, pp. 1233–1257.
Pollack, R. E., and M. M. Burger. 1969. Surface-specific characteristics of a contact-inhibited cell line containing the SV40 viral genome. Proc. Natl. Acad. Sci. USA 62:1074–1076.
Pollack, R., and D. Rifkin. 1975. Actin containing cables within anchorage-dependent rat embryo cells are dissociated by plasmin and trypsin. Cell 6:495–506.
Porter, K. R. 1966. Cytoplasmic microtubules and their functions, *in* Principles of Biomolecular Organization, G. E. Wolstenholm and M. O'Conner, eds., Little Brown and Co., Boston, p. 308.
Porter, K. R. 1976. Introduction: Motility in Cells, *in* Cell Motility, R. Goldman, T. Pollard,

and J. Rosenbaum, eds., Cold Spring Harbor Laboratory, Cold Spring Harbor, New York, pp. 1–28.

Poste, G., D. Papahadjopoulos, K. Jacobson, and G. Shepherd. 1975a. Effects of local anesthetics on membrane properties. I. Changes in the fluidity of phospholipid bilayers. Biochim. Biophys. Acta 394:504–519.

Poste, G., D. Papahadjopoulos, and G. L. Nicolson. 1975b. Local anesthetics affect transmembrane cytoskeletal control of mobility and distribution of cell surface receptors. Proc. Natl. Acad. Sci. USA 72:4430–4434.

Redman, C. M., D. Banerjee, K. Howell, and G. E. Palade. 1975. The step at which colchicine blocks the secretion of plasma proteins by rat liver. Ann. NY Acad. Sci. 253:780–788.

Revel, J. P., P. Hoch, and D. Ho. 1974. Adhesion of culture cells to their substration. Exp. Cell Res. 84:207–218.

Risser, R., and R. Pollack. 1974. A nonselective analysis of SV40 transformation of mouse 3T3 cells. Virology 59:477–489.

Sanger, J. W., and J. M. Sanger. 1976. Actin localization during cell division, *in* Cell Motility, R. Goldman, T. Pollard, and J. Rosenbaum, eds., Cold Spring Harbor Laboratory, Cold Spring Harbor, New York, pp. 1295–1316.

Shelanski, M. L., F. Gaskin, and C. R. Cantor. 1973. Microtubule assembly in the absence of added nucleotides. Proc. Natl. Acad. Sci. USA 70:765–768.

Sheppard, T. 1972. Difference in the cyclic adenosine 3′,5′-monophosphate levels in normal and transformed cells. Nature 236:14–16.

Shin, S., V. H. Freedman, R. Risser, and R. Pollack. 1975. Tumorigenicity of virus-transformed cells in *nude* mice is correlated specifically with anchorage independent growth *in vitro*. Proc. Natl. Acad. Sci. USA 72:4435–4439.

Smith, D. S., V. Jalfors, and B. F. Cameron. 1975. Morphological evidence for the participation of microtubules in axonal transport. Ann. NY Acad. Sci. 253:472–506.

Smith, H. S., C. D. Scher, and G. J. Todaro. 1971. Induction of cell division in medium lacking serum growth factor by SV40. Virology 44:359–370.

Solomon, F. 1976. Characterization of the calcium binding activity of tubulin, *in* Cell Motility, R. Goldman, T. Pollard, and J. Rosenbaum, eds., Cold Spring Harbor Laboratory, Cold Spring Harbor, New York, pp. 1139–1148.

Stoker, M., and I. MacPherson. 1961. Studies on transformation of hamster cells by polyoma virus *in vitro*. Virology 14:359–370.

Stoker, M., and P. Abel. 1962. Conditions affecting transformation by polyoma virus. Cold Spring Harbor Symp. Quant. Biol. 27:375–386.

Taylor, R. B., W. P. H. Duffus, M. C. Raff, and S. de Petris. 1971. Redistribution and pinocytosis of lymphocyte surface immunoglobulin molecules induced by anti-immunoglobulin antibody. Nature New Biol. 233:225–229.

Temin, H. M., and H. Rubin. 1958. Characteristics of an assay for Rous sarcoma virus and Rous sarcoma cells in tissue culture. Virology 6:669–688.

Todaro, G. J., H. Green, and B. D. Goldberg. 1964. Transformation of properties of an established cell line by SV40 and polyoma virus. Proc. Natl. Acad. Sci. USA 51:66–73.

Weber, K. R., T. Pollack, and T. Bibring. 1975. Antibody against tubulin: The specific visualization of cytoplasmic microtubules in tissue culture cells. Proc. Natl. Acad. Sci. USA 72:459–463.

Weisenberg, R. C. 1972. Microtubule formation *in vitro* in solutions containing low calcium concentrations. Science 177:1104–1105.

Wessells, N. K., B. S. Spooner, J. F. Ash, M. O. Bradley, M. A. Luduena, E. L. Taylor, J. T. Wrenn, and K. M. Yamada. 1971. Microfilaments in cellular and developmental processes. Science 171:135–143.

Wiche, G., V. J. Lundblad, and R. D. Cole. 1977. Competence of soluble cell extracts as microtubule assembly systems. J. Biol. Chem. 252:794–796.

Wolff, J., and B. Bhattacharyya. 1975. Microtubules and thyroid hormone mobilization. Ann. NY Acad. Sci. 253:763–770.

Yahara, I., and G. M. Edelman. 1973. Modulation of lymphocyte receptor redistribution by Concanavalin A, anti-mitotic agents and alterations of pH. Nature 246:152–155.

Yahara, I., and G. M. Edelman. 1975. Electron microscopic analysis of the modulation of lymphocyte receptor mobility. Exp. Cell Res. 91:125–142.

Yin, H. H., T. H. Ukena, and R. D. Berlin. 1972. Effect of colchicine, colcemid and vinblastine on the agglutination by Concanavalin A of transformed cells. Science 178:877–879.

Control of Differentiation and Neoplasia—
Stability of Differentiation

Cell Differentiation and Neoplasia, edited by Grady F. Saunders. Raven Press, New York

Induction of Differentiation of Murine Erythroleukemia Cells

Paul A. Marks, Richard A. Rifkind, Arthur Bank, Masaaki Terada, Roberta Reuben, Eitan Fibach, Uri Nudel, Jane Salmon, and Yair Gazitt

Cancer Research Center, and Departments of Medicine and of Human Genetics and Development, Columbia University, New York, New York 10032

We have investigated the mechanism of eukaryotic cell differentiation employing an erythroid model system, induced erythropoietic differentiation of murine erythroleukemia cells (MELC) transformed by Friend virus complex (Friend 1957, Friend *et al.* 1971, Marks *et al.* 1977). This paper will review studies elucidating events involved in the induction of MELC to differentiate to erythroid cells in culture with dimethylsulfoxide and other chemical inducers. MELC show a low level (< 1%) of spontaneous erythroid differentiation in culture, but when cultured with various chemicals, a high proportion of the cells are induced to express a program of erythroid differentiation. This program of differentiation has many similarities to that observed in the erythropoietin-induced differentiation of fetal mouse liver erythropoiesis (Marks and Rifkind 1972), including morphological changes (Friend *et al.* 1971), accumulation of globin mRNAs (Ross *et al.* 1972, Ostertag *et al.* 1972, Gilmour *et al.* 1974, Singer 1975), α- and β-globin synthesis (Boyer *et al.* 1972), increase in heme synthesis (Ebert and Ikawa 1974), synthesis of proteins characteristic of erythrocytes, such as catalase and spectrin (Kabat *et al.* 1975, Arndt-Jovin *et al.* 1976), limited capacity for cell division (Friend *et al.* 1971, Gusella *et al.* 1976, Fibach *et al.* 1977), and appearance of mature red cell specific antigens (Furusawa *et al.* 1971).

MELC has a number of advantages for genetic and biochemical studies of mechanisms of induction of eukaryotic cells to differentiation. MELC can be maintained in continuous culture. A cell line has been developed in our laboratory that has a very low percentage (< 0.5%) of spontaneous erythroid differentiation (Singer *et al.* 1974); chemicals have been synthesized, such as hexamethylene

* All studies discussed in this paper from the authors' laboratory were performed with an MELC line derived from Charlotte Friend's line 745A and is designated DS-19 (Singer *et al.* 1974), or from a variant cell line developed from DS-19, which is resistant to induction with Me_2SO but sensitive to various other inducers and is designated DR-10 (Ohta *et al.* 1976). Unless otherwise specified, MELC refers to DS-19 strain.

bisacetamide (Reuben *et al.* 1976), that can induce essentially 100% of the cells in culture to differentiate to erythroid cells, and variant cell lines have been developed that are resistant to induction by one or another inducer (Ohta *et al.* 1976, Gusella and Housman 1976, Harrison et al. 1974). MELC differ from normal mammalian erythropoiesis in not being dependent on or sensitive to erythropoietin (Marks, Rifkind, and Cantor, unpublished observations) and rarely proceed to a nonnucleated stage of differentiation, which is characteristic of definitive mammalian erythropoiesis (Friend *et al.* 1971, Reuben *et al.* 1976).

MELC* differentiation will be considered with respect to: (1) accumulation of globin mRNA and globin synthesis during differentiation, (2) structure of chemicals active as inducing agents, (3) relation of the action of inducing agents and the cell division cycle, (4) early metabolic events in the induction to differentiation, (5) requirements for stable commitment to differentiation of MELC, and (6) differential effects of various inducers on the expression of globin genes in MELC.

ACCUMULATION OF mRNA FOR GLOBIN AND GLOBIN SYNTHESIS

Before MELC could be used to study the mechanisms regulating induction of differentiation, it was necessary to assess: (1) the uniformity of the cultures with respect to inducibility of cells to erythroid differentiation, and (2) the effect of culture conditions on the proliferation and differentiation of MELC, as characterized by the pattern of accumulation of mRNA for globin and of the synthesis of globin. Studies of the effect of Me_2SO (dimethylsulfoxide) under conditions of clonal growth indicate that MELC are an essentially homogeneous population of transformed erythroid cells (Singer *et al.* 1974, Paul and Hickey 1974). There is evidence which suggests that the transformed cells correspond to proerythroblasts (Tambourin and Wendling 1971).

In MELC cultured with 280 mM Me_2SO, accumulation of mRNA for globin, assayed by cDNA:RNA hybridization, is first detectable at about 24 hours (Singer 1975). An increase in the rate of globin synthesis is also first detectable at approximately 24 hours (Singer 1975). These findings suggest that there is a slight lag in the accumulation of mRNA for globin and the onset of translation of globin mRNAs. In nuclear RNA from MELC cultured with inducers, species of RNA molecules containing globin sequences have been found that are larger than the mature form of globin mRNA (Ross 1976, Curtis and Weissmann 1976, Bastos and Aviv, submitted for publication). These studies suggest that the biosynthesis of mRNA in induced MELC is a multistep process in which an early event is the synthesis of a large precursor molecule, which is subsequently cleaved to intermediate species and then processed to 10S, "functional" globin mRNA in polyribosomes.

CHEMICAL STRUCTURE OF COMPOUNDS ACTIVE AS INDUCING AGENTS

The induction by Me_2SO of the erythropoietic program of differentiation in MELC prompted investigations on the chemical structural requirements for compounds active as inducing agents. It has been demonstrated (Tanaka *et al.* 1975) that a series of planar-polar compounds, such as N-methyl acetamide, induce differentiation in MELC, as assayed by the proportion of cells becoming benzidine reactive at concentrations significantly below that required for Me_2SO. Further, it was shown that dimerization of the inducing agent N-methyl acetamide by linkage at nitrogen through varying numbers of methylenes resulted in a group of inducing agents, polymethylene bisacetamides, which were even more active than the simple monomer (Reuben *et al.* 1976). A variety of other chemicals have been shown to act as inducers, including fatty acids, purine and purine derivatives, heme, and inhibitors of DNA and RNA synthesis (Scher *et al.* 1973, Dube *et al.* 1973, Leder *et al.* 1975, Gusella and Housman 1976, Ross 1976, Ebert *et al.* 1976).

Among the compounds tested in our laboratory, the polar compound hexamethylene bisacetamide (HMBA) (Table 1) has been found to be most effective by the criteria: (a) essentially the entire population of MELC is induced to differentiate, (b) a greater proportion of the total protein synthesized is hemoglobin, and (c) a relatively low concentration is effective (Reuben *et al.* 1976). For example, HMBA is an effective inducer at concentrations of 1 mM, and, at an optimal concentration of 5 mM, induces essentially all cells in culture to differentiate.

There is a direct relationship between the optimal concentration for induction and the concentration at which these compounds inhibit growth and cause cell death in culture. This is a consistent observation for the entire series of polar compounds effective as inducers (Tanaka *et al.* 1975, Reuben *et al.* 1976). These

TABLE 1. *Structure of compounds active as inducing agents*

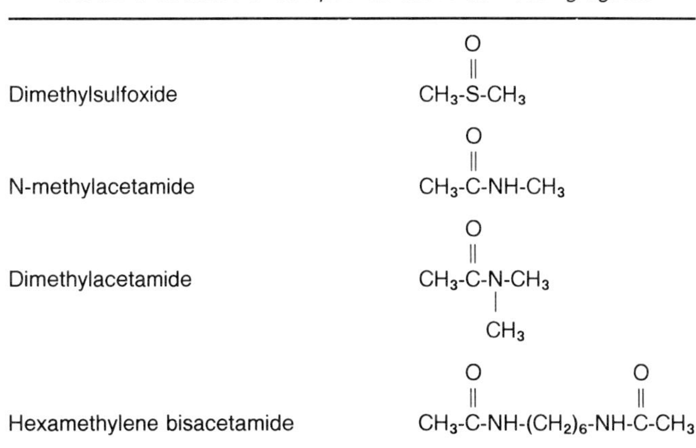

Compound	Structure
Dimethylsulfoxide	$CH_3\text{-}S(=O)\text{-}CH_3$
N-methylacetamide	$CH_3\text{-}C(=O)\text{-}NH\text{-}CH_3$
Dimethylacetamide	$CH_3\text{-}C(=O)\text{-}N(CH_3)\text{-}CH_3$
Hexamethylene bisacetamide	$CH_3\text{-}C(=O)\text{-}NH\text{-}(CH_2)_6\text{-}NH\text{-}C(=O)\text{-}CH_3$

TABLE 2. *Relation of structure and inducing activity of polymethylene compounds*

	Reversal of functional groups (amino and carbonyl)			
	... $(CH_2)_4$ ...			
	CH_3-C(=O)-N ...	H_2N-C(=O) ...	CH_3-NH-C(=O) ...	$(CH_3)_2$-N-C(=O) ...
	(% Benzidine-reactive cells)			
2mM	21	0	27	5
5mM	64	0	71	29
10mM	90	0	(76)	75
20mM	dead	0	dead	99
	... $(CH_2)_6$			
	CH_3-C(=O)-NH ...	NH_2-C(=O) ...	CH_3-NH-C(=O) ...	$(CH_3)_2$-N-C(=O) ...
	(% Benzidine-reactive cells)			
2mM	65	0	81	94
5mM	99	0	96	99
10mM	dead	0	dead	dead
20mM	dead	0	dead	dead

observations suggest that the mechanisms of induction may involve a change in the structure of a cellular component, which, if too extensive, is incompatible with cell survival. Alternatively, the inducers may have toxic effects on the cells, which are independent of the action causing differentiation, but which is apparent at a concentration closely related to that which is optimal for induction. The data at present do not permit one to distinguish between these two possibilities.

The relationship between chemical structure and inducing activity has been further explored by determining if the order of amino and carbonyl groups is important for the activity of the bisacetamides. A series of compounds with either four or six methylene groups was synthesized, in which the carbonyl was internal to the amino group (Reuben *et al.,* unpublished observations). The compounds bearing either one or two terminal methyl groups were found to be comparable in activity to the reverse configuration. The compounds bearing terminal amino groups were inactive as inducers (Table 2).

The nature of the cellular site of the primary action of these polar compounds has not been established. Evidence will be summarized below which suggests that inducing agents cause a series of changes which may initially involve an effect at the level of the membrane and, subsequently, changes in chromatin and product of transcription of the chromatin associated with differentiation.

INDUCTION OF MELC AND DNA SYNTHESIS (S PHASE)

There is accumulating evidence that the transition to the synthesis of a differentiated protein requires DNA synthesis (Marks and Rifkind 1972, Holtzer *et al.* 1972, Rutter *et al.* 1973, Dworkin *et al.* 1972, McClintock and Papaconstantinou 1974, Levy *et al.* 1975, Marks *et al.* 1977). In MELC synchronized with respect to cell division cycle by exposure to high levels of thymidine (Levy *et al.* 1975), it was shown that the inducer must be present during DNA synthesis and, possibly, shortly thereafter, for differentiation to erythroid cells to occur in culture. Thus, it was found that cells which were synchronized by culture for 44 hours in 2 mM thymidine and 280 mM Me_2SO and released from cell division block by transfer to fresh medium without Me_2SO or thymidine proliferated but did not differentiate. Cells that were removed from cell division block and transferred to fresh media with 280 mM Me_2SO and cultured for 4 hours (a time sufficient for essentially all cells to enter S and, some cells, a portion of G_2), and then transferred to fresh media without inducer, proceed to differentiate over the ensuing five days. An additional condition for the induction of erythroid differentiation in these studies was that Me_2SO be present for at least 20 hours prior to the release from cell division block. The preincubation period may, in part, be related to effects of the inducer, which must occur prior to the critical S phase (see below).

These studies indicating that the inducing agent must be present during a critical S phase for MELC to be induced to erythroid differentiation are consistent with findings in other differentiating systems, including chick erythroid cells (Holtzer *et al.* 1972), fetal liver erythroid cells (Marks and Rifkind 1972), mammary tissue (Stockdale *et al.* 1966), uterus and oviduct (O'Malley and Means 1974), melanoma (Varga *et al.* 1974), muscle (Holtzer *et al.* 1972), myeloma (Byars and Kidson 1970). All of these studies suggested, but did not directly demonstrate, that DNA synthesis is a critical step in the sequence of events required for the initiation of synthesis of proteins characteristic of the differentiated state.

TRANSIENT INHIBITION OF INITIATION OF S PHASE (PROLONGATION OF G_1) ASSOCIATED WITH INDUCTION OF MELC

MELC cultured with Me_2SO develop a transient block in initiation of DNA synthesis, or a prolongation of G_1, which appears to be most marked at about 20 hours in nonsynchronous cultures. This was demonstrated by examining the rate of DNA synthesis, proportion of cells in S phase, and pattern of DNA accumulation in MELC grown in the presence and absence of Me_2SO and other inducing agents, such as butyric acid and dimethylacetamide (Terada *et al.* 1977).

In MELC cultured without inducer, there is an initial rise in the rate of thymidine incorporation into DNA, with a maximum value achieved by about 10 hours (Figure 1). The initial increase in the rate of thymidine incorporation probably reflects entry into S phase of cells partially synchronized in the post-log growth phase of the previous culture passage. In comparison, in cells cultured with Me_2SO, although there is an initial rise in the rate of thymidine incorporation, a difference between cultures with and without 280 mM Me_2SO is observed

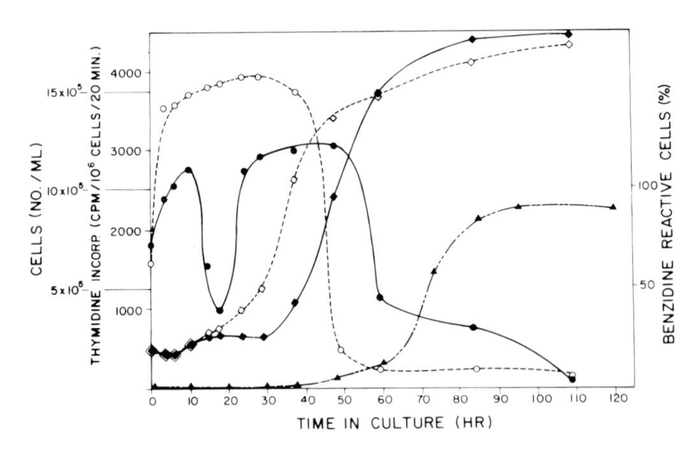

FIG. 1. Cell growth and cells in S phase of MELC cultured with and without 280 mM Me_2SO. MELC after 60 hours of culture were transferred to medium with and without Me_2SO at the initial concentration of 2×10^5 cells/ml, and at each time, an aliquot of cell suspension was removed to determine cell number (◇——◇, without inducer; ◆——◆, with Me_2SO), benzidine-reactive cells (▲-...-▲, Me_2SO), and ^3H-thymidine incorporation (20-minute incubation) into trichloracetic acid-insoluble material (○- - -○, without inducer; ●——●, with Me_2SO). For details of these studies, see Terada *et al.* (1977).

as early as six hours (Figure 1). The population of cells cultured with Me_2SO shows a decrease in thymidine incorporation between 10 and 20 hours. At 20 hours, the rate of thymidine incorporation is at its lowest, about 25% of the rate in control cultures. In the induced cultures, the rate of thymidine incorporation rises between 20 and 30 hours to reach a peak value of about 75% of the highest rate in control cultures.

These findings suggest that there is a transient inhibition of entry of MELC into the S phase when cultured with Me_2SO. This conclusion was supported by analysis of the proportion of cells in S phase (cells that become labeled during 20-minute exposure to ^3H-thymidine) during culture with and without inducing agent (Figure 2) and by determining the relative DNA content per cell, using the propidium iodide binding measured with flow microfluorometry (Krishan 1975, Terada *et al.* 1977) (Figure 3). There is a decrease in the proportion of cells in S phase, which occurs during the early period of culture with inducing agents, and is most marked, in comparison with uninduced cultures, at 20 hours.

In addition, it has been found (Terada *et al.* 1977) that there is a difference in the pattern of binding of the intercalating agent, propidium iodide, to chromatin in cells in G_1 phase of MELC cultured with or without Me_2SO. In control cells cultured without inducer, there is an increase in the binding of propidium iodide as early as 10 hours; this is not observed in cells cultured with Me_2SO. The Me_2SO-resistant cell line, DR-10, behaves like control DS-19, even when cultured with Me_2SO.

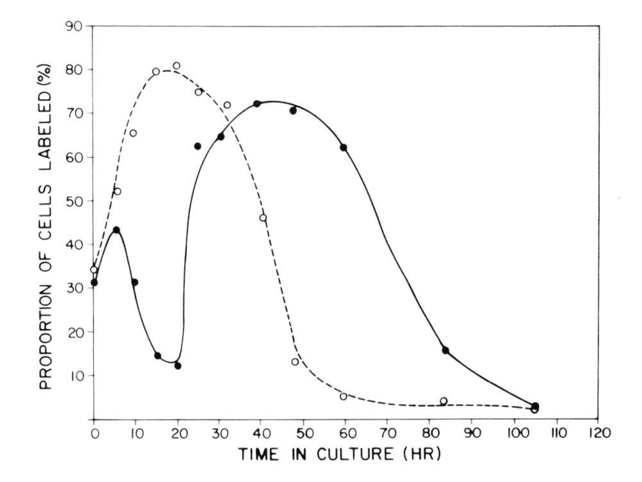

FIG. 2. Thymidine-labeling index during erythroid cell differentiation of MELC induced by Me_2SO. MELC were transferred to medium with and without Me_2SO as described in Figure 1. At each time indicated, an aliquot of cell suspension was removed for incubation with ^3H-thymidine for 20 minutes to obtain the thymidine-labeling index by autoradiography (○- - -○, control; ●——●, Me_2SO). For details of these studies, see Terada *et al.* (1977).

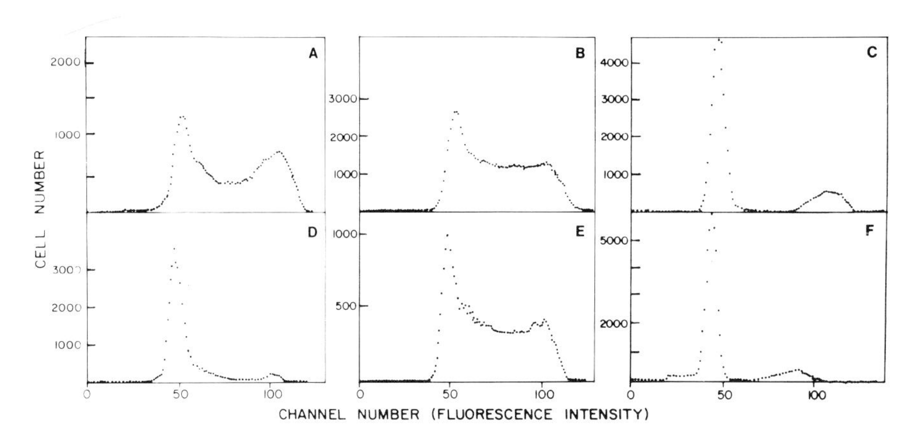

FIG. 3. Distributions of DNA content per cell measured by propidium iodide staining and flow microflurometry (FMF). Peak corresponding to G_1 cells was adjusted to be about channel 50. At the indicated times after beginning of culture with and without Me_2SO, an aliquot of cell suspension was removed to determine the DNA distribution histogram by FMF, according to the method described in Terada *et al.* (1977). (A to C, cells cultured without Me_2SO; and D to F, cells cultured with Me_2SO, for the following periods of time: A and D, 19 hours; B and E, 38 hours; C and F, 88 hours.) For details of these studies, see Terada *et al.* (1977).

These observations suggest that during prolongation of the G_1 phase of the cell cycle, there is an associated alteration in chromatin structure. These changes in G_1 may be related to the induction of MELC to erythroid differentiation.

CHANGES IN DNA ASSOCIATED WITH INDUCTION OF MELC

Further definition of the effects of inducing agents on MELC has been obtained from studies of DNA structure in cells cultured with and without inducers. Analysis of preparations of DNA on alkaline sucrose gradient shows a decrease in the rate of sedimentation of DNA prepared from MELC cultured with Me_2SO, as early as 27 hours after onset of culture (Terada *et al.*, in press) (Figure 4). This is approximately the time at which MELC begin to accumulate mRNA for globin (Singer 1975). No change in the rate of sedimentation of DNA is observed when analyzed on neutral sucrose gradients. The decrease in rate of sedimentation of DNA in alkaline sucrose gradients is observed with DNA from MELC cultured with other inducing agents, such as butyric acid and dimethylacetamide. On the other hand, Me_2SO does not induce a change in

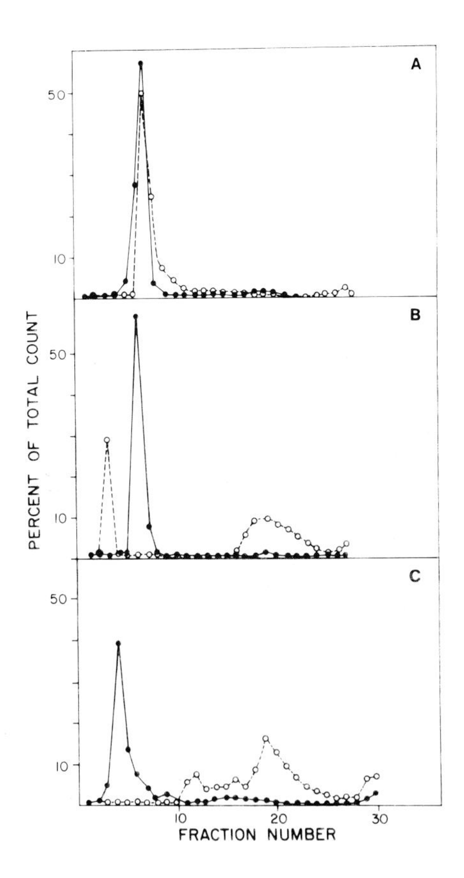

FIG. 4. Alkaline sucrose density gradient sedimentation pattern of DNA from MELC after culture with and without 280 mM Me_2SO. MELC were cultured with [2-^{14}C]thymidine, without (●——●) and with (○- - -○) Me_2SO for (A) 19 hours, (B) 27 hours, and (C) 64 hours. For details of these studies, see Terada *et al.* (in press).

the sedimentation properties of DNA from Me_2SO-resistant DR-10 cells cultured with Me_2SO (Terada *et al.,* in press). Analysis of the slowly sedimenting DNA recovered from the alkali sucrose gradient analysis of material from MELC cultured with inducer suggests that it represents single strand DNA of relatively small size (corresponding to 120S to 150S).

The alkaline sucrose gradient has been used extensively for detection of single-strand breaks introduced in DNA by chemical carcinogens, X-ray irradiation, UV irradiation, and virus infection (Parkhurst *et al.* 1973, Cleaver 1975, Ormerod 1976). One explanation for the change in alkaline sucrose gradient sedimentation pattern of DNA prepared from induced MELC is that single-strand breaks accumulate in DNA of MELC cultured with inducers. Such single-strand breaks may result in rapid unwinding of complementary strands during exposure to alkali. Alternatively, Me_2SO may affect the integrity of association of DNA with other cellular components, thereby sensitizing the DNA to the effects of alkali. Similar changes in DNA sedimentation patterns in alkaline sucrose gradients were observed with human fibroblasts exposed to low levels of UV irradiation (Cleaver 1974). The alteration of DNA associated with induction of MELC may prove to be a critical event leading to the expression of the program of differentiation. Alternatively, the changes in the properties of DNA could be a feature or product of differentiation, subsequent to the events that are critical to the commitment of MELC to differentiate.

COMMITMENT OF MELC TO DIFFERENTIATE

To evaluate the relationship of alterations in cell division cycle, changes in DNA and/or chromatin structure and the action of the inducing agent during the S phase of the cell cycle to the commitment of MELC to differentiate, it is necessary to be able to study the kinetics of recruitment to differentiation at the single cell level, and to study this in a quantitative manner. A technique for the assay of commitment of individual cells to differentiate has been developed (Fibach *et al.* 1977). In this assay, following suspension culture with inducing agent, MELC are cloned in a semisolid medium in the absence of the inducer, and clones are scored for proliferative capacity (number of cells in a colony) and differentiation (number of colonies containing benzidine-reactive cells). Employing this assay, it has been shown that the proportion of cells in the culture which are committed to differentiate is dependent both on the concentration of the inducing agent and the duration of exposure to the inducing agent in the precloning suspension culture. For example, with HMBA, stabilized differentiation of MELC is essentially complete when the cells are cultured with 5 mM HMBA for 50 hours. HMBA-induced erythroid differentiation is associated with a limitation in proliferative capacity, as is characteristic of normal erythropoiesis. By contrast, under conditions less than optimal for induction of differentiation with respect to either the concentration of HMBA or the duration of precloning culture with HMBA, induction to differentiation is not fully stabilized.

TABLE 3. *Heterogeneity of colonies derived from MELC cultured with various concentrations of HMBA and for various times*

HMBA	Undifferentiated colonies*	Differentiated colonies: Uniformly benzidine reactive	Differentiated colonies: "Mixed colonies"†	Differentiated colonies: Proportion of "mixed colonies"‡
Studies with various concentrations of HMBA in precloning culture for 53 hours				
mM	(%)	(%)	(%)	(%)
0	100	0	0	—
0.5	95	2	3	60
1.0	65	28	7	20
2.0	19	76	5	6
3.0	8	90	2	2
4.0	5	94	1	1
5.0	3	96	1	1
Studies with various times of precloning culture with 5 mM HMBA				
Time (hours)				
8	100	0	0	—
13	95	2	3	60
16	86	5	9	64
20	75	14	11	55
26	57	28	15	35
36	26	69	5	7
40	8	90	2	2
50	0	99	1	1

Cells were exposed in suspension culture to various concentrations of HMBA for 53 hours or to 5mM HMBA for various periods before cloning in semisolid medium in the absence of the inducer. Colonies were scored, five days from initiation of the experiments, by benzidine staining. See Fibach *et al.* 1977 for details.

* The criterion for an undifferentiated colony is the absence of any benzidine-reactive cells.

† The criterion for a "mixed colony" is the presence of benzidine-reactive and benzidine-unreactive cells in the same colony.

‡ Proportion of "mixed colonies" $= \frac{\text{"mixed colonies"}}{\text{mixed colonies} + \text{uniformly benzidine reactive}} \times 100$

This is indicated by the fact that a single cell can give rise to a colony containing both differentiated and undifferentiated cells (Table 3).

Employing this assay, committed cells are detected after 12 to 13 hours of culture with HMBA. Colonies derived from cells that had been in suspension culture with HMBA are of three types, as assayed by the benzidine reaction: (a) uniformly benzidine-negative cells, (b) uniformly benzidine-reactive cells, (c) a mixture of benzidine-reactive and nonreactive cells ("mixed colonies"). Precloning culture with 5 mM HMBA for 53 hours yielded colonies of which 96% were uniformly benzidine reactive; 1% were "mixed colonies," and 3% were benzidine nonreactive (see Table 3). Exposure to 1 mM HMBA produced 7% "mixed colonies" and 28% benzidine reactive. Thus, the contribution of "mixed colonies" to the total population of differentiated (benzidine-reactive) colonies is highest at suboptimal concentrations of HMBA or after short periods of exposure to HMBA. Careful inspection of the cultures immediately after inoculation indicated that "mixed colonies" were very unlikely to be due to colony formation by more than one cell. This conclusion was confirmed by scoring "mixed colonies" in plates inoculated with a range of concentrations from 5×10^2 to 5×10^3 per ml; under these conditions, the proportion of "mixed colonies" was independent of inoculant concentration. These observations suggest that under these conditions a committed cell may give rise to both differentiated and undifferentiated progeny.

As discussed above, inducer must be present during DNA synthesis and possibly during a portion of cell cycle thereafter in order to induce differentiation of MELC. If exposure to inducer during S phase of the cell cycle is sufficient to commit a cell to differentiate, then the rate at which cells enter S phase will equal the rate of commitment to differentiation. Alternatively, if exposure to inducer during S phase is required but not sufficient for commitment, then the rate of entry into S will exceed the rate of commitment. These alternatives were examined by scoring cells for DNA synthesis (by ^{3}H-thymidine uptake and autoradiography) during HMBA-induced differentiation. The rate of entry into S phase, measured in this fashion, was greater than the rate of commitment to differentiation, as measured by "transfer out" from medium with HMBA to medium without inducer (Fibach *et al.* 1977) (Figure 5). Taken together, the present data suggest that DNA synthesis (passage through S phase of the cell cycle) is a necessary but not a sufficient condition leading to induced differentiation. The fact that the rate of entry to MELC into S phase is greater than the rate of commitment suggests the possibility that additional steps or events with different kinetics are required to achieve stable commitment of MELC to differentiation. Gusella *et al.* (1976) reported on a quantitative analysis of the kinetics of induction of MELC to differentiation, employing a similar technique for analysis of commitment, and suggested that commitment for each cell is made in a stochastic manner. Irreversible commitment to expression of differentiated functions occurs with a discrete probability per cell generation for many cell generations. In agreement with the observations in our laboratory

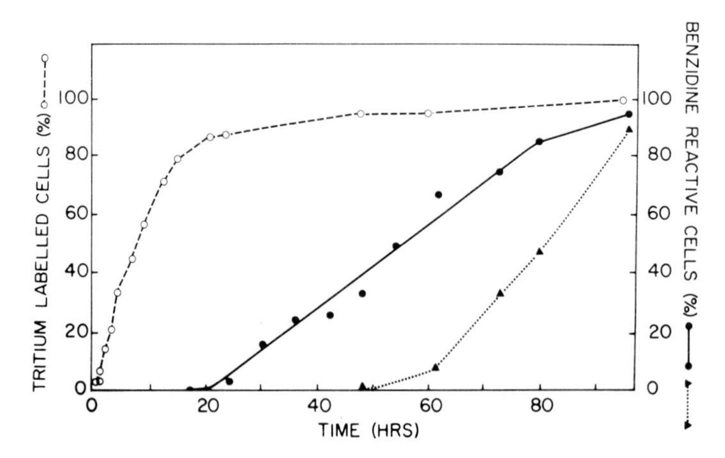

FIG. 5. Effect of HMBA on induction of MELC to differentiate. MELC were maintained in suspension culture with 5mM HMBA and 0.3 μCi of ^{3}H-thymidine added prior to inoculation of cultures with 10^5 cells/ml. Aliquots were taken at the times indicated to determine (a) the proportion of tritium-labeled cells (0), (b) the proportion of benzidine-reactive cells (▲), and (c) the rate of recruitment of MELC committed to differentiate, as assayed by transfer out of cells to fresh medium without HMBA and determination of the proportion of benzidine-reactive cells 120 hours after inoculation of the original culture (●) For details of these studies, see Fibach *et al.* 1977.

(Fibach *et al.* 1977), these workers concluded that the value for this probability is a function of the concentration of inducer, which in their experiments was Me_2SO. In both the studies of Gusella *et al.* (1976) and Fibach *et al.* (1977), the temporal relationship between the kinetics of commitment and development of biochemical characteristics of differentiated erythroid cells suggest that an irreversible commitment to differentiation may precede or at least accompany the first detectable accumulation of globin mRNA and globin synthesis.

DIFFERENTIAL EFFECTS OF CHEMICAL INDUCERS ON THE EXPRESSION OF β-GLOBIN GENES

The proportion of MELC induced to differentiate and the average hemoglobin content achieved per cell varies with the strain of MELC and with the inducing agent used. For example, with MELC (DS-19), 5 mM HMBA induces essentially the entire population (over 99%) of MELC, with marked hemoglobinization of all cells (Reuben *et al.* 1976). Butyric acid, on the other hand, induces a lower proportion of cells, approximately 60%. It is not yet known whether the different agents induce the identical program of erythroid differentiation or whether there are different patterns of differentiation characteristic of the various inducers. This question has been explored in experiments designed to examine one aspect of the problem, namely, the effects of different agents on the relative rates of synthesis of hemoglobin major (Hb^{maj}) and hemoglobin minor (Hb^{min}) (Nudel *et al.* 1977). Hb^{maj} and Hb^{min} differ in amino acid sequence of the β-chain (Hutton *et al.* 1962, Popp and Bailiff 1973, Gilman 1976a, b). On the basis of the available genetic and biochemical evidence (Popp and Bailiff 1973, Gilman 1976a, b), β^{maj} and β^{min} genes appear to be closely linked, and β^{maj}- and β^{min}-globins appear to differ by a relatively few (six to nine) amino acids in the primary structure.

The rate of synthesis of globin was measured with ^{35}S-methionine as the radioactive amino acid precursor. Relative rates of synthesis of Hb^{maj} and Hb^{min} were calculated assuming one methionine in α and two methionines in each of the β-chains. The planar-polar compounds tested, such as HMBA, Me_2SO, and methyl pyrolidinone, induced two to three times more Hb^{maj} than Hb^{min} in both the Me_2SO-sensitive cell line (DS-19) and Me_2SO-resistant cell line (DR-10). Fatty acids, such as butyric acid and propionic acid, induce approximately equal amounts of both hemoglobins in DS-19 and induce relatively more Hb^{min} in DR-10 (Table 4). These differences in relative accumulation of Hb^{maj} and Hb^{min} are associated with similar differences in the rates of synthesis of β^{maj}- and β^{min}-globin and of accumulation of mRNA for β^{maj}- and for β^{min}-globins (Table 5) (Nudel *et al.* 1977). The ratio of the rates of synthesis of the two β-globins in both DS-19 and DR-10 remain similar from day 2 through the duration of the period of culture (day 6) (Table 4). These findings suggest that the differences in the ratio of β^{maj} and β^{min} are not due to changing rates of synthesis at different stages of erythroid differentiation.

TABLE 4. *Ratio of synthesis of β^{maj}-globin to β^{min}-globin in MELC cell lines DS-19 and DR-10 cultured with different inducing agents*

		Inducing agent added			
Cell line	Day	Me_2SO	HMBA	MPL*	BA†
DS-19	2	2.1			1.1
	3	2.7	2.7		1.3
	4	2.1	2.5	2.4	1.1
	5	1.8	2.2		1.3
	6	1.9	2.4		1.2
DR-10	2			3.1	1.1
	3			2.5	1.0
	4		2.3	2.5	0.9
	5		2.1	2.2	0.8
	6		2.6	2.1	0.5

Cell extracts were prepared from cultures incubated with inducing agents for the indicated number of days. Incubation with ^{35}S-methionine, separation and quantitation of globins were as described in Nudel *et al.* (1977). The ratio of synthesis of β^{maj}-to β^{min}-globin was determined by comparing the amount of radioactivity in each globin, assuming that each β-globin chain contains two methionine residues.

* MPL: methyl pyrolidinone

† BA: butyric acid

TABLE 5. *Ratio of synthesis of β^{maj}-globin to β^{min}-globin in a wheat germ cell-free system with MELC poly(A)-containing RNA as template*

	Inducing agent added				
Source of RNA	Me_2SO	HMBA	MPL	BA	None
DS-19 cells	1.9	2.9	—	1.4	—
DR-10 cells	—	3.1	2.4	0.9	—
DBA/2 reticulocytes	—	—	—	—	3.7

MELC were cultured for four days with different inducing agents. Preparation and translation of RNA and quantitation of β-globins were as described in Nudel *et al.* 1977. Ratio of rates of synthesis of β^{maj}-globin to that of β^{min} is in a wheat germ cell-free system to which was added poly(A)-containing RNA purified from the reticulocytes of DBA/2 adult mice.

The finding that the ratio of the relative rates of synthesis of β^{maj}- to β^{min}-globin reflects the relative content of poly(A)-containing mRNA in induced MELC (Nudel *et al.* 1977) and suggests that the differences in the relative amounts of the β-globins observed with different types of inducing agents are not attributable to translational or posttranslational regulatory mechanisms. The inducers appear to act directly or indirectly to control the rates of synthesis of β^{maj}- and β^{min}-globins by affecting the transcription or processing of mRNA for these proteins.

EFFECT OF INDUCERS ON cAMP IN MELC

Inducers, such as Me_2SO, HMBA, and butyric acid, cause changes in MELC including prolonging G_1 and transiently inhibiting entry into S, altering growth rate and development of biochemical and morphological characteristics of differentiation. These effects suggested the possibility that these agents may be affecting intracellular levels of 3′:5′-cyclic adenosine monophosphate (cAMP), since similar changes have been produced by agents that are known to increase cAMP concentrations in other mammalian cells (see review by Prasad and Sinha 1976). The levels of cAMP in MELC during the course of induction with HMBA, Me_2SO, or butyric acid has been measured (Rifkind *et al.*, in press). In MELC cultured with 280 mM Me_2SO, 5 mM HMBA, or 1.5 mM butyric acid, there is a three to fivefold increase in cAMP concentration, the peak value occurring at approximately three to seven hours after addition of inducer. In cultures of MELC without inducing agents, there is a less than twofold increase in cAMP concentration. The levels of cAMP in these cultures return to zero time concentration by 12 to 18 hours after onset of culture. In cultures of MELC cell line (DR-10) resistant to induction with Me_2SO, culture with 5 mM HMBA or with 1.5 mM butyric acid is associated with a three to fivefold increase in cAMP levels. On the other hand, culture of MELC (DR-10) with 280 mM Me_2SO is not associated with an increase in cAMP concentrations. These data suggest that inducing agents cause an early increase in cAMP levels, and this may be a necessary effect for the cells to proceed toward the expression of the program of erythroid differentiation.

MECHANISM OF ACTION OF INDUCING AGENTS

On the basis of the observations summarized above, one can suggest a sequence of events that ensue in culture of MELC with inducing agents causing differentiation. It appears that the virus-transformed cell is an erythroid precursor cell. We assume that the cell has developed to a point where it is "set" to express a coordinated program of protein synthesis characteristic of erythroid cells. Virus transformation causes a block in the responsiveness of these cells to erythropoietin or, alternatively, blocks the cells at a stage beyond which they are normally responsive to erythropoietin (Marks and Rifkind 1972).

When one considers the chemical nature of compounds that have been demonstrated to be active as inducers of differentiation in MELC, it is, at present, difficult to make a single generalizing statement as to the property that makes these compounds active. Several of the most active compounds are relatively highly polar and planar. It is assumed that the specificity of the effects of the inducing agent are due to the properties of the target cell.

One of the earliest detectable biochemical events characteristic of the expression of the erythroid program is the accumulation of mRNA for globin. This suggests that the inducers, either directly or indirectly, act to affect the transcription or processing of the product of the structural genes whose expression is characteristic of erythroid cells. (While this seems to be generally true for inducers with various lines of MELC, Harrison *et al.* [1974] have presented evidence which they interpret as suggesting that a Friend leukemia line may be induced at a level involving translational control of expression of the mRNA for globin.)

The evidence that the inducer must be present during a critical S phase (DNA synthesis) for the transition to expression of the differentiated program suggests DNA replication has a critical role in the onset of expression of globin and other associated structural genes. Weintraub (1975) has provided evidence that the stability of the protein-DNA complexes may be altered as the replication apparatus copies DNA. There is evidence (Terada *et al.*, in press) that changes in DNA protein complex occur in induced cells at a time just prior to or accompanying the first detectable accumulation of mRNA for globin. The presence of an inducer during DNA synthesis is not sufficient for the stable expression of the differentiated program in all progeny of a cell exposed to inducer (Fibach *et al.* 1977).

The finding that an early and transient increase in cAMP levels occurs during culture with the inducer suggests that an initial event in the induction may occur at the cell membrane and that effects on the regulation of transcription are secondary. That chemicals active as inducers of MELC do affect the cell membrane has been suggested by evidence that topical anesthetics, which increase the fluidity of membranes in vitro, interfere with the effects of the inducers

(Bernstein *et al.* 1976) and that there is a correlation between the ability of inducers to cryoprotect erythrocytes and their effectiveness as inducing agents (Priesler 1976).

One cannot exclude a direct effect of the inducing agents on the cell membrane as well as at the level of chromatin (Marks *et al.* 1977). Of interest in this regard are the findings of Nakanishi, Pastan and co-workers (Nakanishi *et al.* 1974a, b) that Me_2SO and other DNA-denaturing agents stimulate the synthesis of specific *gal* and *lambda* RNA transcripts from wild-type and promoter-defective templates of *E. coli* in vivo and in vitro. Me_2SO can directly affect the stability of chromatin and of DNA (Lapeyre and Bekhor 1974).

As a working hypothesis, it is suggested that the induction to erythroid differentiation involves a series of events which include: (1) an effect on the cell membrane that occurs early and leads to a transient increase in cAMP levels, (2) a prolongation of G_1 that is associated with a structural alteration in chromatin, as evidenced by the restricted binding of intercalating dyes and the increased susceptibility to alkali degradation of the DNA of induced cells, (3) "reprogramming" of the chromatin for transcription occurring during the subsequent S phase of the cell cycle, and (4) additional stabilizing events occurring following this S phase, which are required to render differentiation irreversible. It has been suggested (Stein *et al.* 1974) that the activity of genes is dependent on the association of regulatory proteins with particular regions of the chromosomes. It is presumed that these proteins are at least partially dissociated from chromosomes during the process of DNA replication. These regulatory proteins may be synthesized during the previous interphase. Prolongation of the duration of G_1 may permit accumulation or degradation of a regulatory substance to a concentration that can alter the pattern of chromosomal regulatory proteins and, as a consequence, the program of gene transcription of the subsequent S phase. The period of DNA synthesis (S phase) that follows the prolonged G_1 is then essential to the reconstitution of chromatin with the new program of transcription.

ACKNOWLEDGMENTS

Studies reported in this paper were supported, in part, by grants from the National Institute of General Medical Sciences (GM-14552), National Cancer Institute (CA-13696, CA-18314) and contract (NO1-CP-6-1008) and grant (PCM 75-08696) from the National Science Foundation.

M. T. is a Hirschl Trust Scholar; R. R. and E. F. are Schultz Foundation Scholars.

REFERENCES

Arndt-Jovin, D. J., W. Ostertag, H. Eisen, and T. M. Jovin. 1976. Analysis by computer-controlled cell sorter of Friend virus-transformed cells in different stages of differentiation, *in* Modern Trends in Human Leukemia II, R. Neth, R. C. Gallo, K. Mannweiler, and W. C. Moloney, eds., J. F. Lehmanns Verlag, Munich, (Series: Hamatologie and Bluttransfusion, *Band* 19).

Bernstein, A., A. S. Boyd, V. Crickley, and V. Lamb. 1976. Induction and inhibition of Friend leukemic cell differentiation: The role of membrane-active compounds, *in* Biogenesis and Turnover of Membrane Macromolecules, J. S. Cook, ed., Raven Press, New York, pp. 93–103.

Boyer, S. H., K. D. Wuu, A. N. Noyes, R. Young, W. Scher, C. Friend, H. D. Priesler, and A. Bank. 1972. Hemoglobin biosynthesis in murine virus-induced leukemic cells in vitro: Structure and amounts of globin chains produced. Blood 40:823–835.

Byars, N., and C. Kidson. 1970. Programmed synthesis and export of immunoglobulin by synchronized myeloma cells. Nature 226:648–650.

Cleaver, J. E. 1974. Sedimentation of DNA from human fibroblasts irradiated with ultraviolet light: Possible detection of excision breaks in normal and repair-deficient xeroderma pigmentosum cells. Radiat. Res. 57:207–227.

Cleaver, J. E. 1975. Methods for studying repair of DNA damaged by physical and chemical carcinogenesis, *in* Methods in Cancer Research, H. Busch, ed., Vol. 11. Academic Press, Inc., New York, pp. 123–165.

Curtis, P. J., and C. Weissman. 1976. Purification of globin mRNA from dimethylsulfoxide-induced Friend cells and detection of a putative globin mRNA precursor. J. Mol. Biol. 106:1061–1075.

Dube, S. K., G. Gaedicke, N. Kluge, B. J. Weismann, H. Melderis, G. Steinheider, T. Crozier, H. Beckmann, and W. Ostertag 1973. Hemoglobin synthesizing mouse and human erythroleukemic cell lines as model systems for the study of differentiation and control of gene expression, *in* Proceedings of the 4th International Symposium of the Princess Takamatsu Cancer Research Fund, University Park Press, Tokyo, pp. 99–132.

Dworkin, M., J. Higgins, A. Glenn, and J. Mandelstam. 1972. Synchronization of the growth of *Bacillus subtilis* and its effect on sporulation. Spores 5:233–237.

Ebert, P. S., and Y. Ikawa. 1974. Induction of δ-aminolevulinic acid synthetase during erythroid differentiation of cultured leukemia cells. Proc. Soc. Exp. Biol. Med. 146:601–604.

Ebert, P. S., I. Wars, and D. N. Buell. 1976. Erythroid differentiation in cultured Friend leukemia cells treated with metabolic inhibitors. Cancer Res. 36:1809–1813.

Fibach, E., R. Reuben, R. A. Rifkind, and P. A. Marks. 1977. Effect of hexamethylene bisacetamide on the commitment to differentiation of murine erythroleukemia cells. Cancer Res. 37:440–444.

Friend, C. 1957. Cell-free transmission in adult Swiss mice of a disease having the character of a leukemia. J. Exp. Med. 105:307–318.

Friend, C., W. Scher, J. G. Holland, and T. Sato. 1971. Hemoglobin synthesis in murine virus induced leukemic cells *in vitro:* Stimulation of erythroid differentiation by dimethyl sulfoxide. Proc. Natl. Acad. Sci. USA 68:378–382.

Furusawa, M., Y. Ikawa, and H. Sugano. 1971. Development of erythrocyte membrane-specific antigen(s) in clonal cultured cells of Friend virus-induced tumor. Proc. Japan Acad. 47:220–224.

Gilman, J. 1976a. Mouse haemoglobin beta chains: Sequence data on embryonic y chain and genetic linkage of the y-chain locus to the adult β-chain locus Hbb. Biochem. J. 155:231–241.

Gilman, J. 1976b. Mouse haemoglobin beta chains: Comparative sequence data on adult major and minor beta chains from two species, *Mus musculus* and *Mus cervicolor.* Biochem. J. 159:43–53.

Gilmour, R. S., P. R. Harrison, J. W. Windass, M. Affara, and J. Paul. 1974. Globin mRNA synthesis and processing during haemoglobin induction in Friend cells. I. Evidence for transcriptional control in clone M2. Cell Differentiation 3:23–30.

Gusella, J., and D. Housman. 1976. Differentiation in vitro by purine and purine analogs. Cell 8:263–269.

Gusella, J., R. Geller, B. Clarke, V. Weeks, and D. Housman. 1976. Commitment to erythroid differentiation by Friend erythroid leukemia cells: A stochastic analysis. Cell 9:221–229.

Harrison, P. R., R. S. Gilmour, N. A. Affara, D. Conkie, and J. Paul. 1974. Globin messenger RNA synthesis and processing during haemoglobin induction in Friend cells. II. Evidence for post-transcriptional control in clone 707. Cell Differentiation 3:23–30.

Holtzer, H., H. Weintraub, R. Mayne, and B. Mochan. 1972. The cell cycle, cell lineages and cell differentiation. Curr. Top. Dev. Biol. 7:229–256.

Hutton, J. J., J. Bishop, R. Schweet, and E. S. Russell. 1962. Hemoglobin inheritance in inbred mouse strains. I. Structural differences. Proc. Natl. Acad. Sci. USA 48:1505–1513.

Kabat, D., C. C. Sherton, L. H. Evans, R. Bigley, and R. D. Koler. 1975. Synthesis of erythrocyte-specific proteins in cultured Friend leukemia cells. Cell 5:331–338.

Krishan, A. 1975. Rapid flow cytofluorometric analysis of mammalian cell cycle by propidium iodide staining. J. Cell. Biol. 66:188–193.

Lapeyre, J. and I. Bekhor. 1974. Effects of 5-bromo-2'deoxyuridine and dimethylsulfoxide on properties and structure of chromatin. J. Mol. Biol. 89:137–162.

Leder, A., S. Orkin, and P. Leder. 1975. Differentiation of erythroleukemic cells in the presence of inhibitors of DNA synthesis. Science 190:893–894.

Levy, J., M. Terada, R. A. Rifkind, and P. A. Marks. 1975. Induction of erythroid differentiation by dimethylsulfoxide in cells infected with Friend virus: Relationship to the cell cycle. Proc. Natl. Acad. Sci. USA 72:28–32.

Marks, P. A., and R. A. Rifkind. 1972. Protein synthesis: Its control in erythropoiesis. Science 175:955–961.

Marks, P. A., R. A. Rifkind, A. Bank, M. Terada, G. M. Maniatis, R. C. Reuben, and E. Fibach. 1977. Erythroid differentiation and the cell cycle, *in* Growth Kinetics and Biochemical Regulation of Normal and Malignant Cells (The University of Texas System Cancer Center 29th Annual Symposium on Fundamental Cancer Research, 1976), B. Drewinko and R. M. Humphreys, eds., Williams and Wilkins Co., Baltimore, pp. 329–345.

McClintock, P. R., and J. Papaconstantinou. 1974. Regulation of hemoglobin synthesis in a murine erythroblastic leukemic cell: The requirement for replication to induce hemoglobin synthesis. Proc. Natl. Acad. Sci. USA 71:4551–4555.

Nakanishi, S., S. Adhya, M. Gottesman, and I. Pastan. 1974a. Activation of transcription at specific promoters by glycerol. J. Biol. Chem. 249:4050–4056.

Nakanishi, S., S. Adhya, M. Gottesman, and I. Pastan. 1974b. Effects of dimethylsulfoxide on the *E. coli gal* operon and on bacteriophage lambda *in vitro.* Cell 3:39–46.

Nudel, U., J. E. Salmon, M. Terada, A. Bank, R. A. Rifkind, and P. A. Marks. 1977. Differential effects of chemical inducers on expression of β globin genes in murine erythroleukemia cells. Proc. Natl. Acad. Sci. USA 74:1100–1104.

Ohta, Y., M. Tanaka, M. Terada, O. J. Miller, A. Bank, P. A. Marks, and R. A. Rifkind. 1976. Erythroid cell differentiation: Murine erythroleukemia cell variant with unique pattern of induction by polar compounds. Proc. Natl. Acad. Sci. USA 73:1232–1326.

O'Malley, B. W., and A. R. Means. 1974. Female steroid hormones and target cell nuclei. Science 183:610–620.

Ormerod, M. G. 1976. Radiation-induced strand breaks in the DNA of mammalian cells, *in* Biology of Radiation Carcinogenesis, J. M. Yuhas, R. W. Tennant, and J. D. Regan, eds., Raven Press, New York, pp. 67–92.

Ostertag, W., H. Melderis, G. Steinheider, N. Kluge, and S. Dube. 1972. Synthesis of mouse hemoglobin and globin mRNA in leukemic cell cultures. Nature New Biol. 239:231–234.

Parkhurst, J. R., A. R. Peterson, and C. Heidelberger. 1973. Breakdown of HeLa cell DNA mediated by vaccinia virus (viral DNA/alkaline sucrose gradients). Proc. Natl. Acad. Sci. USA 70:3200–3204.

Paul, J., and I. Hickey. 1974. Haemoglobin synthesis in inducible, uninducible and hybrid Friend cell clones. Exp. Cell Res. 87:20–30.

Popp, R. A., and E. G. Bailiff. 1973. Sequence of amino acids in the major and minor beta chains of the diffuse hemoglobin from Balb/c mice. Biochim. Biophys. Acta 303:61–67.

Prasad, K. N., and P. K. Sinha. 1976. Effect of sodium butyrate on cells in culture: A review. In Vitro 12:125–132.

Preisler, H. D. 1976. *In vitro* and preliminary in vivo studies of compounds which induce the differentiation of Friend leukemia cells, *in* Modern Trends in Human Leukemia II, R. Neth, R. C. Gallo, K. Mannweiler, and W. C. Moloney, eds. J. F. Lehmanns Verlag, Munich (Series: Hamatologie and Bluttransfusion, *Band* 19).

Reuben, R. C., R. L. Wife, R. Breslow, R. A. Rifkind, and P. A. Marks. 1976. A new group of potent inducers of differentiation in murine erythroleukemia cells. Proc. Natl. Acad. Sci. USA 73:862–866.

Rifkind, R. A., E. Fibach, R. C. Reuben, Y. Gazitt, H. Yamasaki, I. B. Weinstein, U. Nudel, I. Sumida, M. Terada, and P. A. Marks. 1978. Erythroleukemia cells: Commitment to differentiate and the role of the cell surface, *in* Differentiation of Normal and Neoplastic Hematopoietic Cells, B. Clarkson and P. A. Marks, eds., Cold Spring Harbor Laboratory, Cold Spring Harbor, New York. (In press).

Ross, J., Y. Ikawa, and P. Leder. 1972. Globin mRNA induction during erythroid differentiation of cultured leukemia cells. Proc. Natl. Acad. Sci. USA 69:3620–3623.

Ross, J. 1976. A precursor of globin mRNA. J. Mol. Biol. 106:403–420.

Rutter, W. J., R. L. Pictet, and P. W. Morris. 1973. Toward molecular mechanisms of developmental processes. Ann. Rev. Biochem. 42:601–646.

Scher, W., H. Preisler, and C. Friend. 1973. Hemoglobin synthesis in murine virus-induced leukemic cells in vitro. J. Cell. Physiol. 81:63–70.

Singer, D. 1975. Differentiation of murine erythroleukemia cells. Ph.D. dissertation, Columbia University.

Singer, D., M. Cooper, G. M. Maniatis, P. A. Marks, and R. A. Rifkind. 1974. Erythropoietic differentiation in colonies of Friend virus transformed cells. Proc. Natl. Acad. Sci. USA 71:2668–2670.

Stein, G. S., T. C. Spelsbert, and L. J. Klunsmith. 1974. Non-histone chromosomal proteins and gene regulation. Science 183:817–824.

Stockdale, F. E., W. G. Juergens, and Y. J. Topper. 1966. A histological and biochemical study of hormone-dependent differentiation of mammary gland tissue *in vitro*. Dev. Biol. 13:266–281.

Tambourin, P., and F. Wendling. 1971. Malignant transformation and erythroid differentiation by polycythaemia-inducing Friend virus. Nature New Biol. 234:230–233.

Tanaka, M., J. Levy, M. Terada, R. Breslow, R. A. Rifkind, and P. A. Marks. 1975. Induction of erythroid differentiation in murine virus infected erythroleukemia cells by highly polar compounds. Proc. Natl. Acad. Sci. USA 72:1003–1006.

Terada, M., J. Fried, U. Nudel, R. A. Rifkind, and P. A. Marks. 1977. Transient inhibition of initiation of S-phase associated with dimethylsulfoxide induction of murine erythroleukemia cells to erythroid differentiation. Proc. Natl. Acad. Sci. USA 74:248–252.

Terada, M., U. Nudel, E. Fibach, R. A. Rifkind, and P. A. Marks. 1978. Changes in DNA associated with induction of erythroid differentiation by dimethylsulfoxide in murine erythroleukemia cells. Cancer Res. (In press).

Varga, J. M., A. DiPasquale, J. Pawelek, J. McGuire, and A. B. Lerner. 1974. Regulation of melanocyte stimulating hormone action at the receptor level: Discontinuous binding of hormone to synchronized mouse melanoma cells during the cell cycle. Proc. Natl. Acad. Sci. USA 71:1590–1593.

Weintraub, J. 1975. The organization of red cell differentiation, *in* Cell Cycle and Cell Differentiation (Results and Problems—Cell Differentiation), J. Reinhert and H. Holtzer, eds., Vol. 7. Springer-Verlag, New York, pp. 27–42.

Cell Differentiation and Neoplasia, edited by Grady F. Saunders. Raven Press, New York

Regulation of Gene Expression in Eukaryotes

Bert W. O'Malley, Sophia Y. Tsai, Ming-Jer Tsai, and Howard Towle

Department of Cell Biology, Baylor College of Medicine, Houston, Texas 77030

Estrogen-mediated growth and differentiation of the chick oviduct, and the concomitant induction of ovalbumin, has proved to be an excellent system for studying the mechanism of steroid hormone action (O'Malley and Means 1974, Jensen *et al.* 1974, O'Malley and Schrader 1976). Several lines of evidence indicate that steroid hormones act primarily to regulate oviduct gene expression (O'Malley and McGuire 1968, Cox *et al.* 1973, O'Malley *et al.* 1972, Schwartz *et al.* 1975, Tsai *et al.* 1975). During primary stimulation of immature chicks, estrogen has a dramatic effect on the level of endogenous RNA polymerase activity (O'Malley *et al.* 1969, Cox *et al.* 1973), nuclear RNA synthesis (O'Malley and McGuire 1968, Means *et al.* 1972, Harris *et al.* 1975), chromatin template activity (O'Malley *et al.* 1969, Cox *et al.* 1973, Spelsberg *et al.* 1973), and the number of initiation sites on chromatin available for in vitro RNA synthesis (Schwartz *et al.* 1975). In addition, the appearance of messenger RNA for a specific induced protein, ovalbumin, precedes the accumulation of that protein during estrogen-mediated growth (Means *et al.* 1972, Comstock *et al.* 1972, Rhoads *et al.* 1973). Withdrawal of estrogen leads to a decrease in the number of initiation sites, which temporarily correlates well with the decline in the level of nuclear-bound estrogen receptors (Tsai *et al.* 1975). Readministration of estrogen to these chicks results first in an increase in the level of nuclear-bound receptors, followed by a twofold increase in the level of chromatin initiation sites (Tsai *et al.* 1975), and, finally, by the appearance of specific ovalbumin mRNA sequences ($mRNA_{ov}$) (Chan *et al.* 1973, Cox *et al.* 1974, Palmiter 1973, Harris *et al.* 1975). These combined results indicate that regulation of RNA synthesis by estrogen occurs at the level of transcription.

To date, considerable evidence has implicated nonhistone proteins as the potential candidates for regulating differential gene expression in eukaryotic cells (Stein and Kleinsmith 1975). However, the mechanism of action of the presumptive nonhistone proteins that regulate the transcription of a specific gene is largely unknown. In order to elucidate the mechanism of gene restriction and expression in eukaryotes, it is necessary to develop a sensitive assay system

Present address: H. Towle, Department of Medicine, University of Minnesota, Minneapolis, Minnesota 55455 that can assess the biological function of the nonhistone proteins. In this regard, several groups have used reconstituted chromatin coupled with an in vitro chromatin transcription system to demonstrate the biological significance of the nonhistone proteins. For instance, Paul and Gilmour (1968) have demonstrated that a substantial amount of the RNA transcribed in vitro from reiterated DNA sequences in reconstituted chromatin is tissue specific. The specificity of the in vitro RNA synthesis was determined by the nonhistone proteins. Subsequently, Gilmour and Paul (1973), Barrett *et al.* (1974), and Chiu *et al.* (1975b) demonstrated that globin-specific sequences were detected in in vitro transcripts from reconstituted chromatin containing nonhistones from erythroid chromatin but not from nonhistones of nonerythroid tissues. Similarly, Stein *et al.* (1975) reported that histone mRNA sequences were transcribed from S phase but not G_1 phase HeLa cells, and Park *et al.* (1976) further demonstrated that nonhistone proteins from S phase cells were capable of rendering histone genes available for transcription. These results indicate that the nonhistone proteins of chromatin play a central role in the control of differential gene expression.

The above studies, however, are subject to criticism for the following reasons. First, various amounts of endogenous RNA that associated with the chromatin preparations could not be distinguished from the newly synthesized in vitro RNA products. Thus, contamination with endogenous mRNA sequences during reconstitution could significantly affect interpretation of the results. Second, the fidelity of chromatin reconstitution was not well established. To circumvent these problems, we have examined the in vitro transcription of the ovalbumin gene as well as transcription of total poly(A) RNA sequences in native and reconstituted chromatins. All in vitro transcription experiments were carried out with mercurated uridine triphosphate (UTP) as substrate. The newly synthesized RNA containing HgUTP residues was efficiently separated from the endogenous RNA by sulfhydryl-Sepharose column chromatography under appropriate conditions. Our results demonstrate that reconstitution is faithful with respect to transcription of the above genes.

Recently, we have observed that preferential transcription of certain chromatin genes relative to total RNA synthesis was significantly affected by the condition of in vitro RNA synthesis (Towle *et al.* 1977). For both *Escherichia coli* RNA polymerase and wheat germ RNA polymerase II, lowering the enzyme-to-DNA ratio resulted in an increase in the percentage of ovalbumin mRNA sequences transcribed from chick oviduct chromatin. On the other hand, transcription of the globin gene from oviduct chromatin or ovalbumin gene from reticulocyte chromatin or deproteinized chick DNA was not significantly affected by varying the enzyme-to-DNA ratio. This intrinsic property was further used as a criterion to assess the fidelity of reconstitution. Our results demonstrate that this specific characteristic inherent to native chromatin was not altered after reconstitution. Thus, we have used reconstitution as a tool to study the effect of chromatin protein subfractions on in vitro transcription of the ovalbumin gene. From this study, we hope to learn more about the mode of action by which chromatin proteins regulate the specific genes.

USE OF HgUTP AS A SUBSTRATE FOR CHROMATIN TRANSCRIPTION

In order to separate RNA sequences transcribed from chromatin during the course of incubation from those endogenously bound to the chromatin, RNA was synthesized using an equimolar mixture of HgUTP and UTP, in addition to the other three ribonucleotides. The presence of HgUTP inhibited the synthesis of RNA by about 10% to 15% throughout a two-hour incubation (data not shown), in accordance with published reports (Smith and Huang 1976, Biessman *et al.* 1976). To test the separation of HgRNA from RNA not containing mercury, a double-label experiment was performed. RNA was synthesized from chick oviduct chromatin using either ^{14}C-labeled UTP alone or a mixture of ^{3}H-labeled UTP and HgUTP. After extraction, 50 μg samples of each RNA were mixed before subjecting to sulfhydryl (SH)-Sepharose chromatography (Figure 1). ^{14}C-RNA, containing no mercury substitutions, failed to bind to the column and was quantitatively recovered in the material that flowed through the column. The HgRNA, on the other hand, was largely bound to the column and could be eluted with buffer containing 0.1 M 2-mercaptoethanol. About 80% to 90% of the HgRNA input was recovered in the 2-mercaptoethanol wash, and the level of contamination with ^{14}C-unsubstituted RNA was approximately 1.4%. This low level of contamination might be expected, since no precautions were taken to dissociate RNA-RNA interactions.

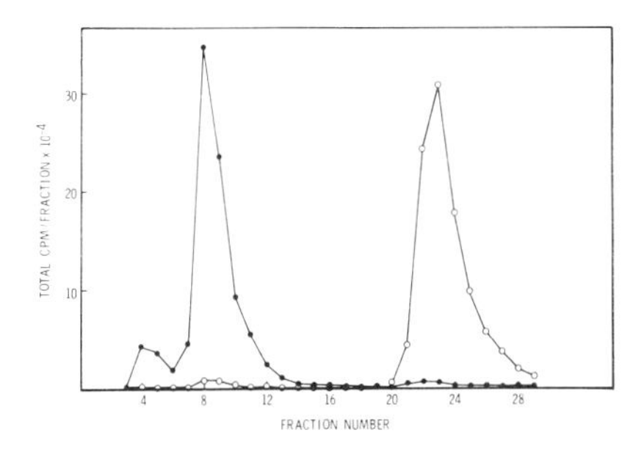

FIG. 1. Separation of mercurated RNA from nonmercurated RNA by SH-Sepharose affinity chromatography. RNA was normally synthesized in 25 ml reactions containing 500 mM Tris-HCl, pH 7.9, 2mM $MnCl_2$, 10 mM 2-mercaptoethanol, 50mM $(NH_4)_2SO_4$, 1 mM each of ATP, CTP, and GTP, 0.5 mM UTP and 0.5 mM HgUTP, 800 μg *E. coli* RNA polymerase, and 200 μg chick oviduct chromatin. In this experiment, 200 μCi of ^{3}H-UTP [with HgUTP (○)] or 30 μCi of ^{14}C-UTP and 1 mM UTP [no HgUTP (●)] were used. After extraction, 50 μg of each RNA sample was mixed and subjected to SH-Sepharose chromatography. The total cpm loaded onto the column was 1.1 × 10^6 cpm for ^{3}H-HgRNA (○) and 9.0 × 10^5 cpm for ^{14}C–non-HgRNA (●). The SH-columns were run according to the procedure of Towle *et al.* (1977).

TABLE 1. *Removal of $mRNA_{ov}$ from in vitro Hg-RNA by SH-Sepharose column chromatography*

Conditions	% Bound to SH column	
	A	B
SH*	73	83
Heated for 2 min at 100° C SH*	31	7
HgRNA†	92	95

* 37.5 ng $mRNA_{ov}$ were added to the reaction mixture containing 200 μg estrogen-withdrawn chromatin, 400 μg *E. coli* RNA polymerase, and HgUTP, as described in Figure 1. The reaction was carried out at 37°C for two hours. RNA was then extracted and loaded first onto Sephadex G-50 columns and then SH columns in 50 mM Tris-HCl, pH 7.5, 0.1 N NaCl (TN), as in A, or in 10 mM Tris-HCl, pH 7.5, 1 mM EDTA (TE), as in B. After loading, the SH columns were washed with TN and TN + 0.1 M 2-mercaptoethanol, as described by Towle *et al.* (1977).

† 50 μg in vitro HgRNA was loaded onto SH columns under conditions A and B.

To more accurately assess the effectiveness of the SH-Sepharose column for separating HgRNA from non-Hg-RNA, a further control was employed. A known amount of ovalbumin messenger RNA was added to an incubation mixture containing estrogen-withdrawn chick oviduct chromatin, and HgRNA was synthesized using an equimolar mixture of HgUTP and UTP. The RNA samples were then extracted and divided into three equal aliquots. Two of the three samples were subjected to SH-Sepharose chromatography to remove the ovalbumin mRNA from the HgRNA transcripts under various conditions. Hybridization was performed using $cDNA_{ov}$ to quantitate the amount of $mRNA_{ov}$ present. By these means, it should be possible to determine the level of nonspecific trapping of $mRNA_{ov}$ containing no mercurated substitutions in the HgRNA sample. As shown in Table 1, 100% of ovalbumin sequences was detected in the control sample. Direct passage of the RNA sample through the SH-Sepharose column either at high-salt or low-salt conditions only removed about 25% to 35% of the ovalbumin mRNA sequences. However, denaturation of the RNA samples in 0.1 M NaCl, 50 mM Tris-HCl before loading on the SH-Sepharose column removed 70% of the ovalbumin mRNA sequences. When the RNA sample was heated at 100°C for two minutes in 10 mM Tris-HCl, 1 mM EDTA, at least 90% of the ovalbumin mRNA sequences was consistently removed. Under the same condition, 95% of the HgRNA was retained by the column. Thus, the level of contaminating endogenous mRNA sequences in the HgRNA samples could be effectively removed under appropriate conditions and would not contribute significantly to the level of $mRNA_{ov}$ sequences in the in vitro transcripts.

TRANSCRIPTION OF OVALBUMIN FROM NATIVE AND RECONSTITUTED CHROMATIN

In order to investigate the transcriptional integrity of reconstituted chromatins, we examined the transcription of two specific genes, ovalbumin and globin. Since ovalbumin and globin mRNAs are highly tissue specific, ovalbumin mRNA should not be synthesized to a significant degree in chick reticulocyte chromatin, and globin mRNA should not be synthesized from oviduct chromatin. To ensure that tissue specificity was maintained in the reconstituted chromatins, HgRNA was synthesized from both native and reconstituted oviduct (estrogen-stimulated) and reticulocyte chromatins. Specific cDNA probes synthesized from pure ovalbumin and globin messenger RNA were then used to quantitate the amount of globin and ovalbumin mRNA sequences present in the in vitro RNA products. As shown in Figure 2, ovalbumin mRNA sequences were transcribed from native oviduct chromatin to a much greater extent than globin mRNA sequences. The concentration of ovalbumin mRNA and globin mRNA sequences transcribed from native oviduct chromatin was calculated to be 0.021% and 0.0013%, respectively (see Table 1). Using reconstituted oviduct chromatin, the titration curves closely follow those obtained from the native oviduct chromatin (see Figure 2). The concentration of ovalbumin and globin mRNAs present in the in vitro transcripts of reconstituted oviduct chromatin was very similar to that of the native chromatin, 0.020% and 0.0016%, respectively (Table 2). Thus, following reconstitution, the oviduct chromatin maintained the expected in vivo tissue specificity seen with native chromatin.

To ascertain that the $mRNA_{ov}$ sequences indeed reflect newly synthesized HgRNA products, in vitro HgRNA from spleen tissue was added to oviduct chromatin, and further incubation was carried out in the absence of RNA

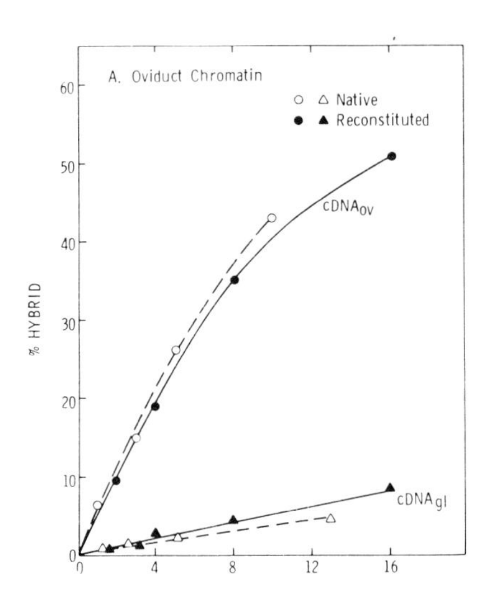

FIG. 2. Tissue specificity of in vitro transcription of ovalbumin and globin genes from native and reconstituted chromatins. RNA was synthesized in a 5-ml reaction mixture containing 400 μg RNA polymerase and 400 μg oviduct chromatin, as described in Figure 1. After extraction, varying amounts of RNA samples from native chromatin (○, △) and reconstituted chromatin (●, ▲) were hybridized to 1.5 ng ovalbumin cDNA (○, ●) or globin cDNA (△, ▲) for 68 hours at 68°C as described by Harris *et al.* (1976).

TABLE 2. *Transcription of specific gene product from native and reconstituted chromatins*

Chromatin	% $mRNA_{ov} \times 10^2$	% $mRNA_{gl} \times 10^2$
Native oviduct	2.1	0.11
Reconstituted oviduct	2.0	0.06

polymerase. The resultant RNA was purified through an SH-Sepharose column and then analyzed for ovalbumin mRNA sequences. Only a very low level (0.001%) of ovalbumin mRNA sequences was detected. Therefore, endogenous contaminating nonmercurated ovalbumin mRNA sequences from oviduct chromatin were effectively removed by chromatography on SH-Sepharose affinity columns.

EFFECT OF VARYING RNA POLYMERASE-TO-CHROMATIN RATIOS

Recently, we have demonstrated that preferential transcription of the ovalbumin gene relative to total RNA synthesis could be obtained by varying the RNA polymerase-to-chromatin ratio (Towle *et al.* 1977). This process is dependent on the presence of specific chromatin proteins, since transcription of the ovalbumin gene from either reticulocyte chromatin or deproteinized DNA was not significantly affected by changing the enzyme-to-chromatin ratio (Towle *et al.* 1977). To examine whether reconstituted chromatin still retained this specific property, HgRNA was transcribed from native and reconstituted stimulated chick oviduct chromatin at two different enzyme-to-chromatin ratios, 4:1 and 1:1. In these experiments, the concentration of chromatin was held constant. As shown in Table 3, the level of $mRNA_{ov}$ sequences present in the in vitro transcripts of both native and reconstituted chromatins increased in a similar manner when the enzyme-to-chromatin ratio was decreased. Over the fourfold range of enzyme-to-chromatin ratios tested, an increase of 2.5-fold in the percentage of $mRNA_{ov}$ sequences was detected in the in vitro transcripts from both native and reconstituted chromatins. The increase in ovalbumin mRNA sequences represented an increase in the absolute amount of $mRNA_{ov}$ synthesized, since the total amount of RNA synthesized using an equivalent amount of enzyme was roughly equal for the two chromatins. Thus, a preferential transcription of the ovalbumin gene was maintained in reconstituted chromatin when the enzyme-to-chromatin ratio was lowered.

HYBRIDIZATION OF NATIVE AND RECONSTITUTED CHROMATIN TRANSCRIPTS

To further provide a general test for the fidelity of reconstitution, a cDNA probe was synthesized using poly(A)-RNA from oviduct tissue as a template.

TABLE 3. *Effect of varying enzyme-to-chromatin ratios on the in vitro transcription of ovalbumin mRNA from native and reconstituted oviduct chromatin*

Enzyme/chromatin ($\mu g/\mu g$)	Total enzyme	Total chromatin (μg DNA)	Amount RNA synthesized (μg)		$\% mRNA_{ov}$	
			Native chromatin	Reconstituted chromatin	Native chromatin	Reconstituted chromatin
4.0	800	200	298	294	0.80	0.86
1.0	800	800	245	282	2.1	2.0

It was thought that this probe, which represented a major fraction of the RNA transcribed in vivo, would enable us to analyze a substantial fraction of the in vitro transcription products. A vast excess of in vitro RNA transcribed from native and reconstituted oviduct chromatin or reticulocyte chromatin was hybridized to cDNA complementary to chick oviduct (estrogen-stimulated) poly(A)-RNA. Figure 3 shows the kinetics of the hybridization reaction. The data reveal that RNA transcribed in vitro from native or reconstituted chromatin reacted with the cDNA probe with strikingly similar kinetics. The rate and the extent of hybridization of in vitro RNA from native and reconstituted oviduct chromatin were virtually indistinguishable and implied that a similarity of sequence complexity and frequency distribution existed in the two in vitro RNA populations. Since the hybridization reaction was not carried out to completion in this experiment, it was not certain that the in vitro RNA transcripts from oviduct chromatin contained all the sequences represented by the probe. However, when the hybridization reaction was carried out to a higher R_0t value, the extent of hybridization generally reached more than 85% (Tsai *et al.*, manuscript in prcparation). This led us to conclude that a major fraction of the in vitro sequences was indeed represented in the in vitro RNA. The rate of hybridization of RNA synthesized in vitro from reticulocyte chromatin to $cDNA_{polyA}$ of oviduct tissue was slower and, to a lesser extent, indicated that

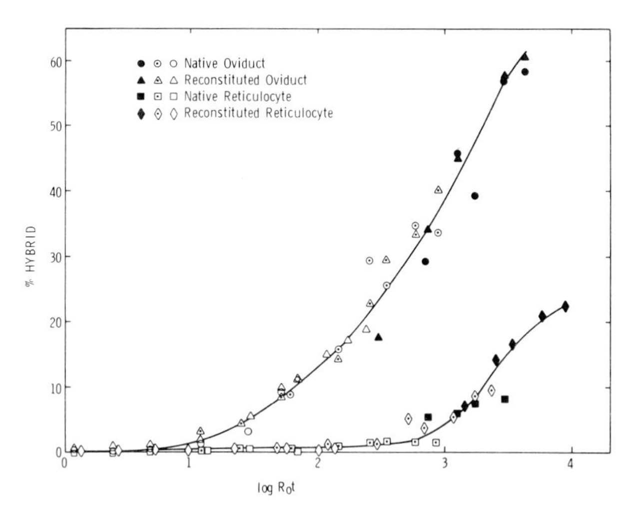

FIG. 3. Hybridization of native and reconstituted chromatin transcripts to cDNA of total cellular poly(A)-containing RNA. RNA was synthesized from native oviduct (●, ⊙, ○) and reticulocyte (■, ⊡, □) chromatins or reconstituted oviduct (▲, ◬, △) and reticulocyte (◆, ◈, ◇) chromatins and extracted as described in Figure 1. Varying amounts of these transcripts were hybridized to 0.05 ng of [^{3}H]-cDNA prepared from total cellular poly(A)-RNA of 14-day diethylstilbestrol-stimulated chick oviducts. The hybridization was carried out to the indicated R_0t values, and hybrids were assayed according to the methods of Monahan *et al.* (1976). The concentrations of chromatin transcripts used were 2.4 mg/ml (◆), 1.2 mg/ml (●, ■, ▲), 0.48 mg/ml (◈), 0.24 mg/ml (⊙, ⊡, ◬), 0.096 mg/ml (◇), and 0.048 mg/ml (○, □, △).

TABLE 4. *Effect of chromatin proteins on in vitro RNA synthesis*

Reconstituted chromatin			% $mRNA_{ov} \times 10^2$
NH_{DES}	H_{DES}	$TBP \cdot DNA_{DES}$	0.88
NH_{DES}	H_{DES}	$TBP \cdot DNA_W$	0.88
NH_{DES}	H_W	$TBP \cdot DNA_{DES}$	0.96
NH_W	H_{DES}	$TBP \cdot DNA_W$	0.11
NH_W	H_W	$TBP \cdot DNA_{DES}$	0.19
NH_W	H_W	$TBP \cdot DNA_W$	0.18

a smaller fraction of common sequences was shared by the in vivo oviduct poly(A)-RNA and the in vitro transcripts from reticulocyte chromatin. Nevertheless, the rate of hybridization was identical in the native and reconstituted chromatin.

EFFECT OF CHROMATIN PROTEINS ON IN VITRO TRANSCRIPTION OF THE OVALBUMIN GENE

Recently, Harris *et al.* (1976) observed that a substantial concentration of ovalbumin sequences was found in the in vitro RNA transcripts of oviduct chromatin isolated from estrogen-stimulated chicks. However, only a negligible amount of ovalbumin messenger RNA sequences was detected in the RNA transcripts of oviduct chromatin isolated from chicks stimulated with estrogen and, subsequently, withdrawn from the hormone. The substantial difference in the concentration of ovalbumin mRNA sequences in the in vitro transcripts of chromatins from the two hormonal states provides an approach for studying the role of chromatin proteins in the regulation of gene expression.

Chromatin proteins were dissociated from oviduct chromatin isolated from estrogen-stimulated or estrogen-withdrawn chicks and subsequently fractionated into three major classes by the modified procedure of Chiu *et al.* (1975a): (1) extractable nonhistone (NH), (2) histone (H), and (3) nonhistone tightly bound to DNA (TBP•DNA). Chromatins were then reconstituted with various protein components from heterologous origins by gradient dialysis, as described previously (Tsai *et al.* 1976a). The various reconstituted chromatins were transcribed using *E. coli* RNA polymerase, and the RNA products were analyzed for ovalbumin mRNA sequences. As shown in Table 4, 0.008% of ovalbumin mRNA sequences was detected in the transcripts of chromatin reconstituted from homologous components of stimulated chromatin. Interchange of TBP•DNA of histone between stimulated and withdrawn chromatin during reconstitution did not affect the synthesis of ovalbumin mRNA as long as the NH was obtained from stimulated chromatin. Similarly, reconstituted chromatin containing NH fraction from withdrawn chromatin could only support the synthesis of a low level (0.001% to 0.002%) of ovalbumin mRNA, regardless of the source of histone or TBP•DNA.

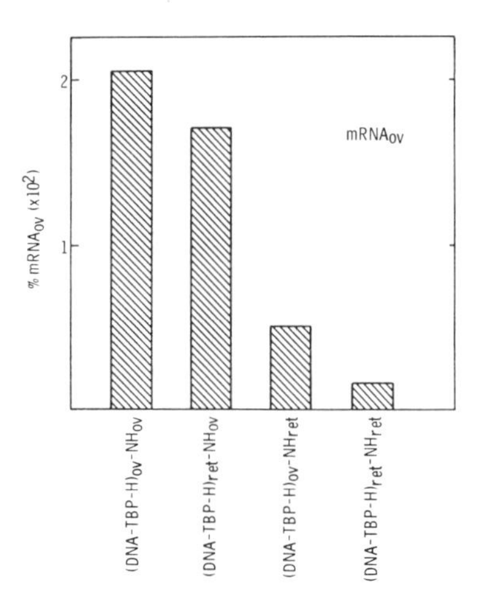

FIG 4. Effect of nonhistones on the in vitro transcription of the ovalbumin gene. Extractable NH proteins were isolated from both oviduct and reticulocyte chromatins and reconstituted to their homologous or heterologous histones and TBP•DNA complex at a ratio equivalent to the amount present in the native chromatins. In the case of native oviduct chromatin, the ratio of NH to DNA on a weight basis was 1.2:1, while in reticulocyte chromatins, the ratio was 0.4:1. RNA was synthesized from the reconstituted chromatin and hybridized to ovalbumin cDNA. The amount of ovalbumin mRNA sequences in the in vitro transcripts was determined as described in Figure 2.

Similar experiments were carried out with oviduct and reticulocyte chromatins. NH proteins were isolated from both oviduct and reticulocyte chromatin and reconstituted to their homologous or heterologous H•TBP•DNA complexes. Analyses of $mRNA_{ov}$ sequences in the in vitro transcripts from these reconstituted chromatins were carried out as shown in Figure 4. When NH from oviduct chromatin was reconstituted with H•TBP•DNA complex from reticulocyte chromatin, the ovalbumin gene was expressed almost to the same extent as that noted in the homologous reconstituted oviduct chromatin. In a parallel experiment, when NH protein from reticulocytes was reconstituted to an H•TBP•DNA complex from oviduct chromatin, the level of ovalbumin mRNA sequences in this heterologous, reconstituted chromatin was slightly higher than that of the homologous reticulocyte chromatin control. This difference can result from incomplete extraction of the functional NH regulators from the oviduct chromatin. Nevertheless, it is clear that the presence of NH proteins from oviduct chromatin can direct synthesis of ovalbumin mRNA, regardless of the origin of the H•TBP•DNA complex in the reconstituted chromatin.

SUMMARY

To date, considerable evidence has implicated NH proteins as the potential candidates for regulating differential gene expression in eukaryotic cells. The estrogen-mediated alteration of gene expression in the chick oviduct provides an excellent system for studying the mechanism of action by which the presumptive NH proteins regulate the specific gene expression. To elucidate the mechanism of gene restriction and gene expression, we developed a sensitive assay system that could assess the biological function of the NH proteins. We used reconstituted chromatin coupled with an in vitro chromatin transcription system to study the significance of the NH proteins in control of gene expression.

Chromatin proteins from oviduct and reticulocyte chromatins were fractionated and then reconstituted with their homologous constituents. DNA complementary to pure ovalbumin mRNA, globin mRNA, or total cellular poly(A)-containing RNA of estrogen-primed chick oviducts was used to examine the fidelity of reconstitution. Our results demonstrate that reconstitution is faithful with respect to transcription of the above genes.

We then examined the effect of chromatin proteins on in vitro transcription of the ovalbumin and globin genes. Reconstituted chromatins containing NH from stimulated oviduct chromatin were capable of synthesizing a high level of ovalbumin mRNA sequences, regardless of the source of histone and TBP•DNA complexes. On the other hand, if NH proteins were obtained from either estrogen-withdrawn oviduct chromatin or reticulocyte chromatin, only a background level of ovalbumin mRNA sequences could be detected in the in vitro RNA products. These results indicate that the control elements regulating the expression of the ovalbumin gene reside in the NH proteins. The regulatory proteins could act as an activator (De Crombrugghe *et al.* 1971, Eron and Block 1971), as an antirepressor (Dickson *et al.* 1975), as an antiterminator (Herskowitz 1973), or as a repressor. However, the exact mode of action of this positive regulator can not be deduced at the present time. It is hoped that the NH regulatory proteins of interest can be identified and purified, thus aiding our understanding of the mechanism involved in control of the ovalbumin gene.

ADDENDUM

Since this article was written, more detailed information relative to the Hg-nucleotide technique for in vitro chromatin transcription has been obtained. Two major problems are inherent in this technique. First, *E. coli* RNA polymerase can utilize endogenous RNA as template and synthesize complementary sequences that remain base-paired to the template. The resulting double-stranded or partially double-stranded RNA containing HgRNA in one of the strands can bind to SH-Sepharose column and thus co-purify with the newly synthesized HgRNA. Second, nonmercurated endogenous RNA can bind to the SH-Sepharose through aggregation with HgRNA and thus be retained in the final RNA preparation. These two problems associated with the Hg-nucleotide technique can be minimized by modifying the conditions for RNA synthesis and SH-Sepharose chromatography. Using the modified procedure, the Hg-nucleotide and SH-Sepharose technique can remove more than 90% of endogenous RNA contaminants.

In order to directly demonstrate that the $mRNA_{ov}$ sequences detected in vitro result from de novo transcription of oviduct chromatin, experiments were also carried out which show that the hybridizable RNA sequences contain Hg element (by way of rebinding to SH-Sepharose after hybridization and S_1 nuclease digestion) and that the synthesis of these RNA sequences is sensitive to low concentrations of actinomycin D. These combined results strongly suggest that the majority of $mRNA_{ov}$ sequences detected by hybridization to $cDNA_{ov}$ is indeed due to DNA-dependent RNA synthesis by *E. coli* RNA polymerase and not due to an artifact of endogenous RNA contamination (Tsai *et al.,* submitted for publication).

In addition, we have developed recently a method in which ^{3}H-RNA synthesized from chromatin can be directly hybridized to immobilized plasmid DNA (pOV_{230}) containing the ovalbumin gene. We found that the ^3H-$mRNA_{ov}$ could be specifically hybridized to the filter-bound ovalbumin gene. These hybridizable sequences were also competed out by excess cold $mRNA_{ov}$. These results once again indicated that *E. coli* RNA polymerase can transcribe the ovalbumin gene in oviduct chromatin.

REFERENCES

Barrett, T., D. Maryanka, P. H. Hamlyn, and H. T. Gould, 1974. Nonhistone proteins control gene expression in reconstituted chromatin. Proc. Natl. Acad. Sci. USA 71:5057–5061.

Biessmann, H., R. A. Gjerset, B. W. Levy, and B. J. McCarthy. 1976. Fidelity of chromatin transcription in vitro. Biochemistry 15:4356–4362.

Chan, L., A. R. Means, and B. W. O'Malley. 1973. Induction rates of specific translatable mRNAs for ovalbumin and avidin by steroid hormones. Proc. Natl. Acad. Sci. USA 70:1870–1874.

Chiu, J. F., M. Hunt, and L. S. Hnilica. 1975a. Tissue specific DNA-protein complexes during azo dye hepatocarcinogenesis. Cancer Res. 35:913–919.

Chiu, J. F., Y. H. Tsai, D. Sakuma, and L. S. Hnilica. 1975b. Regulation of in vitro mRNA transcription by a fraction of chromosomal proteins. J. Biol. Chem. 250:9431–9433.

Comstock, J. P., G. C. Rosenfeld, B. W. O'Malley, and A. R. Means. 1972. Estrogen-induced changes in translation and specific messenger RNA levels during oviduct differentiation. Proc. Natl. Acad. Sci. USA 69:2377–2380.

Cox, R., M. Haines, and N. Carey. 1973. Modification of the template capacity of chick oviduct chromatin for form B polymerase by estradiol. Eur. J. Biochem. 32:513–524.

Cox, R. F., M. E. Haines, and J. S. Emtage. 1974. Quantitation of ovalbumin mRNA in hen and chick oviduct by hybridization to complementary DNA. Eur. J. Biochem. 49:225–236.

De Crombrugghe, B., B. Chen, W. Anderson, P. Nisley, M. Gottesman, I. Pastan, and R. L. Petle. 1971. Lac DNA, RNA polymerase and cAMP receptor protein, cAMP, Lac repressor and inducer are the essential elements for controlled Lac transcription. Nature New Biol. 231:139.

Dickson, R. E., J. Abelson, W. N. Barnes, and W. S. Reznikoff. 1975. Genetic regulation: The Lac control region. Science 187:27–34.

Eron, L., and R. Block, 1971. Mechanism of initiation and repression of in vitro transcription of the *Lac* operon of *Escherichia coli.* Proc. Natl. Acad. Sci. USA 68:1828.

Gilmour, R. S., and J. Paul. 1973. Tissue-specific transcription of the globin gene in isolated chromatin. Proc. Natl. Acad. Sci. USA 70:3440.

Harris, S. E., J. M. Rosen, A. R. Means, and B. W. O'Malley. 1975. Use of a specific probe for ovalbumin messenger RNA to quantitate estrogen-induced gene transcripts. Biochemistry 14:2072–2081.

Harris, S. E., R. J. Schwartz, M. J. Tsai, A. K. Roy, and B. W. O'Malley. 1976. Effect of estrogen on gene expression in the chick oviduct: In vitro transcription of the ovalbumin gene in chromatin. J. Biol. Chem. 251:524–529.

Herskowitz, I. 1973. Control of gene expression in bacteriophage lambda. Annu. Rev. Genet. 7:289–324.

Jensen, E. V., S. Mohla, T. A. Gorell, and E. R. De Sombre. 1974. The role of Estrophilin in estrogen action. Vitam. Horm. 32:89–127.

Means, A. R., J. P. Comstock, G. C. Rosenfeld, and B. W. O'Malley. 1972. Ovalbumin messenger RNA of chick oviducts: Partial characterization, estrogen dependence and translation in vitro. Proc. Natl. Acad. Sci. USA 69:1146–1150.

Monahan, J. J., S. E. Harris, and B. W. O'Malley. 1976. Effect of estrogen on gene expression in the chick oviduct: Effect of estrogen on the sequence and population complexity of chick oviduct poly(A)-containing RNA. J. Biol. Chem. 251:3738–3748.

O'Malley, B. W., and W. L. McGuire. 1968. Studies on the mechanism of estrogen-mediated tissue differentiation: Regulation of nuclear transcription and induction of new RNA species. Proc. Natl. Acad. Sci. USA 60:1527–1534.

O'Malley, B. W., W. L. McGuire, P. O. Kohler, and S. G. Korenman. 1969. Studies on the mechanism of steroid hormone regulation of synthesis of specific proteins. Recent Prog. Horm. Res. 25:105–160.

O'Malley, B. W., and A. R. Means. 1974. Female steroid hormones and target cell nuclei. Science 183:610–620.

O'Malley, B. W., G. C. Rosenfeld, J. P. Comstock, and A. R. Means. 1972. Steroid hormone induction of a specific translatable messenger RNA. Nature New Biol. 240:45–48.

O'Malley, B. W., and W. T. Schrader. 1976. The receptors of steroid hormones. Sci. Am. 234:32.

Palmiter, R. D. 1973. Rate of ovalbumin messenger RNA synthesis in the oviduct of estrogen-primed chicks. J. Biol. Chem. 248:8260–8270.

Park, W. D., J. L. Stein, and G. S. Stein. 1976. Activation of in vitro histone gene transcription for HeLa S_3 chromatin by S-phase nonhistone chromosomal proteins. Biochemistry 15:3296–3300.

Paul, J., and R. S. Gilmour. 1968. Organ-specific restriction of transcription in mammalian chromatin. J. Mol. Biol. 34:305–316.

Rhoads, R. E., G. S. McKnight, and R. T. Schimke. 1973. Quantitative measurement of ovalbumin messenger RNA activity. J. Biol. Chem. 248:2031–2039.

Schwartz, R. J., M. J. Tsai, S. Y. Tsai, and B. W. O'Malley. 1975. Effect of estrogen and gene expression in chick oviduct: Changes in the number of RNA polymerase binding and initiation sites in chromatin. J. Biol. Chem. 250:517–551.

Smith, M. M., and R. C. C. Huang. 1976. Transcription in vitro of immunoglobin kappa light chain genes in isolated mouse myeloma nuclei and chromatin. Proc. Natl. Acad. Sci. USA 73:775–779.

Spelsberg, T. C., W. M. Mitchell, F. Chytil, E. M. Wilson, and B. W. O'Malley. 1973. Chromatin of the developing chick oviduct: Changes in the acidic proteins. Biochim. Biophys. Acta 312:765–768.

Stein, G., W. Park, C. Thrall, R. Means, and J. Stein. 1975. Regulation of cell cycle stage-specific transcription of histone genes from chromatin by nonhistone chromosomal proteins. Nature 257:764.

Stein, G. S., and L. J. Kleinsmith. 1975. Chromosomal Proteins and Their Roles in the Regulation of Gene Expression. Academic Press, New York and London.

Towle, H. C., M. J. Tsai, S. Y. Tsai, and B. W. O'Malley. 1977. Effect of estrogen on gene expression in chick oviduct: Preferential initiation and asymmetrical transcription of specific chromatin gene. J. Biol. Chem. 252:2396.

Tsai, S. Y., M. J. Tsai, R. J. Schwartz, M. Kalimi, J. Clark, and B. W. O'Malley. 1975. Effect of estrogen on gene expression in chick oviduct: Nuclear receptor levels and initiation of transcription. Proc. Natl. Acad. Sci. USA 72:4228.

Tsai, S. Y., S. E. Harris, M. J. Tsai, and B. W. O'Malley. 1976a. Effect of estrogen on gene expression in the chick oviduct: The role of chromatin proteins in regulating transcription of the ovalbumin gene. J. Biol. Chem. 251:4713–4721.

Tsai, S. Y., M. J. Tsai, S. E. Harris, and B. W. O'Malley 1976b. Effect of estrogen on gene expression in the chick oviduct: Control of ovalbumin gene expression by nonhistone proteins. J. Biol. Chem. 251:6475–6478.

Cell Differentiation and Neoplasia, edited by Grady F. Saunders. Raven Press, New York

Pancreas Development: An Analysis of Differentiation at the Transcriptional Level

William J. Rutter, Alan E. Przybyla, Raymond J. MacDonald, John D. Harding, John M. Chirgwin, and Raymond L. Pictet

Department of Biochemistry and Biophysics, University of California at San Francisco, California 94143

The contention that neoplasia is an aberrant aspect of differentiation has considerable merit. The processes of tissue morphogenesis, regulation of growth and gene expression that are fundamental to development, appear to be altered during oncogenesis. However, a definitive test of these putative relationships requires that the molecular events involved in differentiation be known in considerable detail so that divergence from the normal path of regulation can be detected.

Our laboratory is engaged in an analysis of regulatory mechanisms involved in organogenesis, using the pancreas as a paradigm. We reported on our initial studies in the 1965 M. D. Anderson Symposium on Developmental and Metabolic Control Mechanisms and Neoplasia (Rutter and Weber 1965). During the ensuing period, we have characterized the changing patterns of gene expression in the differentiated cells of the pancreas (Rutter *et al.* 1967, 1968, Rall *et al.* 1973, Kemp *et al.* 1972) and identified factors involved in the growth and morphogenesis of this tissue (Ronzio and Rutter 1973, Pictet *et al.* 1975b). We will briefly summarize these results and describe here our recent analysis of the differentiative process at the level of transcription.

CONTROL OF PANCREATIC GROWTH AND DIFFERENTIATION

The pancreas is composed predominantly of exocrine cells. The acinar cells synthesize the digestive enzymes or their precursors and contain typical zymogen granules. The duct cells produce the fluid of the pancreatic juice, as well as line the tubules connecting the acinar structures with the gut. In addition, the pancreas contains endocrine cells located in the islets of Langerhans, which are interspersed within the exocrine gland. In order of decreasing frequency, these are the B-cells that produce insulin, the A-cells that produce glucagon, and the D-cells that produce somatostatin. The differentiated endocrine cells can be distinguished not only by their specific products but also by the distinctive morphology of their secretory granules.

The pancreas appears as a diverticulum on the gut approximately midway through gestation at about 20 to 25 somites (approximately the 11th day of gestation in the rat). All the differentiated endocrine and exocrine cells are derived from the epithelial cells. Nevertheless, as in other organ systems, the mesenchymal cells, which form a cap over the epithelial cells in the early rudiment, are required for normal growth and development of the various cell types.

Mesenchymal Factor(s) Control the Growth and Differentiative Program of Pancreatic Epithelia

Mesenchymal cells influence the morphogenesis and terminal differentiation of many epithelial organ systems (Grobstein 1967, Wolff 1968). A pancreatic epithelial anlage separated from its mesenchymal cap fails to grow and develop normally. Recombination of the tissues leads to a normal developmental pattern. An extract of embryos rich in mesenchymal tissues can replace the mesenchymal requirement for pancreas development (Ronzio and Rutter 1973, Pictet *et al.* 1975a,b). This mesenchymal factor (MF) is neither species nor tissue specific. Cell proliferation, as measured by the incorporation of tritiated thymidine into DNA, is virtually dependent upon MF. Furthermore, MF affects the proportion of differentiated cells in the rudiment. In its absence, a remarkably high proportion of differentiated endocrine cells (largely A-cells) can be observed; the rudiment essentially becomes an islet. In the presence of MF there is rapid cell division and the normal high proportion of acinar cells relative to endocrine cells obtains. More recent studies indicate that MF is a complex material. Two activities can be distinguished. One is destroyed by periodate and is more labile under usual conditions of extraction. This component can be replaced by dibutyryl cyclic AMP or 8-hydroxy cyclic AMP, cyclic AMP derivatives that can penetrate cell membranes. Another component is also required for activity. This second component is not replaced by cyclic GMP or its derivatives, by increased concentrations of calcium ion, or by calcium ionophores (Filosa *et al.* 1975, Pictet *et al.* 1975a).

MF acts largely, if not exclusively, at the epithelial cell surface (Levine *et al.* 1973). Purified MF can be covalently linked to Sepharose beads that are larger than the pancreatic cells. Pancreatic epithelial buds devoid of mesenchymal tissues adhere tightly to the MF-Sepharose beads. Binding is restricted to the basal surface of the cells. As a consequence, the primordium is everted, and the lumen is in direct contact with the culture medium. Cells in direct contact with MF-Sepharose beads rapidly incorporate precursors into DNA and later differentiate into zymogen-containing acinar cells. These results suggest that mesenchymal-epithelial interactions are mediated by molecules that act at the cell membrane. This morphogenetic effect plus the stimulation of MF on DNA synthesis and its influence on the proportion of endocrine and exocrine cells in the differentiated rudiment emphasize the range of controls regulated through the MF system. It thus seems conceivable that alterations in this system may be involved in oncogenesis. For example, modification of the membrane receptors could lead to an altered proliferative signal in cells no longer able to respond normally to effectors of differentiation. Also, internal mechanisms bypassing the MF control system could be instituted by oncogenic viruses or by chemical carcinogens. The net result might be proliferatively competent cells that exhibit varying degrees of differentiative and morphological constraints. The purification of MF to homogeneity and determination of its receptor on the cell surface may allow a direct test of these hypotheses.

Developmental Plasticity of Early Pancreatic Cells

Cells in the process of differentiation appear to have considerable developmental flexibility. This is indicated in part by the previously discussed effects of MF on the proportion of endocrine and exocrine cells in the developing rudiment in vitro. Another demonstration of this plasticity results from studies with the thymidine analogue 5′-bromodeoxyuridine (BrdU). BrdU is known to block cytodifferentiation in many systems; in addition, it induces virus expression (for a review see Rutter *et al.* 1973, Wilt and Anderson 1972). The inhibition of acinar cytodifferentiation is one of the earliest effects ascribed to BrdU (Wessells 1964). More recently, we have shown that BrdU at low concentrations blocks the accumulation of acinar cell secretory products and insulin without significantly altering the synthesis of other macromolecules or the proliferation of the cells (Walther *et al.* 1974). In contrast to normal development in vitro, 14-day pancreatic rudiments cultured six to seven days in the presence of BrdU lack zymogen granules. Instead, the tissue contains large fluid-filled vacuoles surrounded by cells that contain enhanced levels of alkaline phosphatase and carbonic anhydrase (Githens *et al.* 1976), enzymes typically associated with the duct cell population (Gomori 1941, Churg and Richter 1972). Thus, it appears that BrdU may alter the developmental program by enhancing the formation of duct cells at the expense of acinar and endocrine cells. One explanation for these phenomena is that when cells are blocked from progression along their natural developmental course, they can assume another developmental alternative. In this instance, the duct cells may be developmentally related to the cells in the protodifferentiated state. It has been observed that early protodifferentiated cells have the capability of secreting fluid, once stimulated by cyclic AMP (that is, small vacuoles are produced) (Filosa *et al.* 1975). Thus, the duct cells may be a compartment from which other differentiated cells arise. Regeneration of the pancreas, for example, is thought to involve the duct cell compartments. Furthermore, most pancreatic cell tumors are believed to be duct cell tumors. This is consistent with the postulate that many tumors may arise from cells with further developmental potential and which have an inherent capability to grow. This population of developmentally plastic cells may be the prime source of neoplastic activity.

Phased Regulation of Gene Expression during Development

The pancreatic A-cells containing glucagon differentiate coincidently with the formation of the pancreatic diverticulum (Pictet *et al.* 1972, Rall *et al.*

1973). During the following two to three days, there is rapid growth, with continuous formation of new acinar structures and islets but no evidence of cytodifferentiation of acinar or endocrine B-cells. The cytodifferentiation of B-cells and acinar cells begins at 14 to 15 days of gestation and is followed about a day later by the appearance of the β-granules and zymogen granules, respectively. We have shown that there is a biphasic accumulation of the exocrine proteins and insulin. The early embryonic rudiment contains very low levels of these cell-specific proteins; these levels are maintained for several days and increase dramatically just prior to the appearance of zymogen granules. The differentiated level of exocrine proteins, which is 1,000- to 10,000-fold higher than the 12 to 14 day levels, is reached at about 20 days, just prior to birth. During the neonatal period there are also significant changes in the profiles of the exocrine proteins. From these combined studies, we have inferred that there are at least two differentiative transitions. The first results in the formation of the pancreatic diverticulum and the capacity to form pancreas-specific structures and to synthesize low levels of pancreas-specific products. We term this phase the "protodifferentiated state." The second differentiative transition involves the singular dedication of the cells to synthesize and secrete large quantities of their specific products. This transition leads to the differentiated state. The various exocrine enzymes do not accumulate coordinately, and the levels can be selectively modified during development in vitro. The synthesis of the enzymes thus appears to be regulated independently or in small subsets.

The accumulation of specific exocrine proteins in vitro can be dramatically affected by the nutritional quality of the culture medium and also by hormones. Under conditions in which the concentrations of amino acids are limiting, cell-specific protein synthesis appears to be selectively inhibited (Rutter *et al.* 1975). This implies a mechanism for discrimination between differentiative functions as opposed to general physiological functions. The accumulation of specific proteins is also selectively modified by glucocorticoids. The potent synthetic glucocorticoid dexamethasone enhances the synthesis of certain pancreas-specific proteins and inhibits cell proliferation. Thirteen-day embryonic pancreases cultured for seven days in the presence of dexamethasone undergo approximately one less cell division than controls, but the average cell contains twice as much protein, due largely to an enhanced synthesis of amylase and, to a lesser degree, carboxypeptidase B. The specific secretory proteins represent about 40% of the total cell protein in rudiments cultured in the absence of dexamethasone, and about 70% in those cultured in the presence of dexamethasone (Rall *et al.*, submitted for publication).

CHARACTERIZATION OF THE ADULT PANCREAS mRNA POPULATION

We have begun a comprehensive analysis of transcription in adult and embryonic pancreases. For these studies, we have employed in vitro translation and cDNA-RNA hybridization analyses to examine the levels of specific mRNAs. These studies require the isolation of intact RNA, which has been difficult because of the high levels of ribonuclease in the pancreas of many species. Therefore, our first experiments to characterize pancreatic mRNA utilized dog pancreas, since others have shown that some species, including the dog, do not produce substantial amounts of pancreatic ribonuclease (Zendzian and Barnard 1967) and that intact polysomes capable of incorporating amino acids into proteins can be readily prepared (Breillatt and Dickman 1969). We have found that intact RNA can be isolated by phenol-chloroform extraction of dog pancreas polysomes or microsomes and that this RNA is capable of directing the synthesis of pancreas-specific proteins in heterologous cell-free protein synthesizing systems.

In Vitro Synthesis of Dog Pancreas Secretory Proteins

The array of proteins synthesized in response to dog pancreas mRNA in the reticulocyte cell-free system is illustrated in Figure 1 (slot B). Several lines of evidence indicate that the polypeptides synthesized in vitro represent the primary translation products of the mRNAs for pancreatic secretory proteins:

1) The polypeptide distributions of isolated secretory proteins and in vitro translation products resolved by SDS (sodium dodecyl sulfate) polyacrylamide gel electrophoresis are strikingly similar (Figure 1). Fifteen and sixteen major bands are present in the secretory proteins (slot A) and translation products (slot B), respectively. Fourteen have obvious counterparts of similar but not identical size. Twelve translation products migrate significantly slower and are therefore larger, by 500–2,000 daltons, than their putative secretory counterparts. Larger in vitro translation products have been reported for other secretory proteins synthesized in cell-free systems that lack a membrane-bound processing activity (Milstein *et al.* 1972, Kemper *et al.* 1974, Schmeckpeper *et al.* 1974, Boime *et al.* 1975, Devillers-Thiery *et al.* 1975). The two largest in vitro products, designated by asterisks, migrate faster than their presumed secretory counterparts seen in slot A of Figure 1. Since glycosylation decreases electrophoretic mobility in SDS (Segrest *et al.* 1971) and since pancreatic secretory proteins of several species are known to be glycosylated, it is possible that these two polypeptides synthesized in vitro are also larger than the secretory protein standards.

2) A monospecific antibody prepared against purified dog pancreatic amylase selectively precipitates an in vitro translation product approximately the same size as amylase.

3) We have also analyzed the patterns of ^{35}S-methionine-labeled tryptic peptides from these sets of proteins. Identification of many of the in vitro products is complicated by relatively high resistance to trypsin digestion and the paucity of methionine-containing peptides produced. For example, bovine trypsinogen and chymotrypsinogen sequences predict only a single methionine-containing peptide (Mikes *et al.* 1966, Brown and Hartley 1966). As a result, many of

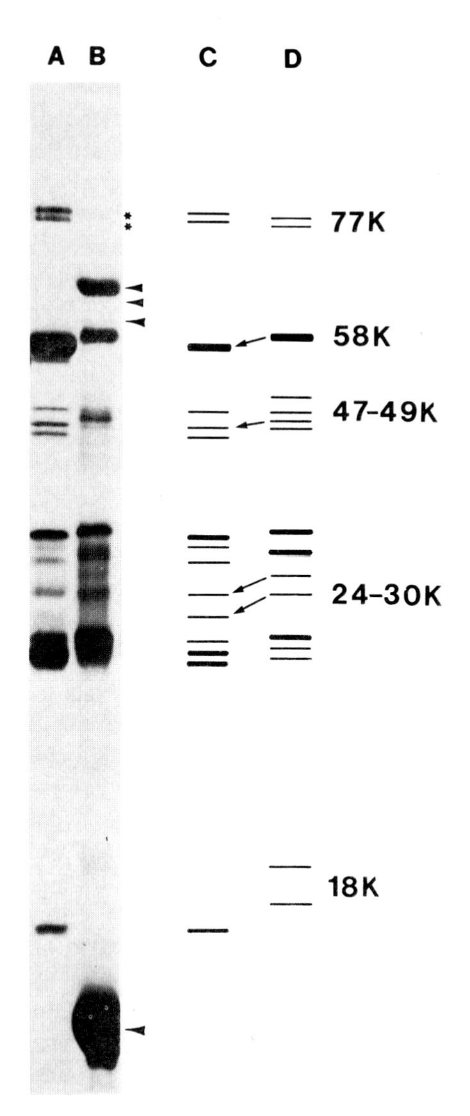

FIG. 1. Comparison of the electrophoretic patterns of pancreatic secretory proteins and pancreatic polypeptides synthesized in vitro. ^{35}S-methionine-labeled secretory proteins (slot A) and reticulocyte lysate translation products of pancreas mRNA (slot B) were carboxymethylated by treatment with iodoacetate prior to electrophoresis in adjacent slots of a 40 cm SDS slab gel. Arrows indicate endogenous reticulocyte polypeptides. The two asterisks indicate two in vitro pancreas products not evident at this autoradiographic exposure time. C and D are schematic representations of slots A and B, respectively. Bands corresponding to endogenous reticulocyte synthesis are omitted in D. Arrows indicate presumptive precursor relationships based on tryptic peptide analyses and specific amylase immunoprecipitation. Numbers at right indicate approximate polypeptide molecular weights in thousands.

the analyses are not diagnostic. However, in three instances, in addition to amylase, the analyses revealed the relationships between authentic and in vitro synthesized polypeptides, indicated by arrows in Figure 1 (slots C and D).

4) Unlike most proteins, the electrophoretic mobilities of authentic secretory proteins and the putative pancreatic secretory proteins synthesized in the reticulocyte cell-free system can be altered extensively by modifying intrachain disulfide bridges (Scheele 1975). This unusual phenomenon may be a result of extensive disulfide bridge cross-linking, which adds stability to these extracellular enzymes. Unreduced, the proteins have relatively high mobilities and have a tendency

to aggregate. Significantly lower mobilities are observed after heating in solutions containing high concentrations of 2-mercaptoethanol. Carboxymethylation, presumably preventing disulfide bond reformation during electrophoresis, further decreases their electrophoretic mobilities. This effect is relatively modest for most of the polypeptides, but four additional species of apparent molecular weights of 30,000–40,000 are resolved in both the standard pancreatic secretory proteins and the in vitro translation products. The similarity in the response to carboxymethylation of the secretory proteins and those synthesized in vitro is consistent with the postulate that these sets of proteins are related. The multiple bands that appear are not modifications of a single polypeptide, since the tryptic peptide maps of each are distinguishable.

5) It has previously been demonstrated that the synthesis of fewer than 19 secretory proteins comprises at least 80% of the total protein synthesis of the pancreas (Jamieson and Palade 1967, Scheele 1975). In vitro translation of a relatively few polypeptides, therefore, appears qualitatively to reflect the synthesis in the gland.

In Vitro Translation Products Generally Appear Larger than Isolated Secretory Proteins

The larger size of most of the translation products compared to the size of the secretory proteins of the gland is consistent with the thesis that these proteins are synthesized as precursors of somewhat higher molecular weight. This issue has been investigated in greater detail for four translation products, and the results support this contention.

Figure 2 shows the results of amylase immunoprecipitation of a mixture of ^{14}C-labeled purified amylase and ^{3}H-labeled pancreas translation products, and

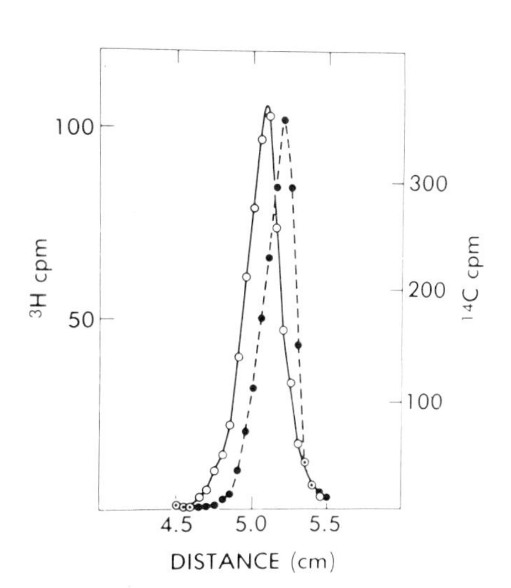

FIG. 2. Comparison of the relative sizes of in vitro and in vivo synthesized amylase. Following the in vitro synthesis of ^{3}H-labeled polypeptides in a reticulocyte lysate primed with pancreatic mRNA; authentic ^{14}C-amylase was added to the reaction and immunoprecipitated with anti-dog amylase immunoglobulin. Electrophoresis of the immunoprecipitate was performed for twice the normal time in a cylindrical SDS polyacrylamide gel to enhance the separation of in vivo and in vitro synthesized amylase. The gel was sectioned into 0.5 mm slices, and the radioactivity of each slice was determined by scintillation counting. Only the amylase region of the gel is shown. (○——○), ^{3}H-labeled in vitro synthesized amylase; (●——●), ^{14}C-labeled authentic amylase.

subsequent electrophoresis in an SDS polyacrylamide gel. The amylase product synthesized in the reticulocyte lysate (molecular weight ~58,000) was clearly separated from authentic amylase (molecular weight 56,000).

Four translation products (including amylase) have been correlated with secretory proteins 1,000–1,500 daltons smaller, by comparison of their tryptic peptide patterns. These relationships are indicated in Figure 1 (slots C and D). Analyses of the methionine-labeled tryptic peptides for authentic amylase and the in vitro amylase product were identical, whereas leucine-labeled in vitro synthesized and authentic amylases digested with protease *S. aureus* VI differed by a single band. Thus, the additional sequence of the in vitro product is likely to contain leucine but not methionine. Devillers-Thiery *et al.* (1975) have demonstrated that leucine (and not methionine) is a major constituent of the additional N-terminal sequence of several dog pancreas proteins synthesized in vitro.

The in vitro synthesis of putative dog pancreas secretory protein precursors has recently been reported by Devillers-Thiery *et al.* (1975). Other secretory proteins, such as human placental lactogen (Hubert and Cedard 1975, Boime *et al.* 1975), immunoglobulin heavy (Cowan and Milstein 1973, Schmeckpeper *et al.* 1974) and light chains (Schmeckpeper *et al.* 1974, Blobel and Dobberstein 1975a), proparathyroid hormone (Kemper *et al.* 1974) and proinsulin synthesized in vivo by membrane-bound polysomes (Permutt and Kipnis 1972, Yip *et al.* 1975), are synthesized in vitro as precursors containing additional amino acid sequences. Blobel and co-workers have proposed (Blobel and Sabatini 1971) and recently presented confirmatory evidence (Devillers-Thiery *et al.* 1975, Blobel and Dobberstein 1975a,b) that the additional amino-terminal sequence commits the polypeptide to transport into the cisternae of the endoplasmic reticulum as a first step to secretion. The presence of a secretion-specific peptide segment on the nascent chain may explain the specificity of binding to the endoplasmic reticulum those polysomes synthesizing proteins for export.

Correlation of the Size of Pancreatic mRNA with the Size of the Translated Proteins

Separation of pancreatic mRNAs was performed in polyacrylamide gels containing formamide to reduce artifacts caused by aggregation and secondary structure. To test for mRNA activities, the gel was sliced and the RNA extracted and translated in the reticulocyte system. The protein products of each reticulocyte translation were analyzed by SDS polyacrylamide gel electrophoresis. Message activity for the pancreatic polypeptides correlated with the RNA absorbance profile (Figure 3). Five size classes of mRNAs, each coding for a different set of polypeptides, are resolved by this procedure: mRNAs coding for (1) the two unidentified polypeptides (mol. wts. 76,000 and 78,000) at 2.1 cm (not shown in Figure 3); (2) amylase (mol. wt. 58,000) at 2.4 cm; (3) putative procarboxypeptidases A and B (mol. wts. 47,000 and 49,000) at 2.4 to 2.6 cm; (4) polypeptides between 24,000 and 30,000 molecular weight at 3.0 to 3.4 cm; and (5) polypeptides ~18,000 mol. wt. at 3.4 cm (not shown in Figure 3).

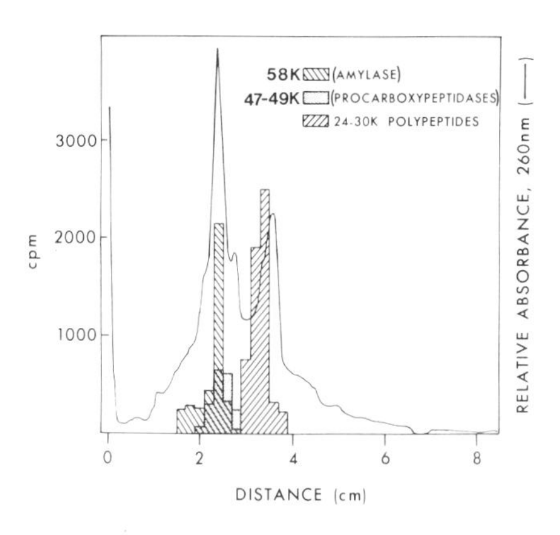

FIG. 3. Fractionation of pancreas mRNA activities. mRNA was resolved in a polyacrylamide gel containing formamide, and the gel was sliced into 2 mm fractions. RNA from each gel fraction was translated in the reticulocyte system. The translation products from each RNA fraction were analyzed by SDS polyacrylamide gel electrophoresis. The radioactivity incorporated into three sets of translation products is displayed. The identities of these products are indicated in Figure 1.

The difficulty of accurately determining the length of nucleotide polymers has been considered in assigning molecular weight estimates for the pancreatic mRNAs. Woo *et al.* (1975) have pointed out the disparities in the estimation of ovalbumin message molecular weight by electrophoresis and other techniques. In contrast, size estimates of globin mRNA (the only other mRNA whose size has been extensively studied) are nearly identical when determined by either electrophoresis in formamide (Berns *et al.* 1974) or electron microscopy (Williamson and Wellauer 1973). When correlating measurements made by formamide gel electrophoresis and another technique, the major source of error is the use of ribosomal RNA molecular weight standards that may retain significant secondary structure under conditions in which the mRNA does not.

Table 1 compares the size estimates of pancreatic mRNAs determined by formamide polyacrylamide gel electrophoresis (as shown in Figure 3) using rRNA molecular weight standards and nondenaturing sucrose gradient sedimentation using both rRNA and mRNA standards. The relative size of the pancreatic mRNAs in denaturing polyacrylamide gels generally correlates with the sizes of the secreted proteins they encode, but each mRNA size estimate is 400–800 nucleotides longer than necessary to code solely for the amino acid sequence. This is consistent with the observation that other eukaryotic mRNAs contain extensive nontranslated nucleotide sequences (Woo *et al.* 1975, Proudfoot and Brownlee 1974, Milstein *et al.* 1974). The mRNA size estimates determined by sucrose gradient centrifugation differ significantly from those obtained by electrophoresis in formamide and are largely dependent upon whether rRNA or mRNA weight standards are employed (Table 1). Messenger RNA standards consistently give lower estimates of pancreatic mRNA sizes; each mRNA contains an estimated 50–300 noncoding nucleotides. The estimated number of nucleotides in excess of those required to code for the amino acid sequence ranges from 170–800 nucleotides, when rRNA standards are used. The uncertainty of the molecular weight estimations does not permit an accurate assessment of the length of nontranslated sequences. It remains to be determined whether the nontranslated sequences are similar in this group of functionally and developmentally related messages.

TABLE 1. *Size determinations of dog pancreas mRNAs*

Size of in vitro polypeptide product* (mol. wt.)	mRNA length required to code for polypeptide	Estimated mRNA sizes (nucleotides)		
		Formamide gel analysis (rRNA standards)†	Sucrose gradient analysis (rRNA standards)†	Sucrose gradient analysis (mRNA standards)†
77,000	1900	2400	2700	2200
58,000	1440	2200	1900	1600
47–49,000	~1200	1800	1700	1400
24–30,000	600–750	~1150	~950	~800
18,000	450	950	620	530

* Pancreatic translation products are identified by their approximate molecular weights as shown in Figure 1.

† RNA molecular weight calibration was performed in the same formamide polyacrylamide gel or sucrose gradient containing the pancreatic mRNAs. The molecular weight standards were as follows (length in nucleotides): rRNAs—28S, 5400; 18S, 2200; 5S, 120; and 4S tRNA, 80. mRNAs—MS2 RNA, 3570; ovalbumin mRNA (a gift of S. Woo), 1890; globin mRNA, 670; and 4S tRNA.

Analysis of Adult Rat Pancreas mRNA Sequences

In contrast to the relative ease of isolation of translatable mRNA from the dog, the isolation of rat pancreas RNA is hampered by the large amounts of ribonuclease present in this tissue. When standard phenol extraction procedures using sodium dodecyl sulfate and ribonuclease inhibitors are employed, the isolated RNA is degraded. We have developed techniques employing rapid homogenization in powerful protein denaturants (4 M guanidine thiocyanate or 6 M guanidine HC1) containing 2-mercaptoethanol or diethylpyrocarbonate. The RNA isolated from adult and embryonic rat pancreases is intact by several criteria, including analysis of unfractionated and poly(A+) RNA by polyacrylamide gel electrophoresis under denaturing conditions, analysis of the polypeptide products after translation in the reticulocyte cell-free system, and efficient template activity with RNA-dependent DNA polymerase.

The polypeptides synthesized in a reticulocyte lysate system in response to adult rat pancreas RNA are shown in Figure 4. Slot 1 depicts the peptides from a control lysate; slot 2 shows, in addition, the bands produced in response to pancreas mRNA. The number, size, and relative intensity of these polypeptides is similar to that produced by dog pancreas RNA in the same system (c.f. Figure 1). The polypeptides synthesized in response to rat pancreas mRNA are also related to the number and size of the in vivo labeled rat pancreas secretory proteins as displayed in Figure 4 (slots 3 and 5). These experiments indicate that rat pancreas mRNA codes primarily for the synthesis of about 15 polypeptides. When the polypeptides produced in a reticulocyte lysate system primed with rat pancreas mRNA (Figure 4, slot 2) were precipitated with antibodies against purified rat amylase, a single polypeptide was evident in the immunoprecipitate (Figure 4, slots 4 and 6). The in vitro synthesized amylase is slightly larger than the authentic amylase isolated from zymogen granules. Thus, the rat amylase mRNA, like the dog mRNA, produces an amylase precursor. This conclusion has been substantiated by peptide mapping of authentic and in vitro synthesized amylase.

The sequence complexity and the frequency distribution of adult rat pancreas poly(A+) RNA were analyzed by means of cDNA-RNA hybridization. Adult pancreas poly(A+) RNA was isolated by oligo(dT) cellulose chromatography and the cDNA synthesized using avian myeloblastosis virus RNA-dependent DNA polymerase in the presence of an oligo(dT) primer (Verma *et al.* 1974). Pancreas cDNA was hybridized with an excess of the poly(A+) RNA from which it was copied, as shown in Figure 5. The great majority of the cDNA hybridized between R_0t values of 5×10^{-4} and 5×10^{-1} mol s 1^{-1}. These R_0t values, as well as the range of the reaction, indicate that a small population of different RNA species is involved. Computer analysis indicates that the data can be described most simply by two ideal pseudo-first order kinetic components. The sequence complexity (total length of different RNA sequences in nucleotides) of the RNA was calculated by comparing the $R_0t_{1/2}$ values of each of the

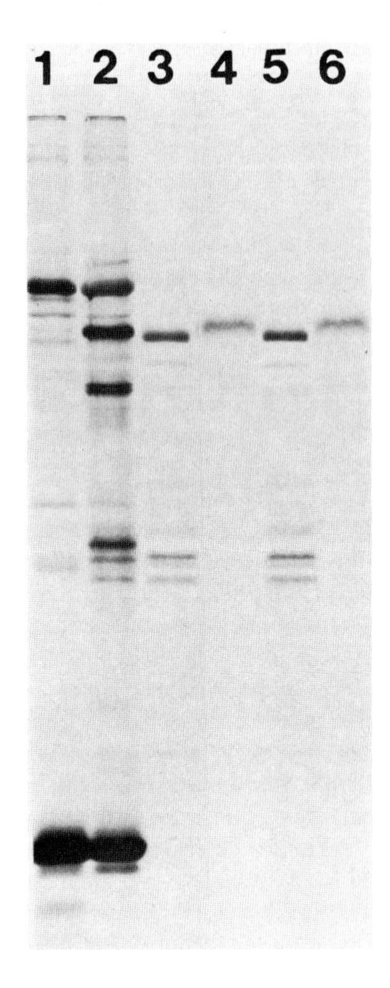

FIG. 4. Translation of guanidine thiocyanate-extracted adult rat pancreas RNA in a reticulocyte lysate. ^{35}S-methionine-labeled in vitro translation products were fractionated by SDS polyacrylamide gel electrophoresis and visualized by autoradiography. *Slot 1:* Control—polypeptides synthesized by the reticulocyte lysate in the absence of exogenous RNA; *Slot 2:* Polypeptides synthesized in response to adult rat pancreas RNA; *Slots 3 and 5:* In vivo labeled secretion products of adult rat pancreas; *Slots 4 and 6:* Polypeptides immunoprecipitated from a reticulocyte lysate containing rat pancreas RNA with rat anti-amylase antibodies.

components with a standard globin cDNA-RNA hybridization curve (Bishop *et al.* 1974, Longacre and Rutter 1977). The first component (0 to 50% hybridization) has a complexity of about 1.6×10^3 nucleotides. The second component (50 to 90% hybridization) has a complexity of about 1.8×10^4 nucleotides. The RNA that hybridizes with approximately 90% of pancreas cDNA therefore has a total complexity of about 2×10^4 nucleotides. These RNA sequences comprise about 2% of the total pancreas RNA, as judged by the relative kinetics of the poly(A+) and total RNA hybridizations (Figure 5). This is an acceptable approximation of mRNA concentrations. It seems probable that the pancreas poly(A+) RNA that hybridizes with pancreas cDNA codes primarily for pancreas-specific secretory proteins, for the following reasons: (1) mRNAs coding for the synthesis of the pancreatic exocrine proteins bind to oligo(dT) cellulose (see above) are therefore polyadenylated and can be copied by RNA-dependent

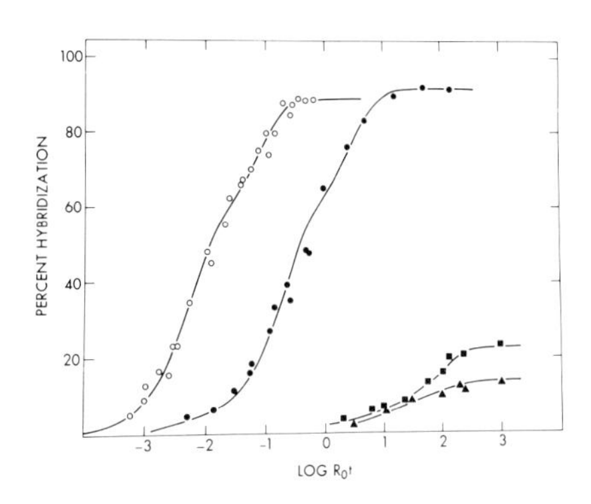

FIG. 5. Hybridization of adult rat pancreas cDNA with rat tissue RNAs. cDNA was synthesized from adult rat pancreas poly(A+) RNA isolated by oligo(dT) cellulose chromatography. Hybridizations were performed at 68°C in 20 mM Tris-HCl, 0.3 M NaCl, 1 mM EDTA, 0.5% SDS, pH 7.4. The extent of hybridization was measured after S1 nuclease digestion of unhybridized nucleic acid. 2.4% of the cDNA is resistant to S1 nuclease in the absence of hybridization; this background is subtracted from each point. Hybridization with pancreas poly(A+) RNA, ○——○ (the two component curve was fit to the data points by computer analysis); hybridization with total adult rat pancreas RNA, ●——● (the curve fit to the open circles has been shifted on the abscissa to fit these data points); hybridization with total liver RNA, ■——■; total brain RNA, ▲——▲.

DNA polymerase. (2) The pancreatic secretory proteins comprise 50 to 90% of the total pancreatic protein (Kemp *et al.* 1972); thus their mRNAs should be most abundant. (3) About 80% of the pancreas cDNA does not hybridize with RNA isolated from other rat tissues or a rat cell line (HTC) at high R_0t values. This is demonstrated for liver and brain RNA in Figure 5. Most of the pancreas cDNA is, therefore, tissue specific, within the limitations of the analysis. Consistent with these data, significant amounts of pancreatic secretory proteins (for example, carboxypeptidase A and lipase) are not found in nonpancreatic tissues (Sanders and Rutter 1974, Bradshaw and Rutter 1972). Pancreatic amylase has a different electrophoretic mobility and different antigenic properties from both parotid and liver amylases and a different amino acid composition and peptide map from the parotid enzyme (Malacinski and Rutter 1969, Sanders and Rutter 1972, Takeuchi *et al.* 1975a,b, Messer and Dean 1975). (4) The abundant pancreas polyadenylated RNA, which has a complexity of about 2×10^4 nucleotides, can code for about 15 average size mRNA sequences. As demonstrated above, rat pancreas poly(A+) RNA codes for a similar number of major proteins in the reticulocyte lysate system. (5) The low sequence complexity of the pancreas polyadenylated RNA preparations is not an artifact of the isolation procedure, since HTC cell poly(A+) RNA, isolated and copied into cDNA by the same techniques, has a sequence complexity of approximately 7×10^6 nucleotides. Furthermore, this result is not unique to the rat pancreas, since hybridization of dog pancreas cDNA with dog poly(A+) RNA exhibits similar kinetics.

The cellular specificity of pancreas cDNA is illustrated by its hybridization in situ with cellular RNA in fixed tissue sections of the pancreas. As shown

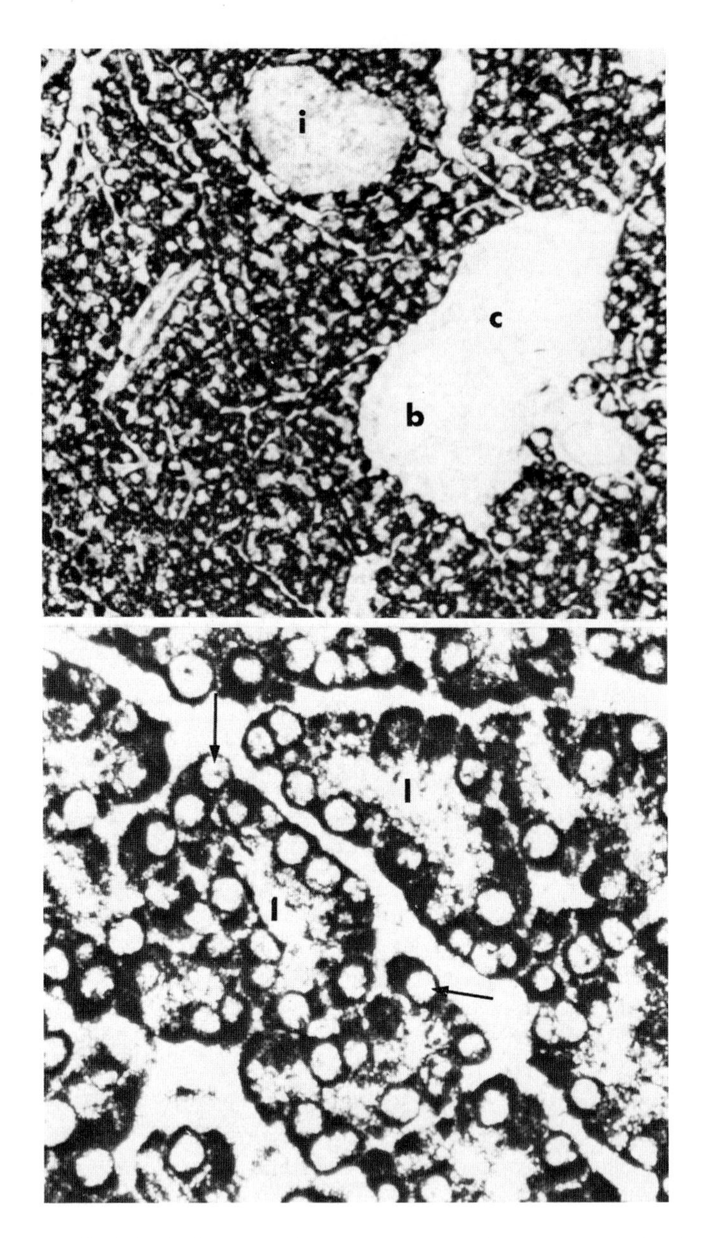

FIG. 6, top. In situ hybridization of adult pancreas tissue with pancreas cDNA. Frozen adult rat pancreas was sectioned in a cryostat at −10°C, collected on glass slides coated with chrome-alum (0.1%)-gelation (1%), and hybridized with 200,000 cpm of pancreas cDNA, as described by Harrison *et al.* (1973). The slide was dried, prepared for autoradiography, and exposed for two weeks. The acinar tissue is heavily labeled. The lumen of a blood vessel (b), endocrine cells of the islets of Langerhans (i), and connective tissue cells (c) are relatively unlabeled. (Magnification, ×100.)

FIG. 7, bottom. A portion of Figure 6 at higher magnification. The basal and perinuclear regions of the acinar cells are heavily labeled. Nuclei (arrows), cell apices, and the lumens of acini (l) are relatively unlabeled. (Magnification, ×550.)

in Figure 6, the acinar tissue is heavily labeled; whereas the lumen of a blood vessel (b), endocrine cells of the islets of Langerhans (i), and connective tissue cells (c) are weakly labeled. Examination of the autoradiograph at higher magnification (Figure 7) shows that the basal and perinuclear regions of the cells (which contain the majority of the rough endoplasmic reticulum) are very heavily labeled; whereas the nuclei (arrows), cell apices, and lumens of the acini (1) are poorly labeled. Thus, RNAs complementary to the cDNA are primarily localized in the regions of the acinar cells actively synthesizing secretory proteins.

QUALITATIVE AND QUANTITATIVE CHANGES IN PANCREATIC mRNAs DURING EMBRYONIC DEVELOPMENT

Pancreas cDNA was hybridized with RNAs isolated from fetal rat pancreases of increasing gestational age, in order to determine the changes in levels of specific pancreatic transcripts during development. The results are depicted in Figure 8. There is a gradual shift of the embryonic hybridization curves toward that of the adult, as development proceeds. Thus, the concentrations of RNA sequences complementary to pancreas cDNA increase as a function of gestational age. The complex shapes of the hybridization curves, however, preclude a precise calculation of changes in RNA concentration for a particular component. Nevertheless, the relative positions on the R_0t scale of the 14-day and adult curves imply about a 500-fold increase in concentration of some of the transcripts during the developmental period. This result is consistent with our previous calculations of the increase in the rate of synthesis of exocrine proteins such as amylase and chymotrypsinogen during development (Sanders and Rutter

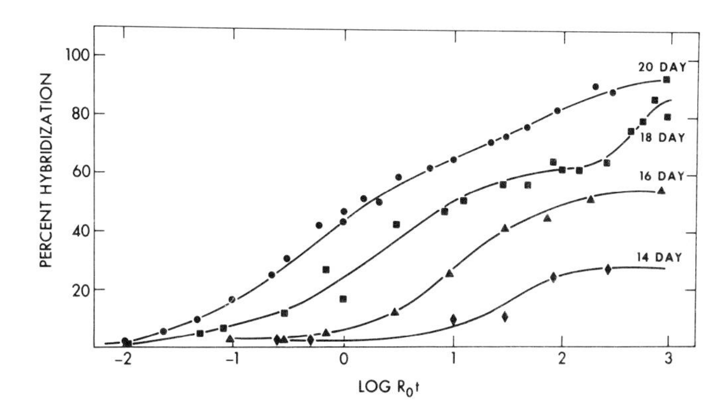

FIG. 8. Hybridization of adult rat pancreas cDNA with embryonic pancreas RNAs. Total RNA was extracted from embryonic pancreas of a given gestational age and hybridized with pancreas cDNA as in Figure 5. Hybridization with 20-day embryonic pancreas RNA, ●——●; 18-day RNA, ■——■; 16-day RNA, ▲——▲; 14-day RNA, ◆——◆.

1974). The hybridization curves in Figure 8 assume different shapes. This indicates that the transcripts being assayed accumulate at different relative rates during development. This result is also consistent with the noncoordinate increase in synthetic rates of the exocrine proteins.

Not all the cDNA is protected by 16- and 14-day RNA preparations at the R_0t values used in these studies. This implies that some RNA sequences are present at very low concentrations or are not present in the cells at these stages. The size of the early pancreatic rudiments limits the amount of RNA that can be isolated practically and thus establishes the maximum R_0t value of the reactions. Nevertheless, the protodifferentiated (14-day) pancreas RNA hybridizes with 10 to 20% more pancreas cDNA at high R_0t values than do nonpancreatic RNAs. This indicates the presence of at least some pancreas-specific transcripts during this period. The distribution of the specific transcripts within cells and the time of the initiation of synthesis may be defined by in situ hybridization techniques.

The distinctive size of amylase and its relative enrichment among pancreatic proteins facilitated the isolation of its mRNA. Resolution of pancreatic poly (A+) RNA by electrophoresis in agarose under denaturing conditions (acid-urea) and subsequent translation of the eluted RNA in a reticulocyte lysate system yielded a fraction highly enriched in amylase mRNA activity. Amylase cDNA was prepared from this material and hybridized with adult and embryonic pancreas RNAs (Table 2). Comparison of the $R_0t_{1/2}$ values of the reactions suggests strong qualitative resemblance to the developmental profile of accumulation of amylase mRNA measured by translation analysis (see below).

The changes in pancreatic mRNA populations during development were also analyzed by in vitro translation. Polypeptides synthesized in the reticulocyte

TABLE 2. *Amylase mRNA accumulation during pancreatic development*

Embryonic age (days)	Amylase accumulation in vivo*	Relative amylase mRNA levels (% 20 day) Translation Analysis†	Hybridization‡
14d	0.15	<1.0	—
15d	0.27	2.8	—
16d	1.8	9.4	4.4
17d	7.8	17	—
18d	33	78	54
19d	70	100	—
20d	100	100	100
Adult	74	560	630

* Data taken from Sanders and Rutter (1974).

† Obtained by translation of RNAs in a message-dependent reticulocyte lysate, immunoprecipitation with anti-amylase immunoglobulin, polyacrylamide gel electrophoresis of the immunoprecipitates, and determination of the radioactivity in the amylase band.

‡ Obtained from hybridization of amylase cDNA with total RNA. The $R_0t_{1/2}$ values of the reactions are as follows: 16-day, 8.0×10^0; 18-day, 6.5×10^{-1}; 20-day, 3.5×10^{-1}; adult, 5.5×10^{-2}.

lysate in response to adult and embryonic pancreas RNAs were resolved by electrophoresis in polyacrylamide gels containing SDS. Also, amylase immunoprecipitates were collected and resolved by SDS gel electrophoresis. Amylase mRNA levels were estimated from the radioactivity in the amylase peak. This procedure eliminates the contribution of nonspecific radioactivity in immunoprecipitates and allows precise measurements of the synthesis of even very low

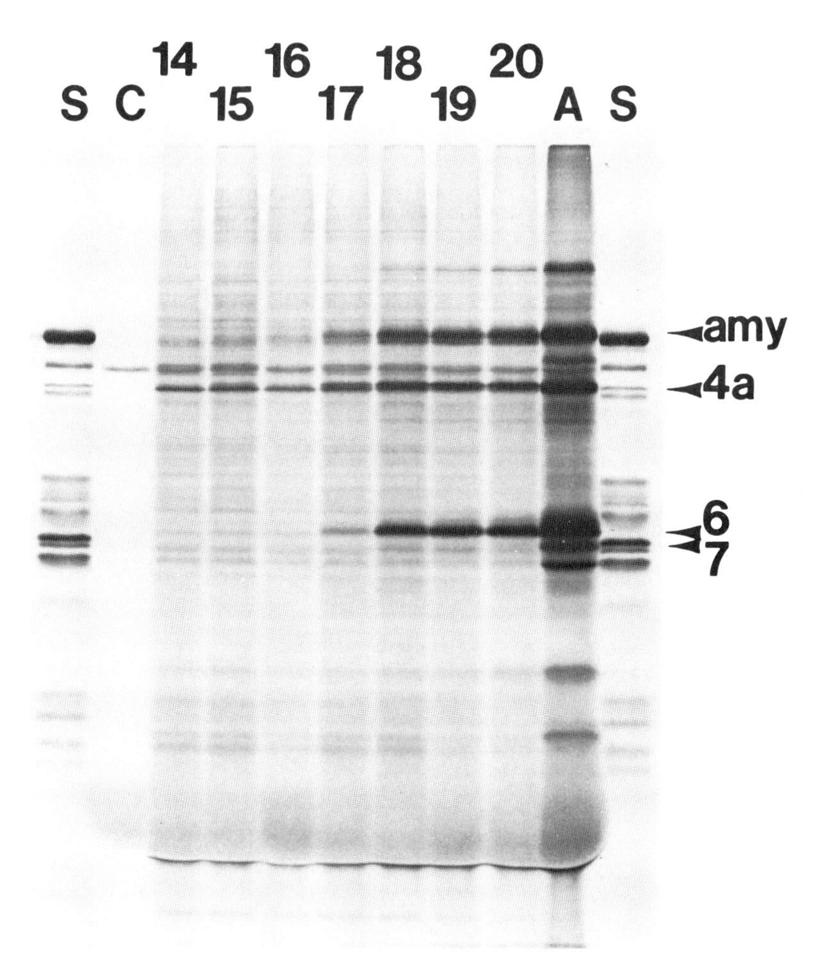

FIG. 9. Translation of guanidine thiocyanate-extracted embryonic and adult rat pancreas RNAs in RNA-dependent reticulocyte lysates. ^{35}S-methionine-labeled in vitro translation products were fractionated by SDS polyacrylamide gel electrophoresis and visualized by autoradiography. *Slot S:* In vivo labeled secretion products of adult rat pancreas; *Slot C:* Control—polypeptides synthesized by the mRNA-dependent reticulocyte lysate in the absence of exogenous RNA; *Slots 14–20:* Polypeptides synthesized in response to RNAs extracted from embryonic pancreases on the day of development indicated by the slot number; *Slot A:* Polypeptides synthesized in response to adult rat pancreas RNA. The arrows refer to polypeptide bands in slots 14–20 and slot A, which were subjected to densitometric analysis as described in Figure 10.

levels of amylase. The results are reported in Table 2 as percent of the 20-day (differentiated state) level of amylase mRNA. Amylase mRNA activity was less than 0.2% of adult levels in 14-day pancreatic rudiments. Thus, amylase mRNA levels increased at least 500-fold during development to adult levels. The levels of amylase mRNA assayed in the translational system are in good agreement with the data obtained by hybridization with amylase cDNA (Table 2). The collective data indicate that there is a dramatic increase in the steady-state concentration of amylase mRNA corresponding with or slightly preceding the increased synthesis of amylase. The experimental results suggest, therefore, that the synthesis of amylase in this period is primarily regulated at the transcriptional level.

Changes in the levels of other mRNAs can be inferred from the polypeptide profiles of the translation system, as shown in Figure 9. The similarity in the fine structure of the background is evident throughout the whole range of the analysis and suggests that a class of pancreatic mRNAs is typical of all develop-

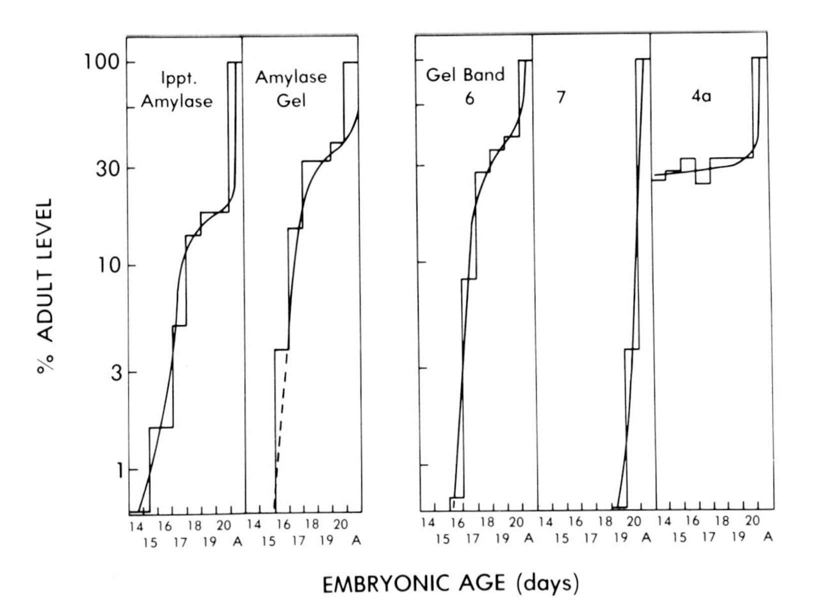

FIG. 10. Characteristic patterns of mRNA accumulation during development of the rat pancreas. mRNA accumulation patterns were derived by densitometric analysis of the bands indicated by arrows in Figure 9, slots 14–20, and slot A. Percent of adult mRNA levels were obtained by scanning identical bands in slots 14–20 and comparing their relative intensities to the bands in slot A. The first two panels of this figure demonstrate the validity of this procedure by comparing relative amylase mRNA levels estimated by immunoprecipitation of amylase synthesized in translation assays with amylase mRNA levels estimated by densitometric scanning. The last three panels illustrate three distinct patterns of mRNA accumulation: Band 6 mRNA is absent at 14 days (the protodifferentiated state) and accumulates throughout development; Band 7 mRNA accumulates only later in development, particularly in the adult stage; Band 4a mRNA is present in the protodifferentiated state and remains relatively constant throughout development.

mental stages, including the protodifferentiated state. Imposed on these minor bands are 15 to 20 major bands, especially evident in the later stages of embryonic development and in the adult. The increasing enrichment of the mRNAs coding for these polypeptides is evident from the translation of the RNAs isolated from 17- to 20-day fetal pancreases. This agrees qualitatively with previous measurements of specific enzymes and with hybridization analysis. Translation analysis does not have the sensitivity of these other methods; thus, the lower range of the accumulation profiles is not evident. Some of the bands characteristic of the adult are of relatively strong intensity, even in the early stages, and appear to remain during the entire course of development. Although the mobilities of these bands correspond closely to those of putative secretory proteins, it is quite possible that they represent other proteins of similar molecular weight. More discriminating resolution techniques, such as analysis on two-dimensional polyacrylamide gels, should distinguish between these possibilities, and, in addition, increase the resolution of the method to the lower levels of mRNA enrichment. An unexpected finding is the qualitative difference in the polypeptides produced from 20-day mRNA as compared with adult mRNA. There are clearly some major components in the adult that are not present at significant levels in the differentiated embryonic pancreas. Since there has never been any indication from measurements of enzyme activity of qualitative changes in the neonatal period, some of the differences may be due to the accumulation of isozymes.

Three characteristic patterns of messenger RNA accumulation discussed above are represented in Figure 10. The data were obtained by densitometric analysis of the bands in Figure 9. A comparison of the results for amylase obtained with this approach with those obtained by immunoprecipitation demonstrates the validity of this procedure (Figure 10).

CONCLUSIONS

The present studies support the existence of three differentiative transitions of the exocrine pancreas. The primary transition initiates pancreatic morphogenesis and specific gene expression. The secondary transition implements overt cytodifferentiation with a dramatic increase in the synthesis of pancreas-specific gene products. The tertiary transition selectively modulates gene expression in the mature organ. Each developmental state is characterized by a distinctive pattern of gene expression apparently unique to the pancreas. Our studies suggest that the major pancreatic messenger RNAs shift qualitatively and quantitatively during each transition. For example, the noncoordinate pattern of gene expression in the secondary transition, inferred from our previous characterization of enzymatic activities, is evident at the mRNA level. The present investigation considers steady-state levels and does not reveal the mechanism of accumulation (mRNA synthesis versus degradation). However, it is clear that the control of the secondary transition must involve more than a single regulatory locus. A more thorough study of the tertiary transition now seems feasible, for example, the determination of the role of glucocorticoids and other regulatory molecules during late embryonic development and in the adult. Clearly, more discriminating studies (for example, with two-dimensional polacrylamide gels and cDNAs synthesized from purified mRNAs) are needed to fully characterize the domains of gene expression within each differentiated state.

By methods similar to those employed here it should be possible to define the pattern of gene expression following the primary transition. Using the in situ hybridization procedure described, we can now perceive the range of transition of individual cells in the population and thus identify the degree of heterogeneity and possible precursor cells. Of course, these studies should be extended to include endocrine and duct cells. With a more comprehensive analysis of gene expression, it may be possible to distinguish oncogenic target cells. If neoplasia involves developmentally primitive populations, this should be apparent from the range of transcription in the transformed cells and perhaps through their residual developmental potential.

ACKNOWLEDGMENTS

This research was supported by grants from the National Science Foundation (BMS72-02222), Juvenile Diabetes Foundation, National Foundation-March of Dimes, and a National Institutes of Health Genetic Center Grant (GM19527). A.E.P. is a Helen Hay Whitney Foundation Postdoctoral Fellow; R.J.M. is an American Cancer Society Postdoctoral Fellow (PF992); J.D.H. is an NIH Postdoctoral Fellow (GM05374); J.M.C. is an NIH Postdoctoral Fellow (GM05385); and R.L.P. is a recipient of an NIH Career Development Award. We would like to thank Jennifer Meek for assistance with embryo dissections, G. Swift for computer analysis of cDNA hybridization data, William Nikovits for preparing tissue sections for in situ hybridization, and Kate Harps for preparation of the manuscript.

REFERENCES

Berns, A., P. Janson, and H. Bloemendal. 1974. The separation of α- and β-rabbit globin mRNA by polyacrylamide gel electrophoresis. FEBS Lett. 47:343–347.

Bishop, J. O., J. G. Morton, M. Rosbash, and M. Richardson. 1974. Three abundance classes in HeLa cell messenger RNA. Nature 250:199–204.

Blobel, G., and B. Dobberstein. 1975a. Transfer of proteins across membranes. I. Presence of proteolytically processed and unprocessed nascent immunoglobulin light chains on membrane-bound ribosomes of murine myeloma. J. Cell Biol. 67:835–851.

Blobel, G., and B. Dobberstein. 1975b. Transfer of proteins across membranes. II. Reconstitution of functional rough microsomes from heterologous components. J. Cell Biol. 67:852–862.

Blobel, G., and D. Sabatini. 1971. Ribosome-membrane interaction in eukaryotic cells, *in* Biomembranes, L. A. Manson, ed., Vol. 2. Plenum Press, New York, pp. 193–195.

Boime, I., S. Boguslawski, and J. Caine. 1975. The translation of a human placental lactogen mRNA fraction in heterologous cell-free systems: The synthesis of a possible precursor. Biochem. Biophys. Res. Commun. 62:103–109.

Bradshaw, W. S., and W. J. Rutter. 1972. Multiple pancreatic lipases. Tissue distribution and pattern of accumulation during embryological development. Biochemistry 11:1517–1528.

Breillatt, J., and S. R. Dickman. 1969. Detachment of ribosomes from dog pancreas polysomes without concurrent polypeptide synthesis. Biochim. Biophys. Acta 195:531–548.

Brown, J. R., and B. S. Hartley. 1966. Location of disulphide bridges by diagonal paper electrophoresis. The disulphide bridges of bovine chymotrypsinogen A. Biochem. J. 101:214–228.

Churg, A., and W. R. Richter. 1972. Histochemical distribution of carbonic anhydrase after ligation of the pancreatic duct. Am. J. Pathol. 68:23–30.

Cowan, J. J., and C. Milstein. 1973. The translation in vitro of mRNA for immunoglobin heavy chains. Eur. J. Biochem. 36:1–7.

Devillers-Thiery, A., T. Kindt, G. Scheele, and G. Blobel. 1975. Homology in amino-terminal sequence of precursors to pancreatic secretory proteins. Proc. Natl. Acad. Sci. USA 72:5016–5020.

Filosa, S., R. Pictet, and W. J. Rutter. 1975. Positive control of cyclic AMP on mesencymal factor controlled DNA synthesis in embryonic pancreas. Nature 257:702–705.

Githens, S., R. Pictet, P. Phelps, and W. J. Rutter. 1976. 5-bromodeoxyuridine may alter the differentiative program of the embryonic pancreas. J. Cell Biol. 71:341–356.

Gomori, G. 1941. The distribution of phosphatase in normal organs and tissues. J. Cell. Comp. Physiol. 17:71–84.

Grobstein, C. 1967. Mechanisms of organogenetic tissue interaction. Natl. Cancer Inst. Monogr. 26:279–282.

Harrison, P. R., D. Conkie, J. Paul, and K. Jones. 1973. Localisation of cellular globin messenger RNA by in situ hybridisation to complementary DNA. FEBS Lett. 32:109–112.

Hubert, C., and L. Cedard. 1975. Isolation and in vitro translation of human placental lactogen messenger RNA from human term placenta. Nucleic Acids Research 2:1903–1910.

Jamieson, J. D., and G. E. Palade. 1967. Intracellular transport of secretory proteins in the pancreatic exocrine cell. II. Transport to condensing vacuoles and zymogen granules. J. Cell Biol. 34:597–615.

Kemp, J. D., B. T. Walther, and W. J. Rutter. 1972. Protein synthesis during the secondary developmental transition of the embryonic rat pancreas. J. Biol. Chem. 247:3941–3952.

Kemper, B., J. F. Habener, R. C. Mulligan, J. T. Potts, Jr., and A. Rich. 1974. Pre-proparathyroid hormone: A direct translation product of parathyroid messenger RNA. Proc. Natl. Acad. Sci. USA 71:3731–3735.

Levine, S., R. L. Pictet, and W. J. Rutter. 1973. Control of cell proliferation and cytodifferentiation by a factor reacting with the cell surface. Nature New Biol. 246:49–52.

Longacre, S. S., and W. J. Rutter. 1977. Isolation of chicken hemoglobin mRNA and synthesis of complementary DNA. J. Biol. Chem. 252:2742–2752.

Malacinski, G. M., and W. J. Rutter. 1969. Multiple forms of α-amylase from the rabbit. Biochemistry 8:4382–4390.

Messer, M., and R. T. Dean. 1975. Immunochemical relationship between α-amylases of rat liver, serum, pancreas and parotid gland. Biochem. J. 151:17–22.

Mikes, O., V. Holeysousky, V. Tomasek, and F. Sorm. 1966. Covalent structure of bovine trypsinogen. The position of the remaining amides. Biochem. Biophys. Res. Commun. 24:346–352.

Milstein, C., G. G. Brownlee, E. M. Cartwright, J. M. Jarvis, and N. J. Proudfoot. 1974. Sequence analysis of immunoglobulin light chain messenger RNA. Nature 252:354–359.

Milstein, C., G. G. Brownlee, T. M. Harrison, and M. B. Matthews. 1972. A possible precursor of immunoglobulin light chains. Nature New Biol. 239:117–120.

Palmiter, R. D. 1974. Differential rates of initiation on conalbumin and ovalbumin messenger ribonucleic acid in reticulocyte lysates. J. Biol. Chem. 249:6779–6787.

Permutt, M. A., and D. M. Kipnis. 1972. Insulin biosynthesis: Studies of islet polyribosomes. Proc. Natl. Acad. Sci. USA 69:505–509.

Pictet, R. L., W. R. Clark, R. H. Williams, and W. J. Rutter. 1972. An ultrastructural analysis of the developing embryonic pancreas. Dev. Biol. 29:436–467.

Pictet, R., S. Filosa, P. Phelps, and W. J. Rutter. 1975a. Control of DNA synthesis in the embryonic pancreas: Interaction of the mesenchymal factor and cyclic AMP, *in* Extracellular Matrix Influences on Gene Expression, H. C. Slavkin and R. C. Greulich, eds., Academic Press, New York, pp. 531–540.

Pictet, R. L., L. Rall, M. de Gasparo, and W. J. Rutter. 1975b. Regulation of differentiation of endocrine cells during pancreatic development in vitro, *in* Early Diabetes in Early Life, R. A. Camerini-Davalos and H. S. Cole, eds., Academic Press, New York, pp. 25–39.

Proudfoot, N. J., and G. G. Brownlee. 1974. Sequence at the 3′ end of globin mRNA shows homology with immunoglobulin light chain mRNA. Nature 252:359–362.
Rall, L. B., R. L. Pictet, R. H. Williams, and W. J. Rutter. 1973. Early differentiation of glucagon-producing cells in embryonic pancreas: A possible developmental role for glucagon. Proc. Natl. Acad. Sci. USA 70:3478–3482.
Ronzio, R. A., and W. J. Rutter. 1973. Effects of a partially purified factor from chick embryos on macromolecular synthesis of embryonic pancreatic epithelia. Dev. Biol. 30:307–320.
Rutter, W. J., W. D. Ball, W. S. Bradshaw, W. R. Clark, and T. G. Sanders. 1967. Levels of regulation in cytodifferentiation, *in* Experimental Biology and Medicine, S. Karger, ed., Vol. 1. Basel, New York, pp. 110–124.
Rutter, W. J., J. D. Kemp, W. S. Bradshaw, W. R. Clark, R. A. Ronzio, and T. G. Sanders. 1968. Regulation of specific protein synthesis in cytodifferentiation. J. Cell. Physiol. 72:1–18.
Rutter, W. J., R. L. Pictet, and P. W. Morris. 1973. Towards molecular mechanisms of developmental processes. Ann. Rev. Biochem. 42:601–646.
Rutter, W. J., R. Pictet, and L. Rall. 1975. Modulations of differentiative competence during embryological development, *in* Normal and Pathological Development of Energy Metabolism, F. A. Hommes and C. J. Van den Berg, eds., Academic Press, New York, pp. 47–53.
Rutter, W. J., and C. S. Weber. 1965. Specific proteins in cytodifferentiation, *in* Developmental and Metabolic Control Mechanisms in Neoplasia (The University of Texas System Cancer Center 19th Annual Symposium on Fundamental Cancer Research, 1965), Williams and Wilkins Co., Baltimore, pp. 195–218.
Sanders, T. G., and W. J. Rutter. 1972. Molecular properties of rat pancreatic and parotid α-amylase. Biochemistry 11:130–136.
Sanders, T. G., and W. J. Rutter. 1974. The developmental regulation amylolytic and proteolytic enzymes in the embryonic rat pancreas. J. Biol. Chem. 249:3500–3509.
Scheele, G. A. 1975. Two-dimensional gel analysis of soluble proteins. Characterization of guinea pig exocrine pancreatic proteins. J. Biol. Chem. 250:5375–5385.
Schmeckpeper, B. J., S. Cory, and J. M. Adams. 1974. Translation of immunoglobulin mRNAs in a wheat germ cell-free system. Molecular Biology Reports 1:355–363.
Segrest, J. P., R. L. Jackson, E. P. Andrews, and V. T. Marchesi. 1971. Human erythrocyte membrane glycoprotein: A re-evaluation of the molecular weight as determined by SDS polyacrylamide gel electrophoresis. Biochem. Biophys. Res. Commun. 44:390–395.
Takeuchi, T., M. Mura, R. Sasaki, T. Matsushima, and T. Sugimura. 1975a. Comparative studies of the electrophoretic mobility and immunogenicity of pancreatic and parotid amylases of the rat. Biochim. Biophys. Acta 403:456–460.
Takeuchi, T., T. Matsushima, and T. Sugimura. 1975b. Electrophoretic and immunological properties of liver α-amylase of well-fed and fasted rats. Biochim. Biophys. Acta 403:122–130.
Verma, I. M., R. A. Firtel, H. F. Lodish, and D. Baltimore. 1974. Synthesis of DNA complementary to cellular slime mold messenger RNA by reverse transcriptase. Biochemistry 13:3917–3922.
Walther, B. T., R. L. Pictet, J. D. David, and W. J. Rutter. 1974. On the mechanism of 5-bromodeoxyuridine inhibition of exocrine pancreas differentiation. J. Biol. Chem. 249:1953–1964.
Wessells, N. K. 1964. DNA synthesis, mitosis and differentiation in pancreatic acinar cells in vitro. J. Cell Biol. 20:415–433.
Williamson, R., and P. Wellauer. 1973. The molecular weight of mouse globin messenger RNA. Carnegie Institution Year Book 72:24–25.
Wilt, F., and M. Anderson. 1972. The action of 5-bromodeoxyuridine on differentiation. Dev. Biol. 28:443–447.
Wolff, E. 1968. Specific interactions between tissues during organogenesis. Curr. Top. Dev. Biol. 3:65–94.
Woo, S. L. C., J. M. Rosen, C. K. Liarakos, U. C. Choi, H. Busch, A. R. Meens, B. W. O'Malley, and D. L. Robberson. 1975. Physical and chemical characterization of purified ovalbumin messenger RNA. J. Biol. Chem. 250:7027–7039.
Yip, C. C., C. L. Hew, and H. Hsu. 1975. Translation of messenger ribonucleic acid from isolated pancreatic islets and human insulinomas. Proc. Natl. Acad. Sci. USA 72:4777–4779.
Zendzian, E. N., and E. A. Barnard. 1967. Distributions of pancreatic ribonuclease, chymotrypsin, and trypsin in vertebrates. Arch. Biochem. Biophys. 122:699–713.

Cell Differentiation and Neoplasia, edited by
Grady F. Saunders. Raven Press, New York
© 1978.

Genetics of Regulation in Cultured Mammalian Cells

Michael J. Siciliano,* Mary R. Bordelon,† and
Ronald M. Humphrey*

** Departments of Biology and Physics, The University of Texas System Cancer Center M. D. Anderson Hospital and Tumor Institute, Houston, Texas 77030; †Department of Cell Biology, Baylor College of Medicine, Houston, Texas 77030*

There are two levels to our study of the genetics of regulation in mammalian somatic cells in culture. First, as in any genetic study, we need to establish or find cell lines that are inherently variant from each other for the trait under study, i.e., gene regulation. Secondly, we wish to combine the two variant genomes so that we can observe the interaction and segregation of gene regulatory factors.

In this presentation we shall first describe the types of biochemical genetic markers that may be used in approaching the problem of gene control in cells. We shall then describe our efforts to generate genetic variants for these markers in a cloned mammalian cell line. Finally, we shall discuss the inheritance and interaction of regulatory factors in somatic cell hybridization studies.

MARKERS FOR STUDYING GENE REGULATION IN CELLS IN CULTURE

The types of regulatory events that one can look for in somatic cells in culture are somewhat limited. We focus our attention on those events that modify the expression of enzymes responsible for cellular metabolism—the so-called "housekeeping enzymes." Considerable evidence exists that many of these enzymes are under tissue-specific regulatory control. This is most obviously seen with enzymes that are coded-for by more than one genetic locus. For most multiple locus-coded enzymes the product of one of the loci predominates in a specific tissue. This is readily observable in animal as well as human biopsy or autopsy material when the products of the different loci are separable by electrophoresis. The classic example of this phenomenon, as well as its observation after electrophoresis, has been demonstrated for the enzyme lactate dehydrogenase (LDH-EC 1.1.1.27).

Markert and Møller (1959) applied a histochemical stain for LDH onto starch gel slices after electrophoresis of homogenates from several different mammalian tissues sampled at different stages of development. Examination of the "zymograms" (stained gel slices) revealed multiple molecular, tissue, and stage-specific

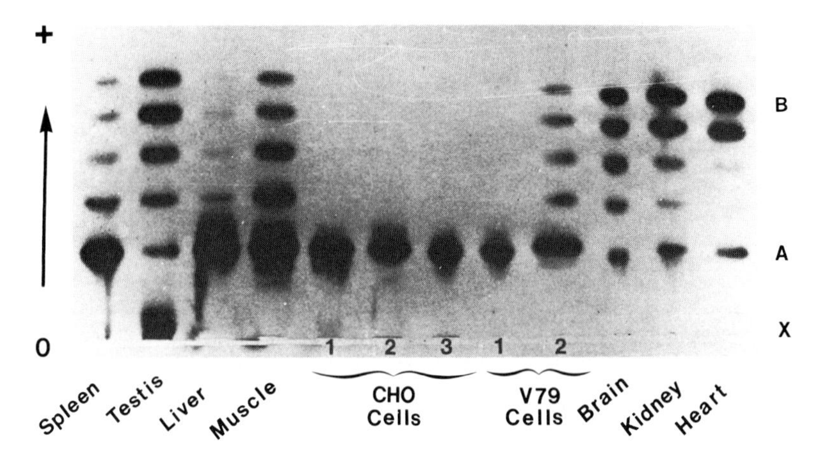

FIG. 1. LDH zymogram. The origin (0) and direction of electrophoresis toward the anode (+) are shown in the left-side margin. Homogenates from Chinese hamster cell lines (CHO and V79) were run along side various tissues from freshly sacrificed Chinese hamsters. Migration of the homopolymers produced by the three LDH loci (A, B, and X) are indicated in the right-hand margin. Bands intermediate in migration between the A and B homopolymers (AAAA and BBBB) are the heteropolymers formed (AAAB, AABB, and ABBB) when both A and B subunits are produced in the same cells. As can be seen, only testis produces LDH-X. In other tissues, either LDH-A or LDH-B is the major form of the enzyme produced. Both cell lines produce LDH-A only (the additional faster bands in the second V79 sample represent leakage from the brain sample).

forms of LDH. Subsequent work studying genetic variants (Shaw and Barto 1963, Nance *et al.* 1963) and additional tissue-specific forms (Blanco and Zinkham 1963) of LDH revealed that there are at least three genetic loci for this enzyme which are under tissue-specific regulatory control (for summary see Markert 1968). An example of the products of these loci differentially expressed in the tissues of Chinese hamsters is shown in Figure 1.

Figure 1 also demonstrates the appropriateness of using these enzymes for studying the expression of genes in cultured cells. The Chinese hamster ovary cell line (CHO: originally derived from ovarian epithelium of the Chinese hamster by Tijo and Puck in 1958) expresses only the products of the LDH-A locus. Since Deaven and Peterson (1973) indicated that CHO cells, although slightly hypodiploid, contain at least one complete set of chromosomes, the genes coding for the unexpressed products (*LDH-B* and *LDH-X*) exist in the cells in a repressed state. Therefore, expression of enzyme gene products in somatic cells appears to be a useful model for studying genetic regulation and the zymogram technique an appropriate technology for such a study.

INDUCTION OF ENZYME GENE VARIANTS

Without inherited variation, there can be no genetics. One could sample the great and diverse array of available mammalian cell material to find lines that are different with respect to the expression of particular enzyme loci and use these in subsequent genetic experiments. While at present this is the most practical approach to the problem, the fact that two such cell lines would be variant at many more loci than just the one under study makes it difficult to evaluate the results of such an effort.

Induction of enzyme variants at specific loci in a cloned cell line would, on the other hand, provide us with variant isogenic lines. Several years ago we began development of a system in which we could induce, identify, and isolate CHO cells that were variant for the products of enzyme loci, as identified by the zymogram approach. Our purposes were to provide not only a somatic cell system in which mutational change could be identified with a specific modified protein but also to establish cell lines with genetic markers for somatic cell genetic studies. In addition, and relevent to the subject of this Symposium, we have found clones that are mutant in gene expression.

Our procedures flow as follows (for details see Siciliano *et al.* 1975):

1) CHO cells are exposed to ultraviolet light (UV) or nitrosoguanidine (MNNG) such that survivals in individual experiments range from 50% to 1% of the plating efficiency of untreated controls.
2) Cells are allowed to divide twice to allow segregation of mutant strands, and then they are cloned in microtest plate wells.
3) After one week, cells in wells containing a single clone are picked and grown up to 50×10^6 cells.
4) 2 to 3×10^6 cells of a clone are kept in culture while the remainder of the clone is washed and homogenized in two volumes of homogenizing medium.
5) Homogenates are cleared by centrifugation and subjected to vertical starch gel electrophoresis.
6) After electrophoresis, gels are sliced and slices stained histochemically for the products of ~40 enzyme loci. Specific procedures for all steps in homogenization, electrophoresis, and histochemical staining are contained in Siciliano and Shaw (1976).
7) Any clone shown to be electrophoretically variant for any enzyme is then subcloned from the 2 to 3×10^6 cells held back. Persistance of the variation in the subclones shows it to be inherited and, therefore, a mutation.

Two types of variants may be expected—electrophoretic shift and variations in isozyme activity.

Electrophoretic Shift

Electrophoretic shift mutants, as demonstrated in Figure 2, are exactly what we expect of a mutation in a structural gene. Such as mutation would cause an amino acid substitution, which results in a net charge difference in the coded protein. These are of interest in mutation frequency studies as well as clonal markers but offer little in terms of the direct study of genetic regulation.

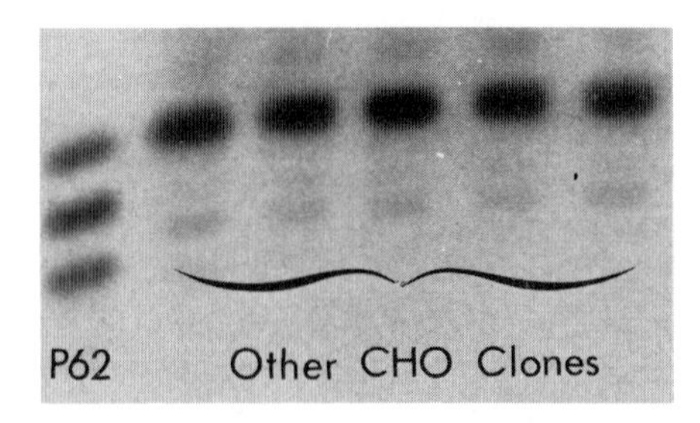

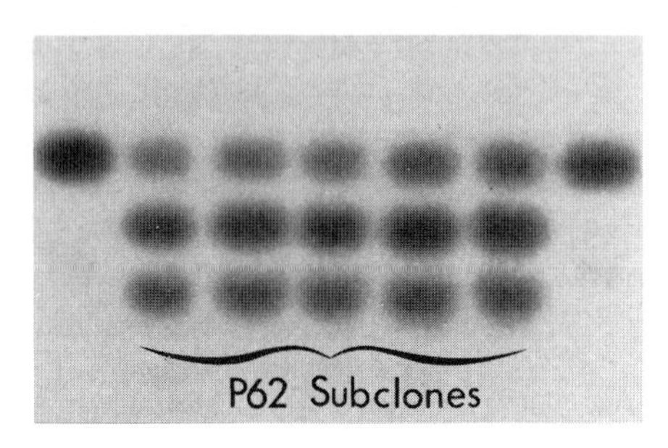

FIG. 2. Zymograms showing a variant (clone P62) at the *malate dehydrogenase-1 (MDH-1)* locus after mutagenesis, cloning, and electrophoresis. Upper panel shows the variant when first detected. The three-banded pattern for a heterozygote is consistent with the known dimer structure of MDH (Davidson and Cortner 1967). The lower panel demonstrates the variant as a true mutant by showing the inheritance of the variant pattern in subclones.

Variations in Isozyme Activity

Two types of isozyme activity mutants have been seen. The first type, exemplified in Figure 3, is one in which a form of the enzyme normally seen in CHO cells is no longer expressed. We are not too surprised to find such mutants surviving for enzymes coded-for by multiple loci, since the loss of one isozymic form apparently can be tolerated by a cell having other forms of the same enzyme. Whether such a variant is due to a mutation in the structural gene coding for the isozyme or in genetic material regulating the activity of the locus or its products is not immediately determinable. Somatic cell fusion studies would be helpful in resolving these possibilities. Fusion of the cells lacking the isozyme to a cell line in which it is expressed and has a different electrophoretic mobility (some other rodent or primate line) should be informative. If the hybrids express both or neither forms of the isozyme, it indicates that the mutation was possibly regulational, and one gets information on the dominance or recessiveness of the factors. If hybrids show only the isozyme of the non-CHO cells, the mutation was possibly structural.

The second type of variant in isozyme activity is the most exciting in terms of identifying regulational mutants. This is one in which forms of the enzyme normally seen in Chinese hamster tissues, but unexpressed in CHO cells, become expressed. Our best example of this is an esterase mutation reported previously (Siciliano *et al.* 1975). The mutant clone had a phenotype for esterases that

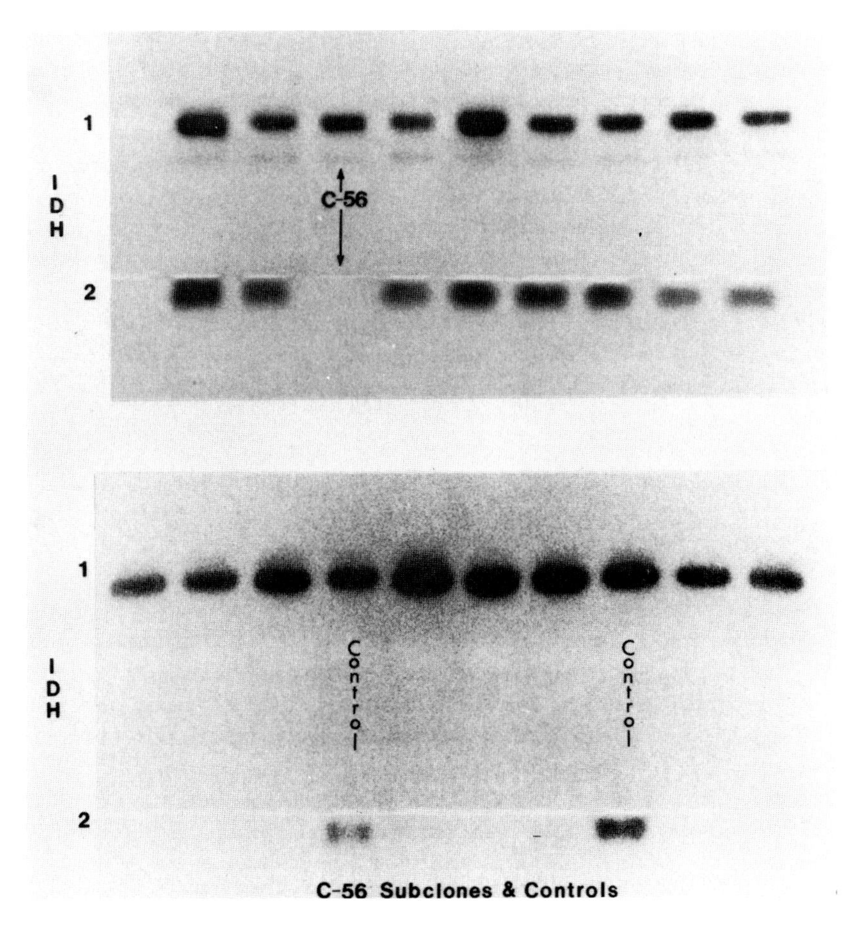

FIG. 3. Zymogram showing variant CHO clone (C-56) picked after mutagenesis, failing to express product of the *isocitrate dehydrogenase (IDH) 2* locus. This was first seen on the upper panel, which shows C-56 compared with other CHO clones exhibiting normal IDH expression. In the lower panel, subclones of C-56 are seen along side of two normal CHO controls, proving the inheritance of the variant pattern.

was completely different from all other clones, or CHO cells in general. Instead of the usual two bands of esterase activity, it has eight (Figure 4a). Suspecting that these extra bands did not represent electrophoretic variants of the two bands seen in the parental CHO cells, the supernatant from the clone was run on the same gel, along with extracts of various Chinese hamster tissues (Figure 4b). As can be seen, the extra esterase bands in this clone all have counterparts in the Chinese hamster tissue. This fact, and the finding that no atypical bands were seen for the products of the other 34 enzyme loci screened in this clone, rules out the possibility that these bands were produced by some

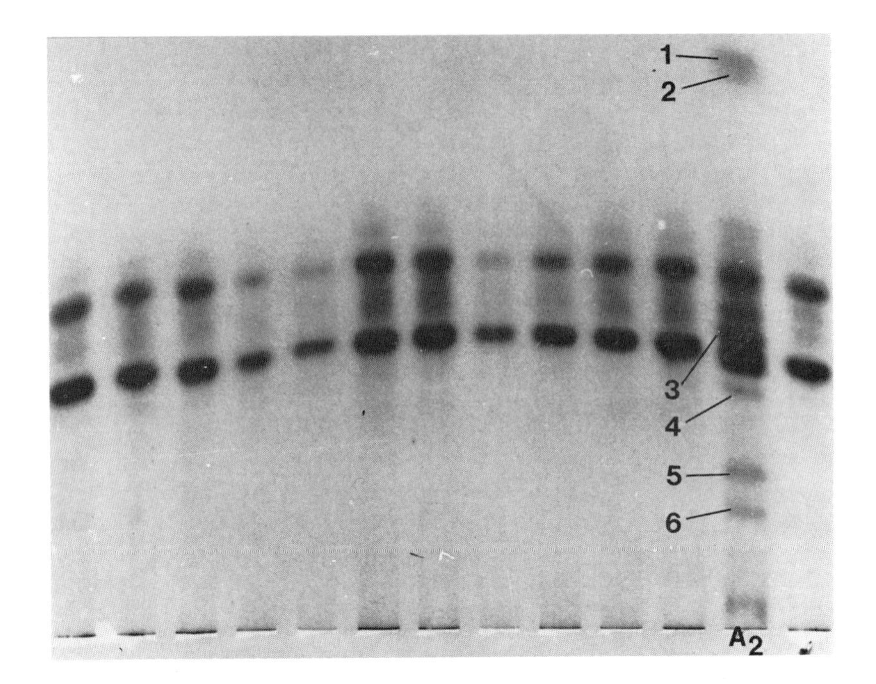
1
2
3
4
5
6
A2

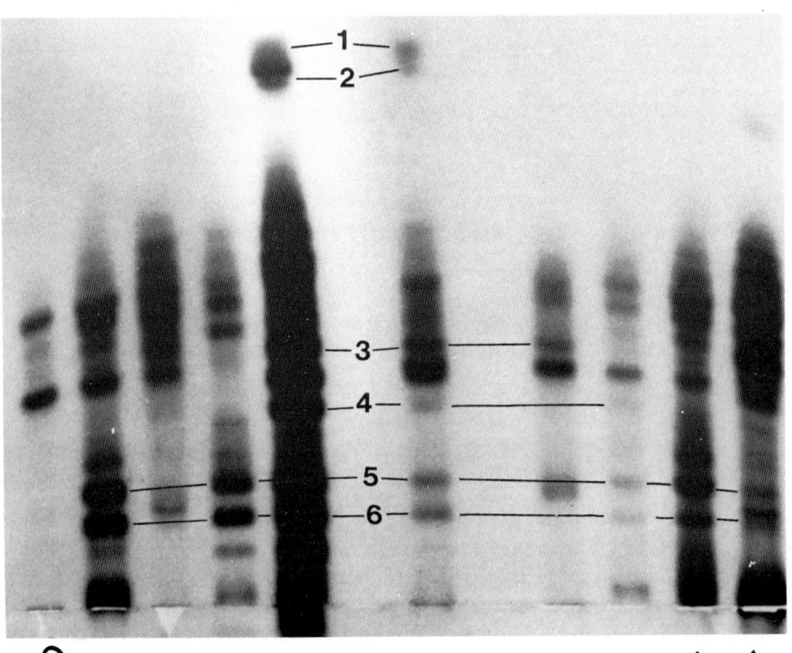
1
2
3
4
5
6
CHO
Testis
Spleen
Muscle
Liver
A2
Brain
Eye
Lung
Kidney

contaminant in the culture. The appearance of Chinese hamster gene products not normally expressed in CHO cells is suggested.

At this point in our studies, we can only speculate about the meaning of the expression of the extra bands. Of particular interest are the two closely spaced, most anodal bands. These correspond in mobility with fast-migrating bands from liver and kidney. These bands appear to occupy the same relative position on the gel and to have the same tissue specificity as *est-2* in mice, as described by Klebe *et al.* (1970). These workers, using somatic cell hybridization, have described a genetic control system for the expression of that locus in certain mouse cell lines. While no genetic data are available indicating that the extra bands in our mutant are truly products of different genetic loci, the tissue specificity of the expression of the corresponding bands in the Chinese hamster material appears to indicate so. The simultaneous re-expression of several genetic loci with shared physiological function in response to a single event seems to be a reasonable interpretation of our findings. Whether or not the fact that many esterase loci appear to be linked on the same chromosome (Petras and Biddle 1967) in mammals has any bearing on this phenomenon is something that future work in this system may elucidate.

Thus far we have screened 266 untreated control clones for a total of 10,376 enzyme loci products (the sum of all the loci over all the clones). In these, no variants were detected. Six hundred and fifty-eight mutagen-exposed clones were isolated and screened, for a total of 24,061 enzyme loci products. Twenty-four electrophoretic enzyme mutants of different types were detected. These data indicate that the mutants were, in fact, induced by the mutagen treatment and not selected from a heterogeneous population of CHO cells—a conclusion further supported by the fact that newly cloned CHO cells are used for the experiments.

So, in terms of the study of the genetics of regulation, the mutation work has

1) produced clones that may be evaluated via somatic cell hybridization procedures to determine if they arose by a structural gene or regulatory gene mutation;
2) produced clones that are most likely regulatory mutants, the genetics of which can be studied by fusion experiments;
3) indicated that regulatory events may occur in response to mutagens.

GENETICS OF REGULATION IN SOMATIC CELL HYBRIDS

While these mutant lines were being developed, we had an opportunity to study the genetics of regulation in fusion experiments between cell lines derived

←

FIG. 4. a, top panel. Esterase zymogram of CHO clones following mutagenesis. Clone A_2 can be seen to have six additional bands of esterase activity not normally seen in CHO cells. b, bottom panel. Clone A_2 (middle) run along side of normal CHO cells (far left) and homogenates from various Chinese hamster tissues. All the extra bands produced in A_2 can be seen in the different Chinese hamster tissues.

from different species. A series of hybrids were produced between human choriocarcinoma cells and established mouse fibroblast cell lines. The following specific cell lines were used:

JEG-3—a cloned human choriocarcinoma cell line established by Kohler and Bridson (1971).
HCG—a sergregant population of JEG cells carried separately through animals, then cultured.
LMTK⁻—mouse fibroblastic cell line deficient in the enzyme thymidine kinase (TK), first described by Dubbs and Kit (1964).
A9—mouse fibroblastic cell line deficient in hypoxanthine phosphoribosyl transferase (HGPRT), as reported by Littlefield (1964).

Four different fusion experiments were conducted: JEG-3 × LMTK⁻, HCG × LMTK⁻, JEG-3 × A9, and HCG × A9. Cells were fused in suspension using Sendai virus inactivated by β-propiolactone or 1,000 MW polyethylene glycol. The cells were then plated two hours later into F-12 medium with additional glucose and salts (GIBCO, F-12, modified by Bordelon). Two days later, F-12 mod. + hypoxanthine-aminopterin-thymidine (HAT) medium was used as selection against the mouse cells (Littlefield 1964). Clones were selected on the basis of morphology, removed into new dishes and grown in HAT medium.

To confirm the hybrid nature of the clones, starch gel electrophoresis, followed by histochemical staining to detect the products of 28 enzyme loci having different electrophoretic mobilities between the mouse and human forms, was conducted on the homogenates of clones surviving hybrid selection conditions. Enzyme loci screened were glucose phosphate isomerase, glucose-6-phosphate dehydrogenase, acid phosphatase *1,* esterases *A* and *D,* malate dehydrogenases *1* and *2,* adenosine deaminase, nucleoside phosphorylase, lactate dehydrogenases *A* and *B,* mannose phosphate isomerase, phosphoglucomutases *1, 2,* and *3,* isocitrate dehydrogenases *1* and *2,* tetrazolium oxidase *1,* glyceraldehyde-3-phosphate dehydrogenase, triose phosphate isomerase, glutamate oxaloacetate transaminase *1,* peptidases *A, B,* and *C,* adenylate kinase *1,* enolase, and hexoseaminidases *A* and *B.* A clone was confirmed as a hybrid if it had for at least one enzyme: the mouse form, the human form, and expected number of intermediately migrating hybrid bands consistent with the subunit structure of a multimeric enzyme (Figure 5).

Interesting results, in terms of genetic regulation, were obtained for the enzyme adenosine deaminase (ADA, EC 3.5.4.4). ADA has attracted recent attention with the discoveries of abnormal levels of this enzyme in patients with genetic diseases. Low levels are associated with severe combined immunodeficiency disease (SCID: Giblett *et al.* 1972, Parkman *et al.* 1975) and elevated ADA with hereditary hemolytic anemia (Valentine *et al.* 1977). Approximately 10% of the human population is heterozygous for ADA, as detected by starch gel electrophoresis, that is, the product of allele *ADA-1* migrates further during electrophoresis than the *ADA-2* product (Spenser *et al.* 1968). Since the enzyme is composed

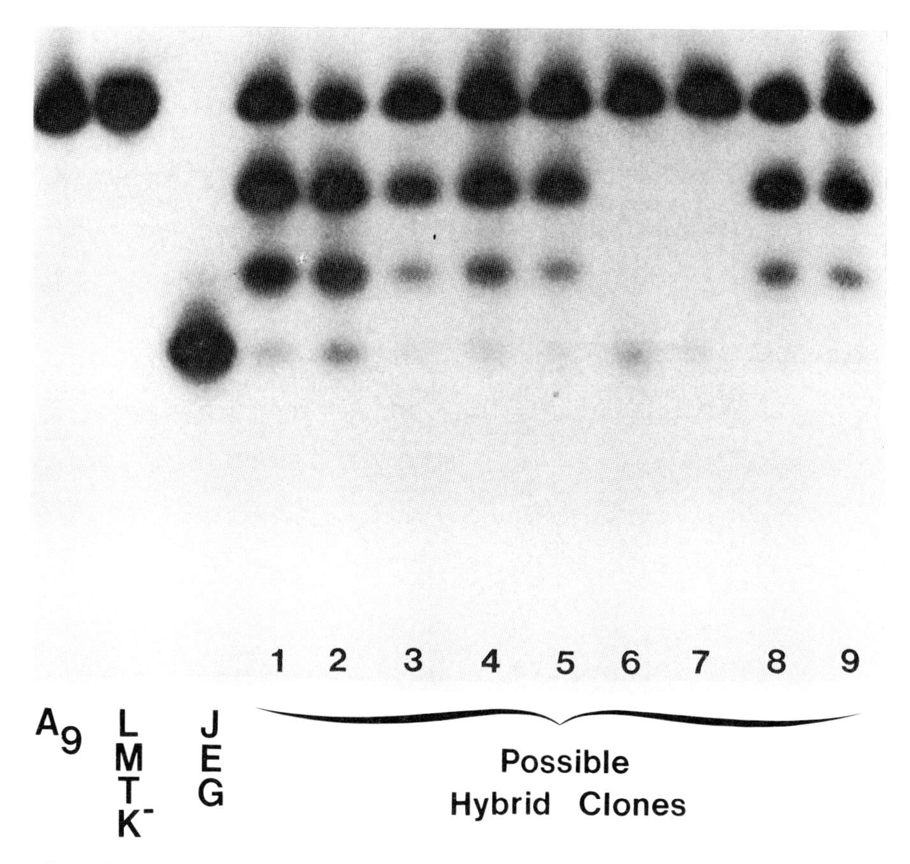

FIG. 5. Zymogram showing the forms of nucleoside phosphorylase (NP) produced in two mouse cell lines (A9 and LMTK$^-$), the human choriocarcinoma cell line (JEG), and nine putative hybrid clones produced by fusion between LMTK$^-$ and JEG. Since NP is a trimeric enzyme producing a four-banded pattern in heterozygous cells (Edwards *et al.* 1971), true hybrids should express the typical mouse band, human band, and two intermediately migrating heteropolymeric hybrid bands. In hybrids, the intensity of activity over the four-banded pattern might be expected to be skewed in the direction of the mouse form, since in hybrid clones many cells may be expected to have lost the human structural gene. By these criteria, clones 1, 2, 3, 4, 5, 8, and 9 are clearly true hybrids because in order for the heteropolymeric hybrid bands to be produced, both mouse and human NP subunits need to be produced in the same cell. Clones 6 and 7, although producing both mouse and human homopolymers, are likely merely mixed cultures and not true hybrids, since the intermediately migrating hybrid bands are not seen.

of a single subunit (monomeric), there are no bands of intermediate migration. Except for some rare types, the remainder of us are *ADA 1–1* homozygotes, producing only the faster band.

We found that the human choriocarcinoma cells used in our fusion experiments are like the SCID cells in that they do not produce ADA bands on starch gels following electrophoresis. As seen in Figure 6, the mouse cells have substan-

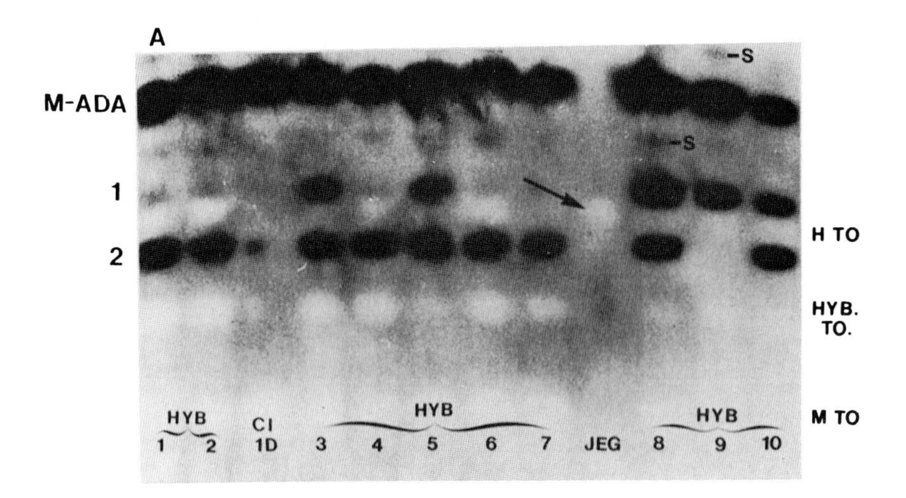

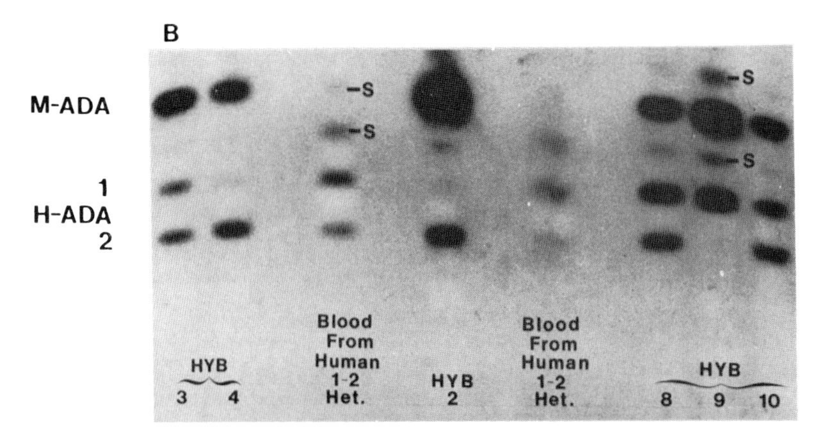

FIG. 6. A, Starch gel slice stained for ADA after electrophoresis (zymogram) of mouse LMTK$^-$-clone 1D cells (Cl 1D), human JEG-3 cells (JEG), and 10 independent hybrid clones derived from them. Direction of electrophoresis is from bottom (origin) toward the top. The dark staining bands represent ADA activity. The light bands represent tetrazolium oxidase (TO). The right margin identifies the mobility of the mouse form of TO (M-TO), human form (H-TO), and the intermediate hybrid molecule (HYB.TO.) produced only in cells in which both human and mouse subunits are made. The TO results indicate the hybrids are true hybrids. Presence of human TO in JEG-3 (arrow) indicates active human sample. This is pointed out since we see no active ADA in JEG-3. Position of mouse form of ADA (M-ADA, left-hand margin) is indicated. As expected, it is present in Cl 1D and in all hybrids. Hybrids also have one, another, or both of two additional bands labeled 1 and 2 (left-hand margin). Typical satellites or secondary bands produced off primary ADA bands (Edwards *et al.* 1971b) are indicated as "S." (The smudge of type "2" ADA and hybrid TO on the left side of the Cl 1D channel represents a small amount of leakage from the adjacent slot containing homogenate from hybrid #2.) B, ADA zymogram on which various hybrids from above are run alongside hemolysates from human *ADA 1–2* heterozygotes. Bands 1 and 2 in hybrids are shown to be identical with the primary products of human *ADA 1* and *2* (H-ADA).

tial ADA activity, with electrophoretic mobility different from any human form. After fusion, different, independently derived hybrid clones had not only mouse ADA but also the products of both human *ADA-1* and *2* alleles (Figure 6). As shown, some hybrid clones had both *1* and *2*, some only *1*, and others only *2*.

We have constructed the following working hypothesis to explain these data (Siciliano *et al.*, in press). The human cells used are virtually absent in ADA activity but have the genetic information for *ADA-1* and *2*. The locus is therefore under negative regulational control. Fusion with mouse cells having activity results in the activation of both human gene products coded-for on separate homologous chromosomes. Presence of *1–2, 1, 2,* or neither ADA allele products depends on the segregation of the structural alleles.

In addition to the observation reported above, there are additional data (Siciliano *et al.*, in press) that support this hypothesis: Radiometric assay of ADA in JEG-3 reveals the specific activity for the enzyme to be ~2.6% that of HeLa controls, and this was shown to be a level too low to be detected on gels. Human ADA turn-on was observed in 30 of 45 independently derived hybrid clones from all four fusion experiments. Segregation information gathered by examining higher passages of hybrid clones (which tend to lose human chromosomes) revealed the same human ADA phenotype as an earlier passage, or the loss of a type. A new form is never gained (Table 1). The conditions of hybridization and selection do not induce human ADA expression in choriocarcinoma cells, and actual heterokaryon formation with hybrid cells is necessary. Our human cell line was shown to be homogeneous for its low level of ADA activity and not a mixture of many cells with low activity and few cells with

TABLE 1. *Human ADA phenotypes expressed in subsequent passages of hybrid clones*

Clone	Passages sampled	Respective human ADA phenotype
112b	7 → 21	1–2 → 2
114b–3	9 → 26	1–2 → 2
114c	4 → 16	0 → 0
117c–1	4 → 7 → 21	1–2 → 1–2 → 1–2
119a	7 → 16	0 → 0
122a	8 → 11	1–2 → 1*–2
123b	2 → 5 → 9 → 23	0 → 0 → 0 → 0
124c	6 → 8 → 27 → 34	1 → 1 → 1 → 1
124c–1	7 → 23 → 26	1 → 1 → 1
130e	4 → 21	1–2 → 2
130g	9 → 14	1–2 → 1–2
134b	6 → 24	2 → 2
137a	7 → 12	0 → 0
222a–1	8 → 12	1–2 → 2
222a–2	8 → 19	1–2 → 1–2
222b	6 → 9	2* → 0

* Extremely weak activity—barely visible on gels.

high activity. Consequently, our results cannot be explained by fusion with the latter type. It should be noted, however, that in experiments such as these, in which one is monitoring for a regulational event after fusion, the purity of the population of cells used in the fusion needs to be established.

Thus, it is clear that ADA is under genetic regulation in these cultured cells. This finding offers an alternative conceptual basis to structural gene mutation for the abnormal levels of ADA associated with SCID and inherited hemolytic anemia.

Actually, the finding of widely varying ADA-specific activity in different mouse tissues and in the same tissue in different ontogenetic stages (Lee 1973) indicates that the enzyme is probably under tissue-specific regulation. Therefore, the low ADA activity of the human choriocarcinoma cells used in these experiments is possibly an extension of the differentiated state of the tissue from which they were derived. Choriocarcinoma cells are trophoblastic or fetal in origin (they contain a Y chromosome). Maternal components of placenta contain up to 20-fold the specific activity of ADA than do fetal components in mammals as diverse as man, guinea pig, cow, and rat (Brady and O'Donavan 1965, Hayashi 1965, Sim and Maguire 1970). Considering this, it appears that we have qualitatively demonstrated an alteration of that differentiated state in human cells as a result of the interaction with the mouse cell genome.

What caused the activation of human enzyme after fusion of ADA-deficient human cells with mouse cells having activity? Has a factor responsible for suppression of ADA activity in the human cells become lost from the hybrid cells, or is a factor for positive activity from the mouse genome responsible for the activation of the human enzyme? We can elucidate these points. In the four hybridization experiments we conducted, two involved fusion of the human choriocarcinoma cells with mouse $LMTK^-$ cells and two with mouse A9 cells. We have demonstrated (Siciliano *et al.,* in press) that human chromosomes are retained to a significantly higher degree in the $LMTK^-$ fusions than in the A9 experiments. $LMTK^-$ hybrids retain ~60% of the 28 human enzyme loci products screened electrophoretically, whereas A9 hybrids retain only ~25%. We also noted that 27 of the 35 hybrids from $LMTK^-$ expressed human ADA, while only three of the ten A9 hybrids had it. In both sets of experiments, therefore, the frequency of hybrid clones that lack human ADA conforms to the frequency of hybrids which might be expected to lose the structural ADA locus.

This result is not consistent with the hypothesis that human ADA activity is due to the loss (segregation) of a human ADA inhibitor. Under that hypothesis, human ADA expression would be possible only in those hybrid clones that lost the human regulatory gene preventing expression. The percentage of such hybrid clones should approximate the percentage expected to have lost a certain particular chromosome. Assuming human chromosome loss is random, that percentage should be equal to the mean percentage of human enzyme loci lost per hybrid clone, i.e., ~40% from the $LMTK^-$ fusion hybrids and ~75% from the A9 fusions. Of those clones losing the supressor, expression of human ADA would be possible only in the ones retaining the ADA structural gene locus—a percentage approximating the mean frequency of human enzyme loci retained per hybrid clone (~60% from LMTK$^-$ and ~25% from A9 fusions). The expected frequency of hybrid clones expressing human ADA under this hypothesis is, therefore, a product of those two probabilities. For LMTK$^-$ fusions, that would be 0.4×0.6, or 0.24. For A9 fusions, the expected frequency is $.75 \times .25$, or 0.19. While there are not enough numbers from the A9 experiments to compare the expected and obtained frequencies of human ADA expression in hybrids, it is clear that the 27 of 35 LMTK$^-$ hybrids expressing human ADA are significantly in excess of the 24% expected ($x^2 = 54.9$, $p < .001$).

On the other hand, the fact that the frequency of hybrid clones which lack human ADA conforms to the frequency of hybrids expected to lose the structural ADA locus fits squarely with the hypothesis that a factor for positive activity from the mouse genome is responsible for the activation of the human enzyme. In this case, since we do not expect loss of mouse genetic material from hybrids, all hybrid clones should have activator, and expression of human ADA depends only on the presence of the structural genes.

Since the concept of random human chromosome loss from somatic cell hybrids is questionable, at best, and central to the calculations in the above probabilistic argument, additional evidence to support the concept of a positive activation is necessary. For instance, it would be nice to show a correlation on the amount of human ADA expressed in a hybrid clone relative to the percentage of cells in the clone retaining human chromosome number 20 (the chromosome believed to carry the structural gene for ADA, Tischfield *et al.* 1974). This work is in progress. Preliminary data on seven independently derived hybrid clones (five with human ADA and two without) show the correlation (the two clones lacking human ADA also lack chromosome #20, and the five with human enzyme have #20 in at least 50% of the cells).

The data in Table 1, although preliminary, are also consistent with the idea that loss of human suppressor is not responsible for human ADA expression in hybrids. This is seen with respect to hybrid clones 114c, 119a, 123b, and 137a. In each of these, human ADA is not expressed initially nor in subsequent passages of the hybrid cell lines. If loss of a human factor for ADA suppression were responsible for activity, in one of these lines we might have expected human ADA to "pop up" as human chromosomes usually continue to be lost in later passages.

At this time, these data are most consistent with the conclusion that a positive factor from the mouse genome is responsible for the activation of the human enzyme and acts as a dominant trait. While we cannot yet determine whether the regulatory signal is operating at the transcriptional, processing, translational, or posttranslational level, we can say that it crosses species lines and simultaneously affects the products of both alleles at the locus.

There are reports in the literature suggesting the alteration of the expression of human gene loci products after fusing human and rodent cells. One hybrid clone secreting human immunoglobin (IgG) was recovered after fusion of IgG-secreting mouse myeloma cells and human lymphocytes (Schwaber and Cohen 1973). Fusion of serum albumin-producing mouse hepatoma cells with human leukocytes resulted in two hybrid clones with human serum albumin (Darlington *et al.* 1974). Human hemoglobin was detected in two hybrid clones after fusion of mouse erythroleukemia cells and human marrow cells (Deisseroth *et al.* 1975). In these reports the numbers are too small to allow any genetic analysis. Also, the purity of the human cell population for cells absent in the trait "turned on" is too questionable to rule out fusion of the mouse cell to one of them.

The ability of the human genome to influence the expression of rodent genes after somatic cell hybridization has also been observed. A form of esterase (ES-2) normally produced in mouse liver and kidney and also expressed in a renal adenocarcinoma cell line is not seen after fusion with a human fibroblast line (Klebe *et al.* 1970). Since it reappears in hybrid clones that lose a human C10 chromosome, the presence of a human repressor is implied. Human activation of esterase activity has been observed as a result of hybridization of CHO cells with human fibroblasts (Kao and Puck 1972). A form of esterase normally unexpressed in CHO cells becomes expressed in hybrid clones that retain the human chromosome satisfying an adenine *(adeB)* requirement in the hybrids. Loss of the human chromosome results in the loss of the Chinese hamster esterase.

Important data on the quantitative aspects of such studies has also been generated. Fusion of highly pigmented hamster melanoma cells with mouse fibroblasts results in hybrids that lack pigment, as well as the important enzymes responsible for melanin production. However, if tetraploid melanoma cells are used, ~50% of the hybrid clones are pigmented. Subcloning pigmented clones produces pigmented and unpigmented cells. Subcloning unpigmented ones produces only unpigmented cells (Davidson 1972). The mouse cell, therefore, is regarded as producing a negative regulator that can control one but not two genomes.

It is clear that as variant cell stocks at appropriate markers are generated, important questions concerning the genetics of regulation can be approached using cultured mammalian cell systems. Here we have demonstrated the types of markers that should prove useful and how mutation work can be valuable not only in producing isogenic variant cell lines but in providing insight into the relationships between mutagenesis, regulation, and cancer. Finally, the somatic cell hybridization work clearly demonstrates alteration of genetic regulation in cells through fusion and suggests the type of analysis that may be conducted, as the genetic factors responsible for regulation interact and segregate.

ACKNOWLEDGMENTS

The authors gratefully acknowledge the support of National Institutes of Health Research Grants ES 01287, GM 19513, and AM 17307. The technical expertise of the following individuals is also very much appreciated: Mary Sue Chennault for her help in preparing the manuscript and figures, Jeanette Siciliano and Marilyn Buck for the tissue culture and chromosome work, and Billie White and Betty Young for the zymogram preparations.

REFERENCES

Blanco, A., and W. H. Zinkham. 1963. Lactate dehydrogenases in human testes. Science 139:601–602.

Brady, T. G., and C. I. O'Donovan. 1965. A study of the tissue distribution of adenosine deaminase in six mammalian species. Comp. Biochem. Physiol. 14:101–120.

Darlington, G. J., H. P. Bernhard, and F. H. Ruddle. 1974. Human serum albumin phenotype activation in mouse hepatoma-human leukocyte cell hybrids. Science 185:859–862.

Davidson, R. L. 1972. Regulation of melanin synthesis in mammalian cells: Effect of gene dosage on the expression of differentiation. Proc. Natl. Acad. Sci. USA 69:951–955.

Davidson, R. G., and J. A Cortner. 1967. Genetic variant of human erythrocyte malate dehydrogenase. Nature 215:761–762.

Deaven, L. L., and D. F. Peterson. 1973. The chromosomes of CHO, an aneuploid Chinese hamster cell line: G-band, and the autoradiographic analysis. Chromosoma 41:129–144.

Deisseroth, A., J. Barker, W. F. Anderson, and A. Nienhuis. 1975. Hemoglobin synthesis in somatic cell hybrids: Coexpression of mouse with human or Chinese hamster globin genes in interspecific somatic cell hybrids of mouse erythroleukemia cells. Proc. Natl. Acad. Sci. USA 72:2682–2686.

Dubbs, D. R., and S. Kit. 1964. Effect of halogenated pyrimidines and thymidine on growth of L cells and a subline lacking thymidine kinase. Exp. Cell Res. 33:19.

Edwards, Y., D. Hopkinson and H. Harris. 1971a. Inherited variants of human nucleoside phosphorylase. Ann. Hum. Genet. 34:395–407.

Edwards, Y. H., D. A. Hopkinson, and H. Harris. 1971b. Adenosine deaminase isozymes in human tissues. Ann. Hum. Genet. 35:207–219.

Giblett, E. R., J. E. Anderson, F. Cohen, B. Pollara, and H. J. Meuwissen. 1972. Adenosine-deaminase deficiency in two patients with severely impaired cellular immunity. Lancet ii:1067–1069.

Hayashi, T. T. 1965. Studies on placental metabolism. Am. J. Obstet. Gynecol. 93:266–268.

Kao, F.-T., and T. T. Puck. 1972. Genetics of somatic mammalian cells: Demonstration of a human esterase activator gene linked to the Ade B gene. Proc. Natl. Acad. Sci. USA 69:3273–3277.

Klebe, R. J., T. Chen, and F. H. Ruddle. 1970. Mapping of a human genetic regulator element by somatic cell genetic analysis. Proc. Natl. Acad. Sci. USA 66:1220–1227.

Kohler, P. O., and W. E. Bridson. 1971. Isolation of hormone producing clonal lines of choriocarcinoma. J. Clin. Endocrinol. Metab. 32:683–687.

Lee, P. C. 1973. Developmental changes of adenosine deaminase, xanthine oxidase, and uricase in mouse tissues. Dev. Biol. 31:227–233.

Littlefield, J. 1964. Three degrees of guanylic acid-inosinic acid pyrophosphorylase deficiency in mouse fibroblasts. Nature 203:1142–1143.

Markert, C. 1968. The molecular basis for isozymes. Ann. NY Acad. Sci. 151:14–40.

Markert, C. L., and F. Møller. 1959. Multiple forms of enzymes: Tissue, ontogenetic and species specific patterns. Proc. Natl. Acad. Sci. USA 45:753–763.

Nance, W. E., A. Claflin, and O. Smithies. 1963. Lactic dehydrogenase: Genetic control in man. Science 142:1075–1077.

Parkman, R., E. W. Gelfand, F. S. Rosen, A. Sanderson, and R. Hirschhorn. 1975. Severe combined immunodeficiency and adenosine deaminase deficiency. N. Engl. J. Med. 292:714–719.

Petras, M. L., and F. G. Biddle. 1967. Serum esterases in the mouse, *Mus musculus.* Can. J. Genet. Cytol. 9:704–710.

Schwaber, J., and E. P. Cohen. 1973. Human X mouse somatic cell hybrid clone secreting immunoglobins of both parental types. Nature 244:444–447.

Shaw, C., and E. Barto. 1963. Genetic evidence for the subunit structure of lactate dehydrogenase isozymes. Proc. Natl. Acad. Sci. USA 50:211–214.

Siciliano, M. J., M. R. Bordelon, and P. O. Kohler. 1977. Expression of human adenosine deaminase

(ADA) after fusion of ADA deficient cells with mouse fibroblasts. Proc. Natl. Acad. Sci. USA (In press).

Siciliano, M. J., R. Humphrey, E. Murgola, and M. C. Watt. 1975. Induction and isolation of electrophoretic mutations in mammalian cell lines, *in* Isozymes IV, C. Markert, ed., Academic Press, New York, pp. 763–780.

Siciliano, M. J., and C. R. Shaw. 1976. Separation and visualization of enzymes on gels, *in* Chromatographic and Electrophoretic Techniques, I. Smith, ed., Vol. 2. 4th ed. Wm. Heinemann Medical Books Ltd., London, pp. 185–209.

Sim, M. K., and M. H. Maguire. 1970. Variation in placental adenosine deaminase activity during gestation. Biol. Reprod. 2:291–298.

Spencer, N., D. A. Hopkinson, and H. Harris. 1968. Adenosine deaminase polymorphism in man. Ann. Hum. Genet. 32:9–14.

Tijo, J. H., and T. T. Puck. 1958. Genetics of somatic mammalian cells. II. Chromosomal constitution of cells in tissue culture. J. Exp. Med. 108:259–268.

Tischfield, J. A., R. P. Creagan, E. A. Nichols, and F. H. Ruddle. 1974. Assignment of a gene for adenosine deaminase to human chromosome 20. Human Heredity 24:1–11.

Valentine, W. N., D. E. Paglia, A. P. Tartaglia, and F. Gilsanz. 1977. Hereditary hemolytic anemia with increased red cell adenosine deaminase (45 to 70 fold) and decreased adenosine triphosphate. Science 195:783–785.

Cell Differentiation and Neoplasia, edited by Grady F. Saunders. Raven Press, New York

Cell Differentiation and Cancer—A Summary

J. Paul

Beatson Institute for Cancer Research, Wolfson Laboratory for Molecular Pathology, Bearsden, Glasgow G61 1BD Scotland

When I studied pathology, over 30 years ago, there were three theories about the cause of cancer—it was due to viruses, chemicals, or disordered embryogenesis. Nothing has been added during the intervening time, and it appears that there may be truth in all three theories. In the 1930s and '40s, chemical carcinogenesis received the most attention; viral theories held the stage in the '60s, and now we are entering an era of increasing interest in the notion that the common lesion in all cancers may be disrupted cell differentiation. This is the topic of this Symposium, and my objective as summary speaker is not so much to present a résumé of all the papers as to extract a logical thread from them, to point out what seem to me to be the "take-home" lessons.

We must never forget that cancer is essentially an environmental disease. The evidence is overwhelming that in the Western world, 80 to 90% of all cancers are due to factors in the environment. This observation immediately suggests that the basic lesion in cancer may be mutation, a proposition which seems to be borne out by much recent research, notably the Ames test (Ames *et al.* 1973). Indeed, some characteristics of cancer are those of mutation, for cancer represents a relatively stable, inherited change in a cell lineage. Mutation is usually associated with an alteration in DNA, but it should be pointed out that normal differentiation itself exhibits some of the characteristics of a mutation in that, during differentiation, a cell lineage undergoes stable, inherited changes. Quite clearly these are not due to irreversible changes in DNA, as has been amply demonstrated by many experiments, notably those of Gurdon (1962) which demonstrated that nuclei from somatic cells can replace the nuclei of frog eggs, which can then develop into normal breeding adults. Moreover, there are suggestions, as in Dr. Siciliano's paper, that stable phenotypic changes of the type associated with cell differentiation may be induced by mutagens and carcinogens. We must, therefore, bear in mind that some changes identifiable as mutations may not entail irreversible changes in the genome.

In connection with mutations, I would like to draw your attention to a major paradox that has perhaps not attracted sufficient attention. An adult human contains about 3×10^{13} cells. For these to accumulate from a single fertilized egg (without losses) would require 3×10^{13} mitoses. In fact, the number of mitoses occurring during the lifetime of a human is much greater than this because many cells are continually dying and being replaced, for example, in the bone marrow, skin, and intestinal mucosa. At a conservative estimate, the number of cell divisions occurring within the human's lifetime must be of the order of 10^{15}. Now, there are indications that the mutation frequency per cell division at single loci in the human is of the order of 10^{-5} to 10^{-6}. Some of the evidence is presented by Dr. Knudson in his paper, and figures of the same order have been obtained in studies of mutational frequency in cultured mammalian cells. The frequency with which double mutations might occur at allelic loci in diploid cells is therefore likely to be of the order of 10^{-10} to 10^{-12}, and it follows that events of this kind should occur between 100,000 and 1,000 times in the lifetime of an individual (i.e., between five to six times a day and 20 times a year). If cancer arose entirely as a consequence of double mutations at key sites, the incidence would be astronomical. Hence, there must be some protective system that neutralizes the effects of these events.

A popular idea in recent years has been that this protection might be due to immune surveillance. However, I think the evidence against it is very strong. For one thing, there are many large multicellular organisms, such as trees and large invertebrates, that have no known cellular immune system. Moreover, in totally immunosuppressed individuals, there is no consistent increase in common tumors; only an increase in rare tumors arising from the reticuloendothelial system itself has been found. There have also been some direct experiments that present evidence against immune surveillance, and one of these is described by Dr. Mintz in her paper.

A second possibility is that multicellular creatures have particularly efficient DNA repair mechanisms. This possibility could be true, but the evidence that it could account for such a high degree of resistance to mutation is not particularly strong.

Finally, there is the possibility that the developing organism has a mechanism for selecting against abnormal cells. Some possibilities were discussed in a paper by Cairns in 1975. One particularly interesting suggestion arises from the hypothesis that in the early stages of commitment to differentiation, stem cells undergo an asymmetric division. If all the "new" DNA were segregated into the differentiating daughter, while the "old" DNA remained in the undifferentiated parent, this might result in rejection of mutant DNA in the differentiated cells.

We do not know the size of the pool of undifferentiated stem cells in an adult animal, but it could be quite small—of the order of 10^9 cells. If all tumors were derived from this pool, our paradox would be resolved.

In making these remarks, of course, I do not wish to imply that there is no genetic element in any cancer. In some kinds of cancer, both in experimental animals and, as very elegantly demonstrated by Dr. Knudson, in human beings, there is a clear-cut genetic element. As his analyses show, some specific kinds of tumors may arise as a result of homozygous lesions in a single locus.

A major reason for interest in the hypothesis that some cancers result from disordered cell differentiation is that the process might be reversed. That this is more than a remote possibility has been demonstrated conclusively by the very beautiful experiments reported by Dr. Mintz in her Bertner Award lecture. There has been a good deal of evidence in the past that, on occasion, tumor cells may revert. For example, hamster fibroblasts transformed by the Schmidt-Ruppin strain of Rous sarcoma virus readily revert from the transformed state to normal fibroblasts (Macpherson 1965). Many experimental examples of reversion seem to be accompanied by chromosomal rearrangement, and Dr. Sachs, among others, has presented some interesting arguments for this (Bloch-Shtacher *et al.* 1972).

Do these results imply that tumors can occur without genetic damage? Again, there has been previous evidence for this. Undifferentiated tissue cultures can be derived from plant tumors. If treated with the correct balance of hormones, these can develop roots and shoots and form quite normal plants (Sacristan and Melchers 1969). In other experiments, tumor cell nuclei have been injected into frog eggs and have given rise to larvae with normal tissues (McKinnell *et al.* 1969). However, the demonstration that perfectly normal mice can be derived from teratocarcinoma cells, which are injected into blastocysts subsequently incubated by surrogate mothers, gives the clearest and most unambiguous demonstration that normal tissues can be derived from tumor cells.

Several possible mechanisms may explain these results. One is that abnormal cells are forced to perform normally in the presence of a strongly inducing environment. A second possibility, however, is that these results arise by selection of normal cells that have been segregated out in the teratocarcinoma cell line. It may be, for example, that the teratocarcinoma cells have a near-normal karyotype and that, as a consequence of nondisjunction, they produce some cells with an abnormal gene balance but others with a perfectly normal gene balance. In the abnormal environment of tissue culture or an abnormal inoculum site, the more abnormal cells may be favored. In the normal inducing environment of the blastocyst, however, selection may entirely favor the normal segregants. It will require more research to determine which of the possibilities is true, or if there is yet another explanation.

One question that we should ask ourselves about these results is this: How good is teratocarcinoma as a general model for cancer? Do these cells have the general abnormalities characteristic of most cancer cells, or are they more like frustrated stem cells that grow in a haphazard way in an abnormal environment? It is not clear where the answer lies. Perhaps, therefore, we should be a little cautious about generalizing from results with teratocarcinoma until similar results have been obtained using other systems.

Another general point arising from these considerations is whether all cancers may be derived from stem cells. This proposition is discussed in the papers by Drs. Mintz, Pierce, Herman, Prasad, McCulloch, and Sachs. All these authors present some evidence that different kinds of cancers can derive directly from stem cells and that these, in fact, may provide the primary targets. As discussed earlier, this notion has its attractions, as it could resolve the paradox arising from the small number of mutations evident in large multicellular creatures. But before we can gain a clear understanding of this question, we must define stem cells rather carefully. In this connection, we have heard about some elegant studies on the hemopoietic system by Drs. Lajtha, Iscove, and McCulloch. From their studies, it is apparent that a mature tissue does not derive directly from a single kind of stem cell. Rather, maturation proceeds through a series of stem cells, and in each different stem cell compartment, many cell divisions may occur, giving rise to a great amplification in cell numbers. An important aim must be to try to recognize which kind of cell is actually transformed in cancer. Is it the most totipotent kind of stem cell, or is it one of the restricted multipotent, or even unipotent, stem cells? Some of the studies on leukemia, perhaps particularly studies on Friend leukemia, suggest that the transformed cell may actually be a stem cell in one the late transit populations, in this instance possibly even a proerythroblast, blocked in its progression to ultimate maturity. If that kind of observation can be generally substantiated, it weakens the earlier explanation of the mitosis/mutation paradox.

Dr. Pierce, in his paper, presents an ingenious idea to explain tumor progression, particularly in some of the common epithelial tumors like squamous epithelioma and breast cancer. He suggests that the transformation event might simultaneously involve a population of stem cells and that the phenomenon of progression might be explained by the emergence of cohorts of progressively less differentiated stem cells, each outgrowing the former one. This idea suggests a polyclonal origin of most cancers, whereas current prejudice is that many cancers are monoclonal. However, this theory of the monoclonal origin of cancers should not be accepted too readily because more rapidly growing cells can very quickly replace more slowly growing cells in a cell population. On the other hand, this idea might be difficult to reconcile with experimental transformation of fibroblasts in vitro and certain phenomena of tumor progression, notably karyotypic progression.

One of the most intriguing possibilities raised at this meeting is that some cancers might actually be cured by causing them to differentiate. That certain tumors do occasionally differentiate spontaneously is, of course, very well known. The outstanding case is neuroblastoma, described by Dr. Herman, which not uncommonly differentiates to form a nonmalignant ganglioneuroma. As described by Dr. Prasad, the differentiation of neuroblastoma cells is accompanied by phenomena that can be well characterized and that can be promoted by a number of factors. Drs. McCulloch and Sachs described similar phenomena in leukemias. Clearly, it would be of the greatest value to discover factors that normally promote differentiation and might influence the behavior of tumor cells.

Many tumors other than neuroblastoma can apparently revert to more normal tissue by differentiation. Although reversion is very rare for most common malignant tumors, there are some hundreds of carefully described cases in the literature that document spontaneous regression of even such tumors as carcinomas of the stomach and bronchus (Everson and Cole 1966). These cases may not be quite so rare as we think because, unless the histopathology has been studied, the diagnosis tends to be revised if a patient recovers from an apparent malignancy. In my own experience, most doctors have one or two anecdotes of "miracle cures." While I would be very cautious about poorly documented clinical anecdotes, I would exhort clinicians to go to some trouble to obtain and report excellent documentation on cases of this nature that they may encounter.

A serious objection to a strategy of treating cancer by promoting differentiation might appear to be the fact that many tumors have obvious karyotypic and, therefore, genetic abnormalities which one might assume to be irreversible. However, it may be worthwhile to point out that karyotypic abnormalities are quite compatible with normal differentiation even in the human being. Individuals with Kleinfelter's or Down's syndromes have quite normal tissues, despite the chromosomal imbalance, and there are a multitude of similar examples in animal and plant genetics. Moreover, some tumors in which cells are constantly differentiating have well-described karyotypic abnormalities.

In the course of this meeting, the old debate about whether tumor cells are undifferentiated or not was resurrected. I have always felt that, rather than describe tumors as undifferentiated, we should regard them as maldifferentiated. In this connection, I was struck by Dr. Becker's use of the happy phrase "phenotypic schizophrenia" to describe the fact that some tumor cells express functions that are mutually exclusive in normal tissues. Some of the most striking examples are of lung tumors; although derived from bronchial epithelium, they produce a variety of hormones, notably pituitary hormones. This behavior is, so far as I can determine, characteristic only of tumor cells. In a search of the literature, I have never found information to suggest that normal cells ever express two radically different functions of this sort.

Whether "phenotypic schizophrenia" represents a breakdown of absolute differences between tissues, or whether it merely represents an altered level of quantitative differences in proteins is not well established because it is possible that all epithelial cells produce one or two molecules of pituitary hormones which are undetectable by present methods.

An indication that common quantitative differences are characteristic of tumor cells emerged from some of the work Dr. Brinkley presents about differences in microfibrillar and microtubular structure in tumor and normal cells. Whether these differences are absolute or quantitative (and this may be difficult to prove one way or the other), they can certainly be exploited usefully in the diagnosis and management of cancer; perhaps the best example is the miraculous improvement in the outlook for patients with chorion carcinoma, an improvement resulting from our ability to monitor the growth of tumor cells by measuring chorionic gonadotrophin.

It must be clear to all of us that to answer many of the questions that I have raised in this summary, we need to know a great deal more about normal differentiation. The fact that we are making great strides in our understanding has been repeatedly illustrated in this meeting. We have heard extremely interesting reports by Drs. Lajtha, Iscove, Cunningham, Rutter, Wright, Edidin, and Meistrich, dealing with different aspects of the biology of differentiation. I would select two observations from these papers as being of special interest and importance. One is the demonstration by Dr. Lajtha and his colleagues of two growth factors, one that represses and another that stimulates the cycling of hemopoietic stem cells. It has always been obvious that there must be feedback signals to regulate the rate of growth of these cells, and this has led to a great deal of speculation about chalones. To provide a convincing demonstration of such entities has proved remarkably elusive, but Dr. Lajtha's paper promises a system from which we may, at last, be able to make concrete observations. The other communication to which I draw your attention is the work described by Dr. Cunningham. His work seems to be bringing us very close to identifying the nature of surface antigens and the possible way in which they may determine cell recognition.

As for the molecular biology of cell differentiation, we have been entertained to a scintillating series of reviews by Drs. Brown, Roeder, Hnilica, Darnell, Lingrel, Lodish, Marks, O'Malley, and Rutter, each giving some inidcation of the exciting state of this field at present. We can now recognize that the cell is somewhat like a computer, and we have learned a great deal about its hardware and the way in which it operates. We are beginning to acquire detailed information about the structure of eukaryotic genes, about the role of chromosomal proteins and polymerases in transcribing the genes, about the primary transcripts and the way in which they are processed into messenger RNAs and other molecules, and about the way in which messengers are translated into proteins. The amount of information in these areas is now immense. The presentations themselves were masterly summaries, and it would be folly for me to summarize them further. However, there are one or two points I might make.

One has to do with another paradox. Several quite recent studies have indicated that the RNA populations of different kinds of cells are probably not very different and, indeed, that most transcribable sequences may always be transcribed, if at very different rates. The observation is borne out by autoradiographic studies in giant chromosomes. Yet, between different tissues, very striking differences in chromatin can be demonstrated, as described in the papers by Drs. Hnilica and O'Malley. It is not clear why these differences should be so striking.

Besides knowing the structural elements in the regulatory network of the cell, we would very much like to know something about the control of regulation itself. In this regard, we have some very interesting and revealing papers by Drs. Gurdon, Siciliano, and Marks. I would draw your attention to one small point that some of you may have missed in the reports by Dr. Marks and Dr. Gurdon. Dr. Marks presents some convincing evidence that induction of hemoglobin synthesis in the Friend cell is cell cycle dependent. Similar observations have been made by a number of groups, including ourselves, and they seem to support the general notion of a quantal mitosis, an idea that has recurred throughout the literature of cell differentiation. However, you may recollect Dr. Gurdon's experiments in which he injected *Xenopus* nuclei into oocytes of *Pleurodeles* and demonstrated that these eggs then produced proteins typical of adult *Xenopus*. In those experiments, no division of the *Xenopus* nuclei occurred. We must, therefore, entertain doubts about the necessity for a round of DNA replication in the process of commitment.

What of the future in research on cell differentiation? In my view, the next five to ten years will see a series of dazzling discoveries. One of the great problems of studying eukaryotic cells is their complexity. The genome contains enough DNA for a million or more genes. However, the whole situation has been transformed within the past two or three years by the emergence of DNA recombinant studies. They now make it possible to clone fragments of the genome, the size of a single gene or less. Moreover, these discoveries have been accompanied by rapid advances in our ability to sequence DNA. It now becomes possible to read genes almost like a book. The floodgates are open, and it is not out of the question that within the next 30 or 40 years we will have complete sequences for the entire human genome. Perhaps some future publisher will indeed produce a book of a thousand volumes containing nothing but these sequences. The volumes may make quite good bedside reading for some. They should certainly be very soporific!

In summary, then, there is now quite strong evidence, much of which has been presented at this meeting, that cancer is essentially a disturbance of the cell computer and that this process may not necessarily require an irreversible alteration of DNA. We still understand very little of the control mechanisms in eukaryotic cells, but quite recently we have acquired a whole collection of tools of incredible precision to enable us to tackle the problem. We can see a clear way ahead to resolving most of them, and who can deny that within the next five years we will obtain very many detailed answers to the problems of both normal and abnormal differentiation, which recent research has made it possible to formulate clearly.

REFERENCES

Ames, B. N., W. E. Durston, E. Yamasaki, and F. D. Lee. 1973. Carcinogens are mutagens: A simple test system combining liver homogenates for activation and bacteria for detection. Proc. Natl. Acad. Sci. USA 70:2281–2285.

Block-Shtacher, N., Z. Rabinowitz, and L. Sachs. 1972. Chromosomal mechanism for the induction of reversion in transformed cells. Int. J. Cancer 9:632–640.

Cairns, J. 1975. Mutation selection and the natural history of cancer. Nature 255:197–200.

Everson, T. C., and W. H. Cole. 1966. Spontaneous Regression of Cancer. W. B. Saunders, Philadelphia and London.

Gurdon, J. B. 1962. The developmental capacity of nuclei taken from intestinal epithelial cells of feeding tadpoles. J. Embryol. Exp. Morphol. 10:622–640.

Macpherson, I. 1965. Reversion in hamster cells transformed by Rous sarcoma virus. Science 148:1731–1733.
McKinnell, R. G., B. A. Deggins, and D. D. Labat. 1969. Transplantation of pluripotential nuclei from triploid frog tumors. Science 165:394–396.
Sacristan, M. D., and G. Melchers. 1969. The caryological analysis of plants regenerated from tumorous and other callus cultures of tobacco. Mol. Gen. Genet. 105:317–333.

Author Index*

* *See also* List of Contributors, pp. xi–xvi.

Subject Index